IBN AL-HAYTHAM'S THEORY
OF CONICS,
GEOMETRICAL CONSTRUCTIONS
AND PRACTICAL GEOMETRY

This book provides a unique primary source on the history and philosophy of mathematics and science from the medieval Arab world. The present text is complemented by two preceding volumes of *A History of Arabic Sciences and Mathematics*, which focused on founding figures and commentators in the ninth and tenth centuries, and the historical and epistemological development of 'infinitesimal mathematics' as it became clearly articulated in the oeuvre of Ibn al-Haytham.

This volume examines the increasing tendency, after the ninth century, to explain mathematical problems inherited from Greek times using the theory of conics. Roshdi Rashed argues that Ibn al-Haytham completes the transformation of this 'area of activity' into a part of geometry concerned with geometrical constructions, dealing not only with the metrical properties of conic sections but also with ways of drawing them and properties of their position and shape.

Including extensive commentary from one of the world's foremost authorities on the subject, this book contributes a more informed and balanced understanding of the internal currents of the history of mathematics and the exact sciences in Islam, and of their adaptive interpretation and assimilation in the European context. This fundamental text will appeal to historians of ideas, epistemologists and mathematicians at the most advanced levels of research.

Roshdi Rashed is one of the most eminent authorities on Arabic mathematics and the exact sciences. A historian and philosopher of mathematics and science and a highly celebrated epistemologist, he is currently Emeritus Research Director (distinguished class) at the Centre National de la Recherche Scientifique (CNRS) in Paris, and is the Director of the Centre for History of Medieval Science and Philosophy at the University of Paris (Denis Diderot, Paris VII). He also holds an Honorary Professorship at the University of Tokyo and an Emeritus Professorship at the University of Mansourah in Egypt.

J. V. Field is a historian of science and is a Visiting Research Fellow in the Department of History of Art and Screen Media, Birkbeck, University of London, UK.

CULTURE AND CIVILIZATION
IN THE MIDDLE EAST
General Editor: Ian Richard Netton
Professor of Islamic Studies, University of Exeter

This series studies the Middle East through the twin foci of its diverse cultures and civilisations. Comprising original monographs as well as scholarly surveys, it covers topics in the fields of Middle Eastern literature, archaeology, law, history, philosophy, science, folklore, art, architecture and language. While there is a plurality of views, the series presents serious scholarship in a lucid and stimulating fashion.

PREVIOUSLY PUBLISHED BY CURZON

THE ORIGINS OF ISLAMIC LAW
The Qur'an, the *Muwatta'* and Madinan *Amal*
Yasin Dutton

A JEWISH ARCHIVE FROM OLD CAIRO
The history of Cambridge University's Genizah collection
Stefan Reif

THE FORMATIVE PERIOD OF TWELVER SHI'ISM
Hadith as discourse between Qum and Baghdad
Andrew J. Newman

QUR'AN TRANSLATION
Discourse, texture and exegesis
Hussein Abdul-Raof

CHRISTIANS IN AL-ANDALUS 711–1000
Ann Rosemary Christys

FOLKLORE AND FOLKLIFE IN THE UNITED ARAB EMIRATES
Sayyid Hamid Hurriez

THE FORMATION OF HANBALISM
Piety into power
Nimrod Hurvitz

ARABIC LITERATURE
An overview
Pierre Cachia

STRUCTURE AND MEANING IN MEDIEVAL ARABIC AND
PERSIAN LYRIC POETRY
Orient pearls
Julie Scott Meisami

MUSLIMS AND CHRISTIANS IN NORMAN SICILY
Arabic-speakers and the end of Islam
Alexander Metcalfe

MODERN ARAB HISTORIOGRAPHY
Historical discourse and the nation-state
Youssef Choueiri

THE PHILOSOPHICAL POETICS OF ALFARABI, AVICENNA
AND AVERROES
The Aristotelian reception
Salim Kemal

PUBLISHED BY ROUTLEDGE

1. THE EPISTEMOLOGY OF IBN KHALDUN
Zaid Ahmad

2. THE HANBALI SCHOOL OF LAW AND IBN TAYMIYYAH
Conflict or conciliation
Abdul Hakim I Al-Matroudi

3. ARABIC RHETORIC
A pragmatic analysis
Hussein Abdul-Raof

4. ARAB REPRESENTATIONS OF THE OCCIDENT
East-West encounters in Arabic fiction
Rasheed El-Enany

IBN AL-HAYTHAM'S THEORY OF CONICS, GEOMETRICAL CONSTRUCTIONS AND PRACTICAL GEOMETRY

A history of Arabic sciences and mathematics

Volume 3

Roshdi Rashed

Translated by J. V. Field

LONDON AND NEW YORK

مركز دراسات الوحدة العربية
CENTRE FOR ARAB UNITY STUDIES

First published 2013
by Routledge
2 Park Square, Milton Park, Abingdon, OX14 4RN

Simultaneously published in the USA and Canada
by Routledge
711 Third Avenue, New York, NY 10017

First issued in paperback 2017

Routledge is an imprint of the Taylor & Francis Group, an informa business

British Library Cataloguing in Publication Data
A catalogue record for this book is available from the British Library

Library of Congress Cataloging in Publication Data
The Library of Congress has cataloged volume 1 of this title under the
LCCN: 2011016464

ISBN 13: 978-0-8153-4876-4 (pbk)
ISBN 13: 978-0-415-58215-5 (hbk)

Typeset in Times New Roman
by Cenveo Publisher Services

This book was prepared from press-ready files supplied by the editor

CONTENTS

FOREWORD

This book is a translation of *Les Mathématiques infinitésimales du IXe au XIe siècle*, vol. III: *Ibn al-Haytham. Théorie des coniques, constructions géométriques et géométrie pratique*. The French version, published in London in 2000, also included critical editions of all the Arabic mathematical texts that were the subjects of analysis and commentary in the volume.

The whole book has been translated, with great scholarly care, by Dr J. V. Field. The translation of the primary texts was not simply made from the French; I checked a draft English version against the Arabic. This procedure converged to give an agreed translation. The convergence was greatly helped by Dr Field's experience in the history of the mathematical sciences and in translating from primary sources. I should like to take this opportunity of expressing my deep gratitude to Dr Field for this work.

Very special thanks are due to Aline Auger (Centre National de la Recherche Scientifique), who helped me check the English translations against the original Arabic texts, prepared the *camera ready* copy and compiled the indexes.

PREFACE

In the first two volumes, we sought to present an overall picture of a tradition of geometrical research on the mathematics of infinitesimals carried out by Arabic followers of the work of Archimedes. But our intentions went further than this: we also wanted to put together a preliminary nucleus of a *corpus* of Arabic geometry. In one form or another, we pursued the same end: to accumulate material and, by all possible means, to break with a manner of writing history that was at best partial and at worst merely anecdotal. We cannot of course lay claim to providing a complete reconstruction – we are protected from entertaining any such illusion both by the limitations of our own resources and by the fact that many writings are either still missing or definitively lost – but we have simply tried to make our search reasonably complete and, above all, sufficiently systematic to expose the exact significance of the mathematical activity of particular periods.

It seemed sensible to start with the geometrical work of Ibn al-Haytham, almost in its entirety, before going back to his predecessors. The justification for this strategy lies in the unusual position that Ibn al-Haytham occupies: he is heir to two centuries of intense research in geometry, and he undertook to push that research forward as far as the logical possibilities allowed. His purpose, stated several times in unequivocal terms, was to 'complete' the contributions made by his Greek and Arabic predecessors. Ibn al-Haytham's explicit project was to correct the errors made by earlier scholars, to develop their intuitions to the full, and to push their successes as far forward as possible. So his geometrical work naturally afforded us a privileged advanced position from which we could look back on the past of the discipline in an orderly way. The methods employed in carrying out this procedure of moving backwards involve the history of the textual tradition of each piece of writing and the tradition of ideas to which the work belongs. We have explained this elsewhere.[1]

---

[1] 'L'histoire des sciences entre épistémologie et histoire', *Historia scientiarum*, 7.1, 1997, pp. 1–10.

The second volume is completely taken up with Ibn al-Haytham's works on infinitesimal or Archimedean geometry. The first volume contains writings by Arabic followers of Archimedes from the period before Ibn al-Haytham. This research tradition began with the Banū Mūsā in the ninth century, and was continued by Ibn al-Haytham's immediate predecessors, scholars such as al-Qūhī and Ibn Sahl, and included a long series of commentators. Two traits are prominent in the character of these contributions. In the first place, the tradition was shaped and developed by geometers. Even when they were familiar with algebra and influenced by it to various degrees, they understood their work as concerned with geometry. The possibilities inherent in algebra are nevertheless noticeable in the introduction of a certain idea of measure, superimposed on the traditional presentation of comparisons between figures, to determine surface areas and volumes of solids with curved boundaries; algebraic possibilities are also apparent in the more intensive use of sums and arithmetical inequalities. The second trait, which has also escaped notice, as it were runs in counterpoint with the first one. These geometers ceased to regard themselves as simply the intellectual descendants of Archimedes, and instead laid an equal claim to be the heirs of Apollonius of Perga. Never, before this period, had there been such heavy emphasis upon any connection between these two geometrical programmes. These mathematicians, coming after Apollonius, had at their disposal a theory of conics much more elaborate than the one used by Archimedes. The works by Apollonius that they knew included the *Conics*, from which they also acquired an interest in metrical relations and in properties of position and form. All these geometers, without exception, from the Banū Mūsā to Ibn al-Haytham, carried out research work that combined the geometry of Archimedes with that of Apollonius. This union between the traditions was powerful, as a principle both for making discoveries and for organization. It is thus no accident that new areas of mathematics were quickly opened up: point-to-point transformations, the study of certain projections, algebraic geometry and others. It is thus a new landscape that we find reflected in the works of Ibn al-Haytham and that forms the backdrop to his researches in infinitesimal geometry. The sheer bulk of the texts, as well as their high quality, required us to devote the following two volumes to them. The title given to the first two volumes is no doubt not very suitable for these later ones; we decided to keep the same title simply so as not to break the overall continuity of the collection; the subtitle will have to carry the burden of identifying the organizing theme of the book. This solution, though slightly out of kilter with the rest, was the least inconvenient.

Like the other volumes in the collection, this one has been read by Christian Houzel, Directeur de recherche at the CNRS; to whom I offer my hearty thanks. I am grateful also to Aline Auger, Ingénieur d'études at the CNRS, for preparing this volume for printing and constructed the glossary and the index.

Roshdi Rashed
Bourg-la-Reine, 1999

Director of Research, Centre National de la
Recherche Scientifique, Paris
Professor, Department of the History and
Philosophy of Science, University of Tokyo

CONIC SECTIONS AND GEOMETRICAL CONSTRUCTIONS

Greek geometers soon noticed that 'geometrical' constructions were not applicable only to plane problems and that 'constructible' problems were not simply those that could be solved by means of straightedge and compasses. This important discovery led some mathematicians to investigate the properties of curves other than the circle, in particular conic sections. The story of these constructions has been told so many times that it need not detain us here.[1] Let us merely remind ourselves that as early as the fourth century BC conics are called upon in the solution of a three-dimensional problem. Menæchmus made use of a parabola and a hyperbola to solve the problem of the duplication of the cube. Was this something unusual or an established practice? Is this a procedure that lays the foundations for the theory of conics? The answers to questions like these are hidden in the mists surrounding the origins of the theory. What concerns us here is to note that from a very early date conic sections were used to construct solutions to three-dimensional problems in geometry.

A little later, towards the mid third century BC, Conon of Samos – according to the account given by Apollonius – takes up the whole problem once again. Apollonius in fact informs us, in the preamble to the fourth book of his *Conics*, that Conon of Samos had investigated the intersection of conic sections, and had tried to find the number of their points of intersection, 'the greatest number of points in which conic sections that do not completely coincide can meet one another'.[2] This is the only echo, distant as it is, to tell us about the form and purpose of Conon's work. Was it like what Apollonius does in the fourth book of his *Conics*, where he discusses the number of points using an argument by *reductio ad absurdum*? We do not know, but this crucial evidence supplied by Apollonius does allow us to infer that after Menæchmus some

---

[1] T. L. Heath, *A History of Greek Mathematics*, 2 vols, Oxford, 1921; repr. Oxford, 1965; O. Becker, *Grundlagen der Mathematik*, 2nd ed., Munich, 1964.

[2] Apollonius, *Les Coniques d'Apollonius de Perge*, French transl. Paul Ver Eecke, Paris, 1959, p. 281. Cf. *Apollonius: Les Coniques,* tome 2.2: *Livre IV*, edition, French transl. and commentary by R. Rashed, Berlin/New York, 2009, p. 116.

mathematicians – and not only minor ones, since they included Conon – attempted to investigate the intersection of conic sections, or at least the number of their points of intersection. If we read further in Apollonius, we find that Conon's work was, however, challenged on two grounds at once: rigour and utility. A contemporary of Conon, Nicoteles of Cyrene, made a double criticism of his work, a criticism that Apollonius reports as follows:

> Conon of Samos expounded it [the matter concerning intersection] to Trasydoeus, but without having paid attention to the proof, as he should have done; for which reason Nicoteles of Cyrene indeed justly found fault with him.[3]

If Apollonius agrees with this first criticism that Nicoteles levels at Conon, he does not endorse the second one, in which the questions raised by Conon are judged to be of no use:

> Nicoteles, he says, is, however, in error, when, in support of his arguments against Conon, he maintains that the things Conon discovered are not any use for further discussions (diorisms); for if, without these things, problems can generally be expressed in a complete form from the point of view of their discussion, the things nevertheless facilitate insights, either concerning the possibility of multiple solutions, or concerning the number of solutions, or, on the contrary, the impossibility of a solution.[4]

In the mid third century BC, an attempt is made to reduce certain discussions (diorisms) to the problem of finding the points of intersection of two conics. But there is no firm foundation for this attempt and it is vitiated by the absence of proof. This may explain why mathematicians such as Nicoteles think it can be set aside. There remains a question we cannot yet answer: on this matter, what position was taken by Archimedes, a young friend of Conon? Did he follow Conon in adopting this new technique, or did he take Nicoteles' part? Did he continue to prefer the technique of using intercalations[5] or did he begin to employ the new technique – considering intersection of conics – because of its simplicity? In *On the Spiral*, Archimedes reduces certain propositions to intercalations and seems ready to accept that this brings the discussion to a conclusion (see Propositions 5, 6, 7 and 9). Can one infer that the new technique was only beginning to replace the older one, at least in the circle surrounding Conon? To move

---

[3] Apollonius, *Les Coniques*, transl. Paul Ver Eecke, p. 281; see also *Les Coniques, tome 2.2: Livre IV*, ed. R. Rashed, p. 116.

[4] Ed. P. Ver Eecke, p. 282; ed. R. Rashed, p. 118.

[5] Intercalation is a technique based on a form of *neusis*: to place between a straight line and a curve, for instance an arc of a circle, a straight line equal to a given straight line and verging (inclined) towards a given point (in the sense that if produced it would pass through that point).

things forward would have required Conon's invention to be backed up by rigorous arguments, which in turn would have needed a better understanding of the local and asymptotic properties of conic sections. Such understanding was to be characteristic of the future. We shall see that the history of the study of conic sections is linked with the development of exactly this understanding. All the same, Apollonius, unlike Nicoteles, had no intention of throwing the baby out with the bath water: he recognizes that the logic of Conon's questions is flawed but he accepts that they have a heuristic value. That might perhaps explain the attitude of Archimedes, since he is younger than Conon and older than Apollonius by one or two generations. Sometimes – for example in the lemma to Book II Proposition 4 of *The Sphere and the Cylinder* – when he has proposed a three-dimensional problem he is content merely to define it (to give the diorism), without a solution. Maybe, as Ibn al-Haytham was to suggest twelve centuries later, Archimedes wished to avoid appealing to intersection of conics, for the reason we have mentioned. However it was, neither Nicoteles' censure, nor Archimedes' silence, nor Apollonius' prudence prevented mathematicians from looking to conic sections in solving three-dimensional problems. Thus Diocles, a successor to Archimedes and a contemporary of Apollonius,[6] uses the intersection of two parabolas to solve the problem of doubling the cube.

Although its impact was indirect, Apollonius' personal contribution was even more important. I refer specifically to Book V of the *Conics*, which is concerned with normals to the curves, and in this connection with finding a set of points – which much later on will become the evolute (Huygens). Apollonius studies the evolute – if we may use the anachronistic term – by considering points that lie both on an equilateral hyperbola and on the conic in question. Thus in Propositions 51, 52, 55, 56, 58, 59, 62 and 63 of Book V he proceeds by discussing intersection of his equilateral hyperbola and one or other of the three conic sections. His work, the most systematic treatment of the matter that we know from Hellenistic mathematics, is nevertheless, from the point of view of what interests us here, defective in two respects: all cases concern an equilateral hyperbola; and, further, the existence of the points of intersection is not established. This deficiency is not due to

---

[6] *Les Catoptriciens grecs.* I: *Les miroirs ardents,* edition, French translation and commentary by R. Rashed, Collection des Universités de France, Paris, 2000, pp. 78ff. Cf. *Apollonius: Les Coniques,* tome 3: *Livre V,* ed. R. Rashed, Berlin/New York, 2008. In Book V of the *Conics* Apollonius tries to find the number of normals that can be drawn from any point. Huygens defines the evolute as the envelope of the normals in connection with a problem in mechanics to do with the cycloid. Unlike Huygens, Apollonius never refers explicitly to the idea of an evolute.

weakness on the part of the mathematician, or to a lack of the technical means required to provide at least a partial proof. It is simply that such a procedure was not yet considered a necessary component of a demonstration if the matter seemed obvious. One can see hints of it in works such as that of Eutocius, where he provides the demonstrations that are lacking in *The Sphere and the Cylinder*, but a proof of the existence of the points of intersection is not yet regarded as truly necessary.

This completes our account of the development of work on the intersection of conic sections up to the ninth century. It is easily established that we are dealing with sparse and scattered contributions, produced here and there when mathematicians encountered three-dimensional problems, and contributions in which employing intersection of conics is merely sporadic (there is no apparent reluctance to use other curves, even though Pappus seems to give preference to conic sections in Books III and IV of his *Collection*). We shall not repeat here a story that has already been told very well several times. It will be enough merely to supply a brief reminder of what these three-dimensional problems were. We have already mentioned doubling the cube, the lemma to Book II Proposition 4 of Archimedes' *The Sphere and the Cylinder*, Book V of Apollonius' *Conics* and trisection of an angle. We may also note the numerous solutions to some of these problems that accumulated over the course of time. Doubling the cube and finding the two means provides a particularly significant example of this: between Diocles, Pappus and Eutocius, we find no fewer than twenty or so solutions, several of which involve conics. We must also note, and this is of crucial importance for the work carried out from the ninth century onwards, that these problems and their solutions involve new idea about the curves and their behaviour. The working out of these ideas was to be left to later mathematicians. On the other hand, we may take note that in themselves these problems already reflect a desire to extend the geometry of Euclid.

From the ninth century onwards the problems inherited from Greek mathematicians are subjected to huge investigation. Although this is not the place to discuss new results, we must note a change in the context. On the one hand, in a relatively short interval of time, we see an unprecedented growth in the number of writings devoted to these inherited problems. On the other hand, it is almost exclusively intersection of conics that is used to solve them. Not only is it rare for other methods to be employed, but when they are used they seem to be connected with an older memory – as in the work of the Banū Mūsā[7] and later that of al-Bīrūnī. Such methods always

---

[7] R. Rashed, *Founding Figures and Commentators in Arabic Mathematics. A history of Arabic sciences and mathematics*, vol. 1, Culture and Civilization in the Middle East, London, 2012, Chapter I.

avoid any demand for an appeal to transcendental curves. That is to say, we are seeing an increased research effort, which displays a certain unity in restricting its methods to those using intersection of conics. From this point on, we are dealing with an 'area of activity' for conic sections.

This area soon became wider. The widening was at first carried out in what one might call a Hellenistic spirit. The most striking example is that of the regular heptagon. While Greek mathematicians have not left us even a single construction for the figure, at the end of the tenth century we witness a genuine debate on the matter in the mathematical community, with the appearance of no less than a dozen essays. But this widening in a 'Hellenistic' style goes together with development of the same area in a new manner that derives from the earliest attempts to solve certain third-degree algebraic equations by using conics. It was at the end of the ninth century that mathematicians began to consider this matter, and it was to be a continuing topic for research until al-Khayyām formulated his general solution to the problem. The names of al-Māhānī, al-Khāzin, al-Qūhī, Ibn 'Irāq, and particularly of Abū al-Jūd, are associated with the various stages of this programme of research. Their writings change the area of operations in constructing solutions to three-dimensional problems using conic sections: together with the ancient problems, the field now includes problems brought forward by algebra.

As we have pointed out, this heterogeneity in the origins of the problems is in contrast to a unity imposed on methods of solution: thereafter, the only method is to use the intersection of conic sections. This methodology seems to be the leading characteristic of research from the ninth century onwards.

This is the tradition to which Ibn al-Haytham belongs, and it is this tradition his work will affect. What he does is to complete the transformation of the 'area of activity' into a part of geometry concerned with geometrical constructions. But before we describe and analyse developments in the theory, let us look at the works Ibn al-Haytham wrote about geometrical constructions. There are no fewer than ten such essays.

1. *On the Lemma for the Side of the Heptagon*
2. *On the Construction of the Heptagon in the Circle*
3. *On the Division of the Line Used by Archimedes*

These three treatises deal with problems introduced by Ibn al-Haytham's predecessors. We shall see that in each of them he tries to complete the treatment of the problem his predecessors had proposed and had either solved only for a particular case, or without proving the two conics intersected one another.

The second group of three-dimensional problems makes it clear that they are based on algebra.

4. *On a Solid Numerical Problem*
5. *On the Determination of Four Straight Lines*

This last treatise, which is unfortunately lost, dealt with the problem of finding four line segments of lengths between those of two given segments such that the lengths of the six segments shall be in continued proportion. A successor to Ibn al-Haytham, the algebraist al-Khayyām, writes that 'this has been proved by Abū 'Alī ibn al-Haytham; only it is very difficult...'.[8] The problem leads to an equation of the fifth degree, which is solved by considering the intersection of a hyperbola and a generalized parabola (a cubic). What al-Khayyām tells us makes it seem likely that Ibn al-Haytham was in possession of a method similar to the one we find later in Fermat's *Dissertatio Tripartita*.[9]

The third group consists of one book.

6. *On the Completion of the Work on Conics*
As we shall see, this book plays a crucial part in making the study of geometrical constructions a subject in its own right.

The fourth and final group consists of several pieces that, to judge by their titles, deal with questions concerning the construction of conics. Unfortunately, no copies of these writings have yet come to light.

7. *On the Properties of Conic Sections*
The title of this treatise mirrors that of another one, on the circle, whose text and translation appear in the fourth volume of the present work.[10] In the latter treatise, Ibn al-Haytham investigates not only metrical properties

---

[8] R. Rashed and B. Vahabzadeh, *Al-Khayyām mathématicien*, Paris, 1999, *Traité d'algèbre et d'al-muqābala*, p. 222. English translation: *Omar Khayyam. The Mathematician*, Persian Heritage Series no. 40, New York, 2000, p. 158.

[9] *Ibid.*, p. 387; English transl., p. 259.

[10] R. Rashed, *Les Mathématiques infinitésimales du IXe au XIe siècle*, vol. IV: *Méthodes géométriques, transformations ponctuelles et philosophie des mathématiques*, London, 2002.

but also some affine properties, and even certain projective properties. Perhaps in the text on the properties of conic sections (parabola, hyperbola, ellipse) he discussed the same properties, but for conics. This is not unlikely: the analogous titles suggest it, as does the fact that his predecessors – Ibn Sinān, al-Qūhī, Ibn Sahl and others – took an interest in these properties.

8. *On the Construction of Conic Sections*

Unlike the preceding one, this book is not mentioned either by old biobibliographers or by old mathematicians that we know of. On the other hand, Ibn al-Haytham himself refers to it in his treatise on parabolic burning mirrors.[11] The context tells us that in this book Ibn al-Haytham had proved the parabola has the following property: the distance from the focus to the vertex is a quarter of the *latus rectum*. The lost treatise no doubt addressed problems of drawing conics, and in some respects resembled Ibn Sinān's treatise *On Drawing Conic Sections*.[12]

9. *On the Compasses for Conic Sections*

The title, reported by old biobibliographers, suggests the treatise dealt with an instrument like the 'perfect compasses' of al-Qūhī, designed to draw conic sections. The loss of the text has clearly deprived us of a new contribution, in succession to those of al-Qūhī and Ibn Sahl, on the subject of mathematical instruments for drawing conic sections.

10. *On the Determination of all the Conic Sections by Means of an Instrument*

The title is mentioned by Ibn al-Haytham,[13] but no other old author – biobibliographer or mathematician – refers to it. It seems that this treatise addresses the same subject as the previous one, and it is reasonable to ask ourselves whether we may in fact be looking at the same work we have already mentioned under its correct title, but which is described here in terms of its uses.

This variety of titles indicates that Ibn al-Haytham was engaging with the properties of conic sections required in their application to problems. Thus he deals with ways of drawing them, with properties of position and shape, and not only with metrical properties – that is to say he considers

---

[11] See The List of Ibn al-Haytham's Works, no. 9, in R. Rashed, *Ibn al-Haytham and Analytical Mathematics,* London, 2012, p. 394.

[12] Text, translation and commentary in R. Rashed and H. Bellosta, *Ibrāhīm ibn Sinān. Logique et géométrie au Xe siècle*, Leiden, 2000, Chapter III.

[13] See The List of Ibn al-Haytham's Works, no. 37, in R. Rashed, *Ibn al-Haytham and Analytical Mathematics*, p. 402.

everything that appears to be necessary for the geometrical constructions to be carried out when conics are used in solving problems. We shall give a systematic account of the different groups of works – that is, of course, of those whose texts have come down to us.

CHAPTER I

THEORY OF CONICS AND GEOMETRICAL CONSTRUCTIONS: 'COMPLETION OF THE CONICS'

1.1. INTRODUCTION

1.1.1. *Ibn al-Haytham and Apollonius'* Conics

The theory of conics – and this point is worth emphasizing – had never held such an important position before the ninth and tenth centuries. This new importance is not only on account of the properties of the curves, but also because they can they can be applied in areas not foreseen by the original mathematicians, specifically not by Archimedes and Apollonius. That is, the theory of conics has ceased to be simply a powerful instrument in the hands of geometers; it now offers algebraists a means of solving cubic equations.[1] Ibn al-Haytham is no exception to the rule. As a geometer he investigates the geometrical properties of conics – as we may see, for example, in his treatise *The Measurement of the Paraboloid*. As a physicist he is concerned with reflection properties of some of the curves – we may think, for example, of his treatise *Parabolic Burning Mirrors*. He also uses conics to achieve success in problems of geometrical construction. In short, a glance at his works will show that conics and their properties run through them from beginning to end. So it is easy to understand his taking an interest in Apollonius' treatise, the more so since – the works of his predecessors having been lost – Apollonius was the only authority. No other Greek mathematical work, apart from Euclid's *Elements*, was so much consulted, studied and cited by Arabic mathematicians, among them Ibn al-Haytham. Not only did he know the *Conics* down to the smallest details, but he even went so far as to transcribe the work, as we know from the copy in his handwriting that has survived down the centuries.[2]

---

[1] R. Rashed, 'Algebra', in R. Rashed (ed.), *Encyclopedia of the History of Arabic Science*, London, 1996, vol. II: *Mathematics and the Physical Sciences*, pp. 31–54.

[2] Apollonius, *Conics*, photographic reproduction from MS Aya Sofya 2762 by M. Nazim Terzioğlu, Istanbul, 1981.

Ibn al-Haytham, as a mathematician and, this once, a copyist, knew the history of the text of the Arabic version of Apollonius' treatise. This history is recorded by the people who oversaw the search for Greek manuscripts and then their translation, the Banū Mūsā. In a text Ibn al-Haytham knew well (since he made several corrections to it)[3] the Banū Mūsā say that they had a first version made up of the first seven books, which was hard to understand, and that later on, while in Damascus, Aḥmad ibn Mūsā found Eutocius' version of the first four books, which is key for understanding the complete work.[4] As we shall see later, the Banū Mūsā report that of the eight books to which Apollonius refers in his preamble to the *Conics*, they (the Banū Mūsā) had in their hands only the first seven, and the eighth was missing. The absence of this book instigated Ibn al-Haytham's project: to make a conjectural reconstruction of the missing book, so as to 'complete' the *Conics*. Thus we must examine possible traces of this eighth book, so that we can then try to understand what was implied by the term 'completion'. Before we make a detailed examination of the significance of Ibn al-Haytham's project, we need to know whether there were any indications concerning the eighth book that might have offered guidance in his researches.

1.1.2. *The eighth book of the* Conics

There are two passages in the *Conics* in which Apollonius refers to the eighth book: the first is in the Preamble, where he describes the contents of his treatise; and a second time in the seventh book.[5] Commentators on the *Conics* are not very helpful in this respect: Hypatia's commentary is lost, Serenus of Antinoë tells us nothing, and Eutocius says no more than Apollonius did. There remain the lemmas for Apollonius' *Conics* that Pappus presents in his *Mathematical Collection*. To which we should add an allusion by a tenth-century biobibliographer, al-Nadīm, who mentions 'four propositions' of the eighth book. It is this story, already repeated more than once, that we shall need to re-examine a little critically.

In fact, the eighth book of the *Conics* seems to have disappeared a long time ago, though no one can say for certain when: after Pappus, there is no

---

[3] Banū Mūsā, *Muqaddamāt Kitāb al-Makhrūṭāt*, MS Istanbul, Aya Sofya 4832, fols 223ᵛ–226ᵛ. A critical edition and French translation of the book by the Banū Mūsā is given in R. Rashed, *Apollonius: Les Coniques*, tome 1.1: *Livre I*, Berlin/New York, 2008, pp. 483–533.

[4] See below, pp. 23–5.

[5] See below pp. 27ff.

doubt; before him, possibly. In the *Mathematical Collection*, Pappus proposes to prove some lemmas to complete the account given by Apollonius in the *Conics*.[6] Pappus draws up lemmas, in succession, for the first book (eleven lemmas), then for the second (thirteen lemmas), the third (also thirteen lemmas) – no lemmas follow the fourth book – for the fifth (ten lemmas), and finally the sixth (eleven lemmas). Thus far, Pappus distinguishes and clearly separates each group of lemmas according to the book to which it applies. Then he finishes his exposition with the lemmas for the seventh and eighth books, taken together. This anomaly seems incomprehensible: why, after carefully separating the lemmas for each book up to the sixth, should he amalgamate the sets of lemmas for the last two books? One might suggest that Pappus to some extent made a habit of this, since in his *Porisms* he combined the lemmas for three books. But that is not the same sort of thing, since in dealing with the *Conics* he had begun by separating the lemmas, whereas in the *Porisms* he put them all together. But an anomaly, if one may say so, never appears on its own. We should have expected Pappus to have used the plural when he refers to two books. But he does not, and the definite article is in the singular; he writes τοῦ *ZH* '<lemmas> for 7 and 8'. Is this a simple copyist's error, or an indication that the text he was handling initially mentioned only one book? There is no textual argument that provides a reasonable reply. All that can be said is that this text raises the problem of what Pappus may have known about the eighth book of the *Conics*, and that it seems the solution to this problem cannot be found in the history of the text.

There is another possible approach: to match up the lemmas and the propositions in Apollonius' seventh book, which has come down to us. This approach is not as easy as it seems, nor indeed so conclusive, for at least two reasons. After the work done by Heiberg, we know that even for the first three books of the *Conics*, which survive in the edition of Eutocius, Pappus'

---

[6] Pappus drew up seventy lemmas for the books of the *Conics*, which appear in the seventh book of his *Mathematical Collection* (see F. Hultsch, vol. II, pp. 636ff.; *Apollonius Pergaeus*, ed. J. L. Heiberg, 2 vols, Leipzig, 1891–1893; repr. Stuttgart, 1974, vol. II, p. 143ff.; *La Collection mathématique*, French transl. P. Ver Eecke, 2 vols, Paris, 1982, vol. II, p. 718; A. Jones, *Pappus of Alexandria*, Book 7 of the Collection, pp. 296ff.). These lemmas are, obviously, very important for writing the history of the textual tradition of the first four books of the *Conics* (see M. Decorps-Foulquier, *Les Coniques d'Apollonius de Perge*, Thesis, University of Lille, 1994, pp. 51ff.; published as *Recherches sur les* Coniques *d'Apollonios de Pergè*, Paris, 2000, pp. 52–9 and pp. 237–65). More than half these lemmas are intended to fill in missing steps in a demonstration by Apollonius, that is a jump in the exposition made deliberately by the author. A reader familiar with the *Elements*, as he should be, would be perfectly capable of filling in the gap for himself.

lemmas are scarcely helpful in reconstructing the propositions. This state of affairs is in no way peculiar to the *Conics*, but derives from Pappus' editing. Moreover, Paul Tannery experienced similar difficulties in regard to Apollonius' *Plane Loci*. The second difficulty arises from the fact that the seventh book of the *Conics* does not survive in Greek, which means that we cannot be confident in making use of philological arguments. As for the eighth book, not the slightest trace of it has survived in any language: a state of affairs that provides fertile ground for the growth of legends. With these warnings in mind, let us nevertheless examine Pappus' lemmas; perhaps they may yield some elements of a response to the question of what Pappus might have known of the eighth book.

We are considering fourteen lemmas in all, referring to the seventh and eighth book. As we noted, Pappus does not indicate where the lemmas for the seventh book begin or end; nor does he say how many lemmas there are. This leaves us completely and inescapably incapable of determining how many propositions Pappus may have known from the eighth book. So all we can do is to try to identify the lemmas that are proposed for the seventh book.

In the first two of this group of fourteen lemmas, Pappus in fact considers two cases of a figure for a single lemma. At first glance, it appears that these lemmas may apply to Proposition 5 of Book VII of the *Conics*,[7] but closer inspection shows the situation is more complicated. It is in fact a third case of the figure, a case Pappus does not mention, that applies to Proposition VII.5. Further, applying it then gives Pythagoras' theorem, from which the required conclusion can be deduced. So the method is very different from that of Apollonius. But let us begin by proving these assertions.

---

[7] Let there be a parabola $\mathscr{P}$ with axis AH, vertex A, and let BI be any diameter, $A\Gamma = c$ the *latus rectum* with respect to the axis and $BZ \perp AH$. The *latus rectum* with respect to the diameter BI is then equal to $c_1 = A\Gamma + 4AZ$. For the text and a French translation, see *Apollonius: Les Coniques*, tome 4: *Livres VI et VII*, ed. R. Rashed, Berlin/New York, 2009, pp. 362–3.

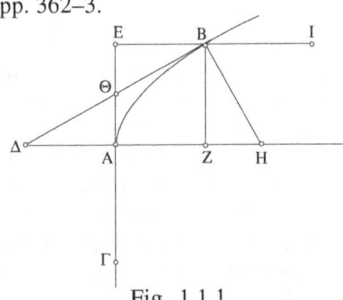

Fig. 1.1.1

In the two lemmas, Pappus introduces a rectangle $AB\Gamma\Delta$ and a straight line from the point A passes through the rectangle in the first lemma, and lies completely outside the rectangle in the second lemma. He shows that

(1) $$AE \cdot AZ = \Delta E \cdot \Delta\Gamma + BZ \cdot B\Gamma.$$

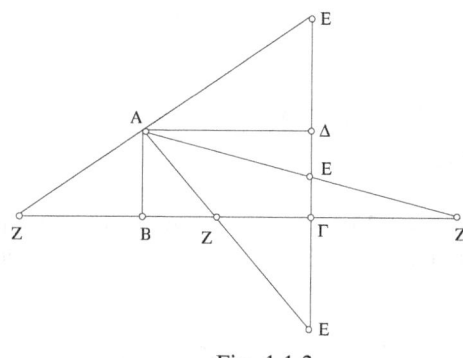

Fig. 1.1.2

The position of the straight line can be defined by E, its point of intersection with the straight line $\Gamma\Delta$, and the possible cases for the figure are as follows:

1. E lies between Δ and Γ; Z then lies beyond Γ on the line $B\Gamma$.
2. E lies beyond Γ; Z then lies between B and Γ, which corresponds to the figure for Pappus' Lemma 1.
3. E lies beyond Δ; Z then lies beyond B, which corresponds to the figure for Pappus' Lemma 2.

So we have three cases for the straight line from A, and in all three figures the reasoning and the conclusion are the same, that is we arrive at equation (1).

The first case of the figure – E between Γ and Δ – is not in Pappus' text, and it is nevertheless the one that corresponds to the figure for Apollonius' Proposition VII.5, where E is the midpoint of $\Gamma\Delta$. However, one can deal with Proposition VII.5 by using Pappus' lemma, in two different ways, without involving the normal, BH, which features in Apollonius' method. but the reasoning is then different from that of Apollonius in his proposition (see the figure for *Conics*, VII.5).

We may consider:

a) The rectangle $AEBZ$ together with the straight line $B\Delta$, the tangent to the parabola at B, and passing through the vertex B of the rectangle (*Conics*, VII.5).

b) The rectangle $AEBZ$ together with the straight line AI from the vertex A of the rectangle and parallel to the line $B\Delta$. The line AI is then the ordinate of A relative to the diameter BI, and we have $AI = B\Delta$; Pappus' lemma gives

a) $B\Theta \cdot B\Delta = E\Theta \cdot EA + ZA \cdot Z\Delta$

b) $AT \cdot AI = EB \cdot EI + ZT \cdot ZB$.

But, from *Conics* I.35, A is the midpoint of $Z\Delta$; so Θ is also the midpoint of $B\Delta$, and T the midpoint of BZ. So both a) and b) are *special cases* of Pappus' lemma:

a) $\Rightarrow B\Delta^2 = EA^2 + Z\Delta^2 \Rightarrow B\Delta^2 = BZ^2 + Z\Delta^2$

b) $\Rightarrow AI^2 = EI^2 + ZB^2 \Rightarrow AI^2 = EI^2 + EA^2$;

that is to say we have Pythagoras' theorem applied to triangle $B\Delta Z$ in case a) and to triangle AEI in case b), results that could have been obtained without Pappus' lemma.

If the *latus rectum* relative to the axis AZ is called c_0 and the *latus rectum* relative to the diameter BI is c, we have

$$B\Delta^2 = AI^2 = c \cdot BI = c \cdot A\Delta = c \cdot AZ,$$
$$BZ^2 = EA^2 = c_0 \cdot AZ,$$
$$Z\Delta^2 = EI^2 = 4 \cdot AZ^2;$$

so from a) and from b) we deduce

$$c \cdot AZ = c_0 \cdot AZ + 4 \cdot AZ^2 \text{ , hence } c = c_0 + 4 \cdot AZ.$$

This rather long discussion shows that the relationship between Pappus' first two lemmas and *Conics* VII.5 is so tenuous that (if one did not already know them) it would be impossible to deduce the statement and content of the proposition from the lemmas alone. Such indeterminacy should, if we care for rigour, prevent us saying this lemma is the one that belongs to that proposition. The most we can say is that Pappus started from the figure for

VII.5, or a figure like it, and proved a general metrical relation that in the case found in Apollonius reduces to Pythagoras' theorem.

Lemma 3 – of which Lemma 4 is only a special case – makes it possible to prove a result concerning the sum of two conjugate diameters of a hyperbola, the subject of Proposition VII.25. When applied to two conjugate diameters of a hyperbola, the hypothesis of Lemma 3 leads to the property proved in Proposition VII.13. So we accept it as a hypothesis and the conclusion of Lemma 3 then directly gives the conclusion of Proposition VII.25. On the other hand, the special case that can be deduced from Lemma 4 is not mentioned by Apollonius. Let us expand these statements a little.

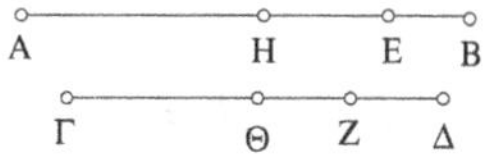

Fig. 1.1.3

In Lemma 3, Pappus considers the case in which $AB > \Gamma\Delta$, $AH = HB$, $\Gamma\Theta = \Theta\Delta$, $E \in [HB]$, $Z \in [\Theta\Delta]$ such that $AE \cdot EB = \Gamma Z \cdot Z\Delta \Rightarrow AE > \Gamma Z$.

Proposition VII.25 investigates the sum of two conjugate diameters of a hyperbola.

Let (d_1, d_1') and (d_2, d_2') be two pairs of conjugate diameters; we know that if $d_1 > d_1'$, we also have $d_2 > d_2'$.

Let us assume $d_2 > d_1 > d_1'$, and let us put $AH = HB = d_2$ and $HE = d_2'$; $\Gamma\Theta = \Theta\Delta = d_1$ and $\Theta Z = d_1'$; we have

$$AE \cdot EB = \Gamma Z \cdot Z\Delta \Leftrightarrow HA^2 - HE^2 = \Theta\Gamma^2 - \Theta Z^2 \Leftrightarrow d_2^2 - d_2'^2 = d_1^2 - d_1'^2,$$

which is the same as the conclusion of Proposition VII.13. We have

$$AE \cdot EB = \Gamma Z \cdot Z\Delta \Rightarrow AE > \Gamma Z.$$

But

$$AE = d_2 + d_2' \text{ and } \Gamma Z = d_1 + d_1'.$$

So we have

$$d_2 + d_2' > d_1 + d_1'.$$

Let the axes of the hyperbola be d_0 and d_0'; we have $d_1 > d_0$, so

$$d_1 + d_1' > d_0 + d_0',$$

and consequently

$$d_0 + d_0' < d_1 + d_1' < d_2 + d_2'.$$

In Lemma 4, Pappus assumes $AH = \Gamma\Theta$, and in this case

$$AE \cdot EB = \Gamma Z \cdot Z\Delta \Rightarrow AE = \Gamma Z.$$

In a hyperbola, two diameters d_1 and d_2 are equal if they are symmetrically placed with respect to one or other of the two axes; if this is so, their conjugates are also symmetrical, and we have $d_1' = d_2'$, hence

$$d_1 + d_1' = d_2 + d_2'.$$

We have $AH = \Gamma\Theta \Rightarrow AE = \Gamma Z \Leftrightarrow d_1 = d_2 \Rightarrow d_1 + d_1' = d_2 + d_2'.$
This special case is not mentioned by Apollonius.

Finally, let us note that if we had $d_1 < d_1'$, we should also have $d_2 < d_2'$, in which case we could apply Lemma 3 by making $AH = HB = d_2'$ and $HE = d_2$; $\Gamma\Theta = \Theta\Delta = d_1'$ and $OZ = d_1$. We should then have

$$AE \cdot EB = \Gamma Z \cdot Z\Delta \Leftrightarrow d_2'^2 - d_2^2 = d_1'^2 - d_1^2,$$

and the remainder of the argument would be the same as before.

So we have now identified the conditions for Lemmas 3 and 4 to be lemmas for Proposition VII.25 of the *Conics*. The first condition is to fulfil the conditions for Proposition VII.13, to which Pappus does not make the slightest reference. The second condition, if we want to speak of perfect correspondence between the lemmas and propositions, would be our possessing a text of the *Conics* in which Apollonius considered the case to which Lemma 4 corresponds, that is the case where there is symmetry with respect to the axes. Now, no such text exists, except in the conjectural circumstances of Pappus perhaps having in his hands a text of the *Conics* that was different from the one we now have available. But it remains impossible to prove that Pappus is indeed referring to this proposition by Apollonius, even if the suggestion is not entirely implausible.

When we turn to Pappus' Lemma 5, we find ourselves in an equally indeterminate situation. One might deduce that the lemma was intended for Proposition VII.27 on the ellipse. But Apollonius does not even give a proof; he apparently regards the result as an obvious consequence of Proposition VII.24, to which his readers are referred.[8] Thus if we agree that Lemma 5 is

---

[8] *Apollonius: Les Coniques*, tome 4: *Livres VI et VII*, pp. 420–1.

intended to apply to that proposition, we shall also have to agree that the same lemma makes it possible to provide a proof that Apollonius never gave.

We can rewrite Pappus' Lemma 5 as follows. Let there be four line segments, respectively of lengths a, b, c, d:

$$a > b, d > c, a > c \text{ and } b > d \Rightarrow a - c > b - d.$$

Let us turn to Proposition VII.27. Let AB and CD be the axes of the ellipse, and let $AB = d_0$, $CD = d_0'$, with $d_0 > d_0'$. Let NM be the diameter that is equal to its conjugate and let $\Gamma\Delta$ be any diameter. Let us make $\Gamma\Delta = d$; and let its conjugate be d'.

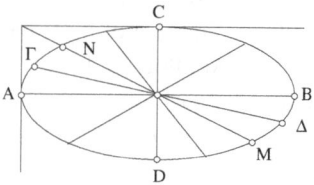

Fig. 1.1.4

– If Γ lies on the arc AN, we have

$$d_0 > d > d' > d_0'.$$

In this case, Lemma 5 gives

$$d_0 - d_0' > d - d'.$$

– If Γ lies on the arc NC, we have $d' > d$; so

$$d_0 > d' > d > d_0',$$

and Lemma 5 gives

$$d_0 - d_0' > d' - d.$$

– If Γ is at point N, we have $d = d'$, $d - d' = 0$. So the general result is

$$d_0 - d_0' > |d - d'|;$$

this proof does not depend on Proposition VII.24.

To say that Lemma 5 is a lemma for Proposition VII.27 is thus to imply either that Pappus, for a reason we do not know, had not seen that the proposition is obvious from VII.24, or that he simply wanted to extract a metrical relationship from the set-up considered in VII.27, without reference to the proposition itself. There is nothing in the text to favour one of these possibilities over the other, and we are more or less faced with just one more lemma, unnecessary if perhaps not without purpose.

Lemma 6 gives an equally trivial result concerning two similar rectangles: if the ratio between corresponding sides is 2, the ratio of the areas is 4. In this lemma Pappus introduces four line segments of lengths a, b, c and d, respectively, such that $c = 2a$, $d = 2b$, and shows that $c \cdot d = 4$ $a \cdot b$.

At no point does Pappus extend this result to the obvious case of parallelograms, still less the general case. Now in Proposition VII.31, from which one might approach the lemma, we find the case of a rectangle, defined by the axes of the ellipse, and that of a parallelogram, defined by the two conjugate diameters. Here is Apollonius' text:

> Four times the parallelogram ΘH, that is the parallelogram HM, is equal to four times the rectangle $A\Theta$ by $\Theta\Gamma$, which is equal to the rectangle enclosed by the axes AB, $\Gamma\Delta$.[9]

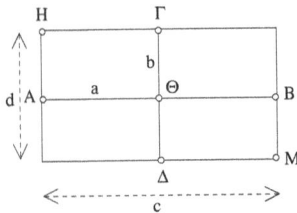 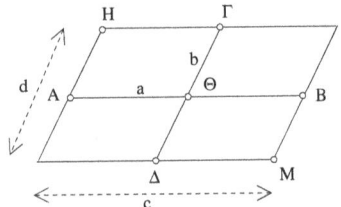

Fig. 1.1.5

So if Lemma 6 were intended for Proposition VII.31, the proposition itself would have had to have partly failed in its purpose.

Lemmas 7, 9 and 11 deal with similar divisions, a matter not treated in the seventh book. Lemma 9 is moreover a special case of Lemma 7; Lemma 11 is the converse of Lemma 9. Lemmas 8 and 10 are inserted between 7 and 9 and 9 and 11, respectively, for no reason, and the properties they establish are completely trivial. That is to say one can see no possible use for Lemmas 8 and 10, nor any reason why they occur where they do. For example, let us consider Lemma 7.

---

[9] *Apollonius*: *Les Coniques*, tome 4: *Livres VI et VII*, p. 432; Arabic text, p. 433, 10–12.

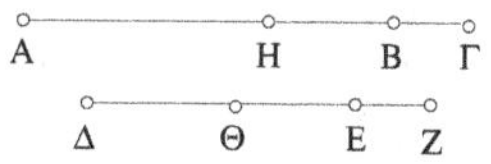

Fig. 1.1.6

Let us assume that

$$\frac{\Delta E}{EZ} = \frac{AB}{B\Gamma} \text{ and } \frac{\Delta E}{E\Theta} = \frac{AB}{BH};$$

we have

(1)
$$\frac{\Delta E}{AB} = \frac{EZ}{B\Gamma} = \frac{E\Theta}{BH} = k;$$

so (A, H, B, Γ) and (Δ, Θ, E, Z) are two similar ranges of points.
From (1) we deduce

(2)
$$\frac{\Delta\Theta}{AH} = \frac{\Theta Z}{H\Gamma} = k.$$

Now
$$(1) \Rightarrow \frac{\Delta E \cdot E\Theta}{AB \cdot BH} = k^2 \text{ and } (2) \Rightarrow \frac{\Delta\Theta \cdot \Theta Z}{AH \cdot H\Gamma} = k^2,$$

hence

$$\frac{\Delta\Theta \cdot \Theta Z}{AH \cdot H\Gamma} = \frac{\Delta E \cdot E\Theta}{AB \cdot BH}.$$

In Lemma 9, B is the midpoint of $A\Gamma$, and E is the midpoint of ΔZ. That is, $AB = B\Gamma$ and $\Delta E = EZ$; moreover $\frac{EZ}{E\Theta} = \frac{B\Gamma}{BH}$. We show that

$$\frac{\Theta\Delta \cdot \Theta E}{ZE \cdot Z\Theta} = \frac{HA \cdot HB}{\Gamma B \cdot \Gamma H}.$$

So we have a special case of Lemma 7.

Lemma 11 is the converse of Lemma 9; it can be written

$$(AB = B\Gamma, \Delta E = EZ \text{ and } \frac{\Theta\Delta \cdot \Theta E}{ZE \cdot Z\Theta} = \frac{HA \cdot HB}{\Gamma B \cdot \Gamma H})$$

$$\Rightarrow \frac{EZ}{E\Theta} = \frac{B\Gamma}{BH} \Rightarrow \frac{\Delta E}{AB} = \frac{EZ}{B\Gamma} = \frac{E\Theta}{BH} = k.$$

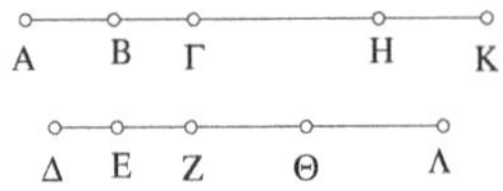

Fig. 1.1.7

In Lemma 9, we can write the relation (1) of Lemma 7, and the proof is immediate. In Lemma 11, we put in a point K on the line AB and a point Λ on the line ΔE such that the ranges (A, B, Γ, K) and $(\Delta, Z, \Theta, \Lambda)$ are similar.

We then deduce from this that the divisions (A, B, Γ, H) and (Δ, E, Z, Θ) are also similar.

Lemma 8, inserted between 7 and 9, reduces to saying that if $AB^2 + B\Gamma^2$ and $AB^2 - B\Gamma^2$ are known, then AB and $B\Gamma$ are known. And Lemma 10, inserted between 9 and 11, can be written

$$\text{if } AB = B\Gamma \text{ and } B\Delta < BE, \text{ then } \frac{A\Delta \cdot \Delta B}{B\Gamma \cdot \Gamma \Delta} < \frac{\Gamma E \cdot EB}{BA \cdot AE}.$$

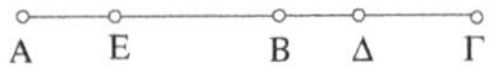

Fig. 1.1.8

It is not clear why Lemmas 8 and 10, which give results that are obvious, should be inserted between lemmas connected with similar divisions.

Let us put these last two lemmas aside and consider all the Lemmas 9, 11, 12, 13 and 14. We may note that all of them involve two line segments and their midpoints – $A\Gamma$ with midpoint B and ΔZ with midpoint E – and a point H on the straight line $A\Gamma$ together with a point Θ on the line ΔZ. In Lemmas 9 and 11, we have similar divisions, and thus a series of equal ratios. In Lemmas 12, 13 and 14 the hypotheses and the conclusions relate to unequal ratios. Thus, from the hypotheses of Lemma 9 we have

$$\Delta E = EZ, AB = \Gamma B \text{ and } \underset{(1)}{\frac{\Delta E}{AB}} = \underset{(2)}{\frac{EZ}{\Gamma B}} = \underset{(3)}{\frac{E\Theta}{BH}} = \underset{(4)}{\frac{Z\Theta}{\Gamma H}} = \underset{(5)}{\frac{\Delta\Theta}{AH}} = k.$$

12) If $\Delta E = EZ$, $AB = \Gamma B$ and $\dfrac{Z\Theta}{\Gamma H} > \dfrac{EZ}{\Gamma B}$, so $\dfrac{\Delta\Theta}{AH} < \dfrac{EZ}{B\Gamma}$ if H lies between B and Γ, and Θ lies between E and Z, or $\dfrac{\Delta\Theta}{AH} > \dfrac{EZ}{B\Gamma}$ if H lies beyond Γ and Θ lies beyond Z.

That is, with the hypotheses of Lemma 9 and the hypothesis (4) > (2), we have (5) < (2) or (5) > (2), according to the position of *H*.

13) If $\Delta E = EZ$, $AB = \Gamma B$ and $\dfrac{\Delta\Theta}{AH} < \dfrac{\Theta E}{HB}$, so $\dfrac{Z\Theta}{\Gamma H} < \dfrac{EZ}{B\Gamma}$, if *H* lies beyond Γ and Θ lies beyond *Z*.

That is, with the hypotheses of Lemma 9 and the hypothesis (5) < (3), we have (4) > (2).

14) If we make the same hypotheses, then $\dfrac{Z\Theta}{\Gamma H} < \dfrac{E\Theta}{BH}$ if *H* lies between *B* and Γ and Θ lies between *E* and *Z*.

We have given summaries of eight lemmas, three of which concern similar divisions, two are obvious, and three deal with divisions in unequal ratios. The five lemmas consider two line segments and their midpoints. We know that when their midpoints coincide, the segments correspond to two diameters of a conic. But it seems that – unless I am mistaken – the seventh book of the *Conics* contains no proposition in which one might make use of any of these lemmas. They could be used, for example, in considering tangents and their intersection with diameters that do not pass through their point of contact with the conic section. But that does not help us much.

To recap: examining the first six lemmas has shown us, if we can identify it from mere hints, a relationship with some propositions of the seventh book of the *Conics*, but the connection is often tenuous and fragile: the case that is relevant to Proposition VII.5 is not considered in the first two lemmas, and, if we decide to ignore that, we have seen that using Lemma 1 gives us Pythagoras' theorem. Lemma 3 can be employed in Proposition VII.25 only on condition that we also use Proposition VII.13, to which no reference is made; Lemma 5 could be useful in a proof of Proposition VII.27, but no proof is given by Apollonius; Lemma 6 is inadequate for VII.31, because it does not consider parallelograms. From Lemma 7 onwards, as we have just seen, it seems there is no relationship with propositions of the seventh book, not even by implication. These lemmas cannot be constructed as components of Apollonius' proofs, instead they appear to be commentaries on the theory of conics, or spin-offs from it, taking up only metrical relationships. On this evidence, we can come to no conclusion about what Pappus may have known of the eighth book of the *Conics*.

Given the inadequacy of the evidence, and the extent of our uncertainty, it would be rash (to say the least) to make a snap judgement about the relationship between Pappus' lemmas and Apollonius' propositions. There is nothing to justify drawing a line dividing the last two books of the *Conics*, passing between Lemmas 6 and 7. Halley's reconstruction of the lost book, for his edition of 1710, could only have been based on pure conjecture, and is of no historical significance. There are also further arguments that should encourage us to exercise prudence.

The first concerns the version of the *Conics* that Pappus knew. He writes:

> The eight books of Apollonius' *Conics* have four hundred and eighty-seven theorems, or figures, and seventy lemmas.[10]

Counting the number of propositions in the seven books, as they survive, in the version due to Eutocius and in the Arabic version, gives a total that differs from this by about a hundred propositions. If Pappus did know the eighth book of the *Conics*, and if his count was true, other things being equal, the eighth book must have contained about a hundred propositions. This would be surprising, because in the other books – including the fifth, which is by far the longest – Apollonius never exceeded 78 propositions. This unexpected feature would at least have been remarked upon by Pappus or by some commentator; but that does not happen. There are various possible, and equally probable, explanations for Pappus' total: a copyist's error, propositions interpolated in Pappus' version that made up this number, a version different from the one that has come down to us, and so on. As for the seventy lemmas, they could equally well be lemmas included in the version of the *Conics* Pappus had in his hands or the seventy-two lemmas found in the *Mathematical Collection*.

The second additional argument concerns the nature of the lemmas. We have not properly understood the guiding principle of Pappus' lemmas for the *Conics*. We have not propositions conceived as parts of the proof that Apollonius may have omitted; rather than that, what we have are commentaries, as it were in counterpoint to Apollonius. In most of the lemmas, Pappus presents commentaries derived from the theory of conics that deal only with metrical relations. Thus we have commentaries on the relations that play a part in Apollonius' proofs, that is, commentaries on isolated properties that are usually intended to draw attention to a metrical property. Almost nothing remains of the properties of conics themselves. Perhaps this is the main reason we find it so difficult to establish secure connections between Pappus' lemmas and Apollonius' propositions.

---

[10] Pappus, *Collection*, Book VII, ed. Hultsch, vol. II, pp. 682, *l.* 21–23 (French transl. P. Ver Eecke, vol. II, p. 512): Ἔχει δὲ τὰ η′ βιϐλία τῶν Ἀπολλωνίου κωνικῶν θεωρήματα, ἤτοι διαγράμματα υπζ′, λήμματα δὲ ἤτοι λαμϐανόμενά ἐστιν εἰς αὐτὰ ο′.

The third argument relates to the content of Lemmas 7 to 14. Lemmas 8 and 10 are not only trivial but also give no clue as to their function, or to why they are on this side of the supposed demarcation line. Lemma 8, for example, tells us that if we have two magnitudes a and b and if $a^2 + b^2$ and $a^2 - b^2$ are known, then a and b are known. So why is Lemma 8 inserted between Lemma 7 and the special case of the same result, that is Lemma 9? It is not clear, either, why Lemma 10, which is trivial, appears between Lemma 9 and its converse, Lemma 11.

Moreover, it should be emphasized that there can be no doubt that Lemmas 7, 9 and 11 relate to propositions in which similar divisions make an appearance, and that the set of lemmas in the group 7, 9, 11, 12, 13, 14 could be useful if one were concerned with two diameters of a conic section.

Thus, however one looks at the evidence provided by Pappus, there is no reasonable justification, that is no sufficiently convincing argument, for extracting from it any useful information about the state of the *Conics* in the copy he used, still less about what Pappus knew of the eighth book. We cannot even be sure that Pappus possessed the whole of this last book. Did he really know it? Did he know some propositions that involved similar divisions? It is, at best, only the last of these possibilities that can be regarded as acceptable, until we have more information. There is no hope of our obtaining it either from the lost commentary by Hypatia, or from the book by Serenus of Antinoë, or even from Eutocius. Unfortunately, everything suggests that Book VIII of the *Conics* was essentially lost, probably in the course of the centuries that separate Apollonius from Pappus. The Arabic tradition provides further confirmation of this conclusion.

We are well informed about the part played by the three brothers Banū Mūsā in searching out Greek manuscripts of the *Conics* and in producing translations. The nature of their contribution has become clearer from the mathematical writings of the youngest brother, al-Ḥasan, which are now better known.[11] In addition to Eutocius' version of the first four books, the Banū Mūsā also possessed another version of seven books of the *Conics*. Thus they had a copy of seven books of the *Conics* that had been made before the ninth century. This copy did not include the eighth book. Here is what they say in an essay they wrote as an introduction for a reader of the *Conics*:

وقد كان وقع إلينا سبع مقالات من الثماني المقالات التي وضعها أبلونيوس.

---

[11] R. Rashed, *Les Mathématiques infinitésimales du IXe au XIe siècle*. Vol. I: *Fondateurs et commentateurs: Banū Mūsā, Thābit ibn Qurra, Ibn Sinān, al-Khāzin, al-Qūhī, Ibn al-Samḥ, Ibn Hūd*, London, 1996.

There have come down to us seven of the eight books composed by Apollonius.[12]

The absence of the eighth book of the *Conics* was confirmed, in more detail, by the tenth-century biobibliographer al-Nadīm. He writes:

وقال بنو موسى: إن الكتاب ثمان مقالات، والموجود منه سبع مقالات وبعض الثامنة. وترجم الأربع المقالات الأولة بين يدي أحمد بن موسى، هلال بن أبي هلال الحمصي، والثلاثة الأواخر، ثابت بن قرة الحراني. والذي يصاب من المقالة الثامنة أربعة أشكال.

The Banū Mūsā said: the work is in eight books; seven and a part of the eighth survive. The first four books were translated in the circle of Aḥmad ibn Mūsā, by Hilāl ibn Abī Hilāl al-Ḥimṣī, and the last three by Thābit ibn Qurra of Ḥarrān. What we possess of the eighth book is four propositions.[13]

Obviously, al-Nadīm is not quoting the Banū Mūsā *verbatim*. In those of their writings that are known today, the Banū Mūsā do not mention these four propositions of the eighth book. On the other hand, the figure is so precise, and the passage we have quoted contains so much that can be substantiated, that we cannot neglect the evidence it offers. The Banū Mūsā in fact speak of only seven books. Thus, Aḥmad ibn Mūsā could have obtained Eutocius' version of the first four books and, they write,[14] 'he managed, with the help of that, to understand the remaining three books of the seven books' (*fa-amkana bi-dhālika fahm al-thalāth al-maqālāt al-bāqiya min al-sab' al-maqālāt*) or again 'The man who undertook the task of translating the remaining three books was Thābit ibn Qurra of Ḥarrān, the geometer' (*wa-kāna al-mutawālī li-tarjamat al-thalāth al-maqālāt al-bāqiya Thābit ibn Qurra al-Ḥarrānī al-muhandis*).

In the writings of the Banū Mūsā, then, it is only a question of seven books, and there is no trace of the four propositions from the eighth book. Where could al-Nadīm have got his information about the four propositions from the eighth book? As he had no direct access to the Greek tradition, his information must come from an Arabic text – possibly one translated from Greek. There is no guarantee this source existed, and we have no idea what it may have been.

The biobibliographers and mathematicians who succeeded al-Nadīm add nothing of substance, except as echoes of the interest taken in the lost book. Al-Qifṭī does no more than pass on what was said by al-Nadīm, but from

---

[12] *Apollonius: Les Coniques*, tome 1.1: *Livre I*, p. 504; Arabic text, p. 505, 1–2.

[13] Al-Nadīm, *Kitāb al-fihrist*, ed. R. Tajaddud, Teheran, 1971, p. 326.

[14] *Apollonius: Les Coniques*, tome 1.1: *Livre I*, p. 506; Arabic text p. 507, 6–7 and 13–14.

his fictionalized account it emerges that at the end of the twelfth century and the beginning of the thirteenth scholars were still looking for the eighth book:

> When books were brought from the Byzantine empire to al-Ma'mūn (the Abbasid caliph), from this book (the *Conics*) they brought the first part, comprising seven books, and no more. When it was translated, the introduction indicated that the work was in eight books and that the eighth book included the concepts of the seven books and more, and that he (Apollonius) had set out useful propositions together with desirable results. From that time until now specialists have been seeking to find this book without hearing anything further of it.[15]

Other old biobibliographers add nothing very important to these stories. They repeat the statements of the Banū Mūsā, that is, that the seven books were passed on and translated into Arabic.

Commentators on the *Conics* had access to only seven books, and provided no information about the eighth. This is true of Naṣīr al-Dīn al-Ṭūsī,[16] al-Iṣfahānī,[17] al-Shīrāzī[18] and al-Yazdī.[19] Only al-Maghribī writes about the eighth book:

أقول: أما هذه المقالة فغير موجودة، بل وجد أشكالها بلا مصادرات، ولم تعلم التراجمة على ماذا
تدل من المسائل، فأهملوها، وبقي من الكتاب سبع مقالات.

> I say: as for this last book, it does not exist <any longer>, but its propositions have been found without their statements, the translators did not know which problems they [the propositions] refer to, so they have neglected it [the book] and only seven books of the work remain.[20]

It is clear that al-Maghribī is contributing no significant information beyond the fact that the eighth book was not translated into Arabic. He does no more than make a conjecture, put forward as an argument, to explain why the book is missing.

So we may state without risk of contradiction that from the ninth century onwards no mathematician ever mentioned any proposition from the eighth book – either commentators on the *Conics* or readers, for

---

[15] Al-Qifṭī, *Ta'rīkh al-ḥukamā'*, ed. J. Lippert, Leipzig, 1903, p. 61.

[16] *Taḥrīr kitāb al-Makhrūtāt*, MS Dublin, Chester Beatty 3076; London, India Office 924.

[17] *Talkhīṣ al-Makhrūtāt,* MS Aya Sofia 2724.

[18] Abū al-Ḥusayn ʿAbd al-Malik ibn Muḥammad al-Shīrāzī, *Kitāb Taṣaffuḥ al-Makhrūtāt*, MS Istanbul, Ahmed III, 3463; Carullah 1507; Yeni Cami 803.

[19] MS Edinburgh, Or. 28.

[20] Ibn Abī al-Shukr al-Maghribī, *Sharḥ Kitāb Abulūniyūs fī al-Makhrūtāt*, MS Teheran, Sepahsalar 556, fol. 2ᵛ.

example Ibn al-Haytham or Ibn Abī Jarrāda. Only al-Nadīm refers to these 'four propositions'. Was he repeating what he had found in an ancient source that had been translated into Arabic?

So, all in all, taking into account both Greek and Arabic traditions, it seems that from the eighth book, by then long lost, some propositions survived, and were probably known to Pappus, and, by one means or another, an echo of them reached into Arabic. The small number of these propositions – four according to al-Nadīm – perhaps explains the anomalies in the main lines of the history of this book. This conjectural explanation has no pretensions to being either definitive or highly plausible. It is one possibility among others. The only certainty, as we have already stated, is that the eighth book was lost, either completely or in great part, some time in the course of the centuries that separate Pappus from Apollonius, or even before Pappus.

But once again we need to discuss the possible contents of the eighth book. And here again, we are reduced to making conjectures. The one most widely accepted is that of T. L. Heath:

> It is probable enough that the book contained a number of problems having for their object the finding of conjugate diameters in a given conic such that Halley attempted a restoration of the Book.[21]

The same explanation is favoured by other eminent historians, G. Loria and H. G. Zeuthen, but it is not the only one that is conceivable. One would have an equal right to defend a completely different opinion. First let us remind ourselves that in the fourth book of his *Mathematical Collection* Pappus seems to indicate that the problem of trisecting an angle, posed in the fourth century BC, was not solved. He writes:

> The earliest geometers could not find a solution to the aforementioned problem concerning the angle, a problem whose nature is three-dimensional (solid), when they sought it by plane methods, for the sections of the cone were not yet familiar to them, and it is because of this that they (the geometers) remained uncertain. They did however achieve trisection of the angle later on, after, in order to find it, having made use of verging which we describe above.[22]

Looking at it from this point of view, we might think the eighth book was concerned with using conics to solve problems involving three-dimensional (solid) *neuseis*, such as those Pappus refers to in the fourth book of the *Collection*. The book on *neuseis* attributed to Apollonius would

---

[21] T. L. Heath, *A History of Greek Mathematics*, vol. II, p. 175.

[22] Pappus d'Alexandrie, *La Collection mathématique*, French transl. P. Ver Eecke, vol. I, p. 209.

then be seen as complementary to *Conics* Book VIII, dealing only with plane problems. In this interpretation we can see similarities between Ibn al-Haytham's reconstruction and Apollonius' eighth book.

In the present state of knowledge, we have no reason to favour either of these conjectural explanations over the other. Lack of information can only give free rein to beliefs. In any case it is in these conditions and with this hypothesis that Ibn al-Haytham conceived his project of 'completing' the *Conics*. But exactly what are we to understand by such 'Completion'?

1.1.3. *The* Completion of the Conics: *the purpose of the enterprise*

It scarcely needs to be said that the *Completion of the Work the Conics* is not a commentary on Apollonius' *Conics*, in any reasonable sense of the term 'commentary'. To convince oneself of this one needs only to compare this book by Ibn al-Haytham with the Greek and Arabic commentaries on the *Conics* – or on several books of the work – that is, the commentaries of Eutocius, of al-Ṭūsī, of al-Maghribī, of al-Shīrāzī, of al-Iṣfahānī, and so on. Moreover, the matter could hardly be otherwise. The *Completion*, unlike commentaries, deals with a book that Ibn al-Haytham has never read. That is, in the absence of a text, there was nothing to comment on.

The *Completion* accordingly belongs to a literary genre distinct from the commentary, that of the reconstruction of a lost text. Finally, we do not know of any attempt to reconstruct this lost book, either in Greek or in Arabic, earlier than the eleventh century. So the *Completion* represents the birth of a literary genre, a new type of mathematical writing whose character we must now examine.

Ibn al-Haytham knew almost nothing about the book he was reconstructing, except for some brief remarks by Apollonius. In the preamble to the first book of the *Conics*, we read that the eighth book τὸ δὲ προβλημάτων κωνικῶν διωρισμένων (ed. Heiberg, p. 4), which in the Arabic translation becomes *al-masā'il allatī taqa'u fī al-makhrūṭāt*.

It is clear that the translator has rendered the verb διορίζειν by the verb *waqa'a*. The immediate and standard translation for the Arabic expression would then be 'the problems that occur in cones', which would, to say the least, appear surprising coming from someone like Hilāl ibn Abī Hilāl al-Ḥimṣī or Isḥāq ibn Ḥunayn. Their knowledge of Greek and of Arabic would have prevented them from abbreviating in this way. So we need to recollect that the senses of the verb *waqa'a* include 'to occur infallibly'.[23]

---

[23] As in the verse in the Koran: ‏.إنَّ عَذَابَ رَبّك لواقع‏

The element of necessity affects the sense of the phrase, which can then be translated 'the problems that necessarily arise (occur) in cones'. A second meaning for this same verb *waqaʿa* is 'firmly established, fixed, determined', as in *waqaʿa al-ḥukm* ('the judgement has been firmly established').[24] Apollonius' phrase would thus mean 'the problems <whose solutions are> firmly established in the conics'. In both cases, we find the sense 'determined' and thus Apollonius' phrase would mean 'the problems <whose solutions are> determined in cones'. The verb διορίζειν means first and foremost 'to set limits on', in the concrete and the abstract senses of the term. Following Aristotle, the dialectico-logical sense ('to determine', 'to define') is extremely widespread, and was not completely displaced by the well-known mathematical sense: 'to find and describe in what and how many ways a question can be resolved'.[25] The Arabic translator thus seems to have opted for the dialectico-logical sense.

The second allusion to the eighth book appears in the preamble to the seventh book. Apollonius first refers to the research in this latter book and writes:

وفي هذه المقالة أشياء كثيرة غريبة حسنة في أمر الأقطار والأشكال التي تعمل عليها مفصلة.

In this book (the seventh), there are many things, surprising and beautiful things, concerning diameters and the figures constructed on them, given in detail.[26]

He continues:

وجميع ذلك عظيم المنفعة في أجناس كثيرة من المسائل، والحاجة إليه شديدة فيما يقع من المسائل في قطوع المخروطات التي ذكرنا مما يجري ذكره وبيانه في المقالة ح من هذا الكتاب.

Here again, there are two possibilities, depending on whether or not the verb *waqaʿa* is still being used to translate the same Greek verb as before. The second hypothesis is, however, the more plausible one, given the method of translation. We would then read:

All this is of great use in many kinds of problems, and we have great need of it in the problems that are determined (lit.: determine themselves) in conic sections, which we have mentioned, as being those that will be set

---

[24] As in the verse of the Koran: فَوَقَعَ الْحَقُّ 'The truth establishes itself firmly'.

[25] Ch. Mugler, *Dictionnaire historique de la terminologie géométrique des Grecs*, Études et commentaires, XXVIII, Paris, 1958, p. 141.

[26] *Apollonius: Les Coniques*, tome 4: *Livres VI et VII*, edited and translated from the Arabic text, and with a historical and mathematical commentary by R. Rashed, Berlin/New York, 2009, p. 350.

out and demonstrated in book VIII of this treatise, which is the last book.[27]

These pieces of information, it must be acknowledged, provide a very slender basis for any kind of reconstitution. Ibn al-Haytham did not have at his disposal any trace or scrap of text on which to exercise the skills of a historian or those of an archaeologist. Moreover, he seems to have known nothing of the four propositions referred to by al-Nadim. So we cannot avoid asking the following question: what, in these circumstances, is the significance of the act of reconstructing a mathematical text about which one knows nothing, and which, furthermore, was written twelve centuries earlier? This enterprise which, at least at first glance, may appear hopeless, for example as a piece of historical research, was nevertheless undertaken by Ibn al-Haytham, and continued to seem attractive to mathematicians who succeeded him, sharpening their imaginations and sometimes even their creative activity. We may think of Maurolico and the *Conics*, of Fermat and Apollonius' *Plane loci*, of Albert Girard and Euclid's *Porisms*, and so on. For all of them, starting with Ibn al-Haytham, the act of 'reconstruction' is in no way an act of restoration. The mathematician is neither an archaeologist nor a historian. Moreover – and this is the second trait that all these attempts have in common – this reconstruction is carried out in accordance with the criteria of rigorous proof. It is not to be equated with the reconstruction of a philosophical system of any kind. In that case the philosopher works up the missing part of a theory so that the theory can be presented in a coherent manner: fundamentally, we have commentary, direct or disguised. For his part, the mathematician must invent, and give rigorous proofs for, the propositions that strengthen the ancient work by extending it. This is so for Ibn al-Haytham as well as for Fermat, for Albert Girard, and, more or less, for Maurolico. We may note that the term 'completion' that Ibn al-Haytham uses has two closely connected senses: 'to complete', to remedy omissions that may be due to Apollonius himself; 'to make perfect', for establishing the internal consistency of the theory of conics. Ibn al-Haytham speaks of *tamām* (perfecting), which carries an implication of a striving for perfection, for accomplishment.[28] He is thus

---

[27] *Ibid.*, p. 350; Arabic p. 351, 7–9.

[28] In the *Kitāb al-'ayn* we read: 'The completion of any thing is what completes it to its limit' ‏. تتمَّة كل شيء، ما يكون تمامًا لغايته‎

In the Koran we read, al-Mā'ida 3: 'Today I have perfected your religion for you, perfected my good deed for you' ‏. اليوم أكملتُ لكم دينكم وأقمت عليكم نعمتي‎

(*Cont. on next page*)

engaged in two forms of approach simultaneously, heuristic and structural. And we cannot, in fact, understand the purpose or the development of his work if we fail to take into account the double nature of the act of 'completing'. Thus, after having pointed out that we lack some 'notions' (*ma'ānī*), that is to say propositions and theorems, 'that should not have been lacking in this work [the *Conics*]', he guesses, of course without any evidence, since there is none, that

> these notions that are missing from these seven books are the notions contained in the eighth book and Apollonius left them until the end because he had not needed to use them for the notions he introduced in the <first> seven books. These ideas we have mentioned are notions required by the notions introduced in the seven books.[29]

Too little attention has been paid to what Ibn al-Haytham says here: he explicitly refers to the set of seven books of the *Conics*, and not to only the seventh book – and, moreover, the examples he chooses are taken from the second book of the work.

The passage we have just quoted gives the clearest possible account of what Ibn al-Haytham intends by reconstructing the eighth book: he wants to find the propositions required by the ones Apollonius proved in the course of the first seven books; to give proofs of them, and thus provide a more solid structure for the *Conics*. We can discern the outlines of the plan that governs the *Completion* and sheds light on the choice of title as well as on the method Ibn al-Haytham employs: to carry out new mathematical research, based on the results obtained in the first seven books, so as to perfect the logical structure of Apollonius' exposition. In this sense the activity of reconstruction is that of active research. But where is the originality to be found, if indeed there is any? At present, all we know is that there is nothing in this undertaking that guarantees that this research, even if it is expressed in terms used by Apollonius, is being carried out precisely within his conceptual world (his *mathesis*), or in his style. In regard to style, Ibn al-Haytham's choice is unambiguous. We know that, throughout the seven books, Apollonius' style is that of pure synthesis. Apart from the problems at the end of Book II, which are all construction problems (44 to 53), for which Apollonius uses analysis and synthesis, we should seek in vain to find anywhere else the slightest allusion to any preliminary analysis. Even

---

(*Cont.*) There are numerous instances in the Koran and in classical poetry that attest to this sense of 'bring something to perfection so that it lacks nothing and has no fault', 'to take it to its limit and perfect it'. It would be mistaken to understand this term *tatimma* as indicating a simple complement.

[29] *Les Mathématiques infinitésimales,* vol. III, Arabic text p. 147, 9–12; English version in this volume p. 173.

the fifth book – the most analytical one – shares the same character. This preference for synthesis makes it impossible for us to speculate about the style of the eighth book. It might have been devoted to construction problems for which Apollonius used analysis and synthesis. But that is not necessarily the case: all the other construction problems, apart from the group at the end of the second book that we have already referred to, were presented in terms of synthesis. Would he, for a reason we do no know, have proceeded in this eighth book to make use of analysis and synthesis, thereby departing from the method he followed in the rest of his work? No serious argument can be put forward in favour of such a conjecture. If something like that had been so, it would have been, above all, book five that would have called for this special treatment. In short, there is no reason to support such a hypothesis in regard to the content of the eighth book and its style. All we can be sure of is that Ibn al-Haytham himself read the first seven books of the *Conics* in the form in which we now know them, that is in their synthetic style of exposition. In fact it is precisely this situation that gives his choice such significance. This is how Ibn al-Haytham himself presents this choice:

> Our procedure is to find these concepts [notions] by the use of analysis, synthesis and discussion [diorism] so as to make this book [the *Completion*] the most perfect of them all [the eight books of the *Conics*] in regard to proofs.[30]

We may ask why Ibn al-Haytham adopted a different approach in his version of a book that was proposed as a continuation of the first seven. This so-to-speak stylistic concern seems to be a response to a new demand.

This demand, as we may note in general terms before setting out to analyse it, springs from a mathematical interest that continued to grow and became important towards the end of the tenth century, notably in the works of Ibn al-Haytham: proving as rigorously as possible the *existence* of points of intersection of the curves of conic sections. Hints of such an interest are no doubt already to be found in some Greek works on geometry – perhaps in Eutocius' commentary on Problem II.4 of Archimedes' *The Sphere and the Cylinder* – but it is not until the tenth century, and in particular with Ibn al-Haytham, that such proof becomes systematic and accordingly takes on the appearance of constituting a standard that must be met.[31] Ibn al-Haytham is in fact concerned with the proof of the existence of

---

[30] *Ibid.*, p. 149, 20–21.

[31] This new demand, which had not been noticed, seemed to us to be so important that we have drawn attention to it more than once: 1) 'La construction de l'heptagone

(*Cont. on next page*)

intersections of two conics, using asymptotic and local properties of conics, and in particular points of contact. Of itself, this new demand in regard to proof makes it impossible, even at the level of exposition, to avoid mentioning analysis and synthesis. The emergence of such research work on the existence of solutions and their number, in accordance with concerns of analysis and synthesis, is closely linked with the emergence of systematic research on geometrical constructions employing the intersection of conics. This research, arising from an interest at once geometrical and algebraic, is no longer driven by questions that happen to be raised and problems that are encountered, as it was in Hellenistic times; it now develops systematically, exploring the field of geometrical problems, the great majority of them solid, but also plane problems.

Thus, in this book about Apollonius' *Conics*, Ibn al-Haytham is especially concerned with geometrical constructions relating to conic sections: the construction of tangents, diameters, *latera recta* and so on, assuming that we know ratios, products, sums and differences of two of these segments. In the course of this work, Ibn al-Haytham employs conic sections to construct solutions not only to solid problems, but also to plane problems. So we meet, in turn, solid problems whose solutions are constructed by means of conics, plane problems constructed by means of conics and plane problems constructed by means of ruler and compasses. Insufficient attention has been paid to this important fact that, moreover, suggests that construction by means of conics had become an acceptable method in geometry, since it is legitimate both for solid problems and for plane problems.

In the problems it considers, in the methods used and in the style it employs, the *Completion of the Work the Conics* belongs to the new area of studies of geometrical construction, in which seeds were undoubtedly sown by Greek mathematicians, to be cultivated by mathematicians of the end of the tenth century, before it became a separate area in its own right, notably in the work of Ibn al-Haytham.

1.1.4. *History of the text*

Ibn al-Haytham's book exists in a single manuscript that is part of an important collection in the Library in Manisa, Turkey – cat. no. 1706. The collection itself contains seventeen treatises, fifteen of which are concerned with mathematics or astronomy. It begins with the commentary on

---

(*Cont.*) régulier par Ibn al-Haytham', *Journal for the History of Arabic Science*, 3, 1979, pp. 309–87; 2) 'La philosophie mathématique d'Ibn al-Haytham. I: L'analyse et la synthèse', *MIDEO*, 20, 1991, pp. 31–231.

Menelaus' *Sphærica* by the thirteenth-century mathematician, Ibn Abī Jarrāda. This commentary is followed by several 'additions' on the same subject. Next comes a short essay (incomplete) on Proposition X.1 of the *Elements*, then a commentary on Apollonius' *Conics*, also incomplete at the beginning and at the end. Most of these texts have been transcribed by the same hand, and the folios are numbered continuously, which shows that they came from a single collection. This first group is immediately followed by another, transcribed by another hand and with the folios numbered differently. The first treatise in this second group is the book by Ibn al-Haytham, which occupies folios 1^v–25^r. Then comes a treatise dictated (but not composed) by Maimonides: *Glosses on Some Propositions of the Conics.*[32] These two treatises are in the same hand, which seems to be more recent than the one that copied the first group of texts. So it seems that Ibn al-Haytham's *Completion* and Maimonides' *Glosses* are derived from another collection. Next comes another treatise, consisting of *Glosses on the Conics*, composed by an anonymous scholar for his personal use, in his own words. The collection continues in the same way, and we find texts transcribed by several hands. For example, one of these treatises was copied at Tabrīz, in Iran, about 699/1300. So everything indicates that this collection was put together from several others, by someone knowledgeable about the mathematical sciences, and who was particularly interested in conic sections.

That is as far as our information goes, in the current state of palaeographic and bibliographic scholarship in Arabic. Thus the history of the text of the *Completion* is very thin: the book was transcribed relatively late, by all indications, the person responsible was someone interested in conics, but the names of the owners do not tell us anything very significant.

So we have a text that has come down to us in a single manuscript, a rather late one. The case is certainly not particularly remarkable, and would not give rise to any problems if the title of the book appeared in one of the lists of Ibn al-Haytham's writings recorded by old biobibliographers, or if the author himself had referred to it in one of the texts we know. Unfortunately, none of this is so, and the situation is clearly favourable for harbouring all kinds of suspicions. That the book is explicitly attributed to al-Ḥasan ibn al-Haytham both in the title and in the colophon is certainly extremely important, but all the same it is not enough to definitively settle the question of attribution. Against that, due weight must be given to the silence of the biobibliographers and that of Ibn al-Haytham himself: a glance

---

[32] R. Rashed, 'Philosophie et mathématiques selon Maïmonide: Le modèle andalou de rencontre philosophique', in *Maïmonide, philosophe et savant (1138–1204)*, Studies collected by Tony Lévy and Roshdi Rashed, Ancient and Classical Sciences and Philosophy, Leuven, 2004, pp. 253–73.

at the tables comparing the lists of works[33] shows that none of the three principal lists is complete – al-Qifṭī, Ibn Abī Uṣaybiʻa, Lahore; and moreover they differ from one another. So the fact that its title does not appear in a list does not *a priori* condemn a work as apocryphal. The same tables also show that Ibn al-Haytham's silence is not a convincing argument for casting doubt on the authenticity of a title; furthermore, the two writings in which he might have referred to the *Completion* are themselves lost: *On the Properties of the Sections* and *On the Construction of the Conic Sections*.

So our question can be narrowed down: here is a text explicitly attributed to al-Ḥasan ibn al-Haytham, without any external evidence that supports or contradicts this attribution. Accordingly, the only approach open to us is to return to the text itself.

The overall structure of the piece and its organization show a style to be found in other works accepted as being by Ibn al-Haytham. He made a habit of beginning by describing what he is aiming to do, the problem he proposes to consider, and then referring to the contribution made by his predecessors, when there was one.

Certainly, Ibn al-Haytham is not the only one to organize his exposition in this way, but there is also the reliable evidence provided by vocabulary: it is indeed that of Ibn al-Haytham. Let us look at some examples:

استقرينا ... وتصفحنا	(*Optics*, p. 62)	استقراء ... وتصفح
تطلع النفوس	(*Analysis and Synthesis*, p. 37, 4)	تسمو النفوس
المعاني التي ذكرناها	(*The Knowns*, p. 151, 2)	المعاني التي لم يذكرها
تمكن هذا المعنى في اعتقادنا وقوي في نفوسنا	(vol. II, p. 83, 7-8)	هذا المعنى هو أحد ما قوى رأي المتفلسفين في اعتقادهم

We can extend our list of examples like these, which are obviously reliable guides to the identity of the author. It suffices to note the very frequent use of *maʻnā/maʻānī*, which is characteristic of the vocabulary and style of Ibn al-Haytham. As for the language, the restricted language, of mathematicians, it is indeed what we find in other writings accepted as being by him, with the important exception of a single expression – *qiṭʻ ṣunūbarī* (section in the shape of a pine cone) – used to designate the parabola, four times, in the *Completion*. Now this term was never used by Ibn al-Haytham in other writings accepted as being his when giving a name to the parabola,

---

[33] R. Rashed, *Les Mathématiques infinitésimales*, vol. II, pp. 512–35.

which is unlike the way the term is used by, for example, his predecessor al-Khāzin. We may note, first, that the word *ṣunūbarī* occurs in the translation of the *Conics* copied out personally by Ibn al-Haytham, in Propositions I.17, I.19, I.20 among others.[34] So it is not impossible that Ibn al-Haytham allowed himself to apply this vocabulary when he composed the *Completion*, it being a book that was meant to follow on from the seventh book of Apollonius and to complete the *Conics*; so that, far from being an argument against the authenticity of the *Completion*, the adoption of the expression 'section in the shape of a pine cone' suggests a possible dating for its composition. The fact that this term *ṣunūbarī* occurs in the *Completion*, and only there, in fact shows a lexical proximity that is closely linked with a proximity in subject matter. It seems as though Ibn al-Haytham the copyist had influenced Ibn al-Haytham the mathematician in his choice of vocabulary. But there is more.

When we examine the copy of the *Conics* that is in the hand of Ibn al-Haytham, in the state in which it has come down to us, we notice that it breaks off towards the end of Proposition XLVIII of Book VII. So we lack the end of this proposition and the four propositions that follow. This loss is by no means a recent one, but dates from before the thirteenth century. This copy in fact belonged to the mathematician Ibn Abī Jarrāda, who made abundant annotations on it; he has written in his hand, in the margin of the last page (fol. 306ᵛ): *baqiya min hadhā al-kitāb al-maqāla al-thāmina* (there remains the eighth book of this work). Now – as is evident from his commentaries on the works of Thābit ibn Qurra[35] – Ibn Abī Jarrāda knew the *Conics* too well not to be aware that the eighth book had not been translated into Arabic. And nevertheless his turn of phrase indicates that the copy he owned indeed included eight books. If our guess is correct, the eighth book might have been none other than the *Completion* by Ibn al-Haytham. Confirmation comes from Maimonides. This twelfth-century philosopher and scholar, who also lived in Cairo, consulted the *Completion*, and even wrote some glosses on certain propositions; now he considered this book to be the last book of the *Conics*. In his glosses Maimonides follows the order in the books of the *Conics* so as to complete some proofs that Apollonius had left to the ingenuity of the reader, this being a matter of

---

[34] MS Aya Sofia 2762. See *Apollonius: Les Coniques,* vol. 1.1: *Livre I,* edited and translated from the Arabic text, and with a historical and mathematical commentary by R. Rashed, Berlin/New York, 2008.

[35] R. Rashed, *Les Mathématiques infinitésimales,* vol. I.

putting in simple intermediate steps.[36] Maimonides makes glosses to the eighth book, which is none other than the *Completion*, and the glosses refer to expressions found in it. So we see that the *Completion* was in circulation between the eleventh and the thirteenth century, passing – at least in the eyes of certain authors – for the eighth book of the *Conics*. However, Ibn al-Haytham cannot be held responsible for this: the introduction to the *Completion* does not admit of any confusion. Where can the misunderstanding have come from? The conjectural explanation that we have just proposed seems to account for all the facts: the exception in vocabulary, the remark made by Ibn Abī Jarrāda, the confusion of Maimonides, the silence of the old biobibliographers.

Ibn al-Haytham may have written the *Completion* immediately after making his copy (or one of his copies) of the *Conics*, and may have put it after the copy. If this was so, he would have written his treatise about 415/1024, that is in the period of his full maturity, which gives a good explanation for the content of this important book on the geometry of conics.

It will be for future research to confirm, correct or disprove this conjecture. For the moment, putting all the evidence together, the arguments we have already set out, to which, as we shall see, we may add considerations of the mathematical content of the book, speak convincingly for the *Completion* having been written by Ibn al-Haytham, and composed while he was copying out the *Conics*.

If we now turn to the copy of the *Completion*, we note that it was transcribed in the elegant *naskhī* script, precise and clear. The figures are drawn with similar care. The few marginal additions are in the hand of the copyist; he made them in the course of his revision, when comparing his copy with the original, since he notes their place in the text with the conventional word *ṣaḥḥ*. The text contains no crossing out and no added glosses.

It was perhaps because of the high quality of this transcription that N. Terzioğlu – the first person to draw attention to this manuscript – provided a photographic reproduction of the text, preceded by a preface and a short introduction. This publication, in 1974,[37] had the great merit of making Ibn al-Haytham's text known and putting it into circulation. In 1981, M. Abdulkabirov[38] provided the first study of the mathematical

---

[36] *Ḥawāshin ʿalā baʿḍ ashkāl kitāb al-Makhrūṭāt*, MS Manisa 1706, fol. 26ᵛ; see R. Rashed, 'Philosophie et mathématiques: Maïmonide et le modèle andalou de rencontre philosophique', in *Maïmonide, philosophe et savant*.

[37] *Das Achte Buch zu den Conica des Apollonius von Perge. Rekonstruiert von Ibn al-Haysam*. Herausgegeben und eingeleitet von N. Terzioğlu, Istanbul, 1974.

[38] In *Matematika i astronomiya v trudakh Ibn Sina, yego sovrenrennikov i posledovatelei*, Tashkent, 1981, pp. 80–94.

content of this book by Ibn al-Haytham, thus making historians of mathematics aware of Ibn al-Haytham's great importance. Three years later, J. P. Hogendijk published a doctoral thesis in which he set out to provide a critical edition, an English translation and an extensive historical and mathematical commentary. This publication had the great advantage of making this book by Ibn al-Haytham known in the West, together with the results he proved.

We have just noted that, perhaps for the very reason that the copy was of such high quality, N. Terzioğlu was content merely to reproduce it. In contrast, J. P. Hogendijk decided it was worth putting out a critical edition.[39] This edition, although it contains errors, is nevertheless an edition. A large number of errors are, moreover, due to a laudable but misguided wish to correct an Arabic text that is, finally, already perfectly correct in the first place.[40]

---

[39] J. P. Hogendijk, *Ibn al-Haytham's Completion of the Conics*, Sources in the History of Mathematics and Physical Sciences 7, New York/Berlin/Heidelberg/Tokyo, 1985.

[40] For the history of the manuscript tradition of the Arabic texts, the critical edition and the French translation, see *Les Mathématiques infinitésimales*, vol. III, pp. 22–6.

1.2. MATHEMATICAL COMMENTARY

The Completion of the Work on Conics has a rather simple overall structure. It starts with a very short introduction, in which the author describes the current state of research and gives a summary of what he intends to do. This introduction, on which we have already commented in the introduction of this chapter, is immediately followed by studies of some geometrical problems, taken in an order that does not seem to reflect didactic considerations. As we shall see later, a difficult problem may be followed by a simpler one. It is in fact the property required that decides the order in the group of problems. Thus in the first problem Ibn al-Haytham requires a point *B* of a parabola with vertex *A* such that the tangent at *B* cuts the axis in *E*, to give *BE/EA* = *k*, a known ratio. The group of problems 2, 3, 4, 5 deal with the same property, but in central conics. Thus is the sense in which the required properties determine the order of the propositions. For each of the problems he considers, Ibn al-Haytham proceeds in the same way: analysis, synthesis and discussion (diorism). But he sometimes leaves the discussion incomplete. This is surprising coming from a mathematician of the stature of Ibn al-Haytham, and it is the less easy to understand for being repeated in other writings – for example in *The Knowns*. Careful examination of the text suggests several explanations, which refer to several orders of incompleteness that, to avoid damaging our argument, we need to distinguish. One of the possible reasons seems to be the objective difficulty of carrying out a discussion of the existence and number of solutions in geometrical terms. The other possible reason, a much more subjective one, is Ibn al-Haytham's not having thought things through before committing himself to using only conic sections, while easier approaches were available, and he knew about them. We should also think about Ibn al-Haytham wanting to make short work in the case of a very easy problem. We shall see these different factors at work later on. But, in order to be able to distinguish all the cases and give the precise conditions for the solution, we have needed to supply two commentaries. There is a geometrical commentary that follows Ibn al-Haytham's mathematics in his own terms; the other commentary is not in his style, but it helps us provide a rigorous discussion when Ibn al-Haytham's is not complete: this commentary is analytic. It need hardly be said that we do not attribute any such treatment to tenth- and eleventh-century mathematicians.

We now turn to an analysis of Ibn al-Haytham's book. The commentary will provide necessary details of the active development of his mathematical ideas. We shall follow the order of exposition in the original.

– **1** – Let there be a parabola *ABC* with axis *AD* and two straight line segments *HI* and *KL* such that $\frac{HI}{KL} = k$ where *k* is a given ratio.

To find the point *B* of the parabola such that the tangent at *B* cuts the axis in a point *E* such that $\frac{BE}{EA} = k$

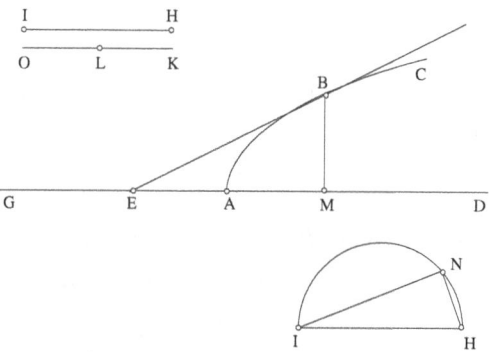

Figs. 1.2.1 and 1.2.2

Analysis: If *BE* is the tangent and *BM* the ordinate, we have *MA* = *AE* (*Conics*, I.35).
We have

$$\frac{BE}{EM} = \frac{BE}{2EA} = \frac{k}{2}$$

and *BM* ⊥ *EM*, so the angle *E* is known. This assumes *k* > 2, so that *HI* > 2*KL*, which is a necessary condition.

– **2** – *Synthesis*: By Proposition 50 of Book II of the *Conics*,[41] we know how to construct a tangent that makes a given angle with the axis.

Let *KO* = 2*LK*, *KO* < *HI*; if in the circle with diameter *HI* we draw a chord *IN* = *KO*, then

$$\frac{IH}{IN} = \frac{k}{2} = \frac{EB}{EM};$$

angle *HIN* is thus the angle made by the required tangent with the axis and we know how to construct this tangent, let it be *BE*. By construction we know that the right-angled triangle *BEM* is similar to triangle *HIN*, so we have

[41] Proposition 50 in Heiberg's edition, Stuttgart, 1974; *Apollonius: Les Coniques*, vol. 1.1 : *Livre I*, ed. R. Rashed.

$$\frac{BE}{EM} = \frac{HI}{IN},$$

hence

$$\frac{BE}{EA} = \frac{HI}{KL} = k.$$

– **3** – Let there be a conic section Γ, an ellipse or a hyperbola, with axis AD, and let G/H be a given ratio, with $G > H$.

To find a point B on Γ such that the tangent at B cuts the axis in the point K and such that

$$\frac{BK}{KA} = \frac{G}{H}.$$

It is clear that, for any point B of the conic, we have $BK > AK$, so the construction of B such that $\frac{BK}{KA} = \frac{G}{H}$ requires that $G > H$.

If Γ is an ellipse, the tangent at the end point B of the axis perpendicular to AB is parallel to AD and K goes to infinity. This can be taken as a limiting case in which $BK = KA$, and the case of the problem in which $G = H$ thus has this solution.

Note: We shall let d denote the length of the diameter in question and let a denote the length of the corresponding *latus rectum*.

Analysis: Let AD be the axis (transverse in the case of the hyperbola), E the centre of the conic Γ, BK the straight line required by the problem and AC the chord parallel to BK; the chord cuts EB in S and we have $SA = SC$.

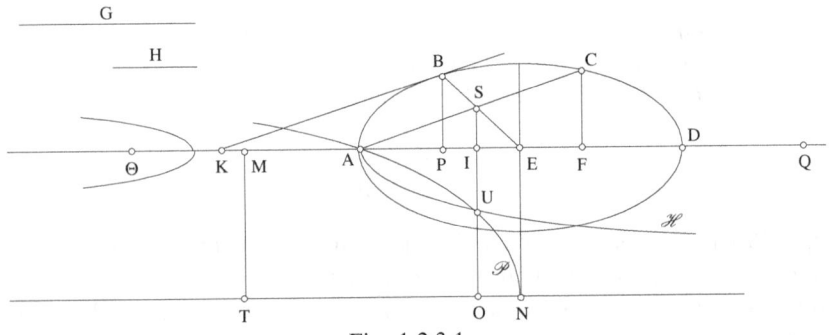

Fig. 1.2.3.1

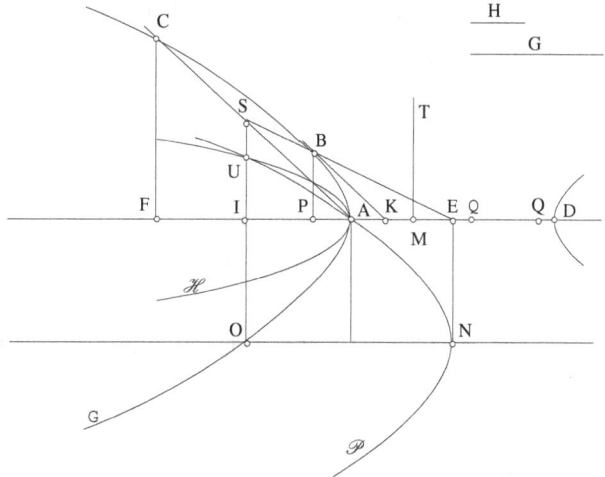

Fig. 1.2.3.2

Let P, I and F be the orthogonal projections of B, S and C on the axis AD. From Proposition 37 of Book I, we have $EA^2 = EK \cdot EP$, hence

(1) $$\frac{PE}{EA} = \frac{EA}{EK} = \frac{PA}{AK},$$

but

$$\frac{AE}{EK} = \frac{SA}{BK},$$

because $SA \parallel BK$.

So we have

$$\frac{SA}{BK} = \frac{PA}{AK},$$

hence

$$\frac{SA}{AP} = \frac{BK}{AK} = \frac{G}{H}.$$

Since SA and BK are parallel, we also have

$$\frac{EA}{EK} = \frac{ES}{EB} = \frac{EI}{EP},$$

so

$$\frac{EI}{EP} = \frac{EP}{EA} \qquad \text{from (1),}$$

hence $EI \cdot EA = EP^2$.

If M is a point on the straight line DA such that $\dfrac{ME}{MA} = \dfrac{d}{a}$, then $\dfrac{ME}{EA}$ is known. We know, from Propositions 2 and 3 of Book VII, that if the point Θ on the straight line AD is such that

$$\frac{\Theta A}{\Theta D} = \frac{a}{d},$$

then

$$\frac{\Theta F \cdot AF}{AC^2} = \frac{\Theta D}{AD}. \;_{42}$$

But S is the midpoint of AC and I is the midpoint of AF and, further, M is the midpoint of ΘA, because

$$\frac{d}{a} = \frac{ME}{MA} = \frac{\Theta D}{\Theta A} \Rightarrow \frac{AE}{AM} = \frac{AD}{\Theta A} = \frac{2AE}{\Theta A} \Rightarrow AM = \frac{1}{2}\Theta A,$$

from which we obtain

$$\frac{\Theta F \cdot AF}{AC^2} = \frac{MI \cdot IA}{AS^2} = \frac{\Theta D}{AD} = \frac{ME}{AE},$$

hence

$$\frac{MI \cdot IA}{AP^2} = \frac{ME}{AE} \cdot \frac{AS^2}{AP^2} = \frac{ME}{AE} \cdot \frac{G^2}{H^2},$$

a known ratio.

Let EN be such that $EN \perp EA$, $EN = EA$ and let NO be parallel to EA. The parabola $\mathscr{P}$ with axis NO, vertex N and *latus rectum* EN passes through A. The straight line SI cuts the parabola in U.

$$OU^2 = EN \cdot ON = EA \cdot ON = EA \cdot EI = EP^2,$$

so we have

---

[42] ΘA is called the homologous line (Apollonius, VII.2–3); ΘD has no name. In the case of the ellipse, Θ is an exterior point of $[AD]$, and we have: if $a < d$, $\Theta D - \Theta A = d$; if $a > d$, $\Theta A - \Theta D = d$. In the case of the hyperbola, Θ lies between A and D, $\Theta A + \Theta D = d$.

(case of the ellipse, $a > d$)

(case of the ellipse, $a < d$)

In all cases, M is the midpoint of $A\Theta$, and $ME = \dfrac{1}{2}\Theta D$.

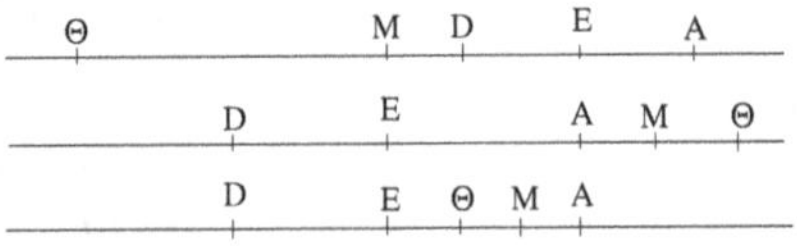

Fig. 1.2.3.3

$$UO = EP,$$

hence

$$UI = AP.^{43}$$

We then have

$$\frac{MI \cdot IA}{UI^2} = \frac{ME}{AE} \cdot \frac{G^2}{H^2}, \text{ a known ratio,}$$

hence U lies on a hyperbola $\mathcal{H}$ whose axis is AM and whose *latus rectum* is known. The point U is thus an intersection of $\mathcal{P}$ and $\mathcal{H}$, so it is known. From U we can find, successively, the points I, F, C, S and B and the line BK parallel to SC.

In the case where AD is the minor axis of the ellipse, the construction is exactly the same, but $\frac{ME}{MA} = \frac{d}{a} < 1$, so the point M (an exterior point of the segment AE) is now on the same side as E. The point I, which lies between A and E, is then between A and M and the auxiliary conic $\mathcal{H}$ that passes through the point U is an ellipse instead of (as before) a hyperbola.

Although Ibn al-Haytham did consider this case, he did not develop it fully, because it is exactly the same as the preceding one.

– **4** – *Synthesis*: Given the conic section (Γ) with axis AD and centre E, and a point M such that $\frac{ME}{MA} = \frac{d}{a}$ is known, MA is known (see Figs. 1.2.3.1 and 1.2.3.2).

As before, we draw the parabola $\mathcal{P}$. If we put

$$\frac{AE}{EQ} = \frac{G^2}{H^2} \quad \text{and} \quad \frac{ME}{EQ} = \frac{AM}{MT},$$

then MT is known, MT is the *latus rectum* of the hyperbola defined above. The equation of $\mathcal{H}$ (Apollonius, *Conics*, I.12) gives

$$\frac{IM \cdot IA}{IU^2} = \frac{ME}{AE} \cdot \frac{G^2}{H^2} = \frac{ME}{AE} \cdot \frac{AE}{EQ} = \frac{ME}{EQ} = \frac{AM}{MT}.$$

If $\mathcal{H}$ and $\mathcal{P}$ cut one another in U (see discussion), we draw $UI \perp AE$; let F be a point such that $AF = 2AI$, let C be the point on the given conic Γ such that $CF \perp AE$, AC cuts the straight line UI in S, we have $AS = SC$, ES is a diameter of Γ and meets Γ in B, and the line through B parallel to AS is a tangent to Γ at B and cuts AE in K. Let us show that

---

[43] We have $UI = OI - OU$ for the ellipse, $UI = OU - OI$ for the hyperbola, if we take U on the semi-parabola NA.

$$\frac{BK}{KA} = \frac{G}{H}.$$

Let $BP \perp EA$, we have $EI \cdot EA = EP^2$ (a property of Γ) and $NO \cdot NE = OU^2$ (the equation of $\mathscr{P}$), hence $UO = EP$ and $UI = AP$. We have

$$\frac{MI \cdot IA}{AS^2} = \frac{ME}{AE} \qquad \text{(a property of } \Gamma\text{)}$$

$$\frac{MI \cdot IA}{IU^2} = \frac{AM}{MT} \qquad \text{(the equation of } \mathscr{H}\text{),}$$

so

$$\frac{AS^2}{IU^2} = \frac{AS^2}{AP^2} = \frac{AM}{MT} \cdot \frac{AE}{ME} = \frac{ME}{EQ} \cdot \frac{AE}{ME} = \frac{AE}{EQ} = \frac{G^2}{H^2},$$

hence

$$\frac{AS}{AP} = \frac{G}{H};$$

but $SA \parallel BK$ implies

$$\frac{AS}{AP} = \frac{BK}{AK},$$

hence

$$\frac{BK}{AK} = \frac{G}{H}.$$

Discussion in the case where Γ is an ellipse with major axis AD. The hyperbola $\mathscr{H}$ has axis AM and *latus rectum MT* (we are considering the branch with vertex A). We have

$$\frac{AM}{MT} = \frac{ME}{EQ} = \frac{ME \cdot EA}{EA \cdot EQ}.$$

We know that

$$\frac{AE}{EQ} = \frac{G^2}{H^2} \quad \text{and} \quad G > H,$$

so $AE > EQ$.
So we have

$$EA \cdot EQ < EA^2 = EN^2,$$

$$\frac{ME \cdot EA}{EN^2} < \frac{AM}{MT}.$$

$\mathscr{H}$ cuts EN between E and N:

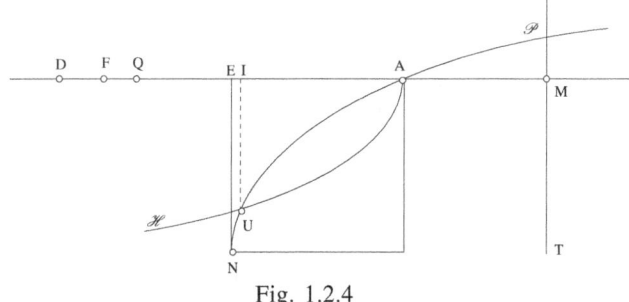

Fig. 1.2.4

$\mathscr{H}$ and $\mathscr{P}$ have their axes parallel, and their concave aspects face one another, $A \in \mathscr{H} \cap \mathscr{P}$, so $\mathscr{H}$ cuts the arc AN of the parabola in U, where U lies between A and N. I, the orthogonal projection of U on the axis lies between A and E, so the point F such that $AF = 2AI$ lies between A and D; there is a point C of the ellipse Γ that corresponds to F, from which we get the point B, the point of contact of a tangent parallel to AC.

The problem can always be solved.

– **5** – *Discussion in the case where Γ is a hyperbola with transverse axis* AD; *we consider the branch with vertex* A, Γ_A. The point M lies between A and D. Let B lie on DA produced and be such that $AB^2 = 2AM \cdot AD$.

Condition for a solution to be possible:

$$\frac{G^2}{H^2} \geq \frac{2AD + 2AM + 3AB}{ME}.$$

On the parabola $\mathscr{P}$, we consider the points: S such that $AS \perp NO$ and U the intersection of SB and $\mathscr{P}$. Now U can be projected orthogonally into the point I on AD, into O on the axis of $\mathscr{P}$ and into V on the line parallel to the axis drawn through S.

We have

$$IO^2 = EA^2$$

$$UO^2 = EA \cdot NO = EA^2 + EA \cdot AI = IO^2 + EA \cdot AI$$

$$EA \cdot AI = UO^2 - IO^2 = (UO + OI) \cdot UI = UV \cdot UI$$

and

$$\frac{UI}{BI} = \frac{UV}{VS} = \frac{UV}{AI} = \frac{EA}{UI},$$

hence

$$UI^2 = EA \cdot BI.$$

We also have

$$\frac{UI}{BI} = \frac{SA}{AB},$$

hence

$$2EA \cdot BI = UI \cdot AB,$$

and in consequence

$$AB = 2UI,$$

hence

$$SA \cdot BI = \frac{1}{2} AB^2 = AM \cdot AD;$$

but

$$SA = AD,$$

hence

$$BI = AM.$$

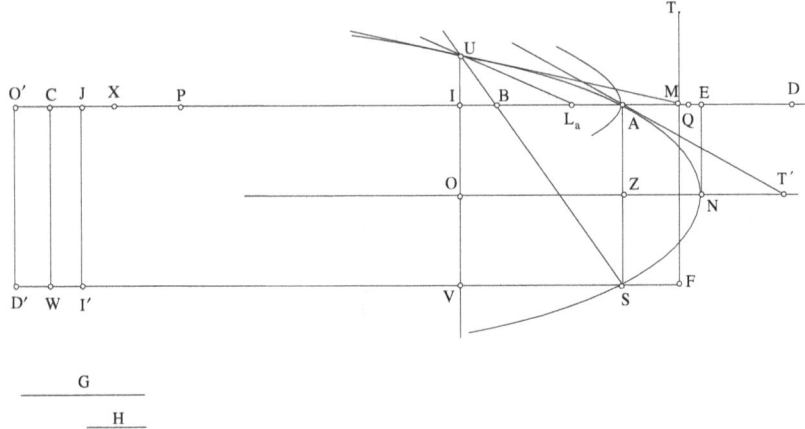

Fig. 1.2.5.1

Let P and C be points such that $IP = 2AD$ and $PC = 2AB$; we then have

$$CM = 2AD + 2AM + 3AB.$$

The condition we imposed becomes

$$\frac{G^2}{H^2} \geq \frac{CM}{ME}.$$

a) Let us assume

$$\frac{G^2}{H^2} = \frac{CM}{ME},$$

so

$$\frac{CM}{ME} = \frac{AE}{EQ}.$$

Let X be the midpoint of PC:

$$PX = AB = 2UI, \ PI = 2AD = 2VI, \ XI = 2UV, \ XI \cdot UI = 2VU \cdot UI.$$

But

$$VU \cdot UI = UO^2 - OI^2 = EA \cdot AI = \frac{1}{2} VI \cdot AI,$$

so

(*) $XI \cdot UI = VI \cdot AI.$

Moreover,

$$CX = AB = 2UI,$$

hence

$$CX \cdot IU = AB \cdot UI = 2UI^2 = 2AE \cdot BI,$$

$$CX \cdot IU = VI \cdot AM.$$

We deduce from this, using (*),

$$CI \cdot IU = VI \cdot MI,$$

hence

$$\frac{CI}{IM} = \frac{VI}{IU} \quad \text{and} \quad \frac{CM}{MI} = \frac{VU}{UI}.$$

But

$$\frac{CM}{AE} = \frac{CM}{MI} \cdot \frac{MI}{AE},$$

and we have

$$\frac{MI}{AE} = \frac{MI \cdot IA}{IA \cdot AE} = \frac{MI \cdot IA}{UV \cdot UI},$$

$$\frac{CM}{MI} = \frac{UV}{UI} = \frac{UV.UI}{UI^2},$$

hence

$$\frac{CM}{AE} = \frac{MI \cdot IA}{UI^2}.$$

We have assumed

$$\frac{CM}{AE} = \frac{ME}{EQ} = \frac{AM}{MT},$$

hence

$$\frac{MI \cdot IA}{UI^2} = \frac{AM}{MT};$$

this relation expresses the fact that U lies on the hyperbola $\mathcal{H}$ with axis AM and *latus rectum MT*. So if $\frac{G^2}{H^2} = \frac{CM}{ME}$, $\mathcal{H}$ and $\mathcal{P}$ have the point U in common, so the problem has at least one solution.

Let us show that there is a second one.

Let AT' be the tangent to the parabola at A and Z the midpoint of AS; we have $ZT' = 2ZN = 2AZ$. Let us draw $UL_a \parallel AT'$, L_a on the straight line AI; triangles UIL_a and AZT' are similar, so $IL_a = 2UI = AB$; we thus have $IB = AL_a = AM$ and consequently the straight line UM is the tangent to the parabola at U.

Let us draw CW and MF perpendicular to VS. From the equality

$$\frac{CM}{MI} = \frac{VU}{UI},$$

we deduce, successively,

$$\frac{MC}{CI} = \frac{UV}{VI}, \quad CM \cdot IV = CI \cdot UV \text{ and } CM \cdot MF = VW \cdot UV;$$

this last equality expresses the fact that the equilateral hyperbola $\mathcal{H}_1$ with asymptotes WF and WC that passes through the point M also passes through U. The segment UM lies inside this hyperbola; but the straight line UM is a tangent to the parabola, so any straight line from U in a position between MU and the tangent to the hyperbola passes inside the parabola; it cuts both the hyperbola $\mathcal{H}_1$ and the parabola $\mathcal{P}$, so $\mathcal{H}_1$ cuts $\mathcal{P}$ in a point situated between A and U. Let that point be U_1.

UM: tangent to $\mathcal{P}$,

Ut: tangent to $\mathcal{H}_1$.

With this point U_1 we associate I_1 and V_1, U_1 being a point of $\mathcal{H}_1$; we have

$$CM \cdot MF = CI_1 \cdot U_1V_1;$$

but

$$MF = I_1V_1,$$

hence

$$\frac{CM}{CI_1} = \frac{U_1V_1}{I_1V_1},$$

from which we deduce

$$\frac{CM}{I_1M} = \frac{U_1V_1}{U_1I_1}.$$

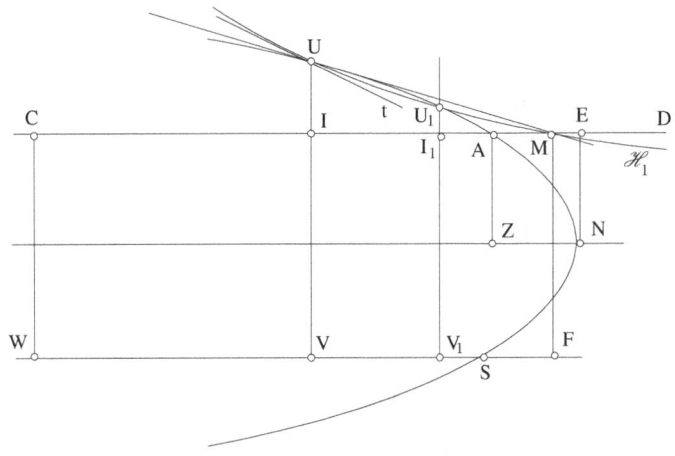

Fig. 1.2.5.2

Proceeding as for point U (pp. 47–8), we show that

$$\frac{MI_1 \cdot AI_1}{U_1I_1^2} = \frac{AM}{MT},$$

so U_1 lies on the hyperbola $\mathcal{H}$.

$\mathcal{H}$ thus cuts $\mathcal{P}$ in two points U and U_1 distinct from A. These two points U and U_1 can be projected onto DA produced beyond the point A; on the branch Γ_A of the given hyperbola there are two points corresponding to U and U_1 (and symmetrical with them with respect to AD), points at which the tangent has the required property.

b) Let us assume $\dfrac{G^2}{H^2} > \dfrac{CM}{ME}$, and let $\dfrac{G^2}{H^2} = \dfrac{O'M}{ME}$, $O'M > CM$. Let $O'D' \perp WF$.

The tangent to the equilateral hyperbola $\mathcal{H}_1$ at M cuts the asymptotes WF and WC in Y and Y_1 and the straight line $D'O$ in Y'; the straight line MU, which joins two points of $\mathcal{H}_1$, cuts these straight lines in R, R_1 and R' respectively; we have

$MY = MY_1$ a property of the tangent,

$MR = UR_1$ a property of the secant MU.
So we have $MY' > MY$ and $UR' > MR$.

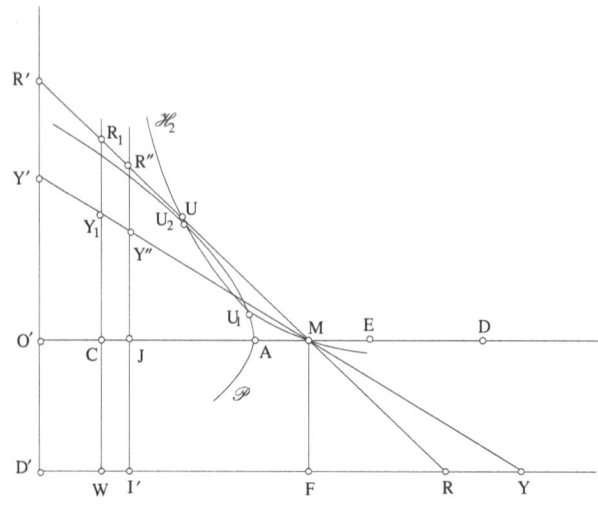

Fig. 1.2.5.3

The equilateral hyperbola $\mathcal{H}_2$ that passes through M and has $D'F$ and $D'O'$ as asymptotes thus cuts the straight line MY in a point of MY' and the straight line MU in a point of UR'. The straight line YY' cuts the parabola $\mathcal{P}$; the hyperbola $\mathcal{H}_2$, which passes through M, thus cuts $\mathcal{P}$ in two points: U_1 lying between A and U, U_2 lying beyond U. We show that U_1 and U_2 are points of the hyperbola $\mathcal{H}$ and we proceed as in the previous case, but replacing CM with $O'M$.

On the branch Γ_A of the given hyperbola there are two points corresponding to U_1 and U_2 (and symmetrical with them with respect to AD), points at which the tangent has the required property.

c) Let us assume $\dfrac{G^2}{H^2} < \dfrac{CM}{ME}$. Let J be such that $\dfrac{G^2}{H^2} = \dfrac{JM}{ME}$; we have $MJ < MC$.

The perpendicular from J to the straight line MC cuts the straight ine WF and the tangent MY in I' and Y'' respectively; we have $MY'' < MY$.

The hyperbola $\mathcal{H}_3$ which passes through M and has $I'J$ and $I'F$ as asymptotes cuts the straight line MY again in a point between M and Y; so it does not cut the parabola $\mathcal{P}$.

Complete analytical study of Problems 3, 4 and 5, independent of the method used by Ibn al-Haytham

I. *The case of the ellipse* with AD as axis. Let us take $AD = d$, $a = latus$ *rectum* corresponding to AD and $k = \dfrac{d}{a}$.

To find B such that the tangent BK satisfies $\dfrac{BK}{KA} = \dfrac{G}{H} > 1$.

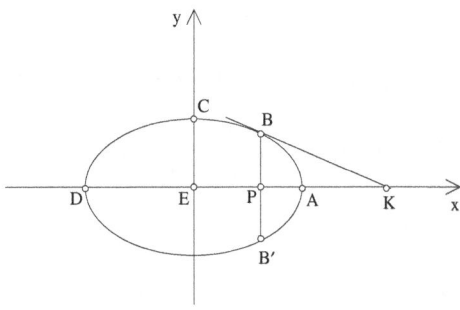

Fig. 1.2.5.4

The equation of the ellipse with respect to its axes is

$$(1) \qquad x^2 + ky^2 = \frac{d^2}{4}.$$

Let $B\,(x, y)$ be such that $0 < x < \dfrac{d}{2}$; from the properties of the tangent, we have

$$\overline{EP} \cdot \overline{EK} = \frac{d^2}{4}, \qquad \text{hence} \qquad \overline{EK} = \frac{d^2}{4x} \qquad (x \neq 0),$$

$$\frac{\overline{EP} \cdot \overline{PK}}{PB^2} = \frac{d}{a}, \qquad \text{hence} \qquad \overline{PK} = \frac{dy^2}{ax},$$

from which we obtain

$$\overline{AK} = \overline{EK} - \overline{EA} = \frac{d}{2x}\left(\frac{d}{2} - x\right)$$

and

$$BK^2 = BP^2 + PK^2 = y^2 + \frac{d^2}{a^2} \cdot \frac{y^4}{x^2} = \frac{y^2}{x^2}\left(x^2 + k^2 y^2\right).$$

So we have

$$\frac{BK^2}{AK^2} = \frac{G^2}{H^2} = \frac{4y^2}{d^2} \cdot \frac{x^2 + k^2 y^2}{\left(\dfrac{d}{2} - x\right)^2}.$$

But from (1)

$$y^2 = \frac{1}{k}\left(\frac{d^2}{4} - x^2\right),$$

so if we assume $B \neq A$, that is $x \neq \frac{d}{2}$, we shall have

$$\frac{G^2}{H^2} = \frac{4}{kd^2} \cdot \frac{\frac{d}{2} + x}{\frac{d}{2} - x}\left[x^2(1-k) + k\frac{d^2}{4}\right].$$

If a point B gives a solution to the problem, its abscissa satisfies the equation

$$4\left(\frac{d}{2} + x\right)\left[x^2(1-k) + k\frac{d^2}{4}\right] - \frac{G^2}{H^2} \cdot kd^2\left(\frac{d}{2} - x\right) = 0.$$

This equation can be written

$$f(x) = 4x^3(1-k) + 2dx^2(1-k) + kd^2x\left(1 + \frac{G^2}{H^2}\right) + k\frac{d^3}{2}\left(1 - \frac{G^2}{H^2}\right) = 0.$$

– If $k = 1$. The ellipse becomes a circle. The equation $f(x) = 0$ is first-degree, and has one root

$$x_0 = \frac{\dfrac{G^2}{H^2} - 1}{\dfrac{G^2}{H^2} + 1} \cdot \frac{d}{2}, \quad \frac{G}{H} > 1 \Rightarrow 0 < x_0 < \frac{d}{2}.$$

The problem has *one* solution. It is clear that the construction can be carried out with straightedge and compasses; a special case that Ibn al-Haytham does not consider, not least because he is dealing with conics. We need to remember that the circle is not considered as a conic until the curves are explicitly defined in terms of their equations.

– If $k \neq 1$.

$$f(0) = \frac{kd^3}{2}\left(1 - \frac{G^2}{H^2}\right) < 0,$$

$$f\left(\frac{d}{2}\right) = d^3 > 0;$$

when $x \to \pm\infty$, $\quad f(x) \cong 4x^3(1-k)$.

– If $k > 1$.

$$\lim_{x \to +\infty} f(x) = -\infty$$

$$\lim_{x \to -\infty} f(x) = +\infty.$$

The equation $f(x) = 0$ then has 3 roots x_1, x_2, x_3

$$x_1 < 0 \qquad 0 < x_2 < \frac{d}{2} \qquad x_3 > \frac{d}{2};$$

the only appropriate root is x_2.

– If $k < 1$.

$$f'(x) = 12(1-k)x^2 + 4(1-k)dx + kd^2\left(\frac{G^2}{H^2} + 1\right).$$

This second-degree polynomial remains positive for $x \geq 0$, so f increases over this domain and $f(x) = 0$ has *only one root* in $\left]0, \frac{d}{2}\right[$.

So, for any given ellipse, for $\frac{G}{H} > 1$, the problem has *one solution* and only one. This reasoning also holds for the case Ibn al-Haytham did not consider, in which AD is the minor axis of the ellipse. In this case, the auxiliary conic involved in the discussion is an ellipse instead of a hyperbola.

The case of the hyperbola with transverse axis AD

$AD = d$, $\dfrac{d}{a} = k$.

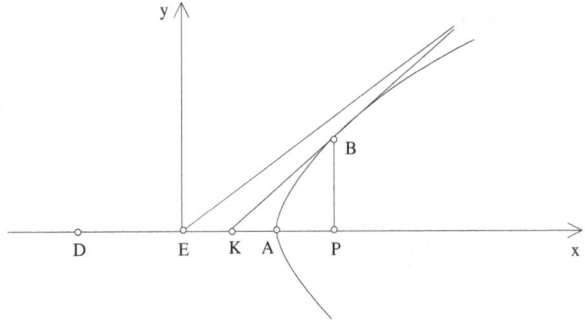

Fig. 1.2.5.5

The equation of the hyperbola is

$$x^2 - ky^2 = \frac{d^2}{4}.$$

Let $B\,(x, y)$, $x > \dfrac{d}{2}$ and let BK be the tangent at B:

$$\overline{EP} \cdot \overline{EK} = \frac{d^2}{4},$$

hence

$$\overline{EK} = \frac{d^2}{4x}, \qquad\qquad 0 < \overline{EK} < \frac{d}{2}$$

and

$$\overline{AK} = \overline{EK} - \overline{EA} = \frac{d}{2x}\left(\frac{d}{2} - x\right) \overline{AK} < 0,$$

$$\frac{\overline{PE} \cdot \overline{PK}}{PB^2} = \frac{d}{a} = k,$$

hence

$$\overline{KP} = k\frac{y^2}{x}$$

and

$$BK^2 = PK^2 + PB^2 = \frac{y^2}{x^2}\left(x^2 + k^2 y^2\right),$$

$$\frac{BK^2}{AK^2} = \frac{G^2}{H^2} = \frac{4y^2}{d^2} \cdot \frac{x^2 + k^2 y^2}{\left(x - \dfrac{d}{2}\right)^2};$$

but

$$y^2 = \frac{1}{k}\left(x^2 - \frac{d^2}{4}\right),$$

hence, assuming $B \neq A$, so $x \neq \dfrac{d}{2}$, we have

$$\frac{G^2}{H^2} = \frac{4}{kd^2} \cdot \frac{x + \dfrac{d}{2}}{x - \dfrac{d}{2}} \cdot \left[x^2(1 + k) - k\frac{d^2}{4}\right],$$

hence

$$f(x) = 4\left(x + \frac{d}{2}\right)\left[x^2(1 + k) - k\frac{d^2}{4}\right] - kd^2 \cdot \frac{G^2}{H^2}\left(x - \frac{d}{2}\right) = 0,$$

$$f(0) = k\frac{d^3}{2}\left(\frac{G^2}{H^2} - 1\right) > 0, \qquad f\left(\frac{d}{2}\right) = d^3 > 0;$$

when $x \to \pm\infty$, $f(x) \cong 4x^3 (1 + k)$, so

$$\lim_{x \to -\infty} f(x) = -\infty$$

$$\lim_{x \to +\infty} f(x) = +\infty.$$

From this we deduce that the equation $f(x) = 0$ has 0 or 2 roots in the interval $\left]\dfrac{d}{2}, +\infty\right[$. The equation can be written:

$$(\alpha) = f(x) = 4x^3(1+k) + 2dx^2(1+k) - kd^2\left(1 + \frac{G^2}{H^2}\right)x + k\frac{d^3}{2}\left(\frac{G^2}{H^2} - 1\right) = 0.$$

We have

$$(\beta) = f'(x) = 12x^2(1+k) + 4dx(1+k) - kd^2\left(1 + \frac{G^2}{H^2}\right).$$

$f'(x) = 0$ has 2 roots x' and x'' with opposite signs, hence

x	$-\infty$		x'	0	x''		$+\infty$
$f'(x)$		$+$	0	$-$	0	$+$	
$f(x)$	$-\infty$	↗	M	↘	m	↗	$+\infty$

The maximum, M, is positive because $M > f(0) > 0$; the equation $f(x) = 0$ thus has one root x_1, $x_1 < x' < 0$. The existence of the other roots depends on the sign of the minima m. For $f(x) = 0$ to have a double root or 2 simple roots in the interval $\left]\dfrac{d}{2}, +\infty\right[$, it is necessary and sufficient that:

1) $x'' > \dfrac{d}{2}$,

2) $m = f(x'') \le 0$.

$x'' > \dfrac{d}{2}$ means that $\dfrac{d}{2}$ lies between the two roots x', x'' of f', that is, that $f'\left(\dfrac{d}{2}\right) < 0$. We have

$$f'\left(\frac{d}{2}\right) = 5d^2(1+k) - kd^2\left(1 + \frac{G^2}{H^2}\right) = d^2\left(5 + 4k - k\frac{G^2}{H^2}\right);$$

the condition can thus be written

(1) $\qquad k\dfrac{G^2}{H^2} > 5 + 4k \iff \dfrac{G^2}{H^2} > \dfrac{5 + 4k}{k}.$

Let $\dfrac{G^2}{H^2} - 4 > \dfrac{5}{k}$. As $M > 0$, the second condition is equivalent to $Mm \le 0$; apart from a positive numerical factor, the first member Mm is the discriminant of the equation. We can find it by the method of remainders for division of f by f' (Euclid, *Elements*, VII); the remainder is

$$-\frac{2d^2x}{9}\left(3k\frac{G^2}{H^2}+4k+1\right)+\frac{kd^3}{9}\left(5\frac{G^2}{H^2}-4\right)=\frac{d^2}{9}(\lambda x+\mu),$$

with

$$\lambda=-2\left(3k\frac{G^2}{H^2}+4k+1\right), \qquad \mu=kd\left(5\frac{G^2}{H^2}-4\right).$$

So we have

$$M=\frac{d^2}{9}(\lambda x'+\mu), \ m=\frac{d^2}{9}(\lambda x''+\mu) \ \text{and} \ Mm=\frac{d^4}{81}\left[\lambda^2 x'x''+\lambda\mu(x'+x'')+\mu^2\right],$$

where

$$x'x''=-\frac{kd^2}{12(1+k)}\left(\frac{G^2}{H^2}+1\right), \ x'+x''=-\frac{d}{3},$$

from equation (β). So

$$\begin{aligned}(\gamma)&=\lambda^2 x'x''+\lambda\mu(x'+x'')+\mu^2\\&=\frac{3kd^2}{1+k}\left(-k^2\frac{G^6}{H^6}+8k^2\frac{G^4}{H^4}-16k^2\frac{G^2}{H^2}+11k\frac{G^4}{H^4}-12k\frac{G^2}{H^2}+\frac{G^2}{H^2}-1\right)\\&=\frac{3kd^2}{1+k}\left[-k^2\frac{G^2}{H^2}\left(\frac{G^2}{H^2}-4\right)^2+k\frac{G^2}{H^2}\left(11\frac{G^2}{H^2}-12\right)+\frac{G^2}{H^2}-1\right].\end{aligned}$$

The figure has been drawn with coordinates $x=\frac{k}{10}$ and $y=\frac{G^2}{H^2}-4$; with these coordinates, the inequality that defines the second condition is

$$(\gamma')=-100x^2y^2(y+4)+10x(y+4)(11y+32)+y+3\le 0,$$

and the first condition can be written $y>\frac{1}{2x}$. If $y>0$, this reduces to $x>\frac{1}{2y}$. Now, putting $x=\frac{1}{2y}$ in the first member of inequality (γ'), we find

$$10(y+4)\left(3+\frac{16}{y}\right)+y+3>0;$$

that is, $\frac{1}{2y}$ lies between the two roots in x of the first member of (γ'), whereas inequality (γ') means that x lies outside the interval defined by the roots. So we see that our two conditions are equivalent to

1) $\dfrac{G}{H} > 2$,

2) $k^2 \dfrac{G^2}{H^2} \left(\dfrac{G^2}{H^2} - 4 \right)^2 - k \dfrac{G^2}{H^2} \left(11 \dfrac{G^2}{H^2} - 12 \right) - \dfrac{G^2}{H^2} + 1 \geq 0$.

The condition given by Ibn al-Haytham can be written

$$\dfrac{G^2}{H^2} \geq 4 + \dfrac{6}{k} \left(1 + \sqrt{1+k} \right),$$

because

$$ME = \dfrac{kd}{2(1+k)}, \quad AM = \dfrac{d}{2(1+k)} \text{ and } AB = \dfrac{d}{\sqrt{1+k}}.$$

In our coordinates, this becomes

$$y \geq \dfrac{3}{5x} \left(1 + \sqrt{1+10x} \right),$$

or

$$x \geq \dfrac{6}{5y} \left(1 + \dfrac{3}{y} \right).$$

Making

$$x = \dfrac{6}{5y} \left(1 + \dfrac{3}{y} \right)$$

in the first member of (γ'), we find

$$-\dfrac{y+3}{y^2} \left(11y^2 + 8 \times 12y + 16 \times 12 \right) < 0.$$

This proves that conditions 1 and 2, which are necessary and sufficient, imply Ibn al-Haytham's condition, which is only sufficient.

We can see, more simply, that the condition for the problem admitting of a solution may be expressed by the fact the ratio $\dfrac{G}{H}$ is equal to or greater than a minimum that depends on k. In fact, the ratio $\dfrac{BK^2}{AK^2}$ is

$$\dfrac{4}{kd^2} \dfrac{x + \dfrac{d}{2}}{x - \dfrac{d}{2}} \left[x^2(1+k) - \dfrac{kd^2}{4} \right].$$

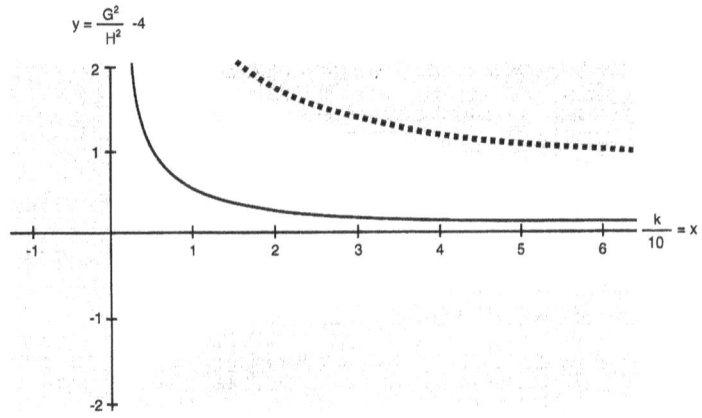

Fig. 1.2.5.6

▭ useful hatched area defined by inequality (γ)

—— curve $\dfrac{G^2}{H^2} - 4 = \dfrac{5}{k}$ or $2xy = 1$

····· curve $\dfrac{G^2}{H^2} - 4 = \dfrac{6}{k}\left(1 + \sqrt{1+k}\right)$;

Ibn al-Haytham's condition means it is above the curve.

Let us examine how this ratio varies when x varies from $\dfrac{d}{2}$ to $+\infty$. Its derivative is

$$\frac{4}{kd^2} \cdot \frac{1}{\left(x - \dfrac{d}{2}\right)^2}\left[-dx^2(1+k) + \frac{kd^2}{4} + 2x(1+k)\left(x^2 - \frac{d^2}{4}\right)\right]$$

$$= \frac{4}{kd^2} \cdot \frac{1}{\left(x - \dfrac{d}{2}\right)^2}\left[2x^3(1+k) - dx^2(1+k) - \frac{d^2 x}{2}(1+k) + \frac{kd^3}{4}\right].$$

The sign of this derivative is that of the third-degree polynomial in the square brackets; the derivative of this polynomial is

$$6x^2(1+k) - 2dx(1+k) - \frac{d^2}{2}(1+k) = 6(1+k)\left(x - \frac{d}{2}\right)\left(x + \frac{d}{6}\right)$$

and it is positive for $x \geq \dfrac{d}{2}$, so the polynomial increases from $-\dfrac{d^3}{4}$ to $+\infty$ and it changes sign once and only once, for a value x_0 of x that gives a minimum of $\dfrac{BK^2}{AK^2}$.

x	$\dfrac{d}{2}$	x_0	$+\infty$
$\dfrac{BK^2}{AK^2}$	$+\infty$ ↘	min.	$+\infty$ ↗

The condition for the problem to have a solution is thus

(*) $$\frac{G^2}{H^2} \geq \frac{4}{kd^2} \cdot \frac{x_0 + \dfrac{d}{2}}{x_0 - \dfrac{d}{2}} \left[x_0^2(1+k) - \frac{kd^2}{4} \right],$$

where x_0 is determined by the conditions

a) $x_0 > \dfrac{d}{2}$,

b) $2x_0^3(1+k) - dx_0^2(1+k) - \dfrac{d^2 x_0}{2}(1+k) + \dfrac{kd^3}{4} = 0.$

If we take account of the last equation, the lower limit on $\dfrac{G^2}{H^2}$ can be written

$$\frac{8(1+k)}{kd^3} x_0 \left(x_0 + \frac{d}{2} \right)^2.$$

We may note that, as $x_0 > \dfrac{d}{2}$, we have

$$\frac{8(1+k)}{kd^3} x_0 \left(x_0 + \frac{d}{2} \right)^2 > 4\frac{1+k}{k} = 4 + \frac{4}{k};$$

in particular, $\dfrac{G}{H} > 2$. Eliminating between (*) and b), we return to inequality (γ).

The discussion by Ibn al-Haytham is based on the idea of replacing the hyperbola ($\mathcal{H}$) by the equilateral hyperbola ($\mathcal{H}_1$), which belongs to the pencil of conics defined by $\mathcal{P}$ and $\mathcal{H}$. This discussion would be complete if it included finding the condition for $\mathcal{P}$ and $\mathcal{H}_1$ to touch one another. As the analytical discussion shows, this condition is difficult to express since it involves terms of the third degree in $\dfrac{G^2}{H^2}$; it is thus at the edge of what can be done in the mathematics of the time. We may suppose that it was on account of this difficulty that Ibn al-Haytham was content to give an incomplete discussion. We may moreover note that Ibn al-Haytham seems to indicate the difficulty: whereas, for the case of the ellipse, he writes that the problem 'can be completed in all cases', for the hyperbola he writes that 'the problem cannot be completed except under one condition and one specification'. He explains the condition, but not the specification.

The incompleteness of Ibn al-Haytham's discussion, which may remind us of a type of error committed by Abū al-Jūd and denounced by al-Khayyām, never the less remains surprising in the work of an author of his calibre. But however matters stand, we need to rule out any interpretation of the ratio $\frac{CM}{CE}$ as being a numerical approximation to the true limit; the mathematical context of this result is purely geometrical so no account can be taken of anything other than exact values.

MC, the tangent to the parabola (which does not depend on $\frac{G}{H}$) makes its appearance naturally; it is impossible to imagine that Ibn al-Haytham confused it with a tangent to the hyperbola.

In general, in these problems, the diorisms are concerned with determining the existence of points of intersection of two conics; the limiting case that separates existence and non-existence of intersection is that given by the point in which the two conics touch one another, as one can see clearly not only in the course of this book by Ibn al-Haytham, but also in other works by him (such as *The Knowns*) and, before him, in the writings of his Arabic and Greek predecessors (for example Eutocius' Commentary on Archimedes' *The Sphere and the Cylinder*, and the fifth book of Apollonius' *Conics*). To make the discussion complete, we need to know how to find the point of contact of the two conics. Finding this point seems a difficult task to carry out geometrically in this case. Let us note what seems to distinguish Ibn al-Haytham from his predecessors in this respect:

– Systematic investigation of constructing solutions to problems using the intersection of conics becomes a separate area of endeavour in geometry. It is no longer a matter of isolated problems appearing sporadically, and being solved using intersection of conics, but rather of a method of exploring the domain of geometrical problems, a large majority of them three-dimensional but also some that are quadratic.

– Within the framework of this new area of investigation, Ibn al-Haytham makes a careful general study concerning the existence and number of solutions, in accordance with his ideas about analysis and synthesis.

This study is based on asymptotic and local properties of conics and in particular tangency.

The algebraic theory developed later by Sharaf al-Dīn al-Ṭūsī is certainly based on this kind of work, but is distinguished from it precisely by the fact that the discussion becomes entirely algebraic. After dealing with the case of the equation $x^3 + c = ax^2$, in which the diorism is the same as that of Eutocius, Sharaf al-Dīn al-Ṭūsī, who is aware of the fact that, in cases near the limiting one, the two solutions are nearly symmetrical with respect to the limiting solution, makes use of his method of affine translation of the unknown[44] to find the derived equation of the second degree, which allows him to find the limiting case for all the other equations under discussion.

Summary of the preceding discussion

We have already noted that Ibn al-Haytham distinguishes the case of the ellipse, in which the problem in question always has a solution, from that of the hyperbola, which requires a condition to be met. The new idea he introduced in this case was to replace the auxiliary hyperbola $\mathcal{H}$ by another hyperbola $\mathcal{H}_1$ belonging to the pencil generated by $\mathcal{P}$ and $\mathcal{H}$. This idea can be exploited in a different way by replacing $\mathcal{H}$ by a second parabola $\mathcal{P}'$ belonging to the same pencil. We shall now summarize what Ibn al-Haytham did by exploiting this idea.

To find the point B of a conic section Γ (an ellipse or hyperbola) with centre E and axis AD, such that the tangent at this point B cuts the axis in K and satisfies the equation $\dfrac{KB}{KA} = \dfrac{G}{H}$, where $\dfrac{G}{H}$ is a given ratio greater than 1.

We note that if a point B gives a solution to the problem, the point symmetric with it with respect to AD, B', also gives a solution. We shall look for B on a semi-conic Γ.

– In Problem 3, Ibn al-Haytham deals with the analysis of this problem. The reasoning is the same whatever the given conic Γ is (an ellipse or a hyperbola), that is, the point M defined by $\dfrac{ME}{MA} = \dfrac{d}{a}$ (where d is the diameter AD and a the corresponding *latus rectum*) will be an exterior point on the segment $[AD]$ in the case of the ellipse, or will lie between A and D in the case of the hyperbola.

In either case, if there exists a tangent BK that gives a solution to the problem, then there exists a point U on a hyperbola $\mathcal{H}$ and on a semi-parabola $\mathcal{P}$ defined as above.

– In Problem 4, Ibn al-Haytham shows that if U is a point common to $\mathcal{H}$ and $\mathcal{P}$, then there is corresponding to it a point B of Γ and a tangent BK that satisfies the relation $\dfrac{KB}{KA} = \dfrac{G}{H}$.

---

[44] That is an affine translation in which $x \rightarrow x \pm a$.

But examining the intersection of $\mathscr{H}$ and $\mathscr{P}$ demands a *discussion* that allows us to separate the two cases:

In Problem 4: the case of the ellipse.

In Problem 5: the case of the hyperbola.

Analytic study of the discussion: d and a designate respectively the axis AD and the corresponding *latus rectum*.

Case of the ellipse. To follow Ibn al-Haytham, we shall assume $d > a$, so $\dfrac{d}{a} = k > 1$.

Let us take A as the origin $A(0, 0)$, and let us have $D(0, d)$, $E\ (0, \dfrac{d}{2})$ N $(\dfrac{d}{2}, -\dfrac{d}{2})$; M is defined by

$$\frac{\overline{ME}}{\overline{MA}} = \frac{d}{a} = k,$$

$$\frac{\overline{ME}}{d} = \frac{\overline{MA}}{a} = \frac{\overline{ME} - \overline{MA}}{d - a} = \frac{\overline{AE}}{d - a},$$

hence

$$\frac{\overline{ME}}{\overline{AE}} = \frac{d}{d-a} = \lambda \text{ and } \overline{MA} = \frac{a \cdot \overline{AE}}{d - a} = \frac{ad}{2(d - a)} = \lambda \cdot \frac{a}{2} = b, \qquad \lambda = \frac{k}{k-1}.$$

The equation of $\mathscr{P}$ is

$$\left(y + \frac{d}{2}\right)^2 = -\frac{d}{2}\left(x - \frac{d}{2}\right) \text{ and } y + \frac{d}{2} > 0$$

or

$$y^2 + dy = -\frac{d}{2}.x \text{ and } y + \frac{d}{2} > 0.$$

The equation of $\mathscr{H}$, vertices $A(0, 0)$, $M\left(-b = -\dfrac{a\lambda}{2}, 0\right)$, is

$$\frac{x(x + b)}{y^2} = \frac{G^2}{H^2} \cdot \lambda \text{ or } y^2 = \frac{H^2}{\lambda G^2} \cdot x(x + b).$$

The abscissa x of a general point of intersection of $\mathscr{P}$ and $\mathscr{H}$ is a root of the equation obtained by eliminating y^2.

To investigate the intersection of $\mathscr{H}$ and $\mathscr{P}$, we can use any two distinct conics from the pencil $\lambda\mathscr{H} + \mu\mathscr{P}$ generated by $\mathscr{H}$ and $\mathscr{P}$. Thus the parabola $\mathscr{P}'$ obtained by eliminating y^2 from the equation

$$y = -\frac{x}{2} - \frac{H^2}{\lambda dG^2}x(x + b),$$

passes through any point U that is an intersection of $\mathscr{H}$ and $\mathscr{P}$.

Examining $\mathcal{H} \cap \mathcal{P}$ can be reduced to examining $\mathcal{P} \cap \mathcal{P}'$:
$\quad$ $\mathcal{P}$ with axis parallel to Ax,
$\quad$ $\mathcal{P}'$ with axis parallel to Ay,
$\quad$ A a point common to $\mathcal{P}$ and $\mathcal{P}'$.
The slope of the tangent to the parabola $\mathcal{P}$ at A has a gradient of $-1/2$.

$$\left[2yy' + dy' = -\frac{d}{2}, \ y = 0 \Rightarrow y' = -\frac{1}{2} \right].$$

The parabola $\mathcal{P}'$ cuts Ox in the point A and in the point with abscissa

$$x = -b - \frac{\lambda d}{2} \cdot \frac{G^2}{H^2} < -b.$$

Its vertex thus has an abscissa < 0 and an ordinate > 0. We have

$$y' = -\frac{1}{2} - \frac{H^2}{d\lambda G^2}(2x + b);$$

for $x = 0$

$$y' = -\frac{1}{2} - \frac{b}{d \cdot \lambda} \cdot \frac{H^2}{G^2} = -\frac{1}{2} - \frac{a \cdot H^2}{2dG^2} < -\frac{1}{2}.$$

The parabola $\mathcal{P}'$ cuts $\mathcal{P}$ in A, and, close to A, we have
$\quad$ for $x > 0$ $\qquad\qquad$ $\mathcal{P}'$ inside $\mathcal{P}$
$\quad$ for $x < 0$ $\qquad\qquad$ $\mathcal{P}'$ outside $\mathcal{P}$.

So $\mathcal{P}'$ necessarily cuts $\mathcal{P}$ in a point U whose abscissa lies between 0 and $\frac{d}{2}$, and in another point whose abscissa is negative. Only point U is to be considered.

Let us note that a point on $\mathcal{P}'$ with abscissa $\frac{d}{2}$ has ordinate

$$y = -\frac{d}{4} - \frac{1}{2}\left(b + \frac{d}{2} \right) \cdot \frac{H^2}{\lambda \cdot G^2}$$

or

$$b + \frac{d}{2} = \frac{d}{d-a} \cdot \frac{a}{2} + \frac{d}{2} = \frac{d}{2}\left(1 + \frac{a}{d-a} \right) = \frac{d}{2} \cdot \lambda,$$

$$y = -\frac{d}{4} - \frac{d}{4} \cdot \frac{H^2}{G^2} = -\frac{d}{4}\left(1 + \frac{H^2}{G^2} \right) > -\frac{d}{2},$$

because $\dfrac{G}{H} > 1$.

The point U thus has an ordinate greater than that of N, U lies on arc NA of the parabola $\mathscr{P}$ and it satisfies the conditions.

So to this point U there corresponds a point B of the semi-ellipse in question, and the tangent at B provides a solution to the problem.

The case of the hyperbola Γ. Ibn al-Haytham considers only the branch Γ_A. The point M lies between A and E:

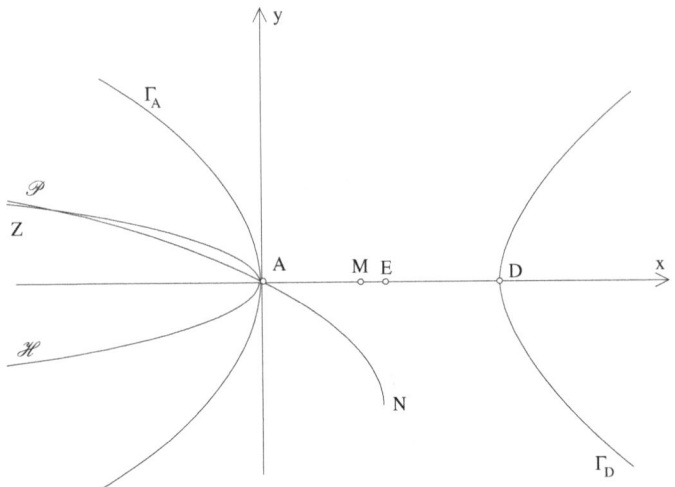

Fig. 1.2.5.7

$$\frac{\overline{ME}}{\overline{MA}} = -\frac{d}{a} = -k,$$

$$\frac{\overline{ME}}{d} = \frac{\overline{MA}}{-a} = \frac{\overline{ME} - \overline{MA}}{d+a} = \frac{\overline{AE}}{d+a},$$

$$\frac{\overline{ME}}{\overline{AE}} = \frac{d}{d+a} = \lambda' < 1, \quad \overline{MA} = -\frac{a \cdot \overline{AE}}{d+a} = -\frac{a}{2} \cdot \lambda' = -b, \quad \overline{AM} = b, \quad \lambda' = \frac{k}{k+1}.$$

With axes as before, we have

equation of $\mathscr{P}$: $y^2 + yd = -\dfrac{d}{2}x$, $[x < 0$ and $y > 0]$ for the arc AZ;

equation of $\mathscr{H}$: $A(0, 0)$, $M(0, b)$;

$$\frac{x(x-b)}{y^2} = \lambda' \cdot \frac{G^2}{H^2} \text{ or } y^2 = \frac{H^2}{\lambda'G^2}x(x-b).$$

Intersection. Any point of intersection satisfies

$$-dy - \frac{dx}{2} = \frac{H^2}{\lambda'G^2}x(x-b), \qquad y = -\frac{x}{2} - \frac{H^2}{d\lambda'G^2}x(x-b),$$

the equation of a parabola $\mathscr{P}'$.

Examining $\mathscr{H} \cap \mathscr{P}$ reduces to examining $\mathscr{P} \cap \mathscr{P}'$.

$\mathscr{P}'$ cuts Ox in the point A ($x = 0$) and in A' with abscissa

$$\alpha = b - \frac{1}{2} \cdot \frac{d\lambda'G^2}{H^2} = b\left[1 - \frac{d}{a}\frac{G^2}{H^2}\right] = b\left(1 - k\frac{G^2}{H^2}\right) < b.$$

The vertex of $\mathscr{P}'$ has abscissa $\frac{\alpha}{2}$ and a positive ordinate.

Tangents at A:

to $\mathscr{P}$ $\qquad y' = -\frac{1}{2}$;

to $\mathscr{P}'$ $\qquad y' = -\frac{1}{2} - \frac{H^2}{d\lambda'G^2}(2x - b)$;

hence, for $x = 0$,

$$y'_{(0)} = -\frac{1}{2} + \frac{bH^2}{d\lambda'G^2} = -\frac{1}{2} + \frac{a.H^2}{2dG^2} = \frac{1}{2}\left(\frac{H^2}{kG^2} - 1\right) > -\frac{1}{2},$$

$$y'_{(0)} = \frac{H^2}{2kG^2}\left(1 - k\frac{G^2}{H^2}\right).$$

α and $y'_{(0)}$ have the same sign, that of $1 - k\frac{G^2}{H^2}$.

For a point of $\mathscr{P} \cap \mathscr{P}'$ to give a solution to the problem, it is necessary that it lies on the arc AZ of $\mathscr{P}$, which requires $\alpha < 0$, $y'_{(0)} < 0$, so $\frac{G^2}{H^2} > \frac{1}{k}$.

Then we have $-\frac{1}{2} < y'_{(0)} < 0$; arc AA' of the parabola $\mathscr{P}'$ thus goes inside $\mathscr{P}$ at the point A.

There are thus three cases:

a) arc AA' lies completely inside $\mathscr{P}$: the problem has no solution.

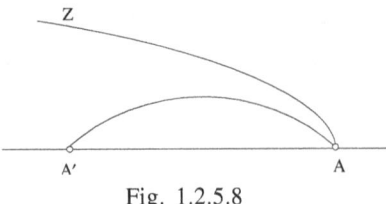

Fig. 1.2.5.8

b) arc AA' is the tangent to $\mathscr{P}$ at a point U_0: we have one solution.

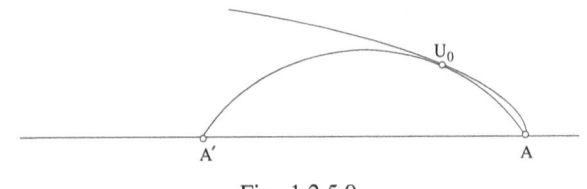

Fig. 1.2.5.9

c) arc AA' cuts $\mathscr{P}$ in two points U_1 and U_2: the problem then has two solutions.

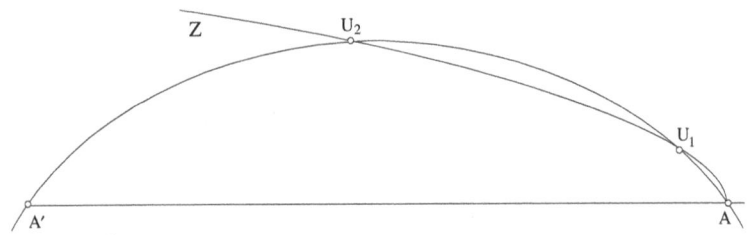

Fig. 1.2.5.10

– **6** – Let Γ be an ellipse or a hyperbola with transverse axis AD and centre E, to find the point B of Γ such that the tangent at this point cuts the axis, on the same side as A, in a point S and satisfies $\dfrac{BS}{DS} = \dfrac{H}{F}$, a given ratio (Figs. 1.2.6.1 and 1.2.6.2). This problem is analogous to the previous one, but we replace the vertex A with the vertex D, the more distant one.

Analysis: Let AI be the ordinate with respect to the diameter EB; we have $AI \parallel BS$ and I the midpoint of AC, C being the point of intersection of AI and Γ. We draw BK, IP, CO perpendicular to AD.

If we put $AD = d$ and $a = $ *latus rectum*, we have

$$\frac{OD \cdot OA}{CO^2} = \frac{d}{a} \text{ and } \frac{KD \cdot KA}{BK^2} = \frac{d}{a}.$$

We have I the midpoint of AC, E the midpoint of AD and $IP \parallel CO$. Therefore

$$AP = \frac{1}{2}\,AO, \quad EP = \frac{1}{2}\,OD, \quad IP = \frac{1}{2}\,CO$$

and we have

$$\frac{PE \cdot PA}{PI^2} = \frac{d}{a}.$$

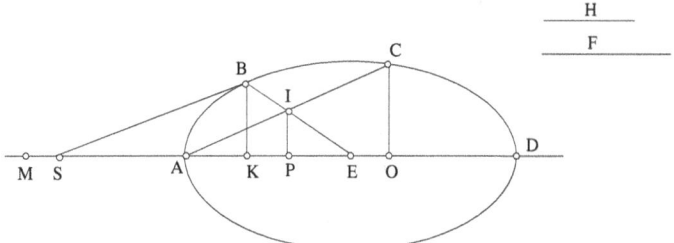

Fig. 1.2.6.1 – Γ is an ellipse

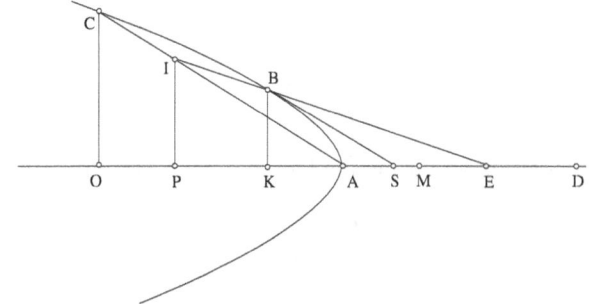

Fig. 1.2.6.2 – Γ is a hyperbola

If we put $\dfrac{EM}{MA} = \dfrac{d}{a}$, then $\dfrac{EM}{EA}$ is known and we have $\dfrac{MP \cdot PA}{AI^2} = \dfrac{ME}{EA}$ (according to *Conics*, VII.2). We have $\dfrac{KA}{AE} = \dfrac{AS}{SE}$ (according to *Conics*, I.37) and $\dfrac{AS}{SE} = \dfrac{IB}{BE} = \dfrac{PK}{KE}$ (because $AI \parallel SB$ and $BK \parallel IP$), so

$$\frac{PK}{KE} = \frac{KA}{AE}.$$

Hence, by composition of ratios in one case and by separation in the other,

$$\frac{EP}{EK} = \frac{EK}{EA},$$

so
$$EK^2 = EP \cdot EA. \qquad (\alpha)$$

From
$$\frac{KA}{AE} = \frac{AS}{SE},$$

we obtain
$$\frac{KA}{AS} = \frac{AE}{SE} = \frac{IA}{BS} \quad \text{and} \quad \frac{IA}{AK} = \frac{BS}{AS}.$$

But
$$\frac{BS}{SD} = \frac{H}{F},$$

hence
$$\frac{DS}{AS} = \frac{IA}{AK} \cdot \frac{F}{H}.$$

But
$$\frac{DS}{SA} = \frac{KD}{KA},$$

because
$$\frac{AE}{ES} = \frac{KE}{AE},$$

which, by composition and using the fact that $AE = DE$, gives us
$$\frac{DS}{ES} = \frac{DK}{AE},$$

so
$$\frac{KD}{KA} = \frac{IA}{AK} \cdot \frac{F}{H},$$

hence
$$\frac{KD}{IA} = \frac{F}{H}.$$

From
$$\frac{KD^2}{IA^2} = \frac{F^2}{H^2} \quad \text{and} \quad \frac{MP \cdot PA}{IA^2} = \frac{ME}{EA},$$

we get
$$\frac{MP \cdot PA}{DK^2} = \frac{ME}{EA} \cdot \frac{H^2}{F^2}. \qquad (\beta)$$

If we start from the equalities (α) and (β), the argument can be developed as in the previous problem – if we make use of a parabola $\mathscr{P}$ and a hyperbola $\mathscr{H}$ that Ibn al-Haytham brings in only in his synthesis.[45] It seems that Ibn al-Haytham wanted to avoid repeating this argument; having proved the equalities (α) and (β) he can affirm that the point P will be known. From the point P we can find, successively, the points O, C, I, then B and the tangent BS.

The synthesis will establish the existence of the point P.

– 7 – *Synthesis of the problem*: let us turn to two figures (Fig. 1.2.7.1 for the ellipse and Fig. 1.2.7.2 for the hyperbola).

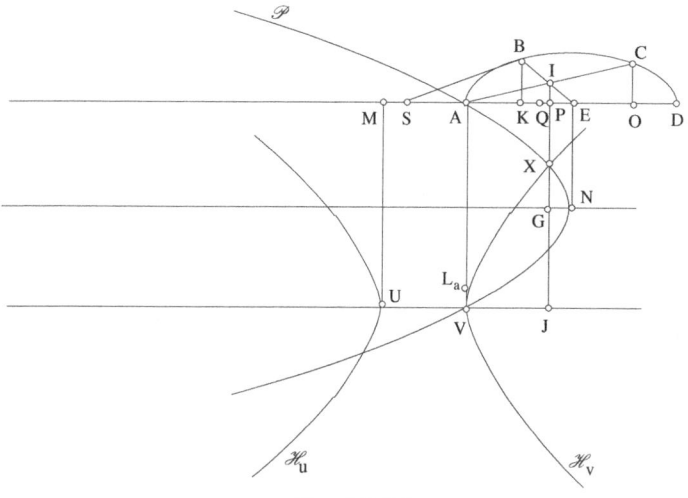

Fig. 1.2.7.1

Let us draw the straight lines EN, AV and MU perpendicular to the straight line AD, with $EN = EA$, $AV = MU = AD$. Let $NG \parallel DA$ and $VJ \parallel DA$. We have $UV = MA$. We put

$$\frac{AE}{EQ} = \frac{H^2}{F^2} \quad \text{and} \quad \frac{UV}{VL_a} = \frac{ME}{EQ}$$

[45] Equation (β) leads to the hyperbola $\mathscr{H}$ without the necessity of using several symmetries to show it; in fact $DK = AK + AD$, hence the need for a downwards translation of the axis of the auxiliary hyperbola $\mathscr{H}$ by a distance AV equal to AD. We can see clearly the link between this problem and the previous one: we move from one construction to the other by a translation of the axis of the auxiliary hyperbola and not by symmetries, as has been mistakenly asserted.

and we draw the hyperbola $\mathscr{H}$ with transverse axis *UV* and *latus rectum* VL_a and the parabola $\mathscr{P}$ with axis *NG* and *latus rectum EA*. The parabola $\mathscr{P}$ passes through *A* and *V*.

In the case of the ellipse, $\mathscr{P}$ and the branch $\mathscr{H}_V$ of the hyperbola have the common point *V* and their concave sides face in opposite directions; $\mathscr{H}_V$ cuts $\mathscr{P}$ again in the point *X*, which can be projected orthogonally on *AD* into the point *P* lying between *A* and *E*. We note that the branch $\mathscr{H}_U$ also cuts $\mathscr{P}$ but in points whose orthogonal projections onto *DA* lie on *DA* produced and do not give solutions.

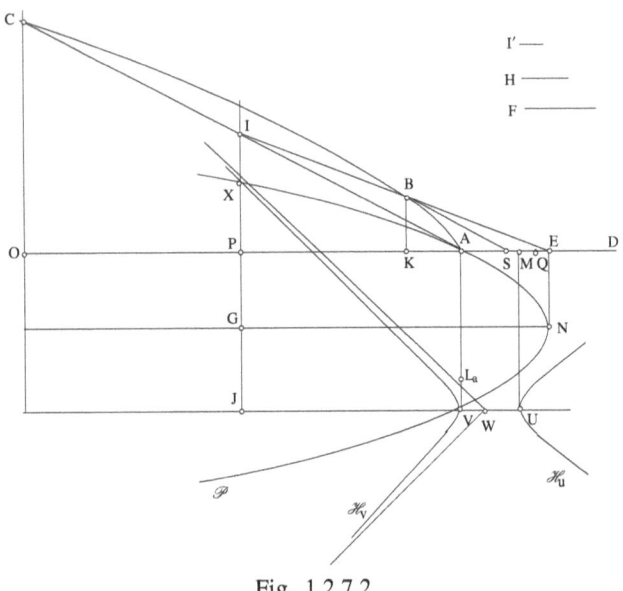

Fig. 1.2.7.2

In the case of the hyperbola, considering the branch $\mathscr{H}_V$ of the hyperbola with axis *UV*, one asymptote cuts the half-line *NG*, which is the axis of the parabola, so the asymptote also cuts the parabola; $\mathscr{H}_V$ goes inside the parabola at the point *V*, it approaches indefinitely close to its asymptote and thus cuts the parabola again in the point *X*.

Let us note that the branch $\mathscr{H}_U$ and $\mathscr{P}$ can have 0, 1 or 2 common points; but these points, if they exist, are projected onto the straight line *AE* between *A* and *E* and do not give any solution for this problem.

The perpendicular from *X* to the straight line *AD* cuts *AD*, *NG* and *UV* in *P*, *G* and *J* respectively.

Let O be a point such that $AP = PO$; the perpendicular to AD at O cuts Γ in C, the straight lines AC and GP cut one another in I, the midpoint of AC, and EI cuts Γ in B. The straight line BS parallel to AI is a tangent to Γ.

We shall show that

$$\frac{BS}{SD} = \frac{H}{F}.$$

We have $EP \cdot EA = EK^2$ (as in the analysis), $PE = NG$ and $NG \cdot EA = GX^2$ (since $X \in \mathscr{P}$), so

$$EK = GX;$$

we have

$$GJ = EN = ED,$$

hence

$$JX = KD.$$

On the other hand,

$$\frac{JU \cdot JV}{JX^2} = \frac{UV}{VL_a} = \frac{ME}{EA} \cdot \frac{EA}{EQ} \quad \text{(since } X \in \mathscr{H}),$$

therefore

$$\frac{JU \cdot JV}{JX^2} = \frac{MP \cdot PA}{AI^2} \cdot \frac{H^2}{F^2};$$

but

$$JU \cdot JV = MP \cdot PA,$$

so

$$\frac{AI^2}{JX^2} = \frac{H^2}{F^2} = \frac{AI^2}{KD^2},$$

hence

$$\frac{AI}{KD} = \frac{H}{F}.$$

Therefore

$$\frac{KD}{KA} = \frac{IA}{KA \cdot \dfrac{H}{F}}.$$

But

$$\frac{IA}{KA} = \frac{BS}{SA}.$$

Let I' be such that

$$\frac{SA}{I'} = \frac{F}{H}.$$

We have

$$\frac{SA \cdot AI}{I' \cdot AI} = \frac{F}{H} = \frac{BS \cdot AK}{I' \cdot AI};$$

therefore

$$\frac{BS}{I'} = \frac{AI}{AK \cdot \dfrac{H}{F}},$$

so

$$\frac{BS}{I'} = \frac{KD}{KA}.$$

On the other hand,

$$\frac{KD}{KA} = \frac{DS}{SA},$$

so

$$\frac{BS}{I'} = \frac{DS}{SA},$$

and consequently

$$\frac{BS}{SD} = \frac{I'}{SA} = \frac{H}{F}.$$

– **8** – Let there be a hyperbola Γ with transverse axis AD and centre E, to find a straight line tangent to Γ in B, cutting the transverse axis in K and such that $\dfrac{BK}{BE} = \dfrac{G}{H}$ ($G < H$). The problem is again of the same kind, but this time the segments AK or DK are replaced by the radius vector EB.

Analysis: Let BK be the solution to the problem. The straight line through A, parallel to BK, cuts the hyperbola and the diameter EB in C and I respectively, points such that $AI = IC$; we have

$$\frac{IE}{IA} = \frac{BE}{BK} = \frac{H}{G}.\,^{46}$$

---

[46] From this equality, we can immediately deduce that I lies on a circle $\mathscr{C}_1$, and that C lies on the circle $\mathscr{C}'_1$ that is the image of $\mathscr{C}_1$ in the homothety $(A, 2)$.

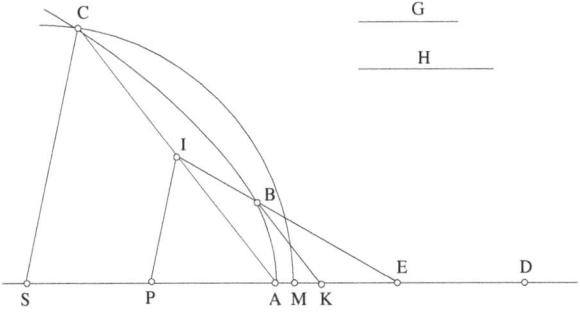

Fig. 1.2.8

Let P be a point of the straight line EA such that $E\hat{I}P = I\hat{A}P$. Triangles EIP and IAP are similar, we have

$$\frac{EP}{PI} = \frac{PI}{PA} = \frac{EI}{IA},$$

hence

$$EP \cdot PA = PI^2 \text{ and } PI > PA.$$

We also have

$$\frac{EP}{PA} = \frac{EP^2}{PI^2} = \frac{EI^2}{IA^2} = \frac{H^2}{G^2},$$

so

$$\frac{EA + AP}{AP} = \frac{H^2}{G^2}, \quad \frac{EA}{AP} = \frac{H^2}{G^2} - 1 = \frac{H^2 - G^2}{G^2} \ (G < H).$$

EA is the semi-transverse axis, so AP is known. Consequently, EP is known and PI also.[47] Let $CS \parallel IP$, we have $AS = 2AP$ and $CS = 2IP$, so $CS > AS$; these lengths are known.

We put $SM = SC$, hence $SM > SA$ and M lies on SA produced. The circle (S, SM) cuts the hyperbola in C; from point C we find I the midpoint of AC, then B and the straight line BK.

$-$ **9** $-$ *Synthesis*: We consider the hyperbola Γ and the ratio $\dfrac{G}{H}$. Let F be a point such that $\dfrac{G}{F} = \dfrac{H}{G}$, then $\dfrac{H}{F} = \dfrac{H^2}{G^2}$.

---

[47] We have $AP = \dfrac{G^2}{H^2 - G^2} \cdot EA$, $EP = \dfrac{H^2}{H^2 - G^2} \cdot EA$, hence $PI = \dfrac{G \cdot H}{H^2 - G^2} \cdot EA$; so we have the centre P and the radius of the circle $\mathscr{C}_1$.

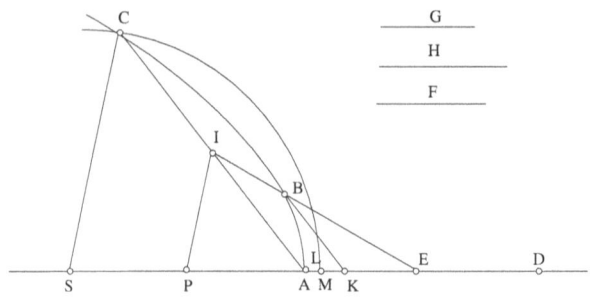

Fig. 1.2.9

Let us put

$$\frac{PE}{PA} = \frac{H}{F} = \frac{H^2}{G^2};$$

this defines a point P on EA produced. Let S be a point such that $PS = PA$ and L a point such that $PE \cdot PA = PL^2$ (so $PA < PL < PE$). Let us draw the circle $(S, 2PL)$, it cuts the axis in M such that $SM = 2PL > SA$, M thus lies outside Γ and S inside it; the circle cuts Γ in a point C.

Let I be the midpoint of AC, the straight line EI cuts the hyperbola in B. Let us draw the tangent BK that is parallel to CA and let us show that

$$\frac{BK}{BE} = \frac{G}{H}.$$

Since P is the midpoint of SA and I is the midpoint of AC, we have $SC \parallel PI$ and $SC = 2PI$, so $PI = PL$ and $PE \cdot PA = PI^2$. Therefore

$$\frac{PE}{PI} = \frac{PI}{PA},$$

hence triangles PEI and PIA are similar to one another. So, as in the analysis, we have

$$\frac{PE}{PA} = \frac{EI^2}{IA^2},$$

so we have

$$\frac{EI^2}{IA^2} = \frac{H^2}{G^2}.$$

However,

$$\frac{EI}{IA} = \frac{EB}{BK},$$

so

$$\frac{BK}{BE} = \frac{G}{H}.$$

– 10 – Let there be a parabola Γ and the tangent CD at point C of that parabola. To find another straight line tangent to the parabola such that, if A is its point of contact and D its point of intersection with CD, we have $\frac{DA}{DC} = \frac{E}{G}$, a known ratio.

Analysis: Let H be the midpoint of AC; DH is a diameter, DH and AC are conjugate directions, DH cuts Γ in B, the midpoint of DH (Fig. 1.2.10.1 or 1.2.10.2).

If $E = G$, then $DA = DC$ (Fig. 1.2.10.2). As $HA = HC$, $A\hat{H}D$ is a right angle; DH is then the axis of Γ, it is known; the point D is thus known. The required tangent is the second tangent from D and it is equal to DC; the points A and C are symmetrical with respect to the axis DH.

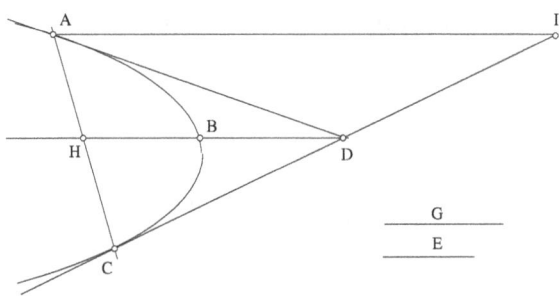

Fig. 1.2.10.1

If $E \neq G$, then $AD \neq DC$, $\hat{H}$ is not a right angle. The straight line CD is known, the direction of DH is known since DH is parallel to the axis of Γ, so $C\hat{D}H$ is a known angle.

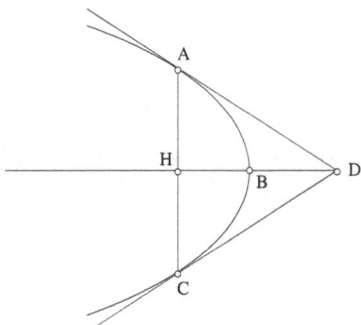

Fig. 1.2.10.2

Let us draw through A a line parallel to the axis of Γ; it cuts CD in I and $DC = DI$, so $\dfrac{DA}{DI} = \dfrac{E}{G}$ and $D\hat{I}A = C\hat{D}H$. Triangle DIA is of known shape, so $A\hat{D}I$ is known, in consequence $A\hat{D}C$ is also known. But $\dfrac{DA}{DC} = \dfrac{E}{G}$, so triangle ADC is of known shape, $D\hat{C}A$ is thus known, the straight line CA is thus known. Its point of intersection with Γ, the point A, is the required point of contact.

 – **11** – *Synthesis*: The axis of Γ meets CD, a known straight line, in the point I. Let K be a point on the axis such that $\dfrac{CI}{CK} = \dfrac{G}{E}$ (see discussion) and let L be the point of the straight line CD such that $CL = CI$. Let us draw through C a line parallel to LK, it cuts the axis and thus cuts Γ at the point A. Let H be the midpoint of CA. The line parallel to the axis drawn through H cuts Γ in B and CD in D, BH is thus a diameter and AH an ordinate; in consequence DA is the second tangent from D.

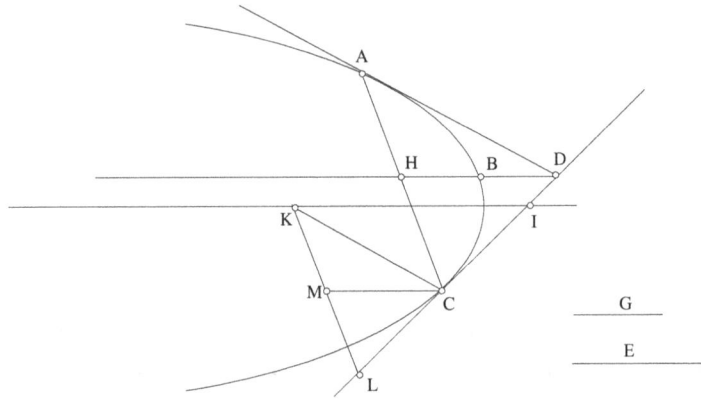

Fig. 1.2.11.1

Let us show that

$$\frac{DA}{DC} = \frac{E}{G}.$$

Let M be the midpoint of KL, we have $CM \parallel IK$, so $CM \parallel DH$; but $CH \parallel KL$, so triangles DCH and CML are similar, with

$$\frac{DC}{CH} = \frac{CL}{LM}.$$

Therefore

$$\frac{DC}{CA} = \frac{CL}{LK};$$

so triangles DCA and CLK are similar and we have

$$\frac{DA}{DC} = \frac{CK}{CL} = \frac{E}{G}.$$

Discussion: The problem has only one solution if the point K exists.

If we call the distance from C to the axis h, K exists if and only if $CK \geq h$; but $\frac{CK}{CI} = \frac{E}{G}$, hence the condition $\frac{E}{G} \geq \frac{h}{CI}$ or again $\frac{E}{G} \geq \sin\Theta$, if Θ is the acute angle between the tangent CD and the axis of the parabola. If this condition is satisfied, there are two points K, K', which coïncide when $\frac{E}{G} = \sin\Theta$; to each of these points there corresponds only one solution of the problem. Ibn al-Haytham does not mention the possibility of there being two solutions. When C is at the vertex of the parabola, I coincides with it and the construction cannot be carried out in the same way. If T denotes the point in which the required tangent AD meets the axis and if P is the projection of A onto the axis, we have $PC = CT$, so $\frac{AT}{AP} = \frac{AD}{DC} = \frac{E}{G}$, which is known, and angle $A\hat{T}P$ is known; the problem thus reduces to Problem 50 of Book II of Apollonius. Ibn al-Haytham does not mention this case; if he thought of it, he may have thought it uninteresting, given that it is so easy.

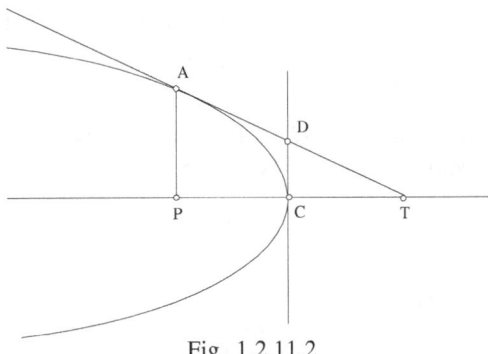

Fig. 1.2.11.2

$- 12 -$ We are given a conic section Γ (an ellipse or a hyperbola) with centre H, BD the tangent at a point B of Γ, a ratio $\frac{E}{G}$.

Problem: To draw a tangent to Γ that cuts BD in the point D such that $\dfrac{DA}{DB} = \dfrac{E}{G}$ (A being the point of contact).

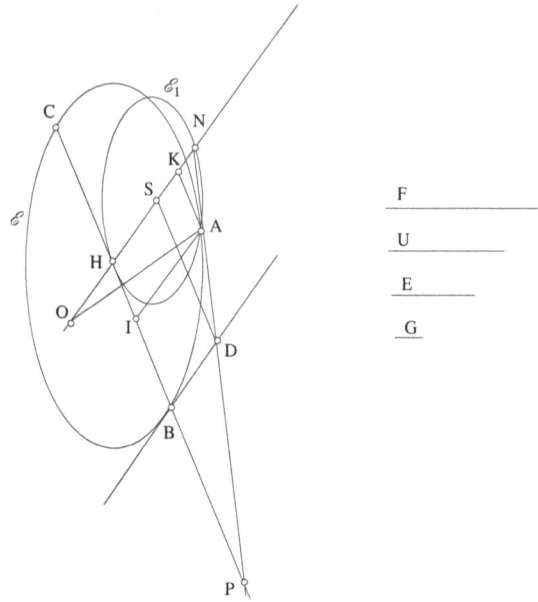

Fig. 1.2.12.1

Analysis: If a is the *latus rectum* relative to the diameter d that passes through B, we have $\dfrac{d}{a} = \dfrac{HB}{\frac{1}{2}a} = k$, a known ratio.

We suppose the tangent AD is known; let P be the point of intersection of AD and HB and let $AI \parallel BD$. We have

$$IH \cdot HP = HB^2 \text{ and } \frac{HI \cdot IP}{AI^2} = k \ (\textit{Conics}, \text{ I.37}),$$

from which we have

$$\frac{HP}{IP} = \frac{HB^2}{k \cdot AI^2}.$$

We draw $HN \parallel AI$ (with N on DA), and $AK \parallel PH$ (with K on HN). We have

$$\frac{HP}{PI} = \frac{HN}{IA} = \frac{HB^2}{k \cdot AI^2},$$

hence

$$NH \cdot AI = \frac{HB^2}{k}$$

and

$$\frac{NH \cdot AI}{HB^2} = \frac{NH \cdot HK}{HB^2} = \frac{1}{k},$$

from which we obtain in succession

$$\frac{NH \cdot HK}{HI \cdot HP} = \frac{1}{k}, \quad \frac{k \cdot KH}{HI} = \frac{PH}{HN}, \quad \frac{k \cdot KH}{AK} = \frac{PH}{HN} = \frac{AK}{KN},$$

hence

$$\frac{KH \cdot KN}{AK^2} = \frac{1}{k}.$$

Let U be the straight line defined by $\dfrac{KH}{U} = \dfrac{1}{k}$, then we have

(1) $$\frac{KH}{U} = \frac{NK \cdot KH}{AK^2},$$

hence

$$U \cdot KN = AK^2.$$

We then have

$$U \cdot HN = KA^2 \cdot \frac{HN}{NK} = HI^2 \cdot \frac{HN}{NK};$$

but

$$\frac{HN}{NK} = \frac{PH}{AK} = \frac{PH}{IH} = \frac{PH.IH}{IH^2} = \frac{HB^2}{HI^2}.$$

Therefore

$$U \cdot HN = HB^2.$$

If Γ is a hyperbola $\mathcal{H}$, we consider the hyperbola $\mathcal{H}_1$ whose transverse axis is NH and *latus rectum* a_1 such that

$$\frac{NH}{a_1} = \frac{KH}{U} = \frac{1}{k};$$

and if Γ is an ellipse $\mathcal{E}$, we consider the ellipse $\mathcal{E}_1$ with diameter NH and *latus rectum* a_1. From equality (1) $\mathcal{H}_1$ (or $\mathcal{E}_1$) passes through the point A of the conic section Γ under consideration.[48]

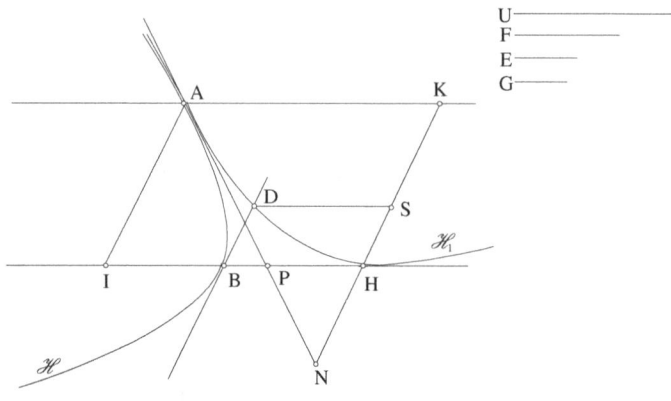

Fig. 1.2.12.2

We have $DB \parallel HN$ and we draw $DS \parallel BH$ (with S on HN); from the equality $HI \cdot HP = HB^2$, we then deduce that

$$AN \cdot NP = ND^2 \quad \text{and} \quad KN \cdot HN = NS^2,$$

hence

$$\frac{AN}{ND} = \frac{ND}{NP} = \frac{AD}{DP} \quad \text{and} \quad AN \cdot DP = AD \cdot DN.$$

Let us put

$$F \cdot NK = AN^2,$$

hence

$$\frac{F}{U} = \frac{AN^2}{AK^2};$$

we can write

$$\frac{F}{AN} = \frac{AN}{NK} = \frac{DN}{NS} = \frac{DP}{DB},$$

hence

$$F \cdot DB = AN \cdot DP = AD \cdot DN.$$

So on the one hand we have

---

[48] From how it has been defined, $\mathcal{H}_1$ (or $\mathcal{E}_1$) is similar to $\mathcal{H}$ (or $\mathcal{E}$), the given conic section (see comment 4, p. 94).

$$\frac{F}{DN} = \frac{AD}{DB},$$

and on the other

$$\frac{F}{U} = \frac{AN^2}{AK^2} = \frac{DN^2}{DS^2},$$

hence

$$F \cdot DS^2 = U \cdot DN^2.$$

But

$$U \cdot HN = HB^2 = DS^2,$$

hence

$$F \cdot HN = DN^2 \text{ and } \frac{F}{DN} = \frac{DN}{HN},$$

from which we have

$$\frac{DN}{HN} = \frac{AD}{DB} = \frac{AD}{HS} = \frac{AN}{NS},$$

but by hypothesis

$$\frac{DA}{DB} = \frac{E}{G},$$

so we have

$$\frac{AN}{NS} = \frac{E}{G}.$$

We know that

$$KN \cdot NH = NS^2,$$

hence

$$\frac{KN \cdot NH}{NA^2} = \frac{NS^2}{NA^2} = \frac{G^2}{E^2};$$

but

$$NA^2 = F \cdot NK,$$

hence

$$\frac{NH}{F} = \frac{G^2}{E^2}.$$

Let $NO = F$, we have $NO \cdot NK = NA^2$, hence

$$\frac{NO}{AN} = \frac{AN}{NK}.$$

So triangles ANK and ONA, which have a common angle N, are similar, hence $N\hat{A}O = A\hat{K}N = B\hat{H}N = P\hat{B}D$, a known angle.

Thus we have

$$\frac{NH}{NO} = \frac{G^2}{E^2} \;^{49} \text{ and } N\hat{A}O = P\hat{B}D = \alpha.$$

The point A lies on the arc $\mathscr{C}$ subtending the angle α constructed on the segment NO; on the other hand it also lies on $\mathscr{H}_1$ in the case of the hyperbola and on $\mathscr{E}_1$ in the case of ellipse.

So if the segment NH was known in magnitude and position, the point A would be known, angles $A\hat{N}H$ and $A\hat{O}N$ also and angle $H\hat{P}N = A\hat{O}N$ as well.

If there exists a tangent that gives a solution to the problem, that tangent makes with the diameter HB an angle $H\hat{P}N = \Theta$, which is determined by the data of the problem, as Ibn al-Haytham will describe at the beginning of his synthesis.

– **13** – *Synthesis*: Outline. a) To construct a figure $MVLXJR$ similar to the figure $NKHOPA$, to find the angle $\Theta = N\hat{O}A = J\hat{L}M$.

b) To show that if a tangent PN is such that $H\hat{P}N = \Theta$, then we have $\frac{DA}{DB} = \frac{E}{G}$.

c) Discussion.

a) Starting with the data of Problem 12:

$$k = \frac{HB}{\frac{a}{2}} = \frac{2HB}{a},$$

α = angle of the ordinates, $\dfrac{E}{G}$ a given ratio,

we take two straight lines LM and LT such that $\dfrac{LM}{LT} = \dfrac{1}{k}$ and we consider two figures – the ellipse $\mathscr{E}_2$ and the branch of the hyperbola $\mathscr{H}_2$ that passes through L – that involves both of them, LM as diameter (a transverse diameter in the case of $\mathscr{H}_2$) and LT as *latus rectum*, with α as the angle of the ordinates. [$\mathscr{H}_2$ and $\mathscr{E}_2$ are thus similar to $\mathscr{H}_1$ and $\mathscr{E}_1$.]

Let MX be the straight line defined by $\dfrac{LM}{MX} = \dfrac{G^2}{E^2}$, X being a point on the straight line LM; we consider the circle $\mathscr{C}$ constructed to have MX as a

---

[49] This equality shows that if $E > G$, then $NO > NH$; if $E = G$, then O is on H; if $E < G$, then $NO < NH$.

chord that subtends the angle α. We should note that Ibn al-Haytham has not specified whether the angle of the ordinates is the acute angle or the obtuse one. One of the arcs MX of circle $\mathscr{C}$ subtends α and the other its supplement. Moreover, there exist two circles $\mathscr{C}$ symmetrical with respect to the straight line LM; we may consider the one that touches the ellipse $\mathscr{E}_2$ (or the second branch of $\mathscr{H}_2$) at M.

Ibn al-Haytham says that the enclosing arc under consideration cuts the conic section ($\mathscr{E}_2$ or $\mathscr{H}_2$) 'in every case'; this is not correct, as Ibn al-Haytham shows later when he examines special cases in his paragraph of discussion.

Let R be the point of intersection and RV the corresponding ordinate. We know that $M\hat{R}X = \alpha = R\hat{V}M$; so triangles MVR and MRX are similar and we have

$$\frac{MR}{MV} = \frac{MX}{MR}$$

or alternatively

$$MR^2 = MV \cdot MX,$$

hence

$$\frac{VM \cdot ML}{MR^2} = \frac{ML}{MX} = \frac{G^2}{E^2}.$$

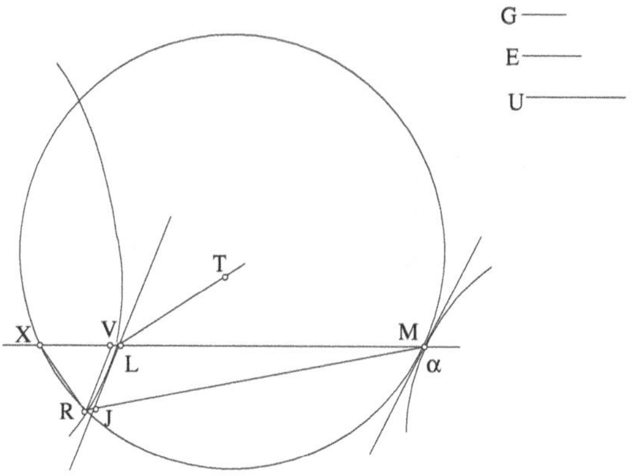

Fig. 1.2.13.1

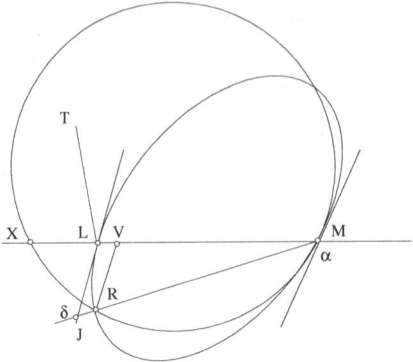

Fig. 1.2.13.2

Another possibility for the enclosing arc is that the tangent at the point X is parallel to the tangent at L to $\mathcal{E}_2$ or to $\mathcal{H}_2$.

Moreover, because R is a point of $\mathcal{H}_2$ or of $\mathcal{E}_2$, we have

$$\frac{MV \cdot VL}{VR^2} = \frac{ML}{LT} = \frac{1}{k} \text{ (equation of the conic section)}.$$

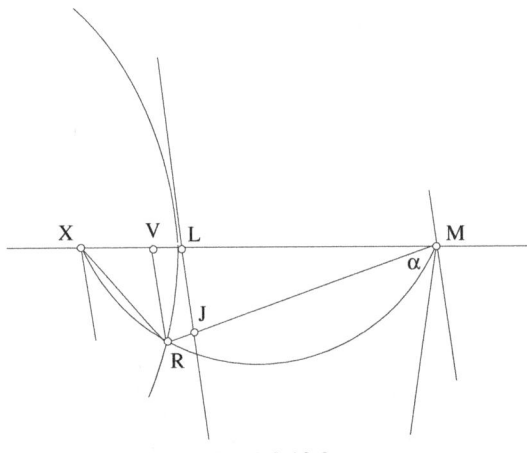

Fig. 1.2.13.3

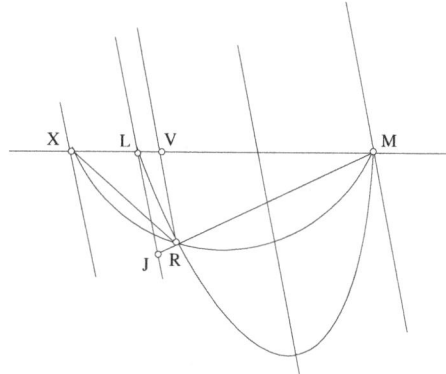

Fig. 1.2.13.4

The line through L parallel to VR cuts MR in J, $L\hat{J}M$ is known because it is given that $J\hat{L}M = \alpha$ and $L\hat{M}J$ is known (because it is defined by the given points L and M and the point R, the intersection of a known conic section $\mathscr{E}_2$ or $\mathscr{H}_2$ with a known arc of a circle). Let $L\hat{J}M = \Theta$.

So, returning to the conic section ABC, we construct a tangent that makes with the diameter HB an angle equal to $L\hat{J}M$; let A be the point of contact, P the point of intersection of the tangent with the diameter HP and AI the ordinate. The points N and K are defined as in Problem 12 and the point O of NH is defined by

$$\frac{ON}{NH} = \frac{XM}{ML} = \frac{E^2}{G^2}, \quad H\hat{P}A = L\hat{J}M = \Theta$$

and

$$A\hat{K}N = P\hat{I}A = P\hat{H}N = J\hat{L}M = \alpha.$$

Triangles NAK, NPH, API, MRV, MJL are similar. We know that

$$\frac{HI \cdot IP}{AI^2} = k \quad \text{(equation of the conic } ABC\text{)}$$

and

$$\frac{MV \cdot VL}{VR^2} = \frac{1}{k},$$

so we have

$$\frac{MV}{VR} \cdot \frac{VL}{VR} = \frac{AI}{IP} \cdot \frac{AI}{HI}.$$

But from the similarity of the triangles it follows that

$$\frac{MV}{VR} = \frac{ML}{LJ} = \frac{NH}{HP} = \frac{AI}{IP},$$

and consequently

$$\frac{VL}{VR} = \frac{AI}{HI} = \frac{HK}{KA}.$$

So we have

$$\frac{NK \cdot KH}{KA^2} = \frac{NK}{KA} \cdot \frac{KH}{KA} = \frac{MV}{RV} \cdot \frac{LV}{VR} = \frac{ML}{LT} = \frac{1}{k}.$$

From

$$\frac{KA}{KN} = \frac{RV}{VM} \text{ and } \frac{HK}{KA} = \frac{LV}{VR},$$

we obtain

$$\frac{HK}{KN} = \frac{LV}{VM} \text{ and } \frac{NH}{KN} = \frac{LM}{VM}.$$

But

$$\frac{HN}{NO} = \frac{ML}{MX},$$

hence

$$\frac{KN}{NO} = \frac{MV}{MX} = \frac{MV^2}{MX \cdot MV} = \frac{MV^2}{MR^2} = \frac{KN^2}{NA^2},$$

from which we get

$$KN \cdot NO = NA^2.$$

Let U be such that $\dfrac{KH}{U} = \dfrac{1}{k}$; we have

$$\frac{KH \cdot NK}{U \cdot NK} = \frac{1}{k}.$$

But

$$\frac{1}{k} = \frac{NK.KH}{KA^2},$$

hence

$$U \cdot NK = KA^2$$

and consequently

$$\frac{NA^2}{KA^2} = \frac{NO}{U}.$$

Moreover

$$\frac{KN}{NH} = \frac{AK}{PH} = \frac{IH}{HP} = \frac{IH^2}{HP \cdot HI} = \frac{IH^2}{HB^2} = \frac{AK^2}{HB^2},$$

so we have

$$\frac{U \cdot KN}{U \cdot NH} = \frac{AK^2}{HB^2},$$

but

$$U \cdot KN = AK^2,$$

so

$$U \cdot NH = HB^2.$$

Let us (as in Problem 12) draw the line DS parallel to AK; from the fact that $HI \cdot HP = HB^2$ we get $NA \cdot NP = ND^2$ and $NK \cdot NH = NS^2$.
Moreover we have

$$\frac{HN}{ON} = \frac{ML}{MX} = \frac{G^2}{E^2} = \frac{HN \cdot NK}{ON \cdot NK} = \frac{NS^2}{NA^2},$$

hence

$$\frac{NA}{NS} = \frac{E}{G}.$$

From

$$ON \cdot NK = NA^2,$$

we obtain

$$\frac{ON}{NA} = \frac{AN}{NK} = \frac{PN}{NH},$$

hence

$$ON \cdot NH = AN \cdot NP = ND^2 \text{ and } \frac{ON}{NH} = \frac{ND^2}{NH^2} = \frac{E^2}{G^2}.$$

Thus we have

$$\frac{AN}{NS} = \frac{ND}{NH} = \frac{AD}{HS} = \frac{AD}{DB},$$

and accordingly

$$\frac{AD}{DB} = \frac{E}{G}.$$

The point A gives a solution to the problem.

c) *Discussion of the problem*:
This discussion is equivalent to considering the existence of the point R, an intersection of $\mathscr{H}_2$ (or $\mathscr{E}_2$) with the arc of $\mathscr{C}$ subtending angle α, both defined by the segments ML and MX. We have put $\frac{ML}{MX} = \frac{G^2}{E^2}$. Ibn al-Haytham says 'Let us take the smaller of the two as a diameter ...', so

if $G < E$, we take ML as a diameter of $\mathcal{H}_2$ (or $\mathcal{E}_2$) and MX as a chord of the arc $\mathcal{C}$; this is what has been done in all our figures;

if $G > E$, we have $ML > MX$, and we should take MX as a diameter, which would come to the same thing as exchanging the letters L and X in the figures. But the argument put forward in the synthesis no longer holds.

Ibn al-Haytham then examines the choice of the arc $\mathcal{C}$ that subtends the angle; there are four possible arcs depending on whether we consider an acute or an obtuse angle α, that is, the half-plane I or the half-plane II in Fig. 1.2.13.5.

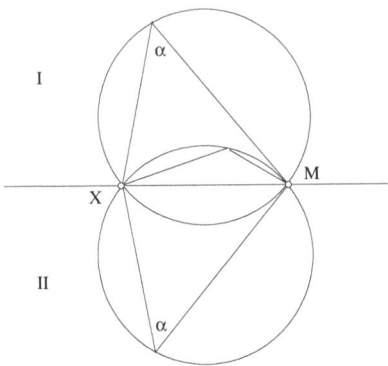

Fig. 1.2.13.5

In the case of the hyperbola $\mathcal{H}_2$, whichever arc we consider, it cuts $\mathcal{H}_2$ in a point because M lies outside and X inside the branch of the hyperbola that passes through L (if we assume $ML < MX$).

In the case of the ellipse $\mathcal{E}_2$, we choose the semi-ellipse for which the tangent LT makes $M\hat{L}T$ acute. The tangent to the ellipse at M is MT_1, and $MT_1 \parallel LT$.

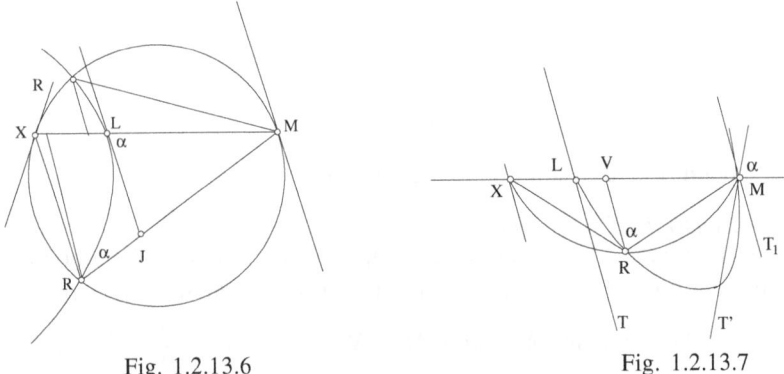

Fig. 1.2.13.6   Fig. 1.2.13.7

Let MT' be such that $L\hat{M}T' = M\hat{L}T$; MT' is a tangent to the arc containing the obtuse angle α. Near the point M, this arc lies inside the ellipse, and the point X of the arc lies outside the ellipse; so the arc in question and the semi-ellipse have one and only one point, R, in common.

If ML is the minor axis of the ellipse $\mathscr{E}_2$ (this assumes $\dfrac{ML}{LT} = \dfrac{1}{k} < 1$, so $k > 1$); then $\mathscr{C}$ has diameter MX, $\mathscr{E}_2$ and $\mathscr{C}$ touch one another at M. In considering the major axis of the ellipse, YZ, and the chord TU cut off in the circle by $Y\hat{M}Z$, an angle that is obtuse because $ML < YZ$, Ibn al-Haytham distinguishes three cases: 1) $YZ = TU$; 2) $TU < YZ$; 3) $TU > YZ$.

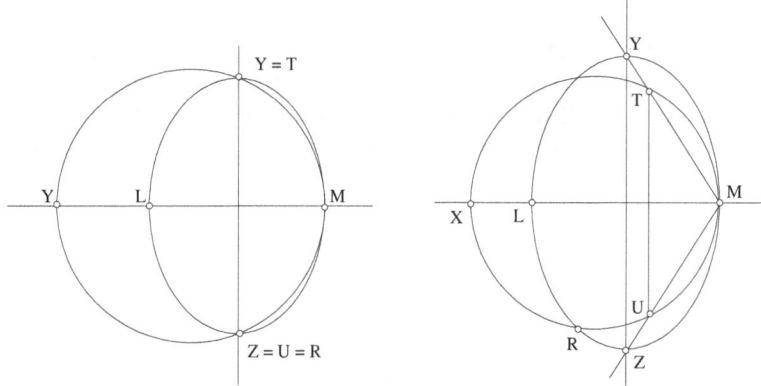

Fig. 1.2.13.8   Fig. 1.2.13.9

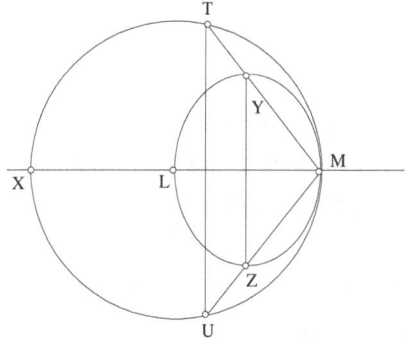

Fig. 1.2.13.10

In the first two cases, $\mathscr{C}$ and $\mathscr{E}_2$ cut one another, that is, R exists.
In the third case, R does not exist.

If the diameter ML of $\mathscr{E}_2$ is its major axis (this assumes $k < 1$), then X lies outside the ellipse.

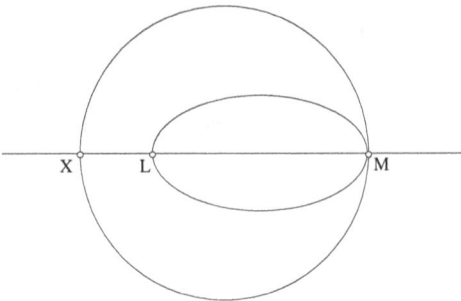

Fig. 1.2.13.11

If $E = G$, we revert to the first conic section Γ ($\mathcal{H}$ or $\mathcal{E}$): if the diameter BH is the transverse axis of $\mathcal{H}$ or an axis of $\mathcal{E}$, the problem has no solution (by *Conics* II.29 and 30). If the diameter BH is not an axis of $\mathcal{E}$, the tangent at B cuts the axis in D, and the second tangent from B, say BA, is equal to BD.

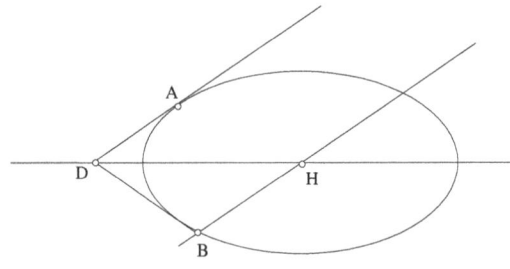

Fig. 1.2.13.12

Investigation of Problems 12 and 13

   Given: A conic section Γ (a hyperbola $\mathcal{H}$ or an ellipse $\mathcal{E}$) with centre H, the tangent BD at a point B of Γ and a ratio $\dfrac{E}{G}$.

   Problem: To draw a tangent to Γ to cut BD in D and to be such that, if A is the point of contact, we have $\dfrac{DA}{DB} = \dfrac{E}{G}$.

   The conic section Γ is specified by the diameter $d = 2HB$, the *latus rectum a* corresponding to it ($k = \dfrac{d}{a}$) and the angle α that HB makes with the conjugate direction, an angle called the 'angle of the ordinates' (there are two possibilities depending on whether α is acute or obtuse).

So in this problem we know HB in position and magnitude, $k = \dfrac{d}{a}$, α and $\dfrac{E}{G}$.

We note that if the conic section Γ is a hyperbola, Ibn al-Haytham considers only one branch, the one that includes the point B.

In Propositions 12 and 13, Ibn al-Haytham proposes to show that the problem concerned reduces to the construction of a tangent that makes a known angle Θ with the diameter HB of the conic section Γ.

1) If the tangent at the point A is a solution to the problem and cuts the diameter HB in P and the diameter conjugate with HB in N, then angle $H\hat{P}N$ is known because triangle HPN is similar to a triangle that can be constructed from what we are given in the problem.

In order to arrive at this result, Ibn al-Haytham, in his analysis, brings in triangle NOA, which is similar to HPN, and at the beginning of the synthesis he gives the construction, from the data provided, of the triangles MXR and MJL, which are similar to the triangles NOA and NPH. We then have $M\hat{J}L = N\hat{P}H = \Theta$, an angle that depends only on what we have been given.

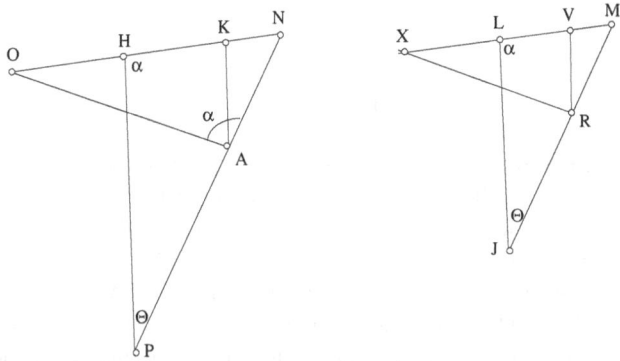

Fig. 1.2.13.13

2) Ibn al-Haytham reminds us that using Propositions 57 and 59 of the second book of *Conics*, we know how to construct a tangent to Γ such that $H\hat{P}N = \Theta$.

Then he shows that if $H\hat{P}N = \Theta$, then the tangent gives a solution to the problem, that is, it satisfies

$$\frac{DA}{DB} = \frac{E}{G}.$$

Comments

1) If Γ is a circle, with the tangent at B known, for any point A, except where A coincides with B' on the diameter HB, the tangent at A cuts the tangent at B in the point D and we have $\dfrac{DA}{DB} = 1$, so $\dfrac{E}{G} = 1$.

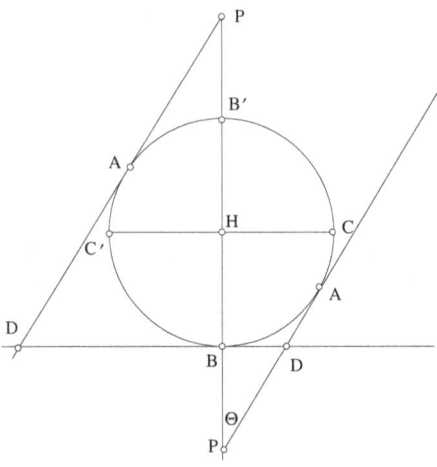

Fig. 13.14

2) Ibn al-Haytham writes: 'we extend AD on the side of D, thus it meets HB; let it meet it at the point P'[50]. This holds true in the case of the hyperbola if we consider only the branch $\mathscr{H}$ that includes the point B. But in the case of the ellipse (as for the circle) several different situations may occur. Let CC' be the diameter conjugate with BHB'. If A is at C or C', the tangent is parallel to HB, and P does not exist. If A lies on the arc BC or the arc BC', we have P beyond D on AD produced; the points concerned occur in the order A, D, P, and H, B, P. But if A lies on one of the arcs CB' or $B'C'$, the points are in the order D, A, P and B, H, P. (See figures below.)

It is thus probable that Ibn al-Haytham is assuming that A lies on the arc CBC'. But we may note that reasoning is valid for any position of A other than C or C' provided we take the 'angle of the ordinates' $\alpha = P\hat{B}D = P\hat{H}N$.

Moreover, if two points A are symmetrical with respect to H, the corresponding tangents are parallel, the points P, as well as the points N associated with them, are symmetrical with respect to H. The two lengths ND associated with them are equal, so the ratio $\dfrac{ND}{NH}$ is the same for the two

[50] See below, p. 198.

points concerned, and the same holds for the ratio $\dfrac{DA}{DB}$, because we can show that $\dfrac{DA}{DB} = \dfrac{DN}{NH}$.

This holds true equally in the case of the hyperbola and in that of the ellipse.

Thus two points of Γ symmetrical with respect to H have associated with them two parallel tangents, and in consequence the same angle $\Theta = H\hat{P}N$ and the same ratio $\dfrac{DA}{DB}$.

This explains why Ibn al-Haytham considers only one branch of a hyperbola or only a semi-ellipse.

3) Starting from the equality $\dfrac{HP}{IP} = \dfrac{NH}{IA} = \dfrac{HB^2}{k \cdot IA^2}$, Ibn al-Haytham brings in:

$\Sigma = k \cdot IA^2$, where Σ is a square bearing a known ratio to the square of IA, but Σ is not known, because IA is not.

$Q = \dfrac{HB^2}{k}$, where Q is a known square, the square of a straight line Δ whose ratio to HB is known. We have $\Delta = \dfrac{HB}{\sqrt{k}}$, and Δ is a known straight line.

Then we have

(1) $$\frac{Q}{IA^2} = \frac{HB^2}{\Sigma} = \frac{NH}{AI}.$$

From (1), we obtain

(2) $$NH \cdot AI = Q \quad \text{(or } NH \cdot HK = Q \text{ since } AI = HK\text{)}$$

and

$$\frac{Q}{HB^2} = \frac{IA^2}{\Sigma},$$

that is to say

(3) $$\frac{NH \cdot HK}{HB^2} = \frac{AI^2}{\Sigma} = \frac{1}{k}.$$

It is not clear what purpose is served by introducing Σ, Q and Δ, to which Ibn al-Haytham does not refer again. The calculation that leads to (2) and (3) need take no more than a few lines without an appeal to Σ or Q, but manipulation of proportions often proceeds by the introduction of such magnitudes.

4) If d denotes the diameter BB' of the conic section Γ ($\mathcal{H}$ or $\mathcal{E}$) and a the *latus rectum* associated with it, and if we denote by d' the diameter CC' conjugate with BB' and by a' the *latus rectum* associated with it, then

$$k = \frac{d}{a} = \frac{a'}{d'},$$

so

$$\frac{d'}{a'} = \frac{1}{k}.$$

Ibn al-Haytham defines a hyperbola $\mathcal{H}_1$ and an ellipse $\mathcal{E}_1$ whose diameter is $d_1 = NH$ and whose *latus rectum* is a_1, such that $\dfrac{d_1}{a_1} = \dfrac{1}{k} = \dfrac{d'}{a'}$.

This is not enough to define $\mathcal{H}_1$ or $\mathcal{E}_1$, but it is clear that Ibn al-Haytham also assumes that the diameter conjugate with HN is parallel to HB, because he takes AK as the ordinate of the point A (meaning $AK \parallel HB$).

In consequence the conic $\mathcal{H}_1$ (or $\mathcal{E}_1$) is similar to the conic section Γ ($\mathcal{H}$ or $\mathcal{E}$) given in the problem.

5) In the part devoted to the discussion, Ibn al-Haytham provides details concerning the auxiliary constructions used in the problem when it is assumed $E \neq G$. He examines a number of special cases where Γ is an ellipse, but he does not consider the problem of establishing the existence of a tangent that satisfies the conditions of the statement: does the proposed problem have a solution for every value of $\dfrac{E}{G}$?

Analytic investigation of the problem
   A) Let Γ be an ellipse with centre H, and BB' a diameter.

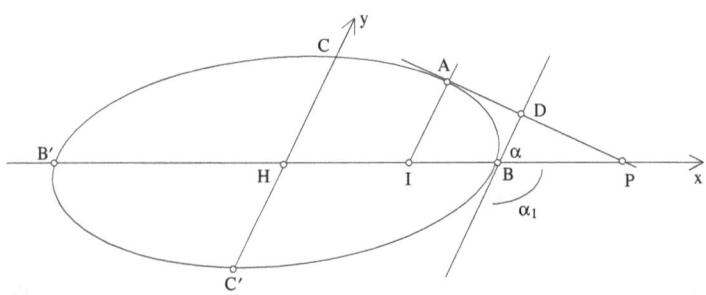

Fig. 13.15

We shall consider A, a variable point on the semi-ellipse CBC'. Let us put $BB' = d$, and $d/a = k$, where a is the *latus rectum* associated with d.

The equation of the ellipse with respect to the axes HB and HC is $x^2 + ky^2 = d^2/4$. If AP is the tangent at A, we have $HI \cdot HP = HB^2$. Let (x_1, y_1) be the coordinates of A; we have, with $x_1 > 0$,

$$HP = \frac{d^2}{4x_1}.$$

We also have $HI \cdot IP = k \cdot AI^2$, hence $IP = \frac{ky_1^2}{x_1}$. The equation of the tangent at A is

$$y - y_1 = (x - x_1) \cdot \frac{-x_1}{ky_1}.$$

The point D of this straight line has abscissa $\frac{d}{2}$, therefore

$$\overline{BD} = y_1 + \left(\frac{d}{2} - x_1\right) \cdot \frac{-x_1}{ky_1} = \frac{ky_1^2 + x_1^2 - \frac{d}{2}x_1}{ky_1},$$

$$\overline{BD} = \frac{d}{2ky_1}\left(\frac{d}{2} - x_1\right)$$

(where $\overline{BD}$ has the same sign as y_1).

We have

$$\frac{AD}{IB} = \frac{AP}{IP},$$

hence

$$AD = AP \cdot \frac{\left(\frac{d}{2} - x_1\right)x_1}{ky_1^2},$$

$$\frac{E}{G} = \frac{AD}{BD} = \frac{2x_1 \cdot AP}{d \cdot |y_1|}.$$

But

$$AP^2 = AI^2 + PI^2 - 2IA \cdot IP \cdot \cos \alpha$$

$$= y_1^2 + \left(\frac{ky_1^2}{x_1}\right)^2 - 2|y_1| \cdot \frac{ky_1^2}{x_1}\cos\alpha$$

$$= \frac{y_1^2}{x_1^2}\left(x_1^2 + k^2y_1^2 - 2k|y_1||x_1|.\cos\alpha\right)$$

where $\alpha < \dfrac{\pi}{2}$ when $A \in BC$ and $\alpha > \dfrac{\pi}{2}$ when $A \in BC'$. So we have

$$\frac{E^2}{G^2} = \frac{4}{d^2}\left(x_1^2 + k^2 y_1^2 - 2k|y_1||x_1| \cdot \cos\alpha\right).$$

But $\dfrac{d^2}{4} = x_1^2 + ky_1^2$; so, for $t > 0$, assuming $x_1 \neq 0$ and putting $\dfrac{|y_1|}{x_1} = t$ we can write

$$\frac{E^2}{G^2} = \frac{x_1^2 + k^2 y_1^2 - 2k|y_1||x_1| \cdot \cos\alpha}{x_1^2 + ky_1^2} = \frac{1 + k^2 t^2 - 2kt\cos\alpha}{1 + kt^2},$$

where $1 + kt^2 = \dfrac{d^2}{4x_1^2}$.

The equation in t can be written

$$(*) \qquad k\left(\frac{E^2}{G^2} - k\right)t^2 + 2kt\,\cos\alpha + \frac{E^2}{G^2} - 1 = 0;$$

its discriminant is

$$\Delta = k\left[k\cos^2\alpha - \left(\frac{E^2}{G^2} - k\right)\left(\frac{E^2}{G^2} - 1\right)\right]$$

and it is positive and equal to $k^2\cos^2\alpha$ for $\dfrac{E^2}{G^2} = 1$ or k. Since

$$\Delta = k\left[-\frac{E^4}{G^4} + \frac{E^2}{G^2}(1 + k) - k\sin^2\alpha\right],$$

we can see that Δ becomes zero for two values of $\dfrac{E^2}{G^2}$, m and M, between which it stays positive; the values 1 and k lie in the interval $[m, M]$. The equation $(*)$ thus has two roots for $\sqrt{m} \leq \dfrac{E}{G} \leq \sqrt{M}$, and these two roots coalesce at the end points of the interval. Their product comes to

$$\frac{1}{k}\frac{\dfrac{E^2}{G^2} - 1}{\dfrac{E^2}{G^2} - k}$$

and it is negative if $\dfrac{E}{G}$ lies between 1 and $\sqrt{k}$; thus equation $(*)$ has one positive root t_0 to which there will correspond one solution with A on the arc BC and another solution A' on the arc BC'. If on the other hand $\dfrac{E}{G}$ is inside the interval $[\sqrt{m}, \sqrt{M}]$, but not between 1 and $\sqrt{k}$, the two roots of (*) have the same sign; half their sum

$$\frac{\cos\alpha}{k - \dfrac{E^2}{G^2}}$$

must be positive for the roots to be acceptable. This condition means that $\dfrac{E}{G} < \sqrt{k}$ if $\alpha < \dfrac{\pi}{2}$ and that $\dfrac{E}{G} > \sqrt{k}$ if $\alpha > \dfrac{\pi}{2}$; in these cases we have two solutions t_0, t_1, which correspond to two points on the same arc BC or BC'.

In the limiting case where $\dfrac{E}{G} = \sqrt{k}$, equation (*) reduces to the first degree $2kt\cos\alpha + k - 1 = 0$, so that $t = \dfrac{1-k}{2k\cos\alpha}$, which corresponds to a solution only if $1 - k$ and $\cos\alpha$ have the same sign, that is, if $k < 1$ for $\alpha < \dfrac{\pi}{2}$ or $k > 1$ for $\alpha > \dfrac{\pi}{2}$. If, in addition, $\alpha = \dfrac{\pi}{2}$, the equation makes $k = 1$ and leaves t indeterminate; in this case the ellipse is a circle and $\dfrac{E}{G} = 1$, the two tangents to the circle from a general point always being equal.

In the limiting case $\dfrac{E}{G} = 1$, equation (*) becomes

$$kt\left[\left(\frac{E^2}{G^2} - k\right)t + 2\cos\alpha\right] = 0$$

and its solution $t = 0$ corresponds to the vertex B of the ellipse.
Calculation gives

$$m = \frac{1 + k - r}{2}, \quad M = \frac{1 + k + r}{2},$$

where $r^2 = (1 + k)^2 - 4k\sin^2\alpha = 1 + 2k\cos2\alpha + k^2$.

B) Γ is one branch of a hyperbola, diameter $BB' = d$, *latus rectum* a, $k = \dfrac{d}{a}$.

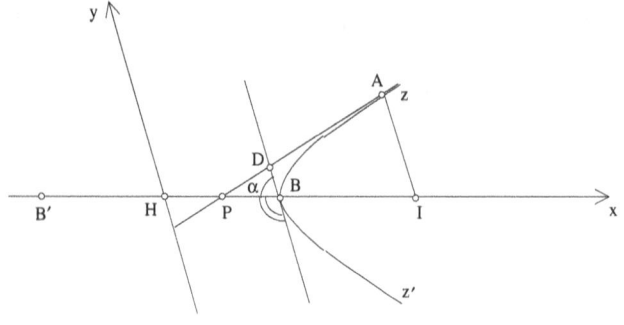

Fig. 1.2.13.16

The equation of the branch of the hyperbola is

$$x^2 - ky^2 = \frac{d^2}{4}, \qquad \text{where } x > \frac{d}{2}.$$

We also have

$$\overline{HI} \cdot \overline{HP} = HB^2, \quad \overline{HI} > 0 \quad \overline{HP} > 0,$$

and

$$\overline{IH} \cdot \overline{IP} = kAI^2, \quad \overline{IH} < 0 \quad \overline{IP} < 0,$$

hence

$$HP = \frac{d^2}{4x_1} \text{ and } IP = \frac{ky_1^2}{x_1}.$$

The equation of the tangent at A is

$$y - y_1 = (x - x_1) \cdot \frac{x_1}{ky_1};$$

$$x = \frac{d}{2} \Rightarrow y = \overline{BD} = y_1 + \left(\frac{d}{2} - x_1\right) \cdot \frac{x_1}{ky_1} = \frac{ky_1^2 - x_1^2 + \frac{d}{2}x_1}{ky_1},$$

so

$$\overline{BD} = \frac{d\left(x_1 - \frac{d}{2}\right)}{2ky_1} \qquad (\overline{BD} \text{ has the same sign as } y_1).$$

We have

$$AI \parallel BD \;\Rightarrow\; \frac{AD}{IB} = \frac{AP}{IP} \Rightarrow AD = \frac{\left(x_1 - \frac{d}{2}\right)x_1}{ky_1^2} \cdot AP,$$

hence

$$\frac{E}{G} = \frac{DA}{DB} = \frac{2x_1}{d} \cdot \frac{AP}{|y_1|}.$$

But

$$AP^2 = AI^2 + IP^2 - 2AI \cdot IP \cos \alpha,$$

$$= y_1^2 + \frac{k^2 y_1^4}{x_1^2} - 2|y_1| \cdot \frac{k y_1^2}{x_1} \cos \alpha,$$

$$= \frac{y_1^2}{x_1^2} \left(x_1^2 + k^2 y_1^2 - 2k x_1 |y_1| \cos \alpha \right),$$

$$\frac{E^2}{G^2} = \frac{4}{d^2} \left(x_1^2 + k^2 y_1^2 - 2k x_1 |y_1| \cos \alpha \right), \text{ with } \frac{d^2}{4} = x_1^2 - k y_1^2;$$

putting $\dfrac{|y_1|}{x_1} = t$ we have

$$\frac{E^2}{G^2} = \frac{1 + k^2 t^2 - 2kt . \cos \alpha}{1 - kt^2} = f(t), \text{ with } kt^2 = 1 - \frac{d^2}{4x_1^2}.$$

The equation in t can be written

(**) $\quad k\left(k + \dfrac{E^2}{G^2} \right) t^2 - 2kt \cos \alpha + 1 - \dfrac{E^2}{G^2} = 0;$

its discriminant is

$$\Delta = k \left[k \cos^2 \alpha - \left(k + \frac{E^2}{G^2} \right)\left(1 - \frac{E^2}{G^2} \right) \right]$$

and this is positive and equal to $k^2 \cos^2 \alpha$ for $\dfrac{E^2}{G^2} = 1$. Since

$$\Delta = k \left[\frac{E^4}{G^4} + (k-1)\frac{E^2}{G^2} - k \sin^2 \alpha \right]$$

is negative for $\dfrac{E^2}{G^2} = 0$, we can see that it becomes zero for a unique value m of $\dfrac{E^2}{G^2} > 0$, a value that lies between 0 and 1. For Δ to be positive, it is necessary and sufficient that $\dfrac{E}{G} \geq \sqrt{m}$; thus equation (**) has two roots, which coalesce when $\dfrac{E}{G} = \sqrt{m}$. The product of these roots has the same sign as $1 - \dfrac{E^2}{G^2}$; when $\dfrac{E}{G} > 1$, only one of the roots is positive and leads to solutions of the problem, A and A', respectively on the arcs BZ and BZ'. If on the contrary $\sqrt{m} < \dfrac{E}{G} < 1$, the two roots have the same sign as $\cos \alpha$; so there are no corresponding solutions except for $\alpha < \dfrac{\pi}{2}$.

In the limiting case $\alpha = \dfrac{\pi}{2}$, the equation requires $\dfrac{E}{G} > 1$. Calculation gives

$$m = \frac{1 - k + r}{2}, \text{ with } r^2 = (k-1)^2 + 4k \sin^2 \alpha = k^2 - 2k \cos 2\alpha + 1.$$

There are two solutions for $\dfrac{E}{G} > \sqrt{m}$; both lie on the arc BZ if $\sqrt{m} < \dfrac{E}{G} < 1$, but for $\dfrac{E}{G} > 1$ one lies on BZ, and the other on BZ'. If $\dfrac{E}{G} = \sqrt{m}$, there is one double solution; if $\dfrac{E}{G} = 1$, there is one solution on the arc BZ; the other solution, which corresponds to $t = 0$, coincides with the point B (a degenerate case).

Ibn al-Haytham's aim was to show that the solution of the problem posed here can be deduced from the solution of Problem 50 of Book II of the *Conics*: Given a conic section, to draw a tangent to it to make with the axis, on the same side as the conic, an angle equal to a given acute angle Θ_1. In the case of the ellipse Problem 50 is possible for any acute angle, and in the case of the hyperbola it is possible only if Θ_1 is greater than the acute angle between an asymptote and the axis.

Here the diameter concerned is a transverse diameter HB assumed to be in general distinct from the axis HX.

The required tangent DA cuts the tangent at B in the point D and the diameter HB in the point P.

In his analysis Ibn al-Haytham shows, using an auxiliary construction considered at the beginning of the synthesis, that if $\dfrac{DA}{DB} = \dfrac{E}{G}$ is known, then angle $A\hat{P}B = \Theta$ is known, and he says in the synthesis that if Θ is known, then Θ_1 is known.

Let β be the angle HX makes with HB; the relationship between Θ (the angle the tangent AP makes with HB), Θ_1 (the angle the tangent AP makes HX) and β will depend on properties of the figure and in particular on $\alpha = P\hat{B}D$.

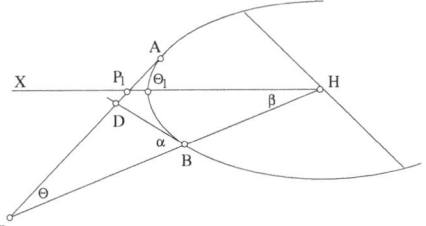

Fig. 1.2.13.17a: $\Theta = \Theta_1 - \beta$, α acute

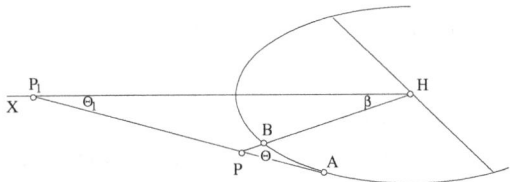

Fig. 1.2.13.17b: $\Theta = \Theta_1 + \beta$, α obtuse

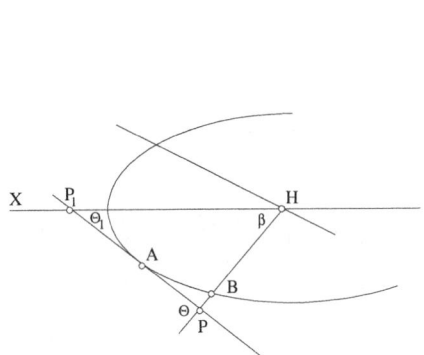

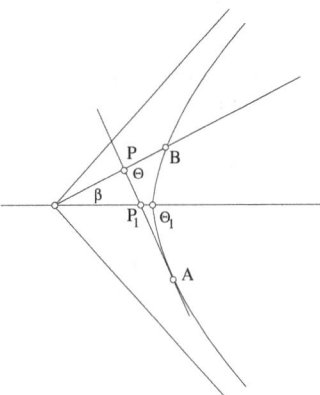

Fig. 1.2.13.17c: $\Theta = \pi - (\Theta_1 + \beta)$, α obtuse   Fig. 1.2.13.17d: $\Theta = \Theta_1 + \beta$, α acute

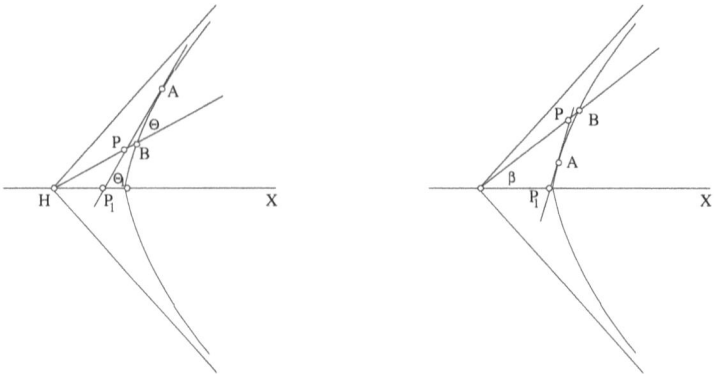

Fig. 1.2.13.17e: $\Theta = \Theta_1 - \beta$, α obtuse   Fig. 1.2.13.17f: $\Theta + \Theta_1 - \beta = \pi$, α acute

Analytical determination of the angle $\Theta = \Theta_1 + \beta$

To determine Θ by an auxiliary construction that uses the given values $\dfrac{d}{a} = k$, a and $\dfrac{E}{G}$ would entail a discussion that Ibn al-Haytham does not supply and whose complexity can be seen in the relationship that connects $\dfrac{E}{G}$, Θ, α and k.

The case of the ellipse
We have

(1) $\dfrac{IB \cdot IB'}{AI^2} = k$, $HI \cdot HP = HB^2$, $\dfrac{HI \cdot IP}{AI^2} = k$.

So

$$\frac{DB}{PB} = \frac{AI}{PI} = \frac{HI}{k \cdot AI} \text{ and } PB = HP - HB = \frac{HB \cdot BI}{HI}.$$

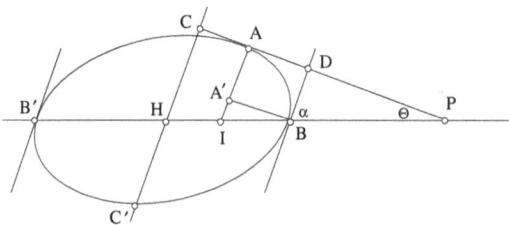

Fig. 1.2.13.18

Moreover

$$\frac{DA}{BI} = \frac{\sin\alpha}{\sin P\hat{A}I},$$

where $P\hat{A}I = P\hat{D}B = \pi - (\alpha + \Theta)$ (α acute or obtuse), hence

$$\sin P\hat{A}I = \sin(\alpha + \Theta)$$

and we have

$$DA = BI \cdot \frac{\sin\alpha}{\sin(\alpha + \Theta)} \text{ and } DB = \frac{HB \cdot BI}{k \cdot AI},$$

hence

$$\frac{E}{G} = \frac{DA}{DB} = \frac{\sin\alpha}{\sin(\alpha + \Theta)} \cdot \frac{k.AI}{HB}.$$

But

$$\frac{HI}{k \cdot AI} = \frac{AI}{PI} = \frac{\sin\Theta}{\sin(\alpha + \Theta)}$$

and (1) can be written

$$HB^2 - HI^2 = k\, AI^2,$$

hence

$$HB^2 = kAI^2 + k^2 AI^2 \cdot \frac{\sin^2\Theta}{\sin^2(\alpha + \Theta)}$$

and

$$\frac{HB^2}{kAI^2} = 1 + k \cdot \frac{\sin^2\Theta}{\sin^2(\alpha + \Theta)}.$$

Thus we have

$$\frac{G^2}{E^2} = \frac{\sin^2(\alpha + \Theta)}{k\sin^2\alpha}\left[1 + \frac{k\sin^2\Theta}{\sin^2(\alpha + \Theta)}\right],$$

$$\frac{G^2}{E^2} = \frac{\sin^2(\alpha + \Theta) + k\sin^2\Theta}{k\sin^2\alpha} = f(\Theta) = \frac{\phi(\Theta)}{k\sin^2\alpha}.$$

This calculation is valid for all positions of the point A on the arc BC $(0 < \alpha < \frac{\pi}{2})$ or on the arc BC' $(\frac{\pi}{2} < \alpha < \pi)$. In both cases we have

$$0 \leq \Theta \leq \pi - \alpha \text{ and } \alpha \leq \alpha + \Theta \leq \pi,$$

where

$$\Theta = 0 \Rightarrow \frac{E}{G} = \sqrt{k} \text{ and } \Theta = \pi - \alpha \Rightarrow \frac{E}{G} = 1.$$

The case of the hyperbola
    We have

(1) $$\frac{IB \cdot IB'}{AI^2} = k \Leftrightarrow HI^2 - HB^2 = k\,AI^2 \Leftrightarrow HB^2 = HI^2 - k\,AI^2$$

and

$$HI \cdot HP = HB^2,$$

$$HI \cdot IP = k\,AI^2,$$

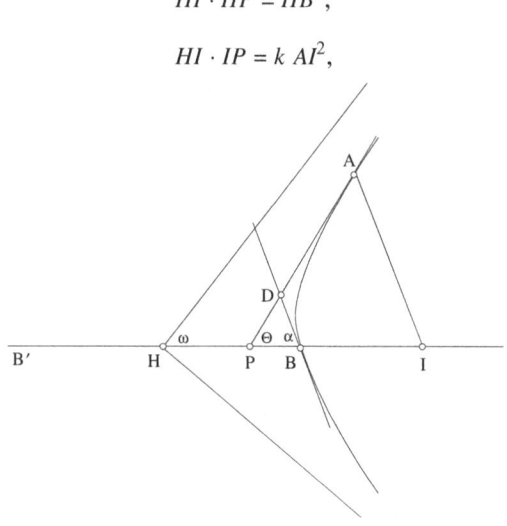

Fig. 13.19

from which we obtain

$$PB = HB - HP = HB - \frac{HB^2}{HI} = \frac{HB \cdot BI}{HI},$$

hence, as in the case of the ellipse,

$$DB = \frac{HB \cdot HI}{kAI}.$$

    We also have

$$DA = BI \cdot \frac{\sin\alpha}{\sin(\alpha + \Theta)},$$

hence

$$\frac{E}{G} = \frac{\sin\alpha}{\sin(\alpha + \Theta)} \cdot \frac{k.AI}{HB}.$$

But

$$\frac{HI}{k \cdot AI} = \frac{AI}{PI} = \frac{\sin\Theta}{\sin(\alpha + \Theta)}.$$

$$(1) \Rightarrow \quad HB^2 = k^2 AI^2 \cdot \frac{\sin^2 \Theta}{\sin^2(\alpha + \Theta)} - k AI^2 = k AI^2 \left[\frac{k \sin^2 \Theta}{\sin^2(\alpha + \Theta)} - 1 \right],$$

hence

$$\frac{G^2}{E^2} = \frac{\sin^2(\alpha + \Theta)}{k \sin^2 \alpha} \left[\frac{k \sin^2 \Theta}{\sin^2(\alpha + \Theta)} - 1 \right],$$

$$\frac{G^2}{E^2} = \frac{k \sin^2 \Theta - \sin^2(\alpha + \Theta)}{k \sin^2 \alpha} = g(\Theta) = \frac{\Psi(\Theta)}{k \sin^2 \alpha},$$

a calculation that is valid for acute and obtuse α.

The relationship between Θ, α, k and $\dfrac{E}{G}$ can thus be written

$$(1) \qquad k \sin^2 \Theta + \varepsilon \sin^2(\alpha + \Theta) = k \frac{G^2}{E^2} \sin^2 \alpha,$$

where $\varepsilon = 1$ in the case of the ellipse and $\varepsilon = -1$ in the case of the hyperbola. The equation becomes simpler if we express it in terms of the angle Θ_1 that the required tangent makes with the principal axis of the conic. Let us denote the ratio of the axes by $c = \dfrac{b}{a}$ ($0 < c \le 1$) and the slope of the diameter HB by p; the equation of the conic with respect to its principal axes HX, HY is $x^2 + \varepsilon \dfrac{y^2}{c^2} = a^2$; the conjugate diameter HC has slope $-\varepsilon \dfrac{c^2}{p}$. We find $k = \dfrac{HB^2}{HC^2}$ by writing

$$HB^2 = (1 + p^2) \frac{a^2}{1 + \dfrac{\varepsilon p^2}{c^2}}, \quad HC^2 = \varepsilon \left(1 + \frac{c^4}{p^2} \right) \frac{a^2}{1 + \varepsilon \dfrac{c^2}{p^2}} ;$$

so

$$k = c^2 \frac{1 + p^2}{p^2 + c^4}.$$

Let us now find the angle α from its tangent:

$$\tan \alpha = \frac{-\varepsilon \dfrac{c^2}{p} - p}{1 - \varepsilon c^2} = -\frac{1}{p} \frac{p^2 + \varepsilon c^2}{1 - \varepsilon c^2}.$$

Equation (1) becomes

$$(2) \qquad (k + \varepsilon \cos 2\alpha) \cos 2\Theta - \varepsilon \sin 2\alpha \sin 2\Theta = k + \varepsilon - k \frac{G^2}{E^2} (1 - \cos 2\alpha)$$

with

$$\sin 2\alpha = -2p\frac{\left(p^2 + \varepsilon c^2\right)\left(1 - \varepsilon c^2\right)}{\left(1 + p^2\right)\left(p^2 + c^4\right)},$$

$$\cos 2\alpha = \frac{\left[p(1-p) - \varepsilon c^2(1+p)\right]\left[p(1+p) + \varepsilon c^2(1-p)\right]}{\left(1 + p^2\right)\left(p^2 + c^4\right)}.$$

We find

$$k + \varepsilon\cos 2\alpha = \frac{\left(1 - \varepsilon c^2\right)\left(c^2 + \varepsilon p^2\right)}{p^2 + c^4}\cdot\frac{1 - p^2}{1 + p^2},$$

$$\varepsilon\sin 2\alpha = -\frac{\left(1 - \varepsilon c^2\right)\left(c^2 + \varepsilon p^2\right)}{p^2 + c^4}\cdot\frac{2p}{1 + p^2}$$

and equation (2) becomes

(3) $$\left(1 - \varepsilon c^2\right)\cos 2(\Theta - \beta) = 1 + \varepsilon c^2 - 2c^2\frac{G^2}{E^2}\frac{c^2 + \varepsilon p^2}{p^2 + c^4},$$

with $\tan\beta = p$ (β is the angle between HX and HB).
     Alternatively

(4) $$\cos 2\Theta_1 = \frac{1 + \varepsilon c^2}{1 - \varepsilon c^2} - \frac{2c^2}{1 - \varepsilon c^2}\frac{G^2}{E^2}\frac{c^2 + \varepsilon p^2}{p^2 + c^4}.$$

The condition for the equation to have a solution is that

$$\left|\frac{1 + \varepsilon c^2}{1 - \varepsilon c^2} - \frac{2c^2}{1 - \varepsilon c^2}\frac{G^2}{E^2}\frac{c^2 + \varepsilon p^2}{p^2 + c^4}\right| \leq 1$$

or

$$\varepsilon c^2 \leq c^2\frac{c^2 + \varepsilon p^2}{p^2 + c^4}\frac{G^2}{E^2} \leq 1.$$

In the case of the ellipse, this condition can be written

(5) $$c^2\frac{c^2 + p^2}{p^2 + c^4} \leq \frac{E^2}{G^2} \leq \frac{c^2 + p^2}{p^2 + c^4};$$

in the case of the hyperbola, it becomes

(5′) $$\frac{E^2}{G^2} \geq c^2\frac{c^2 - p^2}{p^2 + c^4}.$$

We may note that the bounds found in (5) and (5′) coincide with the quantities m and M that appear in the previous discussion.

In the case of the ellipse, if (5) is satisfied, equation (4) defines a unique acute angle Θ_1, which provides the solution to the problem.

In the case of the hyperbola, (4) can also be written

$$(6) \qquad \cos 2\Theta_1 = \cos 2\omega_1 - \frac{G^2}{E^2} c \, \frac{c^2 - p^2}{p^2 + c^4} \sin 2\omega_1,$$

where ω_1 is the angle the asymptote makes with HX; this equation gives a unique acute angle Θ_1 greater than ω_1 because the second member of the equation is less than $\cos 2\omega_1$.

We have just seen, in the course of our discussion of Problem 13, that Ibn al-Haytham did not consider the problem of establishing the existence of a tangent that satisfied the conditions in the statement, namely:

Given the conic section Γ and the point B, so that k and α are known, has the problem got a solution for every value of $\dfrac{E}{G}$?

Let us return again to this same question.

– In the case of the circle, if $\dfrac{E}{G} \neq 1$, the problem has no solution; if $\dfrac{E}{G} = 1$, it is indeterminate, and for every tangent there is a corresponding acute angle $\Theta = B\hat{P}A$. Conversely, $\forall \, \Theta \in \,]0, \frac{\pi}{2}[$, there exists a tangent PA such that $B\hat{P}A = \Theta$ and $\dfrac{DA}{DB} = \dfrac{E}{G} = 1$.

As Θ varies, $\dfrac{E}{G}$ remains constant.

– In the case of a conic section Γ, either an ellipse or a hyperbola, the ratio $\dfrac{E}{G}$ varies as the point A varies.

For example, let us consider an ellipse with major axis BHB'; so $\alpha = \dfrac{\pi}{2}$ and $k > 1$. We know that

$$(1) \qquad \frac{IB \cdot IB'}{AI^2} = k,$$

$$(2) \qquad HI \cdot HP = HB^2,$$

$$(3) \qquad \frac{HI \cdot IP}{AI^2} = k.$$

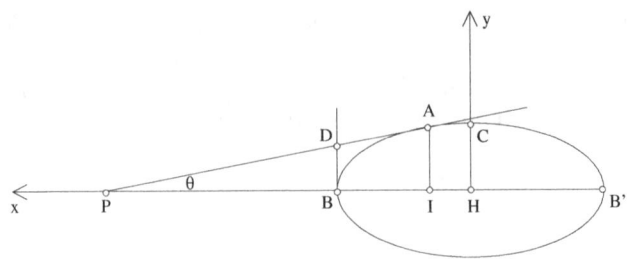

Fig. 1.2.13.20

We assume I lies between H and B

$$HI = x, \qquad\qquad 0 < x < \frac{d}{2}.$$

Taking coordinate axes (Hx, Hy), equality (1) gives

(4) $ky^2 = \left(\frac{d}{2}\right)^2 - x^2,$ the equation of Γ.

From (3) we obtain

$$\frac{HI}{k \cdot AI} = \frac{AI}{IP} = \tan \Theta,$$

hence

(5) $\tan \Theta = \dfrac{x}{ky}.$

We have

$$DA = \frac{BI}{\cos \Theta}, \qquad\qquad DB = PB \cdot \tan \Theta,$$

hence

$$\frac{E}{G} = \frac{DA}{DB} = \frac{BI}{PB \cdot \sin \Theta}.$$

From (2), we obtain

$$HP = \frac{HB^2}{HI} \text{ and } PB = HP - HB = \frac{HB^2}{HI} - HB.$$

$$PB = \frac{HB}{HI}(HB - HI) = \frac{HB \cdot BI}{HI}.$$

Therefore

(6) $\dfrac{E}{G} = \dfrac{HI}{HB \cdot \sin \Theta} = \dfrac{2x}{d \sin \Theta}.$

From (5) and (4), we obtain

$$\left(\frac{d}{2}\right)^2 = x^2 + \frac{1}{k} \cdot \frac{x^2}{\tan^2 \Theta},$$

hence

$$\frac{d^2}{4x^2} = 1 + \frac{1}{k \tan^2 \Theta}.$$

From (6), we then obtain

$$\frac{G^2}{E^2} = \frac{\left(k \tan^2 \Theta + 1\right)\sin^2 \Theta}{k \tan^2 \Theta}$$

$$\frac{G^2}{E^2} = \frac{\left(k \tan^2 \Theta + 1\right) \cos^2 \Theta}{k} = \frac{k \sin^2 \Theta + \cos^2 \Theta}{k} = \frac{(k-1) \sin^2 \Theta + 1}{k}.$$

For given k, $k > 1$, $\dfrac{E}{G}$ is this a function of Θ. As A describes the arc BC, Θ decreases from $\dfrac{\pi}{2}$ to 0; $\sin\Theta$ decreases from 1 to 0, so $\dfrac{G^2}{E^2}$ decreases from 1 to $\dfrac{1}{k}$, and the ratio $\dfrac{E}{G}$ increases from 1 to $\sqrt{k}$.

The problem has no solution unless $1 \le \dfrac{E}{G} \le \sqrt{k}$.

Ellipse with BHB' as any diameter. We shall again assume $k > 1$. The equalities (1), (2) and (3) still hold.

We have

$$\frac{DB}{PB} = \frac{AI}{PI} = \frac{HI}{k \cdot AI} \quad \text{and} \quad PB = HP - HB = \frac{HB \cdot BI}{HI};$$

but

$$\frac{DA}{BI} = \frac{\sin \alpha}{\sin P\hat{A}I}, \qquad P\hat{A}I = P\hat{D}B = \pi - (\alpha + \Theta),$$

$$\sin P\hat{A}I = \sin (\alpha + \Theta),$$

hence

$$DA = BI \cdot \frac{\sin \alpha}{\sin(\alpha + \Theta)} \quad \text{and} \quad DB = \frac{HB \cdot BI}{k \cdot AI},$$

$$\frac{E}{G} = \frac{DA}{DB} = \frac{\sin \alpha}{\sin(\alpha + \Theta)} \cdot \frac{k \cdot AI}{HB},$$

$$\frac{E}{G} = \frac{\sin \alpha}{\sin(\alpha + \Theta)} \cdot \frac{2ky}{d}.$$

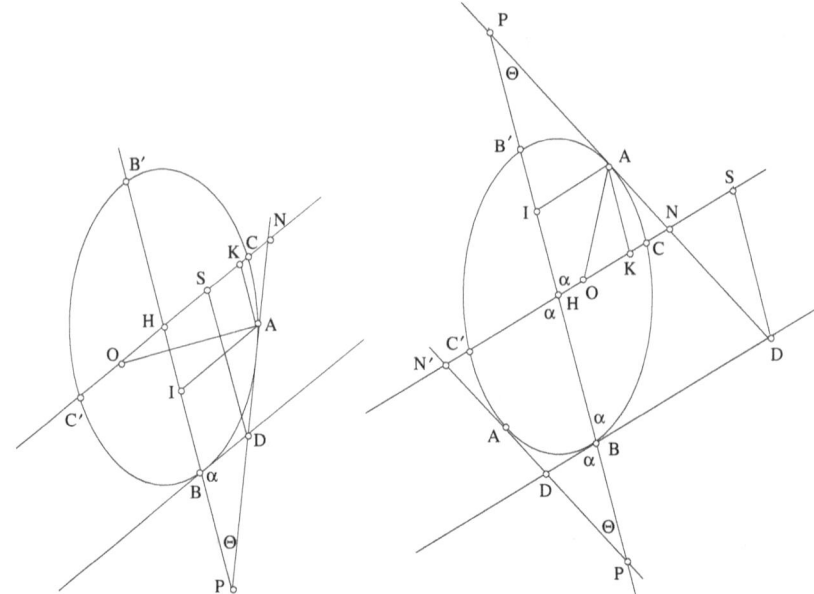

Fig. 1.2.13.21a: $\alpha = P\hat{B}D$ obtuse　　　Fig. 1.2.13.21b: $\alpha = P\hat{B}D$ acute

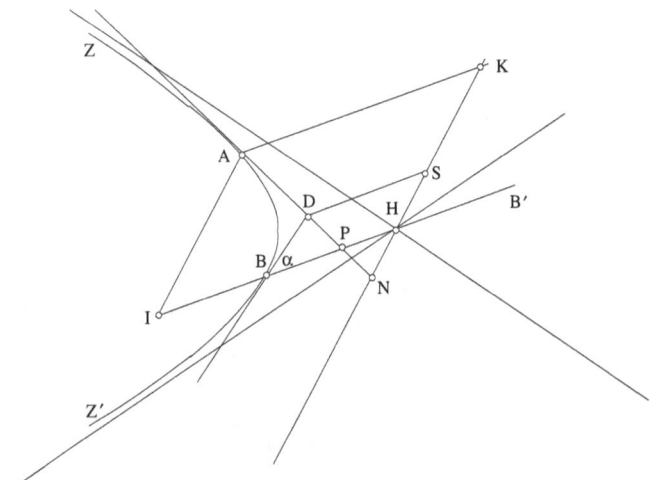

Fig. 1.2.13.21c:
The tangent at any point of the branch of the hyperbola $\mathcal{H}_B$
cuts HB between H and B.
If $A \in BZ$, $\alpha = P\hat{B}D$ is acute.
If $A \in BZ'$, $\alpha = P\hat{B}D$ is obtuse.

But

(4) $$\frac{HI}{k \cdot AI} = \frac{AI}{PI} = \frac{\sin \Theta}{\sin(\alpha + \Theta)} = \frac{x}{ky}$$

and

(5) $$\left(\frac{d}{2}\right)^2 = x^2 + ky^2 \qquad \text{(equation of the ellipse)}.$$

From (4) and (5), we obtain

$$\left(\frac{d}{2}\right)^2 = k^2 y^2 \frac{\sin^2 \Theta}{\sin^2(\alpha + \Theta)} + ky^2,$$

hence

$$\frac{d^2}{4ky^2} = 1 + \frac{k \sin^2 \Theta}{\sin^2(\alpha + \Theta)}.$$

We thus have

$$\frac{G^2}{E^2} = \frac{d^2}{4k^2 y^2} \cdot \frac{\sin^2(\alpha + \Theta)}{\sin^2 \alpha} = \frac{\sin^2(\alpha + \Theta)}{k \sin^2 \alpha} \left[1 + \frac{k \sin^2 \Theta}{\sin^2(\alpha + \Theta)} \right],$$

hence

$$\frac{G^2}{E^2} = \frac{\sin^2(\alpha + \Theta) + k \sin^2 \Theta}{k \sin^2 \alpha}.$$

$\dfrac{E}{G}$ can thus be expressed as a function of Θ, for given k and α.[51] The calculation is valid whether A lies on BC or on BC'. Angle α is acute if A lies on BC', α is obtuse if A lies on BC.

In both cases

$0 \leq \Theta \leq \pi - \alpha$ and $\alpha \leq \alpha + \Theta \leq \pi$.

$\Theta = 0 \Rightarrow \dfrac{E}{G} = \sqrt{k}, \qquad \Theta = \pi - \alpha \Rightarrow \dfrac{E}{G} = 1.$

This discussion concerning the existence of angle Θ thus allows us to come back to the same conditions we obtained earlier by using algebra.

Ibn al-Haytham's discussion only yields cases in which the construction is possible and others in which it is not possible; but the discussion remains incomplete. A complete discussion concerning the existence of the point of intersection R would indeed be difficult to carry out by geometry. As regards the problem as a whole, it is possible to find a solution that is more

---

[51] If $\alpha = \dfrac{\pi}{2}$, we return to $\dfrac{G^2}{E^2} = \dfrac{k \sin^2 \Theta + \cos^2 \Theta}{k} = \dfrac{(k-1)\sin^2 \Theta + 1}{k}$.

elementary than the one developed by Ibn al-Haytham.[52] It is in fact sufficient to consider the intersection of the circle with centre H and radius $\dfrac{E}{G} \cdot \dfrac{\sqrt{ad}}{2}$ with the given ellipse or with the hyperbola that is conjugate with the given hyperbola.

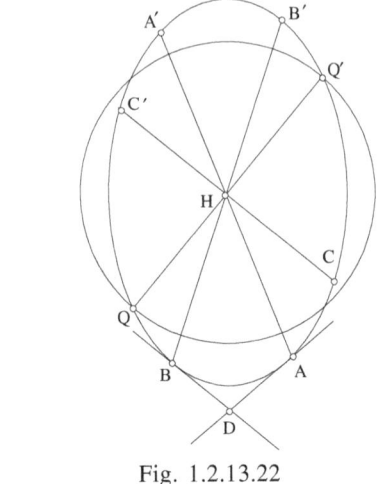

Fig. 1.2.13.22

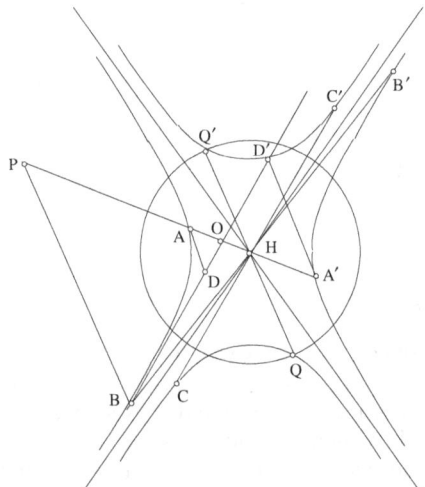

Fig. 1.2.13.23

[52] See J.P. Hogendijk, *Ibn al-Haytham's Completion of the Conics*, Sources in the History of Mathematics and Physical Sciences 7, New York/Berlin/Heidelberg/Tokyo, 1985, p. 383, with a discussion based on an argument from growth and continuity.

The intersection, if there is one, defines two diameters (possibly the same) such as QQ'. We look at AA', the diameter conjugate with QQ' in the conic Γ, and the point A provides a solution to the problem.

For the case of the ellipse, we have $\dfrac{AD^2}{BD^2} = \dfrac{HQ^2}{HC^2}$ from *Conics* III.17; also $HQ^2 = \dfrac{E^2}{G^2} \cdot \dfrac{ad}{4}$ and $HC^2 = \dfrac{ad}{4}$. From this we obtain $\dfrac{AD}{BD} = \dfrac{E}{G}$.

For the case of the hyperbola, we again appeal to the *Conics*, to Proposition 23 of Book III; from this proposition $\dfrac{A'D'^2}{BD'^2} = \dfrac{HQ^2}{HC^2} = \dfrac{E^2}{G^2}$, hence $\dfrac{A'D'}{BD'} = \dfrac{E}{G}$. Moreover, $\dfrac{AD}{BD} = \dfrac{A'D'}{BD'}$ because the triangles $OA'D'$, OAD and OPB are similar to one another, and from Property 36 of Book I of the *Conics*, which gives $\dfrac{AP}{A'P} = \dfrac{AO}{A'O}$.

The number of solutions is given by the number of points of intersection of the circle and the ellipse (respectively with the hyperbola). Calling the axes of the conic α and β, there is no solution if the diameter $\dfrac{E}{G}\sqrt{ad}$ is strictly less than β. In the case of the ellipse, there is no solution either for $\dfrac{E}{G}\sqrt{ad} > \alpha$ and there are two solutions if $\beta < \dfrac{E}{G}\sqrt{ad} < \alpha$. In the case of the hyperbola, there are two solutions if $\dfrac{E}{G}\sqrt{ad} > \beta$. When one of the inequalities is replaced by an equality, the two solutions coincide with one another.

A discussion of the existence and number of solutions, in geometrical terms, requires subtle considerations of the curvature of the conic and seems to be very difficult to carry through without appealing to analytical concerns such as the increase in the radius of the auxiliary circle as a function of the given ratio.

We may wonder why Ibn al-Haytham did not think of such an elementary solution, despite knowing Apollonius' *Conics* better than anyone. In trying to answer this question, we need to remember that elsewhere Ibn al-Haytham uses the intersection of conics to deal with not only three-dimensional problems but also some plane problems, for instance in *The Knowns*. In this latter treatise, he is much less interested in looking for the simplest method of some specific problem than in seeking families of conics that supply solutions to a problem. We have indeed already seen that, in connection with this example, he replaces two conics by two other conics belonging to the same pencil. Let us not forget that Ibn al-Haytham did not usually write to embroider on a known theme, or to propose exercises, but much more to contribute significant advances in mathematics.

– **14** – Let there be a parabola with vertex A and axis AD, and a given straight line E. To find a point B of the parabola such that the tangent at B cuts AD in a point H such that $BH = E$.

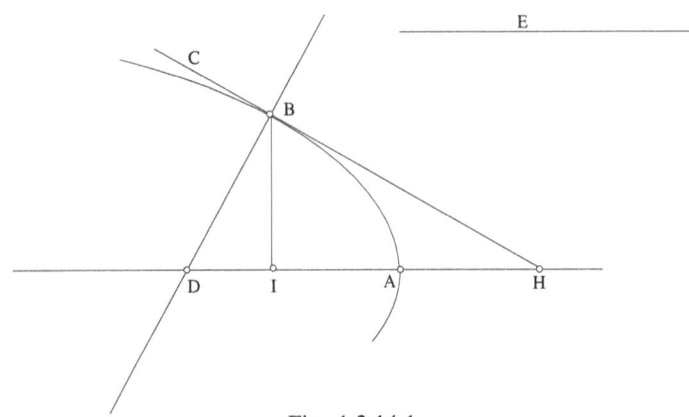

Fig. 1.2.14.1

Analysis: Let us suppose the point B and the straight line BH are known. Let us draw $BI \perp AD$ and $BD \perp BH$. We have $HD \cdot HI = BH^2 = E^2$ (right-angled triangle). In the similar triangles BDH and IBH, we have $\dfrac{DH}{BH} = \dfrac{BH}{HI} \Rightarrow DH \cdot HI = BH^2$.

We know that A is the midpoint of IH, because we have a subtangent, and that $DI = a/2$, because we have a subnormal, where a is the *latus rectum*. From this we find the lengths DH and HI by means of a classical construction;[53] HI is thus known, so the length AI is also known, from which we can find the point I.

---

[53] To construct two lengths x and y when we know that $x - y = l$, $x \cdot y = L^2$, l and L being known lengths.

We draw a circle (O, r), $r = l/2$, we draw a tangent to it TS of length L; the straight line TO cuts the circle in P and Q, we have $TP = x$, $TQ = y$. The construction is always possible.

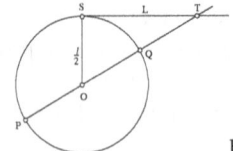

Fig. 1.2.14.2

Note: Let there be a parabola Γ with vertex A and axis AD, and a given straight line E. To draw a tangent to touch Γ at B, and to cut the axis in H and to be such that $BH = E$.

Ibn al-Haytham shows in his analysis that the data are sufficient to determine the point H.

In the synthesis, starting from E and the *latus rectum* of Γ, he says that H is known and simply adds 'we draw from the point H a tangent to the conic section; let it be HB'.[54]

He does not give any indication as to the method of constructing H or B geometrically. Here the two constructions can be carried out with straight-edge and compasses, very simply.

To find the point H, we only need to know AH; now $AH = \dfrac{HI}{2}$, and we have

$$HD \cdot HI = E^2 \text{ and } HD - HI = \frac{1}{2} \text{ latus rectum} = l.$$

Once the point H has been constructed, I can be found from it, because $AH = AI$, then D because $ID = l = \dfrac{1}{2}$ *latus rectum*.

The intersection of the circle with diameter HD with the straight line Δ the perpendicular to the axis of Γ at I is the required point B.

Since he knows H, Ibn al-Haytham knows that the tangent *exists* (Book II.49).

– 15 – *Synthesis*: The synthesis shows that if we are given E and the *latus rectum* a we can find the length HI, and from it deduce H, then I and B. The reasoning then proceeds in the opposite direction from that of the analysis.

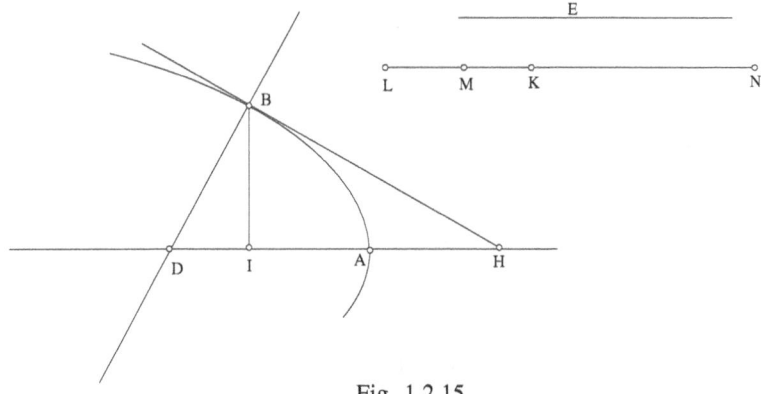

Fig. 1.2.15

[54] See below, p. 208.

Note: The point *B* is the point with abscissa *AI*; it is thus the point of intersection of the parabola and the straight line *Δ* the perpendicular to the axis at *I*; *B* is also the point of intersection of *Δ* with the circle of diameter *HD*, *H* and *D* being defined by what is given.

Finally, we may note that what we have is a problem of intercalation (*neusis*) of the given segment *E* between the parabola and its axis; the condition is that the intercalated line segment shall touch the parabola.

– **16** – Let there be a conic section *Γ*, an ellipse or a hyperbola, with axis *AD* and centre *E*, and a given straight line *F*. To find a point *B* of *Γ* so that the tangent at *B* cuts the axis in a point *H* such that *BH* = *F*.

Analysis: Let *BH* be the required tangent, we draw *BI* ⊥ *AD* and *AC* ∥ *BH*. The straight lines *AC* and *BE* cut one another in *K*, the midpoint of *AC*; we draw *KL* ⊥ *AD*. Let *a* be the *latus rectum*, we define the point *M* by $\dfrac{ME}{MA} = \dfrac{DA}{a}$, *MA* is half the line segment in the same ratio.[55] So we have

$$\frac{ML \cdot LA}{AK^2} = \frac{ME}{AE}.$$

Moreover

$$\frac{HB}{AK} = \frac{HE}{AE},$$

hence

$$AK \cdot HE = AE \cdot BH, \quad \text{a known product.}$$

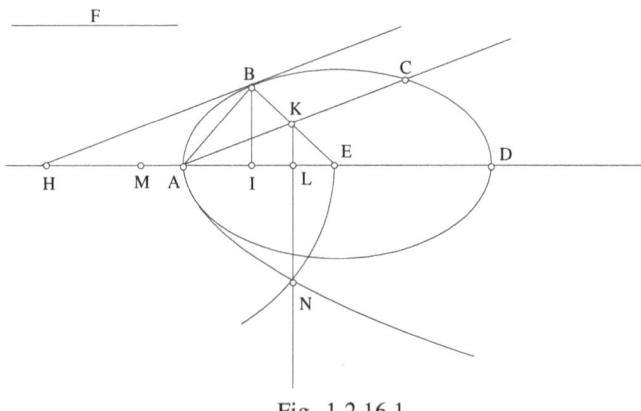

Fig. 1.2.16.1

[55] Apollonius, *Conics*, VII.2-3; see Apollonius de Perge, *Les Coniques*, t. 4: *Livres VI et VII*, pp. 242ff.

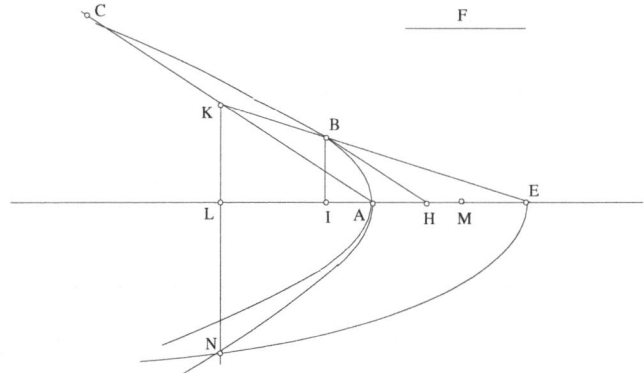

Fig. 1.2.16.2

From a property of the tangent, that (A, D, I, H) is a harmonic range, we have $AE^2 = EI \cdot EH$, and consequently

$$\frac{AK \cdot HE}{EI \cdot EH} = \frac{AE \cdot BH}{AE^2} = \frac{AK}{IE} = \frac{BH}{AE} = \frac{F}{AE} = k, \text{ a known ratio.}$$

From this we obtain

$$\frac{ML \cdot LA}{IE^2} = \frac{ML \cdot LA}{AK^2} \cdot \frac{AK^2}{IE^2} = \frac{ME}{AE} \cdot k^2.$$

But we have seen (Problem 3) that $LE \cdot EA = IE^2$, so

(1) $$\frac{ML \cdot LA}{LE \cdot EA} = \frac{ME}{EA} \cdot k^2.$$

Let $\mathscr{P}$ be the parabola with vertex E, axis EL, *latus rectum* EA, and let $\mathscr{P}$ cut KL in N. Let $\mathscr{H}$ be the hyperbola with vertex A and axis MA, whose *latus rectum* is such that

$$\frac{MA}{a'} = \frac{ME}{EA} \cdot \frac{HB^2}{EA^2} = \frac{ME}{EA} \cdot k^2,$$

where $\mathscr{H}$ cuts KL in N'.

$N \in \mathscr{P}$, so we have $LN^2 = EL \cdot EA$;

$N' \in \mathscr{H}$, so we have $\dfrac{LM \cdot LA}{LN'^2} = \dfrac{MA}{a'} = \dfrac{ME}{EA} \cdot k^2$;

from (1) we thus deduce that $LN'^2 = EL \cdot EA$, so $LN'^2 = LN^2$; the points N and N' thus coincide.

So if the tangent *BH* exists, the hyperbola $\mathcal{H}$ and the parabola $\mathcal{P}$ have a common point *N*. The two auxiliary conics $\mathcal{P}$ and $\mathcal{H}$ meet one another in a point *N* on the vertical through the point *L*, but outside the given conic section Γ. This leads Ibn al-Haytham to give a slightly unusual treatment that can give the misleading impression of resembling an account of a synthesis. It does not help if we try to make a comparison between this passage and the description of the construction of the regular heptagon, in which we can observe a certain vagueness in distinguishing analysis and synthesis.[56]

– 17 – *Synthesis*: As in the analysis we define *M* by $\dfrac{ME}{MA} = \dfrac{DA}{a}$. Let *O* be the straight line defined by $\dfrac{MA}{O} = \dfrac{EM}{EA} \cdot \dfrac{F^2}{EA^2}$; let $\mathcal{H}$ be the hyperbola with transverse axis *MA* and *latus rectum O* and let $\mathcal{P}$ be the parabola with vertex *E*, axis *EA* and *latus rectum AE*. The conic sections $\mathcal{H}$ and $\mathcal{P}$ cut one another in *N* (see the next discussion).

Let *L* and *P* be the points defined by $NL \perp AD$ and $LP = AL$. The points *L* and *P* lie inside Γ, which is an ellipse or a hyperbola. Let *C* be the point of Γ whose projection on the axis is *P*. The straight lines *AC* and *NL* cut one another in *K* and the straight line *EK* cuts Γ in *B*. The tangent at *B* cuts the axis in *H*; we shall show in the next discussion that *BH* = *F*.

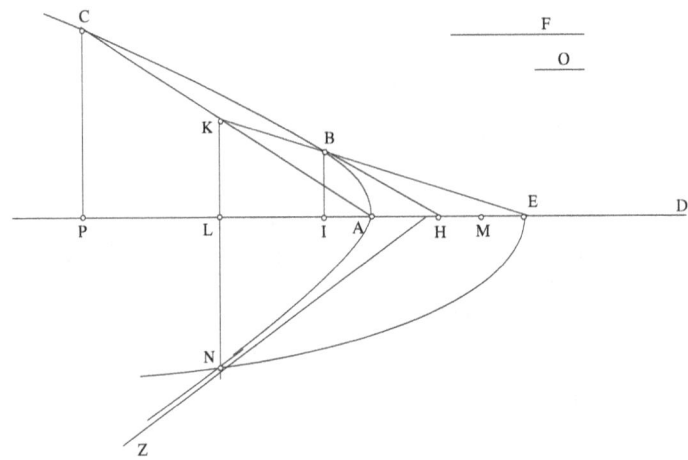

Fig. 1.2.17.1

[56] See below, p. 459.

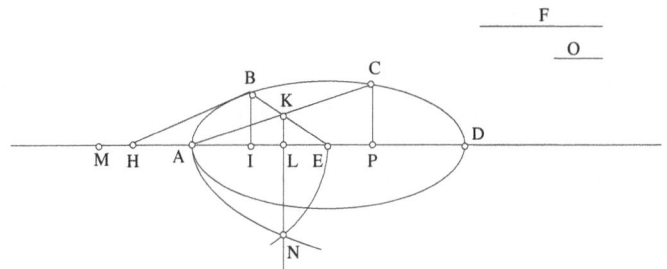

Fig. 1.2.17.2

Let $BI \perp AD$, we have $EL \cdot EA = EI^2$ (see analysis).

Moreover $N \in \mathscr{P}$, hence $EL \cdot EA = LN^2$, so we have $LN = EI$. But $N \in \mathscr{H}$, hence

$$\frac{LA \cdot LM}{LN^2} = \frac{MA}{O} = \frac{ME}{EA} \cdot \frac{F^2}{EA^2} = \frac{LA \cdot LM}{EI^2};$$

now we know that

$$\frac{ML \cdot LA}{AK^2} = \frac{ME}{EA},$$

so we have

$$\frac{AK}{EI} = \frac{F}{EA},$$

and consequently

$$\frac{AK \cdot EH}{EI \cdot EH} = \frac{F}{EA}.$$

But

$$\frac{BH}{AK} = \frac{EH}{EA},$$

hence

$$AK \cdot EH = BH \cdot EA,$$

and moreover

$$EI \cdot EH = EA^2;$$

so we have

$$\frac{F}{EA} = \frac{BH}{EA},$$

hence $BH = F$.

Discussion of the existence of N

If Γ is an ellipse, E the vertex of $\mathscr{P}$ lies inside $\mathscr{H}$ (the branch of a hyperbola in question), and $\mathscr{H}$ and $\mathscr{P}$ have the same axis and are concave in opposite directions; so they cut one another in two points that are symmetrical with respect to the axis AD.

If Γ is a hyperbola, the midpoint of AM, that is, the centre of $\mathscr{H}$, lies inside $\mathscr{P}$, so $\mathscr{P}$ cuts the asymptotes of $\mathscr{H}$ and consequently cuts $\mathscr{H}$.

Notes:

1) As in the previous problem, concerning the parabola, we are making a *neusis*, but this time for central conics.

2) Let us formulate the problem in terms of an equation. The equation of Γ can be written

$$x^2 + ky^2 = \left(\frac{d}{2}\right)^2,$$

where d is the diameter of Γ and $k > 0$ if Γ is an ellipse and $k < 0$ of Γ is a hyperbola. We still have $k\,(k-1) > 0$ because k does not lie between 0 and 1.

The slope of the tangent to Γ at the point B with coordinates (x, y) is $-\dfrac{x}{ky}$, so the equation of the tangent is

$$Y - y = -\frac{x}{ky}(X - x).$$

The tangent meets the axis AD at the point H with abscissa X such that

$$X - x = \frac{ky^2}{x}.$$

So we have

$$BH^2 = y^2 + \frac{k^2 y^4}{x^2} = \frac{y^2}{x^2}\left(x^2 + k^2 y^2\right) = \frac{y^2}{\left(\frac{d}{2}\right)^2 - ky^2}\left[\left(\frac{d}{2}\right)^2 - ky^2 + k^2 y^2\right].$$

The equation that expresses the problem is accordingly

$$y^2\left[1 + k(k-1)\frac{4y^2}{d^2}\right] = F^2\left(1 - \frac{4ky^2}{d^2}\right),$$

or, if we put $\eta = \dfrac{2y^2}{d}$,

$$k(k-1)\eta^2 + \left(\frac{d}{2} + \frac{2kF^2}{d}\right)\eta = F^2.$$

So, to construct the solution to this problem, we first apply the area $\frac{1}{k(k-1)}F^2$ onto the line segment $\frac{1}{k(k-1)}\left(\frac{d}{2} + \frac{2kF^2}{d}\right)$ to fall short by the square η^2. Next, we find y as the side of the square of area equal to $\frac{d}{2} \cdot \eta$.

So, we can see that the problem is again a plane one. However, this is not immediately apparent from a purely geometrical analysis, because we are dealing with a biquadratic equation. In any case, here Ibn al-Haytham again has recourse to intersection of conics and not to a plane construction.

– 18 – Let there be a conic section Γ with vertex A, and let D and E be two points on its axis. To find a point B of Γ such that $\frac{BD}{BE} = \frac{I}{K}$ (where $\frac{I}{K}$ is a given ratio).[57]

Analysis: For $\frac{I}{K} \neq 1$, if there exists a point B that gives a solution to the problem, the point belongs to both Γ and to the circle that is the set of points such that the ratio of their distances from D and E is $\frac{I}{K}$; the end points of the diameter along the straight line DE give a harmonic division of DE.

Ibn al-Haytham finds the centre H and radius of this circle as follows: Let H be such that $D\hat{B}H = B\hat{E}H$. Since triangles HBE and HDB are similar, we have

$$\frac{DH}{HB} = \frac{BH}{HE} = \frac{BD}{BE},$$

hence

$$\frac{DH}{HE} = \frac{DH^2}{HB^2} = \frac{BD^2}{BE^2},$$

so

$$\frac{HD}{HE} = \frac{I^2}{K^2};$$

so the point H is known and the lengths HD and HE are known. Moreover, $HB^2 = HD \cdot HE$, so the length HB is known and B lies on a circle with

---

[57] Ibn al-Haytham does not specify the nature of the conic section Γ. The analysis holds for the ellipse, parabola or hyperbola, but he mentions only one vertex. In the synthesis it is clear, from the start of the discussion, that he is not considering an ellipse, for which the discussion would be very different (see following note).

centre H and known radius $r = HB$; if it exists, the point B is the intersection of the circle (H, r) and Γ.

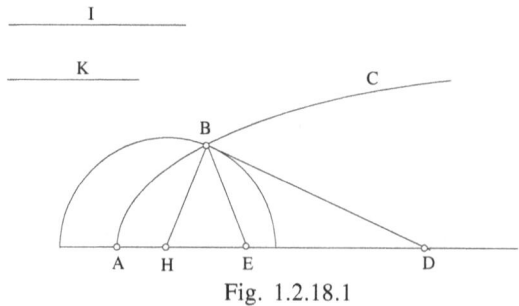

Fig. 1.2.18.1

If $\dfrac{I}{K} = 1$, we have $DB = EB$; if B exists, it lies on Δ, the perpendicular bisector of the segment DE. If B exists, it is at the intersection of Γ and Δ.

Ibn al-Haytham does not envisage the possibility that $\dfrac{I}{K} = 1$ until he comes to the discussion.

Note:

If $\dfrac{I}{K} \neq 1$, H lies outside the segment DE.

If $I > K$, H lies beyond E and $BD > BE$ (BD is called the first segment).

If $I < K$, H lies beyond D and $BD < BE$.

Ibn al-Haytham adopts the hypothesis that $I > K$, which has the consequence that the point H lies outside the segment DE on the same side as E (this is confirmed by both the text and the figures in the manuscript).[58]

---

[58] With the hypothesis $I \neq K$, we can show that the problem does not lose generality if we assume $I/K > 1$, that is, $I > K$.

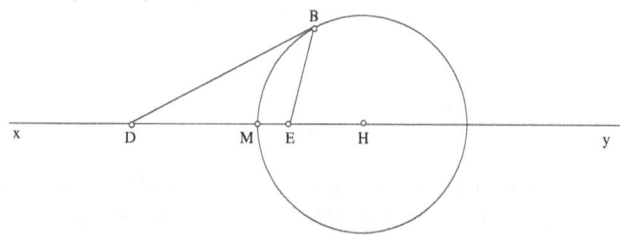

Fig. 1.2.18.2

Let there be a number $\lambda > 1$ and given positions of the points D and E on the axis xy of the conic section Γ, and let us consider Fig. 1.2.18.2, which refers to $BD/BE = \lambda$ and Fig. 1.2.18.3, which refers to $BD/BE = 1/\lambda < 1$.

(*Cont. on next page*)

– 19 – *Synthesis*: We return to Γ and the points D and E. Let H be the point defined by $\dfrac{HD}{HE} = \dfrac{I^2}{K^2}$ and HM the distance defined by $HM^2 = HD \cdot HE$. If the circle (H, HM) cuts Γ in B (see discussion), we shall show that $\dfrac{DB}{DE} = \dfrac{I}{K}$.

We have $HB = HM$, so $HB^2 = HD \cdot HE$, hence $\dfrac{HD}{HB} = \dfrac{HB}{HE}$; the triangles DBH and HBE have a common angle H so they are similar. So we have

$$\frac{DB^2}{BE^2} = \frac{HD^2}{HB^2} = \frac{HD}{HE} = \frac{I^2}{K^2},$$

hence

$$\frac{DB}{BE} = \frac{I}{K}.$$

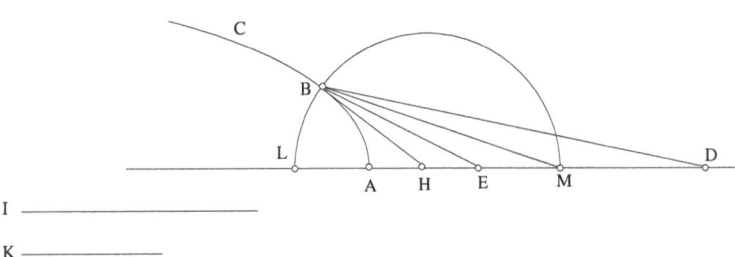

Fig. 1.2.19.1

(*Cont.*) We go from Fig. 1.2.18.2 to Fig. 1.2.18.3 by considering symmetry; so if in Fig. 1.2.18.3 we replace (D, E) by (E', D'), the modified figure will correspond to the case of the problem in which $BD'/BE' = \lambda > 1$.

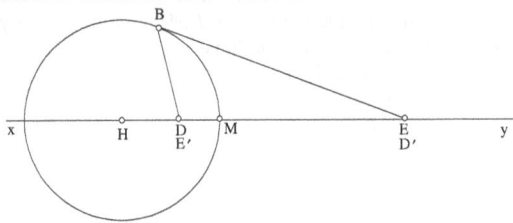

Fig. 1.2.18.3

If we take $\lambda = I/K > 1$, we can thus obtain all the forms of the figure we need to examine in the discussion of the problem.

We may note that from the beginning of the discussion, Ibn al-Haytham assumes $I > K$ and in most cases, after considering a figure with the points A, D, E in a certain order, he then turns to the figure obtained by exchanging the points D and E (see note 3′, p. 126 and note, pp. 128–9).

Note: From the equalities $\dfrac{HD}{HE} = \dfrac{I^2}{K^2}$ and $HM^2 = HD \cdot HE$, we deduce

$$\frac{I^2}{K^2} = \frac{HD^2}{HE \cdot HD} = \frac{HD^2}{HM^2},$$

hence

$$\frac{HD}{HM} = \frac{HM}{HE} = \frac{I}{K}.$$

Discussion: Let us assume $I > K$ (so $BD > BE$).

1) If $E = A$ (the vertex of Γ) and D lies outside Γ, then H lies inside Γ, and M, which is between E and D, lies outside (Fig. 1.2.19.2a); the circle (H, HM) cuts Γ.[59]

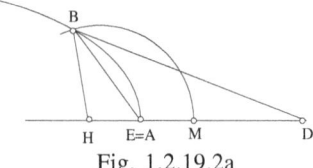

Fig. 1.2.19.2a

2) If $E = A$ and D lies inside Γ, then H lies outside Γ and M lies inside (Fig. 1.2.19.3); the circle cuts Γ.

[59] It is clear that here Ibn al-Haytham assumes that Γ is a parabola or a hyperbola, that is, a conic that is infinite, that the circle (H, HM) must necessarily intersect.

But if Γ is an ellipse with axis AA', we have three possibilities:
1) Γ lies inside the circle (H, HM),
2) Γ touches the circle (H, HM),
3) Γ cuts the circle (H, HM).

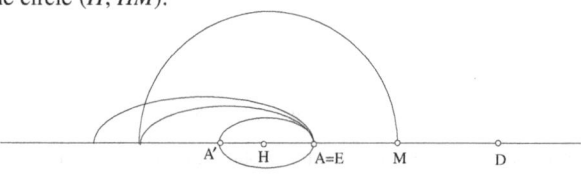

Fig. 1.2.19.2b

The discussion would require us to take account of additional conditions regarding the position of A' the second vertex of the ellipse. We come across difficulties of the same sort in Cases 3, 4 and 5 of the discussion.

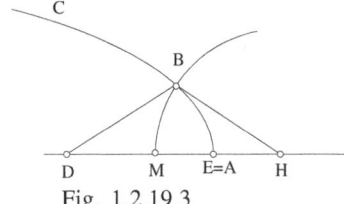

Fig. 1.2.19.3

3) If D and E lie outside Γ, the point H defined by $\dfrac{HD}{HE} = \dfrac{I^2}{K^2}$ may lie inside Γ, on Γ or outside it; M always lies between D and E.

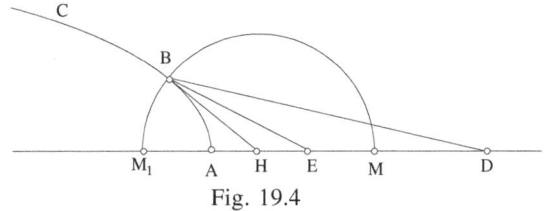

Fig. 19.4

If H lies inside Γ, or at A on Γ, the circle (H, HM) cuts Γ.

If H lies outside Γ, it is between E and A (Fig. 1.2.19.4), the circle (H, HM) will cut Γ if and only if $HA < HM$.

Now

$$HA < HM \Leftrightarrow \frac{DH}{HA} > \frac{DH}{HM} \Leftrightarrow \frac{DH}{HA} > \frac{I}{K}.$$

The condition $\dfrac{ED}{EA} \geq \dfrac{I}{K}$ given by Ibn al-Haytham is a sufficient condition because $\dfrac{DH}{HA} > \dfrac{ED}{EA}$; it is not a necessary condition.

If $HM_1 = HM$, $M_1 \in (H, HM)$, the condition for the intersection becomes

$$HM_1 > HA \Leftrightarrow EM_1 > EA;$$

but

$$\frac{M_1 D}{M_1 E} = \frac{I}{K},$$

hence

$$\frac{DE}{M_1 E} = \frac{I - K}{K} \text{ and } M_1 E = \frac{K.DE}{I - K}.$$

So we have

$$EM_1 > EA \Leftrightarrow \frac{K}{I - K} DE > EA \Leftrightarrow \frac{DE}{EA} > \frac{I - K}{K} \Leftrightarrow \frac{DA}{EA} > \frac{I}{K},$$

which is the necessary and sufficient condition for the existence of a solution when E and D lie outside Γ, with the points in the order A, E, D.

3′) The following paragraph begins 'If the first term of the ratio is on the side of the point E, it is the greater [...]'[60]; it is thus clear that Ibn al-Haytham is retaining the assumption that $I/K > 1$, but wants to say that he is reversing the roles of the points D and E in Figure 1.2.19.4, and that we have Figure 1.2.19.5, with H and M lying outside Γ.

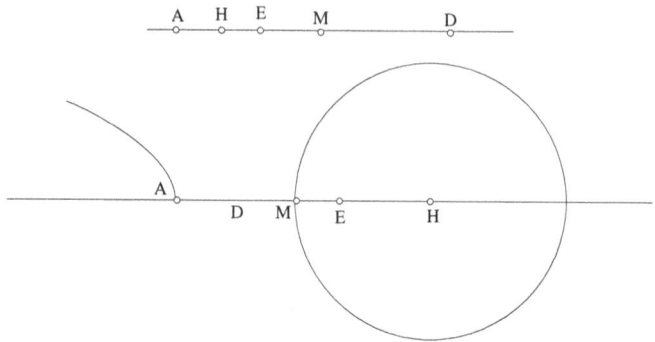

Fig. 1.2.19.5

So we have $HM < HA$; the circle (H, HM) does not cut Γ.

4) If D and E lie inside Γ, and occur in the order A, E, D, the point H may lie outside Γ, at A on Γ or inside Γ between A and E.

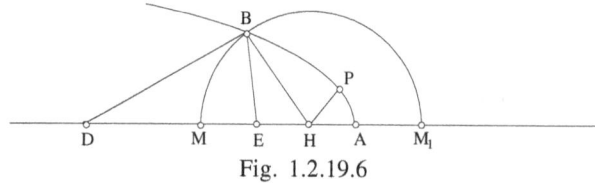

Fig. 1.2.19.6

If H lies outside Γ or at A, since the point M, which is between D and E, lies inside Γ, the circle (H, HM) cuts Γ.

If H is between E and A, if P is the point of Γ such that HP is the *minimum* distance from H to points of Γ, then the necessary and sufficient condition for B to exist is $HM \geq HP$.

The problem can have 0, 1 or 2 solutions.

The condition $\dfrac{DE}{EA} \geq \dfrac{I}{K}$ that is given by Ibn al-Haytham, which implies $HM > HA$, is sufficient, but not necessary.

---

[60] See p. 217.

We may note that on this occasion Ibn al-Haytham does not consider the case where the points D and E lie inside Γ, with the points in the order A, D, E. We should again have H beyond E (Fig. 1.2.19.7), so H and M would lie inside Γ, and we should have $HM < HA$.

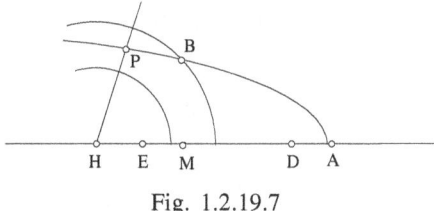

Fig. 1.2.19.7

If HP is the *minimum* distance of the point H from the conic Γ, the necessary and sufficient condition for the point B to exist is again $HM \geq HP$.

5) If D lies outside the conic section and E inside it, still with $l/K > 1$, then the point H lies inside the conic section. But M, which is between E and D can lie inside Γ, at A, or outside it. For the circle (H, HM) to cut Γ, it is necessary and sufficient that $HM \geq HP$, HP being as before the *minimum* distance associated with the point H.

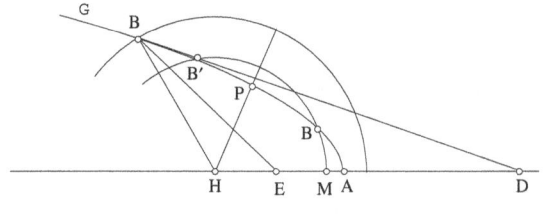

Fig. 1.2.19.8

If D lies inside Γ and E outside it, then H lies outside Γ; M can be inside Γ, at A on Γ, or outside Γ. For the circle (H, HM) to cut Γ, it is necessary and sufficient that $HM > HA$ (Fig. 1.2.19.9).

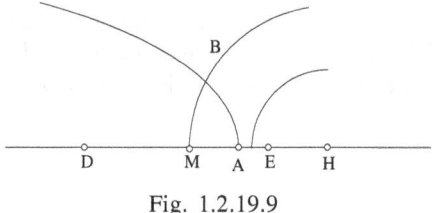

Fig. 1.2.19.9

As in 3 and 4, this condition becomes

$$\frac{DA}{EA} > \frac{I}{K}.$$

6) If $I = K$, we have seen that B lies on Δ, the perpendicular bisector of DE; for Δ to cut Γ, it is necessary and sufficient that M, the midpoint of DE, shall lie inside Γ, and this holds whatever the order of the points A, D, E.

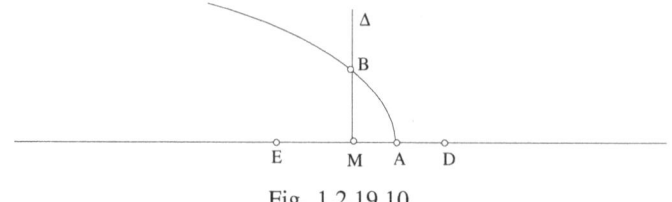

Fig. 1.2.19.10

Note: In the course of this discussion we have found the following arrangements of the points D and E in relation to Γ, in which we may observe that the points D and E can exchange places with one another, as shown in Fig. 1.2.19.11:

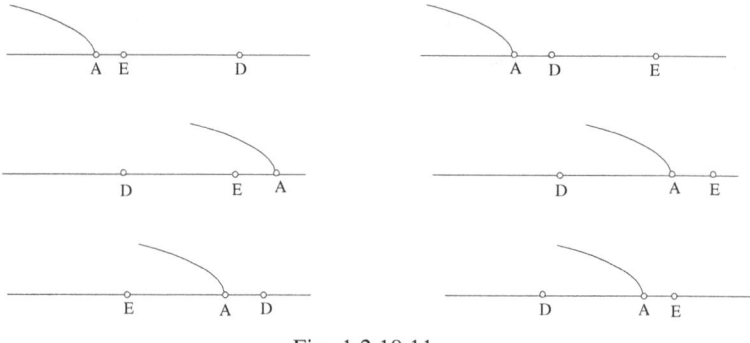

Fig. 1.2.19.11

We may note that in cases 1 and 2, which both give a solution, Ibn al-Haytham has not considered the cases shown in cases 1′ and 2′ in Fig. 1.2.19.12.

In case 1′, it is clear the problem has no solution, since $HM < HA$.

In case 2′, if HP is the shortest distance between H and points of Γ, the condition for the circle (H, HM) and Γ to intersect is $HM \geq HP$.

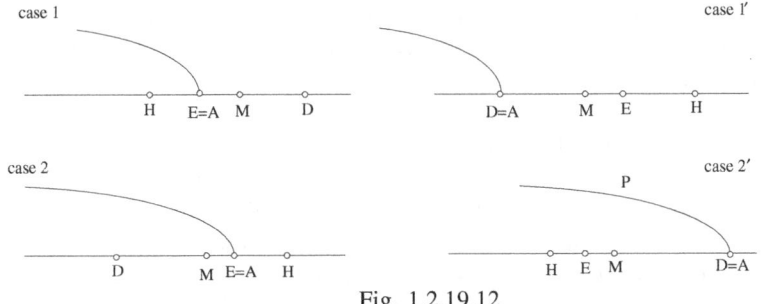

Fig. 1.2.19.12

– 20 – Given a conic section Γ with vertex C, two points D and E on its axis and a straight line G, to find a point B of Γ such that $BD + BE = G$; it is necessary to assume $G > DE$.

Note: The equality $BD + BE = G$ defines an ellipse $\mathscr{E}$ with foci D and E. The problem thus reduces to investigating the intersection of Γ and $\mathscr{E}$.

Analysis: Let H and I be the points of the straight line DE outside the segment DE, and such that $EH = DI = \dfrac{G - DE}{2}$; so we have $HI = G$.

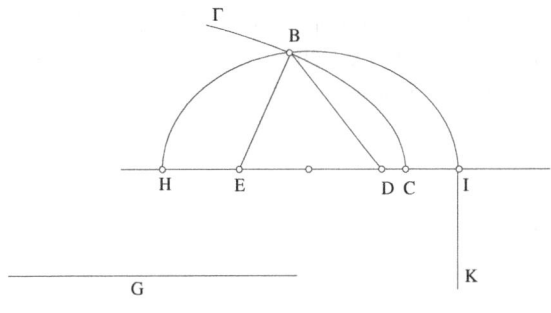

Fig. 1.2.20

We define IK by $HI \cdot IK = 4HD \cdot DI$.

The ellipse $\mathscr{E}$ with axis HI and *latus rectum* IK is thus defined by what has been given and, from Apollonius (*Conics*, III.52), it passes through B.

Throughout this problem, the parts played by D and E are interchangeable, since the perpendicular bisector of DE is an axis of symmetry of the ellipse.

– 21 – *Synthesis*: We construct the ellipse $\mathscr{E}$ as described in the analysis; its axis HI is equal to G.

Let us suppose that $\mathcal{E}$ and Γ cut one another in B. From Apollonius (III.52), $B \in \mathcal{E} \Rightarrow BE + BD = HI$, so $BD + BE = G$. The point B thus gives a solution to the problem.

Discussion: We assume $G > DE$.

1) If one of the points D and E lies outside Γ and the other inside it, or if one of the points is at the vertex C of Γ, where the other may lie either outside or inside Γ, then it follows that $C \in [ED]$; so if Γ is a parabola or a hyperbola, one of the points H and I lies inside Γ and the other outside it (see Fig. 1.2.21.1), so Γ cuts $\mathcal{E}$ in one point and the problem has a solution.

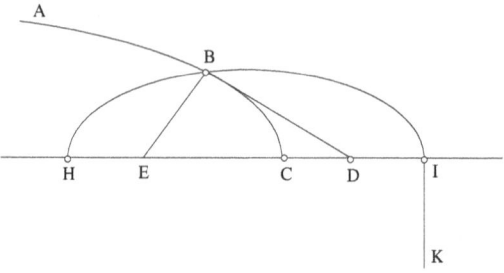

Fig. 1.2.21.1

But if Γ is an ellipse, its second vertex C' may lie between H and E and we thus cannot draw any conclusion. It accordingly seems that Ibn al-Haytham is assuming Γ is a hyperbola or a parabola, as in the previous problem.

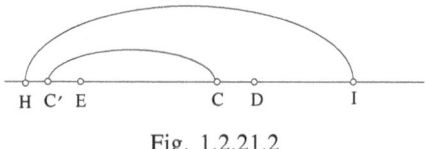

Fig. 1.2.21.2

2) If the two points D and E lies outside Γ in the order C, E, D, we may distinguish two cases:

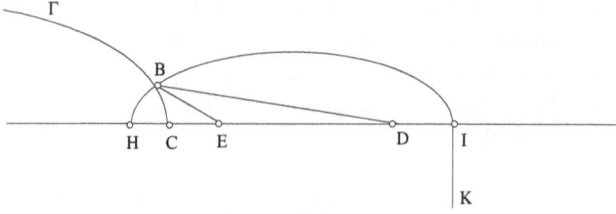

Fig. 1.2.21.3

a) If $EC < \dfrac{G - DE}{2}$, that is, if $EC < EH$, then H lies inside Γ[61] and I lies outside it, so $\mathscr{E}$ cuts Γ in B.

b) If $EC > \dfrac{G - DE}{2}$, then H and I, and consequently all of $\mathscr{E}$, lie outside Γ and there is no solution.

We may note that if $EC = \dfrac{G - DE}{2}$, then H is at C; the point C gives a solution to the problem since $CE + CD = HI = G$.

3) If the two points D and E lie inside Γ in the order C, D, E:

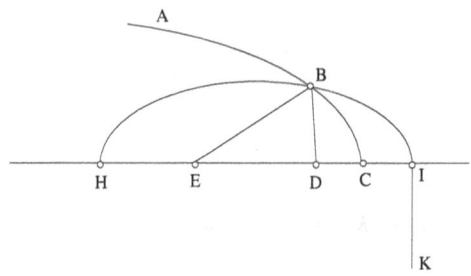

Fig. 1.2.21.4

a) If $CD < \dfrac{G - ED}{2}$, that is, $CD < DI$, then the point I lies outside Γ and the point H lies inside it,[62] so $\mathscr{E}$ cuts Γ.

b) If $CD = \dfrac{G - ED}{2}$, then $I = C$, and C gives a solution to the problem.

c) If $CD > \dfrac{G - ED}{2}$, then I lies inside Γ; the problem requires more substantial treatment, which is provided in 22 for the case where Γ is a parabola and in 23 for the case where Γ is a hyperbola.

– **22** – If the two points D and E lie inside Γ and if $AD > \dfrac{G - DE}{2}$,[63] then H, I, the vertices of the ellipse $\mathscr{E}$, lie inside Γ.

Let us assume that Γ is a *parabola*.

[61] The argument is valid if Γ is a parabola or a hyperbola.
[62] See previous note.
[63] In 22, the letter C is replaced by A.

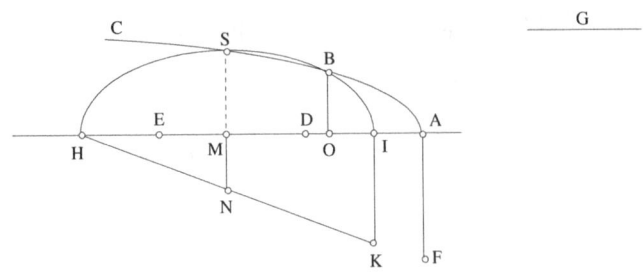

Fig. 1.2.22.1

Let M be the centre of the ellipse and IK its *latus rectum*, A the vertex of the parabola and FA its *latus rectum*. The condition Ibn al-Haytham gives for $\mathscr{E}$ and Γ to intersect is

$$\frac{HM^2}{MA \cdot AF} \geq \frac{HI}{KI}.$$

1) Let us take $\dfrac{HM^2}{MA \cdot AF} = \dfrac{HI}{KI}$ (Fig. 1.2.22.1).

Let $MN \perp HI$, $N \in (HK)$; we have successively

$$\frac{HM^2}{MA \cdot AF} = \frac{HI}{KI} = \frac{HM}{MN} = \frac{HM \cdot MI}{MN \cdot MI} = \frac{HM^2}{MN \cdot MI},$$

from which we obtain

$$MA \cdot AF = MN \cdot MI.$$

Associated with the point M we have a point with ordinate y on the parabola and a point with ordinate Y on the ellipse; we have

$$y^2 = MA \cdot AF, \qquad \text{the equation of } \Gamma,$$

and

$$\frac{MI \cdot MH}{Y^2} = \frac{HI}{IK}, \qquad \text{the equation of } \mathscr{E},$$

so

$$\frac{MH^2}{Y^2} = \frac{MH}{MN},$$

hence

$$Y^2 = MH \cdot MN = MI \cdot MN;$$

so we have $Y = y$, so Γ passes through the vertex S of the ellipse $\mathscr{E}$.

2) Let us take $\dfrac{HM^2}{MA \cdot AF} > \dfrac{HI}{KI}$.

For any point O on the segment HI, we have $0 < HO \cdot OI < HM^2$ and there exists a point O between M and I such that

(1) $$\dfrac{HO \cdot OI}{MA \cdot FA} = \dfrac{HI}{IK} \quad \text{(Fig. 1.2.22.2)}.$$

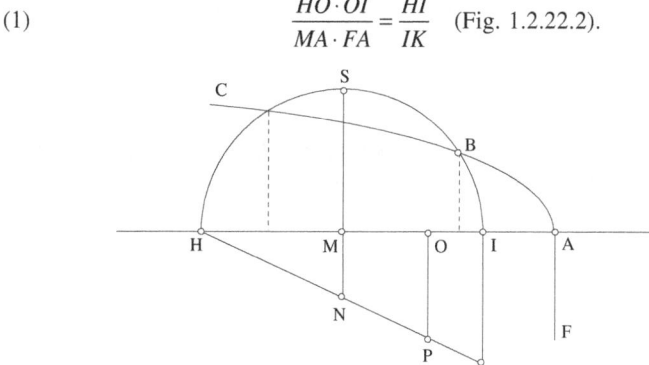

Fig. 1.2.22.2

Let $OP \perp HI$, where P lies on HK; we have

(2) $$\dfrac{HI}{IK} = \dfrac{HO}{OP} = \dfrac{HO \cdot OI}{OP \cdot OI}.$$

From (1) and (2), we obtain $MA \cdot FA = OP \cdot OI$. But $MA \cdot FA > OA \cdot FA$, so $OP \cdot OI > OA \cdot FA$; as before O is the projection on AH of a point of Γ with ordinate y and of a point of $\mathscr{E}$ with ordinate Y, and we have

$$y^2 = OA \cdot FA \text{ and } Y^2 = OP \cdot OI,$$

hence $Y > y$.

So the ellipse has a point that lies outside the parabola, and the two vertices H and I lie inside it; so the ellipse cuts Γ in two points, one whose projection is between I and O, the other with its projection between H and O.

3) Ibn al-Haytham then finishes the first paragraph by showing that when Γ cuts $\mathscr{E}$ at the vertex S, it also cuts it in a second point.

Let us go back to Fig. 1.2.22.1; let O be a point between M and I defined by $MA \cdot MO = MI^2$, that is, O is the harmonic conjugate of A with respect to $\mathscr{E}$; we have

$$\dfrac{MA}{MO} = \dfrac{MI^2}{MO^2},$$

hence

$$\frac{MA}{AO} = \frac{MA}{MA - MO} = \frac{MI^2}{MI^2 - MO^2} = \frac{MI^2}{HO \cdot OI}.$$

Let us use y_M and y_O to designate the ordinates of the points of the parabola Γ that have M and O as their projections on the axis; we have

$$\frac{y_M^2}{y_O^2} = \frac{AM}{AO}.$$

In the same way, let us use Y_M and Y_O to designate the ordinates of the points of $\mathcal{E}$ that have M and O as their projections; we have

$$\frac{Y_M^2}{Y_O^2} = \frac{MI^2}{OH \cdot OI} = \frac{AM}{AO},$$

hence

$$\frac{y_M}{y_O} = \frac{Y_M}{Y_O};$$

but

$$Y_M = y_M = MS,$$

so

$$y_O = Y_O.$$

The parabola Γ and the ellipse $\mathcal{E}$ have a common point whose projection on AH is O.

In conclusion, if $\dfrac{HM^2}{MA \cdot AF} \geq \dfrac{HI}{IK}$, the parabola and the ellipse cut one another in two points, and the problem has two solutions.

Note: The proofs show that the condition that is imposed is *sufficient* to ensure that Γ and $\mathcal{E}$ cut one another in two points. If the condition is satisfied, one of the points is at S, a vertex of the ellipse, or on the arc HS, and the other is on the arc IS.

But it is possible that Γ cuts $\mathcal{E}$ in two points of the arc IS or that Γ is a tangent to $\mathcal{E}$, possibilities that do not emerge in the discussion.

The condition that is imposed is not *necessary*.

Investigation of the intersection of Γ and $\mathcal{E}$

When Γ is a parabola with vertex A and *latus rectum* AF, and D and E are two points on its axis, to find $B \in \Gamma$ such that $BD + BE = G$.

Let us put $AF = c$, $DE = l$. Let M be the midpoint of DE, I and H the points on the axis such that $MI = MH = \dfrac{G}{2}$.

In Problem 22, Ibn al-Haytham assumes the points D and E lie inside Γ and $AE > AD > \dfrac{G - DE}{2} \Leftrightarrow AD > DI$. So the points are in the order A, I, D, M, E, H. We have $BD + BE = G$, so $B \in \mathscr{E}$, an ellipse with foci D and E and major axis HI. Let $IK = c'$, its *latus rectum*. We have $Gc' = G^2 - l^2$, so

$$\frac{HI}{IK} = \frac{G}{c'} = \frac{G^2}{G^2 - l^2} = k.$$

Let us put $AM = m$.

The condition $AD > DI$ gives $m > \dfrac{G}{2}$.

The equation of Γ: $y^2 = c(m - x)$, $\qquad\qquad x \leq m$;

the equation of $\mathscr{E}$: $x^2 + ky^2 = \dfrac{G^2}{4}$, $\qquad -\dfrac{G}{2} \leq x \leq \dfrac{G}{2}$,

the equation for the real-number abscissae of $\Gamma \cap \mathscr{E}$: $x^2 + kc(m - x) = \dfrac{G^2}{4}$,

(1) $\quad f(x) = x^2 - kcx + kcm - \dfrac{G^2}{4} = 0.$

The condition given by Ibn al-Haytham,

$$\frac{HI}{HK} \leq \frac{HM^2}{AM \cdot AF},$$

can be written

$$k \leq \frac{G^2}{4mc} \Leftrightarrow kcm - \frac{G^2}{4} \leq 0.$$

This condition is *sufficient* to make equation (1) have 2 roots, x and x', that satisfy $-\dfrac{G}{2} < x' < 0 \leq x'' < \dfrac{G}{2}$, since the condition implies $f\left(\dfrac{G}{2}\right) > 0, f\left(-\dfrac{G}{2}\right) > 0, f(0) \leq 0.$

But the condition is not *necessary*. If $\Delta = k^2c^2 - 4\left(kcm - \dfrac{G^2}{4}\right) \geq 0$ with $kcm > \dfrac{G^2}{4}$, the equation has 2 roots > 0.

Now, the condition $kcm - \dfrac{G^2}{4} \leq \dfrac{k^2 c^2}{4}$ is necessary and sufficient to make the roots real numbers. We also need to ensure that the roots lie in the interval $\left[-\dfrac{G}{2}, \dfrac{G}{2}\right]$. Since $f\left(\pm\dfrac{G}{2}\right) = kc\left(m \mp \dfrac{G}{2}\right) \geq 0$, the two roots (if they exist) lie on the same side of $\dfrac{G}{2}$ and $-\dfrac{G}{2}$ and it is sufficient to impose the condition that their mean, $\dfrac{kc}{2}$, which is positive, shall be $\leq \dfrac{G}{2}$, that is, that $k \leq \dfrac{G}{c}$. The necessary and sufficient conditions may be written $\lambda^2 - 4m\lambda + G^2 \geq 0$ and $\lambda \leq G$, where $\lambda = kc$. The polynomial $\lambda^2 - 4m\lambda + G^2$ has two positive roots $2m \pm \sqrt{4m^2 - G^2}$ and G lies between them because $G^2 - 4mG + G^2 = 4G\left(\dfrac{G}{2} - m\right) < 0$. The inequality $\lambda^2 - 4m\lambda + G^2 \geq 0$ is equivalent to stating that λ lies outside the interval between the roots; the two conditions are this equivalent to the single inequality $\lambda \leq 2m - \sqrt{4m^2 - G^2}$, or

$$\frac{HI}{IK} = k \leq 2\left(\frac{MA - \sqrt{AI \cdot AH}}{AF}\right).$$

Note: The equation for the real-number abscissae of the points of intersection may also be written

$$\frac{MH^2 - x^2}{\overline{AF} \cdot (\overline{MA} - x)} = \frac{HI}{IK},$$

where the first member becomes zero at the points $x = \pm\dfrac{G}{2}$ and goes through a maximum μ between these two abscissae. The condition for the problem to have a solution is thus that

$$\frac{HI}{IK} \leq \mu;$$

the one proposed by Ibn al-Haytham can be written

$$\frac{HI}{IK} \leq \frac{MH^2}{\overline{AF} \cdot \overline{MA}},$$

the value of the first member for $x = 0$; this condition obviously implies the preceding one. So it is clear that Ibn al-Haytham's condition is too strong.

In fact, his discussion consists of finding the value of k that makes the ellipse $\mathscr{E}$ tangent to the parabola Γ. Finding this by geometry is far from trivial.

– **23** – Let us take Γ to be a *branch of a hyperbola*. Let us return to the assumptions of 22, that is, D and E lie inside Γ and $AD > \dfrac{G-DE}{2}$. The points H and I, the vertices of the ellipse $\mathcal{E}$, thus lie inside Γ.

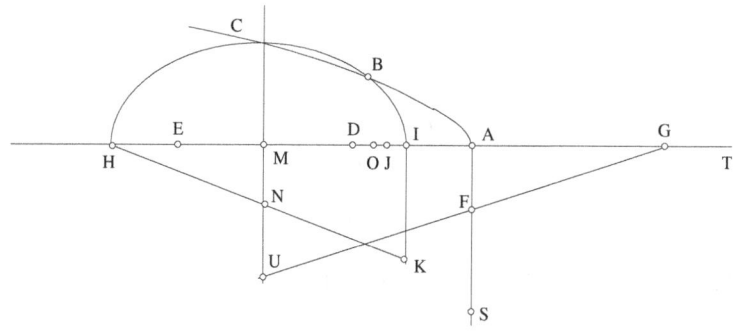

Fig. 1.2.23.1

Let AG be the axis of Γ and AF its *latus rectum*. The straight line MN ($MN \perp AD$) cuts HK in N and GF in U. If S is the point defined by

$$\frac{SA}{AF} = \frac{HI}{IK},$$

we have

$$\frac{SA}{AG} = \frac{HI}{IK} \cdot \frac{AF}{AG}.$$

The condition Ibn al-Haytham gives for $\mathcal{E}$ and Γ to intersect is

$$\frac{HM^2}{MA \cdot MG} \geq \frac{SA}{AG}.$$

1) Let us take $\dfrac{HM^2}{MA \cdot MG} = \dfrac{SA}{AG}$. We have

$$\frac{HM^2}{MA \cdot MG} = \frac{HM \cdot MI}{NM \cdot MI} \cdot \frac{NM \cdot MI}{MA \cdot MG} = \frac{HI}{IK} \cdot \frac{NM \cdot MI}{MA \cdot MG} = \frac{SA}{AF} \cdot \frac{NM \cdot MI}{MA \cdot MG},$$

from which we obtain

$$\frac{NM \cdot MI}{MA \cdot MG} = \frac{AF}{AG} = \frac{MU}{MG} = \frac{MU \cdot MA}{MA \cdot MG},$$

hence

$$NM \cdot MI = MU \cdot MA \quad \text{(Fig. 1.2.23.1)}.$$

With the point M there are associated a point on the ellipse $\mathscr{E}$ with ordinate Y and a point on the hyperbola Γ with ordinate y; we have on $\mathscr{E}$

$$\frac{MH \cdot MI}{Y^2} = \frac{HI}{IK} = \frac{HM}{MN},$$

hence

$$Y^2 = MN \cdot MI,$$

and on Γ

$$\frac{MA \cdot MG}{y^2} = \frac{AG}{AF} = \frac{MG}{MU},$$

hence

$$y^2 = MA \cdot MU;$$

so we have $y^2 = Y^2$, and Γ cuts $\mathscr{E}$ at the vertex C.

2) Let us take $\dfrac{HM^2}{MA \cdot MG} > \dfrac{SA}{AG}$ (Fig. 1.2.23.2).

There exists a point O between M and I such that $\dfrac{HO \cdot OI}{MA \cdot MG} = \dfrac{SA}{AG}.$

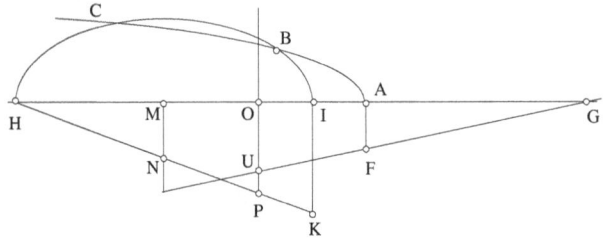

Fig. 1.2.23.2

But

$$MA \cdot MG > OA \cdot OG,$$

hence

$$\frac{HO \cdot OI}{OA \cdot OG} > \frac{SA}{AG}.$$

Let $OP \parallel IK$, where P lies on HK; we have

$$\frac{HO}{OP} = \frac{HI}{IK} = \frac{SA}{AF},$$

hence

$$\frac{HO \cdot OI}{OA \cdot OG} = \frac{HO \cdot OI}{OP \cdot OI} \cdot \frac{OP \cdot OI}{OA \cdot OG} = \frac{SA}{AF} \cdot \frac{OP \cdot OI}{OA \cdot OG},$$

and therefore

$$\frac{OP \cdot OI}{OA \cdot OG} > \frac{AF}{AG}.$$

OP cuts *GF* in *U*, we have

$$\frac{AF}{AG} = \frac{OU}{OG} = \frac{OU \cdot OA}{OA \cdot OG},$$

and consequently

$$OP \cdot OI > OU \cdot OA.$$

With point *O* there are associated a point on the ellipse $\mathscr{E}$ with ordinate *Y* and a point on the hyperbola Γ with ordinate *y*, and we have $Y^2 = OP \cdot OI$ and $y^2 = OU \cdot OA$, hence $y^2 < Y^2$. So Γ cuts the straight line *OP* inside $\mathscr{E}$. The vertex *A* of Γ lies outside $\mathscr{E}$; consequently Γ cuts $\mathscr{E}$ in two points which are one on either side of the straight line *OP*.

3) Ibn al-Haytham then finishes paragraph 1 by showing that if Γ cuts $\mathscr{E}$ at the vertex *C*, it cuts it in another point, which lies on the arc *CI*.

Let us return to Fig. 1.2.23.1; let *O*, between *M* and *I*, be defined by $\frac{MO}{MI} = \frac{MI}{MA}$, and let *J* be defined by $\frac{OJ}{JA} = \frac{AS}{AG}$. We have

$$\frac{OI}{IA} = \frac{MI}{MA} > \frac{IM^2}{MA \cdot MG}.$$

But

$$\frac{IM^2}{MA \cdot MG} = \frac{AS}{AG} = \frac{OJ}{JA},$$

and further

$$\frac{OM}{MG} = \frac{OM}{MI} \cdot \frac{MI}{MG} = \frac{MI}{MA} \cdot \frac{MI}{MG},$$

so

$$\frac{OM}{MG} = \frac{OJ}{JA}.$$

Let us put $GT = JA$; we obtain

$$\frac{OM}{MG} = \frac{OJ}{JA} = \frac{MJ}{MG + JA} = \frac{MJ}{MG + GT} = \frac{MJ}{MT} = \frac{IM^2}{MA \cdot MG} = \frac{MJ^2}{MJ \cdot MT}.$$

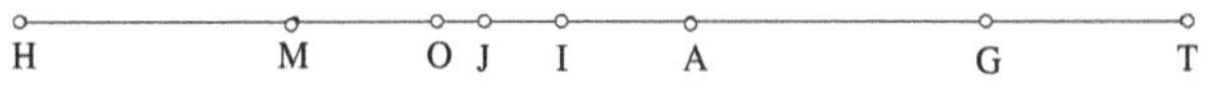

Fig. 1.2.23.3

Now

$$MA \cdot MG = MJ \cdot MT + JA \cdot JG.$$

In fact

$$MA \cdot MG = (MJ + JA)\, MG = MJ \cdot MG + JA \cdot (MJ + JG)$$

$$= MJ \cdot (MG + GT) + JA \cdot JG = MJ \cdot MT + JA \cdot JG,$$

and moreover

$$MI^2 - MJ^2 = (HM + MJ)\,(MI - MJ) = HJ \cdot JI;$$

so

$$\frac{IM^2}{MA \cdot MG} = \frac{IM^2 - MJ^2}{MA \cdot MG - MJ \cdot MT} = \frac{JH \cdot JI}{JA \cdot JG},$$

hence

$$\frac{MH \cdot MI}{JH \cdot JI} = \frac{MA \cdot MG}{JA \cdot JG}.$$

If we use Y_M and Y_J to designate the ordinates of the points of Γ and y_M and y_J for the ordinates of the points of $\mathscr{E}$ that are projected, respectively, into M and J on the axis that is common to $\mathscr{E}$ and $\mathscr{H}$, we have

$$\frac{Y_M}{Y_J} = \frac{y_M}{y_J}.$$

But $Y_M = y_M = MC$, so $Y_J = y_J$; the hyperbola Γ and the ellipse $\mathscr{E}$ have a common point that is projected into J on the axis AH.

In conclusion, if $\dfrac{HM^2}{MA \cdot MG} \geq \dfrac{SA}{AG}$, the two conic sections Γ and $\mathscr{E}$ have two common points, and the problem has two solutions.

Note: As in Problem 22, the condition given by Ibn al-Haytham is sufficient, but it is not necessary.

Analytic investigation of the intersection of Γ and $\mathscr{E}$

Γ is a hyperbola with axis $AG = d$ and *latus rectum* $AF = c$; what we are given is the same as in Problem 22. We again assume that $AD > \dfrac{G - DE}{2}$, which gives $m > \dfrac{G}{2}$.

Taking coordinates $M\,(x, y)$, let us put $\overline{AM} = m$, $\overline{GM} = m + d$. So we have:

– Equation of Γ

$$\frac{(m + x)(m + d + x)}{y^2} = \frac{d}{c}, \qquad\qquad x \geq -m.$$

– Equation of $\mathscr{E}$

$$x^2 + ky^2 = \frac{G^2}{4}, \qquad -\frac{G}{2} \le x \le \frac{G}{2}.$$

– Equation for the abscissae of $\mathscr{E} \cap \Gamma$

$$\frac{G^2}{4} - x^2 = k\frac{c}{d}(m+x)(m+d+x), \qquad -\frac{G}{2} \le x \le \frac{G}{2}.$$

(1) $\qquad f(x) = x^2\left(k\frac{c}{d}+1\right) + k\frac{c}{d}(2m+d)x + k\frac{c}{d}m(m+d) - \frac{G^2}{4} = 0.$

The condition given by Ibn al-Haytham

$$\frac{HM^2}{MA \cdot MG} \ge \frac{HI}{IK} \cdot \frac{AF}{AG}$$

can be written

$$\frac{G^2}{4} \ge \frac{kc}{d}\,m(m+d).$$

This is a *sufficient* condition for equation (1) to have two roots, x' and x'', that satisfy $-\frac{G}{2} < x' < 0 \le x'' < \frac{G}{2}$ because it implies

$$f\!\left(\frac{G}{2}\right) > 0,\ f\!\left(-\frac{G}{2}\right) > 0,\ f(0) \le 0.$$

Indeed

$$f\!\left(\frac{G}{2}\right) = \frac{kc}{d}\left(m+\frac{G}{2}\right)\left(m+d+\frac{G}{2}\right),$$

$$f\!\left(-\frac{G}{2}\right) = \frac{kc}{d}\left(m-\frac{G}{2}\right)\left(m+d-\frac{G}{2}\right), \quad \text{with } m > \frac{G}{2}.$$

But the condition is not necessary.
The discriminant of equation (1) is

$$\Delta = \frac{k^2c^2}{d^2}(2m+d)^2 - \frac{4(kc+d)}{d}\left[\frac{kc}{d}m(m+d) - \frac{G^2}{4}\right].$$

The *necessary and sufficient* condition for the equation to have two roots is thus

$$\frac{G^2}{4} - \frac{kc}{d}m(m+d) \ge -\frac{k^2c^2(2m+d)^2}{4d(kc+d)}.$$

If we put $\lambda = kc$, this condition becomes

$$\varphi(\lambda) = \lambda^2 + \frac{\lambda}{d}\left[G^2 - 4m(m+d)\right] + G^2 \geq 0.$$

We need to add to it the condition expressing the requirement that the roots of the equation in x lie between $-\dfrac{G}{2}$ and $\dfrac{G}{2}$. As half the sum of the roots is $-\dfrac{\lambda(2m+d)}{2(\lambda+d)} < 0$, this last condition becomes

$$\frac{\lambda(2m+d)}{\lambda+d} \leq G,$$

or

$$\lambda \leq \frac{Gd}{2m+d-G} = \alpha.$$

The discriminant of $\varphi(\lambda)$ is

$$\delta = \frac{1}{d^2}\left(4m^2 - G^2\right)\left[4(m+d)^2 - G^2\right] \geq 0,$$

so $\varphi(\lambda)$ has two roots λ_0 and λ_1 and the first condition is that λ must lie outside the interval $[\lambda_0, \lambda_1]$.

Calculation gives us

$$\varphi(\alpha) = -\frac{(2m+d)G\,(2m-G)[2(m+d)-G]}{(2m+d-G)^2} < 0,$$

so α lies between λ_0 and λ_1 and the two conditions finally reduce to the single inequality $\lambda \leq \lambda_0$, that is,

$$\frac{HI}{IK} = k \leq \frac{2m(m+d) - G^2}{cd} - \frac{1}{2cd}\sqrt{\left(4m^2 - G^2\right)\left[4(m+d)^2 - G^2\right]}$$

$$= 2\frac{AM \cdot MG - 2MI^2 - \sqrt{AH \cdot GH \cdot AI \cdot GI}}{AF \cdot AG}.$$

If Γ is an *ellipse*, and assuming D and E lie inside Γ and $AD > \dfrac{G - DE}{2}$, that is, $AD > DI$, the vertex A of Γ lies outside $\mathscr{E}$ and its second vertex G is beyond E; it can lie between E and H, at H or beyond H.

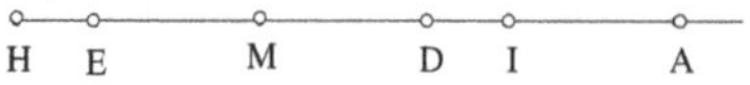

Fig. 1.2.23.4

So the discussion of the intersection of Γ and $\mathscr{E}$ might be different (see below).

Note: $\dfrac{x^2 - MH^2}{(\overline{AM} + x)(\overline{GM} + x)} = \dfrac{SA}{AG}$.

As in the preceding cases, the limiting value given by Ibn al-Haytham corresponds to the case in which the two conic sections have one common point at the vertex of the minor axis of $\mathscr{E}$.

Investigation of Problems 20, 21, 22 and 23

In the statement of the problem, Ibn al-Haytham does not specify the nature of the conic section Γ; this causes no trouble in the analysis in 20. But in 21, the synthesis and the beginning of the discussion, in paragraphs 1, 2, 3, we need to distinguish between the parabola and hyperbola on the one hand and the ellipse on the other.

In stating that if I lies outside Γ, H must lie inside it, Ibn al-Haytham is assuming that Γ is an infinite conic, either a parabola or a branch of a hyperbola. He does not consider the case of the ellipse.

Ibn al-Haytham continues his discussion in 22, where Γ is a parabola, and in 23, where Γ is a hyperbola, and he ends 23 by saying that if Γ is an ellipse, 'the discussion is the same as for the hyperbola, without needing to add or remove anything'.[64] Now, if Γ is an ellipse, we need to distinguish several cases for different positions of G, the other end of the axis of Γ; however, the condition Ibn al-Haytham gives, $\dfrac{SA}{AG} \leq \dfrac{HM^2}{MA \cdot MG}$, is, in all cases, sufficient for the problem to have at least one solution (that is, one or two solutions).

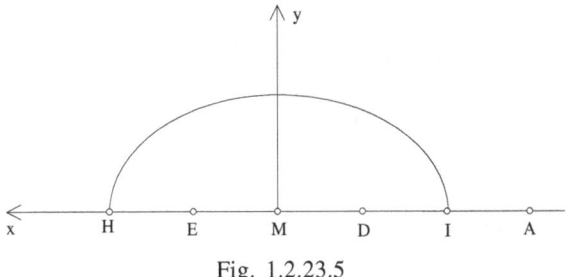

Fig. 1.2.23.5

---

[64] See p. 228.

Intersection of the ellipses $\mathscr{E}$ and Γ with the assumptions of 23.

Equation of $\mathscr{E}$: $\dfrac{MH^2 - x^2}{y^2} = \dfrac{HI}{IK}$ $\qquad$ $\left(\overline{MH} = \dfrac{G}{2}\right)$.

Equation of Γ: $-\dfrac{\left(\overline{AM} + x\right)\left(\overline{GM} + x\right)}{y^2} = \dfrac{AG}{AF}$ $\quad$ $\left(\text{with } \overline{GM} < 0, \ \overline{AM} > \dfrac{G}{2}\right)$.

We must distinguish three cases depending on whether H is beyond G (outside Γ), at G or between M and G (inside Γ). In the first case, $\mathscr{E}$ must cut Γ since its other vertex I lies inside Γ. In the second case, $\mathscr{E}$ touches Γ at their common vertex G. It remains to discuss the intersection in the third case, where $MG > MH$.

The equation for the abscissae of the points of intersection is

$$x^2 - HM^2 = \frac{SA}{AG}\left(\overline{AM} + x\right)\left(\overline{GM} + x\right),$$

or

$$f(x) = x^2\left(\frac{SA}{AG} - 1\right) + \frac{SA}{AG}\left(\overline{AM} + \overline{GM}\right)x + \frac{SA}{AG}\,\overline{AM} \cdot \overline{GM} + HM^2 = 0.$$

We have

$$f\left(\pm\frac{G}{2}\right) = \frac{SA}{AG}\left(\overline{AM} \pm \overline{MH}\right)\left(\overline{GM} \pm \overline{MH}\right) \leq 0$$

because

$$0 < \overline{MH} < \overline{AM} \text{ and } \overline{GM} < -\overline{MH};$$

also

$$f(0) = \frac{SA}{AG}\,\overline{AM} \cdot \overline{GM} + HM^2,$$

and Ibn al-Haytham's condition $\dfrac{SA}{AG} \leq \dfrac{HM^2}{AM \cdot GM}$ thus means that $f(0) \geq 0$, which implies that there exist a root x_1 lying between $-\dfrac{G}{2}$ and 0, and a root x_2 between 0 and $\dfrac{G}{2}$.

If $\dfrac{SA}{AG} > 1$, $f(x)$ becomes zero between $-\infty$ and $-\dfrac{G}{2}$ and between $\dfrac{G}{2}$ and $+\infty$; but these roots do not correspond to points of the ellipse $\mathscr{E}$, so there is no solution. If $\dfrac{SA}{AG} = 1$, $f(x)$ becomes zero only once (the degree of f is 1), at a point that must lie outside the interval $\left[-\dfrac{G}{2}, \dfrac{G}{2}\right]$. So let us assume $\dfrac{SA}{AG} < 1$; the discriminant of f can be written

$$\Delta = \frac{SA^2}{GA^2}\left(\overline{AM} + \overline{GM}\right)^2 - 4\left(\frac{SA}{AG} - 1\right)\left(\frac{SA}{AG}\,\overline{AM}\cdot\overline{GM} + HM^2\right)$$

$$= GA^2 \cdot \frac{SA^2}{AG^2} + 4\left(\overline{AM}.\overline{GM} - HM^2\right)\frac{SA}{AG} + 4HM^2 = \varphi\left(\frac{SA}{AG}\right),$$

a second-degree expression in $\dfrac{SA}{AG}$ whose discriminant is

$$\delta = 4\left[\left(\overline{AM}\cdot\overline{GM} - HM^2\right)^2 - HM^2 \cdot GA^2\right].$$

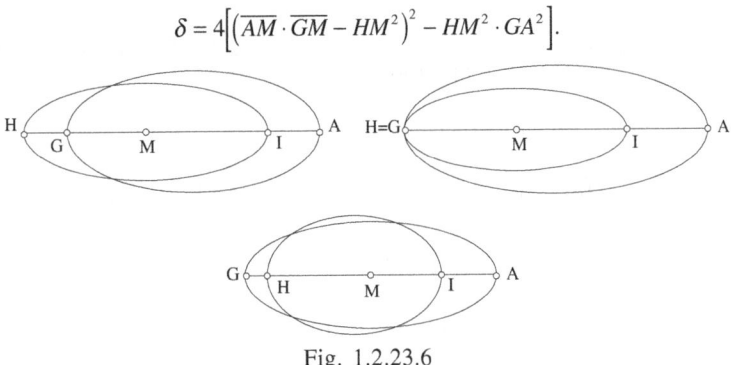

Fig. 1.2.23.6

We note that $\delta \geq 0$ because $AM \cdot MG + HM^2 \geq HM \cdot GA$ on account of the inequality

$$AM \cdot MG \geq HM\,(AM + GH),$$

which can also be written

$$AM \cdot GH \geq HM \cdot GH.$$

Thus Δ becomes zero for two positive values of $\dfrac{SA}{AG}$, α and β, and it is positive if $\dfrac{SA}{AG}$ lies outside the interval $[\alpha, \beta]$. For $\dfrac{SA}{AG} = 1$, Δ becomes

$GA^2 + 4\overline{AM}\cdot\overline{GM} \geq 0$ since $AM \cdot MG \leq \left(\dfrac{1}{2}GA\right)^2$; so we see that 1 lies outside the interval $[\alpha, \beta]$.

We must also impose the condition that the roots of the equation in x shall lie between $-\dfrac{G}{2}$ and $\dfrac{G}{2}$. The absolute value of half the sum of the roots is

$$\frac{\dfrac{SA}{AG}\left|\overline{AM} + \overline{GM}\right|}{2\left(1 - \dfrac{SA}{AG}\right)}$$

and this condition can thus be written

$$\frac{SA}{AG}\left|\overline{AM}+\overline{GM}\right| \le G.\left(1-\frac{SA}{AG}\right),$$

or

$$\frac{SA}{AG} \le \frac{G}{G+\left|\overline{AM}+\overline{GM}\right|}.$$

We have

$$\varphi\left(\frac{G}{G+\left|\overline{AM}+\overline{GM}\right|}\right) = \frac{8HM\cdot\left|\overline{AM}+\overline{GM}\right|}{\left(2HM+\left|\overline{AM}+\overline{GM}\right|\right)^2}\left(HM^2+\left|\overline{AM}+\overline{GM}\right|\cdot HM+\overline{AM}\cdot\overline{GM}\right).$$

This expression is negative because the part in parentheses can be written

$$[HM - \inf (AM, GM)]\ [HM + \sup (AM, GM)],$$

which is negative since HM is less than AM and GM.

Thus $\dfrac{G}{G+\left|\overline{AM}+\overline{GM}\right|}$ is between α and β.

The two conditions accordingly reduce to the single inequality

$$\frac{SA}{AG} \le \alpha = \frac{2HM^2 - \overline{AM}\cdot\overline{GM} - \sqrt{AI\cdot GI\cdot AH\cdot GH}}{AG^2}.$$

The necessary and sufficient condition for the problem to have solutions is finally $\dfrac{SA}{AG} \le \alpha$. When $\dfrac{SA}{AG} = \alpha$, $\Delta = 0$ and the two ellipses touch one another.

– **24** – Let there be a hyperbola Γ with centre H and vertex A; to find a point B on Γ such that the diameter BP from B and the corresponding *latus rectum* PN satisfy

$$BP \cdot PN = EG^2,$$

where EG is a given straight line.

Analysis: Let AD be the transverse axis and AI the *latus rectum* corresponding to AD; then the second diameter Δ associated with AD is such that

$$\Delta^2 = AD \cdot AI.$$

We then have

$$|AD^2 - \Delta^2| = AD \cdot |AD - AI| = AD \cdot DI.$$

Let $EL \perp EG$ with $EL^2 = AD \cdot DI = |AD^2 - \Delta^2|$. If a point B gives a solution to the problem, we have $BP \cdot PN = EG^2$, so the straight line EG is equal to the diameter conjugate with BP. So we have

$$|BP^2 - EG^2| = BP\, |BP - PN| = BP \cdot BN.$$

But

$$|BP^2 - EG^2| = |AD^2 - \Delta^2| \quad (Conics, \text{ VII.13}),$$

so

$$|BP^2 - EG^2\,| = EL^2.$$

1) If $AD > AI$, then $AD > \Delta$ and in this case $BP > EG$ (*Conics*, VII.21), and we have

$$BP^2 - EG^2 = EL^2 \Leftrightarrow BP^2 = EL^2 + EG^2 = LG^2 \text{ (Fig. 1.2.24.1).}$$

The construction shown by Ibn al-Haytham then gives the length BP.

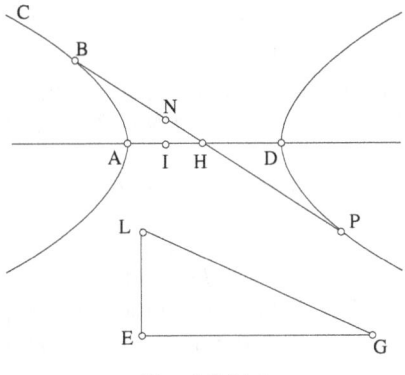

Fig. 1.2.24.1

2) If $AD < AI$, then $AD < \Delta$ and in this case $BP < EG$ (*Conics*, VII.22), we have $EG^2 - BP^2 = EL^2 \Leftrightarrow BP^2 = EG^2 - EL^2$ (Fig. 1.2.24.2).

In this case BP is a side enclosing the right angle in a right-angled triangle whose hypotenuse is EG and the third side EL; this requires $EG > EL$, that is, $EG^2 > AD \cdot DI$.

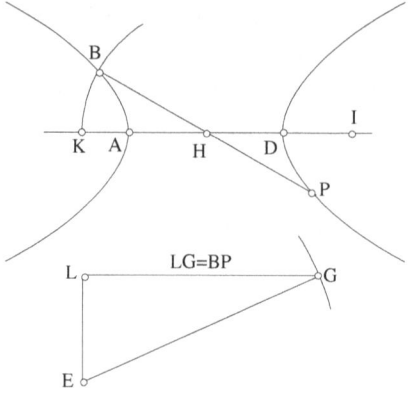

Fig. 1.2.24.2

So in both cases, if the point B exists, the length BP can be found from the equality $|BP^2 - EG^2| = EL^2$.

– **25** – *Synthesis*: Ibn al-Haytham again uses the construction assuming $AD > AI$ (Fig. 1.2.24.1) and considers the circle with centre H and radius $GL/2$; if it cuts Γ in B, this point gives a solution to the problem. Then, we have $BP = 2BH = GL$; and if we define PN by $BP \cdot PN = EG^2$, we have $PN < PB$ because $EG < GL$, and $PB \cdot BN = BP^2 - BP \cdot PN = BP^2 - EG^2 = EL^2$.

PN is in fact the *latus rectum* corresponding to the diameter BP whose conjugate diameter has length EG.

Assuming $AD < AI$ (Fig. 1.2.24.2), if PN is defined by $BP \cdot PN = EG^2$, we have $PN > PB$ because $EG > LG$ and $PB \cdot BN = BP \cdot PN - BP^2 = EG^2 - BP^2 = EL^2$, so BP and PN give solutions to the problem.

The existence of the point B: For the circle $(H, GL/2)$ to cut Γ, it is necessary and sufficient that $GL > AD$. (Any transverse diameter of Γ is greater than the transverse axis.)

$$GL > AD \Leftrightarrow EG > \Delta \Leftrightarrow EG^2 > AD \cdot AI,$$

that is the necessary and sufficient condition for the problem to have a solution (in fact $EG^2 > AD \cdot AI \Rightarrow EG^2 > AD \cdot DI$, the condition known to be required in case 2).

It is only here that Ibn al-Haytham considers the assumption $AD < AI$. So up to now the problem has been treated assuming $AD > AI$. We have seen that construction 1 does not give $BP = LG$ if $AD < AI$. Here Ibn al-Haytham gives a method for obtaining BP from construction 1.

Let the point O of the segment GE be defined by $GO^2 = GE^2 - AD \cdot DI = GE^2 - EL^2$; this assumes $GE^2 > AD \cdot DI$, that is, $GE > EL$; the segment GO is thus equal to the segment LG in construction 2.

If the point M is then defined by $GM \cdot GO = GE^2$, we have $GM > GE > GO$; the length GO is then the same as that of the required diameter and GM is the corresponding *latus rectum*, provided $GO > AD$.

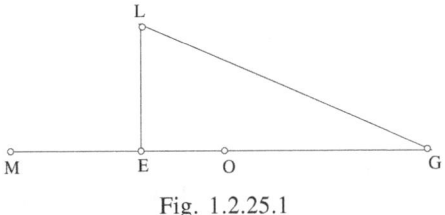

Fig. 1.2.25.1

$$GO > AD \Leftrightarrow GE^2 - AD \cdot DI > AD^2$$
$$\Leftrightarrow GE^2 > AD \, (AD + DI)$$
$$\Leftrightarrow GE^2 > AD \cdot AI.$$

If $AD = AI$, then $AD = \Delta$ and every diameter is equal to its *latus rectum* and to its conjugate diameter (*Conics*, VII.23, equilateral hyperbola), so the required diameter and the corresponding *latus rectum* have the same length EG; the necessary and sufficient condition for B to exist is thus $EG > AD$.

The case where Γ is an ellipse

To find a point B in an ellipse Γ with major axis AD, centre H, such that the diameter BP from B and the corresponding *latus rectum* satisfy $BP \cdot PN = EG^2$, where EG is a given straight line.

Analysis: Let AI be the *latus rectum* corresponding to AD, and the minor axis FK is the diameter conjugate with AD; we have $FK^2 = AD \cdot AI$, $AI < AD$.

If B gives a solution to the problem, we have $BP \cdot PN = EG^2$ and EG is the diameter conjugate with BP; and from Proposition 12 of Book VII, we have

(1) $\qquad AD^2 + FK^2 = BP^2 + EG^2.$

We know the lengths AD, FK and EG, so the length BP is known.

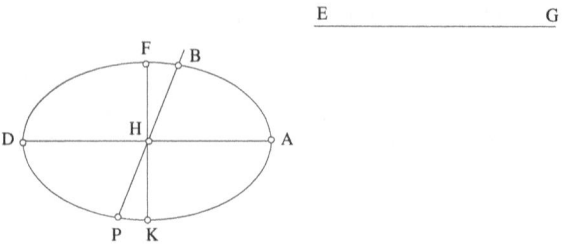

Fig. 1.2.25.2

Synthesis:

a) Construction of the length *BP*.

On the sides enclosing a right angle we make $OM = AD$ and $ON = FK$; we thus have $MN^2 = AD^2 + FK^2$. We draw a semicircle with diameter MN and the circle with centre M and radius EG; they intersect at a point R if and only if $EG < MN$, that is, if $EG^2 < AD^2 + FK^2$. The length RN is equal to BP, the length that was required.

b) The existence of the point *B*.

We draw the circle with centre H and radius $\dfrac{BP}{2} = \dfrac{RN}{2}$. It cuts the ellipse and gives the point B if and only if $FK < BP < AD$. From (1) this condition is equivalent to $FK < EG < AD$. The point B that we find thus gives a solution to the problem, and the *latus rectum* corresponding to BP is PN, which satisfies $BP \cdot PN = EG^2$.

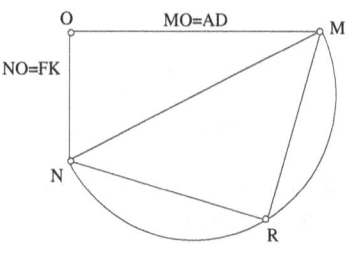

Fig. 1.2.25.3

So, to summarize, in the case of the ellipse, if we are given AD, AI and EG, the condition for the problem to have a solution is $AD \cdot AI < EG^2 < AD^2$.

– **26** – Let there be a hyperbola Γ with axis AD and centre H, and a straight line EG. To find a diameter such that if we add its *latus rectum* to it, we obtain a straight line equal to EG.

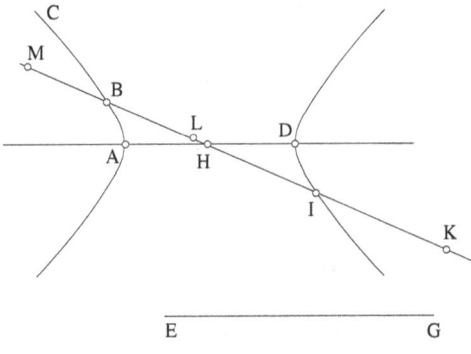

Fig. 1.2.26.1

Analysis: Let *BI* be the diameter that gives a solution to the problem and *IK* its *latus rectum*; we have $BI + IK = BK = EG$.

If Δ is the axis conjugate with *AD*, and Δ' the diameter conjugate with *BI*, we have

$$|AD^2 - \Delta^2| = |BI^2 - \Delta'^2| \text{ and } \Delta'^2 = BI \cdot IK,$$

$$|AD^2 - \Delta^2| = BI \cdot |BI - IK|.$$

Let *L* be a point such that $IL = IK$ and *M* a point such that $BM = BL$. Then we have

$$|BI - IK| = BL = BM, \qquad KM = 2BI.$$

Then we have $BI \cdot BM = |AD^2 - \Delta^2|$, which is known, so the product $KM \cdot BM$ is known and $BK = EG$ is a known length. We thus know the two lengths *KM* and *BM* and we have $BI = KM/2$.

Note: If *C* is the *latus rectum* corresponding to *AD*:

1) $C < AD \Rightarrow \Delta < AD$ and $IK < BI$, so we have the following order of points, which is the same as in Fig. 1.2.26.1:

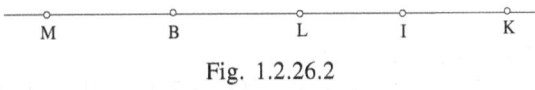

Fig. 1.2.26.2

2) $C > AD \Rightarrow \Delta > AD$ and $IK > BI$, we have the following order:

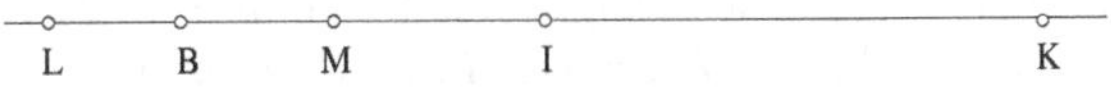

Fig. 1.2.26.3

In both cases, we have $KM = 2BI$, and we thus have

$$KM \cdot BM = 2|\Delta^2 - AD^2| = 2AD \, |AD - C|,$$

which is known.

But in 1) we have

$$BK = KM - BM = EG$$

and in 2) we have

$$BK = KM + BM = EG.$$

The lengths KM and BM are known because we know their geometric mean and their difference (case 1) or their sum (case 2).

In case (1) the construction of KM and BM is always possible and requires no discussion.[65]

In case (2), if we put $BM \cdot KM = 2(\Delta^2 - AD^2) = l^2$, the condition for the construction to be possible is $2l \le EG$, as shown in Fig. 1.2.26.4.

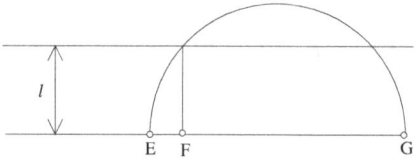

Fig. 1.2.26.4

The lengths EF and FG are the lengths that were required.

– **27** – *Synthesis*: Let $AI = C$ be the *latus rectum* associated with the diameter AD, so we have $DI = |AD - C|$.

1) $AI < AD$ 2) $AI > AD$.

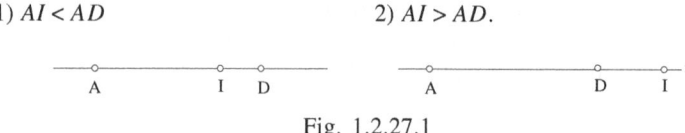

Fig. 1.2.27.1

1) Let us first assume that $AI < AD$.

Let the point K be defined by $KE \cdot KG = 2DA \cdot DI$, and the point M on AD be defined by $HM = EK/4$. The circle (H, HM) cuts Γ at B, the diameter BH gives a solution to the problem.

Proof: Let there be on the straight line BH points U, P, N, L such that $HU = HB$, $BP = EG$ and $BN = BL = KG$ (L, U, P lie on the half-line BH).

---

[65] See above, note 53, p. 114.

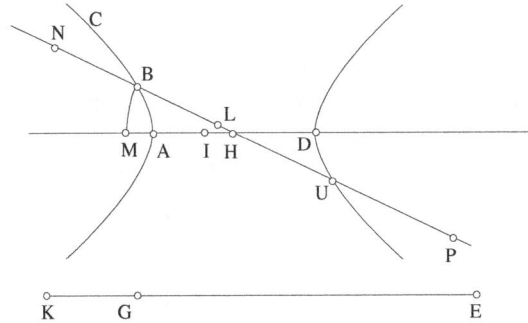

Fig. 1.2.27.2

Thus we have

$$BU = 2HM = \frac{1}{2}\, EK,$$

$$PN = EG + GK = EK,$$

so

$$PU + BN = PU + BL = BU,$$

hence

$$PU = BU - BL,$$

$$PU = LU.$$

We get

$$BU \cdot BL = \frac{1}{2}\, EK \cdot KG = DA \cdot DI,$$

so

$$BU \cdot BL = BU^2 - \Delta'^{\,2},$$

if Δ' is the diameter conjugate with BU, and

$$\Delta'^{\,2} = BU \cdot LU = BU \cdot UP.$$

so UP is the *latus rectum* associated with the diameter BU and we have got $BU + UP = EG$.

Discussion: It is necessary that $BU + UP > AD + AI$ (a property of diameters of a hyperbola and their *latera recta*),[66] so we have

[66] This property holds true only if $3AD \geq AI$. Otherwise, the minimum of the sum (diameter + *latus rectum*) is reached at two positions of the diameter, these positions being symmetrical with respect to AD (see Apollonius, *Conics*, VII.40); see the Note in the main text at the end of this problem.

(1) $EG > AD + AI.$

We need to know whether this condition is sufficient. The point B exists if and only if $HM > HA$, that is $EK > 2DA$, where the point K is defined by $KE \cdot KG = 2DA \cdot DI$, $KE > KG$.

If we assume $AI < AD$, the condition (1) can be written

$$EG > 2AD - DI;$$

so we have

$$EK - KG > 2AD - DI$$

and

$$KE \cdot KG = 2AD \cdot DI,$$

hence

$$EK > 2AD \quad \text{and} \quad KG < DI.$$

Condition (1) is thus sufficient for B to exist.

2) Let us assume $AI > AD$.

Let there be a point U' on the segment EG such that $EU' \cdot U'G = 2AD \cdot DI$, where $U'E > U'G$, and let there be a point M such that $HM = EU'/4$; the circle (H, HM) cuts Γ at B, and the diameter BH gives a solution to the problem.

Discussion: We need to know whether condition (1) $EG > AD + AI$ is sufficient for the points U' and B to exist.

If we put $2AD \cdot DI = l^2$, U' exists if and only if $l < \dfrac{EG}{2}$, that is $EG^2 > 8AD \cdot DI$. Now

$$(1) \Leftrightarrow EG > 2AD + DI \Leftrightarrow EG^2 > 4AD^2 + 4AD \cdot DI + DI^2$$
$$\Leftrightarrow EG^2 > 8AD \cdot DI + 4AD^2 - 4AD \cdot DI + DI^2$$
$$\Leftrightarrow EG^2 > 8AD \cdot DI + (2AD - DI)^2 \qquad (2).$$

So

$$EG > AD + AI \Rightarrow EG^2 > 8AD \cdot DI,$$

and the point U' exists.

We have $HB = HM = EU'/4$; B exists if and only if $HB > AD/2$, that is, $EU' > 2AD$. We know that

$$EU' + U'G = EG \quad \text{and} \quad EU' \cdot U'G = 2AD \cdot DI,$$

hence

$$(EU' - U'G)^2 = EG^2 - 8AD \cdot DI;$$

from (2)

$$(EU' - U'G)^2 > (2AD - DI)^2,$$

so we have

$$EU' + U'G = EG$$

$$EU' - U'G > |2AD - DI| \text{ (since } EU' > U'G),$$

hence

$$EU' > \frac{1}{2} [EG + |2AD - DI|]$$

$$EU' > \frac{1}{2} [2AD + DI + |2AD - DI|].$$

If $2AD > DI$, we have $EU' > 2AD$.

If $2AD < DI$, we have $EU' > DI > 2AD$.

So the condition $EG > AD + DI$ is sufficient for B to exist.

3) Let us assume $AI = AD$.

In this case, any diameter is equal to its *latus rectum*; so the required diameter is equal to $\frac{1}{2}EG$, and the condition for the construction to be possible will be $EG > 2AD$, that is, $EG > AD + AI$.

Note: The condition $BU + UP \geq AD + AI$ is necessary only when $3AD \geq AI$. If this is not so, which means $AI > AD$, the condition must be replaced by

$BU + UP \geq$ minimum of the sum (diameter + *latus rectum*) (see Apollonius, *Conics*, VII.40).

Let us calculate this minimum analytically. The coordinates of a point B of the hyperbola whose equation is $\frac{x^2}{a^2} - \frac{y^2}{b^2} = 1$ are $x = a \cosh t$, $y = b \sinh t$ where t is a parameter that is positive on the branch of the hyperbola we are considering.

The direction of the conjugate of the diameter HB is that of the tangent at B, with steering ratio $a \sinh t$, $b \cosh t$. The corresponding semi-diameters are

$$a' = \sqrt{a^2 \cosh^2 t + b^2 \sinh^2 t}, \quad b' = \sqrt{a^2 \sinh^2 t + b^2 \cosh^2 t}.$$

The associated *latus rectum* is $p' = \dfrac{2b'^2}{a'}$ and the sum $2a' + p'$ is thus

$$2\frac{a'^2 + b'^2}{a'} = 2\sqrt{2}\,\frac{\left(a^2 + b^2\right)\cosh 2t}{\sqrt{\left(a^2 + b^2\right)\cosh 2t + a^2 - b^2}}.$$

Putting $u = (a^2 + b^2)\cosh 2t$ $(u \geq a^2 + b^2)$, we need to find the minimum of

$$v = \frac{u}{\sqrt{u + a^2 - b^2}}.$$

We have

$$v' = \frac{u + 2\left(a^2 - b^2\right)}{2\left(u + a^2 - b^2\right)^{3/2}};$$

this expression is positive for $u \geq 2\,(b^2 - a^2)$.

This condition is always satisfied when $a \geq b$; in this case v is an increasing function of u, and thus of $t \geq 0$, and its minimum is for $t = 0$, and has the value

$$2\frac{a^2 + b^2}{a} = AD + AI \qquad\qquad (B = A).$$

In the case where $b \geq a$, we still have $v' \geq 0$ if $b \leq a\sqrt{3}$ since, in this case, $2(b^2 - a^2) \leq a^2 + b^2$, the initial value of u. So v increases monotonically from its minimum value of $2\dfrac{a^2 + b^2}{a} = AD + AI$. Ibn al-Haytham's condition is still necessary and sufficient. If on the contrary $b > a\sqrt{3}$, v has a local minimum for $u = 2(b^2 - a^2)$. We then have

$$a'^2 = \frac{u + a^2 - b^2}{2} = \frac{b^2 - a^2}{2}, \qquad b'^2 = \frac{u - a^2 + b^2}{2} = 3\frac{b^2 - a^2}{2},$$

that is, the *latus rectum* $\dfrac{2b'^2}{a'} = 6a'$ is equal to three times the transverse diameter; and

$$2a' + p' = 2\frac{a'^2 + b'^2}{a'} = 4\sqrt{2\left(b^2 - a^2\right)} < 2\frac{a^2 + b^2}{a}$$

because

$$(a^2 + b^2)^2 - 8a^2(b^2 - a^2) = (b^2 - 3a^2)^2 > 0.$$

The minimum is thus $4\sqrt{2(b^2 - a^2)} = 2\sqrt{2AD(AI - AD)}$ and the condition for the construction to be possible can be written $EG^2 \geq 8AD\ (AI - AD) = 8AD \cdot DI$. In problems of this type, Ibn al-Haytham tried to find an exact way of obtaining a solution.

The radius a' of the auxiliary circle is greater than $HA = a$ because $\dfrac{b^2 - a^2}{2} > a^2$ is equivalent to $b^2 > 3a^2$.

We may summarize the preceding discussion as follows:

1) If $AI \leq AD$, the point K exists and the condition for B to exist is the inequality $EG \geq AD + AI$ (a necessary and sufficient condition).

2) If $AI > AD$, the condition for the existence of U', which replaces the point K, is the inequality $EG^2 \geq 8AD \cdot AI$ (a necessary and sufficient condition).

This condition ensures the existence of B in the case where $AI \geq 3AD$, but if $AI < 3AD$, the condition needs to be replaced by the stronger one $EG \geq AD + AI$.

Thus, the necessary and sufficient condition for the problem to be soluble is $EG \geq AD + AI$ if $AI \leq 3AD$, and the condition becomes $EG^2 \geq 8AD \cdot DI$ if $AI \geq 3AD$.

When $AI = 3AD$, the two conditions are equivalent.

To complete the discussion, we needed to find the minimum value of the sum of the diameter and the *latus rectum* associated with it, which Apollonius does not do in the case where $AI \geq 3AD$; in this case, where the minimum is not $AD + AI$, Apollonius only states that the minimum in question exists. So it is understandable that Ibn al-Haytham broke off the discussion and omitted to deal with precisely this case.

Here too, we can see that Ibn al-Haytham abandons the attempt to give a complete discussion when he realizes that he cannot provide a precise solution to the diorism. It is as if, having recognized that he could not achieve success in the case where $AI \geq 3AD$, he left that case out, together with what was said about it in Proposition VII.40 of the *Conics*.

– **27a** – The same problem but with Γ an ellipse. Ibn al-Haytham gives only a single indication regarding the method to employ.

Let there be an ellipse Γ with major axis AD and centre H, and a given length EG. To find a diameter BU such that if UP is the *latus rectum* associated with BU, we have $BU + UP = EG$.

Let AI be the *latus rectum* associated with AD ($AI < AD$); we have

(1) $$BU^2 + BU \cdot UP = AD^2 + AD \cdot AI,$$

hence

(2) $\qquad BU \cdot EG = AD\,(AD + AI)$;

EG, AD, AI are known, so *BU* is known.

For the length *BU* to provide a solution to the problem, it is necessary and sufficient that $\Delta < BU < AD$, Δ being the minor axis of the ellipse $(\Delta^2 = AD \cdot AI)$.

From (2), we have $BU < AD$ if and only if $EG > AD + AI$.

$$BU > \Delta \qquad \Leftrightarrow BU \cdot EG > \Delta \cdot EG$$
$$\Leftrightarrow AD\,(AD + AI) > \Delta \cdot EG$$
$$\Leftrightarrow EG < (AD/\Delta)\,(AD + AI).$$

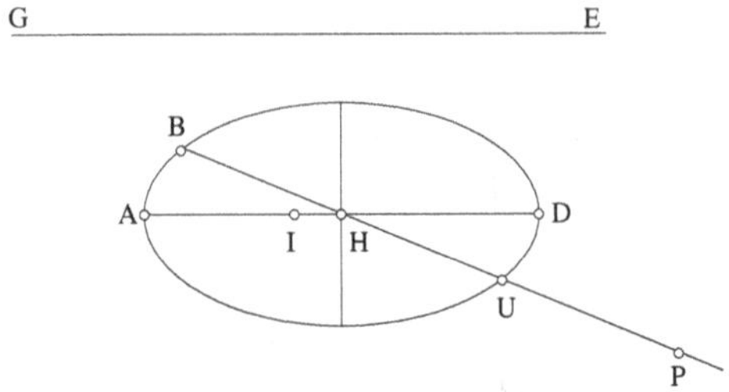

Fig. 1.2.27.3

$$\frac{AD}{\Delta} = \sqrt{\frac{AD}{AI}} > 1,$$

$$BU > \Delta \Leftrightarrow EG < \sqrt{\frac{AD}{AI}}(AD + AI).$$

The problem has a solution if and only if

$$AD + AI < EG < (AD + AI) \cdot \sqrt{\frac{AD}{AI}}.$$

Ibn al-Haytham gives the condition $EG > AD + AI$.

– 27b – Let there be a hyperbola with transverse axis *AD* and centre *H*. To find a point *B* such that the diameter *BE* and the associated *latus rectum* *BF* shall be such that $\dfrac{BF}{BE} = k$, a known ratio.

Let *AI* be the *latus rectum* associated with the diameter *AD* and let $k_0 = \dfrac{AI}{AD}$.

If *BE* gives a solution to the problem, we know that

$$|\,BE^2 - BE \cdot BF\,| = |\,AD^2 - AD \cdot AI\,|,$$
$$BE\,|\,BE - BF\,| = AD\,|\,AD - AI\,|.$$

Let us choose points F and I on the semi-infinite straight lines $[BE)$ and $[AD)$ respectively; we have

(1) $$BE \cdot EF = AD \cdot DI.$$

If $AD > AI$, it is necessary to assume $k_0 < k < 1$ (from Proposition 21, Book VII); we then have $BE > BF$ and

$$\frac{BF}{BE} = k \Rightarrow \frac{BE - BF}{BE} = 1 - k = \frac{EF}{BE};$$

similarly

$$\frac{DI}{AD} = 1 - k_0.$$

(1) $$\Rightarrow BE^2 (1 - k) = AD^2 (1 - k_0).$$

AD, k and k_0 are known, so BE is known. Assuming $k_0 < k < 1$ implies $1 - k < 1 - k_0$ and consequently $BE > AD$, and BE gives a solution to the problem.

If $AD < AI$, it is necessary to assume $1 < k < k_0$ (from Proposition 22, Book VII); we then have

$$BE < BF, \quad \frac{BF - BE}{BE} = \frac{EF}{BE} = k - 1 \text{ and } \frac{DI}{DA} = k_0 - 1$$

(1) $$\Rightarrow BE^2 (k - 1) = AD^2 (k_0 - 1),$$

so BE is known.

$$1 < k < k_0 => k - 1 < k_0 - 1 => BE > AD.$$

For the problem to be soluble it is necessary and sufficient that the given ratio k shall lie between 1 and k_0. $[k_0 < k < 1$ or $1 < k < k_0.]$

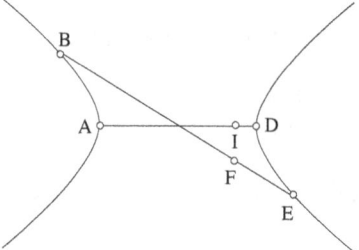

Fig. 1.2.27.4

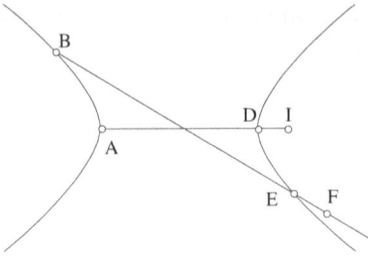

Fig. 1.2.27.5

If $AD = AI$ (an equilateral hyperbola, a case not discussed by Ibn al-Haytham), we have $k_0 = 1$ and the ratio k must also be equal to 1. In this case all the points of the hyperbola give solutions.[67]

– **27c** – The same problem but with Γ an ellipse with major axis AD, *latus rectum AI*, $\dfrac{AI}{AD} = k_0, k_0 < 1$.

If BE is a diameter that gives a solution to the problem, we have

$$\frac{BF}{BE} = k,$$

$$AD\,(AD + AI) = BE\,(BE + BF);$$

but

$$\frac{BF}{BE} = k \Rightarrow \frac{BE + BF}{BE} = k + 1 \Rightarrow BE + BF = (k+1)BE;$$

similarly

$$AD + AI = (k_0 + 1) \cdot AD;$$

hence

(1) $\qquad\qquad AD^2\,(k_0 + 1) = BE^2\,(1 + k),$

so BE is known.

BE gives a solution to the problem if and only if

(2) $\qquad\qquad k_0 \cdot AD^2 < BE^2 < AD^2.$

(1) $\qquad$ gives $\quad BE^2 = \dfrac{AD^2(k_0 + 1)}{1 + k}$

(2) $\Leftrightarrow k_0 < \dfrac{k_0 + 1}{1 + k} < 1 \Leftrightarrow k\,k_0 < 1$ and $k_0 < k \Leftrightarrow k_0 < k < 1/k_0$.

[67] Accordingly, no construction need be carried out. Perhaps this is why Ibn al-Haytham did not think it necessary to consider this case.

The condition for the problem to have a solution is thus $k_0 < k < 1/k_0$. For Problems 27b and 27c Ibn al-Haytham has indicated the method to employ, but he has not investigated the conditions for the problem to be soluble, which are, however, obvious.

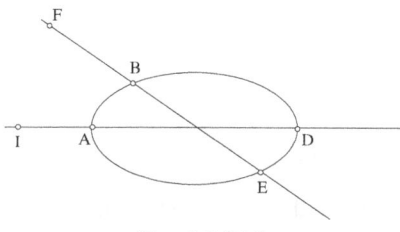

Fig. 1.2.27.6

Note: We construct solutions to Problems 24 to 27 by looking at the intersections of the given conic with a circle having the same centre and a radius that can be constructed with straightedge and compasses.

Thus the problems do not involve any auxiliary conic.

– **28** – Through a given point D on the axis of a given parabola Γ with vertex A, to draw a straight line that cuts Γ in B and C and is such that $BC = F$, where F is a given straight line.

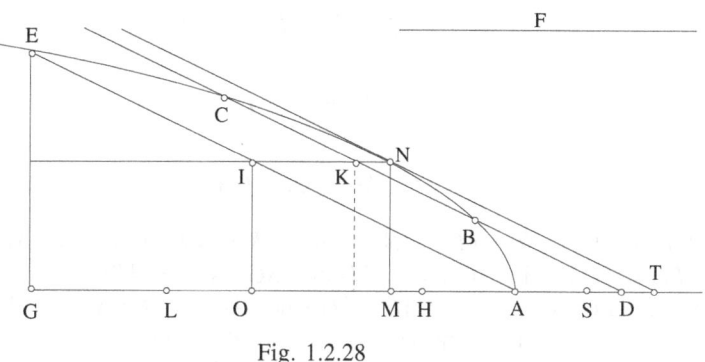

Fig. 1.2.28

Analysis: if the straight line DBC is a solution to the problem, then $BC = F$.

If $AE \parallel BC$, the straight line parallel to AD through the midpoint I of AE cuts BC in its midpoint K and cuts Γ in N. The tangent to Γ at N is parallel to AE, it cuts the axis in T and we have $AT > AD$. If $MN \perp AD$ and $IO \perp AD$, A is the midpoint of TM, so we have $MA = AT = NI = MO$ and $OA = 2AM$. Moreover, $KI = AD$.

Let H be such that $AH = AD = KI$; then $MH = DT = KN$.

If AS is the *latus rectum* with respect to the axis, we have

$$EG^2 = SA \cdot AG,$$

hence

$$EG^2 + GA^2 = EA^2 = SG \cdot GA.$$

But

$$GA = 2OA = 4AM = 4NI,$$

hence

$$SG \cdot NI = IA^2 = EI^2.$$

SG is the *latus rectum* with respect to the diameter NI, K is the midpoint of BC, so

$$DC \cdot DB = DK^2 - BK^2;$$

but

$$DK = IA,$$

so

$$SG \cdot NI = BK^2 + DC \cdot DB;$$

but on the other hand $B \in \Gamma$, so

$$SG \cdot NK = BK^2,$$

hence

$$SG \cdot IK = DC \cdot DB$$

and

$$SG \cdot AH = DC \cdot DB.$$

We also have $SG \cdot MH = BK^2$ since $NK = MH$. If we put $AL = 4AH$, then $LG = 4HM$ (since $GA = 4AM$) and $SG \cdot GL = 4BK^2 = BC^2 = F^2$, a known square. We note that $AL < AG$ since $AH = AD < AT = AM$.

AS and AD were given; AH, AL, SL are thus known segments, and the segment SG and, consequently, the point G are known.

From the point G, we find E on Γ, I the midpoint of AE, then K (since $IK \parallel AD$ and $IK = AD$); we join KD, from this we find B and C since $BK = KC = F/2$.

– **29** – The problem proposed is soluble for all positions of D on the axis of the parabola Γ lying outside the curve (that is, on the side of the vertex opposite to that of the focus), and for any length of the given straight line F.

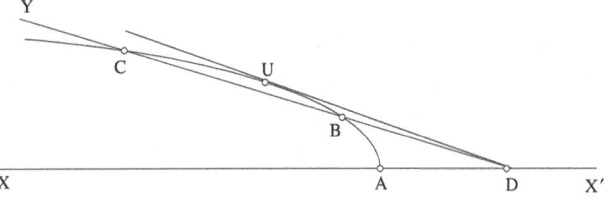

Fig. 1.2.29.1

Indeed, from any point D on AX' we can draw a tangent DU to Γ, and $U\hat{D}X$ will be acute. Any straight line DY lying inside angle $U\hat{D}X$ cuts the parabola in two points B and C and the length BC increases from 0 to $+\infty$ as angle $U\hat{D}Y$ increases from 0 to $U\hat{D}X$; so there exists a value of $U\hat{D}Y$, and only one such value, for which $BC = F$, and thus there is a unique straight line DY that gives a solution to the problem.

We shall show that the construction deduced from the analysis does indeed give the straight line DY.

Let L be a point on the axis such that $AL = 4AD$, G the point lying beyond L such that $SG \cdot GL = F^2$, and E the point of Γ that is projected into G and I the midpoint of AE.

We show that the line DY parallel to the line AE is the required straight line, that is:

a) DY cuts Γ in 2 points B and C,
b) $BC = F$,

a) DY cuts Γ

The straight line parallel to AG passing through I the midpoint of AE cuts Γ in N, and the tangent at N, which we shall call NT, is parallel to AE.

So DY cuts Γ if and only if $AD < AT$. This condition is satisfied since $AL < AG$ and $AL = 4AD$, $AG = 4AT$.

Let B and C be the points of intersection of DY and Γ.

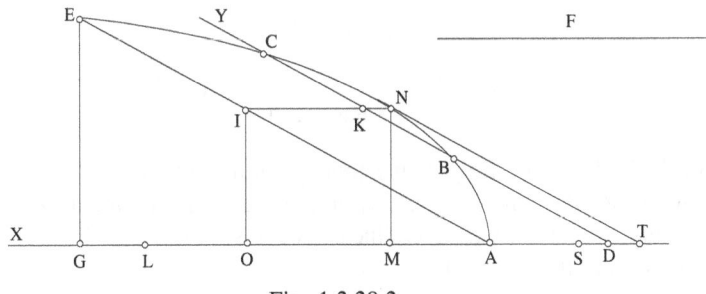

Fig. 1.2.29.2

b) $BC = F$

As in the analysis, we have $SG \cdot GA = EA^2$, hence $SG \cdot NI = AI^2$ (since $NI = \frac{1}{4} GA$); SG is the *latus rectum* with respect to the diameter NI, which cuts BC in its midpoint K. We have

$$BK^2 = SG \cdot KN,$$

$$KN = DT = AT - AD = \frac{1}{4} LG,$$

hence

$$BK^2 = \frac{1}{4} SG \cdot GL,$$

and consequently

$$BC^2 = SG \cdot GL = F^2,$$

so

$$BC = F.$$

We may note that this problem, which presents a clear case of a νεῦσις, is solved solely by plane geometry.

– **30** – We are given a hyperbola Γ with axis AD and centre E, a point H between A and E and a straight line F. To draw through H a straight line that cuts Γ in B and C and such that $BC = F$.

Let the *latus rectum* with respect to AD be called a, let us put $\dfrac{EA}{EH} = k$.

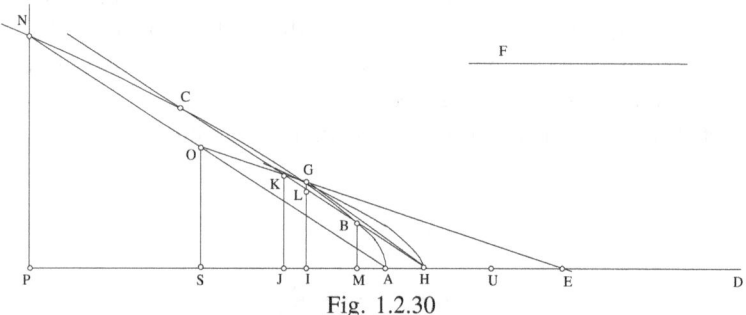

Fig. 1.2.30

Analysis: Let the straight line that gives a solution to the problem be HBC. $AN \parallel BC$, AN cuts Γ in N. If K and O are the midpoints of BC and AN respectively, we have O, K, E collinear. We draw the lines NP, OS, KJ, BM perpendicular to the axis AD.

We have

$$\frac{OS}{KJ} = \frac{SA}{JH} = \frac{OA}{KH} = \frac{EA}{EH} = k \text{ a known ratio,}$$

$$\frac{SA}{JH} = \frac{EA}{EH} = \frac{SE}{JE} = k.$$

Therefore

$$k^2 = \frac{OS^2}{KJ^2} = \frac{ES \cdot SA}{JE \cdot JH} \Rightarrow \frac{ES \cdot SA}{OS^2} = \frac{JE \cdot JH}{JK^2},$$

but

$$PA = 2SA \text{ and } DA = 2AE,$$

hence

$$DP = 2ES;$$

we also have

$$NP = 2OS,$$

so we have

$$\frac{JE \cdot JH}{JK^2} = \frac{DP \cdot PA}{PN^2} = \frac{AD}{a} \text{ (since } N \in \Gamma\text{)}.$$

Let us draw the line HG to be tangent to Γ and $GI \perp AD$; GI cuts the straight line BC in L (if we produce GI, it passes through G', the point of contact of the second tangent to Γ from H); (C, B, L, H) is a harmonic range, so $\frac{CH}{HB} = \frac{CL}{LB}$ (*Conics*, III.37).

We assume $HC > HB$, so $LC > LB$ and L lies between B and K.

We have $\frac{EI \cdot IH}{IG^2} = \frac{AD}{a}$, a property of the tangent (*Conics*, I.37).

If a' is the segment defined by $\frac{EH}{a'} = \frac{AD}{a}$, then the hyperbola $\mathcal{H}$ with axis EH and *latus rectum a'* passes through the points K and G because

$$\frac{JE \cdot JH}{JK^2} = \frac{IE \cdot IH}{IG^2} = \frac{EH}{a'}.$$

Let the point U be defined by $\frac{UE}{UH} = \frac{EH}{a'} = \frac{AD}{a}$, where U lies between E and H and UH is a segment in the same ratio; then $\frac{UJ \cdot HJ}{HK^2} = \frac{UE}{EH}$, which is a known ratio (*Conics*, VII.2).

But

$$\frac{CH}{HB} = \frac{CL}{LB} \Rightarrow \frac{CH + HB}{HB} = \frac{CL + LB}{LB} = \frac{CB}{LB} \Rightarrow \frac{HK}{HB} = \frac{BK}{LB}$$

$$\Rightarrow \frac{HK}{KB} = \frac{BK}{LK} \Rightarrow BK^2 = HK \cdot LK.$$

By orthogonal projection onto the axis AD, we find that $JM^2 = JH \cdot JI$, hence

$$\frac{HJ}{JM} = \frac{JM}{JI} \quad \text{and} \quad \frac{HJ^2}{JM^2} = \frac{HJ}{JI}.$$

But

$$\frac{HJ}{JM} = \frac{HK}{KB},$$

hence

$$\frac{HK^2}{KB^2} = \frac{HJ}{JI} = \frac{HJ \cdot UJ}{JI \cdot UJ} \Rightarrow \frac{JI \cdot UJ}{KB^2} = \frac{HJ \cdot UJ}{HK^2} = \frac{UE}{EH}.$$

$KB^2 = \dfrac{F^2}{4}$ and $\dfrac{UE}{EH}$ is a known ratio, hence $JI \cdot JU$ is a known product and the points I and U are known, so the point J is known.[68]

The perpendicular to AD at J cuts $\mathscr{H}$ in K, the straight line HK cuts Γ in B and C.

– 31 – *Synthesis*: We again have a conic section Γ with vertex A, axis AD and centre E; we draw the tangent HG. We have

$$\frac{EI \cdot IH}{IG^2} = \frac{AD}{a} \quad (Conics, \text{I.37}).$$

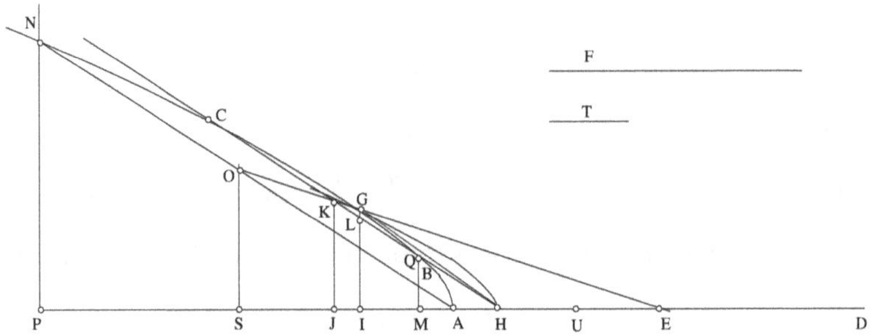

Fig. 31.1

We draw $\mathscr{H}$ (axis EH, *latus rectum* $a' / \dfrac{EH}{a'} = \dfrac{AD}{a}$), $G \in \mathscr{H}$, and let U be a point such that $\dfrac{EU}{UH} = \dfrac{EH}{a'}$, where U lies between H and E (UH is a a segment in the same ratio).

---

[68] The lengths JI and JU are constructed by the method indicated earlier.

Let us put

$$\frac{T^2}{\frac{1}{4}F^2} = \frac{UE}{EH} \text{ and } JU \cdot JI = T^2;$$

the point J lies inside Γ beyond I, the perpendicular to AJ at J cuts $\mathcal{H}$ in K, inside Γ. The straight line HK cuts GI in L and cuts Γ in two points, one on either side of K.

From what is given, we have successively defined the points G, I, H, U, J, K, and hence the straight line HK. We need to show that HK is the required straight line.

We have

$$\frac{UJ.JH}{HK^2} = \frac{UE}{EH} = \frac{T^2}{\frac{1}{4}F^2} = \frac{JU.JI}{\frac{1}{4}F^2} \quad (Conics, \text{ VII.2}),$$

hence we obtain

$$\frac{HJ}{JI} = \frac{HK^2}{\frac{1}{4}F^2}.$$

We define the point M by $HJ \cdot JI = JM^2$, hence

$$\frac{HJ}{JI} = \frac{JM^2}{JI^2} = \frac{JH^2}{JM^2};$$

the perpendicular to the axis at M cuts the straight line HK in B and

$$HJ \cdot JI = JM^2 \Rightarrow KH \cdot KL = KB^2 \Rightarrow \frac{HK}{KL} = \frac{KB^2}{KL^2} = \frac{KH^2}{KB^2}.$$

But

$$\frac{HK}{KL} = \frac{HJ}{JI},$$

hence

$$\frac{HK^2}{\frac{1}{4}F^2} = \frac{HJ^2}{JM^2} = \frac{KH^2}{KB^2},$$

hence

$$KB = \frac{1}{2}F.$$

Through A we draw a line parallel to HK; it cuts EK in O. Let us produce AO to a point N such that $AO = ON$, let NP and OS be perpendicular to AD.

We have

$$\frac{OS}{KJ} = \frac{SA}{JH} = \frac{OA}{KH} = \frac{AE}{EH} = \frac{SA+AE}{JH+EH} = \frac{SE}{JE},$$

$$\frac{AS}{SO} = \frac{JH}{JK}$$

and

$$\frac{ES}{SO} = \frac{JE}{JK} \Rightarrow \frac{ES \cdot SA}{SO^2} = \frac{JE \cdot JH}{JK^2} = \frac{EH}{a'} = \frac{AD}{a},$$

$$NP = 2OS, \ AP = 2SA, \ PD = 2ES,$$

so we have

$$\frac{ES \cdot SA}{SO^2} = \frac{PA \cdot PD}{PN^2} = \frac{AD}{a} \Rightarrow N \in \Gamma.$$

O is the midpoint of AN, so EO is a diameter and, consequently, K is the midpoint of the segment CQ if we use the letters C and Q to designate the points of intersection of HK with Γ, where $HC > HK > HQ$.

We shall show that B and Q are coincident.

The points (C, Q, L, H) form a harmonic range, so $KC^2 = KQ^2 = KL \cdot KH$. But we have seen that $KB^2 = KL \cdot KH$, so $KQ = KB$, the points Q and B, being on the segment BH, are coincident; so we have $KC = KQ = KB = \frac{1}{2}F$, $BC = F$.

Existence of the solution

Throughout this problem, Γ is a branch of a hyperbola, or even half a branch since the investigation takes place within only one of the halves into which the plane is divided by the axis.

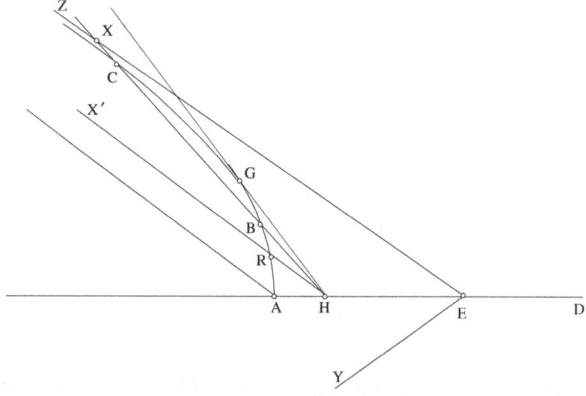

Fig. 1.2.31.2

We shall show that, whatever the given straight line F, the problem has a solution. To prove this Ibn al-Haytham makes use of the asymptotes EX and EY. The straight line drawn through A parallel to EX cuts Γ only at the point A; similarly the straight line HX' parallel to EX cuts Γ at only one point R. Let HG be the tangent to Γ from H; then any semi-infinite straight line HZ starting at H, and lying inside the angle $G\hat{H}X'$, cuts the hyperbola at B between R and G and at the point C beyond G, because such a line necessarily cuts the asymptote EX.

$0 < G\hat{H}Z < G\hat{H}X'$; when $G\hat{H}Z$ increases from 0 to $G\hat{H}X'$, the length BC increases monotonically from 0 to $+\infty$, so it once and only once takes the given value F.

Note: The homothety that transforms A into H and D into E obviously transforms the given hyperbola into the auxiliary hyperbola $\mathcal{H}$. Its centre is the point X such that $EX = \dfrac{1}{2k-1} EA$ and it lies beyond A; the ratio of the homothety is $\dfrac{1}{2k}$.

This homothety, which underlies Ibn al-Haytham's method, transforms the given hyperbola into the locus of the midpoints of the chords from the point H. As H varies, we obtain a complete family of homothetic auxiliary hyperbolae from Γ.

Once again we may observe that we have obtained a linear pencil of conics (hyperbolae that are all homothetic with one another). The points that form the basis of the pencil are the double point E and the two points at infinity on the asymptotes of Γ.

The point J is found by means of a construction involving the application of areas, that is, a plane construction. The point K, the point of intersection of the hyperbola $\mathcal{H}$ known from its *latus rectum* and transverse diameter, and JK the perpendicular to AD, could have been found by another plane construction. Ibn al-Haytham does not mention it because here he is interested specifically in constructions using auxiliary conics.

APPENDIX

Trisection of an angle

Let there be a hyperbola $\mathscr{H}$ with diameter BC, whose *latus rectum* is equal to BC, and the tangent at whose vertex is such that $T\hat{B}C = \alpha$ a given angle.

Let A be a point on $\mathscr{H}$ such that $BA = BC$; then $D\hat{A}B = \dfrac{\alpha}{3}$.

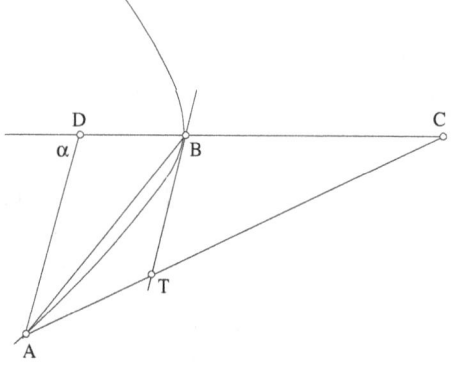

Fig. 1.2.32

Proof: The hyperbola $\mathscr{H}$ is equilateral, so we have $DB \cdot DC = AD^2$, hence

$$\frac{DB}{DA} = \frac{DA}{DC}.$$

So triangles DBA and DAC are similar, hence $D\hat{A}B = B\hat{C}A = B\hat{A}C$ and $D\hat{B}A = 2D\hat{A}B$; accordingly

$$\alpha = D\hat{A}B + D\hat{B}A = 3D\hat{A}B, \quad \alpha = 3D\hat{A}B.$$

This text, which is not part of Ibn al-Haytham's book, but which the manuscript tradition attaches to the end of it, presents a simple application of the same kind of νεῦσις in the problem of trisecting an angle. Here, we insert into an equilateral hyperbola a chord AB equal to a diameter BC and passing through the vertex B of this diameter. The diameter in question is chosen so that the angle that is to be divided is the angle between the ordinate direction and the diameter.

1.3. TRANSLATED TEXT

Al-Ḥasan ibn al-Haytham

On the Completion of the Conics

TREATISE BY AL-ḤASAN IBN AL-ḤASAN IBN AL-HAYTHAM

On the Completion of the *Conics*

Apollonius remarked in the preamble to his work the *Conics* that he had divided his work into eight books, and he set out what each of them contains in terms of notions he had himself discovered. He pointed out that the eighth book deals with problems that arise in the *Conics*. Only seven of the books of this work have been translated into Arabic and the eighth one has not been found.

When we examined this work, considering its ideas one by one and reading through the seven books several times, we found that they were lacking some notions that should not have been lacking in this work. So we thought that these notions that are missing from these seven books are the notions contained in the eighth book and Apollonius left them until the end because he had not needed to use them for the notions he introduced in the <first> seven books. These ideas we have mentioned are notions implied by the notions introduced in the seven books.

Thus he found the ratio in which a tangent divides the axis of a conic section and he showed how to draw a straight line that is a tangent to the conic section and makes with the axis an angle equal to a known angle.[1] These two notions require us to show how to draw a straight line that is a tangent to the conic section and such that its ratio to what it cuts off on the axis is a known ratio; and <how> to draw a straight line to be a tangent to the conic section and such that the part cut off from it between the conic section and the axis is equal to a known straight line. However, these notions are among those which the mind <merely> aspires to know.

---

[1] Problem 50 of Book II for all three conics.

For instance, when he says how to draw a straight line that is a tangent to a conic section and makes with the diameter drawn from the point of contact an acute angle equal to a known angle.[2] This concept too requires us to draw a straight line that is tangent to the conic section and ends on the axis, and is such that its ratio to the diameter drawn from the point of contact is a known ratio.[3]

Thus in the preamble to book seven he spoke of the diameters of the conic sections, mentioning each kind and distinguishing one from another, and he pointed out that they have properties that depend on the *latera recta* of the conic sections concerned. However he says in the preamble to this book that the concepts that follow them in this book are strictly necessary for those of the proposed problems that will be discussed in book eight, <the book> which contains problems concerning diameters and their properties.

For instance, when he says how to draw from a given point a straight line to be a tangent to the conic section and to meet it in a single point. This concept requires us to show how to draw from a given point a straight line that meets the conic section in two points and such that the part of it that lies inside the conic section is equal to a given straight line; and how to draw a straight line that cuts the conic section and is such that the ratio of the part of it that lies outside <the conic> to the part that lies inside is equal to a given ratio.

It is not fitting that these concepts to which we have referred and which we have noted should not be present in this work. These are concepts we consider beautiful and their beauty is no less than the beauty of what the seven books do contain, but indeed on the contrary there are some among them that are more beautiful and more striking than the earlier concepts contained in the seven books. So it is very likely that these concepts are those that we included in the eighth book and if Apollonius did not refer to them before the eighth book, that is because he did not need to use them in the earlier books.

As this concept carried conviction in our mind and strengthened its hold in our thoughts because of the good opinion we have of the author of the work, this good opinion prevailed with us and we came to the conclusion that these concepts and ones like them are those which the

---

[2] Problem 51 of Book II (parabola and hyperbola), Problems 52 and 53 (ellipse).

[3] This is the problem he deals with here for the ellipse and the hyperbola (Problems 13 and 14).

eighth book contained. When we had come to our decision on this matter, we began by working out what these ideas were, proving them and grouping them together in a book that included them all and could thus fulfil the function of the eighth book and could represent the completion of the work the *Conics*. Our procedure is to find these concepts by the use of analysis, synthesis and discussion[4] so as to make this book the most perfect of them all in regard to proofs.

We now begin this book and we ask God for His help.

<1> If we have a given conic section such that the axis of the conic section is extended outside the conic section, <the problem is> how to draw a straight line that is a tangent to the conic section and such that its ratio to what it cuts off from the axis on the side nearer the conic section is equal to a given ratio?[5]

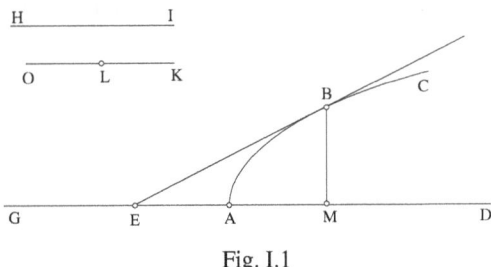

Fig. I.1

First, let the conic section be a parabola. Let the conic be *ABC* and let *AD* be its axis; let us extend *DA* to *G*; let the ratio of *HI* to *KL* be given. We wish to draw a straight line that is a tangent to the conic and ends on the axis, and is such that its ratio to what it cuts off from the axis <on the side nearer the conic section> is equal to the ratio of *HI* to *KL*. <Let us suppose this has been done by analysis; and let the line be *BE*>. Let us draw *BM* as an ordinate, then *MA* will be equal to *AE*, as has been shown in Proposition 35 of Book I. But since the ratio of *BE* to *EM* is equal to the ratio of *HI* to twice *KL* and the ratio of *HI* to *KL* is known, the ratio of *HI* to twice *KL* is known, so the ratio of *BE* to *EM* is known. But angle *M* is a right angle, so the angle *E* is known, the straight line *BE* is thus a tangent to the conic section and the angle <it> encloses with the axis is known; this is possible

---

[4] Diorism.
[5] See Supplementary note [1].

because of what has been shown in Proposition 56 of Book II.[6] So by analysis the problem has been reduced to something that is possible: we draw a straight line that is a tangent to the conic section and that makes with the axis an angle equal to a known angle.

But by this we have also shown that *HI* is greater than twice *KL*, because *BE* is greater than twice *EA*, so it is necessary that *HI* shall be greater than twice *KL*; this is the discussion of the problem.

– **2** – Let us now proceed to the synthesis of the problem: let the conic section be *ABC*, its axis *DAG*; let the given ratio be the ratio of *HI* to *KL*. We make *LO* equal to *LK*, then *KO* will be smaller than *HI*. On the straight line *HI* we construct a semicircle; let it be *HNI*. We draw a chord equal to *KO*; let it be *IN*. Let us draw a straight line that is a tangent to the conic section and makes with the axis an angle equal to the angle *HIN*, as has been shown in Proposition 56 of Book II. Let the tangent be *BE*.

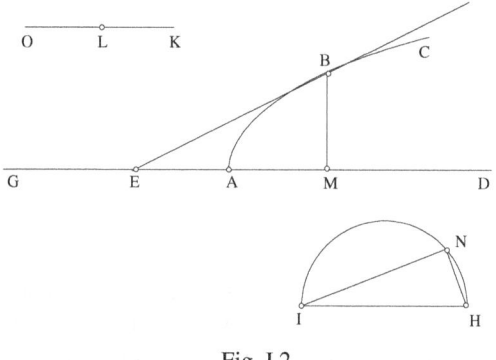

Fig. I.2

I say that the ratio of *BE* to *EA* is equal to the ratio of *HI* to *KL*.

Proof: We take *BM* as an ordinate; angle *M* will be a right angle. We join *HN*, angle *N* will be a right angle. But angle *E* is equal to angle *I*, so triangle *BEM* is similar to triangle *HIN*. The ratio of *BE* to *EM* is thus equal to the ratio of *HI* to *IN*, that is to *KO*. The ratio of *BE* to *EA* – which is half of *EM* – is equal to the ratio of *HI* to *KL* – which is half of *KO*. That is what we wanted to do.

---

[6] This is Proposition 50 in Heiberg's edition.

<3> Let the conic section *ABC* be a hyperbola or an ellipse; let *FAD* be its axis; let the ratio of *G* to *H* be given and let *G* be greater than *H*. We wish to draw a tangent to the conic section, that ends on the axis and is such that its ratio to the segment it cuts off from the axis on the side nearer the vertex of the conic section is equal to the ratio of *G* to *H*.

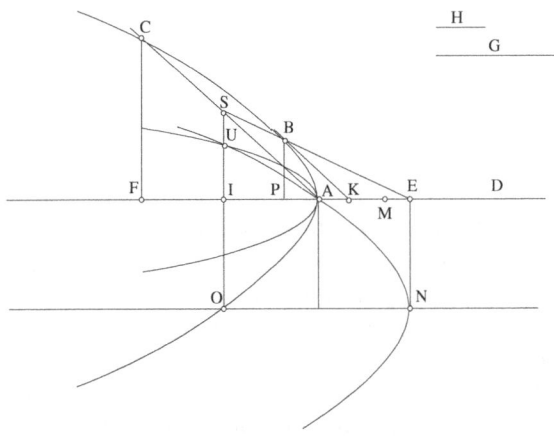

Fig. I.3.1

We suppose this has been done by analysis, let <the line> be *BK*. We draw *AC* parallel to the tangent, let the centre of the conic section be *E*. We join *EB*; and let the line cut *AC* at the point *S*. Thus *ES* will be a diameter of the conic section because *ES* is drawn from the centre and *AS* will be an ordinate because it is parallel to the tangent, which follows from what was shown in the converse of Proposition 32 of the first book. So the straight line *ES* divides *AC* into two equal parts at the point *S*. Let us draw the straight lines *CF*, *SI* and *BP* to be ordinates; the ratio of *PE* to *EA* is equal to the ratio of *AE* to *EK* and to the ratio of *PA* to *AK*, from what has been shown in Proposition 37 of the first book. But the ratio of *AE* to *EK* is equal to the ratio of *SA* to *BK*, so the ratio of *SA* to *BK* is equal to the ratio of *PA* to *AK*, and the ratio of *SA* to *AP* is equal to the ratio of *BK* to *KA*. But the ratio of *BK* to *KA* is equal to the ratio of *G* to *H*, so the ratio of *SA* to *AP* is equal to the ratio of *G* to *H*.

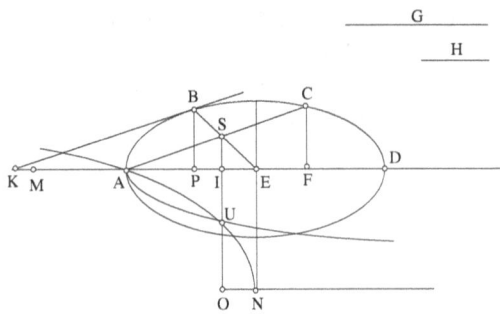

Fig. I.3.2

We put the ratio of *EM* to *MA* equal to the ratio of the transverse diameter[7] to the *latus rectum*; then the straight line *AM* will be equal to half the line that has the homologous ratio.[8] Consequently, the ratio of the product of *MI* and *IA* to the square of *AS* will be equal to the ratio of the homologous straight line plus *AD* to *AD*, which is equal to the ratio of *ME* to *EA*, which is known.[9] But the ratio of the square of *AS* to the square of *AP* is equal to the ratio of the square of *G* to the square of *H* which is known, so the ratio of the product of *MI* and *IA* to the square of *AP* is known. On the other hand, the product of *IE* and *EA* is equal to the square of *EP*.[10] From the point *E* we draw a straight line at a right angle; let it be *EN*. We put *EN* equal to *EA*, from the point *N* we draw a straight line parallel to the straight line *EI*; let it be *NO*. We draw through the point *N* the parabola with axis *NO* and *latus rectum EN*; let the conic section be *NU*. We extend *SI* to *O*, so *UO* will be equal to *EP*; but *IO* is equal to *AE* because it is equal to *EN*, and finally *UI* is equal to *AP*. But the ratio of the product of *MI* and *IA* to the square of *AP* is known, so the ratio of the product of *MI* and *IA* to the square of *IU* is known, so the point *U* lies on the outline of a hyperbola whose axis is *AM* and whose *latus rectum* is known; let this conic section be the conic section *AU*. The conic section *AU* is known in position and the conic section *NU* is known in position, so the point *U* is known. But *UI* is a perpendicular, so the point *I* is known. But

---

[7] We are concerned with the axis *AD*. To judge by the figures that appear in the text, Ibn al-Haytham is considering the major axis in the case of the ellipse.

[8] See Supplementary note [2].

[9] See Mathematical commentary, note 42, p. 42.

[10] This follows from *Conics*, I.37.

the straight line *IS* is known in position and *AS* is equal to *SC*, so the point *C* is known. So the straight line *AC* is known in position, the point *S* is thus known <in position>. But the point *E* is known, so the diameter *EBS* is known in position, and the point *B* is thus known. But the straight line *BK* is parallel to the straight line *AC* which is known in position, so the straight line *BK* is known in position and its ratio to *KA* is equal to the ratio of *G* to *H* which is given. That is what we sought.

<4> The synthesis of this problem is as follows.

We return to the two conic sections and we draw from the point *E* a straight line at right angles; let it be *EN*. We put *EN* equal to *EA*, let us draw *NO* parallel to the axis *EA* and we construct through the point *N* the parabola whose axis is *NO* and *latus rectum NE*; let the conic section be *NU*. We put the ratio of *EM* to *MA* equal to the ratio of the axis *AD* to its *latus rectum*; so the straight line *MA* is half the homologous straight line. We put the ratio of *AE* to *EQ* equal to the ratio of the square of *G* to the square of *H* and we put the ratio of *AM* to *MT* equal to the ratio of *ME* to *EQ*. We construct through the point *A* the hyperbola whose axis is *AM* and *latus rectum MT*, let the conic section be *AU*; let it cut the conic section *NU* at the point *U*. Whether it cuts it or does not cut it <is something that> we shall show later, when we come in our discussion of the problem <as a whole>.

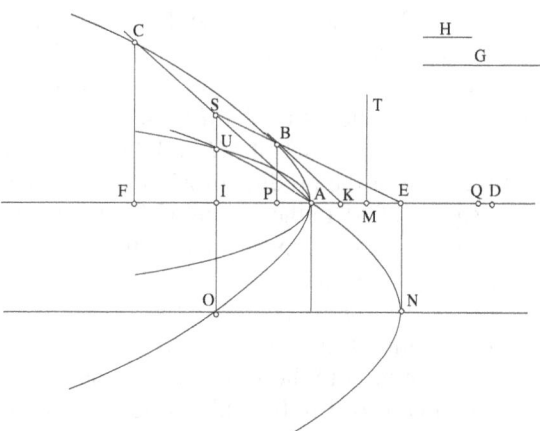

Fig. I.4.1

We draw from the point *U* the perpendicular *UI* and we extend it on both sides; let it cut *NO* at the point *O*. We make *IF* equal to *AI*, we draw the perpendicular *FC* as far as the outline of the conic section and we join *AC*; let it cut *UO* at the point *S*; so *AS* will be equal to *SC*. We join *ES*, it will be a diameter of the conic section *AC*; let it cut the conic section *AC* at the point *B*. We draw *BK* to be a tangent to the conic section.

I say that the ratio of *BK* to *KA* is equal to the ratio of *G* to *H*.

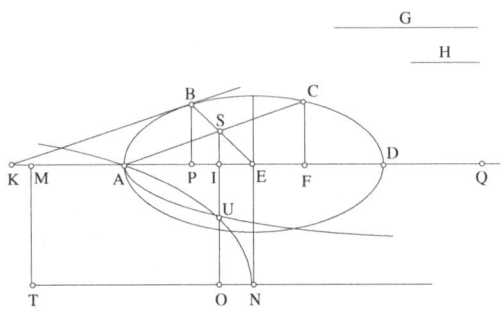

Fig. I.4.2

Proof: We draw *BP* as an ordinate. The product of *IE* and *EA* will be equal to the square of *EP* and the product of *NO* and *NE* is equal to the square of *OU*; but *NO* is equal to *EI* and *NE* is equal to *EA*, so the square of *OU* is equal to the square of *EP*, so the straight line *OU* is equal to the straight line *EP*; but *OI* is equal to *EA*, so the straight line *IU* is equal to the straight line *AP*. The ratio of the product of *MI* and *IA* to the square of *AS* is equal to the ratio of *ME* to *EA*; the ratio of the product of *MI* and *IA* to the square of *AP* is equal to the ratio of *AM* to *MT* which is the ratio of *ME* to *EQ* and the ratio of the product of *MI* and *IA* to the square of *AP* is the compound of the ratio of the product of *MI* and *IA* to the square of *AS* and the ratio of the square of *AS* to the square of *AP*; but the ratio of the product of *MI* and *IA* to the square of *AS* is equal to the ratio of *ME* to *EA*, so the ratio of the square of *AS* to the square of *AP* is equal to the ratio of *AE* to *EQ*. But the ratio of *AE* to *EQ* is equal to the ratio of the square of *G* to the square of *H*, so the ratio of the square of *AS* to the square of *AP* is equal to the ratio of the square of *G* to the square of *H*, and the ratio of *AS* to *AP* is equal to the ratio of *G* to *H*. But the ratio of *AS* to *AP* is equal to the ratio of *BK* to *KA*, so the ratio of *BK* to *KA* is equal to the ratio of *G* to *H*. That is what we wanted to prove.

The discussion of this problem is as follows.

For the ellipse, this solution of the problem is valid[11] in all cases without <imposing any> condition; there is a single solution for each side of the ellipse. Indeed, the hyperbola whose axis is *AM* and its *latus rectum MT*, the ratio of its axis to its *latus rectum* is equal to the ratio of *ME* to *EQ*, which is the ratio of the product of *ME* and *EA* to the product of *AE* and *EQ*, which is smaller than the square of *AE*; so it is smaller than the square of *EN*. The hyperbola whose axis is *AM* cuts the straight line *EN* between the two points *E* and *N*, so it cuts the outline of the parabola between the two points *A* and *N* because the parabola passes through the point *A*. Indeed the perpendicular drawn from the point *A* to the axis *NO* is equal to the straight line *EN* which is the *latus rectum*, and cuts off on the axis a straight line equal to the straight line *EA* which also is equal to the *latus rectum*. The parabola passes through the point *A* and the hyperbola, whose transverse axis is *AM*, has the point *A* as its vertex and its concave side faces the concave side of the parabola; so it cuts the parabola in two points in every case. One of these points is the point *A*, it (the hyperbola) thus <also> cuts it (the parabola) in another point. The conic section (*i.e.* the hyperbola) cuts the straight line *EN*, so it cuts the outline of the parabola in a point between the two points *N* and *A* and does not cut it in any other point except these two points. The problem is solved for all cases and has only a single solution, because the two conic sections, the hyperbola and the parabola, do not cut one another apart from at the point *A* except in one other point.

<5> In the case of the hyperbola, the problem is solved only after imposing a condition, <that is by> taking a special case.

The condition for this conic section is that the ratio of the square of *G* to the square of *H* shall not be smaller than the ratio of the straight line that is compounded from twice the transverse diameter, that is *AD*, from the straight line that has the homologous ratio, which is twice *AM*, and from three times the straight line whose square is equal to the product of the transverse diameter and the straight line with the homologous ratio, to the straight line *ME*.

Let us return to the hyperbola and the parabola and let us complete the parabola. We extend the straight line *DA* in the direction of *A*. We put the

[11] Lit.: is completed.

product of *DA* and twice *AM* equal to the square of *AB*, we draw from the point *A* a perpendicular to the axis of the parabola and let us extend it as far as the outline of the parabola; let it be *AS*. This straight line cuts off from the axis a straight line equal to the straight line *AE* which is the *latus rectum*. We join *SB* and extend it so that it ends on the outline of the parabola; let the line cut the parabola at the point *U*. We draw from the point *U* a straight line that is an ordinate, let it be *UIV*; let it cut the axis of the conic section at the point *O*. The product of the *latus rectum* and the straight line that *AS* cuts off from the axis is equal to the square of *IO*. But the product of the *latus rectum* and *NO* is equal to the square of *UO*, so the product of the *latus rectum* and *BI* is equal to the square of *UI*. Similarly, the ratio of *UI* to *IB* is equal to the ratio of *SA* to *AB*, so the product of *AB* and *UI* is equal to the product of *SA* and *BI*. But the product of *SA* and *BI* is twice the product of the *latus rectum* and *BI*, because *SA* is twice the *latus rectum*. So the product of *AB* and *UI* is twice the square of *UI*, so *AB* is twice *UI*, thus *UI* is half *AB*. But since the product of *SA* and *BI* is equal to the product of *AB* and *UI*, the product of *SA* and *BI* is half the square of *AB*; but the square of *AB* is equal to the product of *DA* and twice *AM*, so the product of *DA* and *AM* is half the square of *AB*; but *DA* is equal to *SA*, so the product of *SA* and *BI* is equal to the product of *SA* and *AM*, so *BI* is equal to *AM* and the sum of *AM* and *BI* is equal to the homologous straight line. We make *IP* twice *AD* and we make *PC* twice *AB*. So the straight line *CM* will be twice *AD*, plus twice *AM*, which is the homologous straight line, plus three times the straight line *AB* whose square is equal to the product of *DA* and the homologous straight line. So the straight line *CM* is the straight line that we defined earlier.

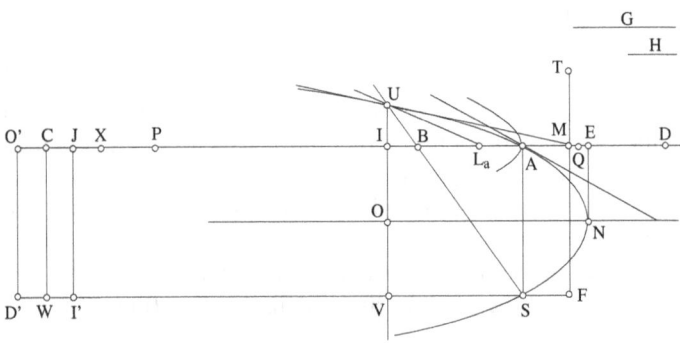

Fig. I.5

I say that if the ratio of the square of *G* to the square of *H* is equal to the ratio of *CM* to *ME* or greater than the ratio of *CM* to *ME*, then the problem is solved and has two solutions. And if the ratio of the square of *G* to the square of *H* is smaller than the ratio of *CM* to *ME*, then the problem is not to be solved in any way. Let us first prove what we have just stated.

Let the ratio of the square of *G* to the square of *H*, which is equal to the ratio of *AE* to *EQ*, be first of all equal to the ratio of *CM* to *ME*. We divide *CP* into two equal parts at the point *X*, then *PX* is equal to *AB* and *AB* is twice *UI*, so the straight line *PX* is twice *UI*. But *IP* is twice *VI*, so the straight line *XI* is twice the straight line *UV*. The product of *XI* and *IU* is thus equal to twice the product of *VU* and *UI*. But *VI* is twice the *latus rectum* and the product of the *latus rectum* and *AI* is equal to the product of *VU* and *UI*, so the product of *VI* and *IA* is equal to the product of *XI* and *IU*. But *CX* is equal to *AB* and *AB* is twice *UI*, so the product of *CX* and *IU* is equal to the product of *VI* and *AM*. The product of *CI* and *IU* is thus equal to the product of *VI* and *IM*, the ratio of *CI* to *IM* is equal to the ratio of *VI* to *IU*, and the ratio of *CM* to *MI* is equal to the ratio of *VU* to *UI*. But the ratio of *CM* to *AE* is compounded from the ratio of *CM* to *MI* and the ratio of *MI* to *AE*, and the ratio of *CM* to *MI* is equal to the ratio of *VU* to *UI*, so the ratio of *CM* to *AE* is compounded from the ratio of *VU* to *UI* and the ratio of *IM* to *AE*, and this ratio is equal to the ratio of the product of *MI* and *IA* to the product of *IA* and *AE*. But *AE* is the *latus rectum* and the product of *IA* and the *latus rectum* is equal to the product of *VU* and *UI*, so the ratio of *IM* to *AE* is equal to the ratio of the product of *MI* and *IA* to the product of *VU* and *UI*. The ratio of *CM* to *AE* is thus compounded from the ratio of *VU* to *UI* and the ratio of the product of *MI* and *IA* to the product of *VU* and *UI*. But the ratio of *VU* to *UI* is equal to the ratio of the product of *VU* and *UI* to the square of *UI*, the ratio of *CM* to *AE* is thus compounded from the ratio of the product of *MI* and *IA* to the product of *VU* and *UI* and the ratio of the product of *VU* and *UI* to the square of *UI*, and this ratio is the ratio of the product of *MI* and *IA* to the square of *IU*. The ratio of *CM* to *AE* is thus equal to the ratio of the product of *MI* and *IA* to the square of *IU*. But the ratio of *CM* to *AE* is equal to the ratio of *ME* to *EQ*, so the ratio of the product of *MI* and *IA* to the square of *IU* is equal to the ratio of *ME* to *EQ*. But the ratio of *AM* to *MT* is equal to the ratio of *ME* to *EQ*, so the ratio of the product of *MI* and *IA* to the square of *IU* is equal to the ratio of *AM* to *MT*.

The hyperbola whose transverse axis is *AM* and *latus rectum MT* passes through the point *U*, but the point *U* lies on the outline of the parabola. So if the ratio of the square of *G* to the square of *H* is equal to the ratio of *CM* to *ME*, then the hyperbola with axis *AM* cuts the parabola and the problem is solved, as we have shown in the synthesis of this problem.

I also say that the problem has two solutions. We draw from the point *U* to the diameter *AI* a straight line that is an ordinate; let the straight line be UL_a. So L_aI is twice *UI*; in fact UL_a is parallel to the straight line that is a tangent to the parabola at the point *A* and the tangent cuts off from the axis outside the conic section a straight line equal to the straight line cut off on the axis by the perpendicular to the axis drawn from the point *A*. But the perpendicular to the axis drawn from the point *A* is equal to what it cuts off from the axis, because this perpendicular is equal to the *latus rectum*. Thus the tangent drawn from the point *A*, together with the axis, gives rise to a triangle whose base is twice the perpendicular drawn from the point *A*. This triangle is similar to the triangle UL_aI, so the straight line L_aI is twice the straight line *IU*, so it is equal to the straight line *AB*. The straight line AL_a is thus equal to the straight line *BI* and *BI* is equal to *AM*, so the straight line AL_a is equal to the straight line *AM*; the straight line *MU* is thus a tangent to the conic section.

We draw from the two points *M* and *C* two perpendiculars to the straight line *SV*; let them be *MF* and *CW*. Since the ratio of *CM* to *MI* is equal to the ratio of *VU* to *UI*, the ratio of *MC* to *CI* is equal to the ratio of *UV* to *VI*, so the product of *CM* and *IV*, that is *MF*, is equal to the product of *UV* and *CI*, that is *VW*. The hyperbola constructed through the point *M*, and which has as its asymptotes the straight lines *FW* and *WC*, passes through the point *U*. The straight line *MU* will thus lie inside this conic section and *MU* is a tangent to the parabola, so there is no straight line between that straight line and the parabola. A straight line drawn from the point *U* between the tangent to the hyperbola and the straight line *MU*, lies inside the parabola. And each of these straight lines lies inside the parabola. So the hyperbola that passes through the two points *M* and *U* cuts the parabola in a point between the points *A* and *U*. Thus if we draw from the point of intersection a perpendicular to the straight line *SW*, the perpendicular cuts the straight line *MC* in such a way that the ratio of *CM* to the part of it cut off by the perpendicular on the same side as *M* is equal to the ratio of the perpendicular to the part of it cut off between the conic section and the straight line *MC*. If we undertake the proof using the method we have indicated, this will allow us to show that the hyperbola whose transverse axis is *AM* and *latus rectum MT* passes through the other point that lies between the two points *A* and *U*. In this way we show that the hyperbola whose axis is *AM* and *latus rectum MT* cuts the parabola in two points distinct from the point *A*. Thus it has been shown that the problem has two solutions.

Now that we have proved this for that ratio, we prove it for a ratio greater than the ratio of *CM* to *ME*.[12] We make *O'M* greater than *CM* and let us draw from the point *O'* the perpendicular *O'D'*. It is clear that if we draw from the point *M* a tangent to the hyperbola that passes through the two points *M* and *U*, and if we then extend it (the straight line) in both directions, it reaches the straight line *WF* and the straight line *WC* – if we extend these two straight lines. The tangent is divided into two equal parts at the point *M*. This tangent will lie between the two straight lines *MU* and *MC*. If we extend this tangent on the other side of *C*, it meets *D'O'* on the side of *O'* and the part of it that lies between the point *M* and the straight line *D'O'* will be greater than the part that lies between the point *M* and the straight line *WF*. If we extend *MU* on the side of *F*, the part of the straight line added to the straight line *MU* that lies between the point *U* and the straight line *D'O'*, the part extended on the side of *O'*, will be greater than the part of the straight line between the point *M* and the straight line *D'F*.[13] The hyperbola constructed through the point *M* and such that its two asymptotes are the straight lines *D'F* and *D'O'* cuts the tangent and cuts it on the side of *O'*; it also cuts the part of the straight line added to the straight line *MU* on the side of *O'* and cuts it beyond the point *U*. The tangent cuts the parabola because it (the line) lies between the two straight lines *MU* and *MC*. The hyperbola constructed through the point *M* and such that the two straight lines *D'O'* and *D'F* are its two asymptotes, thus cuts the parabola and cuts it in two points; one is before the point *U* and the other after the point *U*. So if we draw from the two points of intersection two perpendiculars to the straight line *D'F* and if we continue the proof using the same method as before, it becomes clear that the problem is

---

[12] $\dfrac{CM}{ME} < \dfrac{G^2}{H^2}$, $\dfrac{G^2}{H^2} = \dfrac{O'M}{ME}$, hence $O'M > CM$.

[13] Ibn al-Haytham is using the property of a secant in relation to the asymptotes of the hyperbola.

solved and it has two solutions, as we have shown also for the straight line *MC*.

So we have shown that if the ratio of the square of *G* to the square of *H* is greater than the ratio of *CM* to *ME*, then the problem is solved and it has two solutions. But if the ratio of the square of *G* to the square of *H* is less than the ratio of *CM* to *ME*, then the problem is not solved. This is clear, as we describe it.

We put this ratio equal to the ratio of *JM* to *ME* and we draw the perpendicular *JI'*, which, when extended in the direction of *J*, cuts the tangent that we have described earlier, that is the tangent to the hyperbola that passes through the two points *M* and *U* and a part of which is cut off between the straight line *JI'* and the point *M*, <a part that is> smaller than the part between the point *M* and the straight line *I'F*. If we construct through the point *M* a hyperbola whose asymptotes are the straight lines *I'J* and *I'F*, this hyperbola cuts the tangent that was mentioned above in a point between the point *M* and the straight line *I'F*. So this hyperbola does not cut the parabola. If it does not cut the parabola, the hyperbola whose axis is *AM* and *latus rectum MT* does not cut the parabola; then the problem is not solved.

It is clear, from all that we have proved, that the discussion (diorism) of the problem for the hyperbola is that the ratio of the square of *G* to the square of *H* must not be less than the ratio of *CM* to *ME*, and if the ratio is like this, then the problem has two solutions. That is what we wanted to prove.

– **6** – The conic section *ABC* is a hyperbola or an ellipse whose transverse axis is *AD* and centre *E*; the ratio of *H* to *F* is given. We wish to draw a tangent to the conic section, <a line> ending on the axis and such that its ratio to what it cuts off from the axis on the side towards the further end is equal to the ratio of *H* to *F*.

We suppose this has been done by analysis: let the tangent be *BS*, then the ratio of *BS* to *SD* will be equal to the ratio of *H* to *F*. We join *EB*, we extend it on the side of *B* and we draw from the point *A* and towards the diameter *EB* a straight line that is an ordinate; let it be *AIC*. So *AI* will be half *AC*. We draw from the points *B*, *I* and *C* perpendiculars to the axis; let them be *BK*, *IP* and *CO*.

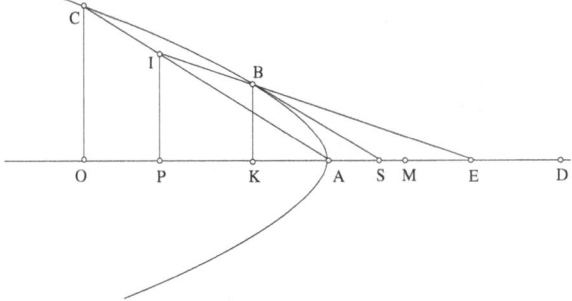

Fig. I.6.1

The ratio of the product of *DO* and *OA* to the square of *OC* is equal to the ratio of the transverse diameter *AD* to the *latus rectum*; the same holds for the ratio of the product of *DK* and *KA* to the square of *KB*, as has been shown in Proposition 21 of Book I. The ratio of the product of *EP* and *PA* to the square of *PI* is equal to this same ratio, since these straight lines are half the other straight lines. We put the ratio of *EM* to *MA* equal to the ratio of the transverse diameter *AD* to its *latus rectum*; the straight line *AM* will thus be equal to half the homologous straight line. The ratio of the product of *MP* and *PA* to the square of *AI* will be equal to the known ratio of *ME* to *EA*, as has been shown in the second proposition of Book VII. But the ratio of *KA* to *AE* is equal to the ratio of *AS* to *SE* and the ratio of *AS* to *SE* is equal to the ratio of *IB* to *BE* which is the ratio of *PK* to *KE*, so the ratio of *PK* to *KE* is equal to the ratio of *KA* to *AE*. The ratio of *PE* to *EK* is equal to the ratio of *KE* to *EA*, so the product of *PE* and *EA* is equal to the square of *EK*. But since the ratio of *KA* to *AE* is equal to the ratio of *AS* to *SE*, the ratio of *KA* to *AS* will be equal to the ratio of *AE* to *ES* which is the ratio of *IA* to *BS*. The ratio of *IA* to *BS* is thus equal to the ratio of *KA* to *AS*, the ratio of *IA* to *AK* is accordingly equal to the ratio of *BS* to *SA*. Since the ratio of *BS* to *SD* is equal to the ratio of *H* to *F*, which is known, the ratio of *DS* to *SA* is equal to the ratio of *BS* to a straight line whose ratio to *SA* is known. The ratio of *DS* to *SA* is thus equal to the ratio of *IA* to a straight line whose ratio to *KA* is known. But the ratio of *DS* to *SA* is equal to the ratio of *DK* to *KA*, as has been shown in Proposition 36 of Book I. The ratio of *DK* to *KA* is thus equal to the ratio of *IA* to a straight line whose ratio to *AK* is a known ratio. The ratio of *DK* to *IA* is thus known; similarly the ratio of the square of *DK* to the square of *AI* is known. But the ratio of the product of *MP* and *PA* to the square of *AI* is known, the ratio of the product of *MP* and *PA* to the square of *DK* is thus known. But it has been shown that the product of *PE* and *EA* is equal to the square of *EK*. But since the ratio of the product of *MP* and *PA* to the square of *KD* is known and the product of *PE* and *EA* is equal to the square of *EK*, the ratio of *EA* to *AP* is known, *AP* will accordingly be known because it is possible to find it; we shall show how to find it in the synthesis of this problem. But if *AP* is known, *PI* will be known in position and *AIC* will be known in magnitude and position; since *AI* is equal to *IC*, the point *I* will be known, so *EBI* will be known in position, the point *B* will be known, the straight line *BS* that is a tangent will be the line that is the solution to the problem and it is known in position.

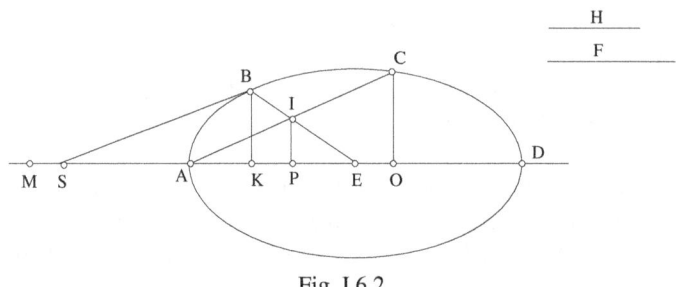

Fig. I.6.2

– **7** – The synthesis of this problem is as follows.

We return to the two conic sections and we put the ratio of *AE* to *EQ* equal to the ratio of the square of *H* to the square of *F*. We draw from the points *E*, *M*, *A* perpendiculars to the axis; let them be *EN*, *MU* and *AV*. We put *EN* equal to *EA* and *MU* equal to *DA*. Let us draw from the two points *N* and *U* two straight lines parallel to the straight line *EA*; let them be *NG* and *UVJ*. *UV* will thus be equal to *MA*. We put the ratio of *UV* to *VL_a* equal to the ratio of *ME* to *EQ*. We construct through the point *V* the hyperbola whose transverse axis is *UV* and *latus rectum VL_a* and we construct through the point *N* the parabola whose vertex is the point *N*, its axis *NG* and *latus rectum EA*; these two conic sections always intersect:

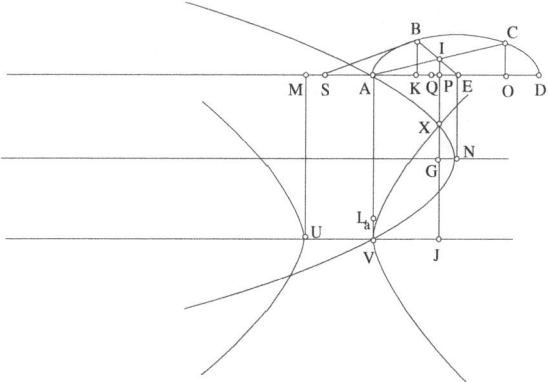

Fig. I.7.1

In the case of the ellipse, it is because the parabola passes through the point *A*, because *EA* is equal to the *latus rectum*; as the parabola is traced out, it then passes through the point *V*, a part of the hyperbola lies inside the parabola and their concave sides face towards one another; so they cut one another in every case and they intersect only on one side in a single point.[14]

In the case of the hyperbola, it is because an asymptote of the hyperbola with axis *UV* cuts the straight line *NG* which is the axis of the parabola and cuts it in a point beyond the point *N*, because it (the asymptote) is drawn from the midpoint of the straight line *UV* and encloses an acute angle with the straight line *UV* beyond the point *V*; if it cuts the straight line *NG* beyond the point *N*, it then falls inside the parabola, so it cuts the outline of the parabola. If the outline of the parabola cuts the asymptote of the hyperbola, it must cut the hyperbola, because the hyperbola approaches indefinitely closely to its asymptote and the parabola becomes indefinitely distant from a straight line that cuts it.

The parabola and the hyperbola thus cut one another in every case; let them cut one another at the point *X* and let the parabola be *NX* and the hyperbola *VX*.

We draw from the point *X* the perpendicular *XPGJ*, we put *PO* equal to *PA*, we extend the perpendicular *OC* to reach the given conic section, we join *AC* and we draw *XP* so far as to meet *AC*; let it meet it at the point *I*.

---

[14] See Supplementary note [3].

Thus *AI* will be equal to *IC*. We join *EI*, it will be a diameter of the conic section; let it cut the outline of the conic section at the point *B*. We draw *BS* parallel to *IA*, it will be a tangent to the conic section, as has been shown in Proposition 17 of Book I.

I say that the ratio of *BS* to *SD* is equal to the ratio of *H* to *F*.

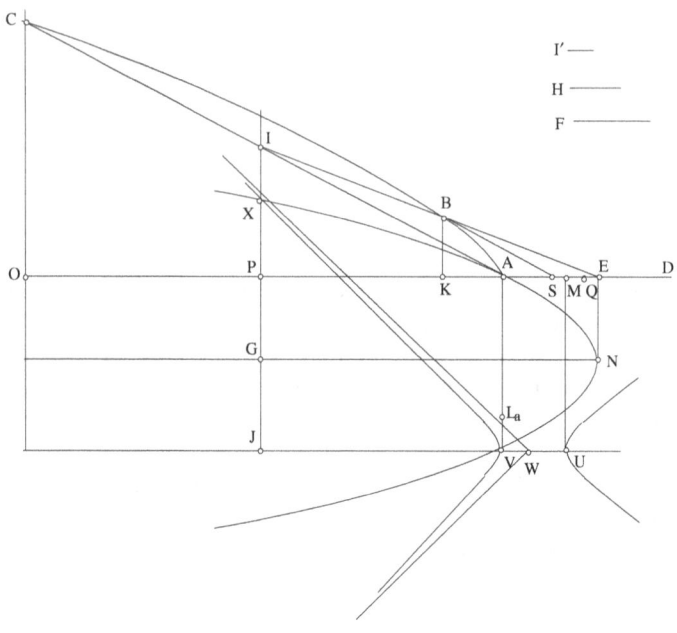

Fig. I.7.2

Proof: We draw the perpendicular *BK*. The product of *PE* and *EA* is equal to the square of *EK*, as has been shown in the analysis. But *PE* is equal to *NG* and the product of *NG* and *EA* is equal to the square of *GX*, because *EA* is the *latus rectum* of the parabola, so the straight line *EK* is equal to *GX*. But *GJ* is equal to *ED* because *PJ* is equal to *AD*, and *GJ* is half *PJ*, the straight line *JX* is thus equal to the straight line *KD*. But the ratio of the product of *UJ* and *JV* to the square of *JX* is equal to the ratio of *UV* to *VL_a* which is the ratio of *ME* to *EQ*; the ratio of the product of *UJ* and *JV* to the square of *JX* is thus a compound of the ratio of *ME* to *EA* and the ratio of *AE* to *EQ*. But the ratio of *ME* to *EA* is equal to the ratio of the product of *MP* and *PA* to the square of *AI*, as has been shown in the analysis, and the ratio of *AE* to *EQ* is equal to the ratio of the square of *H* to the square of *F*, so the ratio of the product of *MP* and *PA* to the square of *JX* is a compound of the ratio of the product of *MP* and *PA* to the square of *AI* and the ratio of the square of *H* to the square of *F*. But the ratio of the product of *MP* and *PA* to the square of *JX* is a compound of the ratio of the product of *MP* and *PA* to the square of *AI* and the ratio of the square of *AI* to the square of *JX*, so the ratio of the square of *AI* to the square of *JX* is equal to the ratio of the square of *H* to the square of *F*. But *JX* is equal to *KD*, so the ratio of the square of *AI* to the square of *KD* is equal to the ratio of the square of *H* to the square of *F* and the ratio of *AI* to *KD* is equal to the ratio of *H* to *F*. So the ratio of *DK* to *KA* is equal to the ratio of *IA* to a straight line whose ratio to *AK* is equal to the ratio of *IA* to *KD*, which is the ratio of *H* to *F*. But the ratio of *IA* to *AK* is equal to the ratio of *BS* to *SA*. We put the ratio of the straight line *I'* to *SA* equal to the ratio of *H* to *F*, the ratio of *SA* to *I'* will then be equal to the ratio of *F* to *H*. But the ratio of *SA* to *I'* is equal to the ratio of *SA* multiplied by *AI* to *AI* multiplied by *I'*, so the ratio of *SA* multiplied by *AI* to *AI* multiplied by *I'* is equal to the ratio of *F* to *H*. But *SA* multiplied by *AI* is equal to *BS* multiplied by *AK*, so the ratio of *BS* multiplied by *AK* to *AI* multiplied by *I'* is equal to the ratio of *F* to *H*. That is why the ratio of *BS* to *I'* is equal to the ratio of *IA* to the straight line whose ratio to *AK* is equal to the ratio of *H* to *F*, a ratio equal to the ratio of *DK* to *KA*. The ratio of *BS* to *SD* is thus equal to the ratio of *I'* to *SA* and the ratio of *I'* to *SA* is equal to the ratio of *H* to *F*, so the ratio of *BS* to *SD* is equal to the ratio of *H* to *F*. Which is what was to be proved.

There is no need for a discussion of this problem, because it has been shown that the two conic sections *NX* and *VX* always intersect. The problem is thus solved in all cases without <imposing> any conditions and there is a solution for each side, because the conic section *NX* cuts the conic section *VX* only on one side and in a single point:

In the case of the ellipse, it is because the concave sides of the two conic sections face towards one another. In the case of the hyperbola, it is because any straight line that cuts a hyperbola in two points, if it is extended in both directions, cuts the two asymptotes and <thus> a part of it is intercepted in the angle between the two asymptotes, an angle which is outside the conic section by what has been shown in Proposition 8 of Book

II. So if the parabola *NX* cuts the hyperbola *VX*,[15] then the two points lie within the angle formed by the two asymptotes, that is the angle that encloses the conic section, beyond the point where the parabola cuts the asymptote of the hyperbola. If we join the two points with a straight line, then this straight line cuts the two conic sections at the same time. So if we extend this straight line, it then cuts the asymptote of the hyperbola beyond the point in which that straight line and the parabola cut one another. The straight line that passes through the two points does not cut the other asymptote of the hyperbola, that is it is no part of it is intercepted in the angle formed by the hyperbola. The parabola and the hyperbola whose <common> axis is *UV* thus do not cut one another in two points on one side. But it has been shown that they cut one another in all cases, thus they cut one another on only one side, in a single point only. That is what we wanted to prove.

It has been shown how, starting from what we have proved in the synthesis, one can find the straight line *AP*, which we stated in the analysis would be shown in some way in the synthesis.

– **8** – The conic section *ABC* is a known hyperbola, its axis is *AD*, its centre *E*, the ratio of *G* to *H* is a given ratio and *G* is smaller than *H*. We wish to find a tangent to the conic section, with one end on its axis, and that is such that its ratio to the <semi>diameter drawn from the point of contact is equal to the ratio of *G* to *H*.

We suppose this has been done by analysis: let the straight line be *BK*. We join *EB*; the ratio of *EB* to *BK* is thus known. We extend *EB* and we draw from the point *A* an ordinate to meet *EB*; let it be *AIC*. *AI* will be equal to *IC* and the ratio of *EI* to *IA* equal to the known ratio of *EB* to *BK*. We make the angle *EIP* equal to the angle *IAP*, then the triangle *AIP* will be similar to the triangle *EIP*, so the ratio of *EP* to *PI* is equal to the ratio of *PI* to *PA*, so the product of *EP* and *PA* is equal to the square of *PI*, the straight line *PI* is thus greater than the straight line *PA*. But the ratio of *EP* to *PA* is equal to the ratio of the square of *EP* to the square of *PI* which is equal to the ratio of the square of *EI* to the square of *IA* which is known; the ratio of *EP* to *PA* is thus known and *EA* is known, so the straight line *AP* is known and the straight line *PI* is known.

---

[15] To be understood: in two points.

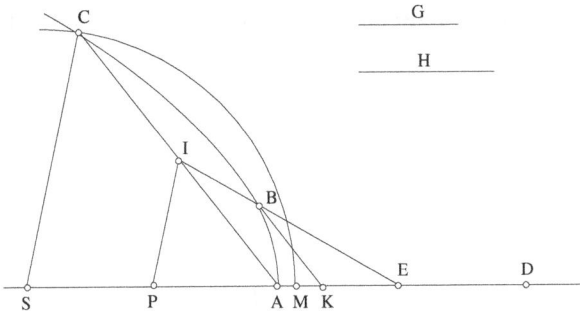

Fig. I.8

We draw from the point C a straight line parallel to the straight line IP; let it be CS. CS is thus twice IP and SA is twice AP. But IP is known and AP is known, so CS is known and SA is known, so the point S is known. But SC is greater than SA, because IP is greater than PA. We make SM equal to SC, so the point M will lie beyond the point A.[16] We take the point S as centre and with distance SM construct a circle;[17] let it be CM. This circle is thus known in size and position, because its centre is of known position; so the point C is known, the straight line AC is known in size and position, so the point I is known, the straight line EI is thus known in position, the point B is thus known and it is the point that provides a solution to the problem.

– **9** – For the synthesis of this problem, we return to the conic section and we put the ratio of G to F equal to the ratio of H to G. The ratio of H to F will thus be equal to the ratio of the square of H to the square of G. We put the ratio of EP to PA equal to the ratio of H to F and we make SP equal to PA; we put the product of EP and PA equal to the square of PL. We take the point S as centre and with a distance equal to twice the straight line PL we construct an arc of a circle; let the arc be MC. We join AC and we divide it into two equal parts at the point I. We join EI; it cuts the conic section because the point I lies inside the conic section; let EI cut the conic section at the point B. We draw BK to be a tangent to the conic section.

[16] Outside the hyperbola, because $SM > SA$.

[17] Ibn al-Haytham does not point out that the circle defined in this way must cut the hyperbola at the point C, but this follows from the inequality $SM > SA$ which has been established in the course of the analysis.

I say that that the ratio of *KB* to *BE* is equal to the ratio of *G* to *H*.

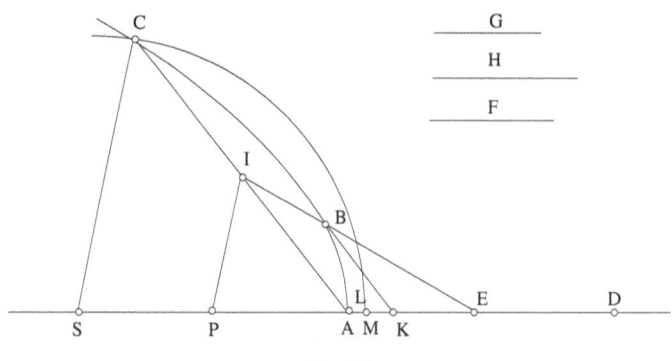

Fig. I.9

Proof: Let us join *SC* and *PI*; they will be parallel, so *CS* will be twice *IP*. The product of *EP* and *PA* is thus equal to the square of *PI*, so the ratio of *EP* to *PI* is equal to the ratio of *IP* to *PA*. Thus the two triangles *EPI* and *API* are similar and the ratio of *EP* to *PI* is equal to the ratio of *IP* to *PA* and to the ratio of *EI* to *IA*. So the ratio of *EP* to *PA* is equal to the ratio of the square of *EP* to the square of *PI*. But the ratio of *EP* to *PA* is equal to the ratio of *H* to *F* which is equal to the ratio of the square of *H* to the square of *G*; the ratio of the square of *EP* to the square of *PI* is thus equal to the ratio of the square of *H* to the square of *G* and the ratio of *EP* to *PI* is equal to the ratio of *H* to *G*. But the ratio of *EP* to *PI* is equal to the ratio of *EI* to *IA* and the ratio of *EI* to *IA* is equal to the ratio of *EB* to *BK*; so the ratio of *EB* to *BK* is equal to the ratio of *H* to *G*, so the ratio of *KB*, the tangent, to *BE*, is equal to the ratio of *G* to *H*. That is what we wanted to prove.

There is no need for a discussion for this problem because it can be solved in all cases.

– 10 – *ABC* is a conic section, the straight line *CD* is a tangent to it and the ratio of *E* to *G* is known. We wish to draw another tangent to the conic section, which meets *CD* and is such that its ratio to what it cuts off from the straight line *CD* is equal to the ratio of *E* to *G*.

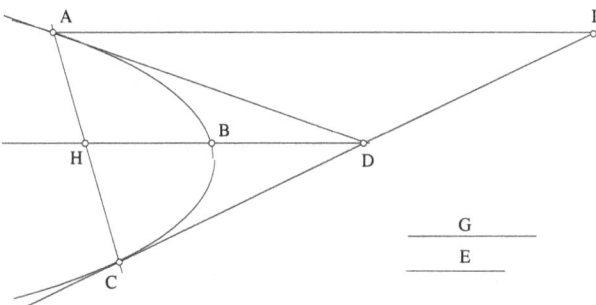

Fig. I.10

We suppose this has been done by analysis: let the required straight line be *AD*, let the conic section *AB* be, first, a parabola. We join *AC*, we divide it into two equal parts at the point *H* and we join *DH*; let it cut the conic section at the point *B*. So *DH* will be a diameter of the conic section because it cuts *CA* into two equal parts and tangents meet it [*DH*] in a single point. So *AC* is an ordinate and *DB* is equal to *BH*. If *E* is equal to *G*, then *AD* is equal to *DC*. But *AH* is equal to *HC*, so angle *H* is a right angle, *DH* is thus the axis of the conic section; it is known in position. But by hypothesis the straight line *CD* is known in position, so the point *D* is known. If we draw from the point *D* a tangent to the conic section, it will be equal to *DC*. That is what we sought.

But if *E* is not equal to *G*, then *AD* is not equal to *DC*, so angle *H* is not a right angle; but since the conic section is a parabola, all its diameters are parallel and parallel to its axis. But its axis is known in position and the straight line *CD* is known in position, so it meets the axis at a known angle. So the straight line *CD* meets each of the diameters of the conic section at that angle; but the straight line *DH* is a diameter, so the angle *CDH* is known. We draw from the point *A* a straight line parallel to the straight line *HD*; let it be *AI*. We extend *CD* until it meets it; let it meet it at the point *I*. The angle *I* will thus be equal to the known angle *HDC*, and *ID* will be equal to *DC*; but the ratio of *DC* to *DA* is known, so the ratio of *ID* to *DA* is known. But the angle *I* is known, so the triangle *AID* is known in shape and angle *ADI* is known, so angle *ADC* is known. But the ratio of *AD* to *DC* is known, so the triangle *ADC* is known in shape, angle *DCA* is known and the straight line *CD* is known in position; so the straight line *CA* is known in position, the point *A* is thus known and the straight line *DA* is the tangent.

– **11** – The synthesis of this problem is as follows.

We draw the axis of the conic section; let it be *IK*. It meets the straight line *CD*; let it meet it at the point *I*.[18] We put the ratio of *IC* to a general straight line equal to the ratio of *G* to *E*; we draw *C K*[19] equal to that <general> straight line, we extend *IC* on the side of *C*, we cut off *CL* equal to *CI*, we join *LK* and we draw from the point *C* a straight line parallel to the straight line *LK*; let it be *CA*. This straight line thus cuts the conic section in another point, because it cuts the axis and in consequence cuts all the diameters; let it cut the conic section at the point *A*. We divide *CA* into two equal parts at the point *H* and we draw from the point *H* a straight line parallel to the axis, so it meets the straight line *CD*; let it be the straight line *HBD*. This straight line will thus be a diameter and *AC* is an ordinate of that diameter, so *DB* is equal to *BH* because *CD* is a tangent and *CH* an ordinate. We join *AD*, it will thus be a tangent because *DB* is equal to *BH*.

I say that the ratio of *AD* to *DC* is equal to the ratio of *E* to *G*.

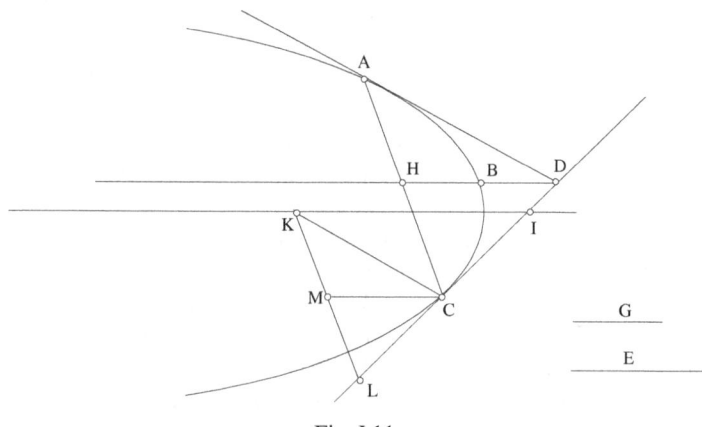

Fig. I.11

Proof: We divide *LK* into two equal parts at the point *M* and we join *CM*; it will be parallel to *IK* because *IC* is equal to *CL*, so the straight line *CM* is parallel to the straight line *DH* and the straight line *CH* is parallel to the straight line *LM*. So triangle *DCH* is similar to triangle *CLM*. So the ratio of *DC* to *CH* is equal to the ratio of *CL* to *LM*. But the ratio of *CH* to

---

[18] So the point *I* here is not the same as the point *I* in the analysis.

[19] The question of the existence of the point *K* is discussed later.

CA is equal to the ratio of *ML* to *LK*, so the ratio of *DC* to *CA* is equal to the ratio of *CL* to *LK*, thus triangle *DCA* is similar to triangle *CLK*. So the ratio of *CD* to *DA* is equal to the ratio of *LC* to *CK*, that is to the ratio of *IC* to *CK*, which is equal to the ratio of *G* to *E*. So the ratio of *AD* to *DC* is equal to the ratio of *E* to *G*. That is what we wanted to prove.

The discussion of this problem is that the ratio of *E* to *G* shall not be less than the ratio of the perpendicular dropped from the point *C* onto the axis to the straight line *CI*, because the ratio of *E* to *G* is equal to the ratio of *KC* to *CI* and the angle *CIK* is acute because it is the angle formed by the tangent and the axis, so it cannot but be acute. That is why the ratio of *KC* to *CI* is not less than the ratio of the perpendicular dropped from the point *C* onto the axis to the straight line *CI*. That is why the ratio of *E* to *G* is not less than the ratio of the perpendicular dropped from the point *C* onto the axis to the straight line *CI*. That is what we wanted to prove.

– **12** – Let the conic section *ABC* be a hyperbola or an ellipse; the straight line *BD* is a tangent and the ratio of *E* to *G* is known. We wish to draw another straight line to be a tangent to the conic section, to meet the straight line *BD* and to be such that its ratio to what it cuts off from the straight line *BD* is equal to the ratio of *E* to *G*.

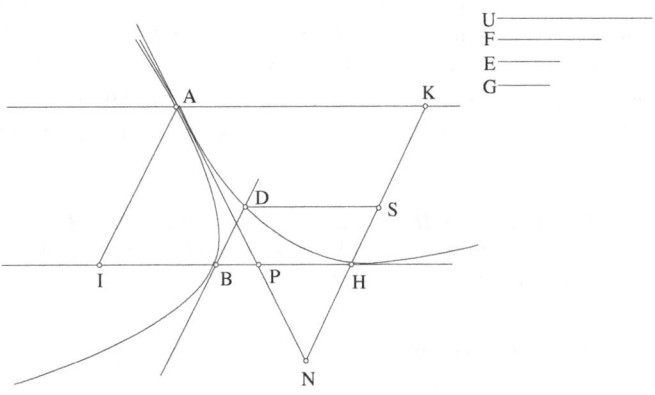

Fig. I.12.1

We suppose this is to be done by means of analysis: let the required tangent be the straight line *AD*. So the ratio of *AD* to *DB* will be known. Let the centre of the conic section be the point *H*. We join *HB*, thus it will be a diameter[20] of known position. Let us draw from the point *A* a straight line that is an ordinate; let it be *AI*. We extend *AD* on the side of *D*, thus it meets *HB*; let it meet it at the point *P*.[21] The product of *IH* and *HP* is equal to the square of *HB*, from what has been shown in Proposition 37 of the first book. We draw from the point *H* a straight line parallel to the straight lines that are ordinates; let it be *HKN*. We extend it in both directions. Let us draw from the point *A* a straight line parallel to *HI*; let it be *AK*. Thus it meets *HK*; let it meet it at the point *K*. We extend *AP* until it meets *KH*; let it meet it at the point *N*. Since *AI* is an ordinate and *AP* is a tangent, the ratio of the product of *HI* and *IP* to the square of *AI* is equal to the ratio of *HB* to half of the *latus rectum*[22] which is a known ratio, as has been shown in Proposition 37 of the first book. The ratio of the product of *HI* and *IP* to the square of *AI* is thus known. But the product of *HI* and *HP* is equal to the square of *HB*, so the ratio of *HP* to *PI* is equal to the ratio of the square of *HB*, which is known, to a square[23] whose ratio to the square of *AI* is known, because the ratio of *HP* to *PI* is equal to the ratio of the product of *HP* and *HI*, which is equal to the square of *BH*, to the product of *HI* and *IP*; but the ratio of the product of *HI* and *IP* to the square of *AI* is known. The ratio of *HP* to *PI* is equal to the ratio of *NH* to *AI*, so the ratio of *NH* to *AI* is equal to the ratio of the square of *HB* to a square[24] whose ratio to the square of *AI* is known; it is thus equal to the ratio of a known square[25] to the square of *AI*. So the product of *NH* and *AI* is equal to the square of a known straight line,[26] whose ratio to *HB* is known. The ratio of the product of *NH* and *HK* to the square of *HB* is thus known because the ratio of *NH* to

---

[20] The straight line *HB* is a diameter, but the segment *HB* is a semidiameter, that is why Ibn al-Haytham associates with it half the *latus rectum* for the diameter.

[21] See Mathematical commentary, p. 92.

[22] See note 17.

[23] If we put the square $\Sigma = K \cdot AI^2$, we have $\dfrac{HB^2}{\Sigma} = \dfrac{NH}{IA}$, the square Σ is not known.

[24] See preceding note.

[25] If we put the square $Q = \Delta^2 = \dfrac{HB^2}{k}$, Q and Δ are known and we have $NH \cdot AI = \dfrac{Q}{AI^2}$, hence $NH \cdot AI = Q$.

[26] See preceding note.

AI is equal to the ratio of a specified area to the square of *AI*, so the ratio of the square of *HB* to the square whose ratio to the square of *AI* is known is equal to the ratio of that area to the square of *AI*. By permutation we can find what we were looking for.[27]

The ratio of the product of *NH* and *HK* to the product of *IH* and *HP* is thus known, the ratio of *KH* to a straight line whose ratio to *HI* is known is thus equal to the ratio of *PH* to *HN*, so the ratio of *HK* to a straight line whose ratio to *AK* is known is equal to the ratio of *AK* to *KN*, so the ratio of the product of *NK* and *KH* to the square of *KA* is known. We put the ratio of *KH* to *U* equal to the ratio of the product of *NK* and *KH* to the square of *KA*, the ratio of *KH* to *U* is thus known, the product of *U* and *KN* is equal to the square of *KA* and the product of *U* and *HN* is equal to the square of *HB*, because *AK* is equal to *HI*.

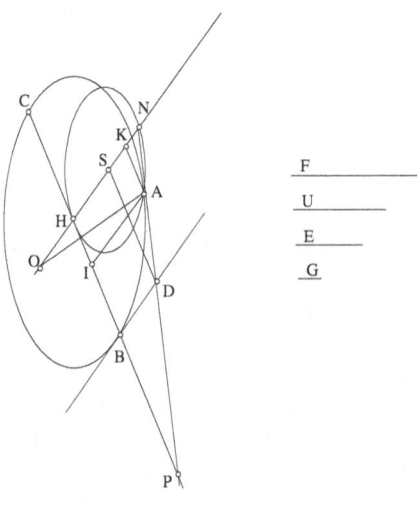

Fig. I.12.2

We construct through the point *H* the hyperbola whose transverse diameter is *NH* and its *latus rectum* a straight line such that the ratio of *NH* to that straight line is equal to the ratio of *KH* to *U*; it accordingly passes through the point *A* of the hyperbola *ABC*. We construct on the straight line

---

[27] From notes 23 and 25, we deduce $\dfrac{NH}{AI} = \dfrac{Q}{AI^2} = \dfrac{HB^2}{\Sigma}$, hence $\dfrac{NH \cdot AI}{AI^2} = \dfrac{HB^2}{\Sigma}$; but $AI = HK$, consequently, by permutation, we can obtain $\dfrac{NH \cdot HK}{HB^2} = \dfrac{AI^2}{\Sigma} = \dfrac{1}{k}$.

NH of the ellipse *ABC* an ellipse such that its diameter is *NH* and its *latus rectum* a straight line such that the ratio of *NH* to that straight line is equal to the ratio of *KH* to *U*; it accordingly passes through the point *A* which lies on the outline of the ellipse *ABC*.[28] We put the product of *F* and *NK* equal to the square of *AN*, the ratio of *F* to *U* will thus be equal to the ratio of the square of *NA* to the square of *AK*. We draw from the point *D* a straight line *DS* parallel to *KA*. But *DB* is a tangent and it is parallel to *HK*; the product of *IH* and *HP* is equal to the square of *HB*, so the product of *AN* and *NP* is equal to the square of *ND* and the product of *KN* and *NH* is equal to the square of *NS*, so the ratio of *AN* to *ND* is equal to the ratio of *AD* to *DP*, the product of *AN* and *DP* is thus equal to the product of *AD* and *DN*. But the ratio of *F* to *AN* is equal to the ratio of *AN* to *NK* which is equal to the ratio of *DN* to *NS* and which is equal to the ratio of *DP* to *SH*, so the ratio of *F* to *AN* is equal to the ratio of *DP* to *SH*, the product of *AN* and *DP* is thus equal to the product of *F* and *SH*, so the product of *AD* and *DN* is equal to the product of *F* and *SH*, that is *DB*. The ratio of *AD* to *DB* is thus equal to the ratio of *F* to *DN*, so the ratio of *F* to *DN* is known. But the ratio of *F* to *U* is equal to the ratio of the square of *NA* to the square of *AK*, thus it is equal to the ratio of the square of *ND* to the square of *DS*. But the product of *U* and *HN* is equal to the square of *DS* because it is equal to the square of *HB*, so the product of *F* and *HN* is equal to the square of *ND*; the ratio of *F* to *ND* is thus equal to the ratio of *ND* to *NH*; but the ratio of *F* to *ND* is equal to the ratio of *AD* to *DB*, so the ratio of *DN* to *NH* is equal to the ratio of *AD* to *DB*, that is the ratio of *AD* to *HS*, and is equal to the ratio of *AN* to *NS*; so the ratio of *AN* to *NS* is equal to the ratio of *AD* to *DB* which is equal to the known ratio of *E* to *G*, and the ratio of *AN* to *NS* is known. But the product of *KN* and *NH* is equal to the square of *NS*, so the ratio of the product of *KN* and *NH* to the square of *NA* is known and it is equal to the ratio of *HN* to *F*. We make *NO* equal to *F*, then the product of *KN* and *NO* is equal to the square of *NA*. We join *AO*, triangle *ANO* will be similar to triangle *ANK*, the angle *NAO* will be equal to the angle *AKN*; but the angle *AKN* is known because it is the angle of an ordinate associated with the diameter *HB*, so the angle *NAO* is known. If the straight line *NH* is known in magnitude and position and if the conic section *HA* is known in position, then the straight line *NO* is of known magnitude, and the point *A* lies on the circumference of a segment of a circle known in position; the point *A* will

---

[28] See the Mathematical commentary, p. 94.

be known, the straight line *AN* will be known in position, the angle *ANH* will then be known and the angle *NPH* will be known. The straight line *AP* is a tangent to the conic section, so it makes a known angle with the diameter *HP*, so the point *A* is known.

– **13** – The synthesis of this problem is as follows.

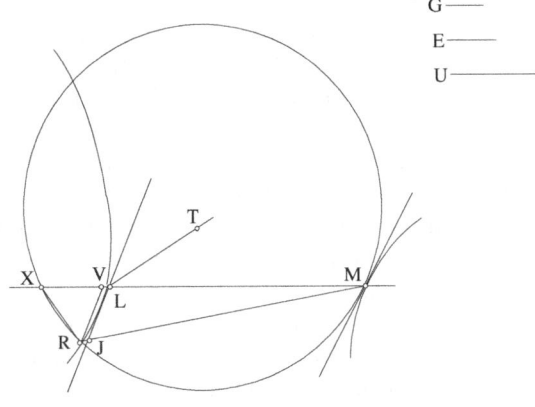

Fig. I.13.1

We return to the conic section and the ratio, we suppose there are two known straight lines[29] *LM* and *LT* and we put the ratio of *LM* to *LT* equal to the ratio of half the *latus rectum* of the diameter *HB* to the diameter *HB*.[30] We construct through the point *L* of one of the two straight lines[31] the hyperbola whose transverse diameter is *LM*, whose *latus rectum* is *LT* and the angle of whose ordinates is the angle of the ordinates of the diameter *HB* of the hyperbola *ABC*; let this conic section be *LR*. We construct on the other straight line *LM* the ellipse whose diameter is *LM* and its *latus rectum* *LT* and the angle of whose ordinates is the angle <of the ordinates> of the diameter *HB* of the ellipse *ABC*; let the conic section be *LRM*. We put the ratio of *LM* to *MX* equal to the ratio of the square of *G* to the square of *E* and we construct on the straight line *XM* a segment that subtends an angle equal to the angle of the ordinate of the diameter *HB*; let the arc be *MRX*.

---

[29] Ibn al-Haytham considers two straight lines *LM* and *LT* and constructs two figures.

[30] As before, *HB* is the semidiameter.

[31] See note 25.

This arc cuts the conic section *LR* in every case; let it cut it at the point *R*. We join *RM*, the angle *M* will thus be known; we join *RX* and we draw *RV* as an ordinate, so the angle *RVM* will be equal to the angle *MRX* and triangle *MRX* will be similar to triangle *MRV*; their sides are therefore proportional. The product of *VM* and *MX* is equal to the square of *MR*, so the ratio of the product of *VM* and *ML* to the square of *MR* is equal to the ratio of *LM* to *MX* which is equal to the ratio of the square of *G* to the square of *E*. But the ratio of the product of *MV* and *VL* to the square of *VR* is equal to the ratio of *ML* to *LT* which is equal to the ratio of the *latus rectum* of the diameter *HB* to the diameter *HB*.[32] We draw from the point *L* a straight line that is an ordinate; let it be *LJ*. The angle *LJM* is known because the two angles *L* and *M* are known. But it has been shown in proposition fifty-seven and in proposition fifty-nine of the second book of the *Conics* how to draw a straight line that is a tangent to the conic section and which makes with the axis of the conic section an angle equal to a known angle. Starting from this case, we shall show how to draw a straight line that is a tangent to the conic section and which makes a known angle with a diameter of known position, because the axis in fact makes a known angle with any diameter of known position. So we draw a straight line that is a tangent to the conic section *ABC* and which makes with the diameter *HB* an angle equal to the angle *LJM*; that is the straight line *AP*. We draw *AI* as an ordinate; we draw *AK* parallel to the diameter *IH* and we draw from the point *H* a straight line parallel to the straight lines that are ordinates, let it be *HKN*; let it cut the straight line *AK* at the point *K* and meet *AP* at the point *N*. We put the ratio of *ON* to *NH* equal to the ratio of *XM* to *ML*, which is equal to the ratio of the square of *E* to the square of *G*; we join *AO*. Since *KH* is parallel to the ordinates and *AK* is parallel to *IH*, the angle *K* will be equal to the angle *I* and to the angle *PHN*. But the angle *PHN* is equal to the angle *JLM* and the angle *P* is equal to the angle *J*, <so> there remains the angle *N* <which is> equal to the angle *M*; triangle *NAK* will thus be similar to triangle *MRV* and triangle *NPH* will be similar to triangle *MJL*. But the ratio of the product of *MV* and *VL* to the square of *VR* is equal to the ratio of *ML* to *LT* which is equal to the ratio of the *latus rectum* relative to the diameter *HB* to the diameter *HB*; so it is equal to the ratio of the square of *AI* to the product of *HI* and *IP*, as has been shown in Proposition 37 of Book I. The ratio compounded from the ratio of *MV* to

---

[32] The diameter is 2*HB* and Ibn al-Haytham considers half of the *latus rectum*. He abbreviates his wording in this same way in all that follows.

VR and the ratio of *LV* to *VR* is the ratio compounded from the ratio of *AI* to *IP* and the ratio of *AI* to *IH*; but the ratio of *MV* to *VR* is equal to the ratio of *ML* to *LJ* which is equal to the ratio of *NH* to *HP* which is equal to the ratio of *AI* to *IP*. There remains the ratio of *LV* to *VR* <which is> equal to the ratio of *AI* to *IH* which is equal to the ratio of *HK* to *KA*; so the ratio of *HK* to *KA* is equal to the ratio of *LV* to *VR*.

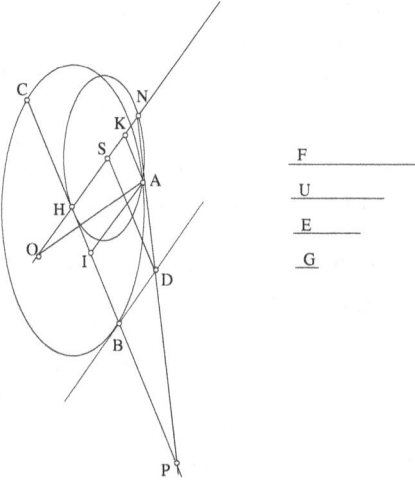

Fig. I.13.2

So the ratio of the product of *NK* and *KH* to the square of *KA* is equal to the ratio of the product of *MV* and *VL* to the square of *VR*, which is equal to the ratio of *ML* to *LT*, which is equal to the ratio of the *latus rectum* associated with the diameter *HB* to the diameter *HB*. But the ratio of *AK* to *KN* is equal to the ratio of *RV* to *VM*, so the ratio of *HK* to *KN* is equal to the ratio of *LV* to *VM*. Thus the ratio of *KN* to *NH* is equal to the ratio of *VM* to *ML*. But the ratio of *HN* to *NO* is equal to the ratio of *LM* to *MX*, so the ratio of *KN* to *NO* is equal to the ratio of *VM* to *MX*. But the ratio of *VM* to *MX* is equal to the ratio of the square of *VM* to the square of *MR*, so the ratio of *KN* to *NO* is equal to the ratio of the square of *VM* to the square of *MR*, which is equal to the ratio of the square of *KN* to the square of *NA*. Thus the product of *KN* and *NO* is equal to the square of *NA*. Let us put the ratio of *HK* to *U* equal to the ratio of the *latus rectum* associated with the diameter *HB* to the diameter *HB*, so the ratio of the product of *NK* and *KH* to the product of *NK* and *U* is equal to the ratio of the *latus rectum* associated with the diameter *HB* to the diameter *HB*. But the ratio of the product of *NK* and *KH* to the square of *AK* is equal to the ratio of the *latus rectum* associated with the diameter *HB* to the diameter *HB*, so the product of *NK* and *U* is equal to the square of *KA*, so the ratio of the square of *NA* to the square of *AK* is equal to the ratio of *NO* to *U*. But the ratio of *KN* to *NH* is equal to the ratio of *AK* to *PH*, which is equal to the ratio of *IH* to *HP*, which is equal to the ratio of the square of *IH* to the square of *HB*, that is the ratio of the square of *AK* to the square of *BH*. So the ratio of the product of *NK* and *U* to the product of *NH* and *U* is equal to the ratio of the square of *AK* to the square of *HB*. But the product of *NK* and *U* is equal to the square of *AK*, so the product of *NH* and *U* is equal to the square of *HB*.

We draw *DS* parallel to *KA*; *DS* will thus be parallel to *HB*. But since the product of *IH* and *HP* is equal to the square of *HB*, the product of *AN* and *NP* will be equal to the square of *ND*, and the product of *KN* and *NH* is equal to the square of *NS*. But the ratio of *ON* to *NH* is equal to the ratio of *XM* to *ML* which is equal to the ratio of the square of *E* to the square of *G*, so the ratio of *ON* to *NH* is equal to the ratio of the square of *E* to the square of *G*; the ratio of *HN* to *NO* is thus equal to the ratio of the square of *G* to the square *E*. Now the ratio of *HN* to *NO* is equal to the ratio of the product of *HN* and *NK* to the product of *NK* and *NO*, which is equal to the square of *NA*. So the ratio of the product of *KN* and *NH* to the square of *NA* is equal to the ratio of the square of *G* to the square of *E*. But the product of *KN* and *NH* is equal to the square of *NS*, so the ratio of the square of *NS* to the square of *NA* is equal to the ratio of the square of *G* to the square of *E*; the ratio of the square of *AN* to the square of *NS* is thus equal to the ratio of the square of *E* to the square of *G*, so the ratio of *AN* to *NS* is equal to the ratio of *E* to *G*. But since the product of *ON* and *NK* is equal to the square of *NA*, the ratio of *ON* to *NA* will be equal to the ratio of *AN* to *NK*, which is equal to the ratio of *PN* to *NH*. So the ratio of *ON* to *NA* is equal to the ratio of *PN* to *NH*, the product of *ON* and *NH* is thus equal to the product of *AN* and *NP* which is the square of *ND*. So the product of *ON* and *NH* is equal to the square of *ND*, so the ratio of *ON* to *NH* is equal to the ratio of the square of *DN* to the square of *NH*. But the ratio of *ON* to *NH* is equal to the ratio of the square of *E* to the square of *G*, so the ratio of the square of *DN* to the square of *NH* is equal to the ratio of the square of *E* to the square of *G*. But the ratio of the square of *AN* to the square of *NS* is equal to the ratio of the square of *E* to the square of *G*, so the ratio of the square of *AN* to the square of *NS* is equal to the ratio of the square of *DN* to the square of

NH; so the ratio of *AN* to *NS* is equal to the ratio of *DN* to *NH* and is equal to the ratio of the remainder, which is *AD*, to the remainder, which is *HS*; so the ratio of *AD* to *DB* is equal to the ratio of *AN* to *NS* which is equal to the ratio of *E* to *G*. Consequently the ratio of *AD* to *DB* is equal to the ratio of *E* to *G*. That is what we wanted to prove – and it is known only to God.

The discussion of this problem – if the two straight lines *E* and *G* are different – consists of supposing we have two straight lines whose ratio to one another is equal to the ratio of the square of *E* to the square of *G* like <the ratio between> the two straight lines *XM* and *LM* which are in the figure. We take the smaller of the two as a diameter of the hyperbola homologous to the conic section *LR*. We do not need to add any other condition in order to construct the hyperbola.

For the ellipse, we must similarly make the smaller of the two straight lines a diameter of the ellipse homologous to the conic section *LRM*. If the diameter of the ellipse is a diameter but not an axis, the excess <length> of the greater straight line over the shorter line — an excess which is homologous to the straight line *LX* — must be placed on the side of the end of the diameter, starting from which the tangent to the conic section makes an acute angle with the diameter of the conic section, so that the other end of the diameter is the end of the base of a segment of the circle, such that the tangent to that segment of the circle at the other end falls inside the ellipse, because the tangent and the base of the segment enclose an acute angle, since in the case of the ellipse the segment subtends an obtuse angle and, in the case of the hyperbola, the segment subtends an acute angle.[33] The tangent to the ellipse makes an obtuse angle with the base of the arc at this end. It follows that this end[34] of the arc lies inside the ellipse and that the other end of the arc that is the end of the greater straight line lies outside the conic section. So the arc cuts the outline of the conic section. This clearly results in the arc cutting the perimeter of the ellipse in a single point and the problem having only a single solution.

If the diameter is an axis and is the smaller axis, then the arc will be a semicircle and the tangent to the arc will be a tangent to the conic section as well. If the chord that cuts off from this circle a segment that subtends an angle equal to the angle made at the end of the smaller axis and enclosed

---

[33] Ibn al-Haytham restricts the scope of the discussion by making the assumption that α is obtuse for the ellipse and α is acute for the hyperbola.

[34] Here what Ibn al-Haytham means by 'end' is not the point itself, which is a point of the ellipse, but a point near this point on the circular arc he is considering.

between the two straight lines drawn from the two ends of the greater axis, <which> is equal to the greater axis, then the outline of the segment passes through the end of the greater axis and the other end of the segment lies outside the conic section; the problem is thus solved.

And if this chord is smaller than the greater axis, then the segment cuts <the major> axis of the conic section inside the conic section; so it cuts the perimeter of the conic section on the side of the outer end and does not cut it on the side of the tangent, because any circle constructed on this diameter such that the chord we have defined in it is smaller than the major axis, and such that this circle is greater than the first circle, is thus a tangent to the first circle, <at a point> outside it, and a tangent to the conic section, <at a point> inside it; the segment that is a semicircle cuts the conic section in a single point; the problem thus has a single solution.

If the chord is greater than the greater axis, then the problem has no solution for the ellipse, because the segment of a circle touches the conic section at the end of the axis and cuts the major axis outside the conic section, and its other end lies outside the conic section; consequently the complete segment lies outside the conic section and does not cut the perimeter of the conic section; the problem thus has no solution.

If the diameter of the conic section is the greater axis, the problem again has no solution, because the complete segment lies outside the conic section.

If the two straight lines E and G are equal, the procedure for the construction is to draw the axis of the hyperbola of the ellipse; it will cut the straight line BD which is the tangent. If it cuts it in a point other than the point B, we draw from the point of intersection another straight line that is a tangent to the conic section, it will accordingly be equal to the first one; the problem is thus solved.

If the axis passes through the point B, then we cannot draw another straight line that is a tangent to the conic section and is equal to the part it cuts off from the straight line BD. In fact the diameter drawn as far as the position where the two straight lines that are tangents meet <one another> divides the straight line that joins the two points of contact into two equal parts – this can be shown from the converse of propositions twenty-nine and thirty of the second book of the *Conics* – and this diameter is not perpendicular to the straight line that joins the two points of contact because it is not the axis, so the two straight lines that are tangents are not equal. If the two straight lines E and G are equal, the problem has a solution only if the point *B* does not lie on the axis. So we have completed the discussion of this problem. That is what we wanted to prove.

<14> If we have a known parabola, <the problem is> how to draw a straight line that is a tangent to it, which ends on its axis and which is equal to a given straight line.

Let the conic section be *ABC*, its axis *AD* and the given straight line *E*. We wish to draw a straight line that is a tangent to the conic section and such that <the part of it> that ends on the axis is equal to the straight line *E*.

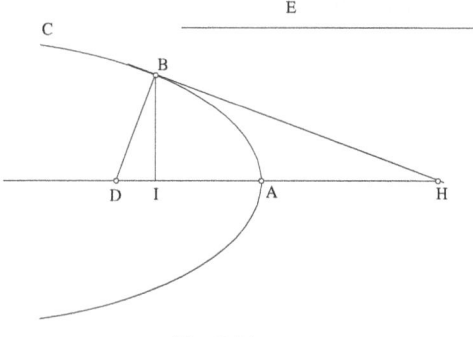

Fig. I.14

We suppose this is to be done by the method of analysis: let the straight line be *BH*; we draw *BI* as an ordinate and we put the angle *HBD* as a right angle. So the product of *DH* and *HI* is equal to the square of *BH*, and the product of *DI* and *IH* is equal to the square of *BI*. But *BI* is an ordinate and *BH* is a tangent, so the straight line *HA* is equal to the straight line *AI*, as has been shown in Proposition 35 of the first book. But the product of the *latus rectum* associated with the axis and *AI* is equal to the square of *BI*, so the straight line *DI* is half the *latus rectum* of the axis. But the *latus rectum* associated with the axis is known, so the straight line *DI* is known and the product of *DH* and *HI* is equal to the square of *HB*. But *HB* is known, so the product of *DH* and *HI* is known.[35] But *DI* is known, so the straight line *HI* is known. But the point *A* is known, so the point *H* is known. Now we have drawn from this point a straight line *HB* that is a tangent to the conic

---

[35] The length of *HI* is obtained by a geometrical construction (see Mathematical commentary).

section, so the point *B* is known, from what has been shown in Proposition 51 of the second book.[36]

<15> The synthesis of this problem is carried out in this way: we return to the conic section and the given straight line; let the *latus rectum* associated with the axis of the conic section be <the straight line> *KL* which we divide into two equal parts at the point *M*. We put the product of *MN* and *NK* equal to the square of *E*.[37] We put *AH* equal to half of *KN* and we draw from the point *H* a tangent to the conic section; let it be *HB*.

I say that *HB* is equal to *E*.

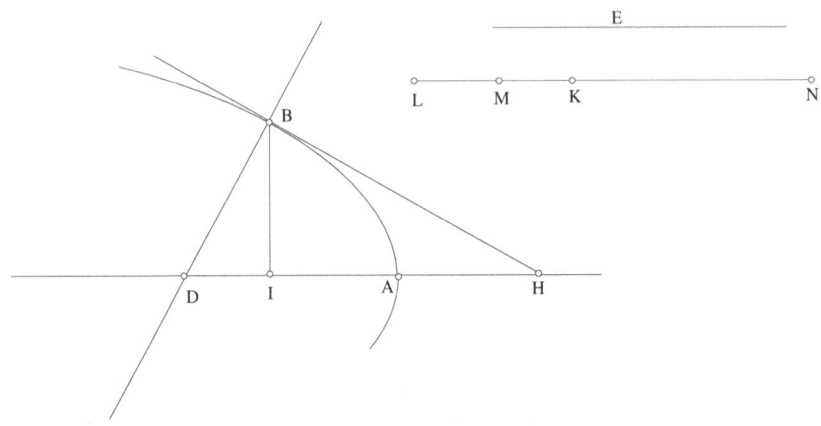

Fig. I.15

Proof: We draw *BI* as an ordinate, then *IA* is equal to *AH* and *IH* is equal to *KN*. We put *ID* equal to *MK*, then the product of *DH* and *HI* is equal to the product of *MN* and *NK*, so the product of *DH* and *HI* is equal to the square of *E*. Since *DI* is equal to half the *latus rectum*, the product of *DI* and twice *IA* is equal to the square of *IB*, so the product of *DI* and *IH* is equal to the square of *IB*. We join *DB*, then the angle *DBH* is a right angle and the product of *DH* and *HI* will thus be equal to the square of *HB*. But the product of *DH* and *HI* is equal to the square of *E*, so the straight line *BH* is equal to the straight line *E* and *BH* is a tangent to the conic section and equal to the straight line *E*. That is what we wanted to construct.

---

[36] Only part of the proposition is used.
[37] The point *N* is found by the same construction.

This problem does not demand a discussion, because it can be solved in all cases.

– 16 – If we have a known hyperbola or a known ellipse, <the problem is> how to draw a straight line that is a tangent to the conic section and ends on its axis and is equal to a known straight line.

Let the conic section be ABC, its axis AD, its centre E and let the straight line F be known. We wish to draw a straight line that is a tangent to the conic section, which ends on the axis and is equal to the known straight line F.

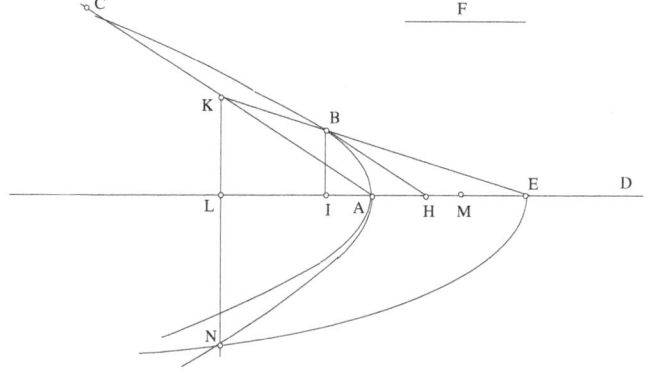

Fig. I.16.1

We suppose this is to be done by means of analysis: let <the tangent> be BH. We draw BI as an ordinate and we draw from the point A the straight line AC parallel to the tangent. We join EB and extend it until it meets AC; let it meet it at the point K. We draw KL as an ordinate and we put the ratio of EM to MA equal to the ratio of DA to its *latus rectum*; so MA is half the homologous straight line and the ratio of the product of ML[38] and LA to the square of AK is equal to the ratio of ME to EA. But since BH is known and AE is known, the product of AE and BH is known. But the product of AE and BH is equal to the product of KA and HE, because the ratio of KA to BH is equal to the ratio of AE to EH; so the product of KA

---

[38] BL in the manuscript (see Mathematical commentary).

and *EH* is known. But we have pointed out earlier[39] that the ratio of *IE* to *EA* is equal to the ratio of *AE* to *EH*, and that the product of *IE* and *EH* is equal to the square of *AE* which is known, so the ratio of the product of *KA* and *EH* to the product of *IE* and *EH* is known, and it is equal to the ratio of *KA* to *IE*. But the ratio of the product of *ML* and *LA* to the square of *KA* is equal to the ratio of *ME* to *EA*, so the ratio of the product of *ML* and *LA* to the square of *IE* is equal to the ratio of *ME* to a straight line whose ratio to *AE* is known. But it has also been shown earlier that the ratio of *LE* to *EI* is equal to the ratio of *IE* to *EA*, so the product of *LE* and *EA* is equal to the square of *EI*.[40] So the ratio of the product of *ML* and *LA* to the product of *LE* and *EA* is a known ratio.

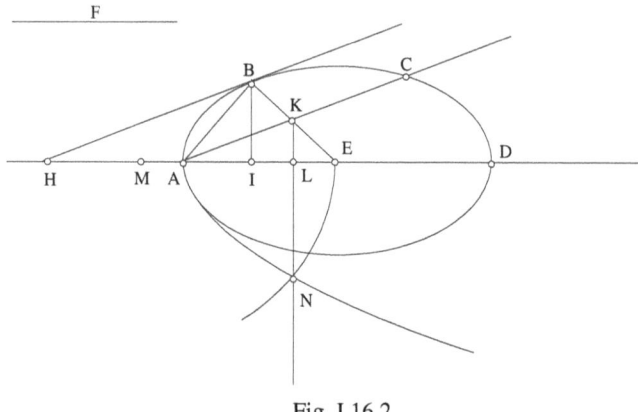

Fig. I.16.2

We construct through the point *E* the parabola whose axis is *EL* and *latus rectum EA*; let the conic section be *EN*. Let us extend *KL* until it meets *EN*; let it meet it at the point *N*. Let us construct through the point *A* the hyperbola whose axis is *MA* and whose *latus rectum* is such that the ratio it has to *MA* is the compound of the ratio of *ME* to *EA* and the ratio of the square of *HB* to the square of *EA*; let it cut the conic section *EN* at the point *N*. Whether it cuts it or does not cut it is a matter we shall deal with

[39] See Mathematical commentary, Problem 2.
[40] *Ibid.*

later.[41] Thus the point N will be the point of intersection of two conic sections known in position, so the point N is known. But NL is a perpendicular, so the point L is known and the straight line NL is known in position; but as it has <merely> been extended, the straight line LK is accordingly known in position and AK is equal to KC, so the straight line AC is known in position, because if we cut off from the axis <a straight line> equal to AL and if from the point of division we draw a straight line as an ordinate, it ends at the point C; so the point C is known, so the straight line AC is known in position. But the straight line LK is known in position, so the point K is known and the point E is known, so the straight line EK is known in position and the conic section ABC is known in position; so the point B is known and the straight line BH is a tangent.

Thus the analysis has led us to draw a tangent from a known point on the outline of the conic section.

– **17** – The synthesis of this problem is as follows.

Let the conic section be ABC; its axis is AD, its centre E, and the given straight line F. We wish to draw a straight line that is a tangent to the conic section, <a line> that ends on the axis and is equal to the straight line F.

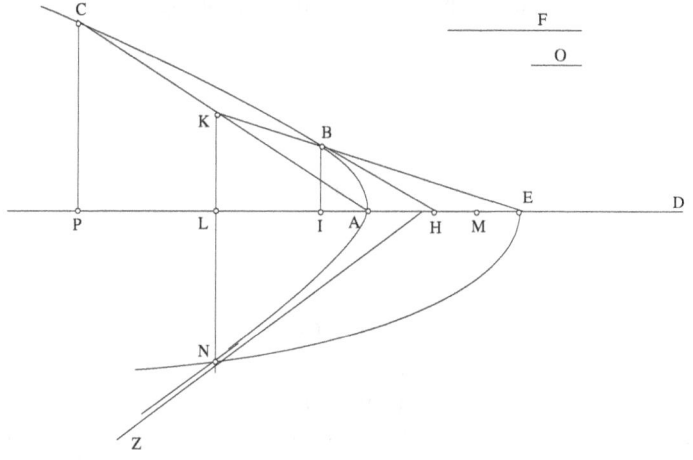

Fig. I.17.1

---

[41] Once the name N has been given to the point of intersection of the parabola and LK, it needs to be shown that the hyperbola cuts LK at the same point, which follows immediately in the analysis and does not require discussion.

We put the ratio of *EM* to *MA* equal to the ratio of *DA* to its *latus rectum*; so the straight line *MA* will be half the homologous straight line. We put the ratio of the straight line *MA* to a straight line *O* equal to the compound of the ratio of *ME* to *EA* and the ratio of the square of *F* to the square of *EA*. We construct through the point *A* the hyperbola whose transverse axis is *MA* and *latus rectum O*; let the conic section be *AN*. We construct through the point *E* the parabola whose axis is *EA* and its extension and whose *latus rectum* is *AE*; it cuts the conic section *AN*; let it cut it at the point *N*; let the conic section be *EN*. Whether it cuts it or not is a matter we shall deal with later. We draw from the point *N* a perpendicular *NL*, the point *L* will thus lie inside the hyperbola *ABC*, because it lies inside the conic section *AN* and the axes of the two sections are <one and the> same straight line. In the case of the ellipse, the point *L* lies between the two points *E* and *A*, because the axis *AE* is common to the two conic sections *AN* and *EN*. We put *LP* equal to *LA* and we draw *PC* as an ordinate, we join *AC* and we extend *NL*; let it meet the straight line *AC* at the point *K*. We join *EK*; let it cut the outline of the conic section at the point *B*. We draw from the point *B* the straight line *BH* to be a tangent to the conic section.

I say that the straight line *BH* is equal to the straight line *F*.

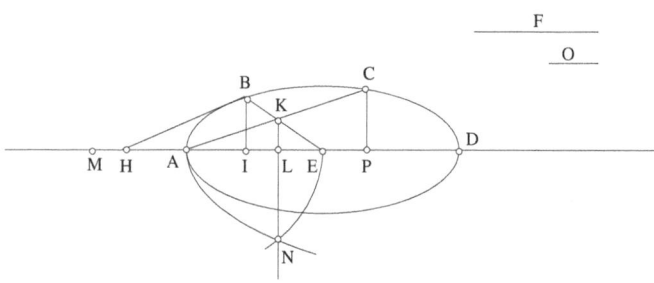

Fig. I.17.2

Proof: We draw *BI* as an ordinate, then the product of *LE* and *EA* will be equal to the square of *EI* and the product of *LE* and *EA* is equal to the square of *NL*; so the straight line *NL* is equal to the straight line *EI*. But the ratio of the product of *ML* and *LA* to the square of *LN* is equal to the ratio of *MA* to the straight line *O* which is a ratio compounded from the ratio of *ME* to *EA* and the ratio of the square of *F* to the square of *EA*, so the ratio of the product of *ML* and *LA* to the square of *EI* is compounded from the ratio of *ME* to *EA* and the ratio of the square of *F* to the square of *EA*. But the ratio of the product of *ML* and *LA* to the square of *EI* is compounded from the ratio of the product of *ML* and *LA* to the square of *AK* and the ratio of the square of *AK* to the square of *EI*. Now the ratio of the product of *ML* and *LA* to the square of *AK* is equal to the ratio of *ME* to *EA*, so it comes out that the ratio of the square of *AK* to the square of *EI* is equal to the ratio of the square of *F* to the square of *EA*; so the ratio of *AK* to *EI* is equal to the ratio of *F* to *EA*. So the ratio of the product of *AK* and *EH* to the product of *IE* and *EH* is equal to the ratio of *F* to *EA*. But the product of *AK* and *EH* is equal to the product of *BH* and *AE*, so the ratio of the product of *BH* and *AE* to the product of *IE* and *EH* is equal to the ratio of *F* to *EA*. But the product of *IE* and *EH* is the square of *EA*, so the ratio of <the product of> *BH* and *AE* to the square of *AE* is equal to the ratio of *F* to *EA*. But the ratio of the product of *BH* and *AE* to the square of *AE* is equal to the ratio of *BH* to *AE*, so the ratio of *BH* to *AE* is equal to the ratio of *F* to *AE*; the straight line *BH* is thus equal to the straight line *F*. That is what we wanted to prove.

The discussion of this problem is that the two conic sections *AN* and *EN* cut one another in every case. This is clear for the ellipse since their concave sides face one another and they have an axis in common. For the hyperbola, the straight line that does not meet the conic section,[42] <a line> drawn from the centre of the conic section, which is the midpoint of the straight line *MA*, <this line that does not meet the hyperbola> cuts the parabola *EN*, crosses it and becomes infinitely distant, and the hyperbola *AN* comes indefinitely close to the straight line which the conic section *EN* meets; so it (the hyperbola) cuts the parabola *EN* before coming close to the straight line it does not meet. The two conic sections must necessarily cut one another, and they cut one another on each side in a single point. The problem can thus be solved in all cases without <imposing> any condition, and it has a single solution on each side <of the axis>. That is what we wanted to prove.

– **18** – If a conic section is known and if we take two points on its axis, <the problem is> how to draw from these two points two straight lines that meet one another on the outline of the conic section and are such that the ratio of one to the other is equal to a given ratio.

---

[42] The line concerned is the asymptote to the hyperbola with axis *AM*.

Let *ABC* be the conic section, *DAE* its axis, *D* and *E* the two points; let the given ratio be the ratio of *I* to *K*. We wish to draw from the two points *D* and *E* two straight lines that meet one another on the outline of the conic section and are such that the ratio of one to the other is equal to the ratio of *I* to *K*.[43]

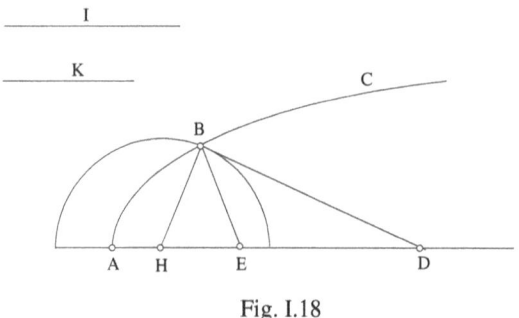

Fig. I.18

We suppose this is to be done by means of analysis: let the two straight lines be *DB* and *EB*. We make the angle *DBH* equal to the angle *BEH*; the two triangles *HBE* and *DHB* are thus similar, so the ratio of *DH* to *HB* is equal to the ratio of *BH* to *HE* and equal to the ratio of *DB* to *BE*. So the ratio of *DH* to *HE* is equal to the ratio of the square of *DH* to the square of *HB* and is equal to the ratio of the square of *DB* to the square of *BE*. So the ratio of *DH* to *HE* is equal to the ratio of the square of *I* to the square of *K*, which is known. So the point *H* is known and the product of *DH* and *HE* is known. But the product of *DH* and *HE* is equal to the square of *HB*, so the straight line *HB* is known. Now the point *H* is known; so the point *B* is known, because the point *B* lies on the circumference of a known circle whose centre is *H* and its semidiameter is known, and *B* lies on the outline of the conic section which is known in position.

Analysis has thus resulted in finding a known point that is the point at which the two straight lines we seek meet one another. Which is what was required.

---

[43] The circle, the set of points *B* such that $\frac{BD}{BE} = IK$ has been investigated in *Analysis and Synthesis*, Problem 1. The centre of the circle is found here as in Problem 1. The similarity of the two proofs is an important argument in favour of *The Completion* being authentic.

– **19** – The synthesis of this problem is carried out in this manner: let the conic section be *ABC*; let the ratio be the ratio of *I* to *K*. We put the ratio of *DH* to *HE* equal to the ratio of the square of *I* to the square of *K*, and we put the product of *DH* and *EH* equal to the square of *HM*. Let us take *H* as centre and construct a circle with distance *HM*; let it cut the outline of the conic section at the point *B*; let the circle be *LBM*. We join *DB* and *BE*.

I say that the ratio of *DB* to *BE* is equal to the ratio of *I* to *K*.

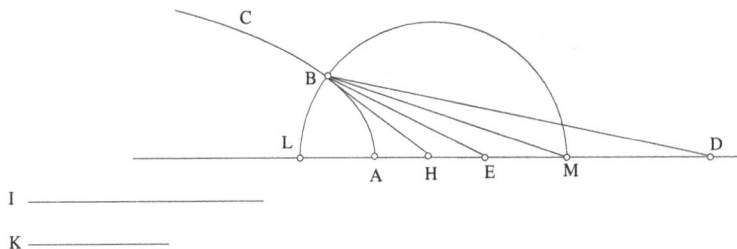

Fig. I.19.1

Proof: Let us join *HB*, the product of *DH* and *EH* is equal to the square of *HB*, so the ratio of *DH* to *HB* is equal to the ratio of *BH* to *HE*, the two triangles *DHB* and *HBE* are thus similar, so the ratio of *DB* to *BE* is equal to the ratio of *DH* to *HB*. So the ratio of the square of *DB* to the square of *BE* is equal to the ratio of the square of *DH* to the square of *BH*. But the ratio of the square of *DH* to the square of *HB* is equal to the ratio of *DH* to *HE*, which is equal to the ratio of the square of *I* to the square of *K*; so the ratio of *DB* to *BE* is equal to the ratio of *I* to *K*. That is what we wanted to prove.

As for the discussion, it will be as follows.

<1> If *I* is greater than *K* and if one of the two points is the vertex of the conic section and the other lies outside the conic section as in the first figure, then the problem can be solved in all cases, <that is> without <imposing any> condition, because the point *H* will lie inside the conic section and the point *M* outside the conic section; the first term of the ratio will thus be on the side exterior to the conic section.

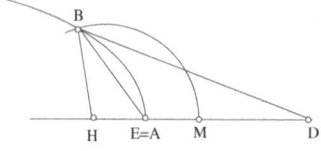

Fig. I.19.2

<2> If one of the two points is the vertex of the conic section and the other lies inside the conic section, as in the second figure, the problem can again be solved in all cases because the point *H* will lie outside the conic section and the point *M* will lie inside the conic section; the first term of the ratio will be on the side of the interior of the conic section.

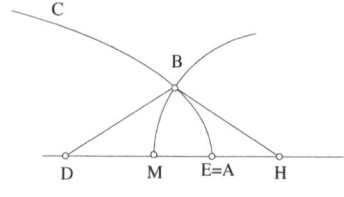

Fig. I.19.3

<3> If the two points lie outside the conic section as in the third figure, the problem can be solved unless we add a condition: the ratio of *DE* to *EA* cannot be smaller than the given ratio, because if we put the ratio of *DH* to *HE* equal to the ratio of the square of *I* to the square of *K*, then the point *H* will lie inside the conic section, or it will lie on the outline of the conic section, or it will lie outside the conic section. But the point *M* always lies between the two points *D* and *E*. If the point *H* lies inside the conic section or on the outline of the conic section, it is clear that the problem can be solved because the centre of the circle will lie inside the conic section, or on the outline of the conic section, and the circumference of the circle <accordingly> lies <partly> outside the conic section, so the circle cuts the conic section. If the point *H* lies outside the conic section, it thus lies between the two points *E* and *A*, if the first term of the ratio is on the side of the point *D*.[44] But the ratio of *DE* to *EA* is not smaller than the ratio of *I* to *K*, so the ratio of *DH* to *HA* is greater than the ratio of *I* to *K*, so the ratio

---

[44] That is, if $\dfrac{I}{K} > 1$ and the points are in the order *A*, *E*, *D*.

of *DH* to *HM* is smaller than the ratio of *DH* to *HA*, so the straight line *HM* is greater than the straight line *HA*; the circle with centre *H* and semi-diameter *HM* thus cuts the conic section and the problem is solved.

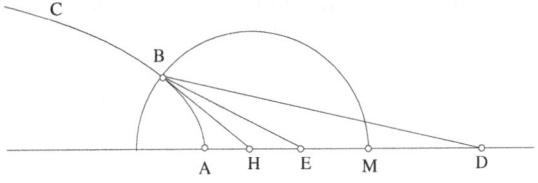

Fig. I.19.4

<3'> If the first term of the ratio is on the side of the point *E*, it is the greater, so the problem cannot be solved in any way because the point *H* will lie further away from the conic section than the point *D* and the point *M* will lie between the two points *D* and *E*.

<4> If the two points lie inside the conic section as in the fourth figure, then the discussion of this figure is like the discussion of the third figure, that is the point *H* lies outside the conic section or it lies on the outline of the conic section, or it lies between the two points *A* and *E*.[45]

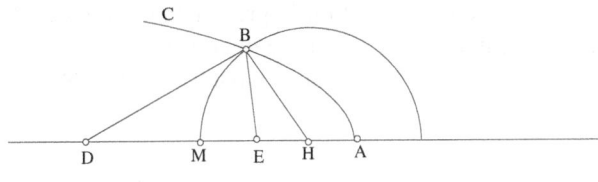

Fig. I.19.5

If the point *H* lies outside the conic section, or on its outline, then the problem can be solved in all cases, and if the point *H* lies between the two points *E* and *A*, the discussion of the problem is that the ratio of *DE* to *EA* must not be smaller than the ratio of *I* to *K* and the first term of the ratio will lie inside the conic section.

---

[45] This implies that, as in the case where the points lay outside the conic section, Ibn al-Haytham supposes that $\dfrac{I}{K} > 1$ and that the points are in the order *A*, *E*, *D*.

<5> If the two points are such that one lies outside the conic section and the other lies inside the conic section as in the fifth figure, then the discussion of the problem is that the straight line *HA* must be smaller than the straight line *HM* and the first term of the ratio lies on the side interior to the section; or that *HM* must not be smaller than the shortest straight line drawn from the point *H* to the conic section and the first term of the ratio lies on the side exterior to the conic section.

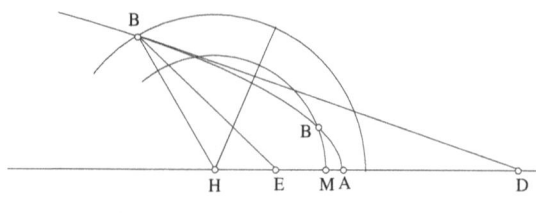

Fig. I.19.6

<6> If *I* and *K* are equal, then the discussion of this problem is that the two points must lie inside the conic section, or that one of them must lie inside the conic section and the other on the outline of the conic section, or that one of them must lie inside the conic section and the other outside the conic section and in such a position that the part of the straight line *DE* which lies inside the conic section is greater than half *DE*. It is thus clear that the problem can be solved in these three ways and this is the discussion of all the cases of this problem. That is what we wanted to prove.

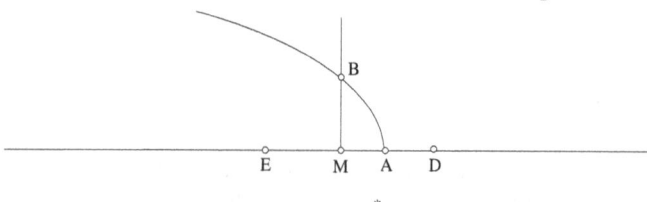

Fig. I.19.7*

– **20** – If we have a given conic section and if we take two points on its axis, <the problem is> to draw from these two points two straight lines which meet on the outline of the conic section and are such that their sum is equal to a given straight line.

---

* The manuscript contains five figures.

Let the conic section be *ABC*, the two points *D* and *E* and the given straight line *G*. We wish to draw from the two points *D* and *E* two straight lines that meet one another on the outline of the conic section and are such that their sum is equal to the straight line *G*.

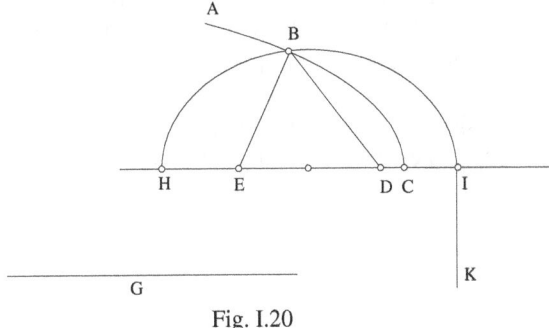

Fig. I.20

We suppose this is to be done by means of analysis: let the two straight lines be *DB* and *EB*; we put each <of the straight lines> *EH* and *DI* equal to half of the amount by which *G* exceeds the straight line *DE*.[46] We put the product of *HI* and *IK* equal to four multiplied by the product of *HD* and *DI*. We construct through the two points *H* and *I* the ellipse whose axis is *HI* and whose *latus rectum* is *IK*; so it passes through the point *B* as has been shown in Proposition 52 of the third book. Let this conic section be the conic section *HIB*. But the straight line *DE* is known in magnitude and position, so the straight line *HI* is of known magnitude and the product of *HD* and *DI* is of known magnitude; so the straight line *IK* is of known magnitude and the conic section *HIB* is known in position; but the conic section *ABC* is known in position, <and> the point *B* is thus known. Analysis has resulted in finding a known point which gives a construction for the problem.

– **21** – The synthesis of this problem is to cut off from *G* a straight line equal to *DE*; we take half of what remains and we add to *DE* an amount that is equal to the two halves, on either end <of the line, that is at> *D* <and at> *E*, and let them (the added halves) be *EH* and *DI*; so *HI* will be equal to *G*. We put the product of *HI* and *IK* equal to four times the product of *HD*

---

[46] Ibn al-Haytham is thus assuming $G > DE$, a condition that is necessary for the triangle *BDE* to exist.

and *DI*, we construct on the straight line *HI* the ellipse whose axis is *HI* and *latus rectum IK*; let it cut the conic section *ABC* at the point *B*. Whether it cuts it or does not cut it <is a matter> we shall examine later. We join *DB* and *BE*, thus their sum will be equal to the straight line *HI* as has been shown in Proposition 52 of the third book. But *HI* is equal to the straight line *G*, so the straight lines *DB* and *BE* have <a sum> equal to the straight line *G*. That is what we wanted to prove.

The discussion of the problem is as follows. In every case the straight line *G* must be greater than the straight line *DE*.

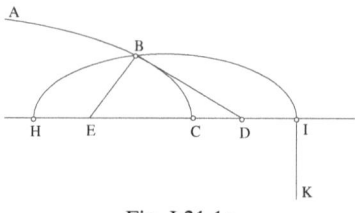

Fig. I.21.1a

<1> If one of the two points *D* and *E* lies outside the conic section and the other lies inside the conic section, or if one of the two points lies outside the conic section and the other is at the vertex of the conic section, or one of the two lies inside the conic section and the other at the vertex of the conic section, then the problem can be solved in all cases without imposing any condition, because in all these cases, it happens that one of the points *H*, *I* lies inside the conic section and the other outside the conic section;[47] so the ellipse cuts the given conic section in every case.

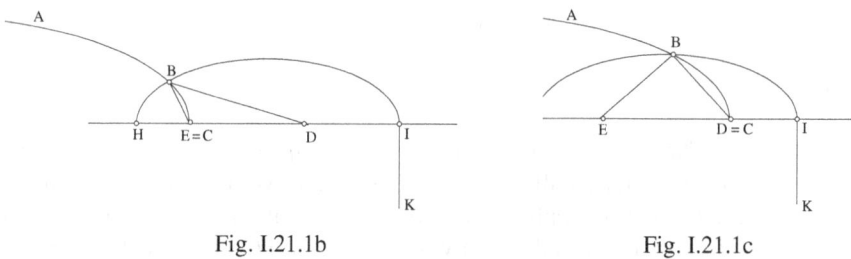

Fig. I.21.1b   Fig. I.21.1c

---

[47] This assumes that the given conic section is a parabola or a hyperbola (see Mathematical commentary).

<2> If the two points *D* and *E* both lie outside the conic section and if the straight line that runs between whichever of the two <points> is closer to the conic section and the vertex of the conic section is smaller than half the amount by which the straight line *G* exceeds the straight line *DE*, then the problem can again be solved without imposing any condition, because one of the two endpoints of the axis of the ellipse will lie inside the conic section and the other endpoint will lie outside. If the straight line that runs between the vertex of the conic section and whichever of the two points is the closer to it (the vertex) is not smaller than half the amount by which the straight line *G* exceeds the straight line *DE*, then the problem cannot be solved because the whole ellipse lies outside the conic section.

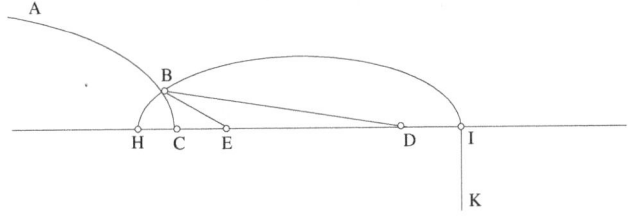

Fig. I.21.2

<3> If the two points *D* and *E* both lie inside the conic section *ABC* and if the straight line that runs between the vertex of the conic section and whichever of the two points is closer to it (the vertex) is smaller than half the amount by which the straight line *G* exceeds the straight line *DE*, then the problem can again be solved without imposing any condition, because one of the endpoints of the axis of the ellipse lies outside the conic section and the other inside.

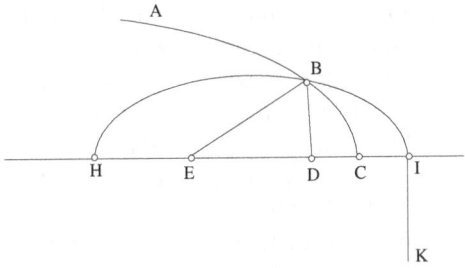

Fig. I.21.3

– **22** – If the two points *D* and *E* lie inside the conic section *ABC* and if the straight line that runs between the vertex of the conic section and whichever of the two points is closer to it (the vertex) is not smaller than half the amount by which the straight line *G* exceeds the straight line *DE*, then the problem cannot be solved unless we impose a condition. This condition is, if the given conic section is a parabola, that the ratio of the square of half the diameter of the ellipse to the product of the straight line that runs between the centre of the ellipse and the vertex of the parabola and the *latus rectum* of the parabola is not smaller than the ratio of the diameter of the ellipse to its *latus rectum*.

Let us return to the figure. We divide the straight line *HI* into two equal parts at the point *M*; we join *HK* and we draw *MN* parallel to *IK*; let the *latus rectum* of the parabola be *FA*.

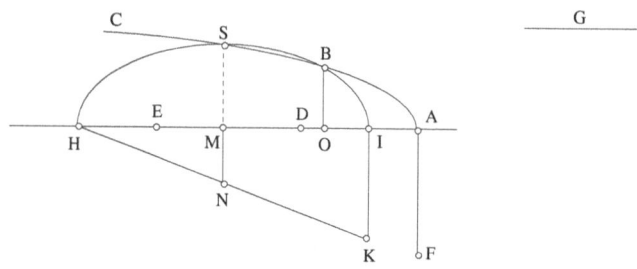

Fig. I.22.1

<1> If the ratio of the square of *HM* to the product of *MA* and *FA* is equal to the ratio of *HI* to *KI*, then the ratio of the square of *HM* to the product of *MA* and *FA* is equal to the ratio of *HM* to *MN*. But the ratio of *HM* to *MN* is equal to the ratio of the product of *HM* and *MI* to the product of *MN* and *MI*, so the ratio of the square of *HM* to the product of *MN* and *MI* is equal to the ratio of the square of *HM* to the product of *MA* and *FA*. The product of *MN* and *MI* is thus equal to the product of *MA* and *FA*. But the product of *MN* and *MI* is the square of the ordinate drawn from the point *M* to the outline of the ellipse and the product of *MA* and *FA* is the square of the ordinate drawn from the point *M* to the outline of the parabola. The ordinate drawn from the point *M* to the outline of the parabola is equal to the ordinate drawn from the point *M* to the outline of the ellipse. So the ellipse cuts the parabola at one end of the ordinate drawn from the point *M*, which is the perpendicular axis.

<2> If the ratio of the square of *HM* to the product of *MA* and *FA* is greater than the ratio of *HI* to *IK*, then the ratio of a part of the square of *HM* to the product of *MA* and *FA* is equal to the ratio of *HI* to *IK*; let this part be the product of *HO* and *OI*. So the ratio of the product of *HO* and *OI* to the product of *MA* and *FA* is equal to the ratio of *HI* to *IK*. We draw the perpendicular *OP*; the ratio of the product of *HO* and *OI* to the product of *OP* and *OI* is thus equal to the ratio of *HI* to *IK*; the ratio of the product of *HO* and *OI* to the product of *MA* and *FA* is equal to the ratio of the product of *HO* and *OI* to the product of *PO* and *OI*, and the product of *PO* and *OI* is equal to the product of *MA* and *FA*. But the product of *MA* and *FA* is greater than the product of *OA* and *FA*, so the product of *PO* and *OI* is greater than the product of *OA* and *FA*. But the product of *PO* and *OI* is the square of the ordinate drawn from the point *O* to the outline of the ellipse and the product of *OA* and *FA* is the square of the ordinate drawn from the point *O* to the outline of the parabola. Consequently, the ordinate drawn from the point *O* to the outline of the ellipse is greater than the ordinate drawn from the point *O* to the outline of the parabola. So the parabola cuts the ordinate of the ellipse drawn from the point *O* <that is> inside the ellipse, so it cuts the outline of the ellipse before cutting the ordinate. If it cuts the outline of the ellipse before cutting the ordinate, then it <also> cuts it in another point after <cutting> the ordinate.

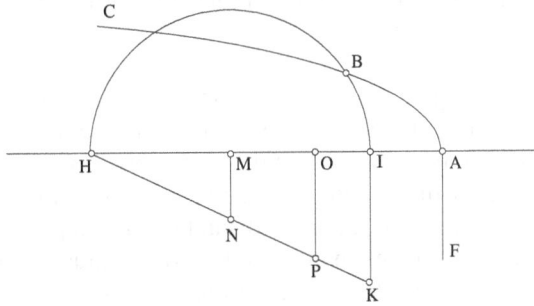

Fig. I.22.2

<3> Similarly, I say that the parabola, if it cuts the ellipse at the end of the perpendicular axis, then it <also> cuts the ellipse in another point before the end of the perpendicular axis.

Proof: We put the product of *AM* and *MO* equal to the square of *MI*, then the ratio of *AM* to *MO* is equal to the ratio of the square of *MI* to the square of *MO*. If we invert and compound the ratio, then the ratio of *MA* to *AO* will be equal to the ratio of the square of *MI*, which is the product of *HM* and *MI*, to the product of *HO* and *OI*. But the ratio of *MA* to *AO* is equal to the ratio of the square of the ordinate drawn from the point *M* to the outline of the parabola, to the square of the ordinate drawn from the point *O* to the outline of the parabola and the ratio of the product of *HM* and *MI* to the product of *HO* and *OI* is equal to the ratio of the square of the ordinate drawn from the point *M* to the outline of the ellipse to the square of the ordinate drawn from the point *O* to the outline of the ellipse. So the ratio of the ordinate drawn from the point *M* to the outline of the ellipse to the ordinate drawn from the point *M* to the outline of the parabola is equal to the ratio of the ordinate drawn from the point *O* to the outline of the ellipse to the ordinate drawn from the point *O* to the outline of the parabola. But the ordinate drawn from the point *M* to the outline of the parabola is equal to the ordinate drawn from the point *M* to the outline of the ellipse, so the ordinate drawn from the point *O* to the outline of the parabola is equal to the ordinate drawn from the point *O* to the outline of the ellipse. So the parabola cuts the ellipse at the endpoint of the ordinate drawn from the point *O*.

So if the ratio of the square of *HM* to the product of *MA* and *FA* is not smaller than the ratio of *HI* to *IK*, then the two conic sections cut one another in all cases and the problem has two solutions in all cases. That is what we wanted to prove.

– 23 – If the given conic section is a hyperbola, then the discussion of the problem is that the ratio of the square of the semidiameter of the ellipse to the product of the straight line that runs between the centre of the ellipse and the vertex of the hyperbola and the straight line that runs between the centre of the ellipse and the more distant of the endpoints of the axis of the hyperbola is not smaller than the ratio compounded from the ratio of the diameter of the ellipse to its *latus rectum* and the ratio of the *latus rectum* associated with the axis of the hyperbola to its transverse diameter.

Let us return to the figure; we divide the straight line *HI* into two equal parts at the point *M*. Let *AG* be the axis of the hyperbola, *AF* its *latus rectum*. We join *HK* and *GF*, we draw *MN* parallel to *IK*, we extend it and we extend *GF* until the two lines meet at the point *U*. We put the ratio of *SA* to *AF* equal to the ratio of *HI* to *IK*; so the ratio of *SA* to *AG* is compounded from the ratio of *HI* to *IK* and the ratio of *FA* to *AG*.

I say that if the ratio of the square of *HM* to the product of *AM* and *MG* is not smaller than the ratio of *SA* to *AG*, then the problem can be solved, and if the ratio of the square of *HM* to the product of *AM* and *MG* is smaller than the ratio of *SA* to *AG*, then the problem cannot be solved.

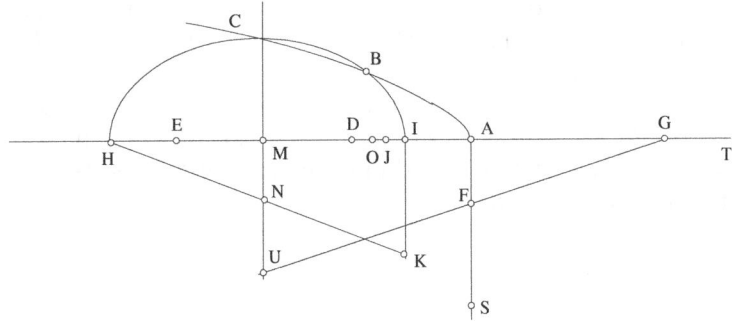

Fig. I.23.1[48]

Proof: <1> If the ratio of the square of *HM* to the product of *AM* and *MG* is equal to the ratio of *SA* to *AG*, then the ratio of the square of *HM* to the product of *MA* and *MG* is compounded from the ratio of the product of *HM* and *MI* to the product of *NM* and *MI* and the ratio of the product of *NM* and *MI* to the product of *MA* and *MG*. But the ratio of the product of *HM* and *MI* to the product of *NM* and *MI* is equal to the ratio of *HI* to *IK* which is equal to the ratio of *SA* to *AF*. So the ratio of the product of *NM* and *MI* to the product of *MA* and *MG* is equal to the ratio of *FA* to *AG*, which is equal to the ratio of *UM* to *MG*, which is equal to the ratio of the product of *UM* and *MA* to the product of *MA* and *MG*. So the ratio of the product of *NM* and *MI* to the product of *MA* and *MG* is equal to the ratio of the product of *UM* and *MA* to the product of *MA* and *MG* and the product of *NM* and *MI* is equal to the product of *UM* and *MA*. But the product of *NM* and *MI* is the square of the ordinate drawn from the point *M* to the outline of the ellipse, and which is the perpendicular axis, and the product of *UM* and *MA* is the square of the ordinate drawn from the point *M* to the outline of the hyperbola. Consequently, the ordinate drawn from the point *M* to the outline of the hyperbola is the ordinate drawn from the point *M* to the

---

[48] The figure has been divided into two separate parts to distinguish between the three cases that are considered and to avoid the confusion that might arise from using the same letters for different points.

outline of the ellipse, the two conic sections thus meet one another at the endpoint of the perpendicular axis.

<2> If the ratio of the square of *HM* to the product of *MA* and *MG* is greater than the ratio of *SA* to *AG*, then the ratio of *SA* to *AG* is equal to the ratio of a part of the square of *HM* to the product of *MA* and *MG*; let this part be the product of *HO* and *OI*. Thus the ratio of the product of *HO* and *OI* to the product of *MA* and *MG* is equal to the ratio of *SA* to *AG*. But the product of *MA* and *MG* is greater than the product of *OA* and *OG*, so the ratio of the product of *HO* and *OI* to the product of *OA* and *OG* is greater than the ratio of *SA* to *AG*.

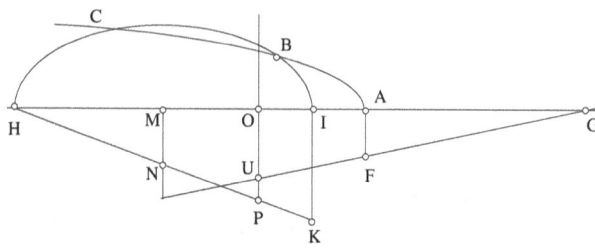

Fig. I.23.2

We draw *OP* parallel to the straight line *IK*, then the ratio of the product of *HO* and *OI* to the product of *OA* and *OG* is equal to the ratio compounded from the ratio of the product of *HO* and *OI* to the product of *PO* and *OI* and the ratio of the product of *PO* and *OI* to the product of *OA* and *OG*. But the ratio of the product of *HO* and *OI* to the product of *PO* and *OI* is equal to the ratio of *HO* to *OP*, which is equal to the ratio of *HI* to *IK*, which is equal to the ratio of *SA* to *AF*. So the ratio of the product of *PO* and *OI* to the product of *OA* and *OG* is greater than the ratio of *FA* to *AG*, which is equal to the ratio of *UO* to *OG*, which is equal to the ratio of the product of *UO* and *OA* to the product of *OA* and *OG*. So the ratio of the product of *PO* and *OI* to the product of *OA* and *OG* is greater than the ratio of the product of *UO* and *OA* to the product of *OA* and *OG*, and the product of *PO* and *OI* is greater than the product of *UO* and *OA*. But the product of *PO* and *OI* is the square of the ordinate drawn from the point *O* to the outline of the ellipse and the product of *UO* and *OA* is the square of the ordinate drawn from the point *O* to the outline of the hyperbola, so the ordinate drawn from the point *O* to the outline of the ellipse is greater than the ordinate drawn from the point O to the outline of the hyperbola; so the hyperbola cuts the ordinate drawn from the point O <at a point> inside the ellipse, and consequently cuts the outline of the ellipse before reaching the ordinate, and if it cuts the outline of the ellipse before reaching the ordinate – such as the ordinate drawn from the point O – then the hyperbola cuts the outline of the ellipse in another point as it exits from the ellipse. Thus, in this case, the hyperbola cuts the ellipse in two points.

<3> Similarly, I say that if the hyperbola cuts the ellipse at the endpoint of the perpendicular axis, it cuts it in another point before the end of the perpendicular axis.

Indeed if we put the ratio of IM to MO equal to the ratio of AM to MI, then the ratio of AI to IO is equal to the ratio of AM to MI; thus the ratio of OI to IA is equal to the ratio of IM to MA. But the ratio of IM to MA is greater than the ratio compounded from the ratio of IM to MA and the ratio of IM to MG which is the ratio of the square of IM – that is HM – to the product of MA and MG, which is the ratio of SA to AG. Let us put the ratio of OJ to JA equal to the ratio of SA to AG. Since the ratio of OM to MI is equal to the ratio of IM to MA, the ratio of OM to MG is compounded from the ratio of IM to MA and the ratio of IM to MG, which is the ratio of SA to AG. So the ratio of OJ to JA is equal to the ratio of OM to MG. Let us put GT equal to JA; so the ratio of JM to MT will be equal to the ratio of SA to AG, which is equal to the ratio of the square of IM to the product of MA and MG. But the ratio of JM to MT is equal to the ratio of the square of JM to the product of JM and MT. So the ratio of the square of JM to the product of JM and MT is equal to the ratio of the square of IM to the product of MA and MG. But the product of JM and MT is the amount by which the product of MA and MG exceeds the product of JA and JG; so the ratio of the square of IM to the product of MA and MG is equal to the ratio of the square of JM to the product of JM and MT and is equal to the ratio of the remainder of the square of IM, which is the product of HJ and JI, to the remainder of the product of MA and MG, which is the product of JA and JG, so the ratio of the square of HM, which is equal to the product of HM and MI, to the product of MA and MG, is equal to the ratio of the product of HJ and JI to the product of JA and JG. The ratio of the square of the ordinate drawn from the point M to the outline of the ellipse to the square of the ordinate drawn from the point J to the outline of the ellipse is accordingly equal to the ratio of the product of MA and MG to the product of *JA* and *JG*, which is the ratio of the square of the ordinate drawn from the point *M* to the outline of the hyperbola to the square of the ordinate drawn from the point *J* to the outline of the hyperbola. The ratio of the two ordinates of the ellipse drawn from the two points *M* and *J*, <their ratio> one to another, is thus equal to the ratio of the two ordinates of the hyperbola drawn from the two points *M* and *J*, <their ratio> one to another. But the ordinate of the ellipse drawn from the point *M* is equal to the ordinate of the hyperbola drawn from the point *M*, so the ordinate of the ellipse drawn from the point *J* is equal to the ordinate of the hyperbola drawn from the point *J*. The two conic sections accordingly meet one another on the ordinate drawn from the point *J*.

It is thus clear from what we have shown that, if the ratio of the square of *HM* to the product of *MA* and *MG* is not smaller than the ratio of *SA* to *AG*, then the two conic sections meet one another in two points, and if the two conic sections meet one another in two points, then the problem has two solutions. That is what we wanted to prove.

If the given conic section is an ellipse, the method of solving the problem is derived from the method we have set out for the hyperbola and the discussion is the same as for the hyperbola, without needing to add or remove anything.[49]

– **24** – The conic section *ABC* is a hyperbola whose centre is *H*. We wish to find a diameter of the conic section that, together with its *latus rectum*, encloses a known area.

We suppose this is to be done by means of analysis: let the required diameter be *BP* and the known area the square of *EG*. The product of *BP* and its *latus rectum* is thus equal to the square of *EG*, and *EG* is the right

---

[49] See Mathematical commentary.

diameter conjugate to the diameter *BP*.[50] We put *AI* as the *latus rectum* of the axis <*AD*>; the product of *AD* and *DI* is thus the difference between the squares of the two axes.

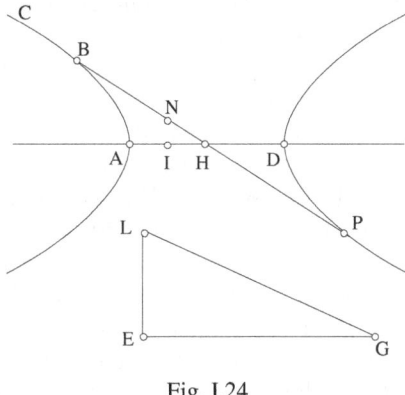

Fig. I.24

At the point *E* we construct the straight line *EL* perpendicular to *EG* and we put its square equal to the product of *AD* and *DI*. We join *LG* and we put *PN* as the *latus rectum* of the diameter *BP*, then the product of *BP* and *PN* is equal to the square of *EG* and the product of *PB* and *BN* is the difference of the squares of the two conjugate diameters; but the difference between the squares of any pair of conjugate diameters of the hyperbola, is the difference between the squares of its two axes, as has been shown in Proposition 13 of the seventh book. So the product of *BP* and *BN* is equal to the square of *EL*, and the square of *BP* is equal to the square of *LG*. But *LG* is known, so the diameter *PB* is known, *HB* is known and the point *B* is known. But the product of *BP* and *PN* is equal to the square of *EG*, which is known; so the straight line *PN* is known and it is the *latus rectum* of the diameter *BP*. That is what we wanted to find.

---

[50] In the case of the hyperbola, the diameter Δ conjugate to a transverse diameter *D* is called a right diameter (see *Les Coniques d'Apollonius de Perge*, transl. P. Ver Eecke, Book VII.6, note 4, p. 557). From Book I, Second Definitions, def. 3, we have $\Delta^2 = D \cdot a$, if *a* is the *latus rectum* of *D*. See our edition, *Les Coniques*, tome 1.1, p. 255, 22–25; tome 1.2, p. 8–7 and tome 1.4, p. 364.

– **25** – The synthesis of this problem is as follows:

Let *ABC* be the hyperbola with axis *AD* and *latus rectum AI* and let there be a given straight line *EG*. We wish to find the diameter of the conic section that, together with the *latus rectum*, encloses an area equal to the square of *EG*. We construct at the point *E* of the straight line *EG* the perpendicular *EL*; we put the square of *EL* equal to the product of *AD* and *DI* and we join *LG*. The square of *LG* is thus equal to <the sum of> the squares of *GE* and *EL*, the square of *LG* thus exceeds the square of *EG* by the area <of the rectangle> enclosed by the two straight lines *AD* and *DI*. We put *HK* equal to half *GL*. We take *H* as centre and we construct with distance *HK* an arc of a circle – let it be *KB*; let it cut the outline of the conic section at the point *B*. We join *HB*, we extend it on the side of *H* as far as the <point> *P* and we put *PH* equal to *HB*, *PB* is thus equal to *GL*. We put the product of *BP* and *PN* equal to the square of *EG*, there remains the product of *PB* and *BN*, <which is> equal to the square of *EL*; so the product of *PB* and *BN* is the difference between the square of *BP* and the square of the right diameter which is conjugate to it. So the straight line *PN* is the *latus rectum* of the diameter *BP*. But the product of *BP* and *PN* is equal to the square of *EG*. That is what we wanted to prove.

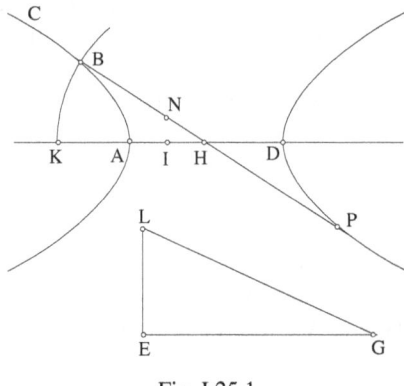

Fig. I.25.1

The discussion of this problem is that the square of *EG* must be greater than the area enclosed by the two straight lines *DA* and *AI*. From this it then follows that the square of *GL* is greater than the square of *DA*, so that half *GL* is greater than *HA*; so the point *K* lies inside the conic section, the arc *KB* cuts the outline of the conic section in every case and the problem can be solved in every case, with the condition for the magnitude of *EG*, that is that its square exceeds the product of *DA* and *AI*, which is the square of the right axis. If the axis <*AD*> is smaller than its *latus rectum*, then each of the diameters is smaller than its *latus rectum*, as has been shown in Proposition 22 of the seventh book. But any diameter is greater than the axis, so the straight line *EG* must be greater than the perpendicular axis. We cut off from the straight line *EG* a straight line such that its square is equal to the amount by which the square of *GE* exceeds the product of *AD* and *DI*,[51] let it be *GO*. We put the product of *MG* and *GO* equal to the square of *GE*; so the ratio of *MG* to *GO* is equal to the ratio of the square of *EG* to the square of *GO*, which is the amount by which the square of *EG* exceeds the product of *AD* and *DI*; so the ratio of *GM* to *MO* is equal to the ratio of the square of *EG* to the product of *AD* and *DI*, so *GO* is the diameter and *GM* is the *latus rectum*, and the construction is completed as before.

The discussion of this problem is that *GO* must be greater than the axis.[52]

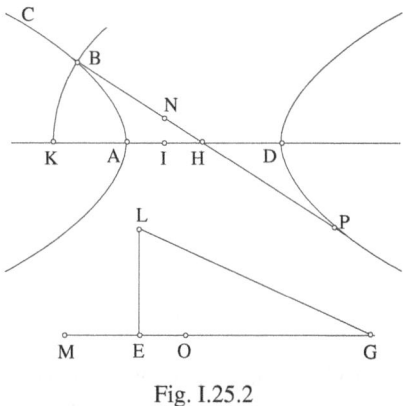

Fig. I.25.2

If the axis is equal to its *latus rectum*, then every diameter is equal to its *latus rectum*, as has been shown in Proposition 23 of the seventh book; *EG* is divided into two equal parts, and <each> half <of it> is half the diameter. The discussion of the problem is that *EG* is greater than the axis.

This idea, that is that the product of the transverse diameter and its *latus rectum* is known,[53] is possible for the ellipse; the method for

---

[51] This assumes that $GE^2 > AD \cdot DI$.
[52] $GO > AD \Leftrightarrow GE^2 > AD \cdot AI$ (see Mathematical commentary).

obtaining it is easier than <the one> for the hyperbola, because the sum of the squares of any pair of conjugate diameters of the ellipse is equal to the sum of the squares of the two axes – this has been shown in Proposition 12 of Book VII. So if the product of the diameter and its *latus rectum* is known or if the square of the right diameter is known, and the sum of the two diameters being known, because the two axes are known, then the result is that the square of the transverse diameter is known, so it is possible and easy to find it.

The discussion of this problem is that the straight line *EG* must be greater than the minor axis.[54]

– **26** – The hyperbola *ABC* is known, <and> has axis *AD*, and the straight line *EG* is known. We wish to find a diameter of this conic section that, <when> added to its *latus rectum*, is equal to the straight line *EG*.

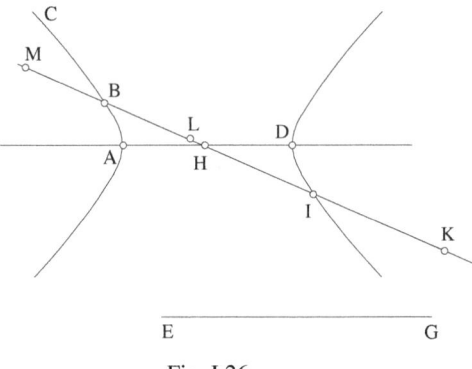

Fig. I.26

We suppose this is to be done by means of analysis: let the diameter be the straight line *BHI* and let its *latus rectum* be *IK*, then *BK* is known. We put *IL* equal to *IK*; the product of *IB* and *BL* will then be the difference between the squares of the two conjugate diameters, because the product of *BI* and *IK* is equal to the square of the conjugate diameter. So the product of *IB* and *BL* is known, because it is equal to the difference between the squares of the two conjugate axes, as has been shown in Proposition 13 of

---

[53] He means that this product is equal to the square of the conjugate diameter.
[54] The length of the straight line *EG* must be intermediate between those of the minor and major axes, a property shared by all diameters of the ellipse.

Book VII. We put BM equal to BL, then KM will be twice IB. But the product of IB and BL is known, so the product of KM and BM is known, because it is twice the product of IB and BL. But KB is known, so the straight line KM is known, so half of it is known, so IB is known, HB is known and the point B is then known.

– **27** – The synthesis of this problem is as follows:

Let the conic section be ABC with axis AD, with centre H and the known straight line EG. We wish to find a diameter of the conic section that, <when> added to its *latus rectum*, is equal to the straight line EG.

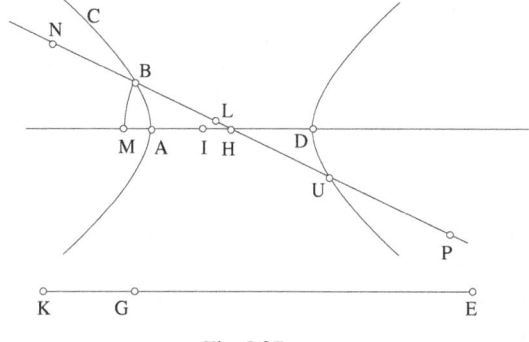

Fig. I.27

We put AI equal to the *latus rectum* of the axis, we put the product of EK and KG[55] equal to twice the product of AD and DI, we put HM equal to a quarter of EK, we take H as centre and with distance HM we construct an arc of a circle; let it be MB. We join HB, we extend HB on both sides and we put BP equal to EG.

I say that BP is equal to the diameter added to the *latus rectum*.

Proof: we cut off BN equal to GK, so PN is equal to EK. We cut off HU equal to HB and we put BL equal to BN, so we are left with LU <which is> equal to UP. So the product of UB and BL is equal to half the product of PN and NB, which is equal to the product of EK and KG, which is twice the product of AD and DI, so the product of UB and BL is equal to the product of AD and DI; so the product of UB and BL is the difference between the square of UB and the square of the right diameter that is conjugate to it, so

---

[55] In this part of his work, Ibn al-Haytham assumes that $AI < AD$ and thus takes K on EG produced, on the side of G.

the product of *BU* and *UL* is the square of the diameter conjugate to the diameter *UB*, so the straight line *UL* is the *latus rectum* of the diameter *UB*. But *UL* is equal to *UP*, so *UP* is the *latus rectum* of the diameter *UB*, so the straight line *BP* is the diameter *BU* added to its *latus rectum*. But *BP* is equal to *EG*, so the diameter *BU* added to its *latus rectum* is equal to the straight line *EG*. That is what we wanted to construct.

The discussion of this problem is that the straight line *EG* must be greater than the sum of the axis and its *latus rectum*, because every diameter of the hyperbola is greater than the transverse axis and its *latus rectum* is greater than the *latus rectum* of the axis. That every diameter is greater than the axis is something that is obvious. That the *latus rectum* of the diameter is greater than the *latus rectum* of the axis is something that follows because it has been proved in Proposition 21 of Book VII that the ratio of any diameter of the hyperbola to its *latus rectum* is smaller than the ratio of the axis to its *latus rectum*.

If the axis is smaller than its *latus rectum*, we divide *EG* into two parts at the point *U'* so that the product of *EU'* and *U'G* is equal to twice the product of *AD* and *DI*. Let us put *HM* equal to a quarter of *EU'*, the construction is completed as before: we shall have *PU* equal to *UN* and *UN* will be the *latus rectum*.

If the axis is equal to the *latus rectum*, we divide *EG* into two equal parts, and one half of it will be the diameter, because if the axis is equal to its *latus rectum*, then every diameter of the conic section is equal to its *latus rectum*.

The discussion of this problem in all cases is that *EG* should be greater than the sum of the axis and its *latus rectum*.

<27a> This idea, that is that the sum of the transverse diameter and its *latus rectum* is equal to a given straight line, is possible and easy <to verify> for the ellipse. Indeed, the sum of the squares of any two conjugate diameters of the ellipse is equal to the sum of the squares of the two axes. But the squares of the two axes are known, so <the sum> of the square of the transverse diameter and its product with its *latus rectum* is known, so the product of the sum of the transverse diameter and its *latus rectum* and the transverse diameter is known. So if the product of the sum of the transverse diameter and its *latus rectum* and the transverse diameter is

known, then the transverse diameter is known, and it will thus be possible and easy to find it.

The discussion of this problem is that the known straight line should be greater than the sum of the greater axis and its *latus rectum* <and smaller than that same sum multiplied by the square root of the ratio of the major axis to the *latus rectum*>.[56]

<27b> We can also easily show how to find a diameter of the hyperbola whose ratio to its *latus rectum* is a known ratio.

Indeed, the difference between the squares of two conjugate diameters in any hyperbola is equal to the difference between the squares of its two axes, as has been shown in Proposition 13 of Book VII. So if the conic section is known, its two axes are known and the difference between their squares is known, so the difference between the square of the diameter and its product with its *latus rectum* is known, because the product of the diameter and the *latus rectum* is equal to the square of the right diameter that is conjugate to it. But the difference between the square of the diameter and its product with its *latus rectum* is the product of the diameter and the amount by which the diameter exceeds its *latus rectum*. If the ratio of the transverse diameter to its *latus rectum* is a known ratio, then the ratio of the transverse diameter to the amount by which that diameter exceeds its *latus rectum* is a known ratio. But its product with this excess amount is known, so the transverse diameter will be known; consequently it is possible and easy to find it.

<27c> Similarly for the ellipse, we <can> easily show how to find a diameter whose ratio to its *latus rectum* is a known ratio.

Indeed the sum of the squares of any two conjugate diameters of the ellipse is known, because it is equal to <the sum> of the two squares of its two axes, as has been proved in Proposition 12 of Book VII. So <the sum> of the square of the transverse diameter and the diameter's product with its *latus rectum* is known. So if the ratio of the transverse diameter to its *latus rectum* is known and if its product with it, plus the square of the transverse diameter, is known, then each of them is known. Thus the diameter whose ratio to its *latus rectum* is known, is <itself> known, <and> in consequence it is possible and easy to find it.

---

[56] This passage at the end of the problem is missing in the manuscript (see Mathematical commentary).

<28> Let there be a known parabola *AB*, with known axis *AD*, the point *D* on its axis outside the conic section and a given straight line *F*. We wish to draw from the point *D* a straight line to cut the conic section in two points and such that the part of it that lies inside the conic section is equal to the given straight line *F*.

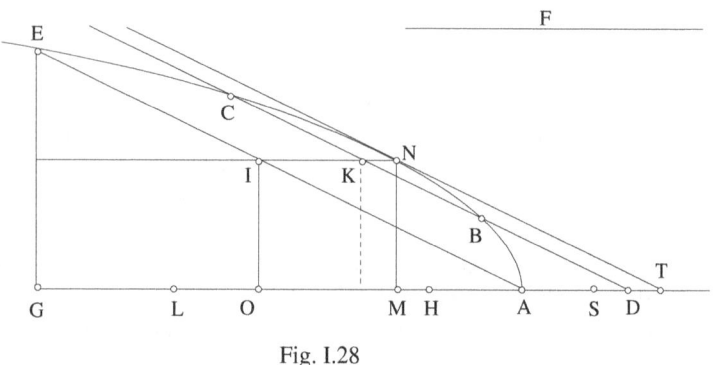

Fig. I.28

We suppose this is to be done by means of analysis: let there be a straight line *DBC*, let *BC* be equal to the straight line *F*. We draw the straight line *AE* parallel to the straight line *DBC*, we draw *EG* as an ordinate and we divide *AE* into two equal parts at the point *I*. We draw from the point *I* a straight line parallel to the axis, let it be *IKN*; thus it will be a diameter of the conic section <and the tangent at the point *N* will be parallel to the straight line *AE*>, as has been shown in Proposition 46 of Book I.[57] We draw *MN* as an ordinate, then *AM* is equal to *NI* because the tangent drawn from the point *N* cuts off from the axis outside the conic section a straight line equal to the straight line *AM*, as has been shown in Proposition 33 of Book I.[58] This straight line cut off by the tangent is equal to the straight line *NI* because the tangent is parallel to the straight line *AE*, so the straight line *AM* is equal to the straight line *NI*; but *KI* is equal to *AD*. We cut off *AH* equal to *AD*, there remains *HM* equal to *NK*. We put *AS* equal to the *latus rectum* of the axis, we have the product of *SG* and *GA* equal to the square of *AE*, as has been shown in Proposition 1 of Book VII. We draw the perpendicular *IO*, so *OM* is equal to *IN*; but *IN* is equal to *AM*,

---

[57] The stated property can be deduced from Proposition I.46.

[58] Propositions 33 and 35 of Book I are converses of one another; the proposition used here is 35.

so *OM* is equal to *MA*, so *OA* is twice *AM*. But *GA* is twice *AO*, so *GA* is four times *AM*, so *GA* is four times *NI*. So the product of *SG* and *NI* is equal to the square of *AI*, so *SG* is the *latus rectum* associated with the diameter *NI*; but *AI* is equal to *DK*, so the product of *SG* and *NI* is equal to the square of *DK*. But since *NI* is a diameter and *BC* is parallel to *AE*, *NI* divides *BC* into two equal parts, the product of *CD* and *DB*, plus the square of *BK*, is thus equal to the square of *DK*, so the product of *SG* and *NI* is equal to the product of *CD* and *DB*, plus the square of *BK*. But the product of *SG* and *NK* is equal to the square of *BK*, because *NK* is a diameter and *SG* is its *latus rectum*, so we are left with the product of *SG* and *IK* equal to the product of *CD* and *DB*; so the product of *SG* and *AH* is equal to the product of *CD* and *DB*. But the product of *SG* and *AM* is equal to the square of *DK*, so we are left with the product of *SG* and *HM* equal to the square of *BK*. We put *AL* equal to four times *AH*, so we have *LG* equal to four times *HM*. But the product of *SG* and *HM* is equal to the square of *BK*, so the product of *SG* and *GL* is equal to the square of *BC* which is known. But *AH* is known, so *AL* is known and *AS* is known, so *SL* is known and the product of *SG* and *GL* is known, so the point *G* is known. But *GE* is <a> perpendicular, so it is known in position. But the conic section is known in position, so the point *E* is known and consequently the straight line *AE* is known in magnitude and in position; so half of it is known and the point *I* is thus known. The straight line *IK* is known in position and in magnitude, so the point *K* is known, so the straight line *DK* is known in position and in magnitude; but <the straight line> *BK* is known in magnitude, so the point *B* is known.

– **29** – The synthesis of this problem is carried out as follows.

We put *AL* equal to four times *AD*, we put *AS* as the *latus rectum* of the axis and we put the product of *SG* and *GL* equal to the square of *F*. We draw the perpendicular *GE*, we join *AE* and we divide it into two equal parts at the point *I*. We draw *IN* parallel to the axis, so *IN* will be a diameter of the conic section. We draw from the point *D* a straight line parallel to the straight line *AE*, then it cuts the conic section in every case, because it makes an acute angle with the axis on the side towards the conic section. But since it cuts the conic section and cuts the axis, it accordingly cuts the conic section in two points, because any straight line that cuts the conic section and that cuts one of the diameters of the conic section, cuts the conic section in two points, as has been shown in Proposition 27 of Book I; let the straight line be *DBC*.

I say that *BC* is equal to *F*.

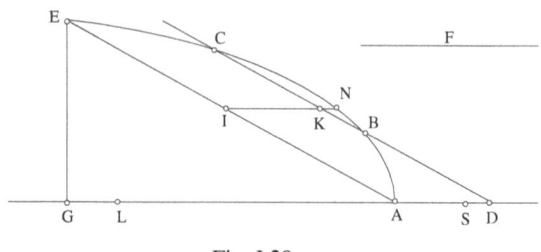

Fig. I.29

Proof: The straight line *BC* cuts the conic section in two points and it is parallel to the straight line *AE*, so the diameter *IN* cuts it into two equal parts; let it cut it at the point *K*. We show, as we showed in the analysis, that the straight line *AG* is four times the straight line *IN*. But the product of *SG* and *GA* is equal to the square of *AE*, so the product of *SG* and *NI* is equal to the square of *AI*. But the straight line *NI* is a diameter, so the straight line *SG* is the *latus rectum* of the diameter *NI*. The product of *SG* and *NK* is equal to the square of *BK*. But *AG* is four times *NI* and *AL* is four times *IK*, because *IK* is equal to *DA*, so *LG* is four times *NK* and the product of *SG* and *GL* is equal to the square of *BC*. But the product of *SG* and *GL* is equal to the square of *F*, so the straight line *BC* is equal to the straight line *F*. That is what we wanted to construct.

In this problem there is no need for a discussion because the product of *SG* and *GL* can be equal to the square of a known straight line, whatever <the magnitude of> that straight line is.

– **30** – The conic section *ABC* is a known hyperbola, with axis *AD* and centre *E*; a straight line *F* is given and a point *H* is given on the axis of the conic section between its centre and its vertex. We wish to draw from the point *H* a straight line that cuts the conic section in two points and <is> such that the part of the straight line that falls inside the conic section is equal to the given straight line *F*.

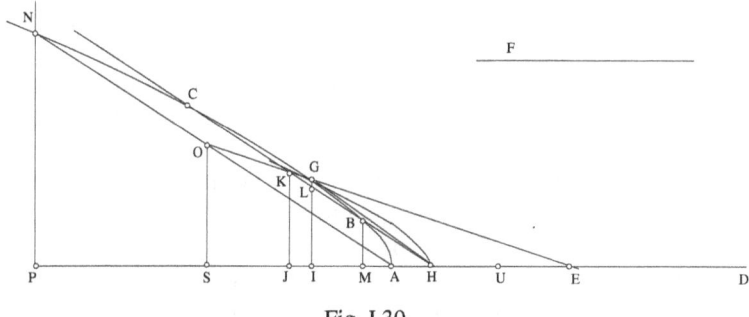

Fig. I.30

We suppose this is to be done by means of analysis: let it (the required line) be the straight line *HBC*; so *BC* will be equal to the given straight line *F*. We divide *BC* into two equal parts at the point *K*, we join *EK* and we extend it. We draw from the point *A* a straight line parallel to the straight line *HK*, so it meets the conic section; let it meet it at the point *N* and let it cut *EK* at the point *O*; so *AO* will be equal to half of *AN*. We draw the straight lines *NP*, *OS*, *KJ*, *BM* as ordinates. So the two triangles *AOS*, *KHJ* are similar, the ratio of *OS* to *KJ* is equal to the ratio of *SA* to *JH* and is equal to the ratio of *OA* to *KH*. But the ratio of *OA* to *KH* is equal to the ratio of *AE* to *EH*, which is a known ratio, because each of the straight lines *AE* and *HE* is known. So the ratio of *OS* to *KJ* is known, the ratio of *SA* to *JH* is known and is equal to the ratio of *AE* to *EH*. So the ratio of *SE* to *EJ* is known, and the ratio of the product of *ES* and *SA* to the square of *SO* is equal to the ratio of the product of *EJ* and *JH* to the square of *JK*. But *PA* is twice *AS*, *DA* is twice *AE* and *PN* is twice *SO*, so the ratio of the product of *ES* and *SA* to the square of *SO* is equal to the ratio of the product of *DP* and *PA* to the square of *PN*, so the ratio of the product of *EJ* and *JH* to the square of *JK* is equal to the ratio of the product of *DP* and *PA* to the square of *PN*, which is equal to the ratio of *DA* to its *latus rectum*, which is a known ratio, so the ratio of the product of *EJ* and *JH* to the square of *JK* is equal to the ratio of *DA* to its *latus rectum*. We draw from the point *H* a tangent to the conic section; let it be *HG*. We draw *GI* as an ordinate, so it cuts the straight line *HC*; let it cut it at the point *L*. Since the point *H* is on the axis, the tangent drawn from the point *H* on the other branch of the conic section is equal to the straight line *HG*, the straight line that joins the two points of contact is perpendicular to the axis and it is an ordinate; so the straight line *GI* is the one that ends at the point of contact on the other side. So the ratio of *CH* to *HB* is equal to the ratio of *CL* to *LB*, as has been shown in Proposition 37 of Book III. But *CH* is greater than *HB*, so *CL* is greater than *LB*, so the point *L* lies between the two points *B* and *K*. But since *GH* is a tangent and *GI* is an ordinate, the ratio of the product of *EI* and *IH* to the square of *IG* is equal to the ratio of *AD* to its *latus rectum*, as has been shown in Proposition 37 of Book I. So the ratio of the product of *EI* and *IH* to the square of *IG* is equal to the ratio of the product of *EJ* and *JH* to the square of *JK*. The hyperbola with axis *EH* and *latus rectum* a straight line such that the ratio of *EH* to that straight line is equal to the ratio of *AD* to its *latus rectum*, thus passes through the points *H*, *G*, *K*; let this conic section be the conic section *HGK*. We put the ratio of *EU* to *HU* equal to the ratio of *AD* to its *latus rectum*, which is the ratio of the axis *EH* to its *latus rectum*; so the point *U* will be known and *UH* will be the straight line with a homologous ratio, as has been shown in the second proposition of the seventh book. So the ratio of the product of *UJ* and *JH* to the square of *HK* is equal to the ratio of *UE* to *EH*, as has been shown in the second proposition of the seventh book. But the ratio of *UE* to *EH* is known, so the ratio of the product of *UJ* and *JH* to the square of *HK* is known. Now, since the ratio of *CH* to *HB* is equal to the ratio of *CL* to *LB*, the ratio of *CH* plus *HB* to *HB* is equal to the ratio of *CB* to *BL* and, similarly, half of one is equal to half of the other.[59] So the ratio of *KH* to *HB* is equal to the ratio of *KB* to *BL*, so the ratio of *HK* to *KB* is equal to the ratio of *BK* to *KL* and consequently the product of *HK* and *KL* is equal to the square of *KB*. The product of *HJ* and *JI* is equal to the square of *JM*. But the ratio of the square of *HK* to the square of *KB* is equal to the ratio of *HJ* to the square of *JM*, and the ratio of the square of *HJ* to the square of *JM* is equal to the ratio of *HJ* to *JI*, so the ratio of the square of *HK* to the square of *KB* is equal to the ratio of *HJ* to *JI*. Now the ratio of *HJ* to *JI* is equal to the ratio of the product of *UJ* and *JH* to the product of *UJ* and *JI*, so the ratio of the product of *UJ* and *JH* to the product of *UJ* and *JI* is equal to the ratio of the square of *HK* to the square of *KB* and the ratio of the product of *UJ* and *JH* to the square of *HK* is equal to the ratio of the product of *UJ* and *JI* to the square of *KB*. But the ratio of the product of *UJ* and *JH* to the square of *HK* is known, because it is equal to the ratio of *UE* to *EH*, so the ratio of the product of *UJ* and *IJ* to the square of *KB* is a known ratio. But *KB* is known because it is half of *F*, so the product of *UJ* and *JI* is known; but the straight line *UI* is known, so the point *J* is known, the straight line *JK* is known in position and the conic section *HK* is known in position, so the point *K* is known in position, the straight line *HK* is thus known in position, so the points *B* and *C* are known and the straight line *BH* is thus known. That is what we wanted to find.

<31> The synthesis of this problem is carried out as follows.

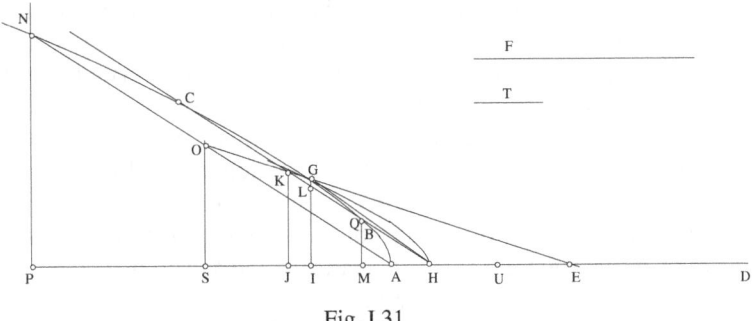

Fig. I.31

We draw from the point *H* a tangent to the conic section *ABC*; let it be *HG*. We draw *GI* as an ordinate, then the ratio of the product of *EI* and *IH* to the square of *IG* is equal to the ratio of the diameter *AD* to its *latus rectum*, as has been shown in Proposition 37 of Book I. We construct through the point *H* a hyperbola with axis *EH* and as *latus rectum* a straight line such that the ratio of *EH* to that straight line is equal to the ratio of *AD* to its *latus rectum*; let the conic section be *HK*. The conic section *HK* passes through the point *G*, because the ratio of the product of *EI* and *IH* to the square of *IG* is equal to the ratio of the diameter *EH* to its *latus rectum*. So the conic section *HK* cuts the conic section *ABC* at the point *G*. We put the ratio of *EU* to *UH* equal to the ratio of the diameter *EH* to its *latus rectum*; the straight line *UH* is the straight line with a homologous ratio. We put the ratio of the square of *T* to a quarter of the square of *F* equal to the ratio of *EU* to *EH* and we put the product of *UJ* and *JI* equal to the square of *T*. We draw from the point *J* a straight line as an ordinate to the outline of the conic section *HK*; let it be *JK*. So the point *K* will lie inside the conic section *ABC*, because the point *G* lies on the outline of *ABC*. We join *HK*; let it cut the straight line *GI* at the point *L*. So the ratio of the product of *UJ* and *JH* to the square of *HK* is equal to the ratio of *UE* to *EH*, as has been shown in Proposition 2 of Book VII; so it is equal to the ratio of the square of T to a quarter of the square of F, so the ratio of the product of UJ and JH to the square of HK is equal to the ratio of the product of UJ and JI to a quarter of the square of F. We put the product of HJ and JI equal to the square of JM and we draw MB as an ordinate;[60] so the product of HK and KL is equal to the square of KB. But since the ratio of the product of UJ and JI to a quarter of the square of F is equal to the ratio of the product of UJ and JH to the square of HK, then the ratio of the product of UJ and JH to the product of UJ and JI is equal to the ratio of the square of HK to a quarter of the square of F. So the ratio of HJ to JI is equal to the ratio of the square of HK to a quarter of the square of F. But the ratio of HJ to JI is equal to the ratio of the square of HJ to the square of JM, so the ratio of the square of HK to a quarter of the square of F is equal to the ratio of the square of HJ to the square of JM, which is equal to the ratio of the square of HK to the square of KB; so the ratio of the square of HK to the square of KB is equal to the ratio of the square of HK to a quarter of the square of F, the square of KB is a quarter of the square of F and the straight line KB is half of the straight line F.

We draw from the point A a straight line parallel to the straight line HK; let it meet the straight line EK at the point O. We extend AO, we put ON equal to OA and we draw NP and OS as ordinates, they will be parallel to the straight line KJ. So triangle AOS will be similar to triangle HKJ, so the ratio of OS to KJ will be equal to the ratio of SA to JH and it is equal to the ratio of OA to KH. But the ratio of OA to KH is equal to the ratio of AE to EH, so the ratio of SE to EJ is equal to the ratio of SA to JH and to the ratio of OS to KJ. But the ratio of AS to SO is equal to the ratio of HJ to JK, so the ratio of the product of ES and SA to the square of SO is equal to the ratio of the product of EJ and JH to the square of JK. But the ratio of the product of EJ and JH to the square of JK is equal to the ratio of EH to its *latus rectum*, which is equal to the ratio of AD to its *latus rectum*, so the ratio of the product of ES and SA to the square of SO is equal to the ratio of AD to its *latus rectum*. But AP is twice AS and DA is twice EA, so DP is twice ES. But NP is twice OS, so the ratio of the product of DP and PA to the square of PN is equal to the ratio of AD to its *latus rectum*, so the point N lies on the outline of the conic section ABC. But since the point K lies inside the conic section ABC, and it lies on the diameter EO and the straight line HK is parallel to the straight line AN, which the diameter EO divides

---

[60] The point B lies on HK. It will be shown later that it also lies on the conic section.

into two equal parts, the straight line *HK*, if extended, cuts the conic section *ABC* in two points and is divided into two equal parts by the straight line *EK*. Let us extend the straight line *HK*; let it cut the conic section at the point *C*.

I say that the point *B* is the second point on the outline of the conic section *ABC*.

If it is not so, let *C* and *Q* be the two points. Then *CK* will be equal to *KQ*. But since *HG* is a tangent, the ratio of *CH* to *HQ* will be equal to the ratio of *CL* to *LQ*; so the ratio of *CH* plus *HQ* to *HQ* is equal to the ratio of *CQ* to *QL* and the ratio of half of one of them to half the other is also equal,[61] so the ratio of *KH* to *HQ* is equal to the ratio of *QK* to *QL* and the product of *HK* and *KL* is equal to the square of *KQ*.[62]

Now it has been shown that the product of *HK* and *KL* is equal to the square of *KB*, so *KB* is equal to *KQ*, which is impossible. So the point *Q* does not lie on the outline of the conic section *ABC* and no point other than the point *B* lies on the outline of the conic section *ABC*. But *EK* is a diameter, and accordingly cuts the straight line *BC* into two equal parts, so *BK* is equal to *KC*; now *BK* is half the straight line *F*, so the straight line *BC* is equal to the given straight line *F* and *BC* lies inside the conic section *ABC*. That is what we wanted to prove.

This problem does not require a discussion,[63] because the straight line drawn from the point *A*, the endpoint of the axis, and which is parallel to the straight line that does not meet the conic section, does not meet the conic section in another point; this is in fact shown from Proposition 13 of the second book. Any straight line drawn from the point *H* between the straight line *HG*, the tangent, and the straight line drawn from the vertex of the conic section parallel to the straight line that does not meet the conic section, cuts the conic section in two points, because it can cut the two straight lines that do not meet the conic section. These straight lines are infinitely many and the parts of these straight lines that fall inside the conic section, those that become more distant from the tangent, increase indefinitely and for those of the straight lines that become closer to the

---

[61] Lit.: and the ratio of half equally.

[62] $\frac{HK}{HQ} = \frac{QK}{QL}$ is written $\frac{HK}{HK - KQ} = \frac{QK}{KQ - KL}$, hence $KQ^2 = KL \cdot KH$.

[63] Ibn al-Haytham means that the problem can be solved without imposing any conditions regarding the given line *F*; but he rightly provides a discussion to prove this, that is he investigates whether a solution exists.

tangent, (the parts that fall inside the conic section) decrease indefinitely. So for any straight line <that is> among the straight lines of finite magnitude, there is a straight line equal to it, which can fall inside the conic section.

So if, to find the straight line from the point *H*, we continue to employ the method we have indicated, then the part of the straight line that falls inside the conic section will be equal to the given straight line. The problem can be solved in all cases and there is no need for a discussion. That is what we wanted to prove.

Thus we have completed what was written by al-Shaykh Abū ʿAlī al-Ḥasan ibn al-Ḥasan ibn al-Haytham on the completion of the work the *Conics*.

Thanks be rendered to God alone. May the blessing of God be upon our Lord Muḥammad, his family and his companions.

Appendix[*]

If from a known angle we wish to take one third, we consider a hyperbola whose *latus rectum* is equal to its transverse diameter, <and> the angle of whose ordinates is equal to the known angle; let the conic section be *AB* and the transverse diameter *BC*. We construct in the conic section a straight line equal to the straight line *BC*, let the straight line be *BA*, and let us draw the straight line *AD* as an ordinate.

I say that the angle *DAB* is the known angle.

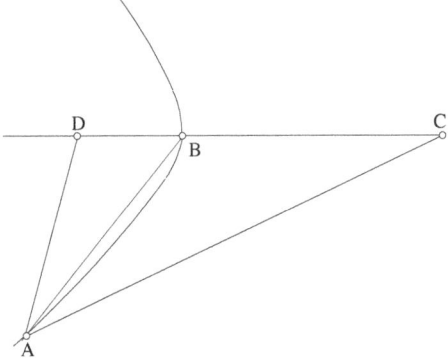

Fig. I.32

Proof: The ratio of the product of *CD* and *DB* to the square of the straight line *AD* is equal to the ratio of the transverse <diameter> to the <latus> *rectum*; but we have supposed the transverse <diameter> to be equal to the <latus> *rectum*, so the product of *CD* and *DB* is equal to the square of *AD*, which is why the triangle *ADB* will be similar to the triangle *ADC*; the angle *DAB* is thus equal to the angle *C* and the angle *ABD* is equal to twice the angle *C*, because *AB* is equal to *BC*, so the angle *ABD* is equal to twice the angle *DAB*. But since <each of the> two external angles of any triangle is equal to <the sum> of the two internal <angles> that are respectively opposite them, so for that reason the angle *DAB* will be one third of the given angle. That is what we wanted to prove.

---

[*] This appendix is anonymous and appears in the manuscript following the treatise of Ibn al-Haytham.

CHAPTER II

CORRECTING THE BANŪ MŪSĀ'S LEMMA
FOR APOLLONIUS' *CONICS*

2.1. INTRODUCTION

In the history of Apollonius' *Conics*, as in that of research on conic sections, the part played by the Banū Mūsā was crucial. They were the ones who had searched for Greek manuscripts, they were also the ones who had employed the translators who made Arabic versions of the seven books of the *Conics* that were found; they were also the ones who had supervised the process of translation; finally, they were the ones who, together with their pupils, reactivated research on conics, which had lain dormant for several centuries. One might point to the works on conics by the youngest brother, al-Ḥasan, and those of their pupil, Thābit ibn Qurra.[1] The tradition on which they set their stamp is certainly the one to which Ibn al-Haytham belongs, as we have seen in the first two volumes. That is to say, it would have been impossible for him to ignore what the Banū Mūsā had written on conics. Indeed, he had good reasons for being interested in their work, first as a student of the theory of conics, second as an expert and serious-minded reader of the *Conics* and finally as one engaged in making a copy of the text.

As an aid to the study of the *Conics*, the Banū Mūsā had composed an essay that supplied the nine lemmas required for Apollonius' proofs.[2] Accordingly, this essay was, understandably, intended to be read before reading the *Conics* and had therefore accompanied the work. The ninth (and last) lemma did not satisfy Ibn al-Haytham, even though it is in accord with the cases considered in the *Conics*. Ibn al-Haytham thought that, in the form in which it was stated, this lemma was not as general as the Banū Mūsā had

---

[1] R. Rashed, *Founding Figures and Commentators in Arabic Mathematics. A history of Arabic sciences and mathematics*, vol. 1, Culture and Civilization in the Middle East, London, 2012, Chapter II.

[2] See *Apollonius: Les Coniques*, tome 1.1: *Livre I*, commentaire historique et mathématique, édition et traduction du texte arabe par R. Rashed, Berlin/New York, 2008.

believed, and he also thought that an 'oversight' in the course of the proof might have reinforced this misunderstanding. Ibn al-Haytham's short treatise has the specific aim of correcting the defects in the Banū Mūsā's essay. It thus belongs to a type of writing that Ibn al-Haytham enjoyed: compositions in which he discusses and corrects his illustrious predecessors: Euclid, Ptolemy, Ibn Sinān, and on this occasion the Banū Mūsā. All the same, in this last case the task is a rather modest one compared with, say, what he undertook in regard to Ptolemy: we are concerned only with a single technical problem with a proof that does not raise any crucial theoretical questions. After all, the importance of the Banū Mūsā's essay itself largely derives from its connection with the *Conics*. That was also the reason that made it of interest to Ibn al-Haytham. The question underlying Ibn al-Haytham's treatise is how to correct the Banū Mūsā's procedure so as to achieve the desired degree of generality. The structure of his work is designed with that end in mind: he begins by setting out the lemma in question and pointing out the difficulty in it; he then returns to the actual problem proposed in the lemma and makes an exhaustive examination of all possible cases. Out of the ten possible cases, he shows that the lemma is true in seven. It is these cases that occur in the *Conics*, so the Banū Mūsā's text, in the state in which Ibn al-Haytham found it, achieves what they had intended. On the other hand, for the three remaining cases – the first, sixth and tenth – the lemma is not always true. But since the tenth reduces to the sixth, we only have two cases to discuss. Ibn al-Haytham accordingly proposes to add a condition to ensure that the lemma always holds true. That is, with this supplementary condition, we obtain the necessary and sufficient conditions for the lemma to always be true. Ibn al-Haytham does not prove the truth of this statement as such. He had established the nature of the seven cases and of the first and sixth cases. That may be why he did not go back to give a general proof. Before we address that possibility, let us first go through Ibn al-Haytham's text point by point.

2.2. MATHEMATICAL COMMENTARY

Lemma of the Banū Mūsā: *Let there be two triangles* ABC *and* DEF, *a point* G *on* BC *and a point* H *on* EF *such that* $\hat{A} = \hat{D}$, $\hat{AGB} = \hat{DHE}$ *and* $\dfrac{BG \cdot CG}{GA^2} = \dfrac{HE \cdot HF}{HD^2}$, *then triangles* ABC *and* DEF *are similar.*

In fact, the lemma will be true, and the triangles proved similar, if we show that the initial assumptions imply $\hat{E} = \hat{B}$. What Ibn al-Haytham will show is, precisely, that the two triangles are not always similar. Let us return to the Banū Mūsā's proof in its original form, as described by Ibn al-Haytham.

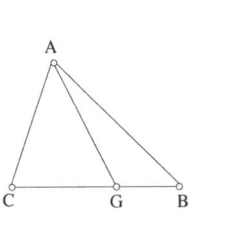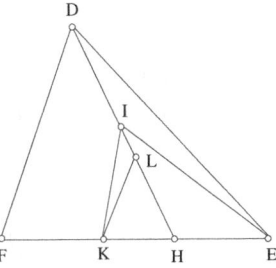

Fig. 2.2.1

Let us assume that the two triangles are not similar, and $\hat{E} \neq \hat{B}$. Let I be a point on DH such that $H\hat{E}I = \hat{B}$, and K a point on EF such that $E\hat{I}K = \hat{A} = \hat{D}$; then triangles EIK and ABC are similar, and so are triangles EIH and BAG. We have

$$\frac{EI}{AB} = \frac{IK}{AC} = \frac{KE}{CB} \quad \text{and} \quad \frac{EI}{AB} = \frac{IH}{AG} = \frac{HE}{BG},$$

hence

$$\frac{IH}{AG} = \frac{HE}{BG} = \frac{KE}{CB} = \frac{KE - HE}{CB - BG} = \frac{KH}{CG};$$

therefore

$$\frac{EH \cdot HK}{IH^2} = \frac{BC \cdot CG}{GA^2}$$

and consequently

$$\frac{EH \cdot HK}{IH^2} = \frac{HE \cdot HF}{HD^2} \Rightarrow \frac{FH}{HK} = \frac{DH^2}{HI^2}.$$

If the point L on DH is such that $\dfrac{DH^2}{HI^2} = \dfrac{DH}{HL}$, we deduce that

(1) $$\frac{DH}{HI} = \frac{HI}{HL}$$

and

(2) $$\frac{DH}{HL} = \frac{FH}{HK}.$$

Now (2) implies that *DF* ∥ *LK*.

We note that the positions of the points *I* and *L* are different according to whether $\hat{E} > \hat{B}$ or $E < B$.

$\hat{E} > \hat{B}$ means that *I* lies between *H* and *D*, so *HI* < *HD*, and, from (1), *HI* < *HD* ⇒ *HL* < *HI* ⇒ *L* lies between *H* and *I* (Fig. 2.2.1).

$E < B$ means that *I* lies beyond *D*, so *HI* > *HD*, so *L* is above *I* (Fig. 2.2.2).

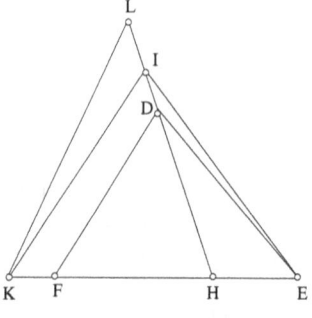

Fig. 2.2.2

The Banū Mūsā's figure (Fig. II.3) is incorrect:[3] *I* lies between *H* and *D* as in Fig. 2.2.1 and *L* is above *I* as in Fig. 2.2.2.

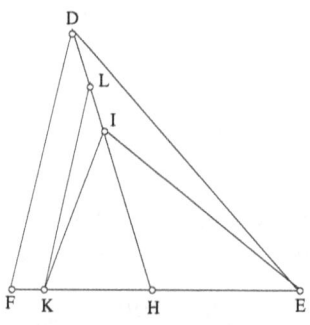

Fig. 2.2.3

[3] See R. Rashed, *Apollonius: Les Coniques*, tome 1.1: *Livre I*, pp. 527–31.

The argument based on this figure gives

$$K\hat{I}H > F\hat{D}H \text{ and } E\hat{I}H > E\hat{D}H,$$

hence $K\hat{I}E < E\hat{D}F$, which is impossible, because by construction $K\hat{I}H = E\hat{D}H$.

But in Fig. 2.2.1 we have

$$K\hat{I}H < F\hat{D}H \text{ and } E\hat{I}H > E\hat{D}H;$$

and in Fig. 2.2.2

$$K\hat{I}H > F\hat{D}H \text{ and } E\hat{I}H < E\hat{D}H;$$

so we cannot draw any conclusions about angles *KIH* and *EDF*.

Faced with this difficulty, Ibn al-Haytham goes back to the problem itself[4] and starts by listing all the possible cases.

Table of the possible cases

$\hat{A} = \hat{D} = 1$ right angle $\hat{G} = \hat{H} = 1$ right angle   (1)

$\hat{G} = \hat{H}$ acute   (2)

$\hat{G} = \hat{H}$ obtuse reduces to (2) (*)

$\hat{A} = \hat{D} > 1$ right angle $\hat{G} = \hat{H} = \hat{A}$ (3)

$\hat{G} = \hat{H}$ obtuse $> \hat{A}$ (4)

$\hat{G} = \hat{H} = 1$ right angle   (5)

$\hat{G} = \hat{H}$ obtuse $< \hat{A}$ (6)

$\hat{G} = \hat{H}$ acute reduces to (4) or (6) (*)

$\hat{A} = \hat{D} < 1$ right angle $\hat{G} = \hat{H} = \hat{A}$ (7)

$\hat{G} = \hat{H} < \hat{A}$ (8)

$\hat{G} = \hat{H} = 1$ right angle   (9)

$\hat{G} = \hat{H}$ acute $> \hat{A}$ (10)

$\hat{G} = \hat{H}$ obtuse reduces to (8) or (10) (*)

---

[4] *Ibid.*

We note (*) that if $A\hat{G}B$ is acute, then $A\hat{G}C$ is obtuse, and if $A\hat{G}B$ is obtuse, then $A\hat{G}C$ is acute. Only one of the cases will be examined, the other can be deduced from it by exchanging the letters B and C (on the one hand) and E and F (on the other). Ibn al-Haytham then investigates the ten cases.

First case: $\hat{A} = \hat{D} = 1$ right angle, $\hat{G} = \hat{H} = 1$ right angle.

In this case, ABC and DEF are right-angled triangles and AG and DH are their heights above their hypotenuses. Whether the triangles are similar or not, we can still write

$$GA^2 = GB \cdot GC \text{ and } HD^2 = HE \cdot HF,$$

hence

(1) $$\frac{GB \cdot GC}{GA^2} = \frac{HE \cdot HF}{HD^2}.$$

Ibn al-Haytham then adds a supplementary condition that is necessary for the two triangles ABC and DEF to be similar:

$$\frac{AG}{DH} = \frac{BC}{EF},$$

hence

(2) $$\frac{GA^2}{HD^2} = \frac{BC^2}{EF^2};$$

from (1) and (2), we get

(3) $$\frac{GB \cdot GC}{BC^2} = \frac{HE \cdot HF}{EF^2}.$$

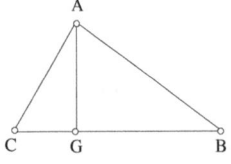 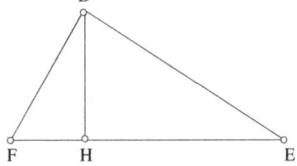

Fig. 2.2.4

Ibn al-Haytham deduces *without offering any justification* that

$$\frac{GB}{GC} = \frac{HE}{HF},$$

and then concludes the triangles are similar.

Second case: $\hat{A} = \hat{D} = 1$ right angle, $A\hat{G}B = D\hat{H}E \neq 1$ right angle and

(1) $$\frac{GB \cdot GC}{GA^2} = \frac{HE \cdot HF}{HD^2}.$$

Ibn al-Haytham first constructs a triangle *DEF* similar to *ABC* ($\hat{E} = \hat{B}$).

If $\hat{G} = \hat{H}$, triangles *AGB* and *AGC* are similar to triangles *DHE* and *DHF* respectively and equation (1) is satisfied.

Let us assume there exists a triangle *EIF*, and a point *K* on *EF*, that satisfy

$\hat{I} = \hat{A} = 1$ right angle, $I\hat{K}E = A\hat{G}B \neq 1$ right angle and $\dfrac{GB \cdot GC}{GA^2} = \dfrac{HE \cdot KF}{KI^2}$.

Then we have *IK* ∥ *DH* and $\dfrac{HE \cdot HF}{HD^2} = \dfrac{KE \cdot KF}{KI^2}$.

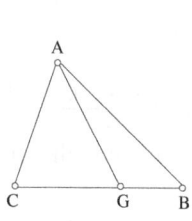
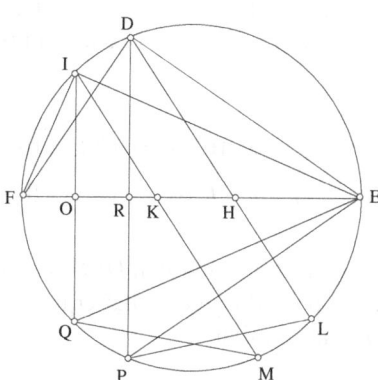

Fig. 2.2.5

DH and *IK* cut the circle of diameter *EF* again in points *L* and *M* respectively; we have

$$HE \cdot HF = HL \cdot HD \text{ and } KI \cdot KM = KE \cdot KF,$$

so

$$\frac{HL}{HD} = \frac{KM}{KI}$$

and, by composition of ratios,

$$\frac{LD}{HD} = \frac{MI}{KI}.$$

But

$$\frac{HD}{DR} = \frac{KI}{IO} \Rightarrow \frac{LD}{DR} = \frac{MI}{IO} \text{ and } \frac{LD}{DP} = \frac{MI}{IQ};$$

on the other hand $L\hat{D}P = M\hat{I}Q$, so triangles *LDP* and *MIQ* are similar, and $D\hat{L}P = I\hat{M}Q$, which implies $D\hat{E}P = I\hat{E}Q$; this is impossible because

$$D\hat{E}F = \frac{1}{2} D\hat{E}P \text{ and } I\hat{E}F = \frac{1}{2} I\hat{E}Q$$

and we have $D\hat{E}F \neq I\hat{E}F$.

Thus the two triangles *ABC* and *DEF* are similar and there exists no other triangle that satisfies the same conditions without being similar to these triangles.

Third case: $\hat{A} = \hat{D}$; $A\hat{G}B = D\hat{H}E > 1$ right angle and $\dfrac{GB \cdot GC}{GA^2} = \dfrac{HE \cdot HF}{HD^2}$.

On a line segment *EF* we construct the arc that subtends the given angle $\hat{A}$.

Let the point *D* be such that $D\hat{E}F = \hat{B}$ and *H* such that $D\hat{H}E = A\hat{G}B$. We show as in the preceding case that equation (1) is satisfied and that triangles *ABC* and *EDF* are similar.

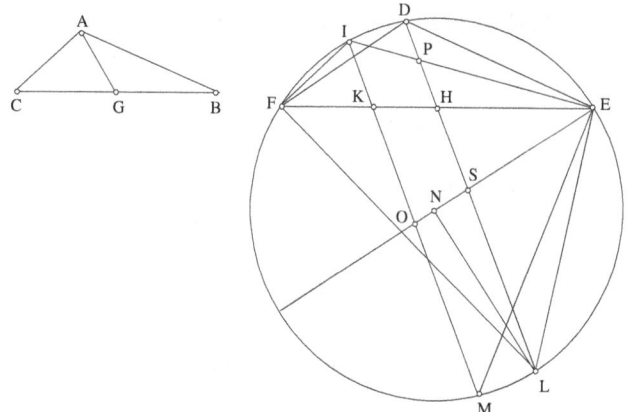

Fig. 2.2.6

Let us assume there exists a triangle EIF, and a point K on EF, such that $E\hat{I}F = \hat{A}$, $I\hat{K}E = A\hat{G}B = \hat{A}$, and

$$\frac{KE \cdot KF}{KI^2} = \frac{GB \cdot GC}{GA^2},$$

then I lies on the arc EDF, $IK \parallel DH$ and $I\hat{E}F \neq \hat{B}$. We have

$$\frac{KE \cdot KF}{KI^2} = \frac{HE \cdot HF}{HD^2}.$$

The straight lines DH and IK cut the circle again in L and M; we have

$$HE \cdot HF = HD \cdot HL \quad \text{and} \quad KE \cdot KF = KI \cdot KM,$$

hence

$$\frac{KM}{KI} = \frac{HL}{HD}$$

and consequently

$$\frac{MI}{IK} = \frac{DL}{HD}.$$

By hypothesis $E\hat{H}D = E\hat{D}F$, hence $ED^2 = EH \cdot EF$.

But $E\hat{H}D = E\hat{D}F$, hence $E\hat{H}L = E\hat{L}F$, hence $EL^2 = EH \cdot EF$, from which we deduce

$$EL = ED; \quad \widehat{EL} = \widehat{ED}.$$

So if N is the centre of the circle, $EN \perp DL$ and $EN \perp IM$. The straight line EN cuts the segments DL and IM at their respective midpoints S and O, so we have

$$\frac{SD}{DH} = \frac{OI}{IK}.$$

The straight line EI cuts DH in P, and we have

$$\frac{OI}{IK} = \frac{SP}{PH}.$$

Therefore

$$\frac{SP}{PH} = \frac{SD}{DH},$$

hence

$$\frac{SH}{HP} = \frac{SH}{HD},$$

which is impossible because $HP \neq HD$.

The reasoning holds whatever the position of the point I on the arc that subtends EDF: if I lies on the arc DF, the point P lies between H and D, $HP < HD$; if I lies on the arc DE, the point P lies beyond D, $HP > HD$. In this case also, the two triangles ABC and DEF are similar and there exists no other triangle that satisfies the same conditions without being similar to these triangles.

Fourth case: 1 right angle $< \hat{A} = \hat{D} < A\hat{G}B = D\hat{H}E$ and $\dfrac{GB \cdot GC}{GA^2} = \dfrac{HE \cdot HF}{DH^2}$.

As in the preceding case, Ibn al-Haytham again constructs on a general line segment EF the arc that subtends the angle A and a triangle EDF similar to ABC with a point H such that $D\hat{H}E = A\hat{G}B = \hat{A}$; relation (1) is satisfied.

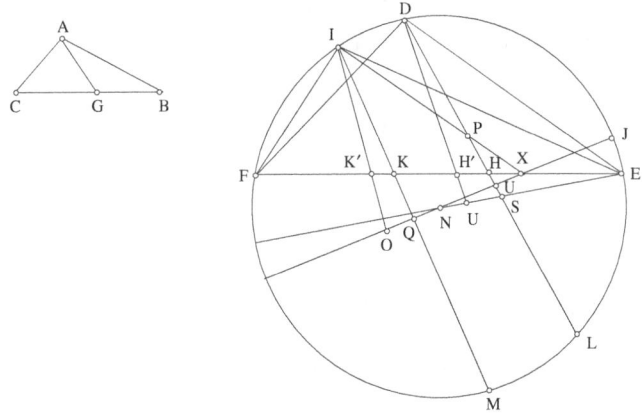

Fig. 2.2.7

Let *EIF* be a triangle that fits the initial assumptions, but is not similar to *ABC*. The notation is the same as in the preceding case and we have

$$\frac{KE \cdot KF}{KI^2} = \frac{HE \cdot HF}{HD^2} \Rightarrow \frac{MI}{IK} = \frac{DL}{HD}.$$

We have $D\hat{H}E > \hat{A}$; if from the point D we draw the straight line DH' such that $D\hat{H}'E > \hat{A}$, then H' lies between H and F and the straight line $DH' \perp EN$. The same is true for $IK' \parallel DH'$.

The perpendicular from N to the straight lines DL and IM cuts the arc DE in J. This straight line cuts DL and IM in their respective midpoints U and Q and cuts EF in X. The straight line IX cuts the straight line DL in P (Fig. 2.2.7); we have

$$\frac{LD}{DH} = \frac{MI}{IK} \Rightarrow \frac{QI}{IK} = \frac{UD}{HD}.$$

The straight lines QI and UD are parallel; (P, H, U) and (I, K, Q) are similar divisions, so

$$\frac{QI}{IK} = \frac{UP}{PH}$$

and consequently

$$\frac{UD}{HD} = \frac{UP}{HP}.$$

If X lies between H and E as Ibn al-Haytham says, and I lies between D and F, then $HD < UD$ and $HP < UP$, and we can write

$$\frac{UD - HD}{HD} = \frac{UP - HP}{HP},$$

or

$$\frac{UH}{HD} = \frac{UH}{HP};$$

which is impossible because $HP \neq HD$.

But there are other possible cases for the figure.

a) It could be that H is the midpoint of DL; then $NH \perp DL$, and U, H, X and P coalesce.

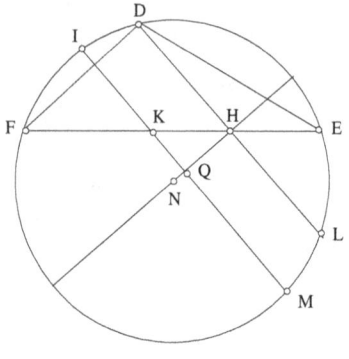

Fig. 2.2.8

In this case $HE \cdot HF = HD^2$. And we should get

$$KE \cdot KF = KI^2;$$

which is impossible, because for any point $I \neq D$, the point K is not the midpoint of MI.

b) It could be that the point U lies between H and D and the point X lies between H and F. If K lies on XF (Fig. 2.2.9), P lies on the semi-infinite straight line HL. We then have $HD > UD$; but $HP < UP$; the equality $\dfrac{UD}{HD} = \dfrac{UP}{HP}$ is thus impossible because $\dfrac{UD}{HD} < 1$.

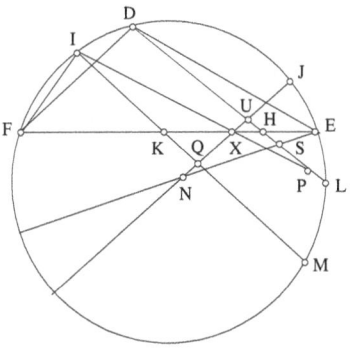

Fig. 2.2.9

c) It could be that U lies between H and D, and the point X lies between H and F; but the point K is at X. In this case, XI does not cut DL (the point P is at infinity), and $Q = K = X$. The point K is thus the midpoint of IM, so $KE \cdot KF = IK^2$; but $HE \cdot HF \neq HD^2$, so it is impossible to have

$$\frac{KE \cdot KF}{IK^2} = \frac{HE \cdot HF}{HD^2}.$$

d) Finally, it could be that U lies between H and D and X lies between H and F; but the point K can lie between X and H, the point P then lies on the semi-infinite straight line HD beyond D. We cannot have

$$\frac{DU}{DH} = \frac{PU}{PH},$$

because $D \neq P$ and D and P both lie outside the segment UH. Thus we conclude that any triangle that conforms with the hypotheses must be similar to the triangle ABC.

Fifth case: $\hat{A} = \hat{D} > 1$ right angle, $\hat{G} = \hat{H} = 1$ right angle and

$$\frac{GB \cdot GC}{GA^2} = \frac{HE \cdot HF}{HD^2}.$$

If $D\hat{E}F = \hat{B}$, triangle DEF is similar to ABC and satisfies (1).

Let us consider triangle EIF, not similar to triangle ABC, with I on the arc EDF, and $I \neq D$; let us assume

$$\frac{KE \cdot KF}{KI^2} = \frac{HE \cdot HF}{HD^2}.$$

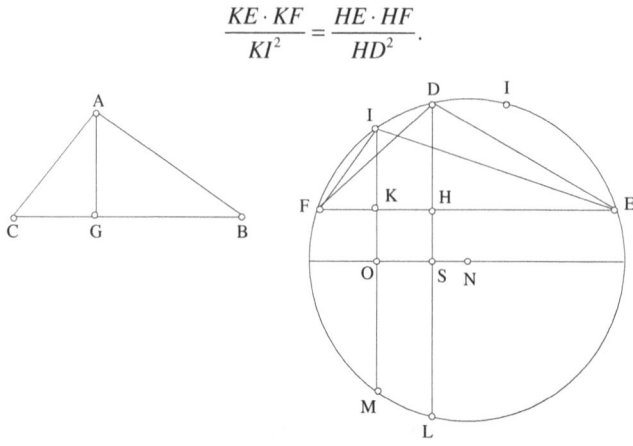

Fig. 2.2.10

Let NO be the diameter parallel to EF. As in the previous cases, we deduce from the last equation that

$$\frac{DS}{DH} = \frac{IO}{IK} \text{ and } \frac{SH}{DH} = \frac{OK}{IK};$$

but $SH = OK$, hence $DH = IK$.

We may note that there exists a point $I \neq D$ for which $IK = DH$; that is, the point symmetrical to D with respect to the perpendicular bisector of EF. Triangles EIF and EDF are thus congruent, so both of them are similar to triangle ABC, but with $I\hat{E}F \neq B$, $I\hat{E}F = \hat{C}$.

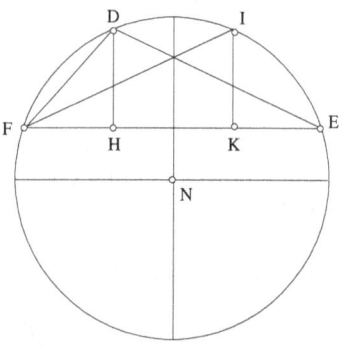

Fig. 2.2.11

In conclusion, any triangle that fits the initial assumption is similar to triangle *ABC*.

Sixth case: 1 right angle $< B\hat{G}A = E\hat{H}D < \hat{A} = \hat{D}$ and

$$\frac{GB \cdot GC}{GA^2} = \frac{HD \cdot HE}{HD^2}.$$

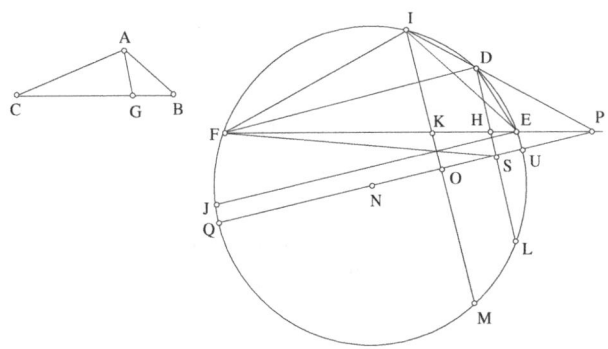

Fig. 2.2.12

Let there be a circle with centre *N*, a chord *EF* of the circle, a point *P* on *FE* produced, a straight line from *P* that cuts the minor arc *EF* in *D* and a point *I* on the half of $\overset{\frown}{EF}$ that lies on the same side as *E*. The straight line *NP* cuts the circle in *U* and *Q*. We draw *DH* and *IK* perpendicular to *NP* at *S* and *O* respectively and *EJ* ∥ *NP* (Fig. 2.2.12). We have

(1) $$D\hat{H}P = I\hat{K}P = 1 \text{ right angle} + F\hat{P}Q$$
$$= 1 \text{ right angle} + F\hat{E}J.$$

Angles *EDF* and *EIF* cut off the arc *EUQJF*, which is equal to

$$\overset{\frown}{EU} + \frac{1}{2} \text{ circle} + \overset{\frown}{QJ} + \overset{\frown}{JF},$$

so

(2) $E\hat{D}F = E\hat{I}F = 1$ right angle $+ F\hat{E}J + 2\alpha$, where α is the inscribed angle that cuts off $\overset{\frown}{EU}$ or $\overset{\frown}{QJ}$.

From (1) and (2), we deduce $\hat{EDF} = \hat{DHE} + 2\alpha$.

We then consider the two triangles EDF and EIF and a triangle ABC similar to EDF, where G is a point on BC such that $A\hat{G}B = D\hat{H}E$; so we have

$$\frac{GB \cdot GC}{GA^2} = \frac{HE \cdot HF}{HD^2}.$$

Moreover, we have

$$\frac{DS}{DH} = \frac{IO}{IK} \quad \text{(similar ranges)},$$

hence

$$\frac{DL}{DH} = \frac{IM}{IK}.$$

Therefore

$$\frac{HL}{HD} = \frac{KM}{KI} \quad \text{and} \quad \frac{HL \cdot HD}{HD^2} = \frac{KM \cdot KI}{KI^2},$$

hence

$$\frac{HE \cdot HF}{HD^2} = \frac{KE \cdot KF}{KI^2}.$$

Triangles DEF and IEF thus conform with the same hypotheses as the similar triangles DEF and ABC; however, triangle IEF is not similar to DEF, because $I\hat{E}F < D\hat{E}F$ and $I\hat{F}E > D\hat{F}E$, so it is not similar to ABC either.

Note: The conditions that are given are not sufficient to make two triangles similar.

Seventh case: $\hat{A} = \hat{D} = A\hat{G}B = D\hat{H}E < 1$ right angle and

(1) $$\frac{GB \cdot GC}{GA^2} = \frac{HE \cdot HF}{HD^2}.$$

Let there be a triangle ABC, a point G on BC lying between B and C such that $\hat{A} = A\hat{G}B < 1$ right angle and a segment EF on which we construct an arc that subtends an angle A. We take a point D on that arc such that $D\hat{E}F = \hat{B}$; the triangle DEF is then similar to ABC, and if a point H on EF is such that $D\hat{H}E = \hat{D} = \hat{A}$, then H lies between E and F and we have the equality (1).

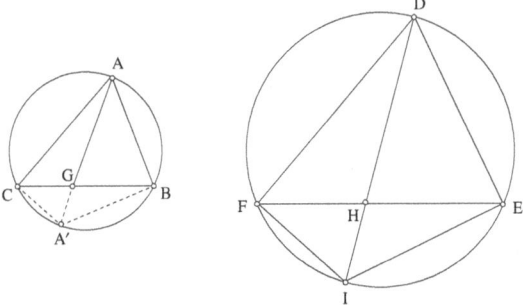

Fig. 2.2.13

The straight line *DH* cuts the circle *DEF* in *I*, so we have

$$E\hat{H}I = E\hat{I}F = 2 \text{ right angles} - \hat{D}.$$

Note: The straight line *AG* cuts the circumcircle of *ABC* again at *A'* and we have $B\hat{A}'C = B\hat{G}A' = 2$ right angles $- \hat{A}$, so the triangles *A'BC* and *IEF* conform with the hypotheses of the third case. But we have seen that there exists no triangle that is not similar to *A'BC* that has the required properties. From this we deduce that there is no triangle that conforms with the hypotheses of the seventh case that is not similar to triangle *ABC*.

Eighth case: $A\hat{G}B = D\hat{H}E < \hat{A} = \hat{D} < 1$ right angle and

$$\frac{GB \cdot GC}{GA^2} = \frac{HE \cdot HF}{HD^2}.$$

The argument proceeds as in the seventh case; we have

$$B\hat{A}'C = E\hat{H}I > A' > 1 \text{ right angle}.$$

Triangles *BA'C* and *EIF* conform with the hypotheses of the fourth case; we have seen that triangle *EIF* is similar to *BA'C* and that there exists no triangle that conforms with these hypotheses that is not similar to *A'BC*.

So there is no triangle that conforms with the hypotheses of the eighth case without also being similar to triangle *ABC*.

Ninth case: $\hat{A} = \hat{D} < 1$ right angle, $A\hat{G}B = D\hat{E}H = 1$ right angle and

$$\frac{GB \cdot GC}{GA^2} = \frac{HE \cdot HF}{HD^2}.$$

By the same reasoning as in the preceding case we reduce this to the fifth case.

Tenth case: $\hat{A} = \hat{D} < A\hat{G}B = D\hat{H}E < 1$ right angle and

$$\frac{GB \cdot GC}{GA^2} = \frac{HE \cdot HF}{HD^2}.$$

By the same process, we reduce this to the sixth case, and we deduce that on a given segment *EF*, we can construct a triangle similar to *ABC* conforming with the hypotheses and another triangle not similar to *ABC* while nevertheless conforming with the hypotheses. So we need to introduce the supplementary condition.

The Banū Mūsā's lemma corrected

1. As we have said, Ibn al-Haytham's stated purpose is to remedy some weaknesses that the Banū Mūsā failed to notice when formulating and proving the ninth lemma. In other words, he needed to find the necessary and sufficient conditions for the lemma to be true in a general form. Let us conclude by returning to this problem.

The Banū Mūsā's proposition can be considered as a converse of the following proposition:

If two triangles *ABC* and *DEF* are similar – *D*, *E*, *F* corresponding respectively to *A*, *B*, *C* – and if $G \in [BC]$ and $H \in [EF]$ are points such that $A\hat{G}B = D\hat{H}E$, then $\frac{GB \cdot GC}{GA^2} = \frac{HE \cdot HF}{DH^2}$ (Fig. 2.2.1).

The proof of this proposition is obvious, so we can write

$$\hat{A} = \hat{D}, \ \hat{B} = \hat{E}, \ A\hat{G}B = D\hat{H}E \implies \frac{GB \cdot GC}{GA^2} = \frac{HE \cdot HF}{DH^2}.$$

Here we propose to show that (1) and (3) imply (2), using the notation

(1) $\hat{A} = \hat{D},$

(2) $A\hat{G}B = D\hat{H}E \Rightarrow \hat{B} = \hat{E},$

(3) $\dfrac{GB \cdot GC}{GA^2} = \dfrac{HE \cdot HF}{DH^2}.$

We have seen that, in his investigation of this converse, Ibn al-Haytham distinguishes ten cases and shows it is false, as a general result, in two principal cases: the first and the sixth. It is also false in the tenth case, but this reduces to the sixth one. To make the conditions necessary and sufficient for the lemma to always be true, Ibn al-Haytham proposes to add the condition

(4) $\dfrac{AG}{BC} = \dfrac{DH}{EF}.$

Taking account of condition (4), let us consider the two problematic cases: the first and the sixth.

In the first case, we have $\hat{A} = \hat{D} = A\hat{G}B = D\hat{H}E = 1$ right angle, triangles ABC and DEF are right-angled at $\hat{A} = \hat{D}$, AG and DH are the heights above the hypotenuse. Hypothesis (3) is a property of any pair of right-angled triangles, whether they are similar or not; its appearance as a condition is redundant. If it is replaced by condition (4), the two triangles ABC and DEF are similar, by Euclid, *Data*, 79. But E may correspond to B and F to C, or E may correspond to C and F to B.

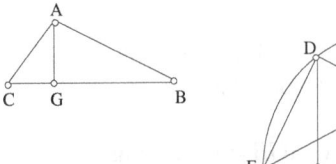

Fig. 2.2.14

In the sixth case, we have $\hat{A} = \hat{D} > A\hat{G}B = D\hat{H}E > 1$ right angle. The straight lines AG and DH are no longer the heights.

Let AK and DL be the heights, let us put $A\hat{G}B = D\hat{H}E = \alpha$. We have $AK = AG \sin \alpha$ and $DL = DH \sin \alpha$. Therefore condition (4) is equivalent to $\dfrac{AK}{BC} = \dfrac{DL}{EC}$. This condition and condition (1) ($\hat{A} = \hat{D}$) together imply that triangles ABC and DEF are similar, and as in the first case there are two possibilities: E and F corresponding respectively to B and C or to C and B.

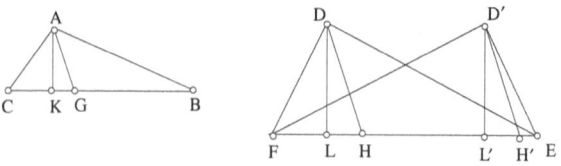

Fig. 2.2.15

Condition (3) is satisfied only if E in triangle EDF corresponds to B; it is not satisfied for the triangle $D'EF$, because we have $DH = D'H'$, but $HE \cdot HF \neq H'E \cdot H'F$.

Here, the two conditions (3) and (4) are complementary and allow us to conclude that $\hat{B} = \hat{E}$.

Thus with condition (4) the Banū Mūsā's lemma is true in all cases.

2. We still need to know how Ibn al-Haytham had found condition (4) and how, when added to those given by the Banū Mūsā, this condition makes it possible to prove the lemma completely generally, without having to distinguish the ten cases, a proof that Ibn al-Haytham does not give. Perhaps he thought it was not necessary to do so after having given corrections for the two defective cases; or perhaps he did not think of it, given how many distinct cases there were.

Let us begin by considering two similar triangles ABC and DEF with a point G on BC and a point H on EF such that

$$\hat{A} = \hat{D}, \ \hat{B} = \hat{E} \text{ and } A\hat{G}B = D\hat{H}E \text{ (Fig. 2.2.16)}.$$

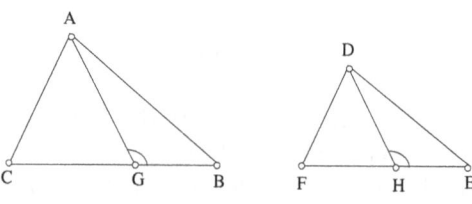

Fig. 2.2.16

Triangles AGB and DHE are similar and so are triangles AGC and DHF. So we have

$$\frac{AB}{DE} = \frac{GB}{HE} = \frac{AG}{DH} = \frac{GC}{FH} = \frac{AC}{DF},$$

hence

$$\frac{AG^2}{DH^2} = \frac{GB \cdot GC}{HE \cdot HF}; \quad \frac{GB + GC}{HE + FH} = \frac{AG}{DH},$$

that is

$$\frac{BC}{EF} = \frac{AG}{DH} \quad \text{and} \quad \frac{GB}{GC} = \frac{HE}{HF},$$

so the divisions (B, G, C) and (E, H, F) are similar.

If, instead of $\hat{B} = \hat{E}$, we were to put $\hat{B} = \hat{F}$ (Fig. 2.2.17), we should need to invert the roles of E and F throughout; we should put $A\hat{G}B = D\hat{H}F$.

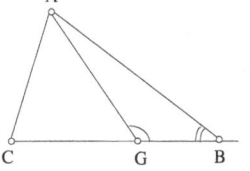 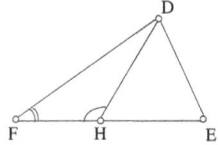

Fig. 2.2.17

The conclusions

$$\frac{AG^2}{DH^2} = \frac{GB \cdot GC}{HE \cdot HF} \quad \text{and} \quad \frac{BC}{EF} = \frac{AG}{DH},$$

would remain the same; but the range (B, G, C) would be similar to the range (F, H, E). Thus we have found conditions (3) and (4) by the previous analysis. Let us now turn our attention to the Banū Mūsā's lemma with the added condition (4); and let us show that $\hat{B} = \hat{E}$.

Lemma of the Banū Mūsā / Ibn al-Haytham
The triangles ABC and DEF are similar if and only if the following conditions are satisfied:

(1) $\hat{A} = \hat{D},$

(2) $A\hat{G}B = D\hat{H}E,$

(3) $\dfrac{GB \cdot GC}{GA^2} = \dfrac{HE \cdot HF}{DH^2},$

(4) $\dfrac{AG}{BC} = \dfrac{DH}{EF}.$

From (3) and (4) we obtain

(5) $\dfrac{GB \cdot GC}{BC^2} = \dfrac{HE \cdot HF}{EF^2},$

(5) $\Leftrightarrow$ $\dfrac{(BC - GC)GC}{BC^2} = \dfrac{(EF - HF)HF}{EF^2} \Leftrightarrow \dfrac{GC}{BC} - \dfrac{GC^2}{BC^2} = \dfrac{HF}{EF} - \dfrac{HF^2}{EF^2}$

$\Leftrightarrow$ $\left(\dfrac{GC}{BC} - \dfrac{HF}{EF} \right)\left[1 - \left(\dfrac{GC}{BC} + \dfrac{HF}{EF} \right) \right] = 0$

$\Leftrightarrow$ $\dfrac{GC}{BC} = \dfrac{HF}{EF}$ or $\dfrac{GC}{BC} = 1 - \dfrac{HF}{EF} = \dfrac{HE}{EF}.$

We then have two cases:

a) If $\dfrac{GC}{BC} = \dfrac{HF}{EF}$, then the ranges (B, G, C) and (E, H, F) are similar and we have

$$\frac{BC}{EF} = \frac{GC}{HF} = \frac{GB}{HE}.$$

b) If $\dfrac{GC}{BC} = \dfrac{HE}{EF}$, then the ranges (B, G, C) and (F, H, E) are similar and we have

$$\frac{BC}{EF} = \frac{GC}{EH} = \frac{GB}{HF}.$$

We can move from a) to b) by inverting the roles of E and F. So let us deal with the two cases.

a) Hypotheses
(1) $\hat{A} = \hat{D},$

(2) $$\frac{BC}{EF} = \frac{GB}{HE},$$

(3) $$A\hat{G}B = D\hat{H}E,$$

(4) $$\frac{AG}{BC} = \frac{DH}{EF}.$$

From (2) and (4), we obtain $\frac{AG}{DH} = \frac{GB}{HE}$; triangles AGB and DHE (Fig. 2.2.16) are thus similar, so $\hat{B} = \hat{E}$ and, consequently, triangles ABC and DEF are similar and points correspond with one another in the order $(A, B, C) \rightarrow (D, E, F)$.

b) Hypotheses

(1) $$\hat{A} = \hat{D},$$

(2) $$\frac{BC}{EF} = \frac{GB}{HF},$$

(3) $$A\hat{G}B = D\hat{H}F,$$

(4) $$\frac{AG}{BC} = \frac{DH}{EF}.$$

From (2) and (4), we obtain $\frac{AG}{DH} = \frac{GB}{HF}$; triangles AGB and DHF (Fig. 2.2.17) are thus similar, so $\hat{B} = \hat{E}$ and, consequently, triangles ABC and DFE are similar and points correspond in the order $(A, B, C) \rightarrow (D, F, E)$. The lemma is thus proved.

2.3. HISTORY OF THE TEXT

The treatise *On a Proposition of the Banū Mūsā* by Ibn al-Haytham exists in five manuscripts.

1) It occurs in an important collection, one part of which is in the Military Museum in Istanbul (Askari Müze 3025, no number), copied by the mathematician Qāḍī Zādeh between 1414 and 1435. Ibn al-Haytham's treatise occupies folios 1 ᵛ–8 ᵛ. Here we shall designate it by the letter S.

We have made a long investigation into the history of this collection.[5] Specifically, we have shown that this collection in the Military Museum is a part of a larger collection, the other part of which is in Berlin, Oct. 2970. Together, the two parts contain numerous treatises by Ibn al-Haytham. We have also shown that the famous collection 'Āṭif 1714 of the Süleymaniye Library in Istanbul was copied from this collection and from it alone.

2) The second manuscript of this treatise is, indeed, a part of the last collection, the one in 'Āṭif 1714, and occupies folios 149^r–157^r. We have designated it by T.

3) The third manuscript is in the Library of the University of Aligarh (India), no. I, fols 28–38, in which the completion of the copy is dated 1072 AH (November 1661) at Jahānābād. The transcription of the whole collection was completed on 26 Rajab 1075 AH, that is, 12 February 1665.[6] The copy is in careful *nasta'līq*. Each folio measures 25.6 × 11.4 cm with 25 lines per page, and about 13 words per line. We designate this collection by the letter A.

4) The fourth manuscript is in the collection of the British Library in London, Add. 14332/2, fols 42–61. In connection with a treatise by Ibrāhīm ibn Sinān,[7] we have shown that this collection was copied exclusively from Aligarh University no. I. A careful examination of Ibn al-Haytham's text confirms this finding, if confirmation is required. All the errors in A are reproduced in B; however, it sometimes happens that B corrects certain grammatical mistakes. This collection was also copied in India and was transferred to the British Museum in the mid nineteenth century. In establishing the text of Ibn al-Haytham's treatise, we might have dispensed with this last manuscript, since B is derived from A and only from A. If we have noted its variants, that is in order to establish the family tree of the manuscripts. Here this manuscript is designated by B.

5) The fifth manuscript is in the collection of the India Office in London, no. 1270 (= Loth 734), fol. 28^{r-v}; this is an incomplete copy that we shall designate by L. An examination of this short fragment shows that it belongs to

---

[5] See Chapter III, below.

[6] The colophon of Ibn Sinān's treatise on *Drawing the Three Sections* says it was transcribed by a certain Muḥammad Akbarābādī. See R. Rashed and H. Bellosta, *Ibrāhīm ibn Sinān. Logique et géométrie au Xe siècle*, Leiden, 2000, p. 261.

[7] *Ibid.*, p. 261.

the same family as A and B. We find five mistakes that are common to A and B, and five individual ones.

So in the end there are only two manuscripts, the one in the Military Museum [S] and the one in Aligarh [A], that are genuinely useful in establishing the text, as is shown in the following *stemma*:

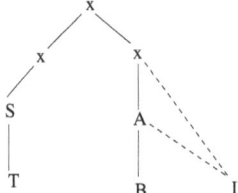

Until now, there has been no critical edition of this treatise. A publication that was not a critical edition appeared in Hyderabad; an examination of the variants shows that this text was printed simply from manuscript B alone. However, this publication had the advantage of attracting attention to this treatise by Ibn al-Haytham.[8]

---

[8] See *Majmū' al-rasā'il*, Osmānia Oriental Publications Bureau, Hyderabad, 1938–1939, no. 6.

2.4. TRANSLATED TEXT

Al-Ḥasan ibn al-Haytham

On a Proposition of the Banū Mūsā

In the name of God the Compassionate the Merciful

TREATISE BY AL-ḤASAN IBN AL-ḤASAN IBN AL-HAYTHAM

On a Proposition of the Banū Mūsā

One of the propositions which the Banū Mūsā put in place before the proofs of the *Conics*, the one that is the last of the lemmas, has not got the property they ascribed to it. They have, in fact, made it <seem to be> a general proposition, whereas it holds only in special cases. Further, they have made a slip in the course of the proof of the proposition and <it is> because of this they have taken it to be general. Now this proposition is required for some proofs of propositions in the *Conics*. That is why we need to explain its form and show that it concerns particular propositions, that it is true in certain cases and false in others, that the use that is made of it in the proofs in the *Conics* involves cases in which it is true and that none of the cases in which it is false is used in the *Conics*. So let us set out what we think about this proposition. We say that the proposition mentioned by the Banū Mūsā and whose property is as we have described above is the following:

Let there be two triangles that have two angles equal. From the two equal angles to the sides opposite them there have been drawn two straight lines that make two equal angles with the two opposite sides, in such a way that the ratios of each of the two rectangles enclosed by the two parts of the two opposite sides to each of the squares of the straight lines drawn to the opposite sides are two equal ratios.

They have claimed that two triangles with this property are similar. It does not follow that the two triangles are always similar. They proved the similarity of the two triangles using a demonstration in which they made a slip.

Let us first indicate where the slip in their proof occurs. They have taken the two triangles, triangles *ABC* and *DEF*; they have drawn in these two triangles the two straight lines *AG* and *DH*; they have taken the two angles *A* and *D* as equal and the two angles *AGB* and *DHE* as also equal; they have made the ratio of the product of *BG* and *GC* to the square of *GA* equal to the ratio of the product of *EH* and *HF* to the square of *HD* and they have claimed in connection with these two triangles that they (triangles) are always similar if they have the property that we have described.

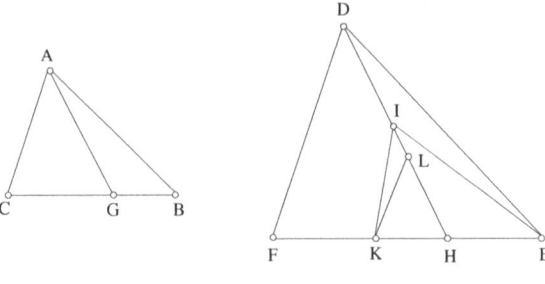

Fig. II.1

They proved this by saying: if angle *DEH* is not equal to angle *ABG*, then let us put angle *HEI* equal to angle *ABG*, and let us put angle *EIK* equal to angle *BAC*; then triangle *EIK* is similar to triangle *BAC* and triangle *EIH* is similar to triangle *BAG*, so the ratio of the product of *EH* and *HK* to the square of *HI* is equal to the ratio of the product of *BG* and *GC* to the square of *GA*, which is equal to the ratio of the product of *EH* and *HF* to the square of *HD*, so the ratio of the product of *EH* and *HF* to the square of *DH* is equal to the ratio of the product of *EH* and *HK* to the square of *HI*. The ratio of *FH* to *HK* is thus equal to the ratio of the square of *DH* to the square of *HI*. They then said: let us put the ratio of the square of *DH* to the square of *HI* equal to the ratio of *DH* to *HL*. They put the point *L* above the point *I*, that is between the two points *D* and *I*. This is the place where they made the slip; if the ratio of *DH* to *HL* is indeed equal to the ratio of the square of *DH* to the square of *HI*, *HL* will be smaller than *HI*, because *HI* is smaller than *HD*. They then joined *LK* which is parallel to the straight line *DF*, because the ratio of *DH* to *HL* becomes equal to ratio of *FH* to *HK*. Next they said: angle *KLH* is equal to angle *FDH* and angle *KLH* is smaller than angle *KIH*, so angle *KIH* is greater than angle *FDH*; but angle *EIH* is greater than angle *EDH*, so angle *EIK* is greater than angle *EDF*; but it is equal to it, which is impossible. However, this impossibility is a necessary consequence of their hypothesis that the point *L* is above the point *I*; now the point *L* cannot but be below the point *I*; if it is below the point *I*, it does not give rise to this impossibility. And if this impossibility does not arise, it does not follow either that these two triangles are similar. It is because of this slip that they thought that the two triangles are always similar; but that is not so.

Having pointed out this slip, let us consider all the <possible> cases[1] for these two triangles[2] and let us show which are the cases where it is necessary that the two triangles be similar, and that there shall exist no other triangle that has the properties that pertain to these two triangles and is not similar to them. We also show which are the parts in which the two triangles are similar and where there also exists another triangle that has the same properties as they do and is not similar to them.

We say that the two triangles that have this property can be divided into several parts; in some parts, it is necessary that the two triangles be similar and that there shall exist no other triangle that has the properties that pertain to these two triangles and that is not similar to them; in other parts, it is necessary that the two triangles be similar and that there shall exist another triangle that has the same properties as they do and that is not similar to them.

So let us examine all the parts for these two triangles. The two triangles can first of all be divided into two parts: in one, the two angles which are at the points *G* and *H* are equal to the angles which are at the points *A* and *D*, and in the other, the two angles which are at the points *G* and *H* are different from the angles which are at the points *A* and *D*. Each of these two parts can then be divided into three parts according to whether the two angles at the points *A* and *D* are right angles, obtuse or acute. So there are six parts. If the two angles *A* and *D* are obtuse and the two angles at the points *G* and *H* are not equal to them, then they will be greater than them or smaller than them. If they are smaller than them, they will be either right angles or obtuse, so the <number of> cases increases by two. In the same way, if the two angles *A* and *D* are acute and the two angles which are at the points *G* and *H* are not equal to them, then they will be greater than them or smaller than them; if they are greater, then they are either right angles or acute and the <number of> cases is increased by two others. So there are ten cases. We shall explain the nature of each of these *cases*.

<First case>. – First of all, let the two angles *A* and *D* be right angles, let the two angles *G* and *H* also be right angles and let the ratio of the product of *BG* and *GH* to the square of *GA* be equal to the ratio of the product of *EH* and *HF* to the square of *HD*. It is possible for there to exist two similar triangles that have this property, as it is possible for there to exist two non-similar triangles that have this property.

---

[1] Lit.: parts.

[2] He means that this holds for all triangles that, two by two, satisfy the given conditions.

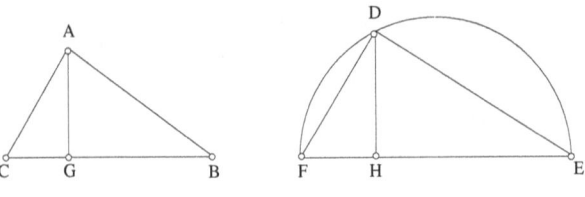

Fig. II.2

Proof: We return to triangle *ABC* and we draw a general straight line whatsoever, let it be *EF*. We construct on this line a semicircle, let it be *EDF*. We make the angle *FED* equal to angle *CBA*, we draw the perpendicular *DH* and we join *FD*; so triangle *EDF* is similar to triangle *ABC* and the two angles at the points *G* and *H* are both right angles. The product of *EH* and *HF* is equal to the square of *HD* and the product of *BG* and *GC* is equal to the square of *GA*, the two triangles thus have the property we have mentioned. However, it is possible for there to exist many triangles, each of which has this property without each of them being similar to the triangle *ABC*. In fact, if from any point taken on the arc *EDF*,[3] a perpendicular is drawn to the diameter *EF* and if this point is joined to the ends of the diameter, we generate a triangle which is not similar to the triangle *ABC*. However, the angle at its vertex is equal to angle *A*, the angle at its base is equal to angle *G* and the ratio of the product of the two parts of the base, which is *EF*, to the square of the perpendicular is equal to the ratio of the product of *BG* and *GC* to the square of *GA*. In this case, it is not necessary the two triangles shall always be similar unless we add to the conditions of the case a further condition, which is: the ratio of *AG* to *DH* is equal to the ratio of *BC* to *EF*, because it necessarily follows that the ratio of the square of *AG* to the square of *DH* is equal to the ratio of the square of *BC* to the square of *EF*; so the ratio of the product of *BG* and *GC* to the square of *BC* is equal to the ratio of the product of *EH* and *HF* to the square of *EF*, and the ratio of *BG* to *GC* is equal to the ratio of *EH* to *HF*.[4] It necessarily follows that triangle *DEH* will be similar to triangle *ABG*, and that triangle *DFH* will be similar to triangle *ACG*. This is the reason why the two triangles *ABC* and *DEF* are similar. If we do not add this condition, it does not necessarily follow that the two triangles *ABC* and *DEF* are similar. That is what we wanted to prove.

Second case. – Let the two angles *A* and *D* be right angles and let the two angles *G* and *H* be equal and not right angles. In this case it necessarily

---

[3] He means: other than the point *D*.
[4] See Mathematical commentary.

follows that the two triangles are similar and there does not exist any other triangle that has the same properties as they do and is not similar to them.

Let us return to the triangle *ABC*. We draw a general straight line; let it be *EF*. We construct a semicircle on this line and we make the angle *FED* equal to the angle *CBA*; we join *FD* and we draw from the point *D* a straight line *DH* to make the angle *DHE* equal to the angle *BGA*; the two triangles that are formed are thus similar to the triangles *ABG* and *AGC*. So the ratio of the product of *BG* and *GC* to the square of *GA* is equal to the ratio of the product of *EH* and *HF* to the square of *HD*. The two triangles *ABC* and *DEF* have the properties we have mentioned and they are also similar.

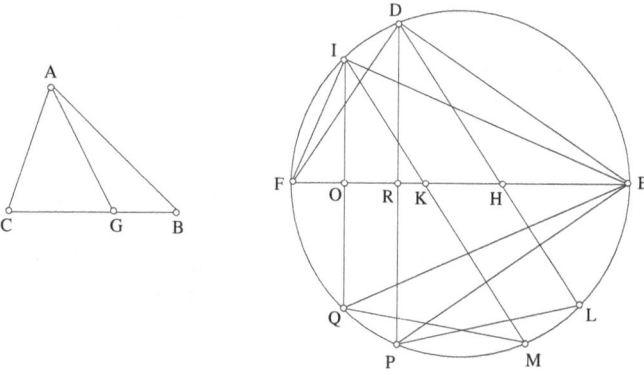

Fig. II.3

I say that it is not possible to find another triangle that has these properties and that is nevertheless not similar to the triangle *ABC*. If that is possible, let it be assumed to be done. It is thus possible to construct on the straight line *EF* a triangle similar to that triangle; the point at its vertex will thus lie on the arc *EDF* and the angle that is homologous to the angle *B* will thus not be equal to the angle *FED*. Let this triangle be the triangle *EIF*. Let the straight line *IK* be the line that makes with the straight line *EF* an angle equal to the angle *DHE*; thus *IK* will be parallel to the straight line *DH* and the ratio of the product of *EK* and *KF* to the square of *KI* is equal to the ratio of the product of *EH* and *HF* to the square of *HD*, if that were possible. We complete the circle *EDF* and we extend the two straight lines *DH* and *IK* to the two points *L* and *M*. We draw the two perpendiculars *DR* and *IO* and we extend them to the points *P* and *Q*; they divide one another into two equal parts at the points *R* and *O*. We join *LP* and *MQ*. Since the ratio of the product of *EH* and *HF* to the square of *HD* is equal to the ratio of the product of *EK* and *KF* to the square of *KI*, the ratio of *LH* to *HD* is equal to the ratio of *MK* to *KI*; so the ratio of *LD* to *DH* is equal to the ratio of *MI* to *KI*. But the two triangles *DHR* and *KIO* are similar, so the ratio of *HD* to *DR* is equal to the ratio of *KI* to *IO*, the ratio of *LD* to *DR* is equal to the ratio of *MI* to *IO*, the ratio of *LD* to *DP* is equal to the ratio of *MI* to *IQ* and the two angles *LDP* and *MIQ* are equal; so the two triangles *LDP* and *MIQ* are similar and the angle *DLP* is equal to the angle *IMQ*; so the sector *DEP* is similar to the sector *IEQ*; which is impossible. This impossibility is a necessary consequence of our assumption that the ratio of the product of *EK* and *KF* to the square of *KI* is equal to the ratio of the product of *EH* and *HF* to the square of *HD*. The triangle *EIF* has not got the same properties as the triangle *ABC*. We can show that the same holds for any triangle that is not similar to the triangle *ABC*. So any triangle that has the same properties as the triangle *ABC* is similar to the triangle *ABC*; and it follows in these two triangles also that the ratio of *AG* to *DH* is equal to the ratio of *BC* to *EF*, because the two triangles *ABG* and *AGC* are similar to the two triangles *DEH* and *DHF*. That is what we wanted to prove.

Third case. – Let the two angles *A* and *D* be obtuse and let the two angles *G* and *H* be equal to them. In this case it is necessary that the two triangles shall be similar and there does not exist any other triangle that has the same properties as they do and is not similar to them.

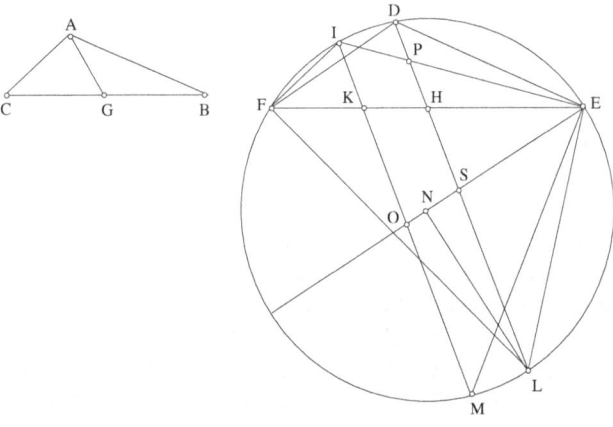

Fig. II.4

Let us return to the triangle *ABC* and let us draw a general straight line, let it be *EF*; on this we construct a portion of a circle that is intercepted by an angle equal to the angle *A*; we make the angle *FED* equal to the angle *CBA* and we join *FD*. The triangle *DEF* is then similar to the triangle *ABC*. Let us draw the straight line *DH* such that the angle *DHE* is equal to the angle *BGA* which is equal to each of the two angles *A* and *D*. The ratio of the product of *EH* and *HF* to the square of *HD* is then equal to the ratio of the product of *BG* and *GC* to the square of *GA*. So the two triangles *ABC* and *DEF* have the properties we mentioned and they are also similar.

I say that it is not possible to find another triangle that has the same properties as these two triangles and which nevertheless is not similar to these two triangles. If that is possible, let it be assumed to be done. We construct on the straight line *EF* a triangle similar to this triangle, the point at its vertex will then lie on the arc *EDF*, so the angle homologous to the angle *B* will not be equal to the angle *E*. Let this triangle be the triangle *EIF*. Let the straight line *IK* be the line that makes with the straight line *EF* an angle equal to angle *DHE*; so the straight line *IK* is parallel to the straight line *DH* and the ratio of the product of *EK* and *KF* to the square of *KI* is equal to the ratio of the product of *EH* and *HF* to the square of *HD*, if that is possible. We complete the circle *EDF* and we extend the two straight lines *DH* and *KI* to the two points *L* and *M*. Let the centre of the circle be the point *N*; we join *NE* and *EL*, the straight line *NE* cuts the two straight lines *DL* and *IM*; let it cut them at the two points *S* and *O*. Since angle *DHE* is equal to angle *FDE*, the product of *FE* and *EH* is equal to the square of *ED*. Since angle *DHE* is equal to angle *FDE*, angle *EHL* is equal to the angle in the sector *ELF*;[5] so the product of *FE* and *EH* is equal to the square of *EL*. So the straight line *EL* is equal to the straight line *ED* and the arc *EL* is thus equal to the arc *ED*, so the straight line *NE* is perpendicular to the two straight lines *DL* and *IM*, so *DS* is equal to *SL* and *IO* is equal to *OM*. But since the ratio of the product of *EH* and *HF* to the square of *HD* is equal to the ratio of the product of *EK* and *KF* to the square of *KI*, the ratio of *LH* to *HD* is equal to the ratio of *MK* to *KI*. The ratio of *LD* to *DH* is thus equal to the ratio of *MI* to *IK*, so the ratio of *SD* to *DH* is equal to the ratio of *OI* to *IK*. But the straight line *EI* cuts the straight line *DH*; let it cut it at the point *P*; then the ratio of *OI* to *IK* is equal to the ratio of *SP* to *PH*, so the ratio of *SP* to *PH* is equal to the ratio of *SD* to *DH*, and the ratio of *SH* to *HP* is equal to the ratio of *SH* to *HD*; which is impossible.

So it is not possible that there exists a triangle that has the same properties as triangle *ABC* and which is not similar to triangle *ABC*. That is what we wanted to prove.

Fourth case. – Let the two angles *A* and *D* be obtuse and let the two angles *G* and *H* also be obtuse and greater than the two angles *A* and *D*,

---

[5] We are concerned with the angle whose vertex is on the arc *ELF* and which intercepts the arc *EDF*.

then the two triangles are similar and there does not exist any other triangle
having the same properties which is not similar to them.

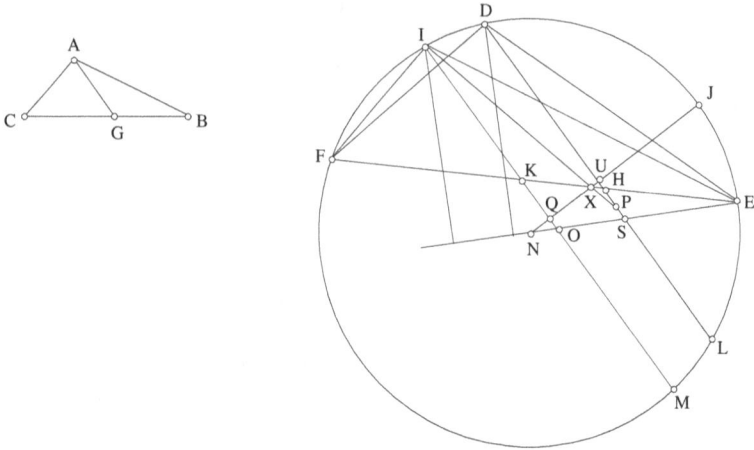

Fig. II.5

Let us return to triangle *ABC* and the previous circle; let triangle *DEF*
be similar to triangle *ABC* and let the properties of the first triangle be
those of the second. Let there be a triangle *EIF* which is not similar to
triangle *ABC* but such that its properties are those of the two triangles *ABC*
and *DEF*, if that is possible. We extend the two straight lines *DH* and *IK* to
L and *M*. So the ratio of *LD* to *DH* is equal to the ratio of *MI* to *IK*. But
since the angle *DHE* is greater than the angle *EDF*, thus the straight line
drawn from the point *D* that makes with the straight line *EF* an angle equal
to the angle *EDF* lies beyond the straight line *DH*, that is in the direction
towards the point *F*; and if we extend it, it meets the straight line *NE* and is
perpendicular to it; the same holds for the straight line that is parallel to it
drawn from the point *I*. Thus it is clear from this that the two angles *DSN*
and *ION* are acute.[6] So the perpendicular drawn from the point *N* to the two
straight lines *DL* and *IM* will be above the straight line *NE*, that is to say it
cuts the arc *ED*; let this perpendicular be the perpendicular *NQUJ*. It cuts
each of the straight lines *DL* and *IM* into two equal parts, so it cuts the
straight line *KE*; let it cut it at the point *X*. We join *XI*. It cuts the straight
line *DH*; let it cut it at the point *P*. Since the ratio of *LD* to *DH* is equal to
the ratio of *MI* to *IK*, the ratio of *QI* to *IK* is equal to the ratio of *UD* to *DH*;
but the ratio of *QI* to *IK* is equal to the ratio of *UP* to *PH*, so the ratio of

---

[6] This is true only when *I* is sufficiently close to *D* for *O* to lie between *N* and *S*.

UD to *DH* is equal to the ratio of *UP* to *PH* and the ratio of *UH* to *HD* is equal to the ratio of *UH* to *HP*. Which is impossible.

If the point *X* falls between the two points *H* and *K* or between the two points *K* and *F*, or at the point *H*, or at the point *K*, then this impossibility is even more unseemly. So it is not possible that there exists a triangle that has the same properties as triangle *ABC* and is not similar to triangle *ABC*. That is what we wanted to prove.

Fifth case. – Let the two angles *A* and *D* be obtuse and let the two angles *G* and *H* be right angles, then the two triangles are similar and there does not exist any other triangle that has the same properties as these and is not similar to them.

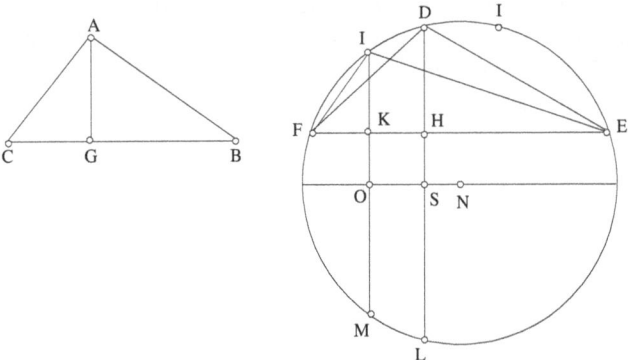

Fig. II.6

Let us return to the triangle *ABC* and the circle; let triangle *DEF* be similar to triangle *ABC* and let its properties be those of triangle *ABC*. Let triangle *EIF* not be similar to triangle *ABC*, but let its properties be those of the two triangles *ABC* and *DEF*, if that is possible. We draw the two straight lines *DH* and *IK* as far as *L* and *M*, then the ratio of *LD* to *DH* is equal to the ratio of *MI* to *IK*; let us draw from the centre of the circle, which is the point *N*, a perpendicular to the two straight lines *DL* and *IM*, let it be *NOS*; *NS* is thus parallel to the straight line *FE*, because the two angles *H* and *K* are right angles; so the ratio of *SD* to *DH* is equal to the ratio of *OI* to *IK*. The ratio of *SH* to *HD* is thus equal to the ratio of *OK* to *KI*; but *SH* is equal to *OK*, so *HD* is equal to *KI*; which is impossible because if *KI* were equal to *DH*, triangle *EIF* would be similar to triangle *EDF*, because the arc *IF* would be equal to the arc *ED* and, in consequence, angle *IEF* would be equal to angle *EFD*, angle *IFE* would be equal to angle *DEF* and triangle *EIF* would thus be similar to triangle *EDF*; but, by hypothesis, it is not similar to it. If triangle *EIF* is not similar to triangle *EDF*, then the straight line *IK* is not equal to the straight line *DH*; so the ratio of *LH* to *HD* is not equal to the ratio of *MK* to *KI*. So the ratio of the product of *EK* and *KF* to the square of *KI* is not equal to the ratio of the product of *EH* and *HF* to the square of *HD* and triangle *EIF* does not have the property found in the two triangles *ABC* and *DEF*. So there does not exist any other triangle that is not similar to the two triangles *ABC* and *DEF*, and has the same properties as they do. That is what we wanted to prove.

Sixth case. – Let the two angles *A* and *D* be obtuse and let the two angles *G* and *H* also be obtuse, but smaller than the two angles *A* and *D*; let the ratio of the product of *BG* and *GC* to the square of *GA* be equal to the ratio of the product of *EH* and *HF* to the square of *HD*.

I say that there can exist two triangles that have this property and are similar and that further there exists another triangle that has this property without being similar to these two similar triangles.

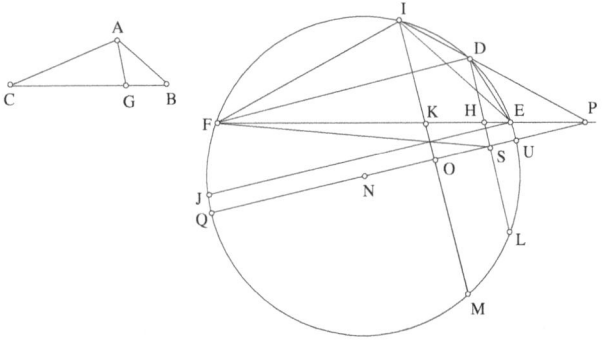

Fig. II.7

Proof: We draw a circle; let it be *EDFM*. Let us cut off on it a part smaller than a semicircle; let the part be *EDF*. We extend *FE* as far as *P* and on this straight line we take a general point; let it be the point *P*. We draw from the point *P* a straight line that cuts the part <of the circle> *EDF* in two points; let the two points lie in the half of the arc *EDF* towards the point *E*; let the straight line be *PDI*; let the centre of the circle be the point *N*. We join *NP*; let it cut the circle at the point *U*. We draw from the two points *D* and *I* two perpendiculars to the straight line *NP*; let the two perpendiculars be *DS* and *IO* which we extend to *L* and *M*; the perpendiculars are thus divided into two equal parts at the points *S* and *O*. We extend *PN* to *Q* and we draw *EJ* parallel to the straight line *PQ*; so angle *FEJ* is equal to angle *FPQ*. Since the two angles *S* and *O* are right angles, the two angles *PHS* and *PKO* are acute, so the two angles *DHE* and *IKE* are obtuse. Since angle *S* is a right angle, the sum of the two angles *SPD* and *SDP* is a right angle, so the two angles *SPD* and *SDP* intercept the arc *UDFQ* which is half of the circle. But angle *JEF*, which is equal to angle *QPF*, intercepts the arc *FJ*; so there remain the two angles *HPD* and *HDP*, that is angle *DHF*, which is the angle that intercepts the arcs *UE*, *ED*, *DF* and *JQ*. So angle *DHF* is smaller than a right angle by an angle that intercepts the arc *FJ*, so angle *DHE* exceeds a right angle by an angle that intercepts the arc *FJ*, so angle *EDF* exceeds a right angle by an angle that intercepts the two arcs *UE* and *FQ* and angle *EDF* exceeds the obtuse angle *DHE* by the angle that intercepts the two arcs *UE* and *FQ*.

Having shown that angle *EDF* is greater than angle *DHE*, let us show that there exist two triangles that have the properties we mentioned and that are nevertheless not similar. We join the straight lines *ED*, *EI*, *FD* and *FI*. Let triangle *ABC* be similar to triangle *DEF*. Since *DS* is parallel to *IO*, the ratio of *SD* to *DH* is equal to the ratio of *OI* to *IK*, so the ratio of *LD* to *DH* is equal to the ratio of *MI* to *IK*. Thus the ratio of *LH* to *HD* is equal to the ratio of *MK* to *KI*. So the ratio of the product of *EH* and *HF* to the square of *HD* is equal to the ratio of the product of *EK* and *KF* to the square of *KI*; so triangle *EIF* has the same properties as the two triangles *ABC* and *DEF*, although it is not similar to them because *IK* is greater than *DH*, since <*D* and *I*> are both on half the arc *EDF*; so the angles <of the triangle *EIF*> are not equal to the angles of the triangle *EDF*.

So if the two angles *A* and *D* are obtuse, the two angles *G* and *H* obtuse and smaller than the two angles *A* and *D* and if the ratio of the product of *BG* and *GC* to the square of *GA* is equal to the ratio of the product of *EH* and *HF* to the square of *HD*, then the two triangles *ABC* and *DEF* are similar and there nevertheless exists a triangle that has these properties but which is not similar to them. That is what we wanted to prove.

Seventh case. – Let the two angles *A* and *D* be acute and let the two angles *G* and *H* be equal to them. In this case it is necessary that the two triangles be similar and there does not exist any other triangle that has the same properties as they do and which is not similar to them.

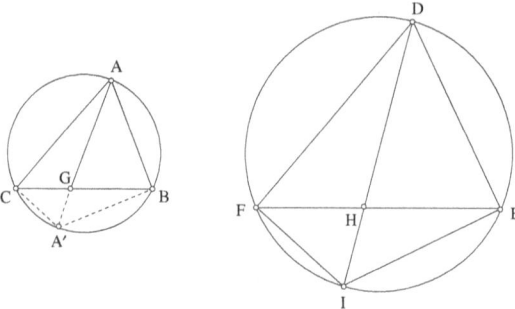

Fig. II.8

Let us return to triangle *ABC* and the circle and let us cut off from the circle a part <that is the arc> that is intercepted by an acute angle equal to angle *BAC*. Let the part be *EDF*. We make angle *FED* equal to angle *CBA* and we join *FD*. So triangle *DEF* is similar to triangle *ABC*. We draw *DH* in such a way that angle *DHE* is equal to angle *EDF*; let it be *DHE*. If the point *G* lies inside the triangle *ABC*, then the point *H* lies inside the triangle *DEF*. We extend *DH* to *I* and we join *EI* and *FI*, so angle *EHI* is equal to angle *EIF*. It necessarily follows that triangle *EIF* has only one triangle that is similar to it and has the properties of triangle *EIF* and there does not exist any other triangle that has the properties of triangle *EIF* and which is not similar to them. If there does not exist any other triangle that has the same properties as triangle *EIF* and which is not similar to it, then there does not exist any other triangle that has the same properties as triangles *ABC* and *DEF* and which is not similar to them. The two triangles *ABC* and *DEF* are similar and there does not exist any other triangle that has the properties of these two triangles without being similar to them. That is what we wanted to prove.

Eighth case. – Let the two angles *A* and *D* be acute and let the two angles *G* and *H* be smaller than them.

In this case, it necessarily follows that the two triangles are similar and there does not exist any other triangle that has their properties and which is not similar to them. In fact, if we make triangle *DEF* similar to triangle *ABC*, if we extend the straight line *DH* as far as *I* and if we complete the triangle *EIF* such that angle *EHI* is greater than angle *EIF*, it is necessary that triangle *EIF* shall have a triangle similar to it that has the properties of triangle *EIF* and there does not exist any other triangle that has their properties and which is not similar to them. So it is necessary that there shall not exist for the two triangles *ABC* and *DEF* any other triangle that has their properties and which is not similar to them. That is what we wanted to prove.

Ninth case. – Let the two angles A and D be acute and let the two angles G and H be right angles. So if we draw DH and complete the triangle EIF, we show <in the same way> as we showed in the fifth case that there exists for the triangle EIF a triangle similar to it that has the same properties as it does and there does not exist any other triangle that has the same properties as it does and which is not similar to it, so it is necessary that there does not exist for the two triangles ABC and DEF any other triangle that has the same properties as them and which is not similar to them.

Tenth case. – Let the two angles A and D be acute and let the two angles G and H be acute, but greater than A and D.

It necessarily follows that angle EHI is smaller than angle EIF. We show <in the same way> as we showed in the sixth case, that there can exist for the triangle EIF a triangle similar to it and which has the same properties as it does, that there exists another triangle that has the same properties as the triangle EIF and which is not similar to it. It necessarily follows that the two triangles ABC and DEF are similar and that there exists another triangle that has the properties of these two triangles and which is not similar to them.

The cases into which this proposition is divided are ten in number: in seven of them, the assertion mentioned by the Banū Mūsā is true, and in three of them this assertion does not hold. The cases in which this assertion mentioned by the Banū Mūsā is true, require that the ratio of the base of the triangle to the base of the triangle shall be equal to the ratio of the straight line drawn to the base of one to the straight line drawn to the base of the other; if, indeed, the two triangles are similar, then their angles are equal. So we suppose that each of the two triangles into which one of the two large triangles is divided is similar to the corresponding figure in the other large triangle, it necessarily follows that the ratio of the two parts of the base of one of the two triangles, one to the other, is equal to the ratio of the two parts of the base of the other triangle, one to the other. It necessarily follows that the ratio of the base of one of the two large triangles to the straight line drawn to that base is equal to the ratio of the base of the other large triangle to the straight line drawn to this base. It necessarily follows that the ratio of the straight line drawn to the other straight line <that has been> drawn is equal to the ratio of the base to the base.

So if we add to the conditions <imposed> for the two triangles that the ratio of the straight line drawn to the other straight line <that has been> drawn is equal to the ratio of the base to the base, the proposition becomes general and will not be defective in any case. All the cases of this proposition that are used in the *Conics* are among the true cases which we have proved and none of the defective cases is used in the *Conics*.

It is thus clearly apparent, from all that we have shown, that the proposition that the Banū Mūsā employed in assessing these two triangles is not a general proposition, that is it is true in certain cases of the two triangles and that it is false in the other cases of the triangles, that is that two triangles that have the properties the Banū Mūsā have mentioned are not always similar, but in certain cases they are similar and in other cases they are not similar. We have also shown the slip they made in the course of the demonstration of this proposition; that is what we wanted to show in this treatise.

The treatise on the proposition of the Banū Mūsā is completed.

CHAPTER III

PROBLEMS OF GEOMETRICAL CONSTRUCTION

3.1. THE REGULAR HEPTAGON

3.1.1. *Introduction*

One of the geometrical construction problems that most interested mathematicians between the ninth and eleventh centuries was that of constructing the regular heptagon. If we are to believe what the Arabic mathematicians themselves say, this problem belongs to a whole group of three-dimensional (solid) problems inherited from Greek mathematicians; but historians have placed too little emphasis upon what makes it different from other such problems. Unlike, for example, investigations of the two means or the trisection of an angle, work on the regular heptagon was resumed rather late. Whereas the first two problems were subjects of intense research from the mid ninth century onwards, from such early scholars as the Banū Mūsā, Thābit ibn Qurra and others, it was not until the second half of the tenth century that study of the regular heptagon was resumed. But once it started, such research became something of a craze among eminent mathematicians: it is as if each of them wanted to leave his mark on it. This is how, a little later on, Ibn al-Haytham presented such research:

> One of the geometrical problems – over which geometers enter into rivalry, \<problems\> in which those who surpass the others take pride, and through which the prowess of those who succeed in solving it is revealed – is the construction of a regular heptagon in a circle.[1]

Why was there such enthusiasm, and why did work start relatively late? The fashion for studying this topic may be explained by the high standing of Archimedes. The mathematicians of the time, including Ibn al-Haytham, refer to the lemma that Archimedes proposed for this construction. It is clear that they are concerned with a text attributed to Archimedes, and one of which no trace has come down to us. If we are to believe the mathematicians, Archimedes had not, however, proved his lemma, but had

---

[1] See below, *On the Construction of the Heptagon in a Circle*, p. 441.

simply assumed it was accepted as true. So the regular heptagon basked in the reflected glory of Archimedes, but did so without having been constructed by him: a state of affairs peculiarly intriguing and exciting. But that does not explain why it was not until the second half of the tenth century that mathematicians revived a problem already known, they say, to Thābit ibn Qurra and his contemporaries.[2] The time that elapsed might be explained by reference to the new interests that emerged during this period. These interests, which have various origins – algebraic and geometrical – encouraged mathematicians to investigate this problem, along with many others previously neglected to which we shall return. Interest in matters of algebra, as we have seen elsewhere,[3] is directed to the development of a theory of algebraic equations of degree less than or equal to three, and to research in algebraic geometry. As we shall see, the interest in geometry is bound up with the changes brought about in work on geometrical construction by the introduction of the use of conic sections.

To construct the regular heptagon, Ibn al-Haytham's predecessors first constructed a triangle in a given circle, the triangle having a particular ratio between its angles. Thus Abū al-Jūd, and similarly al-Sijzī, considered the ratio (1, 3, 3); Ibn Sahl tried to prove Archimedes' lemma;[4] al-Qūhī made separate studies of two cases, (1, 2, 4) – which takes us back to Archimedes' lemma, as we shall see – and (1, 5, 1). As for al-Ṣāghānī, he too discusses the case (1, 2, 4). In a first essay, with the title *Treatise on the*

---

[2] Qusṭā ibn Lūqā mentions the problem of the regular heptagon as one of those that illustrate a special power. That is, in the mid ninth century, mathematicians knew of the existence of this problem and its difficulty. See 'Une correspondance islamo-chrétienne entre Ibn al-Munaǧǧim, Ḥunayn ibn Isḥāq et Qusṭā ibn Lūqā', Introduction, édition, divisions, notes et index par Khalil Samir; Introduction, traduction et notes par Paul Nwyia in F. Graffin, *Patrologia Orientalis*, vol. 40, fasc. 4, no. 185, Turnhout, 1981, pp. 674-6:

وماذا تقولُ .... في مـستخرج خطّين بينَ خطّين على مناسبةٍ، وفي تدوير المُسَبَّع، وفي غير ذلك من الأشياء التي عجز الناسُ عن وجودها ؟ إنْ أخرجَها لك أحدٌ من أهلِ زماننا، أتُنزلُ ذلك من فعْله منزلةَ إحياءِ الموتى وفَلقِ البحر، وتُقرُّ له بالنُبوَّةِ؟

'What do you say [...] about someone who finds two segments between two segments in continuous proportion, <solves the problem of> the inscription of the regular heptagon in the circle and other matters that men were incapables of solving? If one of our contemporaries were to solve these problems, would you place his action on the same level as raising the dead and parting the sea, and would you recognize him as thus fulfilling the prophecy?'.

[3] R. Rashed, 'Algebra', in *id.* (ed.), *Encyclopedia of the History of Arabic Science*, London, 1996, pp. 349–75.

[4] See below, pp. 295ff.

Determination of the Lemma of the Heptagon, Ibn al-Haytham himself tries to prove Archimedes' lemma for constructing the heptagon. In this essay, he considers a triangle whose angles are in the ratio (1, 2, 4). In a second essay, with the title *On the Construction of the Regular Heptagon* – a more substantial piece that was written later – he undertakes the first systematic study of all possible constructions, and thus of all the triangles that can be formed from sides and diagonals of the regular heptagon. In this essay, he deals in succession with the cases (1, 3, 3), (3, 2, 2), (1, 5, 1) and finally (1, 2, 4). We may note that for the case (1, 3, 3), the analysis leads to division of a segment, that is a range of points on a line, not to be found in any of the studies by his predecessors, and that for the case (1, 2, 4) Ibn al-Haytham begins by proving that such a case can be obtained from preceding ones: (1, 3, 3), (3, 2, 2) and (1, 5, 1). He does, however, give an analysis of (1, 2, 4) that leads to the division of the segment proposed in Archimedes' lemma (see below). Here we see the emergence of a characteristic of these constructions that is important for several reasons: an emphasis on the geometrical nature of the problem. Ibn al-Haytham, following the example of his predecessors, reduces the construction of the regular heptagon to that of a triangle and not that of the angle $\frac{\pi}{7}$; from then on, there are in all four possible triangles, and he examined the construction of each of them. To say that the problem reduces to finding the angle $\frac{\pi}{7}$ is to interpret Ibn al-Haytham's text in a framework that is foreign to it. What he does is to construct a triangle whose angles are in the ratio (3, 2, 2). Ibn al-Haytham's view is in fact deliberately geometrical and not trigonometrical. Finally, as in his essay on Archimedes' division of the line, Ibn al-Haytham – unlike some of his predecessors, such as Abū Naṣr ibn 'Irāq[5] – was not tempted by the possibility of making an algebraic version of the problem. His proof is deliberately geometrical, and he is intent on considering all possible cases.

Once we have reconstructed the research tradition to which it belongs, Ibn al-Haytham's contribution allows us to discern two intentions on the part of its author. He obviously wished to give a systematic account of what was known, thereby settling the question of constructing the heptagon, and thus bringing the tradition to a successful conclusion. But if we now examine the construction problems Ibn al-Haytham inherited from Antiquity, we can only find two, both of which carry the name of Archimedes: the regular heptagon and Archimedes' division of the line (the lemma for the fourth proposition of the second book of *The Sphere and the Cylinder*). It is, it

---

[5] R. Rashed and B. Vahabzadeh, *Omar Khayyam. The Mathematician*, Persian Heritage Series no. 40, New York, 2000, p. 174.

seems to me, an important fact, despite its having escaped attention, that Ibn al-Haytham never deals with any other problem inherited from his Greek or Arabic predecessors. Perhaps his purpose was to remedy the inadequacies of this unfinished work by Archimedes, as he had done for Apollonius' *Conics*. Was that his intention, one of the motives for his three pieces on the solid problems that had been passed down from Antiquity? The suggestion is worth some consideration.

First, we shall reconstruct the tradition of research on the construction of the regular heptagon, and then we shall examine each of Ibn al-Haytham's two treatises.

3.1.2. *The traces of a work by Archimedes on the regular heptagon*

It is worth repeating that the history of the problem of constructing the regular heptagon differs from that of the other solid problems: trisecting an angle, the two means, the lemma for the fourth proposition of the second book of Archimedes' *The Sphere and Cylinder*. Apart from some references in archaeological sources (Babylonian tablets dating back to about 1800 BC) and a practical method like that of Heron of Alexandria,[6] we know of no

---

[6] On several mathematical tablets of the Old Babylonian period (about 1800 BC) we find carefully drawn geometrical figures, including the regular heptagon. Heron of Alexandria returns more than once to the regular heptagon, not to explore its geometrical properties, but to give an approximate calculation. See Heron, *Metrica*, ed. E. M. Bruins, *Codex Constantinopolitanus Palatii Veteris n. 1*, Part 2 [Greek Text], Leiden, 1964, pp. 101–2. Heron writes as follows:

'Lemma. If a regular heptagon is inscribed in a circle, the ratio of the radius to the side of the heptagon will be equal to 8/7. Let there be the circle *BC* with centre *A*, and let us construct inside it the side *BC* of the hexagon, that is a side equal to the radius of the circle. Let us draw on this side the height *AD*. *AD* will thus be almost equal (ὡς ἔγγιστα) to the side of the heptagon. Let us join *BA*, *AC*. The triangle *ABC* will thus be equilateral. So the square of *AD* will be three times the square of *DB*. So the ratio of *AD* to *DB* to the power of two will be almost equal to 49/16. And in length, the ratio of *AD* to *DB* will be equal to 7/4. And *BC* is double *BD*. The ratio of *BC* to *DA* is thus equal to 8/7.' E. M. Bruins commented on this lemma as follows: 'In this lemma the ratio of the side of the heptagon to the radius of the circle is *not* deduced from the properties of this regular polygon. Equating the side of the heptagon to the perpendicular from the centre of the circle to the side of the heptagon, i.e. equating the side of the heptagon to one half of the side of the equilateral triangle in the same circle, the ratio 7 to 8 is computed' (Part 3, pp. 230–1).

(*Cont. on next page*)

contribution on the construction of this figure from the ancient world – with one exception: a work attributed to Archimedes, that only Arabic sources take into account. The ancients' complete silence is as surprising as it is intriguing, and prompts reflection. One may wonder why, after all, so much interest was shown, for example, in the trisection of an angle (on the evidence of Pappus' *Mathematical Collection*), whereas there was such apparent indifference to the heptagon. The same question might be asked about the lemma in *The Sphere and Cylinder*, where, this time, the evidence is supplied by Eutocius. What makes the heptagon the exception despite the fact, if we are to believe the Arabic sources, that the problem was raised by Archimedes?

It is even more surprising that this silence from antiquity was not much interrupted in the ninth century, that is, even when, it seems, Archimedes' text was available. The scarce testimony we have from that period leaves no doubt that the problem was known. As we have seen, Qusṭā ibn Lūqā, the famous translator and scholar, mentions it in a religious debate, as a wonderful discovery, without giving the name of Archimedes. On the other hand, he has done no better than any other mathematician of the time by not leaving us any indication of how the figure constructed. Nevertheless the biobibliographer al-Nadīm cites in his list of the works of Archimedes known in Arabic translation 'a book on the regular heptagon (*Kitāb fī tasbī' al-dā'ira)'.[7] But he is niggardly with further information and does not supply either the name of the translator or the date of the translation. However, if the translation did indeed take place, we may suppose this happened in the ninth century. Apart from these bare indications, all we know about the work attributed to Archimedes and its possible translator goes back to a single, and very late, source – it dates from the eighteenth

---

(*Cont.*) Heron also writes: 'Let there be a regular heptagon *ABCDEFG*, with side equal to 10 units. To find its area. Let us take the centre of the circle circumscribed about it, *H*, and let us join *DH*, *HE*. Let *HI* be the perpendicular to *DE*. The ratio of *HD* to *DE* is thus equal to 8/7 and its ratio to *DI* is equal to 8 over 3 ½, or 16/7. So that the ratio of *HI* to *ID* is almost exactly (ὡς ἔγγιστα) that of 14⅓ to 7, or 43/21. From that it follows that the ratio of *DE* to *IH* is equal to 42/43, or 84/86. From that it follows that the ratio of the square of *DE* to *DE. IH* is the same. From that it follows that the ratio of *DE* to the triangle *DHE* is equal to 84/43. Now the ratio of the triangle to the heptagon is 1/7. The ratio of the square of *DE* to the heptagon is thus equal to 12/43. And the square of *DE* is given. The heptagon is thus given. The synthesis is carried out in the following way: multiply 10 by itself; we obtain 100; multiply that by 43; we obtain 4300; divide by 12; we obtain 358 ⅓. That will be the area of the heptagon'.

Here too we have an approximate calculation. See also E. M. Bruins, Part 3, p. 234.

[7] Al-Nadīm, *Kitāb al-fihrist*, ed. R. Tajaddud, Tehran, 1971, p. 326.

century – a well known copyist: Muṣṭafā Ṣidqī.[8] Educated and familiar with mathematics, Muṣṭafā Ṣidqī transcribed a treatise called *Book of the Construction of the Circle Divided into Seven Equal Parts, by Archimedes, Translated by Abū al-Ḥasan Thābit ibn Qurra de Ḥarrān. A Single Book in Eighteen Propositions.*[9] A work by Archimedes translated by Thābit ibn Qurra is an important document, and was considered as such after being translated into German by C. Schoy.[10] However, enthusiasm quickly abated: the copyist, in all honesty, delivers a prompt warning when he writes:

> when I wanted to transcribe this book, I had only obtained one copy, damaged and affected by the ignorance of the copyist and his lack of understanding. I did all I could to force myself to check the problems, to make the synthesis of the analyses and to set out the propositions in terms that are easy and accessible; and I have introduced several proofs by modern mathematicians [...].[11]

Muṣṭafā Ṣidqī's admission puts us in an impossible situation. There are two overlapping questions that preclude us from drawing any firm conclusion: What remains of the original version after this work of editing, commentary and making additions? What evidence have we that this original version was truly the work of Archimedes? As it is, Muṣṭafā Ṣidqī's edition is, at best, an edition that has been rewritten, incorporating proofs by late mathematicians like al-Ḥubūbī and al-Shannī; at worst, the edition is a 'pot-pourri', in which the problem of the heptagon takes up only two of the eighteen propositions. But the text also raises other difficulties. The first sixteen propositions are very elementary: they deal with calculation with segments of a straight line using some notable identities, $(a \pm b)^2$ and $(a + b + c)^2$, they also consider Pythagoras' theorem, and expressions for the area of a right-angled triangle. From Proposition 9 onwards, we meet the circle inscribed in a triangle, and we use the property of there being two tangents from a point outside the circle and the concept of the power of a point. On occasion the editor gives three or four proofs of the same proposition, two such proofs being explicitly attributed to tenth-century mathematicians such as al-Shannī. In short, the principle behind the

---

[8] R. Rashed, *Founding Figures and Commentators in Arabic Mathematics*. A history of Arabic sciences and mathematics, vol. 1, Culture and Civilization in the Middle East, London, 2012, p. 126; R. Rashed, *Geometry and Dioptrics in Classical Islam*, London, 2005, e.g. p. 10.

[9] See Appendix I: *Kitāb 'amal al-dā'ira al-maqsūma bi-sab'a aqsām mutasāwiya li-Arshimīdis*, MS Cairo 41, fols. 105ʳ–110ʳ.

[10] See C. Schoy, *Die trigonometrischen Lehren des persischen Astronomen Abū'l Raiḥān Muḥammad Ibn Aḥmad al-Bīrūnī*, Hanover, 1927, pp. 74ff.

[11] See below, p. 587.

organization of the text is not clear, and we may note that *none* of the first sixteen propositions is of any use for the last two, which concern the heptagon. Which amounts to saying that we cannot expect an analysis of the content to yield any indication that might allow us to ascribe such a work to Archimedes or to Thābit ibn Qurra.[12]

---

[12] This is so for the tenth proposition, the one that is closest to those that relate to Archimedes' lemma. This proposition can be rewritten:

Let I be the point of intersection of HD with the perpendicular to CB at C; then

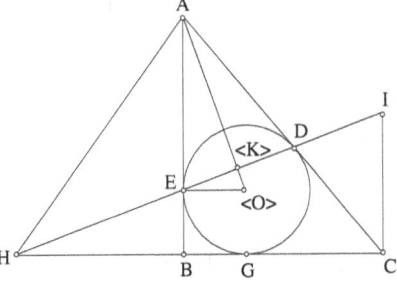

Fig. 3.1

$$\frac{CH}{HB} = \frac{DC}{EB}.$$

We have $CI \perp CB \Rightarrow CI \parallel AE \Rightarrow$ triangles ICD and EAD are similar and

$$\frac{AE}{CI} = \frac{AD}{CD} \Rightarrow \frac{AE}{AD} = \frac{CI}{CD}.$$

Now $AE = AD$, so $CI = CD$.

On the other hand, triangles HBE and HCI are similar, hence

$$\frac{CH}{HB} = \frac{CI}{EB} = \frac{CD}{EB}.$$

Note: Thus we have

$$\frac{CH}{CB} = \frac{CG}{BG} \Rightarrow CB \cdot CG = CH \cdot BG,$$

which gives a property of the range (C, B, G, H), a property that will not be used later when this range is investigated.

What is much more serious and significant is the error that seems to have been made in the ninth proposition, an error that neither Archimedes nor Thābit ibn Qurra could have committed. This proposition can be rewritten:

Let ABC be a triangle right-angled at B, circumscribed about a circle DEG; then $BH = AD$ (see Fig. 3.1).

Here in brief is the proof given in the text:

(1) $HD \cdot HE + EA^2 = AH^2 \Rightarrow HG^2 + EA^2 = AB^2 + BH^2,$

hence

$$EA^2 + HG^2 = HB^2 + AB^2 \Rightarrow EA^2 + BG^2 + 2\,HB \cdot BG = AB^2,$$

hence

$$BG^2 + 2\,HB \cdot BG = AB^2 - EA^2 = 2\,AE \cdot EB + EB^2;$$

but $BG = EB$, hence the result.

(*Cont. on next page*)

When we try to authenticate the work at least partly, with the help of its presumed translator, we quickly run up against a stone wall. For indeed if we turn to the titles of works by Thābit ibn Qurra – original works and translations – we can only be disappointed: there is no source that would allow us to ascribe any such translation to him, and no old biobibliographer suggests anything like it.[13] Even more significantly, the title does not appear in the list of Thābit ibn Qurra's works that was kept by his family. This list is, understandably, to be taken as authoritative, and was provided by his grandson al-Muḥsin Ibrāhīm al-Ṣābi'.[14] We are left what is to be found in the tenth-century mathematicians who were interested in the heptagon. Almost all of them agree in saying that Archimedes wrote a work on the heptagon that included a famous lemma. But whereas the work is described in slightly vague terms, the lemma is well reported; nevertheless, even though the sense is there, the details vary each time the statement of the lemma is presented as appearing in the work designated as by Archimedes. Prudence is required.

From Abū al-Jūd ibn al-Layth, in 968–969, up to Ibn al-Haytham in the following century, there are references to what the former calls 'a treatise by Archimedes on the construction of the heptagon (*Risālat Arshimīdis fī 'amal al-musabba'*)'. Abū al-Jūd's contemporary, al-Qūhī, says of this treatise:

(*Cont.*) Now, the equality given by the author – $HD \cdot HE + EA^2 = AH^2$ – does not follow directly from the hypotheses, but is deduced from the required result

(2) $BH = AE = AD$.

This is an error in the reasoning, because we cannot suppose that the statement of the proposition contains an error that affects the lettering, or an error in regard to the required property. In fact, the results (1) and (2) are correct: we prove (2), and we deduce (1). Moreover, equality (2), together with the equality $HG = AB$ that is deduced from it, are used in Proposition 11 to show that area $(ABC) = AD \cdot AC$.

Note: To show that we have $BH = AD = AE$, let us consider the centre O of the circumscribed circle and the point K, the point of intersection of AO and ED. We have $A\hat{E}O = A\hat{K}E$, one right angle; so $A\hat{O}E = A\hat{E}K$. But $A\hat{E}K = H\hat{E}B$, so $A\hat{O}E = H\hat{E}B$; so triangles AEO and HEB are similar and also have one side equal, because $EB = BG = EO$; thus they are congruent, and $HB = AE = AD$.

Let us show that (2) $\Rightarrow$ (1). If $BH = AD$, we also have $AE = HB$ and $AB = HG$. Moreover, $HG^2 = HE \cdot HD$, hence $AH^2 = HB^2 + AB^2 = HE \cdot HD + AE^2$.

This proof could not have escaped the notice of either Archimedes or Thābit ibn Qurra; it thus rules out any possibility of the text being authentic.

[13] We have reached this conclusion after examining all the old biobibliographical writings.

[14] R. Rashed, *Founding Figures and Commentators in Arabic Mathematics*, Chapter II, p. 121.

It is a subtle book whose purpose he (Archimedes) did not fulfil nor did he achieve his aim in finding the heptagon by a single method, not to speak of more than one.[15]

Although the wording is not very precise, it nevertheless (unless we are determined upon total scepticism) gives us a rational basis for drawing the conclusion that there existed a text attributed to Archimedes, translated into Arabic and consulted by the mathematicians of the mid tenth century. What we need to know is whether that work is the same as the text copied by Muṣṭafā Ṣidqī. Archimedes' lemma, which tenth-century mathematicians found valuable for a variety of reasons, provides a starting point for the discussion.

In the edition by Muṣṭafā Ṣidqī, the lemma is stated as follows:

Let us suppose we have a square *ABCD*. We extend the side *AB* in the direction of *A* as far as *E*; we join the diagonal *BC* and we put one end of the ruler at the point *D* and the other end on the straight line *EA* which is such that it cuts *EA* at the point *G* and makes the triangle *GAH* equal to the triangle *CID*. We draw from the point *I* the straight line *KIL* parallel to *AC*. I say that the product of *AB* and *KB* is equal to the square of *GA*, that the product of *GK* and *AK* is equal to the square of *KB* and that each of the straight lines *BK* and *GA* is longer than the straight line *AK*.[16]

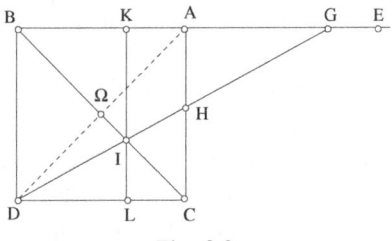

Fig. 3.2

In this edition, the lemma is presented as a kind of *neusis* for carrying out the division of the line segment, which, according to T. L. Heath, is a construction problem Archimedes solved 'by means of a ruler, without troubling to show how it might alternatively be solved by means of conics or otherwise'.[17] If that is so, the proof follows immediately. From the equality of the areas *AHG* and *CID*, we obtain

$$DC \cdot IL = AG \cdot AH \Rightarrow \frac{AB}{AG} = \frac{AH}{IL}.$$

---

[15] See below, p. 651.

[16] See below, p. 602.

[17] T. L. Heath, *A Manual of Greek Mathematics*, New York, 1963, p. 341.

Since triangles *GAH*, *GIK* and *ILD* are similar, we obtain

(1) $\qquad \dfrac{AH}{IL} = \dfrac{GA}{LD} = \dfrac{GA}{KB} \Rightarrow \dfrac{AB}{AG} = \dfrac{GA}{KB} \Rightarrow GA^2 = AB \cdot KB.$

We also have

(2) $IL = KA, \quad KI = BK = DL \Rightarrow \dfrac{KA}{KB} = \dfrac{IL}{KI} = \dfrac{DL}{GK} = \dfrac{BK}{GK} \Rightarrow BK^2 = KA \cdot GK.$

Finally, $GK > AK$; so, from (2), $KB > KA$. But $AB > KB$, and from (1) $GA > KB$, so $GA > AK$. The inequalities $BK > KA$ and $GA > AK$ follow from (1) and (2).

It is enough to look at each of the versions of Archimedes' lemma that appear in the tenth century to recognize that all of them differ from the statement we have just cited, insofar as none of them mentions the movable ruler. Here is how Abū al-Jūd, quoted by al-Sijzī, reports Archimedes' lemma:

> Let us draw the diagonal of the square *ABCD*, say *AC*. Let us extend *AB* to *E*, without an endpoint, and let us draw from a point of *BE*, let it be *E*, a straight line to the corner of the square, at the point *D*, which cuts the diagonal *AC* at the point *G* and the side *BC* at the point *H*, so that the triangle *BHE* outside the square is equal to the triangle *CDG*.[18]

So it is indeed the same figure (almost down to the lettering), but there is absolutely no sign of the movable ruler. The same is true for all the versions of this lemma in the writings of tenth- and eleventh-century mathematicians, and even those of later ones writing in Arabic.

So there is an important and irreducible difference between version of the lemma known to the mathematicians of the tenth and eleventh centuries and that attributed to Archimedes in the eighteenth-century manuscript: the use of the movable ruler. If it had been mentioned, the use of the instrument could not have escaped the notice of geometers like al-Qūhī and Ibn al-Haytham. Although perfectly capable of appreciating the ingenuity of the procedure, they would not have been able to accept it as legitimate, that is as a technique leading to a valid geometrical construction. We are not putting forward only an argument *ex silentio*, because it concerns the very nature of the work of mathematics, above all when we are considering a scholar as

---

[18] See *Book on the Construction of the Heptagon in the Circle and the Trisection of a Rectilinear Angle of Aḥmad ibn Muḥammad ibn 'Abd al-Jalīl al-Sijzī*, below, p. 631.

preoccupied with existence questions as Ibn al-Haytham was. Would he, and his predecessors, have complained about the way Archimedes expressed the matter? That is hard to believe, it is not his style, no more than it is theirs; it is more likely that they would have raised the question of whether the construction was valid. Their books of criticisms indeed show their taste for this.[19]

We might hope that there is a way out of this situation by a judicious deployment of established philological methods. But there is not. A careful and detailed examination of Muṣṭafā Ṣidqī's language – a purely classical Arabic – reveals not even a hint of employing Greek, and does not permit us to identify any vestige of a translation anywhere in the text. Finally, perhaps one might reject the idea that Archimedes ever wrote anything on the heptagon. This solution would be cavalier in several respects.

It does happen that Archimedes proposes problems for which he sketches the diorism, but says nothing about the solution. One such problem is the famous lemma in *The Sphere and Cylinder* mentioned above. The problem of the regular heptagon could belong in the same category.

Moreover, the problem is one that might have interested Archimedes for two reasons. As Heath has pointed out,[20] the problem of the regular heptagon very probably derives from that of inscribing regular polygons in the circle. These questions interested Archimedes. The successful use of straightedge and compasses to construct the triangle, the square, the pentagon and the hexagon inevitably suggest looking for an analogous technique for constructing the heptagon. Archimedes would surely not have been unwilling to adopt approximation procedures to accomplish this task, and such a style can indeed be characterized as Archimedean. Finally, if we look at the figure, if G lies beyond A, we *see* a solution, and only one, because, on one hand, DIC decreases from $D\Omega C$ to zero, and, on the other hand, AHG increases from zero to infinity. This observation leads to a *neusis* that Archimedes could have carried out.

Then, an allusion by Abū al-Jūd refers the reader who wants to solve the problem of the heptagon by a *neusis* to Archimedes' lemma. He writes:

> We obtain the well known triangle constructed by Archimedes and others who have sought to construct the heptagon using an instrument and movement, with the help of the lemma he had assumed, because the angles of this triangle follow one another in double ratio.[21]

---

[19] See the critique of the *Almagest* by Ibn al-Haytham or his commentaries on Euclid's *Elements*.

[20] T. L. Heath, *A History of Greek Mathematics*, vol. I, p. 235.

[21] See p. 619.

One might see this sentence as referring to a work attributed to Archimedes that contained Propositions 17 and 18 of the text copied by Muṣṭafā Ṣidqī.

Putting all this together, we can accordingly say: there was a work on the regular heptagon attributed to Archimedes; such a work was in circulation in the tenth century; that two kinds of vestiges of this work have come down to us: one type in the references by mathematicians of the time, and the other in the edition by Muṣṭafā Ṣidqī (Propositions 17 and 18), to which Abū al-Jūd seems to allude; that between the two types there is a crucial difference that we have no means of reducing. In short, it seems this work comprised two propositions – which are the origin of Muṣṭafā Ṣidqī's version of Propositions 17 and 18 – and that the tenth-century mathematicians ignored the *neusis* and kept only the two results. That is all we can say, and from this point on we shall follow tradition by referring to 'Archimedes' lemma'.

3.1.3. *A priority dispute: al-Sijzī against Abū al-Jūd*

Interest in solid problems had never been so strong as it became in the last third of the tenth century: the number of such problems continued to grow, as did the number of mathematicians working on them. There are two kinds of reason for this. First of all, as we have already emphasized, there is the part played by new geometrical and algebraic interests.[22] Then – and this is indeed the second reason – the new forms of activity in mathematics itself and in related sciences provided a solid basis for these interests. We have already mentioned them but these new forms of activity have not yet received the attention they deserve: what we have is no less than the emergence of a new mathematical community, whose members are linked by exchanges of letters, and face to face conversations in the numerous salons (*majālis*), or even at the courts of Kings, Viziers, and around patrons.[23] Previously unknown, this situation had, among other effects, that of sharpening competition, encouraging challenges and stirring up polemics. To surpass one's rivals was a way to draw closer to the centres of power, even sometimes to reach to the very top.

It is in precisely such communities and such an atmosphere that the problem of the heptagon was revived. The matter is immediately taken up

---

[22] See R. Rashed and B. Vahabzadeh, *Omar Khayyam, The Mathematician*.

[23] See A. Anbouba, 'Tasbiʿ al-Dāʾira (La construction de l'heptagone régulier)', *Journal for the History of Arabic Science*, vol. 1, no. 2, 1977, pp. 352–84; R. Rashed, *Geometry and Dioptrics*, Chapter I.

as a challenge: as witness the writings of eminent correspondents such as Muḥammad ʿAbdallāh al-Ḥāsib, Aḥmad al-Ghādī (?) and others; the essay is dedicated to the King in person, as by al-Ṣāghānī to ʿAḍud al-Dawla, by al-Qūhī to him and to his son, Sharaf al-Dawla; competitors are denounced to the community. In short, the construction of the heptagon acquires a worldly dimension far more striking than that of any other piece of mathematical research of the period. That is to say, the heptagon appears on the scene in an atmosphere of polemics.

The story of what happened with this construction, which we shall recount here, has been presented twice; similarly that of the polemic. The first time, it was near the end of the tenth century that a mathematician who, evidently, did not like Abū al-Jūd composed an essay with the title *The search for the trap laid by Abū al-Jūd*. In this hypercritical essay, al-Shannī – who is the author concerned – even goes so far as to invective. The second time is very recent: A. Anbouba wrote, in Arabic, the first modern history of the regular heptagon in this period, that is, before Ibn al-Haytham.[24] This is a serious piece of work and in it the author rightly deplores the sometimes unacceptable remarks made by al-Shannī, but without managing to completely break free from the influence of the story he tells. So we must, in turn, exercise particular prudence and submit al-Shannī's testimony to the rules of historical criticism.

The story that al-Shannī tells, that is, the first history of the heptagon, is presented in two parts. In his first section, al-Shannī mentions Archimedes' lemma, to remind us that it has remained 'in its state [he means it has not been solved] until it became possible for Abū Sahl Wayjān ibn Rustum al-Qūhī and Abū Ḥāmid al-Ṣāghānī, to each of them, to solve it by using

---

[24] A. Anbouba traces the history of the construction of the regular heptagon from Abū al-Jūd to al-Shannī in his article 'Tasbīʿ al-Dāʾira (The construction of the regular heptagon)' (cited in our preceding note). At the request of the *Journal*, he gave a brief summary of his study in French, entitled 'La construction de l'heptagone régulier' (The construction of the regular heptagon), which appeared in the same *Journal*, vol. 2, no. 2, pp. 264–9. Scarcely a year later, we ourselves published 'La construction de l'heptagone régulier par Ibn al-Haytham' (The construction of the regular heptagon by Ibn al-Haytham), which thus completes A. Anbouba's study, in the *Journal for the History of Arabic Science*, vol. 3, no. 2, (1979), pp. 309–87. Readers can find an account in English, in an article by J. P. Hogendijk, which repeats A. Anbouba's study, for the period from the last third of the tenth century, with the addition of an edition and English translation of the treatise by al-Sijzī and several quotations from various mathematicians ['Greek and Arabic constructions of regular heptagon', *Archive for History of Exact Sciences*, 30, 1984, pp. 197–330]. The texts and translations in this work are, however, far from satisfactory. For example, in a short text by Ibn Sahl – seven lines – we find no less than ten editorial errors, not to mention the translation (p. 324). For this text, compare our *Geometry and Dioptrics*, p. 480.

conic sections'. In other words, the lemma was proved twice, separately: by al-Qūhī in 360/969–970 and by al-Ṣāghānī in 970. The former divided the straight line segment with the help of a parabola and a hyperbola; as for al-Ṣāghānī, his procedure was by means of two branches of a hyperbola (the two opposing sections) and one branch of another hyperbola.

The second part begins with Abū al-Jūd. Al-Shannī, and despite the repulsive picture he draws of the man, who is accused of incompetence and plagiarism, nevertheless attributes to him a division of the straight line segment, the division we designate below[25] as D_2. It gives the range of points that leads to the construction of the regular heptagon. Al-Shannī complains, however, that Abū al-Jūd has made two errors and left a gap in his argument. The two errors are the same ones that al-Sijzī had pointed out at least a quarter of a century before. In the course of the proof, Abū al-Jūd had replaced the ratio obtained at the end of his division – the correct ratio – with another one. Worse still, he had not actually proved such a division was possible. The weakness of which he is accused elsewhere was sufficient to prevent his noticing that his division is equivalent to that of Archimedes.

This essay by Abū al-Jūd, the object of such vehement criticisms by al-Shannī, is dated 358/968–969. So it is the first essay to be mentioned that deals with the heptagon. Al-Shannī does not explicitly situate it in relation to the work by al-Qūhī, which was composed about a year later. But he does suggest that it is in fact the first essay, even though he considers Abū al-Jūd has not really succeeded. So if he does not place the essay in its proper position, that is perhaps because he considered it a semi-failure.

It is at this point that al-Shannī mentions the crucial part played by Ibn Sahl. According to al-Shannī, at the end of the 960s, al-Sijzī, then a young mathematician, had noticed Abū al-Jūd's supposed error in the division of the straight line segment, was himself unable to give a solution, and sent the problem to Ibn Sahl. The latter then gave the analyse by means of a hyperbola and a parabola. Al-Sijzī, after having himself carried out the synthesis corresponding to Ibn Sahl's analysis, then (allegedly) laid claim to having carried out the construction of the heptagon. Ibn Sahl's solution, it is said, fell into the hands of Abū al-Jūd, who in turn committed the same act of plagiarism as al-Sijzī. We are told the latter then became angry and wrote an essay in which he made violent criticisms of Abū al-Jūd.

That is a brief summary of the history of the heptagon as told by al-Shannī, which can be found in full below.[26] We may note that part of the story has been copied from al-Sijzī, that another part does not correspond

---

[25] See below, p. 341.
[26] See Text 1.2.8.

to what we find in surviving documents, and finally that it is the only source that describes Ibn Sahl's intervention in any detail. It is accordingly necessary to make comparisons between the account given by al-Shannī and the various other versions that we have at our disposal.

We know that Abū al-Jūd in person composed three essays on the heptagon. The first, written in 358/968–969, is a letter addressed to a certain Abū al-Ḥusayn ʿUbayd Allāh ibn Aḥmad. This is the essay that attracted the criticism from al-Sijzī and, later on, from al-Shannī. Now, it is precisely this first essay that reactivates research on the heptagon. But all our information about it is indirect, since the essay itself is lost. So we are reduced to taking testimony from Abū al-Jūd and his detractors. We may note, for the present, that all of them agree that in his essay Abū al-Jūd reduced the study of the heptagon to that of a triangle of type (1, 3, 3) – that is, a triangle whose angles are in the ratio 1:3:3 – which has been constructed from the division of a straight line segment that he proposed. We shall see later that Abū al-Jūd explains how he was led to take that step.

In the second essay, in which he mentions the preceding one, Abū al-Jūd writes:

> I learned that a certain geometer arbitrarily attributed this construction to Abū Sahl al-Qūhī, then that he changed part of it and claimed it for his own.[27]

This accusation seems to be aimed at al-Sijzī and looks very like a counterattack on him. Al-Sijzī had indeed just complained of the weakness of Abū al-Jūd's proof, and had claimed the correct solution as his own; but he had made no claim to have proposed the problem himself. Having got under way at the beginning of the 970s, the series of polemical exchanges only became more strident. Later on, al-Sijzī himself was to abandon prudence and address to Abū al-Jūd reproaches that he had not expressed at the time he made the construction. Thus in *A response to the problems posed by the people of Khurāsān*, he chooses to write:

إذ هو يستقبح الأشكال التي لا يتهيأ له معرفتها ... إذ حكيت في صدر كتابي في المسبع ركاكته في هذه الصناعة وهو المعروف بمحمد بن الليث الذي أخذ مقدمات كتابي في المسبع بعد معرفته في الهندسة وأضافها إلى نفسه.

He (Abū al-Jūd) rejects as ugly all the propositions that it is not possible for him to understand [...] I have told the story, in the preamble of my book on the heptagon, of the weakness in this art of the man called Muḥammad

---

[27] Abū al-Jūd, *On the Construction of the Heptagon in the Circle, which he Sent to Abū al-Ḥasan Aḥmad ibn Muḥammad ibn Isḥāq al-Ghādī*, below, p. 605.

<Abū al-Jūd> ibn al-Layth, who has taken the lemmas from my book on the heptagon after having acquired his knowledge in geometry, and has claimed them as his own.[28]

To enable us to identify the elements in this dispute, we return to its beginning, before it took on such breadth, and let us start by listening to Abū al-Jūd. He writes:

Some years, and no small number of them, after what I did, Abū Sahl al-Qūhī composed a treatise on this figure, in which he relies upon Archimedes' lemmas.[29]

He continues:

Abū Ḥāmid al-Ṣāghānī later composed a treatise on this figure, in which he addressed himself to this square (Archimedes' lemma) […], and for that he made use of three hyperbolas, two opposite conic sections and a third one, in a long construction and using many figures and straight lines.[30]

These statements are correct, apart from slight details. Abū al-Jūd's detractors are in tacit agreement on this version of events, and the only point on which they disagree with him is the success of the construction, but Abū al-Jūd seems to be exaggerating the time that elapsed between his work and that of al-Qūhī: it is not a matter of 'some years', but only about one year. Taken as a whole Abū al-Jūd's version thus seems to be correct, at least in providing the succession of events in the story of the construction. It could not of course be otherwise, given al-Qūhī's contribution, undoubtedly authoritative, and that of al-Ṣāghānī, Abū al-Jūd's real teacher. But, all the same, Abū al-Jūd's account underestimates al-Sijzī, and leaves him out of the honourable company of al-Qūhī and al-Ṣāghānī. In short, according to Abū al-Jūd, the discovery developed with contributions from mathematicians in the following order: himself, al-Qūhī and al-Ṣāghānī. Al-Sijzī appears only in the wings, for having surreptitiously 'plagiarized' the contribution by al-Qūhī. As for Ibn Sahl, he is simply not there.

Let us now look at al-Sijzī's version. In a highly rhetorical preamble, he denounced Abū al-Jūd for his irreverence in regard to Archimedes, the Geometer, and then censures him for making two mistakes in his proof, which to al-Sijzī's mind suffice to deprive him of the desired priority. In doing this, al-Sijzī draws a less than flattering portrait of Abū al-Jūd

---

[28] *Jawāb Aḥmad ibn Muḥammad ibn 'Abd al-Jalīl 'an masā'il handasiyya su'ila 'anhu ahl Khurāsān*, MS Chester Beatty 3652, fol. 55ʳ; Reshit 1191, fol. 114ᵛ.

[29] Abū al-Jūd, *On the Construction of the Heptagon in the Circle, which he Sent to Abū al-Ḥasan Aḥmad ibn Muḥammad ibn Isḥāq al-Ghādī*, below, p. 606.

[30] *Ibid.*

– making him a second-rate mathematician, if not a downright mediocre one – and suggests, without saying so explicitly, that he was not familiar with the properties of conic sections. This picture inevitably caused irritation, since it is thoroughly belied not only by Abū al-Jūd's works, but also by the opinions of his successors, and high-ranking ones, since we mean al-Bīrūnī and al-Khayyām.[31]

As for the accusation of disrespect to Archimedes, which is indeed repeated by al-Shannī, it seems to arise from a misrepresentation. This is what al-Sijzī himself cites from Abū al-Jūd concerning Archimedes:

قال : قد قلد أرشميدسُ – في خلال مقدمات كثيرة قدّمها لقسمة الدائرة بسبعة أقسـام متساوية – مقدمة لم

يبين عملها ولم يبرهن عليها ، ولعلها أصعب عملاً وبرهاناً مما له قدمها .

All the criticism of the two authors is directed to the verb قلد . Al-Sijzī had read it as *qallada*, and had preferred the sense 'imitate' (*ḥākā*); al-Shannī had done the same, and the paragraph we have cited will thus be translated:

He (Abū al-Jūd) said: among the many lemmas that he introduced for the division of the circle into seven equal parts, Archimedes imitated a lemma whose construction he has not shown and which he has not proved; and it perhaps involves a more difficult construction and a less accessible proof than the matter for which he introduced it.[32]

One does not need to be a linguist to find the expression 'imitated a lemma' jarring: one cannot imitate a lemma or a proposition. But the verb has other meanings that are obviously more natural here, and fit better with the syntax. There is a choice: either we have the form *qalada*, or the form *qallada*. In the first case, the dictionaries (Ibn Manẓūr, Ibn Fāris, Ibn Durayd al-Zabīdī, etc.) give the sense 'reunite', 'associate', 'bind' one thing with or to another; in the second case, we find the meanings of the word include 'impose' and 'require' (*alzama*). So we might read, in one case: '[…] Archimedes associated with numerous [other] lemmas a lemma […]';

---

[31] Al-Khayyām is ready to speak of 'the eminent geometer' in his *Treatise on Algebra*, English transl. and commentary in R. Rashed and B. Vahabzadeh, *Omar Khayyam. The Mathematician*, pp. 138, 147, 174 (Arabic text in the French edition, pp. 179, 197, 257). Similarly, al-Bīrūnī includes him among 'the distinguished men of our time (*al-mubarrizūna min ahli zamāninā ka-Abī Sahl wa-Abī al-Jūd*)', in *al-Qanūn al-Mas'ūdī*, Hyderabad, 1954, vol. I, p. 297. See R. Rashed, 'Les constructions géométriques entre géométrie et algèbre: l'Épître d'Abū al-Jūd à al-Bīrūnī', *Arabic Sciences and Philosophy*, 20.1, 2010, pp. 1–51.

[32] Cf. below, pp. 630–1.

and in the other case 'Archimedes imposed with numerous [other] lemmas a lemma [...]'. Whichever of the two forms we choose, al-Sijzī's objection no longer applies, and Abū al-Jūd's paragraph becomes clear, both as to syntax and as to sense.

Once we have removed this misrepresentation, which was no doubt driven by the dispute, there remain the two errors Abū al-Jūd is accused of making. Al-Sijzī, as ever followed by al-Shannī, states that in his letter of 358/968–969, Abū al-Jūd carried out the construction of the heptagon with the help of four lemmas, which we can rewrite in the form:

1. If the radius of a circle of centre A is equal to the distance from A to a straight line, then the circle touches this straight line.

2. Draw through a point M of the side AB of a triangle ABC a straight line parallel to the side AC to cut the side BC in N, and such that $MN = CN$.

3. Given a straight line AB and the ratio of C to D, to draw a straight line X such that $\dfrac{X}{AB} = \dfrac{C}{D}$.

4. To divide a straight line AB in a point M such that $AB \cdot AM = X^2$ and $\dfrac{X}{BM} = \dfrac{C}{D}$.

We may observe that the proof of 1 is immediate, and that 2 and 3 are very simple constructions; 4 can be rewritten:

to find on AB a point M such that $\dfrac{\sqrt{AB \cdot BM}}{MB} = \dfrac{C}{D}$ (a given ratio); in the case where $\dfrac{C}{D} = \dfrac{AB}{AB + BM}$, we have the division of a segment – division D_2, see below – which leads to the construction of the heptagon.

Al-Sijzī's criticisms are precise. Instead of using the ratio given in 4 in the course of the proof, Abū al-Jūd replaced it by another ratio. Contemporaries could check the validity of this criticism, even if we, having no text, cannot do so. This criticism leads us to the most hotly contested point of the debate. We shall return to it later.

The second criticism is less clear insofar as al-Sijzī suggests more than he actually states. He writes:

He (Abū al-Jūd) believed it is possible to construct that (the division and then the heptagon) by means of the lemma in the fourth proposition. But the construction is not possible except by means of conic sections and, for someone who, in geometry, knows neither the cone nor the sections, it would be by means of the lemmas presented in the books of the ancients, thanks to which it is possible to construct the heptagon <that is> for someone who adds his lemmas to them. But by means of his lemmas, and by others analogous to them, it is difficult to find the hexagon inscribed in the circle [...], *a fortiori* if it is a matter of finding the heptagon.[33]

What, exactly, is al-Sijzī suggesting by these complicated, not to say slightly confused, remarks? Perhaps that Abū al-Jūd had not proceeded by using conic sections, but by using straightedge and compasses. Abū al-Jūd was replying to this criticism when he wrote: 'It was not possible for me to do this (the division of the straight line), except by using two intersecting conic sections, a hyperbola and a parabola.'[34] unless we are to suppose Abū al-Jūd was not telling the truth about something that was (after all) verifiable, which is difficult to believe. Perhaps al-Sijzī was insinuating that Abū al-Jūd was not familiar with conics. That seems equally extreme. Or perhaps he wanted to imply that Abū al-Jūd's discussion did not lead to the required construction. That criticism is also incorrect. Later, al-Shannī suggests that Abū al-Jūd did not succeed in dividing the straight line. He even goes so far as to say that al-Sijzī tried to cover up this deficiency, but without success. Perhaps it was this incomplete, or incorrect, proof that al-Sijzī meant to complain of in his somewhat Sybilline remarks, when he says that it was by using conic sections that one might succeed, that if not it was enough to return to Archimedes' division with lemmas different from those employed by Abū al-Jūd. Perhaps he was complaining of Abū al-Jūd's initial hesitation before turning to conic sections.

At this moment in the dispute, an entrance is made by one of the most eminent mathematicians of the time, and the most underestimated until recent times: Ibn Sahl.

As we have seen, according to al-Shannī, it was Ibn Sahl who provided the solution for al-Sijzī and Abū al-Jūd then misappropriated it. This is the story as told by al-Shannī, who, of course, prolonged and revived a dispute that by then was already about a quarter of a century old.

This version is not correct; at the very least we can set against it the fact that on two occasions al-Sijzī acknowledges his indebtedness to Ibn Sahl. At the beginning of his treatise on the regular heptagon, he says he profited by 'the mathematical learning of Archimedes and the preliminary parts of Apollonius, and in particular that of moderns such as al-'Alā' ibn Sahl.' More importantly, when he comes to the division, he writes:

---

[33] Cf. below, p. 633.

[34] *Treatise by Abū al-Jūd Muḥammad ibn al-Layth* Addressed to the Eminent Master *Abū Muḥammad 'Abd Allāh ibn 'Alī al-Ḥāsib on the Account of the Two Methods of the Master Abū Sahl al-Qūhī the Geometer, and of his Own Master Abū Ḥāmid al-Ṣāghānī, and on the Route he Himself Took to Construct the Regular Heptagon in the Circle*, below, p. 622.

Abū Sahl al-'Alā' ibn Sahl has proved this proposition by using the method of analysis and our synthesis is a part of his analysis.[35]

What al-Shannī says is thus incorrect. Moreover, if Abū al-Jūd had done what al-Shannī accuses him of doing – if he had laid claim to Ibn Sahl's solution – it could not, in any case, have been in the letter of 968–969, the subject of the dispute. Abū al-Jūd's later versions of the matter will be considered below. However, we may say at once that if they had been under scrutiny, al-Sijzī, as a declared enemy of Abū al-Jūd, could hardly have failed to mention the fact. And he does not mention it.

Finally, still according to al-Shannī, al-Sijzī had addressed another question to Ibn Sahl. This is confirmed by another source: the *Book on the Synthesis of Problems Solved by Analysis by Abū Sa'd al-'Alā' ibn Sahl.*[36] The matter in hand was a sort of generalization of Archimedes' lemma to the case of a parallelogram. Ibn Sahl had given the analysis, and it is al-Shannī himself (not Abū al-Jūd as A. Anbouba inadvertently implies[37]) who set out the synthesis for this analysis.

The moment has at last come for us to examine the first criticism al-Sijzī addresses to Abū al-Jūd, a criticism repeated by al-Shannī. It is important because it allows us to pinpoint a matter of key significance in the construction of the heptagon, but, in the absence of the document (Abū al-Jūd's first essay), it is not possible to decide whether the criticism is justified. We need to try come closer to answering the question raised by this criticism: what relationship is there between Abū al-Jūd's four lemmas and his lemma on the division of form D_2 that he proposes, and consequently the construction of the heptagon? We want to know whether the error that al-Sijzī, followed by al-Shannī, accuses Abū al-Jūd of making lies in the conception of this relationship and in the course of the proof of the lemma on division D_2. Al-Sijzī's attitude, which at the very least is equivocal, complicates the issue, which is already made difficult by our not having Abū al-Jūd's essay: in the passage we have already quoted, he states, in the clearest possible terms, that Abū al-Jūd made use of the four lemmas to construct the heptagon, a procedure for which he is sharply upbraided. But al-Sijzī also states that, in the proof of the lemma on division D_2, he did not use the ratio established by the last of this group of four lemmas, but instead had recourse to another ratio, a complaint that is in contradiction with the

---

[35] Al-Sijzī, *Book on the Construction of the Heptagon in the Circle and the Trisection of a Rectilinear Angle*, below, p. 635.

[36] R. Rashed, *Geometry and Dioptrics*, pp. 7–10, 329ff, 480 ff.

[37] Although it is of no consequence in A. Anbouba's study, this inadvertent mistake, when it happens to be combined with an astonishingly faulty edition of Ibn Sahl's paragraph, has had serious consequences later. See note 24.

previous one. To say the least, al-Sijzī's position requires some clarification. The fourth lemma in the group we mentioned raises a precise question, that of finding the point M on a given segment AB such that

$$AB \cdot AM = X^2 \text{ and } \frac{X}{MB} = \frac{C}{D},$$

where $\frac{C}{D}$ is a given ratio.

Point M can thus be defined by the equation

$$\frac{AB \cdot AM}{MB^2} = \frac{C^2}{D^2}.$$

To understand what Abū al-Jūd conceives to be the relationship between this lemma and the one that gives the division D_2, we need to answer the following questions: Did Abū al-Jūd believe he was dealing with a *neusis* or with an acceptable geometrical construction? Had he noticed the difference between this construction and that of division D_2, or did he regard them as the same? Finally, assuming he did distinguish between them, why did he introduce this fourth lemma?

The first question can be disposed of quickly, thanks to an anonymous text.[38] The unnamed author more or less repeats Abū al-Jūd's lemma. Perhaps he knew Abū al-Jūd's essay. Or perhaps he had come upon the problem independently in the course of his own researches. At present we cannot make any comment at all on these possibilities. But what matters to us here is to see how a mathematician belonging to this tradition dealt with this problem; he does so as follows.

He starts with two segments EG and EH such that

$$\frac{EG^2}{EH^2} = \frac{C'}{D'},$$

and poses the problem of constructing the point M such that

$$\frac{AB \cdot AM}{BM^2} = \frac{C'}{D'}.$$

We may note that this will correspond to Lemma 4 if we put

---

[38] This text was noted by J. P. Hogendijk, who translated a short paragraph from it. He thought this paragraph gave a *neusis*. See 'Greek and Arabic constructions of the regular heptagon', pp. 246–8.

$$\frac{C'}{D'} = \frac{C^2}{D^2},$$

hence

$$\frac{EG}{EH} = \frac{C}{D}.$$

If we are given EH, then EG is the segment defined by Abū al-Jūd's third lemma.

To carry out his construction, the anonymous author begins by giving a lemma that shows that if two ranges (A, M, B) and (G, I, K) are similar, then

$$\frac{AB \cdot AM}{MB^2} = \frac{GK \cdot GI}{IK^2}.$$

The converse of this lemma is easily shown to be true, even if the author neglects to do so. In any case, it is the converse he uses. He constructs the range (G, I, K) such that

$$\frac{GK \cdot GI}{IK^2} = \frac{EG^2}{EH^2},$$

and from this he derives a similar range on the given segment AB. He then carries out the construction. We take $EG \perp EH$ and draw from G a straight line to cut the circle (H, HE) in I and K such that IK is the side of the regular hexagon inscribed in the circle. As for constructing the straight line, the author writes 'that is possible for us', and it is thus within the capabilities of any mathematician of the time.

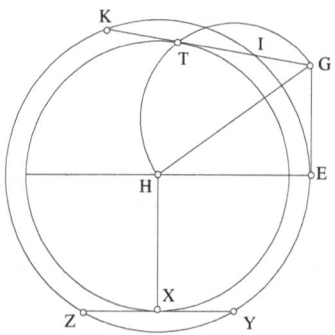

Fig. 3.3

Let $ZY = HE$ be a side of the hexagon and X the midpoint of ZY; the circle (H, HX) is then the circle inscribed in the hexagon. If from G we draw a tangent to this circle, the tangent will cut the circle (H, HE) in I and K and we have $IK = ZY = HE$. This is a classical construction using straightedge and compasses. The point T, the point of contact of the tangent from G, is the point of intersection of the circle (H, HX) and the circle with diameter GH. We have

$$EG^2 = GI \cdot GK \quad \text{and} \quad IK = HE,$$

so

$$\frac{KG \cdot GI}{IK^2} = \frac{EG^2}{EH^2}.$$

It follows, from the converse of the lemma we mentioned above, that the range (G, I, K) is similar to the range (A, M, B).[39] So the construction can be carried out by means of straightedge and compasses, making use of the circle inscribed in the hexagon. This explains why, on the one hand, the

[39] A segment equal to the segment GK is drawn parallel to the given segment AB. There are two possible cases:

1) $AB \neq GK$. The lines AG and BK cut one another in O, and the line OI cuts AB at the required point M (homothety).

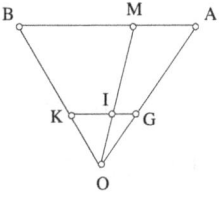

Fig. 3.4

2) $AB = KG$. In this case AG and BK are parallel, and we draw $IM \parallel GA$. The divisions are then equal, and correspond to one another by translation.

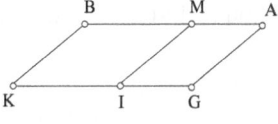

Fig. 3.5

author writes: 'That is possible for us', and on the other hand supposes the construction to be known, as we see in his reference to the hexagon.[40]

It does not seem rash to claim that a mathematician of Abū al-Jūd's stature could have carried out this construction. But that still leaves the difficulty of accounting for al-Sijzī's formulation of the problem, according to which it is difficult to construct the hexagon with Abū al-Jūd's lemmas and analogous results. The remark is astonishing in more than one respect, since al-Sijzī knew better than anyone that these lemmas are not required for the construction in question, for which one merely needs to use the compasses to make a chord equal to the radius of the circle. That is, moreover, what he is talking about when he writes that carpenters carry out this construction 'with one simple opening of the compasses'.[41] Perhaps, in the heat of the conflict, he merely meant to say that Abū al-Jūd's lemmas are not used in constructing the hexagon and they would only complicate something simple.

Let us now return to the difference between the ratio used in this lemma and the one considered in the lemma for Abū al-Jūd's division D_2.

In the fourth lemma, the length X is defined by a given ratio, whereas in the expression for the division D_2 this same length is defined by a ratio that depends on the required point M.

If M is a general point on the given segment AB, and ABC is a triangle such that $BC = BM$, Abū al-Jūd's Lemma 2 allows us to construct a point E such that $EG \parallel BC$ implies $EG = BE$. Then

$$\frac{EG}{BM} = \frac{EG}{BC} = \frac{AE}{AB} = \frac{EG + AE}{AB + BM} = \frac{AE + BE}{AB + BM} = \frac{AB}{AB + BM};$$

so we have

$$\frac{EG}{BM} = \frac{AB}{AB + BM},$$

a relation that holds for any point M on AB.

If we put $EG = X$, we have

$$\frac{X}{BM} = \frac{AB}{AB + BM}.$$

The point M required by Abū al-Jūd must satisfy this equality, but also

---

[40] To say that this construction is a *neusis* (see note 37), rather than an acceptable geometrical construction, is thus obviously untrue, and the construction that uses an arc of 120° follows immediately from the properties of the hexagon.

[41] See below, p. 633.

$$X^2 = AB \cdot AM,$$

where the length X is an auxiliary length defined by a ratio that depends on M.

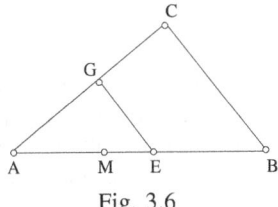

Fig. 3.6

So the construction of point M cannot be carried out by a method analogous to the one used for the point M of the fourth lemma, where X was defined by a *known ratio*. It is accordingly inconceivable that Abū al-Jūd could have used Lemma 4 in constructing the division D_2. Indeed, neither al-Sijzī nor al-Shannī accuses him of making such an error. For example, after quoting Abū al-Jūd's fourth lemma, al-Shannī writes:

> He considered this ratio and later in the construction of the heptagon he used another ratio, different from the one he had introduced.[42]

So Abū al-Jūd seems not to have confused these two lemmas, or the two ratios. Why then did he give the four lemmas? Perhaps, at first, before he had fully realized that constructing the heptagon is a solid problem, he believed his lemmas might allow him to find a construction for the heptagon. An error he then immediately corrected by making use of another ratio, whereas the point M is found by the intersection of two conic sections. This seems to be the only possible explanation we can offer, and if it is correct then it was presumably Abū al-Jūd's uncertainty that attracted blame. This explanation, or something like it, would also account for the somewhat contradictory statements made by al-Sijzī as well as the claims of Abū al-Jūd.

The controversy whose main outlines we have just sketched allows us at least one advantage: it identifies the groups that took part in the work on constructing the heptagon, and shows, behind the sometimes acrimonious exchanges, something more or less like collaboration, forced but productive.

The first grouping is that of Abū al-Jūd and al-Sijzī, with whom, as it were without prior consent, Ibn Sahl was also associated. Abū al-Jūd

---

[42] *Book on the Discovery of the Deceit of Abū al-Jūd concerning the Two Lemmas he Introduced in order to Construct the Heptagon*, see below, p. 675.

proposes a division D_2 of a segment, which seemed to him to be rather better than that proposed by Archimedes (in fact the divisions are equivalent). His proof was undoubtedly imperfect in some respects; accordingly, al-Sijzī turns to Ibn Sahl to request that he provides a rigorous proof. So the latter probably became involved in the controversy without making a deliberate choice in the matter. Thanks to Ibn Sahl, al-Sijzī gave a proof of Abū al-Jūd's division D_2.

The second group consists of mathematicians of the calibre of Ibn Sahl, who are obviously above the controversy. In any case, they took no part in this dispute. Each one of them starts with Archimedes' division D_1, which they all prove, and then construct the heptagon. Notable examples are al-Qūhī and al-Ṣāghānī, with whom Abū al-Jūd later allies himself.

Abū al-Jūd intervenes again to propose another division, D_3. As for al-Shannī, he has a dual role: as a historian who is directly involved in the dispute he reports events and stokes the fire of controversy; as a mathematician he proves several interesting propositions.

This is the varied and polymorphous tradition that Ibn al-Haytham brings to a close by making it systematic. To show how much ground was covered, we shall organize our account round the successive forms of division, together with the constructions they enabled, beginning with Archimedes' division, even though it was in effect taken up again after that of Abū al-Jūd.

3.1.4. *The lemmas for the construction of the heptagon: the division of a segment*

To understand the part played by the various lemmas – Archimedes' lemma and the other ones – that were necessary for the construction of the heptagon, specifically in the last third of the tenth century, we shall first of all take a look at the problem of the heptagon itself.

Let us consider a regular heptagon (*ABCDEFG*) inscribed in a circle. The side of the heptagon subtends at the circumference an angle $\alpha = \frac{\pi}{7}$. If we draw the chords *AB, AC, AD, AE, AF, AG, BC, CD, DE, EF, FG* and *CF*, the angles we obtain are thus all α or a multiple of α. The triangles defined in terms of this measure α are (see Fig. 3.7):

T_1	Triangles *ABC* or *AGF*,	type (1, 5, 1)
T_2	Triangles *ADC* or *AEF*,	type (1, 2, 4)
T_3	Triangle *ADE*,	type (1, 3, 3)
T_4	Triangle *ACF*,	type (3, 2, 2).

As we have already mentioned, Ibn al-Haytham was the first to state that there are only these four types, and he investigated all of them. In contrast, his predecessors had only investigated one or two types at a time. Thus it is these triangles that we shall find in the treatises on the regular heptagon.

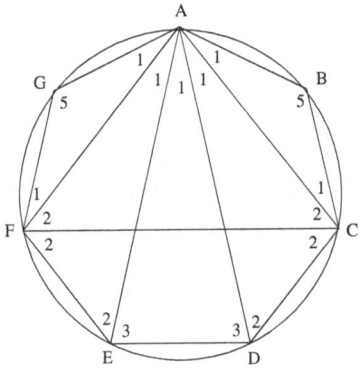

Fig. 3.7

Examining the construction of each of these triangles leads to one or more types of division of a given segment into two or three parts. However, these divisions can only be obtained by using the intersection of conics. Let us go through them systematically. The first division to be proposed is that of Archimedes' lemma.

3.1.4.1. *Archimedes' division* (D_1)

Archimedes' division appears in two textual traditions, that of the mathematical writings of the last third of the tenth century, and that of the eighteenth-century manuscript, that is, the edition by Muṣṭafā Ṣidqī. Happily, these two traditions are in agreement, and allow us to present Archimedes' lemma in the following terms:

Let there be a segment [*AB*]; let us divide it in the points *C* and *D* such that

$$(1) \qquad BD^2 = AD \cdot CD, \qquad\qquad (2) \qquad AC^2 = CB \cdot BD.$$

B ————————— D — C —————— A

Fig. 3.8

If now, starting from that division, we construct a point *E* such that *DE = DB* and *CE = CA*, we obtain four triangles *EDC*, *EDB*, *ECA* and *BAE*, which are respectively of types T_2, T_4, T_1 and T_2.

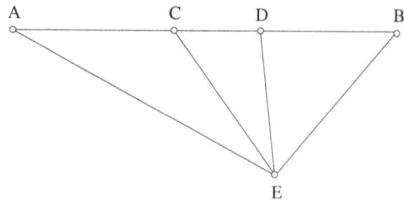

Fig. 3.9

To understand how research on this lemma unfolded from the 960s onwards, we need to distinguish three stages. The first, about which we have only slender information, saw the reception of the text attributed to Archimedes and its translation. Here all we know is contained in the two clues to be found in the edition by Muṣṭafā Ṣidqī. In the second stage, mathematicians set about research, they begin to want to provide a solid basis of proof for the lemma; this period is marked by the intervention of Ibn Sahl. Finally, at the third stage, when Ibn Sahl's competitors come to grips with the lemma, we wish to speak in particular of al-Qūhī and al-Ṣāghānī. That is the overview of our account.

3.1.4.1.1. *First stage: the division in the text attributed to Archimedes*

We have seen that in the lemma (Proposition 17 of Muṣṭafā Ṣidqī's edition), the author gives the range (A, C, D, B) that satisfies equations (1) and (2). In proposition 18, he constructs on the segment DC the triangle DEC such that $DE = DB$ and $CE = CA$, and shows that triangles DEC and EAB (triangles of type (1, 2, 4), although the author does not say so) are similar. He deduces that

$$D\hat{E}C = E\hat{A}B = A\hat{E}C.$$

The straight lines EC and ED cut the circumcircle of triangle AEB at the points G and H; we thus have $\widehat{HG} = \widehat{GA} = \widehat{BE}$. He then shows that each of the arcs AE and BH is double each of the preceding arcs. If M and N are respectively the midpoints of the arcs AE and BH, then $AMEBNHG$ is a regular heptagon.

If we turn back to the earlier summary of Archimedes' procedure, we can see he does indeed start from a given segment AB, on which he supposes the points of the range (A, C, D, B) of type D_1 to be known. He next constructs on AB a triangle of type (1, 2, 4), then its circumcircle, and from that he finds the heptagon.

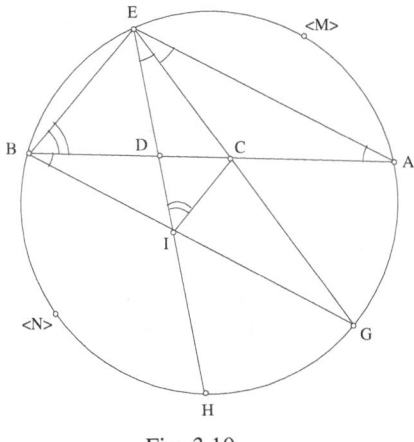

Fig. 3.10

We need to emphasize what distinguishes this procedure from that of all the tenth-century mathematicians. In all the treatises we have been able to examine, the author starts by constructing one of the four triangles T_1, T_2, T_3, T_4, then inscribes in the given circle a triangle similar to the triangle just constructed.

With this difference in mind, let us now present Archimedes' procedure as it appears in Muṣṭafā Ṣidqī's edition.

Let there be the range (A, C, D, B) that satisfies (1) and (2). From the lemma, we have $AC > CD$ and $BD > CD$. Let us construct the triangle ECD such that $CE = CA$ and $DE = DB$. We draw the circumcircle of triangle EAB (a triangle of type T_2, although the author does not mention the fact); the straight lines EC and ED cut the circle in G and H, the chord BG cuts EH in I. We have $E\hat{A}C = A\hat{E}C$, so $\overset{\frown}{AG} = \overset{\frown}{EB}$.

From equation (1), we have $\dfrac{AD}{BD} = \dfrac{BD}{DC} = \dfrac{DE}{DC}$, so triangles AED and CED are similar, and $\overset{\frown}{GH} = \overset{\frown}{EB} = \overset{\frown}{AG}$.

So we have $A\hat{B}G = E\hat{A}B$, hence $GB \parallel AE$. Angles CEI and CBI are equal, points C, E, B, I are concyclic, and $D\hat{C}I = D\hat{I}C$, so $CD = DI$ and $CE = IB$.

From equation (2) and the equalities $EC = CA$, $DB = DE$ and $CB = EI$, we get

$$EI \cdot DE = EC^2 \Rightarrow \frac{EC}{ED} = \frac{IE}{EC},$$

so triangles CED and IEC are similar, and $D\hat{C}E = E\hat{I}C$.

But $D\hat{C}E = 2C\hat{A}E$, so $C\hat{I}E = 2C\hat{A}E$. Moreover, $C\hat{I}E = C\hat{B}E$, so $C\hat{B}E = C\hat{A}E$ and $\widehat{AE} = 2\widehat{BE}$. Similarly, $D\hat{B}E = D\hat{E}B = 2C\hat{A}E$, so $\widehat{HB} = 2\widehat{BE}$.

If we divide the two arcs AE and HB in half at the points M and N, the circle will be divided into seven equal arcs and $AMEBNHG$ will be the required regular heptagon.

3.1.4.1.2. *Second stage: Ibn Sahl*

According to al-Sijzī, and later also al-Shannī, the period of proofs of Archimedes' lemma begins with Ibn Sahl. It is no surprise to find that Ibn Sahl is also the first to attempt to generalize this lemma to the parallelogram.

If we are to believe al-Shannī, al-Sijzī asked Ibn Sahl two questions. The first relates to the proof of Abū al-Jūd's division – a division equivalent to that of Archimedes. Ibn Sahl gave a proof that used a parabola and a hyperbola. It is al-Sijzī who, in a way, preserves this method and Ibn Sahl's result, in carrying out the synthesis of Ibn Sahl's analysis; furthermore, let us remind ourselves, he was scrupulous in acknowledging his debt to Ibn Sahl. The second question al-Sijzī asked Ibn Sahl is reported by al-Shannī, who is (precisely) engaged in carrying out the synthesis of Ibn Sahl's analysis. Here is that second question:

ABCD is a parallelogram in which there has been drawn a diagonal which is *BC*; the side *CD* has been extended indefinitely in the direction of *D*. How may we draw a straight line, say the straight line *AEGH*, in such a way that the ratio of the triangle *BEG* to the triangle *GDH* is a given ratio?[43]

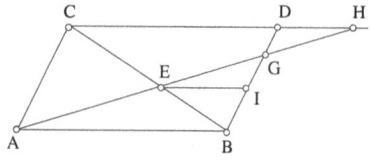

Fig. 3.11

This is the question as reported by al-Shannī in his essay on the heptagon. We find confirmation of the terms of the question in a text that was certainly also written by al-Shannī, the *Book on the Synthesis of the Problems Analysed by Abū Sa'd al-'Alā' ibn Sahl*.[44] From reading this text, we understand that Ibn Sahl's purpose – and very probably that of al-Sijzī when he raised the question – has two parts: to prove Archimedes'

---

[43] See below, p. 678.
[44] R. Rashed, *Geometry and Dioptrics*, p. 472ff.

lemma in the case of the parallelogram; to find the ratio between areas, a ratio that is not unity, by looking at another triangle, that is to say by slightly departing from Archimedes' approach. We have two reasons for this interpretation: the manner in which contemporaries such as al-Shannī understood the problem; and, further, the mathematical analysis of the synthesis given in this book, which we shall not consider here. In fact, for al-Shannī the matter is entirely that of generalizing Archimedes' lemma. However, the question addressed to Ibn Sahl and the construction he proposes, which shows through in the synthesis given by al-Shannī, lead to a solution in the case where we are comparing the areas of triangles *BEG* and *GDH*, whereas Archimedes' problem considered triangles *AEB* and *GDH*. We show that the two problems are not identical.[45] Ibn Sahl himself acknowledged that his construction does not solve the problem of Archimedes' lemma in this case. In a famous text he writes:

> As for the means of extending mathematical knowledge by giving the ratio between the two triangles *DCG* and *LAE* (here *EBG* and *GDH*), there is no method that comes to mind for leading us to determine it by analysis, nor for obtaining a lemma, and if we had found an approach that allowed us to achieve it, we should have held to it so as to have knowledge of that which escaped until it could be shown to follow.[46]

While he acknowledges Ibn Sahl's eminence as a mathematician, al-Shannī never the less launches into a diatribe against him, accusing him of vanity. No doubt al-Shannī had become confused and had failed to notice that this problem was different from that considered by Archimedes. His confusion is the more surprising since he had himself correctly copied out the precise quotation. He actually writes 'the triangle *CGD*' – here *BEA* – instead of 'triangle *CGE*' – here *BEG*. His own words are:

> I say: if the plane figure *ABCD* is a square, and if the ratio of the two triangles is the ratio unity, then it is the same figure introduced by Archimedes for the construction of the heptagon, and for which Abū Sahl Wayjan ibn Rustum al-Qūhī followed the method of dividing a straight line into three parts in a ratio that must occur.[47] If, later, the ratio of the two triangles is a different ratio, then, from the figure constructed by Abū Sahl, the straight line is divided in the ratio we mentioned [...].[48]

Al-Shannī gives a summary of al-Qūhī's solution for the case of the square, using an equilateral hyperbola and a parabola, a solution we shall

[45] See R. Rashed, *Geometry and Dioptrics*, pp. 324–34.
[46] *Ibid.*, p. 480.
[47] The expected ratio is $K/L \neq 1$.
[48] See R. Rashed, *Geometry and Dioptrics*, p. 480.

examine presently. He then adapts al-Qūhī's construction to the case of the parallelogram. Here, briefly, is his solution, using his own lettering.

Let there be a parallelogram *ABEG*, on whose side *AB* we have the range (*A, C, D, B*). The straight line *GD* cuts the diagonal in *I* and the produced part of *EB* in *M*.

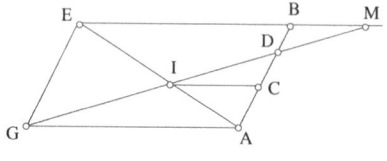

Fig. 3.12

We want

$$\frac{\text{area }(BDM)}{\text{area }(GIA)} = \frac{K}{L} \neq 1.$$

Let there be two segments *CD, DE*, such that $DE \perp CD$ and $DE = DC$; the parabola $\mathscr{P}$ with vertex *C*, and *latus rectum DE*; the hyperbola $\mathscr{H}$ with axis *DE* and *latus rectum H* defined by $\frac{H}{DE} = \frac{K}{L}$. The two curves cut one another in four points; let *G* be the point of intersection that is projected into *I* and *B*, which lie on *DE* and *CD* respectively.

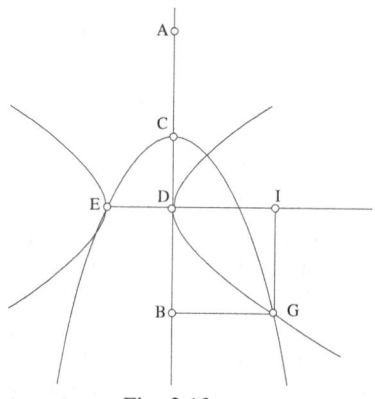

Fig. 3.13

We obtain

$$GB^2 = CB \cdot DE = CB \cdot CD \qquad \text{(from the parabola)}$$

and

$$GI^2 = EI \cdot ID \cdot \frac{H}{ED} = EI \cdot ID \cdot \frac{K}{L} \text{ (from the hyperbola)}.$$

If we produce DC beyond C by a distance $AC = GB$, we have

(1) $AC^2 = CB \cdot CD$,

(2) $BD^2 = AD \cdot AC \cdot \dfrac{K}{L}$.

If the range (A, C, D, B) on the side of the parallelogram satisfies equations (1) and (2), then from (1) we deduce that if we draw from the point C a line parallel to AG, it cuts AE and GD at the same point, that is, in I. From (2) we get

$$\frac{BD^2}{AD \cdot AC} = \frac{BM}{AG} \cdot \frac{DM}{IG} = \frac{K}{L};$$

from which the result follows.[49]

The inspiration for this approach is very much drawn from the work of al-Qūhī, so that it is indeed as if we had a commentary on Ibn Sahl's study written in the light of a reading of al-Qūhī.

3.1.4.1.3. *Third stage: al-Qūhī and al-Ṣāghānī*
3.1.4.1.3.1. *Al-Qūhī: the first treatise*
Round about the 970s, two mathematicians picked up the problem as Archimedes had left it. Nothing in the three texts composed by these mathematicians suggests they had played the slightest part in the famous controversy. Perhaps they did not know of its existence, or perhaps they simply wished to avoid becoming involved in it. We do not know, but what concerns us here is that both of them used Archimedes' range and carried out their construction of the heptagon as Archimedes had wished, starting from the triangle (1, 2, 4). This is what al-Qūhī did in a first essay on the subject.

In this work, which is dedicated to the King himself, ʿAḍud al-Dawla, al-Qūhī proceeds by analysis and synthesis. So he has decided his treatment of the problem shall be complete. His starting point is an analysis of the problem itself.

Let us suppose that the chord BC is the side of the regular heptagon inscribed in the given circle, and A is the point on the circle that satisfies the equation $\widehat{AB} = 2\widehat{BC}$; so A is a vertex of the heptagon. From this we deduce that

---

[49] *Geometry and Dioptrics*, pp. 324–34.

$$A\hat{B}C = 2A\hat{C}B = 4B\hat{A}C.$$

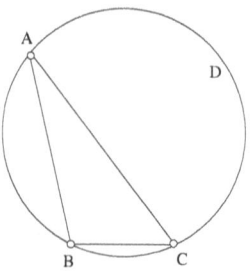

Fig. 3.14

Al-Qūhī naturally goes on to an analysis of the construction of a triangle (1, 2, 4) with apex at the point A and base BC. He shows that this analysis leads to the range used by Archimedes (E, B, C, D).

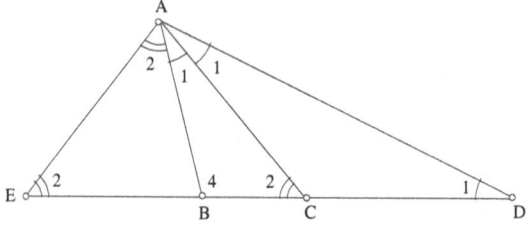

Fig. 3.15

Let us produce BC in both directions, to give points D and E with $BE = BA$ and $CD = CA$. Triangle ACD is isosceles, so $B\hat{C}A = 2C\hat{A}D = 2A\hat{D}B$. Now $B\hat{C}A = 2B\hat{A}C$, hence $B\hat{A}C = A\hat{D}B$; triangles ABD and ABC are similar, hence $\dfrac{DB}{AB} = \dfrac{AB}{BC}$; so we have $AB^2 = BC \cdot BD$, that is,

(1) $BE^2 = BC \cdot BD.$

Similarly, $A\hat{B}C = 2B\hat{A}E = 2A\hat{E}C$. Now $A\hat{B}C = 2A\hat{C}B$, so $A\hat{C}B = B\hat{A}E$, and the two triangles ACE and ABE are similar; hence $\dfrac{CE}{EA} = \dfrac{EA}{EB}$; so we have $AE^2 = CE \cdot EB$. But, $EA = AC = CD$, so

(2) $CD^2 = CE \cdot EB.$

Now (1) and (2) define Archimedes' range.

The above analysis shows that a (1, 2, 4) triangle is associated with that range. It is clear that triangle *ACD* is of the type (1, 5, 1) which leads to the construction of the heptagon. For this reason al-Qūhī returns to the same figure in a second essay in which he considers this triangle.

In the following propositions – numbers 3 to 5 – al-Qūhī uses analysis and synthesis to carry out the construction of a range (*A*, *B*, *C*, *D*) of type D_1. Thus, in Proposition 3 of his treatise, completed by Proposition 4, he gives the analysis that leads to the parabola and hyperbola that allow the range to be constructed. In Proposition 5, he gives the synthesis to which we shall now turn, keeping the lettering and the figure used for the analysis.

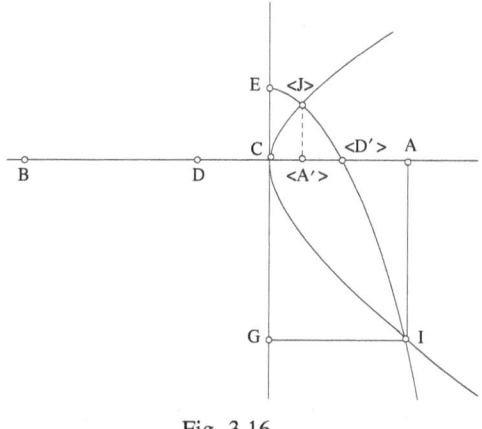

Fig. 3.16

Let *CD* be a given segment and *CE* a segment such that *CE* ⊥ *CD* and *CE* = *CD*. Let us consider the parabola $\mathscr{P}$ with vertex *E*, axis *EC* and *latus rectum DC*, and the hyperbola $\mathscr{H}$ with vertex *C*, transverse axis *DC* and *latus rectum DC*. The two curves cut one another at the point *I*. Let us draw *IA* ∥ *CE* and *IG* ∥ *CD*; let us produce *CD* by a length *DB* equal to *AI*. Then we have *EG* = *CB*, because *CG* = *AI* = *DB*. But, since *I* ∈ $\mathscr{P}$, we have $IG^2 = EG \cdot CD$, so

(1) $AC^2 = CB \cdot CD,$

and since *I* ∈ $\mathscr{H}$, we have $IA^2 = AC \cdot AD$, so

(2) $DB^2 = AC \cdot AD.$

So if we take the segment CD, we can construct $\mathscr{P}$ and $\mathscr{H}$, and from their point of intersection I we can find the points A and B such that (A, C, D, B) is a range of the kind given by Archimedes.

We may note that the parabola $\mathscr{P}$ cuts the straight line DC in a point D' symmetrical with D with respect to the point C, and that $\mathscr{P}$ cuts $\mathscr{H}$ in two points J and I. The point I chosen by al-Qūhī is the appropriate one, because $CA > CD'$, so $CA > CD$, a condition the author's analysis establishes as necessary; whereas J would give us A' with $CA' < CD$.

Once again, the problem has one and only one solution. Let us turn to an analytical method of finding the point I.

Let CA, CE be orthogonal axes (Cx, Cy), with $DC = a$ and $CE = a$. The equation of the parabola $\mathscr{P}$ with vertex E, axis Cy and *latus rectum a* is

$$x^2 = a(a - y).$$

The equation of the hyperbola $\mathscr{H}$ with vertex C, transverse axis DC and *latus rectum a* is
$$y^2 = x(a + x),$$

and the equation for the abscissae of the points of intersection is

$$x^3 - ax^2 - 2a^2x + a^3 = 0,$$

which has three roots $x_1 < 0$, $0 < x_2 < a$ and $x_3 > a$. It is this third root that satisfies the conditions of the problem.

In the following proposition, the sixth, al-Qūhī proves the converse of Proposition 2 (which was an analysis), namely that with a range (A, C, D, B) of type D_1 we can associate a triangle with base CD, of type $(1, 2, 4)$.

We first construct Archimedes' range with points in the order (A, C, D, B), such that equations (1) and (2) are satisfied, using the method of the preceding proposition. Now we know that $BD + AC > DC$; so we can construct the triangle DCE such that $DE = DB$ and $CE = CA$.

We produce EC by $GC = CD$, we then have $EG = AD$. We have

$$BC \cdot CD = AC^2 = EC^2,$$

so

$$\frac{BC}{EC} = \frac{EC}{CD}.$$

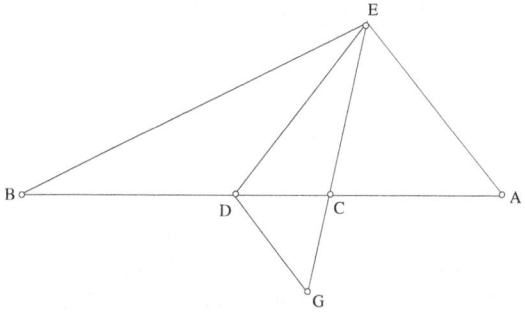

Fig. 3.17

Triangles BCE and DCE are similar, with common angle C. But

(3) $E\hat{D}C = D\hat{B}E + B\hat{E}D = 2E\hat{B}D = 2C\hat{E}D$.

In the same way

$$E\hat{C}D = C\hat{D}G + C\hat{G}D = 2D\hat{G}C;$$

moreover $DA \cdot AC = DB^2$, hence $GE \cdot EC = DE^2$; so we have

$$\frac{GE}{DE} = \frac{DE}{EC}$$

and triangles GED and DEC, which have a common angle E, are similar. In consequence

$$D\hat{G}C = E\hat{D}C,$$

so

(4) $E\hat{C}D = 2E\hat{D}C$.

So we deduce from (3) and (4) that

$$E\hat{C}D = 2E\hat{D}C = 4C\hat{E}D.$$

So triangle ECD gives a solution to the problem.

In Proposition 7 al-Qūhī finally carries out the construction of the heptagon, by drawing in the given circle a triangle ABC similar to the preceding one; the arc *CB* will be a seventh of the circumference and the chord *CB* will be the side of the heptagon.

3.1.4.1.3.2. *Al-Ṣāghānī*

It was scarcely two years after Abū al-Jūd, still closer in time to Ibn Sahl, and less than a year after al-Qūhī, that al-Ṣāghānī attacked the problem of Archimedes' lemma and the construction of the heptagon. He completed a first essay intended for the Library of the King, 'Aḍud al-Dawla. He then returned to this essay and wrote a second one, also dedicated to the King. It is this second composition that has come down to us.

As regards the history of the problem, al-Ṣāghānī is rather discreet. He merely remarks that 'the determination of the chord of the heptagon resisted the efforts of geometers', and that 'Archimedes had stated a lemma whose proof would have made it possible to find the chord of the heptagon'; and that 'this is how the problem has made its way down to our own time'.[50] The most one might detect in this is an allusion to the research being directed to the problem. All the same, this study by al-Ṣāghānī is one of the most detailed.

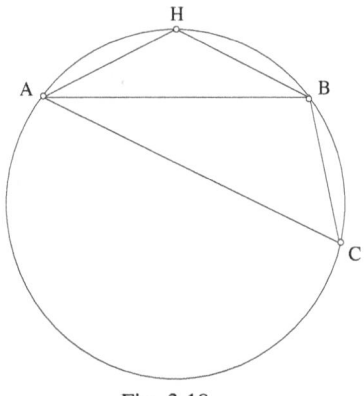

Fig. 3.18

He begins by proving that if *A*, *H*, *B*, *C* are four consecutive vertices of a regular heptagon, then $A\hat{B}C = 2A\hat{C}B = 4B\hat{A}C$. So triangle *ABC* is of type T_2 (1, 2, 4).

We may note that triangle *AHB* is of type T_1 (1, 5, 1). Al-Ṣāghānī then states that, if one knows how to construct T_2, the problem of constructing the heptagon is solved.

---

[50] See below, p. 661.

The propositions that follow are dedicated to investigating this triangle.

First proposition: Let there be a triangle ADU such that $\hat{U} = 2\hat{D} = 4\hat{A}$. We produce the base DU in both directions, making $DH = DA$ and $UB = AU$.

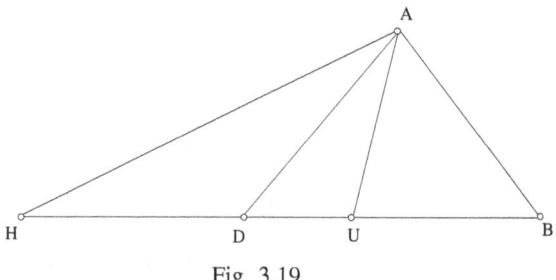

Fig. 3.19

We have

$$A\hat{U}D = 2A\hat{B}U = 2B\hat{A}U,$$

hence $B\hat{A}U = A\hat{D}U$; so triangles ABU and ADB are similar and isosceles, $AD = AB$. Thus we have

$$\frac{DB}{AB} = \frac{AB}{BU},$$

hence $AB^2 = DB \cdot BU$, or again $DH^2 = DB \cdot BU$.

Similarly, $A\hat{D}U = 2D\hat{A}U = 2A\hat{H}U$, so $D\hat{A}U = A\hat{H}U$; so triangles AUD and AUH are similar, and we have

$$\frac{UH}{AU} = \frac{AU}{UD},$$

hence $AU^2 = UD \cdot UH$, or again $BU^2 = UH \cdot UD$.

Thus the analysis shows that the triangle ADU of type $(1, 2, 4)$ is associated with Archimedes' range, that is, D_1, (B, U, D, H).

In the second proposition, al-Ṣāghānī considers such a range, lettered (A, E, D, B), and on AD constructs the square $AFID$; the straight line FB cuts ID in K and IA in H. We draw $HC \parallel ID$. We then show that:
• the points C and E coincide
• area (IFH) = area (KBD).

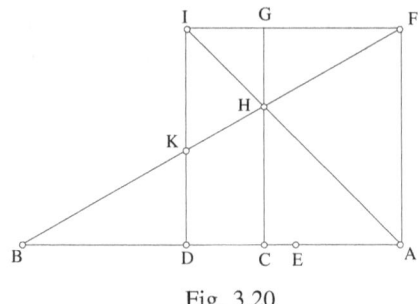

Fig. 3.20

In the third and fourth propositions, al-Ṣāghānī starts with a square *ABDC* with diagonal *BC*, and shows, by analysis and synthesis, how to construct a straight line from *A* to cut *BC* in *E* and *BD* in *H* in order to give us area $(AEC) = $ area (GDH).

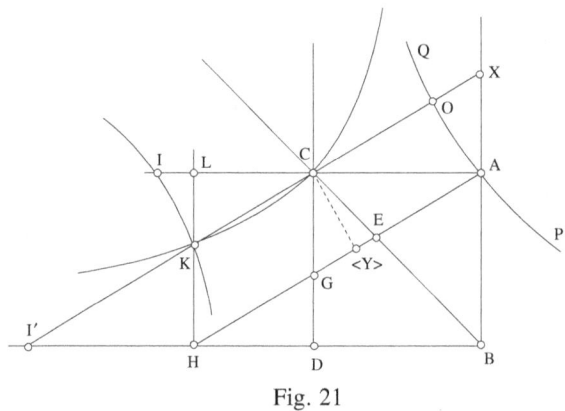

Fig. 21

Let us begin with the analysis and let us suppose that *AH* is the required straight line, *L* the fourth vertex of the rectangle *ABHL* and *XC* the line parallel to *AH* that cuts *HL* in *K* and *BD* in *I'*. We have $AC = I'H$. Since

$$\text{area } (AEC) = \text{area } (GDH) \text{ and } E\hat{A}C = G\hat{H}D,$$

thus

$$AC \cdot AE = HD \cdot HG,$$

hence

$$\frac{AC}{DH} = \frac{GH}{AE};$$

but

$$\frac{AC}{DH} = \frac{AG}{GH},$$

so we have

$$\frac{GH}{AE} = \frac{AG}{GH},$$

hence

$$GH^2 = AG \cdot AE.$$

But in triangle ACG we have $AG \cdot AE < AC^2$ (because $AE < AY$ if $CY \perp AG$), so $GH < AC$, hence $KC < AC < CX$. We mark on CX a point O such that $CO = KC = GH$, so we have $OC^2 = AG \cdot AE$.

The hyperbola $\mathcal{H}_1$ that passes through A and has asymptotes CB and CD also passes through O.[51]

Moreover, the hyperbola $\mathcal{H}_2$ that passes through C and has asymptotes BX and BI' also passes through K, because $XC = KI'$.[52]

We know that $CL < GH < AC$; let I be a point on CL produced such that $CI = CA$. The two points I and K are symmetrical to A and O with respect to C which is the centre of $\mathcal{H}_1$, so the second branch of $\mathcal{H}_1$ passes through K and I. So the point K, the point of intersection of $\mathcal{H}_1$ and $\mathcal{H}_2$, is known, and the straight lines CK and AH are thus known.

In the fourth proposition al-Ṣāghānī carries out the synthesis for this analysis.

We take a square $ABDC$ with diagonal BC, and a point I on AC produced such that $CI = CA$. We draw the hyperbola $\mathcal{H}_1$, with asymptotes CB and CD, one of whose branches passes through A; so the other branch, KI, passes through I. We then draw the hyperbola $\mathcal{H}_2$ that passes through C and has asymptotes BA and BD. The branch KI of $\mathcal{H}_1$ cuts the two parallel lines AI and BD; so it cuts $\mathcal{H}_2$ in a point K lying between the lines AI and CD. The straight line KC cuts AB in X and BD in I'. We draw $KH \perp BD$; we join AH, it cuts CB and CD in E and G respectively.

We then show that area (GDH) = area (AEC).

From Apollonius II.8, we have $KI' = XC$, so triangles AXC and KHI' are congruent; hence $HI' = AC$; so we have $AH \parallel XI'$. If O is the point of intersection of XI' with the hyperbola $\mathcal{H}_1$, we have $CO = CK$,[53] and since

[51] Apollonius, *Conics*, II.11, ed. Heiberg.

[52] Apollonius, *Conics*, II.8, ed. Heiberg.

[53] Apollonius, *Conics*, I.30, ed. Heiberg.

$AG \parallel OC$, we have $AG \cdot AE = OC^2$, from Proposition II.11 of Apollonius. But $OC = CK = GH$, so $GH^2 = AG \cdot AE$, hence

$$\frac{AG}{GH} = \frac{GH}{AE}.$$

Triangles ACG and DGH are similar, so

$$\frac{AC}{DH} = \frac{GH}{AE}.$$

Now $E\hat{A}C = G\hat{H}D$; hence the result.

We have just repeated the analysis and synthesis given by al-Ṣāghānī. Let us go back over his proof using a different formal language – that of algebra.

Let (BD, BA) be coordinate axes (Bx, By) with $BD = BA = a$.

$$\mathcal{H}_1 = \left\{(x,y); y = x - \frac{a^2}{x-a}\right\}.$$
$$\mathcal{H}_2 = \left\{(x,y); xy = a^2\right\}$$

The equation for the abscissae of the points of intersection is

$$x^3 - ax^2 - 2a^2x + a^3 = 0,$$

which has three roots $x_1 < 0$, $0 < x_2 < a$ and $a < x_3 < 2a$. The required point K corresponds to the root x_3.

Note 1: Let U be the orthogonal projection of E on BD. Let us put $BD = a$; we require U and H such that

$$BU^2 = UH \cdot UD \text{ and } HD^2 = BU \cdot BD.$$

Let us put $BH = x$, $BU = y$ and $BD = a$, where $a < x < 2a$ and $0 < y < a$. We have

(1) $y^2 = (x-y)(a-y) \iff y = \dfrac{ax}{a+x}$

(2) $(x-a)^2 = ay.$

From (1) and (2) we have

$$(x-a)^2(a+x) = a^2 x,$$

and, when we reduce this, we have the same cubic equation as before, which gives the abscissa of K, the point of intersection of $\mathscr{H}_1$ and $\mathscr{H}_2$, a point that has the same abscissa as H.

Note 2: In his analysis, al-Ṣāghānī proved the inference

$$C\hat{A}E = G\hat{H}D \text{ and area } (AEC) = \text{area } (GDH) \Rightarrow GH^2 = AG \cdot AE$$

and, in his synthesis

$$GH^2 = AG \cdot AE \text{ and } C\hat{A}E = G\hat{H}D \Rightarrow \text{area } (AEC) = \text{area } (GDH).$$

So the argument has made use of only one of the two equalities that serve to characterize Archimedes' range (A, E, G, H), a range al-Ṣāghānī obtained in his first proposition.

In his fifth proposition, he does indeed show that this range (A, E, G, H) considered in Propositions 3 and 4 also satisfies the second equation that is necessary to define an Archimedean range D_1, and from what he has already proved he derives a construction for a triangle of type $(1, 2, 4)$. Proposition 5 thus gives the synthesis that corresponds to the analysis in Proposition 1. Al-Ṣāghānī proceeds as follows:

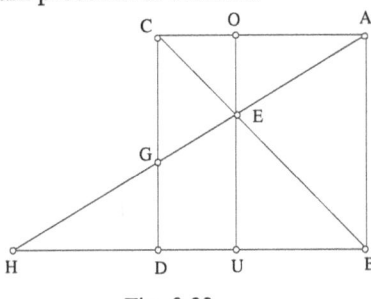

Fig. 3.22

Let us return to the square $ABDC$ and the straight line AH with the points E and G obtained in Proposition 4. From E we draw a perpendicular to BD to cut AC in O and BD in U.

We know that $GH^2 = AG \cdot AE$; the two ranges (A, E, G, H) and (B, U, D, H) are similar, so $DH^2 = DB \cdot BU$. Moreover, triangles AOE and EUH are similar, so

$$\frac{HU}{UE} = \frac{AO}{OE};$$

but

$$OE = OC = UD \text{ and } AO = UB = UE,$$

hence

$$\frac{HU}{UB} = \frac{UB}{UD}.$$

So $BU^2 = HU \cdot UD$ and consequently $AE^2 = EH \cdot EG$. The two ranges are thus both of type D_1.

We have seen in Proposition 3 that $AC > GH$; now $GH > DH$ and $AC = BD$; so $BD > DH$, that is $BU + UD > DH$. Moreover $BU^2 = HU \cdot UD$, where $HU > UD$; so we have $BU + DH > UD$ and $HD + DU > BU$. So we can construct a triangle from the three segments BU, UD, DH.

It is true that al-Ṣāghānī did not state that the ranges (A, E, G, H) and (B, U, D, H) were similar; but it is clear that it is implicit when he concludes:

Similarly, we can construct a triangle whose sides are equal to the straight lines AE, EG, GH starting from the straight line AH and the points E and G.[54]

So, starting from the range (B, U, D, H), he constructs a triangle ADU where $AU = UB$ and $DA = DH$, and proves that

$$A\hat{U}D = 2A\hat{D}U \text{ and } A\hat{D}U = 2U\hat{A}D.$$

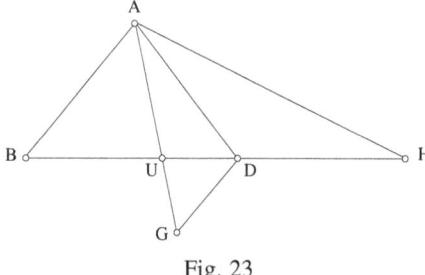

Fig. 23

The proof is immediate: we produce AU by a length $UG = UD$, and use the fact that triangles AUD, AUH and ADG are similar to deduce that they are equiangular. These triangles are of type $(1, 2, 4)$.

[54] See below, p. 668.

To construct the regular heptagon, it is sufficient to inscribe in the circle a triangle ABC similar to the triangle AUD; which is what al-Ṣāghānī does in the fifth and sixth propositions of his treatise.

Al-Ṣāghānī's solution was subjected to careful investigation by Abū al-Jūd in his *Letter to Abū Muḥammad 'Abd Allāh ibn 'Alī al-Ḥāsib*. Abū al-Jūd does not go over al-Ṣāghānī's analysis, but he repeats the whole of the proof of the synthesis following it step by step (only two letters of the figure are changed).

3.1.4.1.3.3. *Al-Qūhī: the second treatise*

Some years after his first treatise, in which he constructed the heptagon with the help of a triangle of type $(1, 2, 4)$, al-Qūhī wrote a second treatise, in which the heptagon is constructed starting from a triangle of type $(1, 1, 5)$, again making use of a range of type D_1. Whereas the first treatise was composed around 970 and dedicated to the King 'Aḍud al-Dawla, the second is dedicated to the King's son, Abū al-Fawāris. The son's youth, and the manner in which al-Qūhī addresses him, leave no doubt about the order in which the texts were composed, as A. Anbouba correctly observed.[55] This second treatise was separated from the first by a period of several years. In any case, it was composed before the death of the King who was the dedicatee's father, in 982, that is before Abū al-Fawāris became the Prince Sharaf al-Dawla of Fārs.

As before, al-Qūhī starts by referring to the contribution made by Archimedes, and, amazingly, this is the only name he mentions in connection with the construction of the heptagon. No other text, not even his own, is referred to in this second treatise.

Al-Qūhī considers three consecutive vertices A, B, C, of a regular heptagon inscribed in a circle and proves that the isosceles triangle ABC is of type $(1, 5, 1)$. In a second proposition, he proves that analysis of the construction of an isosceles triangle ABC in which $AB = BC$ and $A\hat{B}C = 5B\hat{A}C = 5B\hat{C}A$ leads to Archimedes' range (C, B, D, E). Let us follow al-Qūhī's procedure for constructing such a triangle ABC.

Let D and E be points on CB such that $B\hat{A}D = B\hat{A}C$ and $DE = AD$. Triangles ACD and ABD are similar, so we have

$$\frac{CD}{DA} = \frac{DA}{DB},$$

hence

$$DA^2 = DB \cdot DC,$$

[55] A. Anbouba, *'Tasbī' al-Dā'ira'*.

and consequently

(1)
$$DE^2 = DB \cdot DC.$$

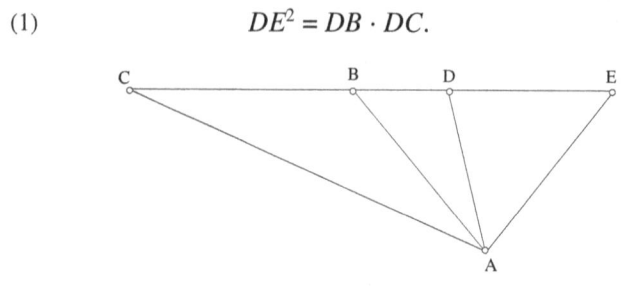

Fig. 3.24

We have
$$A\hat{B}C = 5B\hat{A}D \text{ and } A\hat{B}C = B\hat{A}D + B\hat{D}A;$$

so
$$B\hat{D}A = 4B\hat{A}D.$$

Moreover, $AD = DE$, so $B\hat{D}A = 2D\hat{A}E = 2D\hat{E}A$. Triangles ABE and ADE are thus similar, and we have

$$\frac{EB}{EA} = \frac{EA}{ED},$$

hence
$$EA^2 = EB \cdot ED.$$

But $EA = AB$ and $AB = BC$, hence

(2)
$$BC^2 = EB \cdot ED.$$

So we require to find, on the line BC, points D and E such that (1) and (2) are satisfied. The analysis has thus led to the inference:

[Triangle of type (1, 5, 1)] $\Rightarrow$ Archimedes' range D_1.

Note: This figure includes the triangles *ABD* and *CAE*, which are of type (1, 2, 4] and the triangle *DAE* of type (3, 2, 2). But al-Qūhī pays no attention to them.[56] In the first treatise we find the same figure for considering a triangle (1, 2, 4), and the other types, although present, are not discussed. We may remind ourselves that it was to be Ibn al-Haytham who dealt with all the types of triangle, taking an approach he himself characterized as more complete: considering all the possible cases.

In the third and fifth propositions of his treatise, al-Qūhī uses analysis and synthesis to construct the range (A, B, C, D) of type D_1. In the third proposition, he gives the following analysis:

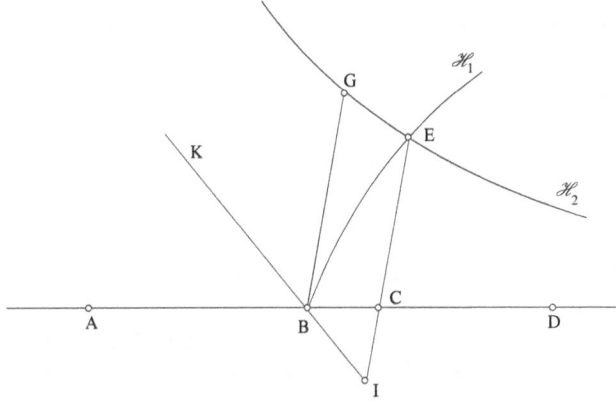

Fig. 3.25

Let $A\hat{B}G = \alpha$ be given and let $CE \parallel BG$, with $BG = BA$ and $CD = CE$ and KB, the bisector of angle α cuts EC in I. By hypothesis we have $CD^2 = CE^2 = AC \cdot AB$; so E lies on the hyperbola $\mathcal{H}_1$, which has diameter AB and *latus rectum* $c = AB$, so the angle of the ordinates is $E\hat{C}B = \alpha$ (so BG is the tangent to the hyperbola at B).

We have $C\hat{B}I = K\hat{B}A$ and $C\hat{I}B = K\hat{B}G$, hence $C\hat{B}I = C\hat{I}B$ and $CB = CI$, and consequently $IE = BD$; so we have $IE \cdot EC = BA^2 = BG^2$; so the point E lies on a hyperbola $\mathcal{H}_2$ that has asymptotes KB and BD and passes through G.

Al-Qūhī then concludes by considering a segment AB of known length and position, and choosing an angle α, which allows him to construct G and the bisector BK. From this we find the hyperbolae $\mathcal{H}_1$ and $\mathcal{H}_2$ and their point of intersection E. The points C and D are obtained from E, and the

---

[56] In a shortened version of al-Qūhī's treatise on the trisection of an angle and the construction of the heptagon (Thurston 3, fol. 130ᵛ; Marsh 720, fol. 264ᵛ), he constructs the triangle (3, 3, 1), and thus recognizes the relationship between the two constructions.

range (A, B, C, D) is thus known. Effectively, this strategy corresponds to the conclusion of the synthesis carried out in the fifth proposition. However, we first need to ensure that certain inequalities are satisfied. These inequalities follow immediately from the properties of Archimedes' range and are necessary for the construction of the triangle.

In Proposition 4, al-Qūhī proves the following inequalities:

$$AB < BC + CD, \quad BC < AB + CD \quad \text{and} \quad CD < AB + BC.$$

We shall now present the synthesis, using the same lettering and figure as in the analysis.

Let AB be a given segment, BG a segment such that $BG = AB$ and $A\hat{B}C = \alpha$, a given angle, and let KB be the bisector of $A\hat{B}G$.

We draw the hyperbola $\mathcal{H}_1$ with transverse diameter AB and *latus rectum* AB, with α as the angle of the ordinates. We then draw the hyperbola $\mathcal{H}_2$ that passes through G and has asymptotes KB and AB. The two hyperbolae must intersect, say in a point E. Let EC be the ordinate of this point, $EC \parallel BG$ and it cuts BK in I.

From *symptoma* of $\mathcal{H}_1$ we have $EC^2 = CA \cdot CB$. We produce AC by $CD = CE$, so we have $CD^2 = CA \cdot CB$. The *symptoma* of $\mathcal{H}_2$ gives $GB^2 = EC \cdot EI$, but $GB = AB$, $EC = CD$ and $EI = BD$; so we have $AB^2 = DB \cdot DC$.

Given AB and the angle α (arbitrary choices), $\mathcal{H}_1$ and $\mathcal{H}_2$ and their point of intersection E are determinate; from the point E we can find C and D, which do not depend on the angle α.

We may make the following comments on al-Qūhī's procedure, employing a type of presentation that is different from his.

Let (BD, BG) be the coordinate axes (Bx, By), with $AB = BG = a$. The hyperbola $\mathcal{H}_1$ that passes through B, and to which By is a tangent, has the equation

(1) $y^2 = x(a + x).$

The hyperbola $\mathcal{H}_2$ has asymptotes BD and BK and passes through $G(0, a)$; it has the equation

(2) $y(y + x) = a^2.$

From (1) and (2) it follows that

(3) $\left(a^2 - x^2 - ax\right)^2 = x^2\left(x^2 + ax\right),$

which after simplification reduces to

(4) $x^3 - ax^2 - 2a^2x + a^3 = 0.$

This equation has three roots, $x_1 < 0$, $0 < x_2 < a$, $x_3 > a$. One of the positive roots is the abscissa of E. Now we know that the ordinate of E must be positive. But

$$y = \frac{a^2 - x^2 - ax}{x} = \frac{a^2}{x} - (x + a),$$

so $x_2 \in\ ]0, a[$, which is the abscissa of the point E (because $x_3 > a$ gives $y < 0$).

So al-Qūhī's choice is understandable. It remains to explain the disappearance of the term x^4 from (3). The reason for this is that one of the asymptotes of $\mathcal{H}_1$ is a parallel to the asymptote GK of $\mathcal{H}_2$, drawn through the midpoint of AB, a transverse diameter of $\mathcal{H}_1$. Let us draw the figure taking $\alpha = \dfrac{\pi}{2}$. The hyperbola $\mathcal{H}_1$ is equilateral and has asymptotes

$$y = x + \frac{a}{2} \quad \text{and} \quad y = -x - \frac{a}{2}.$$

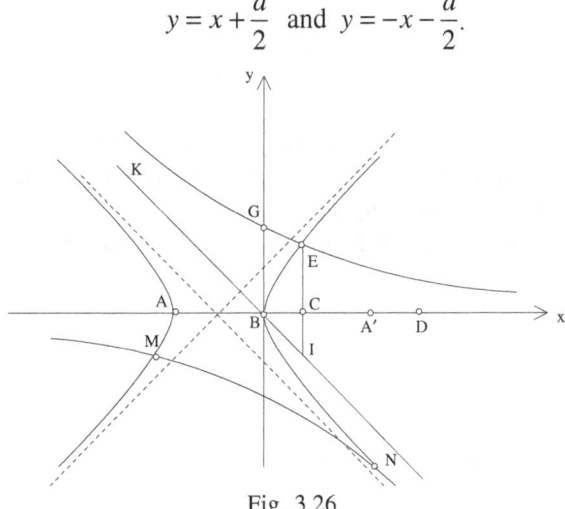

Fig. 3.26

Let A' be such that $AB = BA' = a$. So we have three points of intersection: M where $x_M = x_1$, E (the solution to the problem) where $x_E = x_2 \in \,]0, a[$, and N where $x_N > a$ and $y_N < 0$.

In the sixth proposition al-Qūhī proves the converse of Proposition 2 (which was an analysis); that is, that if we have an Archimedean division D_1, there is an associated triangle of type $T_1 - (1, 5, 1)$.

From Proposition 4, we know it is possible to construct such a triangle. Let DEC be a triangle such that $DE = DB$ and $CE = CA$. The straight line EC cuts the line through D parallel to AE in the point I. We have $CD = CI$ and triangle CDI is isosceles, as is triangle ACE; hence $IE = AD$.

So we have $IE \cdot EC = DA \cdot AC = DB^2 = DE^2$, from which we obtain

$$\frac{IE}{ED} = \frac{ED}{EC},$$

so triangles EDC and EID are similar and we have

$$E\hat{C}D = 2C\hat{I}D = 2E\hat{D}C = 4E\hat{B}D.$$

Moreover,

$$BC \cdot CD = AC^2 = EC^2,$$

hence

$$\frac{BC}{CE} = \frac{EC}{CD};$$

so triangles BCE and ECD are similar, so $E\hat{B}D = D\hat{E}C$, and we thus have $E\hat{D}C = 2D\hat{E}C$ and $E\hat{C}D = 4D\hat{E}C$. As $B\hat{D}E$ is an exterior angle of triangle EDC, $B\hat{D}E = D\hat{C}E + D\hat{E}C = 5D\hat{E}C = 5D\hat{B}E$.

So the isosceles triangle EBD gives a solution to the problem.

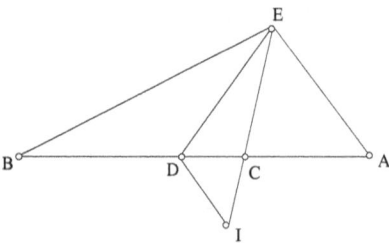

Fig. 3.27

In the following proposition, the seventh, al-Qūhī returns to the construction of the side of the regular heptagon inscribed in a given circle, using the method we have already examined more than once: we construct a triangle similar to that in the previous proposition and we inscribe it in the circle.

For constructing Archimedes' range (D_1), the idea we find for the triangle of type $(1, 2, 4)$ in all three principal authors – Ibn Sahl, al-Qūhī in his two treatises and Ibn al-Haytham, in his two treatises – consists of taking two of the four points as given, and the others as determined by Archimedes' two conditions. In fact these authors read the two conditions as the *symptomata* of two conic sections whose points of intersection determine the two points that are required. In fact, one of these two points is the projection on the line of the range of a suitably chosen point of intersection, while the other point is found by marking off the ordinate of the point of intersection on the line, starting from one of the points that are already known.

$$AC^2 = CB \cdot BD,$$

$$BD^2 = AD \cdot CD.$$

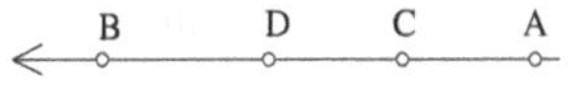

Fig. 3.28

Ibn Sahl and al-Qūhī, in his first treatise, take the two points C and D as given; the point B is the projection of a point of intersection of a parabola and a hyperbola and CA is the ordinate of this point of intersection. We put $CD = a$, $BD = x$ and $AC = y$; Archimedes' conditions can be written

$$y^2 = x(a + x),$$

which is the equation of an equilateral hyperbola with transverse axis CD, and

$$x^2 = a(a + y),$$

which is the equation of a parabola with axis DE perpendicular to CD, vertex E such that $DE = CD = a$ and *latus rectum a*.

As x and y have positive values, only one of the points of intersection can be used.

We may note that in the case of al-Qūhī the sign of y is changed. Ibn al-Haytham, like al-Qūhī in his second treatise, takes the points B and D as given; the point C is the projection of the point of intersection of a parabola and a hyperbola and CA is equal to the ordinate of this point of intersection. We put $BD = a$, $CD = -x$ and $AC = y$; the conditions can be written

$$y^2 = a(a - x)$$

which is the equation of a parabola with axis BD, vertex B and *latus rectum* a; and

$$a^2 = x\,(x - y),$$

which is the equation of a hyperbola that passes through B and has asymptotes $x = 0$ and $y = x$.

Here $x < 0$ and $y > 0$ and this determines which is the relevant point of intersection. To find the equations given in the commentary, we need to change the signs of x and y.

Al-Ṣāghānī's procedure is completely different, although he too starts with two of the four points, taking them as given, and then determines the other two. He assumes we know B and C; he finds A as the projection of the point of intersection of two hyperbolae and he obtains D from A by means of the geometrical construction of the square of side BC, say $BCC'B'$; D is the projection onto BC of the point of intersection of the straight lines $B'A$ and BC'.

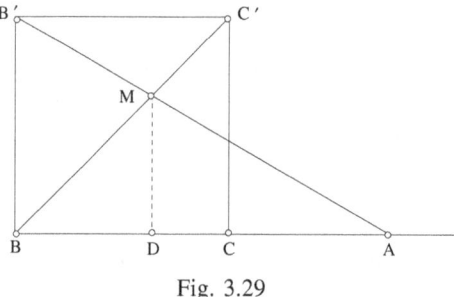

Fig. 3.29

This construction replaces the use of Archimedes' first relation, which does not appear in al-Ṣāghānī. The first of the two hyperbolae that he uses is chosen *a priori*, to have equation $xy = a^2$ ($a = BC$); the second hyperbola is chosen so that it expresses Archimedes' second relation. Thus, as noted in our commentary, the determination of A involves only the use of this second relation. Al-Ṣāghānī's procedure is much less transparent than those of the other authors.

3.1.4.2. *The range studied by Abū al-Jūd and al-Sijzī* (D₂)

We are in fact concerned with Abū al-Jūd's first range: to divide a segment AB at a point C such that

$$\frac{\sqrt{AB \cdot AC}}{BC} = \frac{AB}{AB + BC}.$$

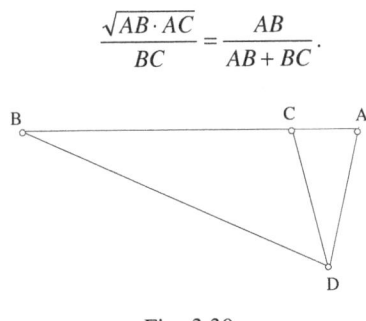

Fig. 3.30

From that range we derive a construction for a point D such that $BD = BA$ and $AD = \sqrt{AB \cdot AC}$. We show that D lies on the perpendicular bisector of AC and that the triangles ABD and ACD are isosceles and similar; they are of type T_3.

According to his own account, Abū al-Jūd had proposed this range in the first letter, that of 968–969. He explains how he was led to the problem and to this range. His model is Euclid and his construction of the regular pentagon. This construction involves an isosceles triangle each of whose base angles is twice the angle at the vertex; thus $\alpha = \frac{\pi}{5}$. An angle α inscribed in a circle stands on an arc whose chord is equal to the side of the regular pentagon. Abū al-Jūd notes that the same principle holds for any polygon with an odd number of sides. Thus, for a polygon with $2n + 1$ sides, we would employ a triangle each of whose base angles is n times the angle α at the vertex, that is the triangle $(1, n, n)$, $\alpha = \frac{\pi}{2n+1}$. This explains why Abū al-Jūd chose the triangle $(1, 3, 3)$ for the heptagon.

The analogy clearly has its limits, and Abū al-Jūd seems to have realized that the heptagon cannot be constructed by means of straightedge and compasses. He had carried out the following analysis for constructing such a triangle.

Let ABC be a triangle such that $AB = AC$ and $A\hat{B}C = B\hat{C}A = 3B\hat{A}C$. Let there be a point D on AB and a point E on AC such that $B\hat{C}D = B\hat{A}C$ and $A\hat{D}E = B\hat{A}C$.

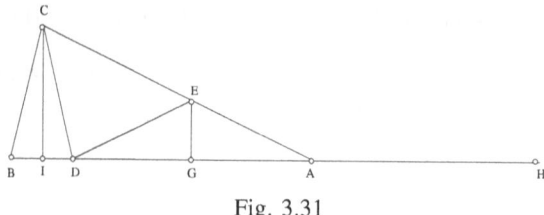

Fig. 3.31

We have $BC = CD$ and $DE = EA$.

Moreover, $D\hat{C}E = 2\hat{A}$, so $DC = DE$, and we have $BC = EA$. Triangles ABC and BCD are isosceles and similar; so

$$\frac{AB}{BC} = \frac{CB}{BD} \Rightarrow BC^2 = AB \cdot BD;$$

hence

(1) $AE^2 = AB \cdot BD.$

We draw $CI \perp BD$ and $GE \perp AD$; we obtain $IB = ID$ and $GA = GD$. We have

$$\frac{AE}{AG} = \frac{AC}{AI} = \frac{AB}{AI}.$$

We produce BA by $AH = AD$; thus $BH = 2\,AI$. Now $AD = 2\,AG$, so

(2) $\dfrac{AE}{AD} = \dfrac{AB}{BH}.$

So the points B, D, A, H satisfy (1) and (2); and $BH = AB + AD$. Thus we can describe the range (A, D, B) by the equality

(3) $\dfrac{AB}{AB + AD} = \dfrac{\sqrt{AB \cdot BD}}{AD}.$

Conversely, starting with a straight line AB and a point D lying on it that satisfies equality (3), we can find an isosceles triangle such that the sum of its angles is seven times the angle α at its vertex, so $\alpha = \dfrac{\pi}{7}$. From this triangle we can go on to construct the heptagon.

In this work Abū al-Jūd does not give either the construction for the range, or that for the heptagon, but merely points out that he has already done all this in his essay of 968–969 using a parabola and one branch of a hyperbola. But that is precisely what is in dispute. He returns to the matters of this earlier essay in another treatise, entitled *Book on the Construction of the Heptagon in the Circle*, where he writes:

> As for my former letter about the construction of the heptagon, in which I was ahead of everyone and I distinguished myself from others by the route I followed, I repeat it here for you in its entirety in a single proposition proved with the aid of God and His assistance.[57]

Abū al-Jūd then gives the following proof:

Let there be a segment *AI*, with midpoint *B*, and the square *BIKL*; let there be a parabola $\mathscr{P}$ with vertex *A*, axis *AI* and *latus rectum AB*, and $\mathscr{H}$ a branch of a hyperbola with vertex *B*, transverse diameter 2 *BK* and *latus rectum* 2 *BK* – so $\mathscr{H}$ has the straight lines *KI* and *KL* as asymptotes. The vertex *B* of $\mathscr{H}$ lies inside $\mathscr{P}$, so $\mathscr{H}$ cuts $\mathscr{P}$ in two points; let *M* be the point of intersection that lies between *A* and *L*. We draw *MD* ⊥ *AB* and construct a point *C* such that *AC = CD = MD*. So we have $AB \cdot AD = AC^2$; the two triangles *ABC* and *ADC* are similar, they are of type (1, 3, 3) and consequently $A\hat{B}C = A\hat{C}D = \dfrac{\pi}{7}$.

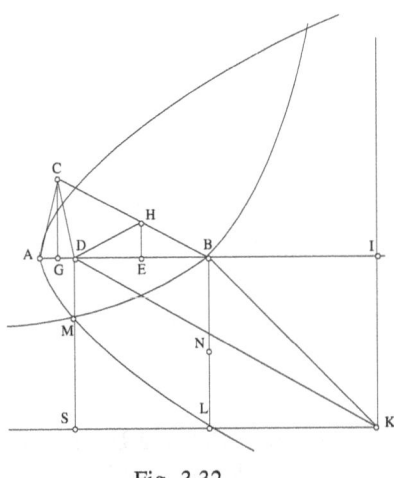

Fig. 3.32

We may note that the construction shows two other triangles: *CBD*, of type T_2, and *DHB*, of type T_1.

So Abū al-Jūd gives the range (A, D, B) of type D_2, which is not mentioned. He uses only the equality $AC^2 = AB \cdot AD$, where $AC = MD$. Finally, he gives only the synthesis.

Is this an abbreviated version of the original letter, in which Abū al-Jūd has, in the process of shortening it, corrected some errors for which he has been reproved? The only justification for such a suspicion is in the very existence of the dispute; because if a proof like this was to be found in that original letter, it is hard to understand the criticisms made by al-Sijzī.

The latter constructs the range D_2 in his treatise, which thus happens to be the oldest *available* document that contains the proof, which he acknowledges is in its essentials due to Ibn Sahl. Al-Sijzī gives the following proof of the lemma:

Lemma 1 of al-Sijzī: *To construct on a given segment* AB *a point* C *such that we obtain the range* D_2.

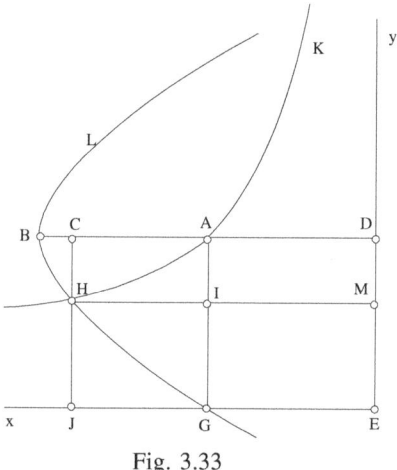

Fig. 3.33

Let D be a point such that $BD = 2BA$, and let $ADEG$ be the square constructed on AD. We consider a hyperbola $\mathscr{H}$ with vertex A, having asymptotes ED and EG; and the parabola $\mathscr{P}$ with vertex B, axis BD and *latus rectum AB*. This parabola passes through G, so it cuts the hyperbola at the point H. From H we draw the perpendicular to AB – let it be CHJ, and the line parallel to AB – let it be HIM. We have

$$\text{area } (HMEJ) = \text{area } (ADEG) \;\Rightarrow\; \text{area } (JHIG) = \text{area } (IADM)$$
$$\Rightarrow \text{area } (JCAG) = \text{area } (HCDM).$$

So we have

$$CA \cdot AG = CH \cdot CD \Rightarrow CA \cdot AB = CH \cdot CD.$$

But, $H \in \mathcal{P} \Rightarrow CH^2 = BC \cdot AB$ and $CD = AB + AC$, hence

$$CA \cdot AB = (AB + AC)\sqrt{BC \cdot AB}.$$

So the problem has one solution; we show it is unique. Let us take up the solution in different terms. We consider orthogonal axes (Ex, Ey) and take points A $(a; a)$ and B $(2a; a)$, where $a > 0$. The problem is to find a point $C \in [A \, B]$, $C(x, a)$, where $a < x < 2a$ such that

$$\frac{\sqrt{a(2a - x)}}{x - a} = \frac{a}{x}.$$

The equations of $\mathcal{H}$ and $\mathcal{P}$ in this system of coordinates are respectively

(1) $xy = a^2$ – we consider only the branch $x > 0$, $y > 0$
(2) $a(2a - x) = (a - y)^2$.

For any point of $\mathcal{H}$ we have

$$xy = a^2 \Leftrightarrow a(x - a) = x(a - y),$$

hence for $x \neq 0$ and $x \neq a$

(3) $\dfrac{a - y}{x - a} = \dfrac{a}{x}.$

So if $H(x_0, y_0) \in \mathcal{H} \cap \mathcal{P}$, from (2) and (3) we have

$$\frac{\sqrt{a(2a - x_0)}}{x_0 - a} = \frac{a}{x_0},$$

and if $a < x_0 < 2a$, the point H gives a solution to the problem.

To prove the point H exists, we may note that A, the vertex of the branch of the hyperbola in question, lies inside $\mathcal{P}$; so $\mathcal{H}$ cuts $\mathcal{P}$ in two points, which lie on either side of A. The one that is nearer the vertex B of the parabola is the one we require. This is how (by implicit considerations of continuity and convexity) the mathematicians of the time – in particular Ibn al-Haytham – establish the existence of a point of intersection for two convex curves. Analytical investigation of $\mathscr{H} \cap \mathscr{P}$ gives, by eliminating y between (2) and (3), the equation

$$x^3 - ax^2 - 2a^2x + a^3 = 0,$$

which has three solutions: $x_1 < 0$, $0 < x_2 < a$ and $a < x_3 < 2a$. So the problem has one and only one solution, which corresponds to the root x_3.

Once the range D_2 has been set up, al-Sijzī gives a second lemma that considers the construction of a triangle ABC of type (1, 3, 3), making use of the range (B, C, A) of the same type (in this lemma, the ordering A, B of the previous lemma becomes B, A). The proof uses the point E on the segment $[B, A]$ such that $BE^2 = AB \cdot AC$; the range (B, E, C, A) is then of type D_3, as put forward by Abū al-Jūd.

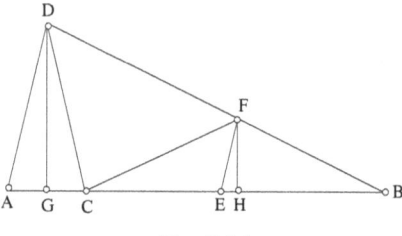

Fig. 3.34

Lemme 2 of al-Sijzī: *Given a segment* AB, *to construct the triangle* ABD *such that* BA = BD *and* Â = 3B̂.

Let C be the point on BA obtained by constructing the range D_2 (taking points in the order B, A). We have

$$\frac{\sqrt{AB \cdot AC}}{BC} = \frac{AB}{AB + BC}.$$

We construct D such that $BA = BD$ and $AD = \sqrt{AB \cdot AC}$; D is the required point.

Let us follow al-Sijzī's proof; we have

$$\frac{AD}{BC} = \frac{AB}{AB + BC},$$

so $AD < BC$.

Let E be such that $BE = AD$, $EF \parallel AD$, $FH \perp AB$, $DG \perp AB$. We have

$$AD^2 = AB \cdot AC,$$

so

$$\frac{AB}{AD} = \frac{AD}{AC},$$

and the two triangles ABD and ADC are similar ($\hat{A}$ is common to both). Then triangle ADC is isosceles and G is the midpoint of AC; hence

$$GB = GC + CB = \frac{1}{2}(AC + 2CB) = \frac{1}{2}(AB + CB).$$

But

$$\frac{AD}{BC} = \frac{AB}{AB + BC} \quad \text{and} \quad AD = EB,$$

hence

$$\frac{EB}{\frac{1}{2}BC} = \frac{AB}{\frac{1}{2}(AB + BC)} = \frac{AB}{GB}.$$

Moreover, the two triangles FHB and DGB are similar, hence

$$\frac{EB}{HB} = \frac{AB}{GB},$$

since $EB = FB$ and $AB = BD$.
Then we have

$$HB = \frac{1}{2}BC,$$

H is the midpoint of BC, so $FC = FB = EB = AD = DC$, hence

$$D\hat{F}C = F\hat{C}B + \hat{B} = 2\hat{B}, \quad C\hat{D}F = 2\hat{B}$$

and

$$A\hat{C}D = C\hat{D}F + \hat{B} = 3\hat{B}.$$

But $\hat{A} = A\hat{C}D$, so finally we have $\hat{A} = 3\hat{B}$, hence $\hat{B} = \frac{\pi}{7}$.

We may note that D is the point of intersection of the circle (B, AB) and the perpendicular bisector of AC.

Al-Sijzī then goes on to construct the regular heptagon inscribed in a given circle $\mathscr{C}$. Using Lemma 2, we construct an isosceles triangle EGD such that $\hat{E} = \hat{G} = 3\hat{D}$, and we construct in $\mathscr{C}$ a triangle ABC similar to triangle DEG. We have $\hat{B} = \hat{C} = 3\hat{A}$, hence $\hat{A} = \dfrac{\pi}{7}$ and BC is the side of the regular heptagon. To construct the vertices of the heptagon, al-Sijzī considers the point H on the circle such that $B\hat{C}H = B\hat{A}C$ and CI is the bisector of $A\hat{C}H$; then $B\hat{A}C = B\hat{C}H = H\hat{C}I = I\hat{C}A$, hence $CB = BH = HI = IA$. In the same way, we obtain the points J and K such that $CJ = JK = KA$, and thus the regular heptagon $AIHBCJK$.

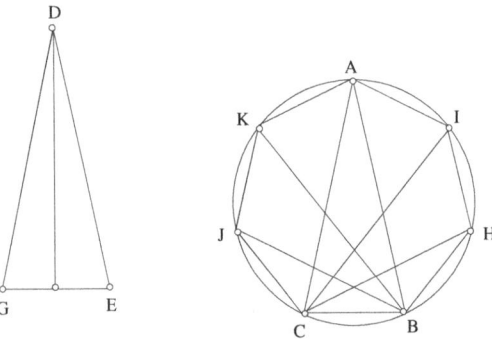

Fig. 3.35

However, we may note that al-Sijzī does not tell us how to construct the triangle ABC similar to triangle DEG. It can be done like this: let R be the radius of the given circle and r the radius of the circumcircle of triangle DEG, we have $\dfrac{BC}{EG} = \dfrac{R}{r}$, which corresponds to the construction mentioned by Abū al-Jūd in his Lemma 3, which al-Sijzī himself cites. In the same way, we construct AB, $\dfrac{AB}{DE} = \dfrac{R}{r}$.

In our examination of the treatises by Abū al-Jūd and al-Sijzī, we have of course remarked on several occasions, and with identical figures, that the ranges of types D_1 and D_2 have been set up by means of the intersection of a parabola $\mathscr{P}$ and a hyperbola $\mathscr{H}$. Let us consider this intersection again, in a slightly different way.

Let $ABCD$ be a square of side a; let us take (AD, AB) as coordinate axes (Ax, Ay) and let I be the point such that $BI = 2\,BC = 2a$. We consider the equilateral hyperbola $\mathscr{H}$ (one branch of it) that passes through C and has asymptotes Ax and Ay; and the parabola $\mathscr{P}$ with vertex I, *latus rectum a* and axis IB parallel to the asymptote Ax. The parabola $\mathscr{P}$ passes through D and the point C lies inside $\mathscr{P}$ (see Fig. 3.36); we have

$$\mathscr{H} = \left\{ (x,y) ; xy = a^2 \right\} \text{ where } x > 0, y > 0,$$

$$\mathscr{P} = \left\{ (x,y); (a-y)^2 = a(2a-x) \right\}.$$

The equation for the abscissae of the points of intersection is

$$x^3 - ax^2 - 2a^2x + a^3 = 0,$$

which has two positive roots $x_2 \in \ ]0, a[$, the abscissa of the point L, and $x_3 \in \ ]a, 2a[$, the abscissa of the point H.

Abū al-Jūd, in his *Book on the Construction of the Heptagon in the Circle*, and al-Sijzī, in his treatise, both use the point H; its projection J on the axis of the parabola gives the range (I, J, C) of type D_2 with $HJ^2 = IJ \cdot IC$. The isosceles triangle constructed with base IJ and $IK = JK = HJ$, is similar to triangle IKC; they are triangles of type $(1, 3, 3)$.

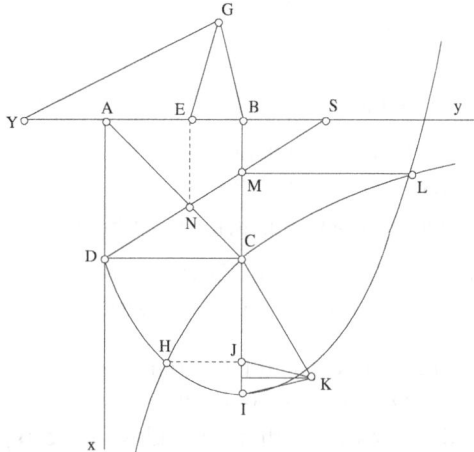

Fig. 3.36

Abū al-Jūd – like another, anonymous, author[58] – uses the point L; its projection M on the side BC of the square $ABCD$ determines the range (D, N, M, S) of type D_1 (Archimedes' range). The range (A, E, B, S) is also of type D_1. From (A, E, B, S) Abū al-Jūd derives (B, E, A, Y) by putting $AY = BS$, $AY^2 = BS^2 = BA \cdot BE$; this is a range of type D_3, so the range (B, E, Y) is of type D_2. The isosceles triangle EBG, with base EB and $BG = EG = BS$, is similar to triangle BGY; these are triangles of type $(1, 3, 3)$.

In conclusion, in using the method employed by Abū al-Jūd and al-Sijzī, we are seeking to divide a segment AB in a point C such that

$$\frac{\sqrt{AB \cdot BC}}{AC} = \frac{AB}{AB + AC}.$$

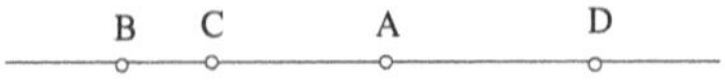

Fig. 3.37

If we take the origin of the abscissae to be the point D, symmetrical to B with respect to A, and put $AB = a$, this condition can be written

$$\frac{\sqrt{a(2a - x)}}{x - a} = \frac{a}{x}.$$

Let us introduce the quantity z such that $\dfrac{z}{x - a} = \dfrac{a}{x}$. We have $z = \sqrt{a(2a - x)}$ or $z^2 = a(2a - x)$, the equation of a parabola with coordinates (x, z); its vertex is the point B, its axis is BD and its *latus rectum* is a. Meanwhile, the relation $\dfrac{z}{x - a} = \dfrac{a}{x}$ is the equation of an equilateral hyperbola that passes through A and has asymptotes $x = 0$ and $z = a$. Observe that z corresponds to $a - y$. We may note that al-Sijzī proves that there exists a point of intersection of the two conics.

Let us remind ourselves that for the range D_1 there were four points and two relations; in D_2 there are only three points and a single relation. Geometrical considerations concerning triangles of the type $(1, 3, 3)$ have made it possible to eliminate one of the unknown points.

---

[58] See below, p. 693.

3.1.4.3. *Abū al-Jūd's range* (D_3)

Let us return to the preceding range D_2 (*A*, *C*, *B*). We had[59] $AD = \sqrt{AB \cdot AC}$; let there be a point *E* on [*AB*] such that $BE^2 = AB \cdot AC$.

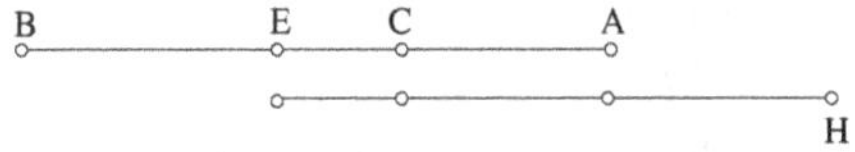

Fig. 3.38

So we have $BE = AD$, and consequently

$$\frac{BE}{BC} = \frac{\sqrt{AB \cdot AC}}{BC} = \frac{AB}{AB + BC} \Rightarrow \frac{BE}{EC} = \frac{AB}{BC} \Rightarrow \frac{AB}{BE} = \frac{BC}{EC} \Rightarrow \frac{AE}{BE} = \frac{BE}{EC} \Rightarrow BE^2 = AE \cdot EC.$$

The range (*A*, *C*, *E*, *B*) can thus be defined by

$$BE^2 = AB \cdot AC$$

and

$$\frac{BE}{BC} = \frac{AB}{AB + BC},$$

or by

$$BE^2 = AB \cdot AC \text{ and } BE^2 = AE \cdot EC.$$

We note that if we produce *BA* by a length *AH* equal to *BE*, then *EH* = *AB* and (*H*, *A*, *C*, *E*) is a range of type D_1, Archimedes' range.

In fact, we have

$$HA^2 = AE \cdot EC$$

and moreover

$$AB \cdot AC = AE \cdot EC \Rightarrow HE \cdot AC = AE \cdot EC \Rightarrow \frac{HE}{EC} = \frac{AE}{AC} \Rightarrow \frac{CH}{EC} = \frac{EC}{CA}$$
$$\Rightarrow CE^2 = AC \cdot CH.$$

So we can pass from a range (*A*, *C*, *E*, *B*) of type D_3 to a range (*H*, *A*, *E*, *C*) of type D_1, and conversely. Types D_1 and D_3 are thus equivalent. Abū al-Jūd makes use of D_3 in his *Letter to Abū Muḥammad 'Abd Allāh ibn 'Alī al-Ḥāsib*. He returns to the analysis that leads him to a triangle of type (1, 3, 3) with which he associates this range D_3. He proceeds as follows:

---

[59] See above, p. 341.

Let E, B, H, G be four points on a circle such that $\overset{\frown}{EG} = \overset{\frown}{GH} = \overset{\frown}{HB} = 2\overset{\frown}{EB}$ then $\overset{\frown}{BE} = \dfrac{1}{7}$ of the complete circumference. GE and HB cut one another in A. From $\overset{\frown}{EG} = \overset{\frown}{BH}$, we get

a) $EG = BH$ and b) $B\hat{H}G = E\hat{G}H$, hence $AG = AH$; from a) and b) it follows that $AE = AB$ and $EB \parallel GH$.

From $\overset{\frown}{GH} = 2\overset{\frown}{EB}$ we get $G\hat{E}H = E\hat{H}B$; but $G\hat{E}H = E\hat{H}B + E\hat{A}H$, hence $E\hat{H}B = E\hat{A}H$, and in consequence $EH = EA = AB$.

If $EI \perp AB$, then $IH = IA$. If we take a point C on IA such that $IB = IC$, then $AC = HB$; but $BH > BE$, so $AC > BE$.

Let D be a point on AC such that $AD = BE$; we have

$$\frac{GH}{AH} = \frac{EB}{AB} = \frac{AD}{AB} = \frac{AC}{AH} \quad \text{(because } AC = BH = GH\text{).}$$

Therefore

$$\frac{AD}{DB} = \frac{AC}{CH} \Rightarrow \frac{AD}{AC} = \frac{DB}{CH} \Rightarrow \frac{AD}{AC} = \frac{DB}{AB} \Rightarrow \frac{AD}{DC} = \frac{DB}{AD} \Rightarrow AD^2 = DC \cdot DB.$$

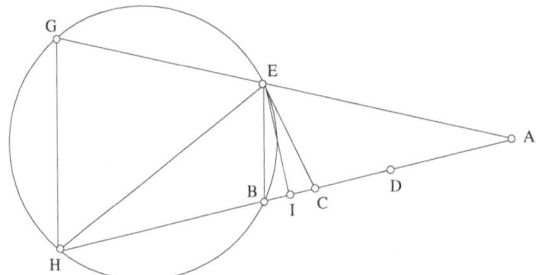

Fig. 3.39

Moreover, the angle at I is a right angle and $IB = IC \Rightarrow EB = EC$; the isosceles triangles ABE and CBE are similar, so $\dfrac{AB}{BE} = \dfrac{EB}{BC}$; but $EB = AD$, so $AD^2 = AB \cdot BC$.

The segment AB has thus been divided into three parts, AD, DC, CB, such that

$$AD^2 = AB \cdot BC \quad \text{and} \quad AD^2 = DC \cdot DB.$$

This is the range (A, D, C, B) of type D_3.

3.1.4.4. *Comparing the ranges: Abū al-Jūd, al-Shannī, Kamāl al-Dīn ibn Yūnus*

Among other questions raised in the course of the dispute there is the question of the choice of range. In fact, a priority dispute, if contingent elements are stripped away, in most cases reduces to arguing over what constitutes the discovery: Here, the matter in contention is the type of range. If we read Abū al-Jūd carefully, we observe that one of his major claims is that the range he uses is new, economical and easy to use. He writes about the range of type D_2 as follows:

> But dividing the given straight line into two parts, as I did it, is more immediate than dividing it into three parts, as they did (Abū Sahl al-Qūhī and our teacher Abū Ḥāmid al-Ṣāghānī).[60]

That is to say, the preference he expresses for D_2 as against Archimedes' range is not defended on grounds of truth – from that point of view the ranges are equal – but on grounds of efficiency. Further, Abū al-Jūd congratulates himself on his discovery of D_2 as he does on that of D_3. Indeed he deploys similar arguments in presenting the range of type D_3. This range, he writes,

> is more accessible and easier than dividing a straight line into three parts such that the product of the sum of the first and the second parts and the first <part> is equal to the square of the third part and the product of the sum of the second and the third parts and the second <part> is equal to the square of the first part, as was proposed by Archimedes and as has been constructed by the Master Abū Sahl and our Master Abū Ḥāmid – may God sustain them – in order to construct the heptagon. It is also easier than dividing the straight line into two parts such that the product of the complete straight line and one of them is equal to the square of a straight line whose ratio to the other part is equal to the ratio of the whole straight line to the sum of the straight line and this other part, which is what I did before, also in order to construct the heptagon.[61]

Continuing to apply only the criterion of efficiency, Abū al-Jūd's order of preference would be $D_3 - D_2 - D_1$; and the comparison between the types of range appeared at the very moment that research was beginning to be done on the construction of the heptagon.

---

[60] See *Treatise by Abū al-Jūd Muḥammad ibn al-Layth Addressed to the Eminent Master Abū Muḥammad 'Abd Allāh ibn 'Alī al-Ḥāsib on the Account of the Two Methods of the Master Abū Sahl al-Qūhī the Geometer, and his own Master Abū Ḥāmid al-Ṣāghānī*, below, p. 623.

[61] *Ibid.*, pp. 624–5.

Without a word of explanation, al-Sijzī adopts the range of type D_2 introduced by Abū al-Jūd. Later, al-Shannī returns to the problem when he makes his criticisms of Abū al-Jūd. There are two criticisms: on the one hand, if we prove the three types of range are equivalent – or at least that they are interdependent – Abū al-Jūd's claim to priority, even if it does not completely collapse, is greatly weakened. On the other hand, Abū al-Jūd claimed to have set up the range D_2 using a single conic section. This assertion, which is taken literally by al-Shannī, is certainly erroneous. Nevertheless there are other possibilities – perhaps he thought he had carried out the construction using a conic section and a circle (which at the time was not considered to be a conic section)[62] – but, since Abū al-Jūd does not explain, further discussion is impossible. So here we shall concern ourselves only with the first criticism.

To dismantle Abū al-Jūd's argument, al-Shannī first of all proves that D_2 implies D_3; but, since D_3 also implies D_2, the two types of range are equivalent. Let there be two points C and E on AB such that

$$AB \cdot AC = BE^2 \ \text{ and } \ \frac{BE}{BC} = \frac{AB}{AB + BC};$$

then

$$BE^2 = AB \cdot AC \ \text{ and } \ BE^2 = AE \cdot EC,$$

since we have

$$\frac{AB}{AB + BC} = \frac{BE}{BC} \Rightarrow \frac{AB}{BC} = \frac{BE}{EC} \Rightarrow \frac{AB}{BE} = \frac{BC}{EC} \Rightarrow \frac{AE}{EB} = \frac{BE}{EC} \Rightarrow EB^2 = AE \cdot EC.$$

But, by hypothesis, we have $BE^2 = AB \cdot AC$; so the result follows.

D_2 and D_3 define the same range on AB, that is the range (A, C, E, B) that we have considered above.

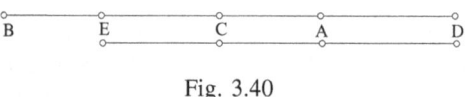

Fig. 3.40

Let us now produce AB by $AD = BE$; then the range (E, C, A, D) is of type D_1 (Archimedes' range), and we have

$$AD^2 = AE \cdot EC \ \text{ and } \ EC^2 = DC \cdot CA.$$

---

[62] We know, from Book III of Descartes' *Géométrie*, that such a construction, using a parabola and a circle, is always possible.

In fact we have

$$AD = BE \Rightarrow AE \cdot EC = EB^2 \Rightarrow AE \cdot EC = AD^2.$$

Moreover, by hypothesis we have $AB \cdot AC = AE \cdot EC$; but $AB = DE$, so $ED \cdot AC = AE \cdot EC$, hence

$$\frac{ED}{EA} = \frac{EC}{AC} \Rightarrow \frac{ED}{EC} = \frac{EA}{AC} \Rightarrow \frac{DC}{EC} = \frac{EC}{CA} \Rightarrow CE^2 = DC \cdot CA;$$

so

$$D_3 \Rightarrow D_1.$$

Al-Shannī has thus proved that the range (A, C, B) of type D_2 implies the range (A, C, E, B) of type D_3, which in turn implies (D, A, C, E) of type D_1.

Conversely, it is easy to proceed from a range of type D_1 to one of type D_3, that is from (E, C, A, D) to (A, C, E, D), which is what Abū al-Jūd did. The three ranges are in fact equivalent.

Finally, we may wonder whether, once the dispute had died down and the whole problem was re-examined by Ibn al-Haytham, in the time of al-Shannī perhaps, or a little later, the ranges were no longer compared with one another. This is far from certain. We can find at least one exception: Kamāl al-Dīn ibn Yūnus, who died in 639/1242. He was a pupil of Sharaf al-Dīn al-Ṭūsī. He returns to this question, and even to some extent to the dispute, although by now it is more than two centuries old. One of his correspondents, Muḥammad ibn al-Ḥusayn, reminds him that al-Sijzī took the question of Archimedes' lemma so seriously

> that at the beginning of his book on the regular heptagon inscribed in the circle he discussed remarks made by the man who affirmed: 'and it perhaps involves a more difficult construction and a less accessible proof than the matter for which he introduced it;   and perhaps these latter are not possible'.[63]

Ibn Yūnus thus returns to the material of the dispute and conjures up the ghosts of Abū al-Jūd and al-Sijzī. He reproves al-Sijzī for having seen only that $D_2 \Rightarrow D_1$. So he in turn wants to give a proof:

---

[63] *Treatise by Kamāl al-Dīn ibn Yūnus to Muḥammad ibn Ḥusayn on the Proof for the Lemma neglected by Archimedes in his Book on the Construction of the Heptagon Inscribed in the Circle*, see below, p. 698.

Let AB be a straight line and (A, C, B) a range of type D_2, defined by

$$\frac{AB}{AB + AC} = \frac{\sqrt{AB \cdot BC}}{AC};$$

if on AB we define the points N and P by $BP = CN = \sqrt{AB \cdot BC}$, then

$$BP^2 = AP \cdot AN \quad \text{and} \quad AN^2 = NB \cdot NP,$$

that is, the range (A, N, B, P) is a range D_1, Archimedes' range.

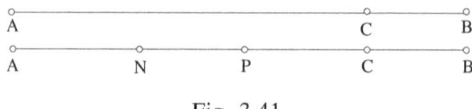

Fig. 3.41

We shall give a brief summary of his proof as a whole, beginning with the proof of the lemma. Let there be a square $ABCD$; on AB let us take the point H such that $AB = AH$ and on the straight line DC let us take the points J and N such that $JD = DC = CN$; let us produce NH to the point K such that $HK = HN$ and let us draw $KL \perp AL$.

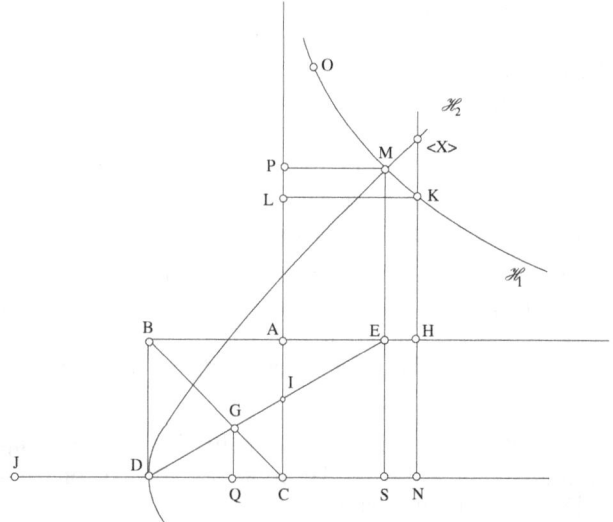

Fig. 3.42

Let $\mathcal{H}_1$ be the hyperbola that passes through K and has asymptotes AH and AL, and let $\mathcal{H}_2$ the hyperbola with vertex D, with DJ as a transverse diameter and *latus rectum*; $\mathcal{H}_2$ cuts the straight line NK in a point X such that

$$NX^2 = JN \cdot ND > NK^2,$$

hence $NX > NK$.

Moreover, the point K is the only point of intersection of $\mathcal{H}_1$ with the straight line HK, which is parallel to the asymptote AL.

The two curves $\mathcal{H}_1$ and $\mathcal{H}_2$ thus cut one another in a point M with ordinate $MS > KN$. The straight line MS cuts BA in a point E, which is the point required for a solution to Archimedes' lemma.

Since triangles ESD and GQD are similar, we have

$$\frac{ES}{DS} = \frac{GQ}{DQ} = \frac{DJ}{DS},$$

hence

$$\frac{JS}{SD} = \frac{CD}{DQ}.$$

But $M \in \mathcal{H}_2$, so

$$\frac{JS}{SD} = \frac{MS^2}{SD^2};$$

and $M \in \mathcal{H}_1$, hence

$$\text{area } (AEMP) = \text{area } (AHKL) = \text{area } (ABDC),$$

from which it follows that

$$\text{area } (EBDS) = \text{area } (MPCS) \Leftrightarrow ES \cdot SD = MS \cdot SC \Leftrightarrow \frac{MS}{SD} = \frac{ES}{SC}.$$

Consequently

$$\frac{CD}{DQ} = \frac{ES^2}{SC^2} = \frac{CD^2}{SC^2},$$

hence

$$\frac{SC}{DQ} = \frac{CD}{SC} \quad \text{or} \quad \frac{EA}{DQ} = \frac{CD}{EA}.$$

But triangles EAI and GDQ are similar, hence

$$\frac{EA}{DQ} = \frac{AI}{GQ};$$

so

$$\frac{CD}{EA} = \frac{AI}{GQ},$$

hence $CD \cdot GQ = AI \cdot EA$; the two areas (CDG) and (AEI) are equal, and the lemma is proved.

Ibn Yūnus here carries out a synthesis: the point E, the orthogonal projection on AB of M, the point of intersection of $\mathcal{H}_1$ and $\mathcal{H}_2$, defines an Archimedean range (E, I, G, D) such that the two areas (AEI) and (CGD) are equal. We may note that he proves the existence of the point of intersection of $\mathcal{H}_1$ and $\mathcal{H}_2$, presenting a proof comparable to that given by Ibn al-Haytham in the first part of his first treatise, *Lemma for the Side of the Heptagon*, regarding the intersection of a parabola and a hyperbola. Both of them in fact make use of Proposition 13 of the second book of the *Conics*: a line parallel to an asymptote cuts the conic section in one point and only one point.

This is when Ibn Yūnus is concerned with the implication $D_2 \Rightarrow D_1$. His proof is as follows:[64]

We put $CN = BP = CH$. We draw through N a line parallel to AG; it cuts EG in S and HM in O, and we have $AI = CH = CN$. But we had $CH^2 = AB \cdot BC$,

$$\frac{AB}{AB + AC} = \frac{CH}{AC} = \frac{CN}{AC},$$

hence

$$\frac{CN}{NA} = \frac{AB}{AC}.$$

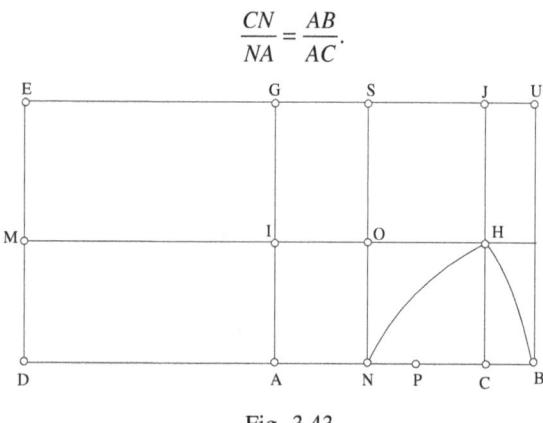

Fig. 3.43

[64] The figure is the same as that in al-Sijzī's treatise (see below, p. 634) for investigating the range (A, C, B), a range of type D_2; we have $AB = AD = AG$ and $CH^2 = AB \cdot BC$.

But

$$\frac{CN}{NA} = \frac{AI}{IO} \quad \text{and} \quad \frac{AB}{AC} = \frac{AG}{HI};$$

so we have

$$\frac{AB}{AC} = \frac{AG}{HI} = \frac{AI}{IO} = \frac{GI}{HO} = \frac{GI}{CN} \Rightarrow GI \cdot IO = AI \cdot CN = CN^2.$$

Now $GI = PA$, so

$$(1) \qquad PA \cdot AN = CN^2 = BP^2.$$

But we have seen[65] that area $[H, G]$ = area $[I, D]$ = area $[B, I]$; by adding the area $[H, U]$ we have area $[U, I]$ = area $[C, U]$ + area $[C, I]$; but area $[C, U] = AB \cdot CB = CH^2 = PA \cdot AN = GI \cdot IO$ = area $[G, O]$, so area $[C, I]$ = area $[U, I]$ – area $[G, O]$ = area $[U, O]$.

Now area $[C, I] = CA \cdot AI = AB \cdot IO = AG \cdot AN$ and area $[U, O] = US \cdot SO$. So

$$\frac{AG}{US} = \frac{AB}{BN} = \frac{SO}{AN} = \frac{AP}{AN} \quad \text{(because } SO = AP\text{)}.$$

Then we have

$$\frac{AB}{BN} = \frac{AP}{AN} \Rightarrow \frac{AN}{BN} = \frac{PN}{AN},$$

hence

$$(2) \qquad AN^2 = BN \cdot NP.$$

From (1) and (2) the range (A, N, P, B) is a range of type D_1, so $D_2 \Rightarrow D_1$.

In other words, after having proved Archimedes' lemma and found the range (D, G, I, E), Ibn Yūnus returns to the figure used by al-Sijzī for the Abū al-Jūd/al-Sijzī range: to divide the segment AB at the point C such that

$$\frac{\sqrt{AB \cdot BC}}{AC} = \frac{AB}{AB + BC}.$$

Al-Sijzī has found the point H such that $CH = \sqrt{AB \cdot BC}$ by using the intersection of the parabola with vertex B, axis AB and *latus rectum* AB, and the equilateral hyperbola that passes through A and has asymptotes ED and EG. He has proved that if $HC \perp AB$, C is the point required.

---

[65] Cf. al-Sijzī's treatise.

Kamāl al-Dīn ibn Yūnus then proves that if we take points N and P on AB such that $CN = BP = CH$, we divide AB at the points P and N in the required way. He proves that we have

$$BN \cdot PN = NA^2 \text{ and } BP^2 = PA \cdot AN,$$

that is, the two equalities that define Archimedes' range.

Although it was composed in the thirteenth century, Ibn Yūnus's study is, as it were, that of a mathematician of the second half of the tenth century; despite being written about two centuries after Ibn al-Haytham, it in fact belongs to a distinctly earlier phase than he does. Was Ibn Yūnus unaware of Ibn al-Haytham's work? Did he know the tradition that lay behind it? These questions give rise to many others, in particular concerning the diffusion of learned treatises in the Islamic world during these centuries. It would be premature to venture onto such ground. However this came about, Ibn Yūnus's contribution finds its true place within a discussion of the types of range and their equivalence.

3.1.4.5. *Ibn al-Haytham's ranges* (D_4 and D_5)

The two pieces that Ibn al-Haytham wrote on the regular heptagon not only belong to two distinct periods, they also form part of two different projects. The *Lemma for the Side of the Heptagon* – the author's first piece of writing on this subject – is from the same mould as the works of his predecessors. In it Ibn al-Haytham seeks to prove Archimedes' lemma, to construct the triangle of type (1, 2, 4) and finally to construct the heptagon. The work is traditional in style, of course, but not in one respect: Ibn al-Haytham is much more concerned than his predecessors with proofs of existence, here the existence of the points of intersection of the conic sections. This difference, which is often overlooked, but seems crucial to the historian, is found in other work as well as in his investigations of the heptagon, but we should stress that in this treatise Ibn al-Haytham does indeed show himself to be a mathematician who belongs to the tenth century, that is, thus far he is thinking like his predecessors.

The later treatise, *On the Construction of the Regular Heptagon*, is very different in scope, and also directed to a different end. There is every indication that what Ibn al-Haytham intends here is to carry through a project in which he has succeeded elsewhere: to 'complete' the tradition, to bring it to fulfilment. In this there is often a reform that demands the starting point be changed.

Ibn al-Haytham seems to have a good knowledge of the tradition. He is in possession of a treatise by al-Qūhī and another one, whose author he does not name, which proves Archimedes' lemma and constructs the heptagon; perhaps he has other texts too, all of which have in common a characteristic that Ibn al-Haytham wants to break with. In fact he writes:

> We have not found a treatise that explains sufficiently fully by any of the ancients or by modern scholars, <that is one> in which there are included all the ways in which the construction of the heptagon can be achieved.[66]

This time, we no longer start from Archimedes' lemma, nor from any equivalent lemma, but from the problem of the heptagon as a whole, so as to open up a route that allows us to find 'all the ways by which one can complete the construction of the heptagon'. We shall in fact need to find all possible triangles that lead to the construction of the heptagon. This search for the 'possible' allows us to speak of the generality of this *exhaustive* procedure (the word *exhaustive* is Ibn al-Haytham's, يستوعب), which has no known precedent. Thus, in this last treatise, Ibn al-Haytham explicitly breaks with tradition. It is no less a difference of project than it is a break in style, because identifying the ranges actually makes it necessary to prove their existence. So if we gloss over this difference between Ibn al-Haytham's two treatises, and consider them simply as following on one from the other, we cannot judge either the new contribution he made, or what links him with the tradition from which he came.

There is a short and simple way of understanding the situation: we shall work through his ranges and triangles, comparing then with those of his predecessors.

3.1.4.5.1. *Triangle (1, 3, 3) and Ibn al-Haytham's range* (D_5)

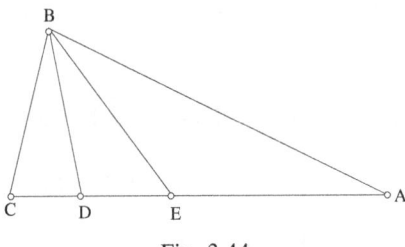

Fig. 3.44

Analysis gave the range (A, E, D, C) such that

(1) $CE^2 = AC \cdot CD$ and $CE^2 = AD \cdot DE.$

---

[66] *On the Construction of the Heptagon in the Circle*, see below, p. 441.

The triangle (1, 3, 3) is found in Abū al-Jūd, al-Sijzī, and al-Qūhī in connection with trisecting an angle; later on we meet it in Naṣr ibn 'Abd Allāh. But Ibn al-Haytham's range D_5 is not found in the work of any of his predecessors that we know about.

We may note that the range (C, D, A) is that of Abū al-Jūd and al-Sijzī, and that the point E is obtained from these three points by making $CE^2 = CA \cdot CD$.

We may also note that this range of Ibn al-Haytham's is different from Abū al-Jūd's range $D_3 - (C, D, F, A)$ – deduced from (C, D, A) by taking a point F defined by $AF^2 = AC \cdot CD$.

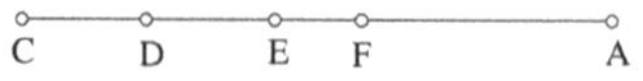

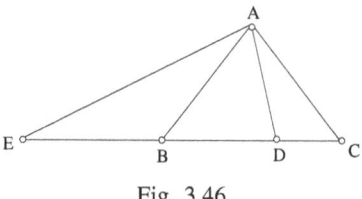

Fig. 3.45

To obtain this range D_5 Ibn al-Haytham starts with two hyperbolae one of which is equilateral.[67]

3.1.4.5.2. *Triangle (3, 2, 2) and the range of type* D_3

This triangle does not occur in the work of any of Ibn al-Haytham's predecessors; which is additional evidence that there was no attempt to find all possible triangles.

Fig. 3.46

Analysis leads to the range (B, E, D, C) such that

$$EB^2 = CE \cdot CD \ \text{ and } \ EB^2 = BD \cdot BC,$$

which is the range D_3 given by Abū al-Jūd.

To obtain this range Ibn al-Haytham again uses two hyperbolae, one of them equilateral; that is, his procedure is not at all like that of Abū al-Jūd.

---

[67] See below.

3.1.4.5.3. *Triangle (1, 5, 1) and Ibn al-Haytham's range (D₄)*

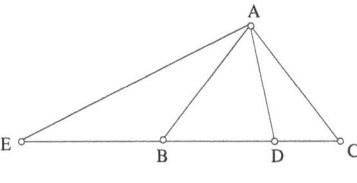

Fig. 3.47

Analysis leads to a range (B, E, D, C) such that

$$BE^2 = BC \cdot CD \text{ and } CD^2 = BD \cdot DE,$$

a range which is not found in any of Ibn al-Haytham's predecessors. He obtains it by using a hyperbola and a parabola.

We may note that Naṣr ibn 'Abd Allāh employs the range D_4 to construct the triangle of type $(1, 3, 3)$ – triangle ABE. We may also note that al-Qūhī had studied the triangle $(1, 5, 1)$.

3.1.4.5.4. *Triangle (1, 2, 4) and the range* D_1

Unlike his predecessors, and departing from the procedure he had adopted himself in his first treatise, this time Ibn al-Haytham begins by proving that a triangle of this type can be obtained from each of the triangles already studied. If we divide the angle ACB into four equal parts, each is equal to $\hat{A} = \dfrac{\pi}{7}$; then on the segment AB we obtain the points $D, E,$ G such that ACD is a triangle of type $(1, 3, 3)$, EBC a triangle of type $(3, 2, 2)$ and AGC a triangle of type $(1, 5, 1)$; and from the construction of one or other of these triangles we can deduce a construction for triangle ABC, which is of type $(1, 2, 4)$. So this last triangle has no logical priority over the others, and its privileged position is due solely to an accident of history.

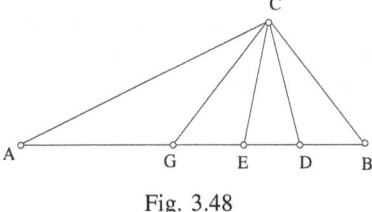

Fig. 3.48

Ibn al-Haytham then gives an analysis of the construction of this triangle that leads to Archimedes' range D_1 such that

$$DC^2 = EC \cdot BC \text{ and } EB^2 = DB \cdot DC.$$

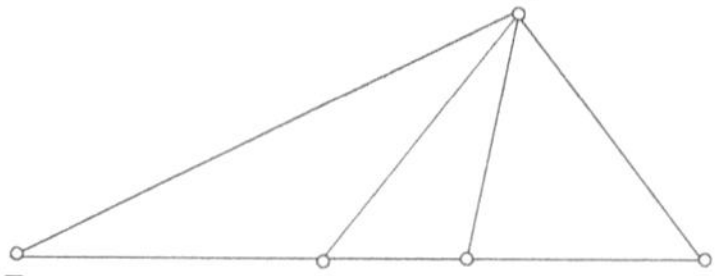

Fig. 3.49

To obtain this range, Ibn al-Haytham uses the same curves as in the preceding case. We have already encountered this triangle in the text attributed to Archimedes, in al-Qūhī and in al-Ṣāghānī, and it was also mentioned by Abū al-Jūd.

So we can see in detail, and in regard to a major element in the construction of the heptagon, how far Ibn al-Haytham has progressed beyond the tradition in which he began. This tradition has not completely dropped out of sight, since Ibn al-Haytham reduces each construction of a triangle to finding two points of a range on a segment. All the same, by introducing a new range in each case, Ibn al-Haytham seems to suggest that the problem of choosing the right range is no longer important. Indeed one can carry out the construction by means of any two conics belonging to a pencil of which only the base points are fixed; each choice of two conics gives two *symptomata* that define a range on the segment.

3.1.5. *Two supplementary constructions: Naṣr ibn 'Abd Allāh and an anonymous author*

With the work of Ibn al-Haytham, the tradition of research on the regular heptagon draws to a close. We have seen that a later contribution – by Ibn Yūnus – in fact adds nothing very substantial. However, if we wish to be exhaustive, there are two further constructions for the heptagon which we should report. The first is the work of Naṣr ibn 'Abd Allāh and the second that of an anonymous author.

3.1.5.1. *Naṣr ibn 'Abd Allāh*

Like al-Qūhī and Ibn al-Haytham (in *The Construction of the Heptagon*) Naṣr does not start with the square given in Archimedes' lemma, but instead directly with a triangle of type (1, 3, 3). Thus, with the chord *BC*, the side of the regular heptagon inscribed in a circle, he associates a point *A*, the midpoint of the major arc *BC*; the isosceles triangle *ABC* is a triangle of type (1, 3, 3).

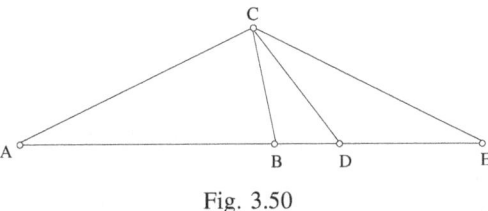

Fig. 3.50

Analysis of a triangle of this type leads to the range (*A*, *B*, *D*, *E*) such that

(1) $AB^2 = AE \cdot ED$ and (2) $DE^2 = AD \cdot DB$.

This we recognize as the range D_4, which is to be found in Ibn al-Haytham – and nowhere else – where it is used in his investigation of the triangle (1, 5, 1). However, in this study of D_4, Naṣr takes the segment *AB* as given whereas Ibn al-Haytham takes as given a segment that corresponds to the segment *AE*. This difference explains the difference that governs the choice of curves: while Ibn al-Haytham opts for a parabola and a hyperbola to show the range, Naṣr proceeds to use two hyperbolae (one branch of each).

Let us examine his analysis and synthesis.

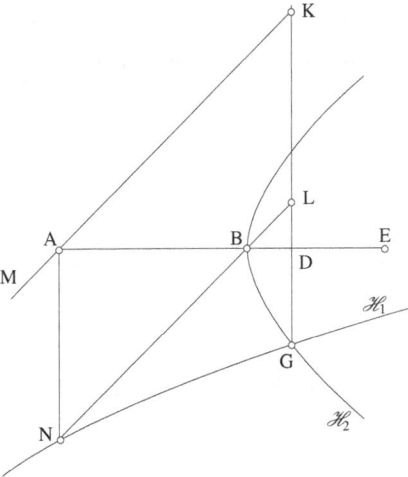

Fig. 3.51

We draw AN and DG perpendicular to AD, with $AN = AB$ and $DG = DE$. The line NB cuts DG in L; so we have $DL = DB$. The line MA, constructed to be parallel to NB cuts DG in K; we have $LK = AN = AB$, and so $KG = AE$ and

$$(1) \Rightarrow KG \cdot GD = KL^2 = AB^2 = AN^2.$$

The points N and G thus lie on a hyperbola $\mathcal{H}_1$, which has asymptotes AM and AE. Similarly, $DG^2 = DE^2 = DA \cdot DB$, so B and G lie on a hyperbola $\mathcal{H}_2$ with vertex B, transverse axis AB and *latus rectum* equal to AB. So G is the point of intersection of $\mathcal{H}_1$ and $\mathcal{H}_2$ ($\mathcal{H}_2$ is an equilateral hyperbola, its asymptotes pass through the midpoint of AB and one of them is parallel to AM).

We may note that the range (K, L, D, G) can be obtained from (A, B, D, E) by a $90°$ rotation about D.

Let us now examine the synthesis, which is stated as follows: given the segment AB, to find the segments BD and DE on AB produced so that the range (A, B, D, E) satisfies (1) and (2).

We draw NA perpendicular to AB, $NA = AB$; we draw NB and MA, where $NB \parallel MA$. Let $\mathcal{H}_1$ be the hyperbola that passes through N and has asymptotes AM and AB, and let $\mathcal{H}_2$ be the hyperbola with vertex B, transverse axis AB and *latus rectum* equal to AB. The two curves must cut one another because $\mathcal{H}_1$ has the axis of $\mathcal{H}_2$ as an asymptote. Let G be the point of intersection. We draw GD perpendicular to AB; GD cuts AM in K. Let E be a point on AD produced such that $DE = DG$. We get

$$G \in \mathcal{H}_1 \Rightarrow GD \cdot GK = NA^2;$$

but $DK = DA$, so

$$GK = GD + DK = DE + AD = AE,$$

hence equality (1).
    Moreover

$$G \in \mathcal{H}_2 \Rightarrow GD^2 = BD \cdot AD,$$

hence equality (2).

Let us express Naṣr's proof in different language. Let us put $AB = a$, $BD = x$, $DE = y$. The equalities (1) and (2) give the equations of $\mathcal{H}_1$ and $\mathcal{H}_2$ (since $DE = DG$). We can write:

$$\mathscr{H}_1 = \{(x, y);\ y(y + x + a) = a^2\},$$
$$\mathscr{H}_2 = \{(x, y);\ x(a + x) = y^2\}.$$

The equation for the abscissae of the points of intersection is written

$$x^3 + 4ax^2 + 3a^2x - a^3 = 0,$$

an equation that has one positive root that corresponds to the point G and two negative roots that correspond to the points of intersection of the second branch of $\mathscr{H}_2$ with each of the branches of $\mathscr{H}_1$.

Starting from this range that he has now set up, Naṣr goes on to construct a triangle of type $(1, 3, 3)$. Let there be circles (A, AB) and (D, DE); they cut one another since $DE > DB$; let C be their point of intersection; we have

$$DE^2 = AD \cdot DB \quad \text{and} \quad DE = DC,$$

hence

$$\frac{AD}{DC} = \frac{DC}{DB};$$

so triangles DCA and DCB are similar, hence $\hat{A} = B\hat{C}D$.

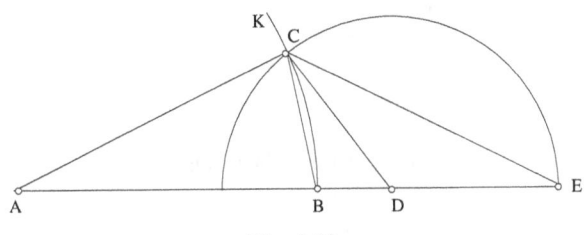

Fig. 3.52

To prove angles A and E are equal, Naṣr replaces AB by EC in the equality $AB^2 = DE \cdot AE$. Now, the equality $AB = AC = CE$, proved in the analysis, is not a hypothesis in the synthesis, because in the latter the hypotheses relate only to the range (A, B, D, E). So we need to prove that $\hat{A} = \hat{E}$, but this point is missing in Naṣr's proof; and it is not true to say that his synthesis is the strict converse of the analysis. So let us prove that $\hat{A} = \hat{E}$.

We have

$$C\hat{D}B = 2A\hat{E}C \quad \text{and} \quad A\hat{C}E = \pi - \hat{A} - \hat{E}.$$

In triangle ACD, we have

$$\frac{AC}{\sin C\hat{D}B} = \frac{CD}{\sin \hat{A}},$$

hence

$$\frac{AC}{CD} = \frac{\sin 2\hat{E}}{\sin \hat{A}}.$$

But $AC = AB$ and $CD = DE$, so

$$\frac{\sin 2\hat{E}}{\sin \hat{A}} = \frac{AB}{DE}.$$

But $AB^2 = DE \cdot AE$. So we have

$$\frac{\sin 2\hat{E}}{\sin \hat{A}} = \frac{AE}{AB} = \frac{AE}{AC}.$$

Now in triangle AEC, we have

$$\frac{AE}{AC} = \frac{\sin A\hat{C}E}{\sin \hat{E}} = \frac{\sin\left(\hat{A} + \hat{E}\right)}{\sin \hat{E}}.$$

So we have

$$\frac{\sin 2\hat{E}}{\sin \hat{A}} = \frac{\sin\left(\hat{A} + \hat{E}\right)}{\sin \hat{E}},$$

hence

$$\sin 2\hat{E} \cdot \sin \hat{E} = \sin \hat{A} \cdot \sin\left(\hat{A} + \hat{E}\right)$$

$$\Rightarrow \cos \hat{E} - \cos 3\hat{E} = \cos \hat{E} - \cos\left(2\hat{A} + \hat{E}\right)$$

$$\Rightarrow 3\hat{E} = 2\hat{A} + \hat{E} \Rightarrow \hat{E} = \hat{A},$$

and it follows that $\hat{A} = E\hat{C}D$; hence

$$B\hat{D}C = 2\hat{A} \quad \text{and} \quad A\hat{B}C = B\hat{D}C + B\hat{C}D = 3\hat{A}.$$

So triangle ABC is of type $(1, 3, 3)$.

Like everyone, Naṣr finishes by inscribing in a given circle an isosceles triangle similar to this last triangle, that is of type $(1, 3, 3)$.

3.1.5.2. *An anonymous text*

The text, which was copied by Muṣṭafā Ṣidqī without the name of its author, contains no hints that provide material for a plausible conjecture about the identity of the author or even his date. As the title indicates, we have a synthesis of the analysis of Archimedes' lemma, but employing a different method: to divide a straight line AB into two parts AC and CB such that

$$AB(AB + AC) = X^2, \text{ where } \frac{X}{AB} = \frac{AC}{BC}.$$

As we have explained above, this range had been obtained by considering the intersection of a parabola and a hyperbola.

The author sets out to find the straight line from the point A, a vertex of the square $ABCD$, that cuts the diagonal BC in E and the line CD in H, and which gives the equality area (AEB) = area (GDH). To do this, he uses the point G of the segment DB, defined by

$$BD(BD + BG) = F^2, \text{ where } \frac{F}{BD} = \frac{BG}{GD}.$$

Authors who deal with this problem generally use the point H on the segment CD. Their analysis takes them to the range (A, E, G, H) or the similar range derived from it by orthogonal projection onto the line CD.

If we make an orthogonal projection of the point E onto the line BD, we obtain the point I, and the range (B, I, G, D) is similar to the range (A, E, G, H), which satisfies

(1) $AE^2 = EH \cdot EG$ and (2) $HG^2 = AG \cdot AE$.

We also have

(1′) $BI^2 = ID \cdot IG$ and (2′) $DG^2 = BG \cdot BI$.

From (1′) we get

$$BI^2 = (BD - BI)(BD - BI - GD),$$

hence

$$BD^2 - BD \cdot GD = BI\,(2BD - GD);$$

hence

$$BD \cdot BG = BI(BD + BG) \Rightarrow BI = \frac{BD \cdot BG}{BD + BG}.$$

From (2′) we get

$$DG^2 = \frac{BG^2 \cdot BD}{BD + BG} \Rightarrow \frac{DG^2}{BG^2} = \frac{BD}{BD + BG} = \frac{BD^2}{BD(BD + BG)};$$

so we have

$$\frac{BG}{DG} = \frac{\sqrt{BD(BD + BG)}}{BD}.$$

If we put $F^2 = BD(BD + BG)$, we come back to the equalities that were given to define the point G and the range.

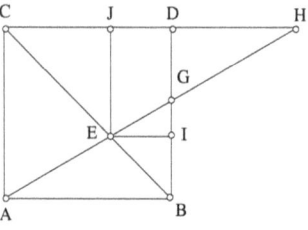

Fig. 3.53

We must note that there is some degree of analogy between this procedure and that of Abū al-Jūd, which we have considered before. To obtain a triangle of type (1, 3, 3), Abū al-Jūd employs a range that is similar to the range (C, J, D, H) that we have here. He obtains this range by finding a point that corresponds to the point G, by considering the intersection of a parabola and a hyperbola, exactly the ones used by the anonymous author here. One might perhaps hazard the conjecture that Abū al-Jūd's study inspired this anonymous author to carry out his synthesis.

3.1.6. *Ibn al-Haytham's two treatises on the construction of the heptagon*

3.1.6.1. *On the determination of the lemma for the side of the heptagon*

1. Ibn al-Haytham is referring to the lemma Archimedes gave for constructing the heptagon.

Lemma:[68] Let *ABCD* be a square and *AC* its diagonal. Let us produce *AD* to *E* and let us draw *BGHE* such that the two triangles *BGC* and *HDE* have equal areas. Let us draw *KGI* parallel to *BA*; we have

(1) $DA \cdot AI = DE^2,$

(2) $EI \cdot ID = IA^2.$

Now, while (1) and (2) can be deduced from the areas of the two triangles *BGC* and *HDE* being equal, the pair of points (E, I) can be constructed only by using conic sections.

So Ibn al-Haytham first tries to prove the lemma. He starts by trying to simplify the problem, and sets about doing so by analysis. So let us join *BD*, it cuts *AC* in its midpoint, *M*. We have

(3)   tr. $(BMC) = $ tr. (AMD)

tr. $(BMC) = $ tr. $(BMG) + $ tr. $(EDH),$

hence, from (3),

tr. $(AMD) = $ tr. $(EDH) + $ tr. $(BMG).$

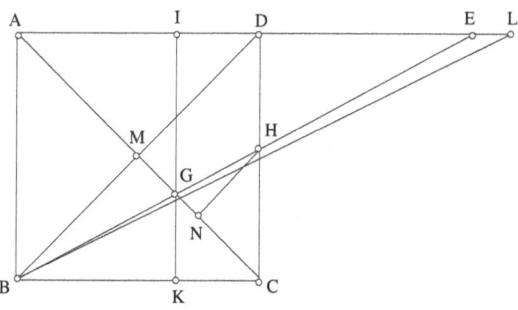

Fig. 3.54

[68] We may note that this lemma can be reduced to: to find on a given segment *AD* a point *I* and on *AD* produced a point *E* such that (1) and (2). These two conditions can be expressed algebraically as follows: If we put $AD = a$, $ID = y$, $DE = x$, we get

$$* \qquad \begin{cases} x^2 = a(a-y), \\ (a-y)^2 = (x+y)y, \end{cases}$$

which give $x^3 + 2ax^2 = a^2x + a^3$, an equation that can be solved by considering the inter-section of the two curves whose equations are given by *. The first curve is a parabola, the second is a hyperbola.

If we add the quadrilateral *MDHG* to both sides, we have

$$\text{tr. } (BDE) = \text{tr. } (ADHG).$$

Let *L* be a point such that tr. $(BEL) = $ tr. (CGH). We have

$$\text{tr. } (BDL) = \text{tr. } (ADC),$$

and they are between two parallel lines. So

(4) $LD = DA$

and we have

$$\frac{\text{tr. } (BDL)}{\text{tr. } (BEL)} = \frac{\text{tr. } (ADC)}{\text{tr. } (CGH)}.$$

Let us draw $HN \perp GC$; we have

$$HN \cdot \frac{1}{2}GC = \text{tr. } (GHC),$$

and similarly

$$DM \cdot \frac{1}{2}AC = \text{tr. } (ADC), \text{ since } DM \perp AM,$$

so

$$\frac{\text{tr. } (ADC)}{\text{tr. } (CGH)} = \frac{DM}{HN} \cdot \frac{AC}{GC} = \frac{DC}{CH} \cdot \frac{AC}{GC};$$

and since

$$\frac{DC}{CH} = \frac{BE}{BH} \text{ and } \frac{AC}{GC} = \frac{EB}{BG},$$

we have

$$\frac{\text{tr. } (ACD)}{\text{tr. } (CGH)} = \frac{EB}{BH} \cdot \frac{EB}{BG} = \frac{EB^2}{BH \cdot BG}.$$

If *ABCD* is a rectangle (see Fig. 3.55), we need to draw *DM′*, the perpendicular to *AC*. The line *DM′* then replaces *DM*, and we obtain the same ratios as before.

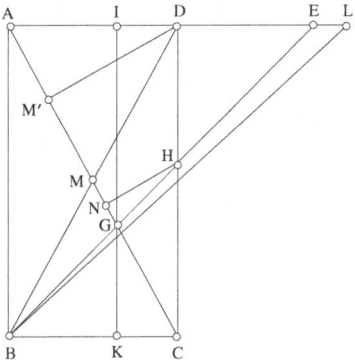

Fig. 3.55

We have

$$\frac{\text{tr. } (ACD)}{\text{tr. } (CGH)} = \frac{\text{tr. } (BDL)}{\text{tr. } (BEL)} = \frac{DL}{LE},$$

and consequently

$$\frac{DL}{LE} = \frac{EB}{BH} \cdot \frac{EB}{BG} = \frac{EA}{AD} \cdot \frac{EA}{AI} = \frac{EA^2}{AD \cdot AI};$$

but

$$DA \cdot AI = DE^{2},{}^{69}$$

hence

(5) $$\frac{DL}{LE} = \frac{AE^2}{DE^2}.$$

---

[69] According to the text attributed to Archimedes, this relation expresses the equivalence of the triangles BGC and DEH. Indeed, the area of the first one is

$$\frac{1}{2} BC \cdot GK = \frac{1}{2} AD \cdot DI \quad \text{or} \quad \frac{1}{2} AD \cdot DI \cdot \frac{AB}{BC},$$

while that of the other one is

$$\frac{1}{2} DH \cdot DE = \frac{1}{2} IG \cdot \frac{DE}{EI} \cdot DE = \frac{1}{2} \frac{AI \cdot DE^2}{EI} \quad \text{or} \quad \frac{1}{2} \frac{AI \cdot DE^2}{EI} \cdot \frac{AB}{BC};$$

so the required relation can be written $AD \cdot DI \cdot EI = AI \cdot DE^2$.

Now $AI^2 = DI \cdot IE$ because triangles KGB and IGE are similar to one another. So we have $AD \cdot AI = DE^2$.

Now, from (4), $AD = DL$. So the construction reduces to dividing AL in a point E such that it satisfies (5). But this division of the segment AL can be carried out only by using conic sections.

So let us continue the analysis and let us suppose that the segment has been divided in this manner. Let us produce CD to O and let us make $DO = AE$. Let us draw from E the line EF perpendicular to AL such that $EF = DE$ (see Fig. 3.56). We have

(6) $$\frac{DL}{EL} = \frac{OD^2}{EF^2}.$$

Let

(7) $$DL \cdot s = OD^2 \quad \text{where } s = \frac{EF^2}{EL}.$$

The parabola with axis DL and *latus rectum s* accordingly passes through O, from (7). Let (L, F, O) be that parabola; it passes through F, because from (6) and (7) we have

(8) $$LE \cdot s = EF^2.$$

Let us put $DQ = DL$ and let us join LQ. It cuts EF in U. We know LDQ [isosceles right-angled triangle], and $O\hat{Q}U$ is known [= 135°]. The ratio $\frac{QU}{DE}$ is also known since $\frac{QU}{DE} = \frac{QL}{DL} = \sqrt{2}$.

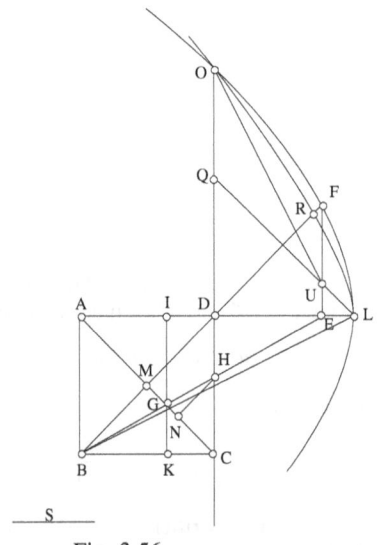

Fig. 3.56

Now $OD = EA$ and $QD = DL = DA$, so $QO = DE$, and consequently $\dfrac{OQ}{QU}$ is known.

It follows that triangle OQU is of known shape and $\dfrac{UO}{OQ}$ is known.

We have $OQ = DE$ and $DE = EF$, so $OQ = EF$ and $\dfrac{OU^2}{FE^2}$ is known.

From (8) we know $\dfrac{LE \cdot s}{OU^2}$ and $\dfrac{EL}{LU}$, so $\dfrac{LU \cdot s}{OU^2}$ is known and $O\hat{U}L$ is known.

So the parabola with diameter LQ, vertex L, angle of its ordinates OUL and having as its *latus rectum* a segment whose ratio to s is known passes through O. Let (L, R, O) be this parabola.

$$\left[\frac{LU}{OU^2} \cdot s = k \Rightarrow LU \cdot \frac{s}{k} = OU^2 \Rightarrow LU \cdot s_1 = OU^2 \Rightarrow s_1 \text{ the } latus\ rectum \right].$$

Unfortunately, knowing the *latus rectum* s assumes we know point E, which is what we want to construct; in fact s is defined by

$$s = \frac{OD^2}{DL} = \frac{EA^2}{DL}.$$

However, from Ibn al-Haytham's analysis, we may imagine a synthesis that allows us to construct the square $ABCD$ in accordance with the conditions impose by Archimedes' lemma. If we suppose the point L, the line LD and the magnitude s are known, then we know the parabola LFO, with axis LD, vertex L and *latus rectum* s. The second auxiliary parabola passes through L and has a diameter LQ such that $D\hat{L}Q = 45°$. The angle OUL, the angle between the ordinates and this diameter, is constructed as follows: let us consider an isosceles triangle UVQ, right-angled at V, let us produce the side VQ by a length $QO = VQ$; let us join OU and let us produce QU to L; the required angle is $O\hat{U}L$.

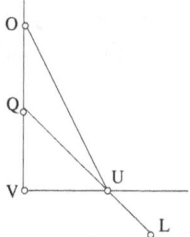

Fig. 3.57

The *latus rectum* s_1 of this second parabola can be found if we note that it is equal to

$$\frac{OU^2}{LU} = \frac{OU^2}{LE} \cdot \frac{1}{\sqrt{2}},$$

where

$$OU^2 = OQ^2 + QU^2 + OQ \cdot QU\sqrt{2} = 5QV^2 = 5DE^2;$$

so

$$s_1 = \frac{5}{\sqrt{2}} \cdot \frac{DE^2}{LE} = \frac{5}{\sqrt{2}} s$$

since

$$s = \frac{EA^2}{DL} \quad \text{and} \quad \frac{EA^2}{DE^2} = \frac{DL}{LE}.$$

So the second auxiliary parabola is also known. The two parabolae intersect in a point O (different from L) whose projection on LD defines the point D. We take the point A on LD produced such that $DA = LD$ and the point E on LD such that $DE = OQ$ where Q is the point where the diameter LQ intersects OD. Archimedes' figure is thus reconstructed, up to a homothety.

Ibn al-Haytham follows this solution with another one, which makes corrections to the range constructed from the initial data, a solution that is, moreover, taken from the work of his predecessors: we may see this procedure as proof that the first solution did not seem to him to be satisfactory.

Moreover, it has been claimed that in the second part of this treatise Ibn al-Haytham employs 'an analysis as well as a synthesis', and that 'the analysis is rendered in a confused manner'.[70] However, a simple examination of the text shows that it contains no analysis and that Ibn al-Haytham employs only synthesis.

This synthesis starts from data that are different from what was given at first and that is in fact the point at issue: it is assumed we know the magnitude s and not segment AD. Perhaps that is why Ibn al-Haytham looked for a different method rather than provide the synthesis we have just given.[71]

[70] J. P. Hogendijk, 'Greek and Arabic constructions of the regular heptagon', p. 234.

[71] We have been criticized for saying that Ibn al-Haytham's analysis was 'erroneous'; we have never written anything of the sort, but only that 'the analysis does

(*Cont. on next page*)

2. There is thus every indication that, recognizing the difficulty we have mentioned, Ibn al-Haytham returns to the problem in the second part of his first treatise. He first notes that constructing the regular heptagon using Archimedes' lemma in fact reduces to dividing a segment AB at points C and D in such a way that

$$DA \cdot AC = DB^2 \text{ and } BC \cdot CD = AC^2,$$

where

$$AC > DC \text{ and } DB > DC.$$

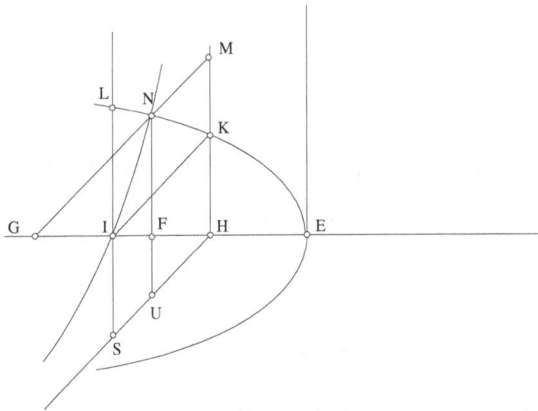

Fig. 3.58

This time he dispenses with the analysis and passes immediately to the synthesis.

Let there be a general segment IE with midpoint H. Let $HK \perp HE$, $HK = HE$ and $HK \parallel IL$.

Let us draw the parabola ($\mathscr{P}$) with axis EG, vertex E and *latus rectum* EH; from I.52 of the *Conics*, we have $K \in (\mathscr{P})$ (see Fig. 3.59a).

Fig. 3.59a

(*Cont.*) not lead to a solution of the problem in the form in which it was stated', that is with a given segment AD; see 'The construction of the heptagon', p. 314. Perhaps we should, moreover, see in this the reason why Ibn al-Haytham did not think it worth his while to give a synthesis, which we may note is not a matter of any difficulty. Other criticisms, concerning the remainder of our edition and commentary, have no more weight than that last one. So we shall not address them here.

Let $L \in (\mathscr{P})$ be such that $LI \perp IE$. Let us produce LI to S such that $IS = IH = KH$. The quadrilateral $KISH$ is thus a parallelogram.

Let us draw the hyperbola $(\mathscr{H})$ passing through I and with asymptotes HK and HS; an $(\mathscr{H})$ with these properties must exist, from II.4 of the *Conics*.

Now $IL \parallel KH$ and KH is an asymptote. So the straight line IL cuts $(\mathscr{H})$ in a single point I. The semi-infinite straight line IL lies inside $(\mathscr{H})$ and meets $(\mathscr{H})$ only at the point I.

Let N and N' be two general points on $(\mathscr{H})$ and F and F' their projections on EG (see Fig. 3.59b):

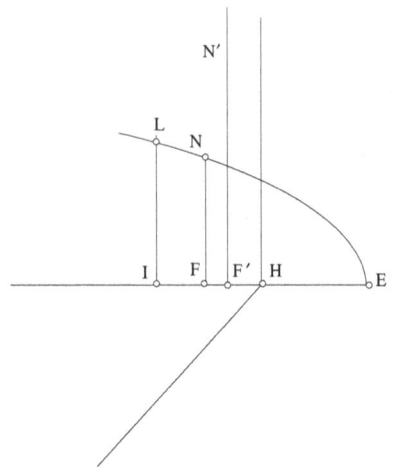

Fig. 3.59b

if $N'F' > NF$, then $d\,(N', HK) < d\,(N, HK)$,

if $N'F' \to +\infty$, then $d\,(N', HK) \to 0$.

So the part of $(\mathscr{H})$ between the lines IL, HK is cut by the arc KL of $(\mathscr{P})$ in a point N.

Let $NM \parallel KI$ and $NU \parallel HK$. We have

$$NM \cdot NU = KI \cdot IS \quad \text{[from the equation of } (\mathscr{H})\text{]}$$

and

$$\text{area } (N, H) = \text{area } (S, K).$$

Now $HF \perp NU$ and $HI \perp IS$, so

$$NU \cdot HF = SI \cdot IH = EH^2.$$

Let us put $FG = NF$. Since $FU = FH$, then $HG = NU$, hence

$$GH \cdot HF = EH^2.$$

Moreover

$$FE \cdot EH = FN^2 \text{ [equation of } (\mathscr{P})].$$

Now $FN = FG$, so $FE \cdot EH = FG^2$.

We then move from the division of EG to that of AB by using a homothety that transforms (E, F, G, H) into (A, D, B, C). So we have

$$DA \cdot AC = DB^2,$$
$$BC \cdot CD = CA^2.$$

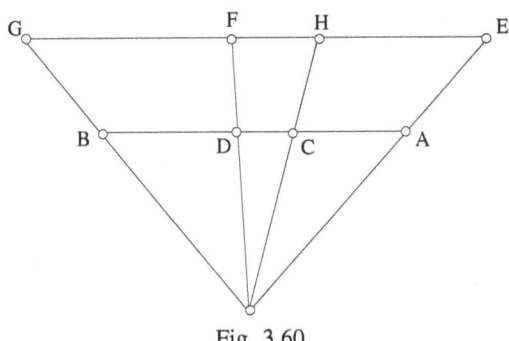

Fig. 3.60

We still need to prove that $AC > CD$ and $DB > CD$. We have $FE \cdot EH = FG^2 = FN^2$; now $FE > EH$; hence $FN > EH$.

But $EH = HI$, hence $FN > HI$ and $FN > HF$, since $HF < HI$. Now $NF = FG$, so $FG > FH$; $EH > HF$ because $EH = HI$. So $EH > HF$ and $FG > HF$, hence by homothety $AC > CD$ and $DB > CD$.

At the end, we have divided AB in points C and D in accordance with the given conditions. Q.E.D.

Constructing the heptagon can finally be reduced to constructing a triangle ECD such that $EC = CA$ and $ED = DB$. In fact, if we draw the circumcircle of triangle AEB, AB is a diagonal of the required heptagon and E a neighbouring vertex to A; EC and ED are two other diagonals of the heptagon because, as we shall show later, triangle ECD has angles $\pi/7$, $4\pi/7$ and $2\pi/7$.

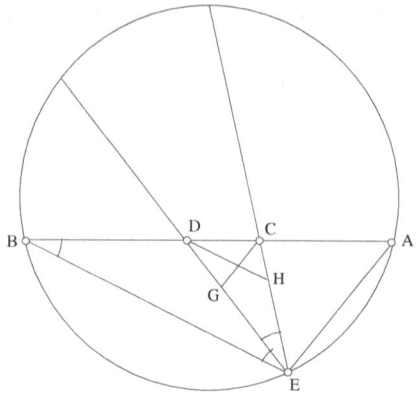

Fig. 3.61

The circumcircle of this triangle leads directly to the regular heptagon inscribed in the circle. Now, Ibn al-Haytham has already established that this triangle can be constructed since

$$AC > CD \text{ and } DB > CD \Rightarrow AC + DB > CD;$$

moreover

$$AC^2 = CD \cdot BC \text{ and } AC > CD \Rightarrow AC < CB,$$

hence

$$AC < CD + DB,$$

so

$$AC - DB < CD,$$

hence

$$AC - DB < CD < AC + DB.$$

Let us prove that triangle ECD is of the type $(1, 2, 4)$, that is,

$$E\hat{D}C = 2C\hat{E}D \text{ and } E\hat{C}D = 4C\hat{E}D.$$

Let DH be the bisector of $C\hat{D}E$ and CG the bisector of $E\hat{C}D$; we have

$$\frac{EH}{HC} = \frac{ED}{DC} = \frac{BD}{DC},$$

hence, by composition of ratios,

$$\frac{EC}{CH} = \frac{BC}{CD}.$$

But

$$\frac{BC}{CD} = \frac{AC^2}{CD^2}$$

because $BC \cdot CD = AC^2$, so

$$\frac{EC}{CH} = \frac{AC^2}{CD^2} = \frac{CE^2}{CD^2};$$

hence

$$CD^2 = CH \cdot EC,$$

hence

$$\frac{CE}{CD} = \frac{CD}{CH}.$$

So triangles DEC and CDH are similar, and in consequence

$$D\hat{H}C = E\hat{D}C \ \text{ and } \ D\hat{H}C = E\hat{D}H + D\hat{E}H$$

$$D\hat{E}H = H\hat{D}C, \ E\hat{D}C = 2H\hat{D}C, \ E\hat{D}C = 2D\hat{E}C.$$

We also have

$$\frac{DG}{EG} = \frac{CD}{CE} = \frac{CD}{AC},$$

hence, by composition of ratios,

$$\frac{DE}{EG} = \frac{AD}{AC} = \frac{BD^2}{AC^2} = \frac{DE^2}{CE^2},$$

and in consequence

$$\frac{DE}{CE} = \frac{CE}{EG},$$

so triangles ECD and ECG are similar, and in consequence

$$C\hat{G}E = E\hat{C}D \ \text{and} \ C\hat{G}E = G\hat{C}D + G\hat{D}C,$$

$$E\hat{D}C = E\hat{C}G. \ E\hat{C}D = 2E\hat{C}G. \ E\hat{C}D = 2E\hat{D}C. \ E\hat{C}D = 4C\hat{E}D.$$

If we now construct in the circle a triangle whose angles are equal to those of triangle ECD and if we divide $E\hat{C}D$ in half, and then divide each half into halves again, and divide $E\hat{D}C$ in half, the straight lines we have drawn divide the circle into 7 equal parts.

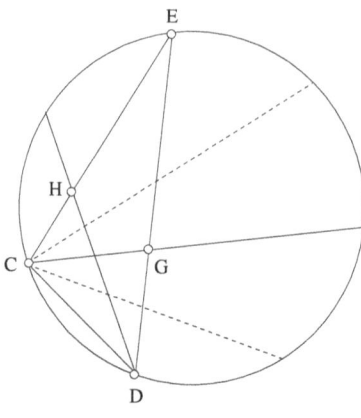

Fig. 3.62

To put it differently, we may say that, to divide the segment in accordance with the given conditions, Ibn al-Haytham considers (HI, HK) as a pair of rectangular coordinate axes, to which he refers the conics used in the construction.

So let $(HI, HK) = (Ox, Oy)$, $EH = a$,

$$(\mathscr{P}) = \{(x, y); \ y^2 = a\,(x + a)\},$$
$$(\mathscr{H}) = \{(x, y); \ y = \frac{a^2}{x} - x\}.$$

He proves that these two curves must necessarily cut one another in a point whose abscissa lies between 0 and a. The equation for the abscissae is of the fourth degree with one obvious root $x = -a$ (the second branch of the hyperbola passes through the point E, symmetrical with I).

$$(a^2 - x^2)^2 = ax^2\,(a + x),$$

hence

$$x^3 - 2ax^2 - a^2x + a^3 = 0.$$

It is clear that this equation has three roots $x_1 < 0$, $0 < x_2 < a$, $x^3 > a$; it is the positive root x_2 that corresponds to the point N (see Fig. 3.59a).

A property of the asymptote to the hyperbola (*Conics*, II.14) allows Ibn al-Haytham to prove the existence of this positive root.

Ibn al-Haytham then constructs a triangle of type (1, 2, 4) to complete the solution of the problem. Apart from the discussion – which has historical importance for its treatment of the intersection of the curves – Ibn al-Haytham's solution, although the procedure differs, is not actually distinct from those given by al-Ṣāghānī or al-Qūhī. It is in his second treatise that he changes the very status of the problem of the heptagon.

3.1.6.2. *On the construction of the heptagon*

In the introduction to this treatise, Ibn al-Haytham first of all notes that until his time all constructions of the regular heptagon have been based on Archimedes' lemma, that is, on dividing one of the diagonals of the heptagon by the two others. For his part, Ibn al-Haytham sets out to find constructions for the heptagon by examining constructions for the triangles that can be made from its sides and diagonals. This change of viewpoint, of which historians have taken too little notice, leads Ibn al-Haytham to work systematically through all possible constructions and thus go beyond the single solutions given by each of his predecessors; and he does this by considering all partitions of the integer 7. It is in precisely this sense that his solution to the problem is more general.

He refers to mathematicians who have already considered this problem; these are al-Qūhī and an anonymous author whose solution is based on Archimedes' lemma, probably al-Ṣāghānī. Ibn al-Haytham thus proceeds with analyses of the problems and states the following proposition:

Let there be a circle ABC; let us suppose the problem is solved. Let $ADEBCGH$ be the regular heptagon obtained. Let BEC, BDC, BAC, BDH be four triangles inscribed in the circle. Any other triangle formed from these 7 points is congruent with one of these four triangles.

In fact, we have

1. ABC: $\hat{A} = \dfrac{\pi}{7}$, $\qquad \hat{B} = \dfrac{3\pi}{7}$, $\qquad \hat{C} = \dfrac{3\pi}{7}$, $\qquad$ (1, 3, 3),

2. BDH: $\hat{B} = \dfrac{2\pi}{7}$, $\qquad \hat{D} = \dfrac{3\pi}{7}$, $\qquad \hat{H} = \dfrac{2\pi}{7}$, $\qquad$ (2, 3, 2),

3. EBC: $\hat{E} = \dfrac{\pi}{7}$, $\hat{B} = \dfrac{5\pi}{7}$, $\hat{C} = \dfrac{\pi}{7}$, $(1, 5, 1)$,

4. DBC: $\hat{D} = \dfrac{\pi}{7}$, $\hat{B} = \dfrac{4\pi}{7}$, $\hat{C} = \dfrac{2\pi}{7}$, $(1, 4, 2)$.

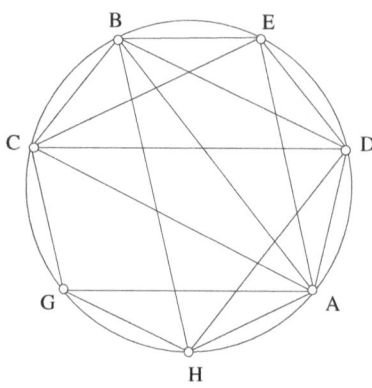

Fig. 3.63

In other words, there exist only four triplets of natural numbers a, b, c, such that $a + b + c = 7$. Ibn al-Haytham does not present a justification for that last assertion, whose proof is immediate. Let us present the proof in the style of the time:

Let us suppose that $a \geq b \geq c$. It is impossible to have $a = b = c$, because that would give $3a = 7$, which is impossible since a is a natural number. Let us put $b + c \geq 2$, we have $a \leq 5$; moreover $7 = a + b + c < 3a$, so we have $a > 2$. So we can take three values:

$a = 5$, $b + c = 2$, $b = 1,\ c = 1$, $(1, 5, 1)$,

$a = 4$, $b + c = 3$, $b = 2,\ c = 1$, $(1, 2, 4)$,
        $b \geq c$,

$a = 3$, $b + c = 4$, $b = 3,\ c = 1$, $(1, 3, 3)$,
        $b \geq c$, $b = 2,\ c = 2$, $(2, 3, 2)$.

The remainder of the treatise is accordingly devoted to the synthesis of the proposition just stated. The purpose is to prove that each of these triangles gives a possible construction for the regular heptagon.

1. *The case (1, 3, 3)*

Analysis: Let us suppose that we have found a triangle ABC (see Fig. 3.64) whose angles A, B, C, are of the type (1, 3, 3). Triangle ABC is isosceles. Let the point D on AC be such that $C\hat{B}D = B\hat{A}C$. So triangles BCD and ABC are similar. We have

$$BD = BC \text{ and } \frac{AC}{CB} = \frac{BC}{CD},$$

hence

(1) $AC \cdot CD = BC^2.$

Let the point E on DA be such that $D\hat{B}E = B\hat{A}C$. Since $A\hat{B}C$ is three times $B\hat{A}C$,

$$A\hat{B}D = B\hat{E}C = C\hat{B}E = \frac{2\pi}{7},$$

so

$$EC = CB.$$

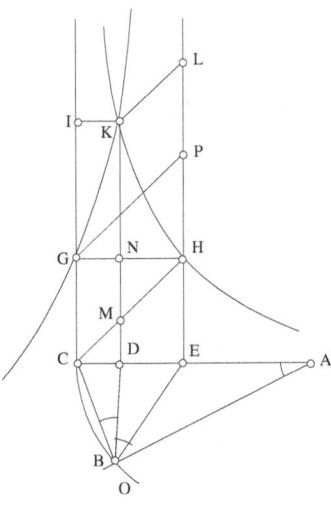

Fig. 3.64

But triangles DBE and ABD are similar, so

(2) $AD \cdot DE = DB^2.$

Now $BD = BC$, so, from (1) and (2),

$$AD \cdot DE = AC \cdot CD.$$

$$BD = BC = CE,$$

and, from (2),

(3) $AD \cdot DE = CE^2,$

(4) $AC \cdot CD = CE^2.$

The range (A, E, D, C) on the segment AC, constructed so as to obtain the triangle whose angles are in the ratio $(1, 3, 3)$, and satisfying equations (3) and (4), is not found in the work of any predecessor known to Ibn al-Haytham.

Then, on CE we construct the square $CEHG$ and the hyperbola $(\mathcal{H}_1)$ that passes through H and has asymptotes CE and CG. The line drawn from D parallel to CG cuts $(\mathcal{H}_1)$ in K and GH in N.

Let P be a point on HE such that $HP = HE$; let us draw PG and HC. The segment HC cuts DN in M.

We have

(5) $CD = DM$ and $DE = HN.$

Let us draw KI parallel to DC; we have

(6) $KD \cdot DC = HE \cdot EC = CE^2$ [equation of $(\mathcal{H}_1)$].

We have, from (4),
$$KD = AC,$$

and from (5),
$$KM = AD.$$

Hence, from (3) and (5),
$$KM \cdot NH = CE^2;$$

since
$$\frac{NH}{MH} = \frac{GH}{CH},$$

we have

$$\frac{KM \cdot NH}{KM \cdot MH} = \frac{GH}{CH} = \frac{GH^2}{GH \cdot CH} = \frac{GH^2}{ND \cdot HC}.$$

Now

$$GH = CE,$$

hence

$$KM \cdot MH = HC \cdot ND = HP \cdot HC.$$

Let us draw KL parallel to HM, where $L \in HE$; we have

$$MK \cdot KL = HP \cdot PG,$$

so the hyperbola ($\mathcal{H}_2$) that passes through G and has asymptotes HC and HL also passes through K. So $K \in (\mathcal{H}_1) \cap (\mathcal{H}_2)$. The projection of K on CE is the point D.

Synthesis: Let CE be a general segment; let us construct on CE the square $EHGC$; let us place P on EH so that $HP = HE$. Let us then draw the hyperbola ($\mathcal{H}_1$) that passes through H and has asymptotes CE and CG, and the hyperbola ($\mathcal{H}_2$) that passes through G and has asymptotes HC and HP. The arcs of ($\mathcal{H}_1$) and ($\mathcal{H}_2$) included in the area bounded by the two parallel asymptotes cut one another in K.

Let $D \in CE$ be such that $KD \parallel GC$, $I \in CG$ be such that $KI \parallel CE$, $L \in EH$ be such that $KL \parallel MH$; $\{M\} = (CH) \cap (DK)$, $A \in CE$ be such that $CA = KD$.

Let us draw the circle ($\mathcal{C}_1$) with centre A and radius AC, and the circle ($\mathcal{C}_2$) with centre C and radius CE. The circles ($\mathcal{C}_1$) and ($\mathcal{C}_2$) cut one another in B, and we have

$$AC \cdot CD = KD \cdot DC = KD \cdot KI = GH \cdot HE = CE^2 \quad \text{[equation of } (\mathcal{H}_1)\text{]}.$$

Since $CB = CE$, we have

(7) $AC \cdot CD = CB^2$;

and since $KD = AC$ and $CD = DM$, we have $AD = KM$. But

$$MK \cdot KL = GC \cdot GP \quad \text{[equation of } (\mathcal{H}_2)\text{]},$$

hence

$$KM \cdot MH = GC \cdot CH;$$

now

$$\frac{MH}{HN} = \frac{CH}{HG} \quad [EH \parallel CG \parallel DK, HG \parallel EC],$$

hence

$$\frac{KM \cdot MH}{KM \cdot HN} = \frac{CH \cdot HG}{HG^2} = \frac{PG \cdot GH}{CG^2};$$

now

$$KM \cdot MH = PG \cdot GH \quad [\text{equation of } (\mathscr{H}_2)],$$

so

$$KM \cdot HN = CG^2 = CE^2,$$

and

$$HN = DE, KM = AD$$

hence

$$AD \cdot DE = CE^2 = CG^2 = CB^2.$$

Triangles ABC and BDC are similar, from (7); so $B\hat{D}C = A\hat{B}C$ and $C\hat{B}D = B\hat{A}C$, and $BD = BC$, because triangle ABC is isosceles, hence

$$AD \cdot DE = BD^2.$$

So triangles ABD and BED are similar.

$$B\hat{E}D = A\hat{B}D \quad \text{and} \quad D\hat{B}E = B\hat{A}D,$$

hence

$$D\hat{B}E = C\hat{B}D.$$

Triangles ABC and CBD are similar, hence

$$\frac{AB}{BC} = \frac{BD}{DC}.$$

Now

$$BC = BD = EC,$$

so

$$\frac{AB}{BD} = \frac{CE}{CD} = \frac{AC}{CE};$$

and, by separation, we find that these ratios are also equal to $\dfrac{AE}{ED}$.

So the point E lies on the bisector of $D\hat{B}A$, so $D\hat{B}E = A\hat{B}E$. Thus the angle B is divided into three equal parts. The construction of the heptagon follows as before.

So we may finally sum up Ibn al-Haytham's solution:

Let (CE, CG) be a pair of coordinate axes (Ox, Oy). Let us put $CE = a$ and consider the two hyperbolae

$$(\mathcal{H}_1) = \{(x, y); xy = a^2\},$$

$$(\mathcal{H}_2) = \left\{(x, y) ; y = x - \frac{a^2}{x-a}\right\};$$

$(\mathcal{H}_1)$ and $(\mathcal{H}_2)$ must cut one another at the point K (x_0, y_0) such that $x_0 \in \,]0, a[$.

So there exists a unique $x_0 \in \,]0, a[$ such that

$$x_0^3 - ax_0^2 - 2a^2x_0 + a^3 = 0.$$

In fact, the equation for the abscissae of the points of intersection has three roots, and x_0 is the one that gives the solution to the problem.

Let D $(x_0, 0)$ be the projection of K (x_0, y_0) on CE.

Let A be a point of CE such that $CA = DK = y_0$. The circles $(\mathcal{C}_1)$ and $(\mathcal{C}_2)$ cut one another. Let B be one of their points of intersection. The triangle ABC that we obtain is of type $(1, 3, 3)$. By a homothety we construct in the given circle a triangle similar to triangle ABC. We may observe that here Ibn al-Haytham's procedure involves trisecting the angle CBA. We may also note that $(\mathcal{C}_1)$ passes through the centre of $(\mathcal{C}_2)$; so the two circles intersect, and we do not need the inequality concerning the distance between the centres and the radii.

2. *Second case* (3, 2, 2)

First, we may note that, as far as we know, this case had not been examined by any of Ibn al-Haytham's predecessors.

Analysis: Let us suppose we have found the required triangle ABC (see Fig. 3.65); so angles A, B, C are of the type $(3, 2, 2)$.

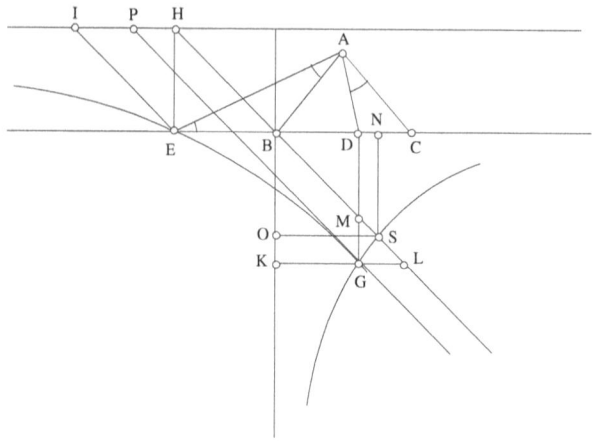

Fig. 3.65

Thus triangle ABC is isosceles, $AB = AC$. Let D be a point on BC such that $B\hat{A}D = \hat{C}$. Let us produce CB to a point E such that $BE = BA$. Then triangles ABD and CBA are similar; we have

$$CB \cdot BD = BE^2.$$

Triangle ABE is isosceles:

$$B\hat{A}E = B\hat{E}A = \frac{1}{2}\hat{B} \quad \text{and} \quad C\hat{A}D = A\hat{E}C,$$

because, by hypothesis, $B\hat{A}C$ is three times $B\hat{E}A$; so triangles ADC and EAC are similar.
    We have

(1) $EC \cdot CD = AC^2 = EB^2,$

so

$$EC \cdot CD = CB \cdot BD.$$

Let there be a segment EH such that $EH \perp BE$ and $EH = BE$, a segment HI such that $HI \parallel BE$ and $HI = BE$, a segment BK such that $BK \perp BE$ and $BK = BC$, a segment KL such that $KL \parallel BC$ and $KL = BC$, and a segment $DG \parallel BK$, where $G \in KL$; DG cuts BL in M.
    Let P be the fourth vertex of the parallelogram $HLGP$.

Let the point N on BC be such that $BN = BE$ and $BNSO$ the square constructed on BN. We have

$$\text{area } (N, O) = BE^2 \text{ and } KB \cdot KG = BE^2,$$

hence

$$DG \cdot GK = NS \cdot SO.$$

So the hyperbola ($\mathscr{H}_1$) that passes through S and has asymptotes DB and BO passes through G.

We have

$$\frac{LB}{BK} = \frac{LB}{BC} = \frac{BH}{BE} = \frac{LH}{CE},$$

hence

$$\frac{LH \cdot CD}{EC \cdot CD} = \frac{HB \cdot BE}{BE^2};$$

and from (1) we have

$$LH \cdot CD = HB \cdot BE;$$

we have

$$LH = PG \text{ and } CD = GL$$

because

$$KL = BC \text{ and } KG = BD,$$

so

$$PG \cdot GL = HB \cdot BE = IE \cdot EB.$$

So the hyperbola ($\mathscr{H}_2$) that passes through E and has asymptotes HL and HI passes through G. So

$$G \in (\mathscr{H}_1) \cap (\mathscr{H}_2).$$

The projection of G on BC is D; Ibn al-Haytham then deduces that $CB \cdot BD = BE^2$; so we know BA and AC. But we are already moving on to the synthesis.

Synthesis: Let BE be a given general segment, N the point symmetrical to E with respect to B, and let the square $BNSO$ be constructed on BN. Let us draw the hyperbola ($\mathcal{H}_1$) that passes through S and has asymptotes BN and BO. Let H be such that $HE \perp EB$ and $HE = EB$, the point I such that $HI \parallel BE$ and $EI \parallel BH$. Let us also draw the hyperbola ($\mathcal{H}_2$) that passes through E and has asymptotes HS and HI. The hyperbolae ($\mathcal{H}_1$) and ($\mathcal{H}_2$) cut one another at the point G because ($\mathcal{H}_2$) approaches indefinitely close to HS.

Let D be the projection of G on EB, L a point on HS such that $GL \parallel EB$, C a point on BE such that $DC = GL$ and K a point on BO such that $GK \parallel EB$. We have

$$BC = KL = KB,$$

so

(1) $CB \cdot BD = BE^2$ [equation of $\mathcal{H}_1$].

Let P be a point on HI such that $GP \parallel HS$; we have

(2) $PG \cdot GL = EI \cdot EB$ [equation of $\mathcal{H}_2$].

But

$$\frac{LB}{BK} = \frac{LB}{BC} = \frac{HB}{HE} = \frac{HB}{BE} = \frac{HL}{EC},$$

hence

$$\frac{HB}{BE} = \frac{IE \cdot BE}{EB^2} = \frac{HL}{EC},$$

so

$$\frac{IE \cdot EB}{EB^2} = \frac{HL \cdot DC}{EC \cdot DC};$$

but

$$CD = LG \quad \text{and} \quad HL = PG,$$

so

$$\frac{PG \cdot GL}{EC \cdot CD} = \frac{IE \cdot EB}{EB^2},$$

hence, from (2)

$$EC \cdot CD = EB^2$$

and from (1)

$$EC \cdot CD = CB \cdot BD,$$

hence

$$\frac{EC}{CB} = \frac{BD}{DC}.$$

But

$$EC > CB,$$

hence

$$DB > DC.$$

So we have

$$BD > DC \Rightarrow 2BD > BD + DC = BC$$

$$2BN > 2BD > BC \Rightarrow BE + BN > BC.$$

We can thus construct triangle ABC such that $BA = AC = BE$, and we have

$$CB \cdot BD = BA^2.$$

So triangles ABD and CBA are similar, and we have

$$B\hat{A}D = A\hat{C}B \text{ and } A\hat{D}B = B\hat{A}C.$$

Since $EC \cdot CD = BE^2 = CA^2$, triangles ADC and AEC are similar, hence

$$C\hat{A}D = A\hat{E}C.$$

Therefore

$$A\hat{B}C = 2A\hat{E}C,$$
$$A\hat{B}C = 2C\hat{A}D,$$
$$A\hat{D}B = 3C\hat{A}D,$$
$$B\hat{A}C = 3C\hat{A}D.$$

So if angle BAC is three parts, then each of the angles ABC and ACE is two parts. We construct in the given circle a triangle similar to ABC, and finally we obtain the heptagon.

Let us take a rapid look at Ibn al-Haytham's solution using algebraic notation.

Let there be a segment EB; let N and E be two points symmetrical with respect to B. Let us construct the square $BNSO$; let (BO, BC) be coordinate axes (Ox, Oy) and let us put $BE = a$.

Let us consider the two hyperbolae

$$(\mathcal{H}_1) = \{(x, y); xy = a^2\},$$

$$(\mathcal{H}_2) = \left\{(x, y)\; ;\; y = x - \frac{a^2}{x+a}\right\}.$$

$(\mathcal{H}_1)$ and $(\mathcal{H}_2)$ must cut one another at $G(x_0, y_0)$ such that $x_0 > 0$. So there exists a unique $x_0 > 0$ such that

$$x_0^3 + ax_0^2 - 2a^2x_0 - a^3 = 0.$$

In fact, the equation for the abscissae of the points of intersection has three roots, of which x_0 is the one that provides the required solution.

From $G(x_0, y_0)$ we can find $D(0, y_0)$, $L(x_0, x_0)$ and $C(0, x_0)$. We construct A as the point of intersection of the two circles $\mathcal{C}_1$ (B, a) and $\mathcal{C}_2$ (C, a). These two circles have the same radius a, and they cut one another if $BC < 2a$, which Ibn al-Haytham proves is true – that is he shows that $BE + BN > BC$.

The triangle we obtain, triangle ABC, is of the type $(2, 3, 2)$. By a homothety, we construct in the given circle a triangle similar to triangle ABC. Finally, we may note that in constructing this solution Ibn al-Haytham has carried out a trisection of the angle BAC.

3. *The case* $(1, 5, 1)$

Analysis: Let us suppose we have found a triangle ABC such that

$$A\hat{B}C = A\hat{C}B = \frac{\pi}{7} \quad \text{and} \quad B\hat{A}C = \frac{5\pi}{7}.$$

Let us put $C\hat{A}D = A\hat{B}C$ and $D\hat{A}E = A\hat{B}C$. Triangles CAD and ABC are similar, and we have

$$\frac{BC}{CA} = \frac{AC}{CD},$$

hence

(1) $BC \cdot CD = AC^2 = AB^2.$

Triangles ADE and ABD are also similar, and we have

$$BD \cdot DE = AD^2.$$

But $AD = CD$, because $\hat{CAD} = \hat{ACD}$, hence

$$BD \cdot DE = CD^2.$$

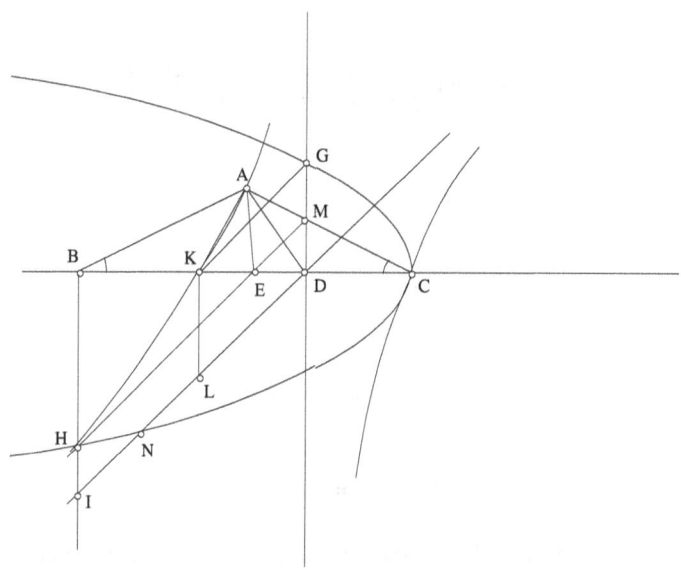

Fig. 3.66

Since

$$\hat{CAD} = \hat{DAE} = \hat{ABD} = \hat{ACD},$$

we have

$$\hat{ADE} = 2A\hat{C}D$$

and

$$\hat{AEB} = 3A\hat{C}B,$$
$$\hat{BAC} = 5A\hat{C}B \qquad \text{(by hypothesis)},$$
$$\hat{EAC} = 2A\hat{C}B,$$
$$\hat{BAE} = 3A\hat{C}B,$$
$$\hat{BAE} = A\hat{E}B, \qquad \text{hence } AB = BE.$$

From (1), we have

(2) $BC \cdot CD = BE^2.$

Let us put $DK = CD$. Let us draw $KL \perp DK$ with $KL = KD$, and at D construct the perpendicular DG such that $DG = DK$. Let us join GK and DL, and at B construct the perpendicular BH such that $BH = BE$. Let us join EH and produce it to M; let us produce DL to meet BH in the point I. Then we have

$$\frac{HE}{EB} = \frac{EM}{DE} = \frac{HM}{BD} \text{ and } \frac{HE}{BE} = \frac{GK}{DK},$$

so

$$\frac{HM}{BD} = \frac{GK}{DK} \text{ and } \frac{HM \cdot DE}{BD \cdot DE} = \frac{GK}{DK} = \frac{GK}{KL} = \frac{GK \cdot KL}{KL^2}.$$

But

$$BD \cdot DE = CD^2 = KL^2,$$

so

$$HM \cdot DE = GK \cdot KL.$$

But

$$DE = DM \text{ and } DM = HI,$$

so

$$HM \cdot HI = GK \cdot DK.$$

The hyperbola $(\mathcal{H})$ that passes through K and has asymptotes GD and DI thus passes through H. But from (2) and the hypothesis that $BH = BE$, the parabola $(\mathcal{P})$ with axis BC, vertex C and *latus rectum DC* passes through H. So

$$H \in (\mathcal{H}) \cap (\mathcal{P}).$$

So if we knew CD, $(\mathcal{H})$ and $(\mathcal{P})$ would be known, point H would also be known, as would points E and B.

Synthesis: Let CK be a given general segment. Let us divide CK into two halves at the point D, and at the points D and K let us construct perpendiculars DG and KL such that $DG = KL = DK$. Let us join GK and DL, and let us produce DL to the point I. Let us draw the hyperbola $(\mathcal{H})$ that passes through K and has asymptotes GD and DI.[72] Let us also draw the parabola $(\mathcal{P})$ with axis CK, vertex C and *latus rectum CD*.

---

[72] The point I will be defined below.

($\mathcal{P}$) cuts *DI*, because any straight line that cuts the axis of ($\mathcal{P}$) cuts ($\mathcal{P}$) in two points on opposite sides of the axis. If, further, ($\mathcal{P}$) extends beyond *DI*, it becomes more distant from it, because the tangent at the point of intersection cuts *DI*. So ($\mathcal{P}$) remains above the tangent. If ($\mathcal{P}$) becomes more distant from the intersection, the curve thus becomes more distant from *DI*. But as we extend ($\mathcal{H}$), it comes closer to *DI*. It must accordingly follow that ($\mathcal{H}$) and ($\mathcal{P}$) intersect; let their point of intersection be *H*.

At the point *H* let us construct *HB* perpendicular to the axis of ($\mathcal{P}$); let *HB* meet *DL* in the point *I*; let us draw *HEM* parallel to *DL*. Triangles *HBE* and *EDM* are then similar to triangle *DKL*, and we have

$$HB = BE \quad \text{and} \quad ED = DM = HI,$$

hence

$$\frac{HE}{BE} = \frac{DL}{DK} = \frac{EM}{DE} = \frac{HM}{BD},$$

so

$$\frac{HM}{BD} = \frac{DL}{DK},$$

hence

$$\frac{HM \cdot DE}{BD \cdot DE} = \frac{GK}{KL} = \frac{GK \cdot KL}{KL^2},$$

hence

$$\frac{HM \cdot HI}{BD \cdot DE} = \frac{GK \cdot KL}{KL^2}.$$

But

$$HM \cdot HI = GK \cdot KL \quad \text{[equation of ($\mathcal{H}$)]},$$

so

$$BD \cdot DE = KL^2,$$

hence

$$BD \cdot DE = CD^2.$$

But

$$KC = 2CD,$$

hence

$$KC \cdot CD = 2KL^2.$$

The point L thus lies inside $(\mathscr{P})$.[73] So the parabola $(\mathscr{P})$ cuts DI beyond L, say at the point N. Thus the straight line HB lies beyond KL, and $BD > DK$. But

$$BD \cdot DE = CD^2 = DK^2,$$

so $DE < DK$ and consequently $DE < CD$ and $EC < 2CD$. But

$$BC \cdot CD = HB^2 \quad \text{[equation of } (\mathscr{P})\text{]}$$

and

$$HB = BE,$$

hence

$$BC \cdot CD = BE^2$$

and we have

$$BC \cdot CE < BC \cdot 2CD \quad \text{and} \quad BC \cdot CE < 2EB^2,$$

so

$$CE < BE$$

[in fact $BC = BE + CE$, hence $(BE + CE) \cdot CE < 2EB^2$, then the hypothesis $CE \geq BE$ gives $(BE + CE) \cdot CE \geq 2EB^2$, and $2\,BE > BC$.]

So it is possible to construct on BC an isosceles triangle such that its base is BC and its other sides are equal to BE. Let the triangle be ABC. Let us join AD, AE.

Since $AC = BE$, we have $BC \cdot CD = AC^2$. Triangles ACD and ABC are similar, hence

$$C\hat{A}D = A\hat{B}C = A\hat{C}B.$$

---

[73] This statement is easily proved: let $L' \in (\mathscr{P})$ be projected to K; we have

$$KL'^2 = KC \cdot CD.$$

But

$$KC \cdot CD = 2KL^2,$$

hence

$$KL'^2 = 2KL^2,$$

so $KL' > KL$ and L lies inside $(\mathscr{P})$.

Then we have

$$AD = CD \text{ and } BD \cdot DE = AC^2,$$

hence triangles ADE and ABD are similar and

$$D\hat{A}E = A\hat{B}D = A\hat{C}D,$$

so

$$A\hat{E}B = 3A\hat{C}B.$$

Since $AB = BE$, we have

$$B\hat{A}E = B\hat{E}A,$$

so

$$B\hat{A}E = 2A\hat{C}B \text{ and } C\hat{A}E = 2A\hat{C}B$$

and we have

$$B\hat{A}C = 5A\hat{C}B.$$

So triangle ABC is of type $(1, 5, 1)$. By a homothety we construct in the given circle a triangle similar to ABC, and we obtain the heptagon.

Let us now retrace Ibn al-Haytham's procedure in a language of algebraic functions that is (obviously) not his, and let us put $(BD, DG) = (Ox, Oy)$ and $CD = a$.

Let us consider

$$(\mathscr{P}) = \{(x, y); y^2 = a(x + a)\},$$

$$(\mathscr{H}) = \left\{(x, y) ; y = x - \frac{a^2}{x}\right\}.$$

$(\mathscr{P})$ and $(\mathscr{H})$ must intersect at the point $H(x_1, y_1)$, such that $x_1 \in \mathbf{R}^*_+$ and $y_1 \in \mathbf{R}^*_+$. In fact,

Let f_1: $[0, \infty[\to \mathbf{R}$ such that $f_1(x) = \sqrt{a(x + a)}$,

f_2: $]0, \infty[\to \mathbf{R}$ such that $f_2(x) = x - \frac{a^2}{x}$.

Let h: $]0, \infty[\to \mathbf{R}$ such that $h(x) = f_2(x) - f_1(x)$; h is defined as increasing monotonically; so we have

$$h(x) = x - \sqrt{a(x+a)} - \frac{a^2}{x},$$

which is the difference of two functions

$$x - f_1(x) = x - \sqrt{a(x+a)} \text{ and } x - f_2(x) = x - \left(x - \frac{a^2}{x}\right),$$

the first of which is increasing and the second decreasing. These two functions represent the respective positions of $\mathscr{P}$ and $\mathscr{H}$ with respect to the asymptote $y = x$; these positions are exactly what Ibn al-Haytham investigates.

We have

$$\lim_{\substack{x \to 0 \\ x > 0}} h(x) = -\infty \text{ and } \lim_{x \to \infty} h(x) = +\infty.$$

So there exists a unique $x_1 \in \,]0, \infty[$ such that $h(x_1) = 0$ where $y_1 > 0$. The root x_1 is one of the two positive roots of the equation for the abscissae, which, after we divide through by $x + a$, can be written

$$x^3 - 2ax^2 - a^2x + a^3 = 0.$$

Ibn al-Haytham then constructs a triangle of type (1, 5, 1), and by a homothety he constructs in the given circle a triangle similar to the first one and, finally, obtains the heptagon.

4. The case (1, 2, 4)

Analysis: Ibn al-Haytham first proves that this case can be reduced to those investigated earlier. Let us suppose that we have found the triangle *ABC* (see Fig. 3.67) such that angles *A*, *B* and *C* are of the type (1, 2, 4). Let us put $B\hat{C}D = \frac{\pi}{7}$, we have $A\hat{C}D = \frac{3\pi}{7}$, and hence

$$A\hat{D}C = A\hat{B}C + B\hat{C}D = \frac{3\pi}{7}.$$

So triangle *ACD* is of type (3, 3, 1). Given triangle *ACD*, we increase $A\hat{C}D$ by $D\hat{C}B = C\hat{A}D$, and we obtain triangle *ABC* of type (1, 2, 4).

If, similarly, we put $B\hat{C}E = \dfrac{2\pi}{7}$, we have $C\hat{E}B = \dfrac{3\pi}{7}$ because $E\hat{B}C = \dfrac{2\pi}{7}$. So triangle BEC is of type $(2, 3, 2)$. Given triangle BEC and $E\hat{C}A = E\hat{C}B$; we have

$$A\hat{C}B = \frac{4\pi}{7} \quad \text{and} \quad C\hat{A}B = \frac{\pi}{7}.$$

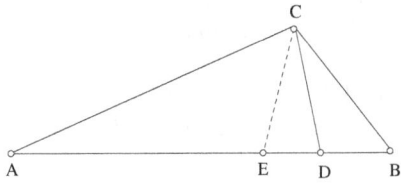

Fig. 3.67

So triangle ABC is of type $(1, 2, 4)$.

Similarly (see Fig. 3.68), if we put $A\hat{C}G = C\hat{A}G = \dfrac{\pi}{7}$, then

$$G\hat{C}B = \frac{3\pi}{7} \quad \text{and} \quad A\hat{G}C = \frac{5\pi}{7},$$

because

$$A\hat{G}C = G\hat{C}B + G\hat{B}C.$$

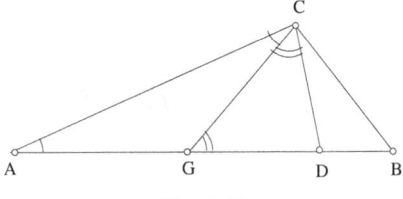

Fig. 3.68

So triangle AGC is of type $(1, 5, 1)$. Let triangle AGC be given and let us take $G\hat{C}D = C\hat{G}D$. Now

$$C\hat{G}D = \frac{2\pi}{7},$$

because

$$C\hat{G}D = A\hat{C}G + G\hat{A}C,$$

so

$$A\hat{C}D = C\hat{D}G = \frac{3\pi}{7}.$$

So if we take $D\hat{C}B = C\hat{A}B = \frac{\pi}{7}$, then

$$A\hat{C}B = \frac{4\pi}{7} \text{ and } C\hat{A}B = \frac{\pi}{7} \text{ and } A\hat{B}C = \frac{2\pi}{7}.$$

So the case (1, 2, 4) can be reduced to the preceding ones.

But it is possible to construct a triangle of type (1, 2, 4) without reducing it to the preceding cases. Analysis shows that we then return to Archimedes' lemma.

Let the triangle be ABC (see Fig. 3.69); let us produce BC in both directions to D and E respectively such that $CD = CA$ and $BE = BA$.

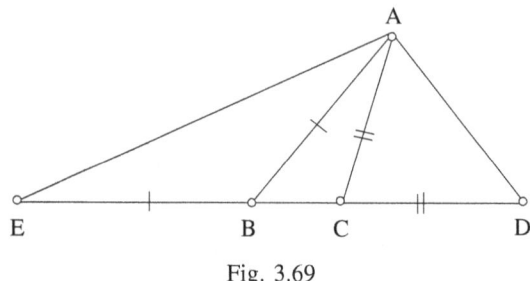

Fig. 3.69

Since

$$A\hat{C}B = \frac{4\pi}{7},$$

then

$$A\hat{D}C = C\hat{A}D = A\hat{B}C = \frac{2\pi}{7},$$

so

$$B\hat{A}D = \frac{3\pi}{7} \text{ and } A\hat{B}D = A\hat{D}B,$$

hence

$$AB = AD \text{ and } AD = BE.$$

But

$$A\hat{B}C = \frac{2\pi}{7},$$

thus

$$A\hat{E}B = \frac{\pi}{7};$$

hence

$$A\hat{E}C = B\hat{A}C.$$

So triangles ABC and AEC are similar. We have

(1) $EC \cdot CB = CA^2 = CD^2.$

Now

$$A\hat{C}B = \frac{4\pi}{7},$$

thus

$$D\hat{A}C = A\hat{B}C = \frac{2\pi}{7}.$$

So triangles ADC and ABD are similar. Thus we have

(2) $BD \cdot DC = DA^2 = BE^2.$

So the segment ED must be divided at points B and C such that we have (1) and (2); which corresponds to Archimedes' lemma.

Ibn al-Haytham then mentions that al-Qūhī had carried out a division of the segment in this ratio to construct the triangle of type (1, 2, 4) and then the regular heptagon. He proposes employing a method different from al-Qūhī's. But before enquiring into this difference, let us continue our account of Ibn al-Haytham's analysis.

To divide the segment ED at points B and C into parts that satisfy the given conditions, let us put $CK = CD$; $KG \perp CD$ such that $KG = KC$; $BH \perp BC$ such that $BH = BE$; $CL \perp BC$ (see Fig. 3.70). Let us draw GI parallel to KC with $GI = GK$ and let us join GC and IK. The straight line HB cuts GC in M.

Let us draw the parabola $(\mathscr{P})$ with vertex D, axis DB and *latus rectum* DC. Since $HB^2 = EB^2$, we have $HB^2 = BD \cdot DC$, so $H \in (\mathscr{P})$.

Let us draw the hyperbola $(\mathscr{H})$ that passes through K and has asymptotes CL and CG. Since $KG = KC$, we have $BM = BC$, hence $HM = EC$.

So the equality $EC \cdot CB = CD^2$ implies $MH \cdot CB = KG \cdot KC$. But

$$\frac{HL}{CB} = \frac{MC}{CB} = \frac{GC}{KC},$$

hence

$$\frac{HL \cdot MH}{CB \cdot MH} = \frac{GC \cdot KG}{KG \cdot KC},$$

so

$$MH \cdot HL = KG \cdot GC,$$

hence $H \in (\mathcal{H})$.

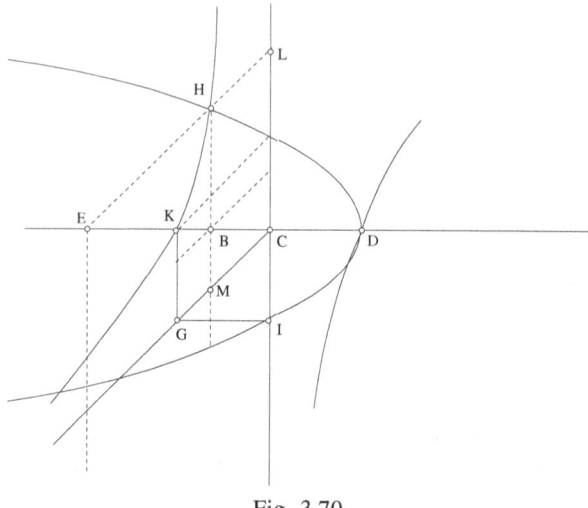

Fig. 3.70

Finally, we have $H \in (\mathcal{P}) \cap (\mathcal{H})$. So if we know points C and D, we know $(\mathcal{P})$ and $(\mathcal{H})$, and consequently H. We also know B, the projection of H, and finally E, because $BH = BE$.

Synthesis: Let KD be a given general segment and C its midpoint; let us draw $KG \perp KD$ such that $KG = KC$, GI parallel to KC such that $GI = KC$, $CL \perp DK$. Let us join GC, IK.

Let us draw the hyperbola $(\mathcal{H})$ that passes through K and has asymptotes GC and CL, and the parabola $(\mathcal{P})$ with vertex D, axis KD and *latus rectum CD*. The parabola $(\mathcal{P})$ cuts $(\mathcal{H})$ in H, for the reasons we gave earlier.

Let us draw $HB \perp DK$, $EH \parallel GC$. Let us produce HB to M and EH to L. We have

$$HB = BE \quad \text{and} \quad BM = BC;$$

so
$$HM = CE \text{ and } HL = MC,$$

$$EC \cdot CM = HM \cdot MC.$$

But
$$HM \cdot MC = IK \cdot KC$$

[since $HM \cdot MC = KG \cdot GC$, the equation of $(\mathscr{H})$].
Now

$$\frac{MC}{CB} = \frac{GC}{CK} = \frac{IK}{CK} = \frac{IK \cdot KC}{KC^2},$$

so

$$\frac{EC \cdot CM}{EC \cdot CB} = \frac{IK \cdot KC}{KC^2}.$$

But
$$EC \cdot CM = IK \cdot KC,$$

hence
$$EC \cdot CB = KC^2 = CD^2.$$

Now
$$BD \cdot DC = HB^2 \text{ [equation of } (\mathscr{P})],$$

hence
$$BD \cdot DC = BE^2.$$

So we have divided ED into three parts such that

(3) $EC \cdot CB = CD^2,$

(4) $BD \cdot DC = BE^2.$

These conditions are those for Archimedes' range (E, B, C, D) on the segment (ED), used to construct the triangle of type $(1, 2, 4)$.

Now, from (3), we have $CD > CB$ [because $EC > CB$] and consequently $EC > CD$. From (4) we have $BE > CD$ [because $BD > CD$] and consequently $BD > BE$. So the sum of any two of the segments EB, BC, CD is greater than the third segment [$EB + CD > BC$, because

CD > BC]. So the triangle *ABC* can be constructed from these segments. The construction of the heptagon is then carried out as before.

We may note that, if we put $(CE, CI) = (Ox, Oy)$ and $CD = a$, we again return to the curves from the previous case, that is,

$$(\mathscr{P}) = \{(x, y); y^2 = a(x + a)\},$$

$$(\mathscr{H}) = \left\{(x, y) \; ; \; y = x - \frac{a^2}{x}\right\}.$$

As before, we show that $(\mathscr{P})$ and $(\mathscr{H})$ intersect at $H(x_0, y_0)$ where $x_0 \in \,]0,\,a[$ and we have the same equation

$$x^3 - 2ax^2 - a^2x + a^3 = 0.$$

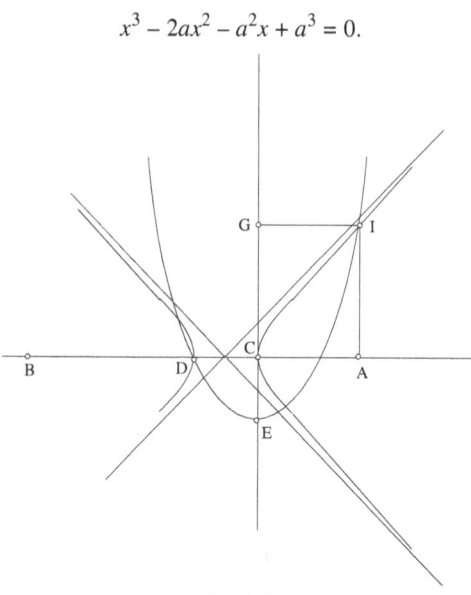

Fig. 3.71

To pinpoint the difference between Ibn al-Haytham's last step and that of al-Qūhī, we need to have a quick look at the latter's text.[74] In his analysis, al-Qūhī supposes we have a segment *AB* (see Fig. 3.71) divided at points *C* and *D* such that

$$AD \cdot AC = DB^2,$$

[74] Al-Qūhī, *On the Determination of the Side of the Heptagon*, p. 639.

$$CB \cdot CD = AC^2.$$

Let us put $ECG \perp AB$; $EC = CD$ and $CG = DB$. Let us draw $GI \parallel BA$ and $AI \parallel CG$. We have

$$IG^2 = AC^2 = CB \cdot CD,$$

hence

$$IG^2 = CE \cdot EG.$$

the point I thus lies on the parabola $(\mathscr{P})$ with axis EG, vertex E and *latus rectum EC*.

Moreover

$$AI^2 = CG^2 = BD^2 = AD \cdot AC.$$

so the point I lies on the hyperbola $(\mathscr{H})$ with vertex C, axis AC and *latus rectum CD*.

So if we put $CD = CE = a$ and $(CA, CG) = (Ox, Oy)$, we have

$$(\mathscr{P}) = \{(x, y); \, ay = x^2 - a^2\},$$

$$(\mathscr{H}) = \{(x, y); \, y^2 = ax + x^2\}.$$

$(\mathscr{H})$ is an equilateral hyperbola whose second vertex is D.

The point of intersection we have investigated here corresponds to the greater of the two positive roots of the equation

$$x^3 - ax^2 - 2a^2x + a^3 = 0.$$

The difference between the procedures adopted by Ibn al-Haytham and al-Qūhī thus lies in their choice of curves. But this difference leads us to another more important one: Ibn al-Haytham has chosen the same curves in this case as in that of (1, 5, 1), so the equation obtained is the same. He not only wanted to solve the problem, but also – calling on as few curves as possible – to provide solutions to the problem of the heptagon in all possible cases. That is why he chose a method different from that of al-Qūhī, who (for his part) was not looking for a solution as exhaustive as Ibn al-Haythm's.

Al-Qūhī's synthesis follows immediately after his analysis:

Let us put $AB = AC$ and $AB \perp AC$ (see Fig. 3.72). Let us draw the parabola ($\mathscr{P}$) with vertex B, axis AB and *latus rectum* AB; the hyperbola ($\mathscr{H}$) with vertex A, axis AC and *latus rectum* $AB = AC$. The curves intersect at E.

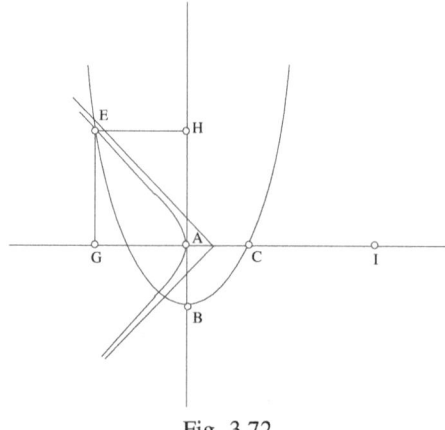

Fig. 3.72

Through E we draw EH parallel to AC and EG parallel to AH. Let us take the point I on AC such that $IC = AH$. We have

$$EG^2 = GA \cdot GC \quad \text{[equation of ($\mathscr{H}$)]},$$

hence

$$IG^2 = GA \cdot GC.$$

Moreover

$$AG^2 = AB \cdot BH \quad \text{[equation of ($\mathscr{P}$)]},$$

hence

$$AG^2 = AC \cdot AI,$$

hence the result.

The remainder of the construction is carried out as usual.

This is the form of solution Ibn al-Haytham gives for the problem of the heptagon. After listing the various possible cases, that is, the different possible triangles, he investigates them all. Beneath the apparent diversity, his investigation essentially reduces to solving three cubic equations. We shall summarize the different cases to give ourselves an overview:

First case:

$$(\mathcal{H}_1) = \{(x, y); xy = a^2\},$$

$$(\mathcal{H}_2) = \left\{(x, y) ; y = x - \frac{a^2}{x - a}\right\},$$

hence

$$x^3 + a^3 = ax^2 + 2a^2x.$$

There are three real roots, two of them positive: $x_0 \in \,]0, a[, x_1 > a$. Ibn al-Haytham uses x_0.

Second case:

$$(\mathcal{H}_1) = \{(x, y); xy = a^2\},$$

$$(\mathcal{H}_2) = \left\{(x, y) ; y = x - \frac{a^2}{x + a}\right\},$$

hence

$$x^3 + ax^2 = 2a^2x + a^3.$$

There are three real roots, one of which is positive x_0; this is the one Ibn al-Haytham uses.

Third and fourth cases:

$$(\mathcal{P}) = \{(x, y); y^2 = a(x + a)\},$$

$$(\mathcal{H}) = \left\{(x, y) ; y = x - \frac{a^2}{x}\right\},$$

hence

$$x^3 + a^3 = 2ax^2 + a^2x.$$

There are three real roots, two of them positive:

$$x_0 \in \,]0, a[\quad \text{and} \quad x_1 > a.$$

In the third case, Ibn al-Haytham uses x_1 and, in the fourth one, x_0.

This summary makes clear that Ibn al-Haytham is concerned with economy and thus shows that his systematization represents an advance, not only in going beyond all the solutions given by his predecessors, but also beyond what he himself had presented in his first essay. Ibn al-Haytham's systematic approach thus shows the history of the problem of the heptagon in Arabic mathematics in a new light, not only because it is systematic but also because he made various investigations of the conic curves. Another reason for giving further consideration to these investigations is that, later on, they seem to have exercised some influence on algebraists (al-Khayyām and Sharaf al-Dīn al-Ṭūsī).

Furthermore, Ibn al-Haytham noted, in the first two cases, that the problem of constructing the regular heptagon is equivalent to that of trisecting an angle. The equivalence of the two problems thus appears explicitly in Ibn al-Haytham's treatise. From work by Viète, it is now known that the solution of all irreducible cubic equations can be expressed as the general problem of trisecting an angle.

We have been examining the beginnings, and then the unfolding and transformation, of the study of the regular heptagon in Greek geometry and in Arabic geometry. Its history opens with Abū al-Jūd – who was the first to try to solve the construction problem – and closes, about half a century later, with Ibn al-Haytham. This does not, of course, mean that after Ibn al-Haytham work on the heptagon abruptly and finally came to a stop. We know this was not so, as witness the example of Ibn Yūnus. We merely wish to emphasize that later contributions – in Arabic as well as in Latin – added nothing of major significance. It is only much later that, with the theory of algebraic numbers and Wantzel's theorem, that new work was done in this area. The story of the regular heptagon is thus a good example of a mathematical investigation that exhausts its subject matter, only for the subject matter to then, as it were, renew itself in a quite different form from its original one. Neither Ibn Yūnus nor Viète makes any difference to this.

We still need to answer the question raised at the beginning of this chapter: why did this problem of constructing the heptagon draw in so many mathematicians of the time, and ones who were highly placed in the learned community? Why did this problem prompt some of them – Abū al-Jūd, al-Qūhī, Ibn al-Haytham – to come back to their essays on it a second time or more? We have turned to the milieu, or even psychological factors, to account for such intense activity, and no doubt these matters are relevant. But the essence of the explanation lies elsewhere: the reasons for this ferment of activity lie in geometry itself, and they are, it seems to us, the same reasons that entitle us to speak of this work as forming a *chapter*: at once an intellectual domain, a unifying element and a style.

Let us return to the tenth century and take note that the problem of the heptagon is only one element in an increasingly large group, that of three-dimensional problems. This group included inherited problems – of which that of the heptagon is one – and others that were new. Unlike that of the heptagon, the other inherited problems came down to mathematicians accompanied by solutions, and all had two claims to excite their interest: on the one hand their connections with the theory of conic sections, and, on the other hand, their undergoing a double process of translation that had never existed in Greek mathematics.

Having a grasp of the theory of conics, mathematicians saw these three-dimensional problems as providing opportunities for making good use of the tools offered by Apollonius' *Conics*. Of course, there were other areas for such applications: for example burning mirrors;[75] but, to confine ourselves to the problem of the heptagon, it can be seen that the authors have made use of no fewer than sixteen propositions from the *Conics*, and that they have drawn no fewer than seven parabolae and hyperbolae, defined by the positions of axes and asymptotes.

There is every indication that such frequent use of conic sections, together with the many studies of points of intersection, eventually changed ideas about what constituted a legitimate geometrical construction. Henceforth, a construction is legitimate if it uses straightedge and compasses, or if it employs the intersection of two conic sections. Constructions are rejected if they involve using a *neusis*[76] or transcendental curves. This principle is applied rigorously and the occasional exceptions confirm the rule. This decision was to dominate Arabic mathematics, and we may ask ourselves about the part algebra played in its formulation and the generalization of its demands. In any case, the message was clear and it was well understood by algebraists.

By the end of the ninth century the algebraists (Thābit ibn Qurra, al-Māhānī, ...) were already beginning to translate geometrical problems into the language of algebra. In the mid tenth century, a new idea appears: that of solving cubic equations by considering intersections of two conic sections – for example in the work of al-Khāzin. So it is in connection with three-dimensional (solid) problems that this idea of double translation takes shape, and, according to al-Khayyām,[77] it is Abū al-Jūd who was the first to make a systematic study of the matter. Al-Khayyām also informs us that it was Ibn 'Irāq who translated the problem of the heptagon into an algebraic equation. The fact that Abū al-Jūd took an interest in both algebra and

---

[75] *Les Catoptriciens grecs*. I: *Les miroirs ardents*, Texts edited and translated and with commentary by R. Rashed, Collection des Universités de France, Paris, 2000.

[76] This seems to explain why there were so few mechanical constructions using a *neusis*. One of these, by an anonymous author, has been preserved in a Latin translation by Gherard of Cremona. It would be very helpful to be able to date the original Arabic text. The Latin text has been edited by Marshall Clagett, *Archimedes in the Middle Ages*, vol. V: *Quasi-Archimedean Geometry in the Thirteenth Century*, Philadelphia, 1984, pp. 596–9.

[77] R. Rashed and B. Vahabzadeh, *Omar Khayyam. The Mathematician*.

geometry can hardly be a matter of pure coincidence.

In short, algebraists led geometers on to open up the new field of geometrical construction by means of conic sections, and encouraged them in their work. The series of studies of the heptagon can be seen as part of this new process, and circumstances favoured such studies. It is accordingly understandable that they aroused such active interest.

3.2. DIVISION OF THE STRAIGHT LINE

One of the famous problems of geometrical construction encountered in classical and late antiquity is Archimedes' division of a segment. The history of the problem from Archimedes to Eutocius is well known.[78] We shall summarize the principal elements to make it clear what was contributed by Ibn al-Haytham.

It all begins with Archimedes' statement in the fourth proposition of the second book of *The Sphere and Cylinder*:

So we need to cut a segment of a straight line ΔZ in a point X and do it so that XZ shall be to the given segment $Z\Theta$ as the given square, on $B\Delta$, is to the square on ΔX.[79]

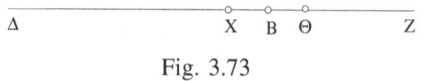

Fig. 3.73

According to the Greek text, Archimedes then went on to determine the special conditions for the problem examined in Proposition II.4, that is with $\Delta B = 2BZ$, and $ZB > Z\Theta$; which leads to the following formulation:

given two straight line segments $B\Delta$ and BZ, $B\Delta$ being twice BZ, and a point Θ on BZ, to cut ΔB in a point X in such a way that the square on $B\Delta$ shall be to the square on ΔX as XZ is to $Z\Theta$; each of these problems will be examined at the end (ἐπὶ τέλει).[80]

The final remark is lacking in the Arabic version, and very probably also in the whole manuscript tradition of Archimedes' text that was translated into

[78] See, for instance, Oskar Becker, *Das mathematische Denken der Antike*, Göttingen, 1966, pp. 89–90; T. L. Heath, *The Works of Archimedes*, Cambridge, 1897; Dover Reprint, 1953, pp. 65–79; E. J. Dijksterhuis, *Archimedes*, transl. by C. Dickshoorn with a new bibliographic essay by Wilbur Knorr, Princeton, 1987, pp. 143–205.

[79] See *Archimède, Commentaires d'Eutocius et fragments*, Texte établi et traduit par Charles Mugler, Collection des Universités de France, Paris, 1972, p. 113.

[80] *Ibid.*, p. 113, *l.* 10–14, translation modified.

Arabic. All we find is:

فينبغي أن نقسم خط ز د المعلوم بقسمين على نقطة ح حتى تكون نسبة ح ز إلى ز ط المعلوم كنسبة مربع
ب د إلى مربع د ح، ونظام ما ذكرنا وتأليفه وتركيبه على ما أصف.

So we must divide the known straight line $Z\Delta$ into two parts at the point X, in such a way that the ratio of XZ to $Z\Theta$, which is known, shall be the same as the ratio of the square of $B\Delta$ to the square of ΔX; the order of what we mentioned, its composition and its synthesis, as I describe.

This is the text that would have been available to Ibn al-Haytham. He also knew Eutocius' commentary on *The Sphere and Cylinder*, also translated into Arabic. In this book, Eutocius writes:

Archimedes did indeed promise to prove, at the end, what precedes, but the promised explanation is not found in any manuscript.[81]

This broken promise appears only in the Greek text, and is not found in the Arabic version. It does not matter much here whether the Greek sentence is authentic or not: the proof is missing, and its absence is attested well before Eutocius. Diocles, only one or two generations after Archimedes, explicitly mentions the fact,[82] and Dionysodorus, when giving a proof, seems to confirm Eutocius' statement. Dionysodorus himself proposes a proof, using a parabola and a hyperbola.

Ibn al-Haytham knew Archimedes' text in the Arabic version, together with Eutocius' commentary. He knew that Archimedes set out this construction, and that a proof could only be given with the help of conic sections. For Ibn al-Haytham, the latter fact explains the former one: the absence of a proof is not an oversight by Archimedes, but springs from a deliberate choice to separate the genera, to avoid using conic sections. This is a very strange style of learned discussion; it shows us how Ibn al-Haytham conceives constructions of solutions to problems by means of intersection of conics: as a special body of results within geometry.

This is the conception that governs the study Ibn al-Haytham devotes to Archimedes' line segment. In any case, it is the conception that characterizes his solution, which an unprepared reader might tend to place in a tradition derived directly from Eutocius, which is not in fact the case. To appreciate the historical position of Ibn al-Haytham's solution, we need to look briefly at its context: Ibn al-Haytham's predecessors, from the end of the ninth century onwards, had chosen to give an algebraic solution to the problem. This was the method adopted by al-Māhānī, al-Khāzin and Abū Naṣr ibn 'Irāq.

---

[81] *Ibid.*, p. 88.
[82] R. Rashed, *Les Catoptriciens grecs. I: Les miroirs ardents*, p. 121, *l.* 20–21.

At this point we should mention the evidence provided by al-Khayyām, who writes:

> As for the moderns, among them it is al-Māhānī who found himself led to carry out an algebraic analysis of the lemma Archimedes used, taking it as accepted, in proposition four of the second book of his work on *The Sphere and the Cylinder*. He then arrived at cubes, squares and numbers in an equation he did not succeed in solving after having reflected on it for a long time; so he broke off, deciding it was impossible, until Abū Ja'far al-Khāzin appeared and solved the equation by means of conic sections.[83]

Al-Khayyām says the same thing in another text.[84]

The problem of Archimedes' line, in which mathematicians began to take a new interest during the tenth century, had thus been translated into a cubic equation before it was solved by considering the intersection of two conic sections. At the same time that progress was being made by the algebraic approach, other mathematicians, who were probably well aware of this work and of course also aware of algebra itself, chose to take a deliberately geometrical path. Among such mathematicians we find al-Qūhī[85] and Ibn al-Haytham.

Al-Qūhī looks simultaneously at two problems according to whether the point of division X lies within ΔZ (to retain Archimedes' lettering) or on ΔZ produced. However, he assumes the special condition $Z\Theta = B\Delta$. In the case where the point X lies on ΔZ produced, the problem always has a solution (the cubic equation always has one positive root). In the other case, we require a condition that al-Qūhī states: the cube of the line segment $Z\Theta$ must be smaller than 4/27 of the cube of $Z\Delta$.

For his part, Ibn al-Haytham makes no special assumptions about the segment $Z\Theta$, but he considers only the case envisaged by Archimedes, in which the point of division X lies within the segment ΔZ. So we may assume that, even if he knew the study by al-Qūhī, Ibn al-Haytham began his work directly from the text of Archimedes.

Let us now look at Ibn al-Haytham's proof, preserving the Greek lettering.

Let $(Z\Delta, Z\Gamma)$ be coordinate axes (Zx, Zy), Δ $(\beta, 0)$ a point on Zx and B $(a, 0)$ another point. Let Θ $(\alpha, 0)$ be such that $\alpha < a$ and $A(\beta, \beta - a)$.

To find a point X $(x, 0)$ such that

---

[83] *Treatise on Algebra*, in R. Rashed and B. Vahabzadeh, *Omar Khayyam. The Mathematician*, p. 111 (Arabic text in the French edition, p. 117, *l.* 11–15).

[84] *Treatise on the Division of a Quadrant of a Circle*, in R. Rashed and B. Vahabzadeh, *Omar Khayyam. The Mathematician*, p. 173 (Arabic text in the French edition, p. 255, *l.* 4–8).

[85] See Supplementary Note.

(1)
$$\frac{x}{\alpha} = \frac{(\beta - a)^2}{(\beta - x)^2},$$

which is a version of Archimedes' condition

$$\frac{ZX}{Z\Theta} = \frac{B\Delta^2}{\Delta X^2}.$$

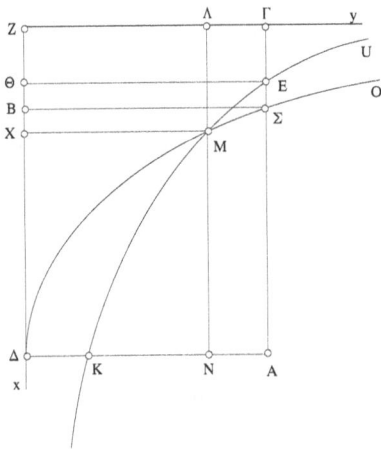

Fig. 3.74

Let there be a hyperbola

$$\mathcal{H} = \{(x, y); xy = \alpha(\beta - a)\},$$

$$E\,(\alpha, \beta - a) \in \mathcal{H}.$$

Since $\mathcal{H}$ tends to infinity, and $Z\Delta$ is an asymptote to $\mathcal{H}$, $\mathcal{H}$ must cut $A\Delta$, say in a point K.

Let there be a parabola

$$\mathcal{P} = \{(x, y); (\beta - x)^2 = y(\beta - a)\}.$$

$\mathcal{P}$ tends to infinity, and $A\Gamma$ is perpendicular to its axis. So $\mathcal{P}$ must cut $A\Gamma$, say in a point Σ, with ordinate $\beta - a$.

$$\Sigma\,(x, \beta - a) \in \mathcal{P} \Rightarrow (\beta - x)^2 = (\beta - a)^2,$$

so $x = a$; but $\alpha < a$ by construction; so $E \in \mathcal{H}$ lies outside $\mathcal{P}$ and $K \in \mathcal{H}$ lies inside $\mathcal{P}$, so (by an implicit argument from continuity) $\mathcal{P}$ cuts $\mathcal{H}$. Let $M(x_0, y_0) \in \mathcal{H} \cap \mathcal{P}$. The point $M(x_0, y_0) \in \mathcal{P}$, so $(\beta - a)y_0 = (\beta - x_0)^2$, hence

$$\frac{\beta - a}{\beta - x_0} = \frac{\beta - x_0}{y_0},$$

hence

$$\frac{\beta - a}{y_0} = \frac{(\beta - a)^2}{(\beta - x_0)^2}.$$

The point $M(x_0, y_0) \in \mathcal{H}$, so $x_0 y_0 = \alpha(\beta - a)$, hence

$$\frac{x_0}{\alpha} = \frac{\beta - a}{y_0},$$

hence

$$\frac{x_0}{\alpha} = \frac{(\beta - a)^2}{(\beta - x_0)^2}.$$

Which is what it was required to prove.

Comparing it with that of Eutocius, we can see that Ibn al-Haytham's geometrical construction is also carried out by means of a parabola and a hyperbola. Translated into the algebraic terms of al-Khayyām, Eutocius' version of the problem reduces to solving the equation

$$x^3 + c = ax^2,$$

that is to say Equation 17 in al-Khayyām,[86] whereas Ibn al-Haytham's version leads to the equation

$$x^3 + bx = ax^2 + c,$$

that is to Equation 24 in al-Khayyām.[87]

---

[86] R. Rashed and B. Vahabzadeh, *Omar Khayyam. The Mathematician*, p. 46ff.
[87] *Ibid.*, pp. 77ff. Al-Khayyām uses the intersection of a semicircle and an equilateral hyperbola

(*Cont. on next page*)

The solution Ibn al-Haytham finds is not unique. His construction gives him a second solution that, as it happens, does not satisfy the conditions for the geometrical problem as proposed, for which we are required to find a point X between Δ and B (in Archimedes' problem ΔB is the diameter of the sphere). Ibn al-Haytham's second solution lies outside ΔB. On the other hand, a division such as that proposed by Archimedes, with Θ between B and Z, which in our notation is written $\alpha < a$, gives Ibn al-Haytham a sufficient condition for the solution to exist. In fact $Z\Theta \cdot B\Delta^2 < ZB \cdot B\Delta^2 = \alpha(\beta - a)^2$, the equation we are required to solve being $x(\beta - x)^2 = \alpha(\beta - a)^2$; this equation has a solution between 0 and β if $\alpha(\beta - a)^2$ is less than the maximum value of $x(\beta - x)^2$ for $0 < x < \beta$, a condition that is satisfied for $0 < \alpha < a < \beta$. The maximum value in question is $\dfrac{4\beta^3}{27}$, so the necessary and sufficient condition for there to be a solution in the interval $]0, \beta[$ is that $\alpha(\beta - a)^2 < \dfrac{4\beta^3}{27}$; this condition is implied by $\alpha < a$.

We may note that Eutocius establishes that this condition is necessary.[88] Al-Qūhī also points out this condition in the special case of the problem that he investigates.

For his part, Ibn al-Haytham abides strictly by the conditions of Archimedes' problem, which means he can dispense with this necessary condition (which is satisfied automatically).

We find another important difference between Eutocius' text and Ibn al-Haytham's, a difference that relates to proving two conic sections do intersect. Eutocius as it were sees the result without pausing to point it out or prove it. In addition, Ibn al-Haytham uses the coordinates more directly by referring the two curves to the same coordinate axes, which decreases the complexity of manipulating the proportions. However, he explicitly set out to give a proof by considering properties of the convexity of the curves, their behaviour at infinity, and their assumed continuity. Finally, we may note that, unlike Eutocius, Ibn al-Haytham shows the *symptoma*, the basic relationships that define the two conic sections, in the form of an equality of products of segments. However, he also notes that the parabola passes

---

(*Cont.*)

$$\mathscr{C} = \left\{ (x,y); \left(b^{\frac{1}{2}} - y \right)^2 = \left(x - \frac{c}{b} \right)(a - x) \right\},$$

$$\mathscr{H} = \left\{ (x,y); xy = b^{\frac{1}{2}} \times \frac{c}{b} \right\}.$$

[88] *Commentaires d'Eutocius et fragments*, ed. and transl. Charles Mugler, pp. 94f.

through a point M that lies inside the hyperbola, and he does not explicitly draw the appropriate conclusions regarding the intersection of the two curves.[89]

3.3. ON A SOLID NUMERICAL PROBLEM

'To divide a known number into two parts such that one is the cube of the other'. This is the problem Ibn al-Haytham considers in a text that, while short, is nonetheless of major significance. The importance of the text derives from the nature of the problem. We are indeed dealing with one of the types of problem that, since the time of Diophantus, number theorists were in the habit of proposing, problems in which the algebraists, following al-Khayyām, took a lively interest. But while the number theorists looked for rational numbers as solutions, the algebraists wanted positive real numbers. Since Ibn al-Haytham is engaging with a three-dimensional (solid) problem, he proposes to employ the techniques of geometrical construction that he has developed, in order to solve the problem without needing to call on algebra: thus the field for geometrical constructions extends beyond the area in which it originated.

The facts that the task was new, and that the means employed could accomplish it, could hardly fail to attract the notice of Ibn al-Haytham's successors: they were to seize upon the essential elements of his work before integrating it into algebra. That integration was carried out by al-Khayyām, who was certainly aware of Ibn al-Haytham's work.[90] So let us begin by analysing Ibn al-Haytham's text[91] using a mathematical language different from his.

Problem:
Let $a > 0$ be a given number. To find $x > 0$ such that $x < a$ and

(1) $x = (a - x)^3$.

To solve this problem, Ibn al-Haytham begins by proving the following lemma:

---

[89] *Ibid.*, pp. 92–3.
[90] R. Rashed and B. Vahabzadeh, *Al-Khayyām mathématicien*, especially pp. 222, 224.
[91] See below.

Lemma: To find four magnitudes a_1, a_2, a_3, a_4 such that

1. $$0 < a_1 < a_2 < a_3 < a_4,$$

2. $$\frac{a_1}{a_2} = \frac{a_2}{a_3} = \frac{a_3}{a_4},$$

3. $$\frac{a_4 - a_3}{a_1} = k = \frac{b}{c}, \text{ a known ratio.}$$

With coordinate axes (Nx, Ny), let us take the points $A(c, b)$, $B(c, 0)$, $D(2c, 0)$ and $E(2c, b)$. Let us then draw

$$\mathcal{H} = \{(x, y); \, y(x - c) = bc\},$$

$$\mathcal{P} = \{x, y); \, y = \frac{x^2}{c}, x > 0\}.$$

Ibn al-Haytham then uses the behaviour of these conics as they tend to infinity to prove that the two curves must cut one another in a unique point with abscissa $x_0 > c$. He notes that when x increases from c to $+\infty$, $y_\mathcal{H}$ decreases from $+\infty$ to zero, since the hyperbola has asymptotes BA and BC; and (for the parabola) $y_\mathcal{P}$ increases from c to $+\infty$, so there exists a unique value $x_0 > c$ such that $G(x_0, y_0) \in \mathcal{H} \cap \mathcal{P}$. Since $x_0 > c$, we have $y_0 > x_0$.

We get $y_0 = \frac{x_0^2}{c}$ since $G(x_0, y_0) \in \mathcal{P}$ and $y_0(x_0 - c) = bc$ since $G(x_0, y_0) \in \mathcal{H}$. From these two relations we obtain respectively

$$\frac{c}{x_0} = \frac{x_0}{y_0} \text{ and } \frac{c}{x_0} = \frac{y_0}{y_0 + b};$$

hence

$$\frac{c}{x_0} = \frac{x_0}{y_0} = \frac{y_0}{y_0 + b},$$

where $c < x_0 < y_0 < y_0 + b$.

It is now enough to put

$$a_1 = c, \, a_2 = x_0, \, a_3 = y_0, \, a_4 = y_0 + b,$$

in order to satisfy the conditions of the lemma, since

$$\frac{a_4 - a_3}{a_1} = \frac{b}{c} = k.$$

Solution of the problem:

Let us now look for x, the solution to (1).

Let $AB = a$, $AI = x$; $x < a$ and the problem is to find x such that

$$x = (a - x)^3 \Leftrightarrow AI = BI^3.$$

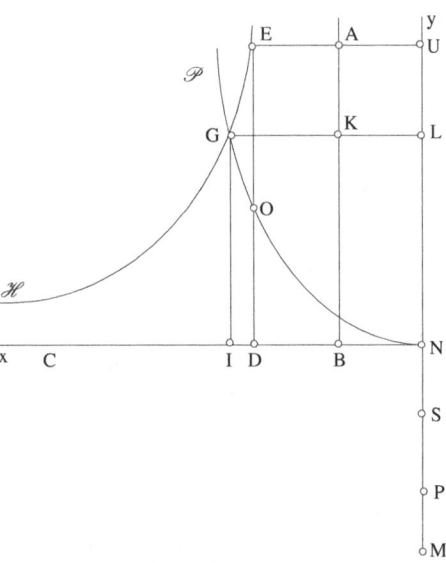

Fig. 3.75

From the lemma, we can find four positive numbers $a_1 < a_2 < a_3 < a_4$ such that

$$\frac{a_1}{a_2} = \frac{a_2}{a_3} = \frac{a_3}{a_4} \text{ and that } \frac{a_4 - a_3}{a_1} = a^2.$$

We prove that, if

$$\frac{x}{a - x} = \frac{a_4 - a_3}{a_3},$$

then $x = (a - x)^3$.

By hypothesis we have $a_4 - a_3 = a_1 \cdot a^2$. The relation $\dfrac{x}{a - x} = \dfrac{a_4 - a_3}{a_3}$

implies

(2) $\dfrac{x}{a-x} = a^2 \cdot \dfrac{a_1}{a_3} = a^2\left(\dfrac{a_3}{a_4}\right)^2.$

But we also have

(3) $\dfrac{a}{a-x} = \dfrac{a_4}{a_3}.$

From (2) and (3), we obtain

$$\frac{x}{a^2(a-x)} = \frac{(a-x)^2}{a^2},$$

hence the result.

Notes:

1. Ibn al-Haytham starts from the range (D, H, G, E, C), which is assumed known with $DC = a_4$, $DE = a_3$, $DG = a_2$, $DH = a_1$, and constructs from another given segment AB the range (B, L, K, I, A) similar to the first one, with AI corresponding to CE.

2. The construction of the point I on AB is equivalent to finding $x = AI$ where $a - x = BI$; which is the same as solving equation (1). If we put $y = a - x$, that is the same as solving $y^3 = a - y$ with $a > 0$. The solution of this problem depends on the lemma. We have seen that the solution is obtained from the intersection of a parabola and a hyperbola. Later, al-Khayyām, after translating the problem into an algebraic form – see Equation 13 of his treatise[92] – solved it by means of the intersection of a parabola and a circle.

If we now reverse Ibn al-Haytham's procedure and start from the magnitudes in continued proportion, we find an equation of the form

$$x^3 = \alpha_1 x^2 + \alpha_3,$$

which is of al-Khayyām's type 18.[93]

If, instead, we start directly from equation (1), we come back to the equation

[92] R. Rashed and B. Vahabzadeh, *Omar Khayyam. The Mathematician*, pp. 35–8. See also Sharaf al-Dīn al-Ṭūsī, *Œuvres mathématiques*, 2 vols, Paris, 1986, vol. I, Eq. 13, pp. CLV–CLVIII.

[93] See R. Rashed and B. Vahabzadeh, *Omar Khayyam. The Mathematician*, pp. 52ff.

$$x^3 + (3a^2 + 1)x = 3ax^2 + a^3$$

of al-Khayyām's type 24, which can have three positive solutions.

3. We may note that if $x = c$, from the equation of the parabola we have $y = c$, and if $x = 2c$, we have $y = 4c$. The parabola cuts the straight line EA between E and A if $b > 4c$, $(k > 4)$, which is the case shown in the figure, and we then have $y_0 < b$ and $x_0 > 2c$. If $k < 4$, we have the situation shown in the following figure, giving $y_0 > b$ and $x_0 < 2c$, but all the reasoning is still valid.

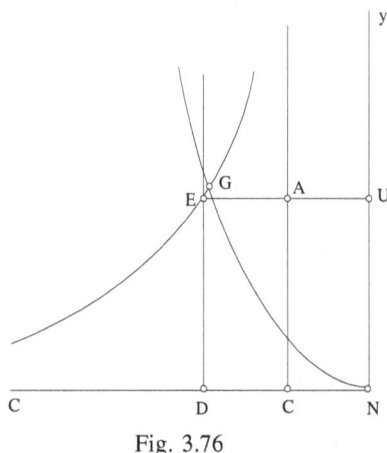

Fig. 3.76

4. The solution to equation (1) constructed in this way is, in general, irrational and consequently beyond the scope of the theory of numbers. The solution thus belongs to geometry or to the algebra developed at this time. Ibn al-Haytham's study, as we have just seen, is strictly geometrical.

3.4. HISTORY OF THE TEXTS OF IBN AL-HAYTHAM

3.4.1. *On the Construction of the Regular Heptagon*

This title is the one adopted by the old biobibliographers, al-Qifṭī and Ibn Abī Uṣaybiʿa, in their lists of the works of Ibn al-Haytham.[94] It has been

---

[94] R. Rashed, *Ibn al-Haytham and Analytical Mathematics*, London, 2012, no. 8, p. 392.

known since 1979[95] that this treatise was preserved in the famous manuscript 'Āṭif 1714 in Istanbul. Apart from twenty treatises by Ibn al-Haytham on mathematics, astronomy and optics, the collection, which is of relatively late date (1158 AH/1745), includes a piece by the mathematician Yaḥyā al-Kāshī and a piece by an anonymous author on the planisphere. The treatise *On the Regular Heptagon* was the nineteenth work passed on in this collection, which until very recently was believed to provide the only surviving example of this text by Ibn al-Haytham.

In the course of our research on the mathematics and optics of Ibn al-Haytham, we have been led to establish several texts included in this collection, such as the *Measurement of the Sphere*,[96] the *Exhaustive Ttreatise on Lunes*[97] and the *Burning Sphere*.[98] In this connection, we have shown that in each case the copyist of manuscript 'Āṭif worked from only a single original, luckily one that is extant, manuscript Oct. 2970 in the Staatsbibliothek in Berlin. In fact we can go further: this conclusion holds for all the treatises – fifteen treatises by Ibn al-Haytham, one by al-Kāshī – contained in the Berlin collection: they are all found, in the form of apographs, in manuscript 'Āṭif.

So, apart from the rather mysterious question of the anonymous text on the planisphere, there remains the problem of the five texts by Ibn al-Haytham that are absent from the Berlin manuscript and present in manuscript 'Āṭif, namely:

– *On a Proposition of the Banū Mūsā*,
– *On the Hour Lines*,
– *On the Construction of the Waterclock*,
– *On the Formation of Shadows*,
– *On the Construction of the Regular Heptagon*.

The question that came to mind was the following: did these five texts once form part of the Berlin collection and were later separated from it, either deliberately or accidentally? A careful examination of this collection confirmed our initial suspicions by revealing numbering in the original volume, which had included not sixteen treatises, as it does today, but ins-

---

[95] R. Rashed, 'La construction de l'heptagone régulier par Ibn al-Haytham', *Journal for the History of Arabic Science*, 3.2, 1979, pp. 309–87.

[96] *Les Mathématiques infinitésimales*, vol. II, p. 294–323; *Ibn al-Haytham and Analytical Mathematics*, Chapter II.

[97] *Les Mathématiques infinitésimales*, vol. II, pp. 102–75; *Ibn al-Haytham and Analytical Mathematics*, Chapter I.

[98] R. Rashed, *Géométrie et dioptrique*, pp. 111–32; *Geometry and Dioptrics*, Chapter III.

tead twenty-two or twenty-three. The same examination also showed that the Berlin collection had been copied out over a period of at least twenty years, between 817/1414 and 839/1435. The hand – contrary to what we said earlier, there is only one[99] – is none other than that of the famous scholar Qāḍī Zādeh, who, in the service of Ulugh Beg, was for some time the director of the observatory at Samarkand. In fact, on fol. 12[r], we read the following note:

> Treatise on the proof two problems, one about what is the basis for the measurement of the surface of the sphere and the other about the area of a parallelogram ... by the hand of Qāḍī Zādeh.

In the colophon to this same text, we read the following information (fol. 21[v]):

> His calligraphy (*tanmīqihi*) was completed on ten Rabīʿ al-ākhir of the year eight hundred and seventeen at Samarkand.

Finally, in the colophon of Ibn al-Haytham's *The Measurement of the Sphere* (fol. 152[r]), the scribe has dated his copy to the year 839/1435. As for the history of this manuscript between when it was written and its incorporation into the German national collection in 1930, we know almost nothing. However, given the transcription dates that appear in it – covering an interval of at least twenty years – the identity of the copyist and the place where the copy was made, it does not seem rash to suggest that what we have here may be a personal working copy belonging to Qāḍī Zādeh. He very probably copied the set of works now in Berlin for his personal use, and with an eye to his own mathematical work – evidence, if it were really needed, that the mathematical, optical and astronomical corpus of work by Ibn al-Haytham was still found interesting in the first half of the fifteenth century. It is now up to the sociologists of science to give us in-depth information about stakes and modalities of scientific activity at the court of Ulugh Beg. More modestly, and to return to the textual question we began with, we for our part need to ask whether, together with the four other texts that are absent from the Berlin manuscript, the treatise *On the Regular Heptagon* may have formed part of this collection made by Qāḍī Zādeh – if the answer to the question were in the affirmative, we should be able to take the history of our treatise forward by about three centuries.

This affirmative answer is possible and, by a lucky conjunction of the disciplines of philology, codicology and palaeography, it can even be rigorously demonstrated. The solution lies in investigating a third piece of evidence, never examined until now: manuscript 3025 in the Military Museum in Istanbul. This collection includes the five treatises by Ibn al-Haytham that

---

[99] *Ibid.*, pp. CXLVIII–CXLIX.

are lacking in the Berlin collection, and a commentary on Apollonius' *Conics* by al-Ḥusayn ibn ʿAbd al-Malik. However, whereas the five works by Ibn al-Haytham are copied in the same hand, this last commentary comes from a different source. More importantly: the hand in which the treatises by Ibn al-Haytham were copied is none other than that of the Berlin manuscript, that is, the hand of Qāḍī Zādeh. As far as the microfilms permit us to judge, the layout of the pages in the two manuscripts – those in Berlin and in the Military Museum – shows an identical pattern of ruled lines: 21 lines to the page, almost equal ratio between the dimensions of the four margins to those of the area of writing.

So we may conclude without the slightest hesitation that the present two manuscripts (in Berlin and in the Military Museum) were originally a single manuscript – and remained so at least until the year 1158/1745, the date when the apograph manuscript ʿĀṭif 1714 was put together.[100] After this date, the huge collection was divided into two separate volumes, and it was perhaps then – unless this did not take place until later still – that the five treatises today in the Military Museum in Istanbul were put together with the commentary on the *Conics* by al-Ḥusayn ibn ʿAbd al-Malik. So the manuscript ʿĀṭif 1714 no longer has any independent value in regard to the stemma.

This conclusion is corroborated in a definitive manner by a philological comparison of the two copies of the five texts (Military Museum 3025 and ʿĀṭif 1714). Here we shall confine our discussion to the case of the *Construction of the Regular Heptagon*. It is clear that the copyist of ʿĀṭif [A] worked from a single original, the text of the Military Museum [M]. Every error or omission in M can in fact be found in A, while A presents numerous distinctive mistakes of its own, errors and omissions, which do not appear in M. Finally, we may note that certain phrases copied in the margin of M are integrated into the text in A. This extends to the letters on the

---

[100] So Qāḍī Zādeh's manuscript, copied at Samarkand, was in Istanbul in the eighteenth century and there it was transcribed by the copyist of ʿĀṭif. What we read in the colophon of Ibn al-Haytham's treatise on the *Direction of the Qibla* (fol. 9ᵛ) is:

« ونقل مما خطه موسى الشهير بقاضي زاده الرومي، ووقع الفراغ في خلال محرم الحرام لسنة ثمان وخمسين
ومائة وألف في البلد الطيبة قسطنطنية المحمية...»

'It has been copied from what was transcribed by Mūsā, known under the name of Qāḍī Zādeh the Byzantine. The copy was completed during the sacred month of Muḥarrām in the year one thousand one hundred and fifty-eight in the good city of Constantinople, protected <by God>.'

The collection was later divided and the greater part of it was acquired by the Staatsbibliothek, while the remainder was taken into the Military Museum in Istanbul.

geometrical figures, which, if they are forgotten in M, are similarly absent from A.

The text of M is transcribed in *nasta'līq*. We find some marginal corrections in the hand of the copyist. The geometrical figures are drawn, but the lettering on the figures has sometimes been forgotten.

In the collection 'Āṭif 1714, the text on the heptagon occupies folios 200ᵛ–210ʳ, the script is *naskhī* and the figures are copies resembling those in M.[101]

Ibn al-Haytham's text on the regular heptagon has so far appeared only in a single edition, the one we published in 1979. This was made from the only manuscript then known: 'Āṭif 1714. The history of the text that we have just recounted compelled us to revise that critical edition, and the French translation that accompanies it. But there is no occasion to expect important modifications. The omissions and mistakes in A do not make any changes in the text. However, it goes without saying that as far as we are concerned this new edition and its translation supersede the former ones. So we decided, without being obliged to do so, to note the variants from A in the apparatus criticus so as to provide access to a certain amount of the evidence we used in support of our argument.

3.4.2. *Treatise on the Determination of the Lemma of the Heptagon*

This title is the one mentioned by the old biobibliographers, al-Qifṭī and Ibn Abī Uṣaybi'a,[102] and it is thus different from the one given in the only manuscript of the whole text. In fact, in manuscript 1270/21 (= Loth 734), fols 122ʳ–123ᵛ of the India Office in London [designated by A], Ibn al-Haytham's treatise is presented with the title: *Faṣl li-al-Ḥasan ibn al-Ḥasan ibn al-Haytham fī muqaddimat al-musabba'* (Chapter of al-Ḥasan ibn al-Ḥasan ibn al-Haytham on the lemma of the heptagon). The author himself refers to this treatise in his second and substantial version *On the Construction of the Regular Heptagon*, in these terms:

> We, for our part, have proved the lemma that Archimedes used, in a separate treatise, not in this <present> treatise.

---

[101] Thus we read in the colophon: 'The drawing of the figures of this treatise was carried out in conformity with the original from which it was copied, in the night that ends the twentieth of the month of sha'bān 1158.'

[102] Ibn Abī Uṣaybi'a, *'Uyūn al-anbā' fī ṭabaqāt al-aṭibbā'*, ed. N. Riḍā, Beirut, 1965, with the title *Qawl fī istikhrāj muqaddimat ḍil' al-musabba'*, p. 559; al-Qifṭī, *Ta'rīkh al-ḥukamā'*, ed. J. Lippert, Leipzig, 1903, with the title *Muqaddimat ḍil' al-musabba'*, p. 167.

This perfect agreement between Ibn al-Haytham and the old biobibliographers in speaking of a *qawl* (treatise or discourse), and not of a *faṣl* (chapter), tells us the true title of this treatise. Moreover, it is surprising to speak of a *faṣl* for a short treatise that deals with a single proposition. Further, we may add that the word *faṣl* could be a corrupt reading introduced in the course of copying the term *qawl*. So here we have opted for the title referred to by the old biobibliographers.

Concerning manuscript 1270 of the India Office, we have already told the little that we know about it. Ibn al-Haytham's text is transcribed without crossing out or additions, certain words are illegible because of damage by damp.

Until now this was the only known manuscript of this treatise. We have been able to identify a fragment of this text in manuscript 678, fol. 27^{r-v}, of the collection 'Abd al-Ḥayy in the University Library in Aligarh [designated by O], in India, copied in 721/1321–22 at al-Sulṭāniyya, in *nasta'līq* script. This fragment is all that remains, after the loss of several pages of this manuscript. Comparison of this 'Extract' with Ibn al-Haytham's text in the manuscript in the India Office shows that the former has been made after an ancestor of the latter. However, this fragment allows us to add a supplementary textual argument in favour of the authenticity of the text.

The textual tradition of this treatise is not confined to the copy in the India Office and the fragment in Aligarh. We also have a shortened version of this text in the Bodleian Library in Oxford, Thurston 3, fol. 131.[103] This version presents itself for what it is, since it has the title: *Min kalām Ibn al-Haytham 'alā muqaddimat Arshimīdis fī ḍil' al-musabba'* (Extract from Ibn al-Haytham's discourse on the lemma of Archimedes for the side of the heptagon). Thus in this version we do not find either the first paragraph or the last one; the remainder is summarized rather briefly.

Finally, this last text was copied, without doubt recently, in the Bodleian Library manuscript, Marsh 720, fol. 259^r.

There has been only a single critical edition of this text by Ibn al-Haytham, our own, published in 1979,[104] based on the single manuscript in the India Office. In *Les Mathématiques infinitésimales*, vol. III, we presented a revised version of that edition based on the same manuscript and the fragment in Aligarh.

---

[103] We gave the edition of this extract in *Les Mathématiques infinitésimales*, vol. III, pp. 914–19.

[104] 'La construction de l'heptagone régulier par Ibn al-Haytham', pp. 385 ff.

This text has been translated into German by C. Schoy.[105] The translation was very useful in making the text known. Here we present an English translation that we hope is as accurate as possible.

3.4.3. The Division of the Straight Line Used by Archimedes in the Second Book of his Work on The Sphere and the Cylinder

Ibn al-Haytham's treatise on *The Division of the Straight Line Used by Archimedes in the Second Book of* The Sphere and the Cylinder exists in numerous manuscripts. It has come down to us with other treatises by Ibn al-Haytham as well as on its own, being included in *Mutawassiṭāt*, intermediate collections. In *Les Mathématiques infinitésimales* we gave the *editio princeps* of this piece based on the eight manuscripts we have been able to obtain:[106]

1. Leiden, Or. 14/16, fols 498–501, designated by L. This is the famous manuscript transcribed in the seventeenth century at the request of the mathematician and Orientalist Golius, as is reported by R.P.A. Dozy.[107] We have discussed the history of this manuscript elsewhere.[108]

2. Istanbul, Topkapi Sarayi, Ahmet III 3453/16, fol. 179ᵛ, designated by D. This manuscript was copied by ʿAbd al-Kāfī ʿAbd al-Majīd ʿAbd Allāh al-Tabrīzī in 677 AH (1278) in Baghdad. Fatḥ Allāh al-Tabrīzī owned this manuscript in 848 AH (1444). The script is *naskhī* (page 17.1×13.2 cm; text 13.9×9.6 cm). The numbering of the folios is modern.

3. Istanbul, Topkapi Sarayi, Ahmet III 3456/18, fols 81ᵛ–82ʳ, designated by E. This text was transcribed on 12 Rabīʿ al-awwal 651 AH (12 May 1253). The script is *nastaʿlīq* (page 25.5×11.3 cm; text 19.4×8.9 cm). The numbering is old.

4. Istanbul, Süleymaniye, Carullah 1502, fol. 222ᵛ–223ʳ, designated by C.

This is a collection transcribed, in 894 AH, from the copy of the celebrated astronomer Quṭb al-Dīn al-Shīrāzī, as is affirmed by the copyist Ibn Maḥmūd ibn Muḥammad al-Kunyānī. The script is *naskhī*; each page has 25 lines (page 25.5×17.9 cm; text 17.2×11.2 cm).

---

[105] *Die trigonometrischen Lehren des persischen Astronomen Abū'l Raiḥān Muḥammad Ibn Aḥmad al-Bīrūnī*, Hanover, 1927.

[106] F. Sezgin mentions that the collection Algiers 1446/9 includes a copy of this text in fols 119–126. On checking, this proves not to be so (*Geschichte des arabischen Schrifttums*, V, Leiden, 1974, p. 372).

[107] *Catalogus Codicum Orientalium Bibliothecae Academiae Lugduno Batavae*, Leiden, 1851, p. XV.

[108] R. Rashed and B. Vahabzadeh, *Al-Khayyām mathématicien*, Paris, 1999, pp. 109–10 and especially below, Chapter IV, pp. 506–7.

5. Istanbul, Beşiraga 440, fol. 275^v, designated by B.

The copy dates from the beginning of Dhū al-Qaʿda 1134 AH (August 1722). The script is *naskhī*, and is very elegant (page 28.2 × 15.7 cm; text 18.5 × 8.6 cm).

6. Istanbul, Haci Selimaga 743, fols 135^r–136^v, designated by S.

This manuscript was copied in 1099 AH. In fact we read: 'The copy was completed on 15 Shaʿbān one thousand and ninety-nine', that is 14 June 1688. This manuscript, made up of two different parts, but on paper from the same batch, is in *naskhī* script (each page is 22.2 × 13.3 cm; 18 × 8.8 cm for the text).

7. Istanbul, Süleymaniye, Atif 1712/17, fol. 147^{r-v}, designated by O. This is an intermediary collection of books.

8. London, India Office 1270/18, fol. 119^v, designated by A. We do not know the date of transcription, which might be in the tenth century of the Hegira.

Since Ibn al-Haytham's text is short, it would be highly arbitrary to establish a *stemma* on the basis of a single page, without investigating the history of the collections of which this page forms part. But such an investigation must await the distant future, given the current state of research into the history of Arabic manuscripts.

The text was translated into French by F. Woepcke with the title 'Mémoire d'Ibn Alhaïtham, c'est-à-dire du Chaïkh Aboûl Haçan Ben Alhaçan Ben Alhaïtham sur la section d'une ligne employée par Archimède dans le second livre' (Essay by Ibn Alhaïtham, that is by Chaïkh Aboûl Haçan Ben Alhaçan Ben Alhaïtham on the section of a line used by Archimedes in the second book). This translation, which is a little free, appeared as a first appendix to the translation of the *Algebra* of al-Khayyām.[109]

3.4.4. *On a Solid Numerical Problem*

This text exists in a single manuscript in the India Office Library, London, no. 1270, fols 118^v–119^r (a manuscript to which we have referred several times),[110] with the title *Fī mas'ala ʿadadiyya mujassama* (*On a*

---

[109] See his translation of the *Algebra*, *L'Algèbre d'Omar Alkhayyâmî*, Paris, 1851, pp. 91–6.

[110] See above and R. Rashed, *Ibn al-Haytham and Analytical Mathematics*, p. 33.

Solid Numerical Problem). This is the title under which it appears in the old biobibliographers' lists of the writings of Ibn al-Haytham.[111]

In *Les Mathématiques infinitésimales* we gave the *editio princeps* of this treatise, together with a French translation that we have made as exact as possible. A French translation of this text has already been published by J. Sesiano, 'Mémoire d'Ibn al-Haytham sur un problème arithmétique solide', *Centaurus*, 20.3, 1976, pp. 189–95. This translation sometimes runs into difficulties in rendering nuances in Ibn al-Haytham's line of thought.

[111] *Ibid.*, p. 412–13.

3.5. TRANSLATED TEXTS

Al-Ḥasan ibn al-Haytham

3.5.1. *A Lemma for the Side of the Heptagon*

3.5.2. *On the Construction of the Regular Heptagon*

3.5.3. *On the Division of the Straight Line Used by Archimedes in Book Two of* The Sphere and the Cylinder

3.5.4. *On a Solid Numerical Problem*

In the Name of God, the Compassionate, the Merciful
Glory to God

TREATISE BY AL-ḤASAN IBN AL-ḤASAN IBN AL-HAYTHAM :

A Lemma for the Side of the Heptagon

Archimedes based <his derivation of> the side of the heptagon on the square he introduced, without showing how to construct the square that has the property he required. In fact, if he did not show this, that is because the construction of the square with the property he required cannot be carried out except by using conic sections. Now, since in the book at the end of which he deals with the heptagon, he has not referred to conic sections, he deliberately did not put into the book something that does not belong in a book of that kind, so he considered this square as an assumption and it was on this basis that he built up the side of the heptagon.

As for the manner of constructing the square with the property he required: we draw the square to which he referred, that is the square $ABCD$. We extend AC as he did, we extend the straight line AD to E, we extend the straight line $BGHE$ and we assume for the purpose of analysis that the triangle HDE is equal to the triangle BGC. We draw the straight line KGI parallel to BA as he did. So the product of DA and AI is equal to the square of DE, as Archimedes showed. We join BD, then it cuts the diameter AC into two equal parts, because the square $ABCD$ is a right-angled parallelogram; let it cut it at the point M. So the triangle BMC is equal to the triangle AMD. But since the triangle EDH is equal to the triangle BGC, the triangle BMC will be equal to the triangle EDH plus the triangle BMG. But the triangle BMC is equal to the triangle AMD, so the triangle AMD is equal to the <sum of the> triangles EDH and BMG. We consider their common quadrilateral $MDHG$, so the triangle BDE is equal to the quadrilateral $ADHG$. Let the triangle BEL be equal to the triangle CGH, so the triangle BDL is equal to the triangle ADC, and they lie between two parallel straight lines. So the straight line *LD* is equal to the straight line *DA* and the ratio of triangle *BDL* to the triangle *BEL* is equal to the ratio of the triangle *ADC* to the triangle *CHG*.

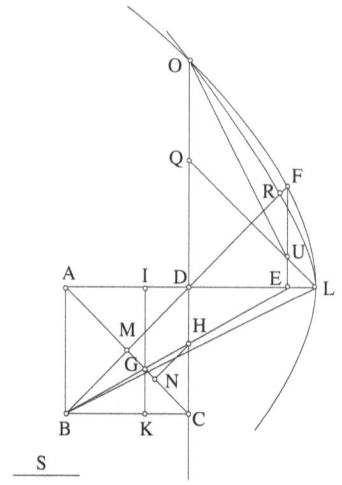

Fig. III.1.1

We draw the straight line *HN* perpendicular to the straight line *GC*. So the product of *HN* and half of *GC* is equal to the triangle *HGC*, and the product of *DM* and half of *AC* is equal to the triangle *ADC* because *DM* is perpendicular to *AM* if the rectangle has equal sides, so the ratio of the triangle *ADC* to the triangle *CGH* is compounded from the ratio of *DM* to *HN* – which is equal to the ratio of *DC* to *CH* – and the ratio of half of *AC* to half of *CG* – which is equal to the ratio of *AC* to *CG*; so the ratio of the triangle *ADC* to the triangle *CGH* is compounded from the ratio of *DC* to *CH* and the ratio of *AC* to *CG*. But the ratio of *DC* to *CH* is equal to the ratio of *EB* to *BH* and the ratio of *AC* to *CG* is equal to the ratio of *EB* to *BG*, so the ratio of the triangle *ACD* to the triangle *CGH* is compounded from the ratio of *EB* to *BH* and the ratio of *EB* to *BG*. In the same way, it necessarily follows that if the rectangle has <sides of> two different lengths, we draw from the point *D* a perpendicular to the straight line *AC*, then the perpendicular takes the place of *DM*, and the case reduces to the two ratios <that we have already> mentioned. And the ratio of the triangle *ACD* to the triangle *CGH* is equal to the ratio of the triangle *BDL* to the triangle *BEL*, which is equal to the ratio of *DL* to *LE*, so the ratio of *DL* to

LE is compounded from the ratio of *EB* to *BH* – which is equal to the ratio of *EA* to *AD* – and the ratio of *EB* to *BG* – which is equal to the ratio of *EA* to *AI* – so the ratio of *DL* to *LE* is compounded from the ratio of *EA* to *AD* and the ratio of *EA* to *AI* – which is equal to the ratio of the square of *EA* to the product of *DA* and *AI*, which is equal to the square of *DE* – so the ratio of *DL* to *LE* is equal to the ratio of the square of *AE* to the square of *ED*, and the straight line *AD* is equal to the straight line *DL*.

Thus, <the construction of the> square has been reduced by analysis to the division of the straight line *AL* – which is twice *AD* – at a point *E* such that the ratio of *DL* to *LE* is equal to the ratio of the square of *AE* to the square of *ED*. But the straight line cannot be divided in this ratio without using conic sections.

By the method of analysis we assume that the straight line has been divided, we extend the straight line *CD* to *O* and we put *DO* equal to *AE*. We draw from the point *E* the perpendicular *EF* and we put *EF* equal to *DE*. So the ratio of *DL* to *LE* is equal to the ratio of the square of *OD* to the square of *FE*. Let the product of *DL* and the straight line *S* be equal to the square of *OD*. The parabola whose axis is *DL*, and *latus rectum* the straight line *S*, passes through the points *O* and *F*. It <necessarily> passes through the point *O* since the square of *DO* is equal to the product of *DL* and the *latus rectum*, and this is the property of the parabola; the parabola <necessarily> passes through the point *F* since the ratio of *DL* to *LE* is equal to the ratio of the square of *OD* to the square of *FE*, as has been shown in Proposition 20 of Book I of the *Conics*; let this conic section be *LFO*. We put the straight line *DQ* equal to <the straight line> *DL* and we join *LQ*; let it cut the straight line *FE* at the point *U*. So the triangle *LDQ* is known in shape, the angle *OQU* is known and the ratio of *QU* to *DE* is known because it is equal to the ratio of *QL* to *LD*, <which is> known. But since *OD* is equal to *EA* and *QD* is equal to *DL* – which is equal to *DA* – we have *QO* equal to *DE*, so the ratio of *OQ* to *QU* is known and the angle *OQU* is known. We join *OU*, so the triangle *OQU* is of known shape and the ratio of *UO* to *OQ* is known; but *OQ* is equal to *DE* and *DE* is equal to *EF*, so the straight line *OQ* is equal to the straight line *FE* and the ratio of the square of *OU* to the square of *FE* is known. But the square of *FE* is equal to the product of *LE* and the straight line *S*, so the ratio of the product of *LE* and *S* to the square of *UO* is known; but the ratio of *EL* to *LU* is known, so the ratio of the product of *LU* and *S* to the square of *UO* is known and the angle *OUL* is known. So the parabola – whose diameter is

LQ, vertex the point *L*, the angle of its ordinate the angle *OUL* and *latus rectum* a straight line whose ratio to the straight line *S* is a known ratio – passes through the point *O*; let this conic section be the conic section *LRO*.

So if the straight line *AD* is known in position, if the point *L* is known and the straight line *S* is of known magnitude, then the conic section *LFO* is known in position, the straight line *LQ* is known in position because the angle *DLQ* is known, the *latus rectum* of the conic section *LRO* is of known magnitude and the angle *OUL* is known, so the conic section *LRO* will be known in position, so the point *O* will be known. But the straight line *OD* is perpendicular to the straight line *LD* which is known in position, so the straight line *OD* is known in magnitude and in position, the point *D* is known and the straight line *DL* is of known magnitude, so the ratio of *OD* to *DL* is known; *OD* is equal to *AE* and *DL* is equal to *AD*, so the ratio of *AE* to *AD* is known. And since we can find two straight lines that are equal to them by the method that we have demonstrated, <lines> which are the straight lines *OD* and *DL*, and since the straight line *AD* is known, so the straight line *DE* is known, and since the point *D* is known, so the point *E* is known, and this is the one that gives the square *ABCD* the property that Archimedes required.

Archimedes similarly assumed this square <can be constructed> and by means of analysis he reduced the problem to the lemma he needed to construct the heptagon, that is that the product of *DA* and *AI* is equal to the square of *DE* and the product of *EI* and *ID* is equal to the square of *AI*, each of the two straight lines *AI* and *ED* being greater than *ID*. So he supposed there was a known straight line and he divided it in this ratio and from it he constructed the heptagon. It is possible to divide a straight line in this ratio by means of conic sections even without recourse to the square.

Let us suppose we have the straight line; let it be *AB*. We wish to divide it into three parts, the parts *AC*, *CD*, *DB* such that the product of *DA* and *AC* is equal to the square of *DB* and the product of *BC* and *CD* is equal to the square of *AC* and each of the straight lines *AC*, *DB* is greater than *DC*.

We suppose we have a general straight line, let it be *EG*; we cut off from it a known arbitrary magnitude, let it be *EH*. We construct a parabola with axis *EG*, vertex the point *E* and *latus rectum* the straight line *EH*, as in Proposition 52 of Book I of the *Conics*; let the conic section be *EKL*. We cut off *HI* equal to *HE* and we draw from the points *H* and *I* two perpendiculars whose endpoints are on the conic section, let them be *HK* and *IL*. So *HK* is equal to *HE*, because the square of *KH* is equal to the product of *HE* and the *latus rectum*, and *HE* is the *latus rectum*, so the square of *KH* is equal to the product of *HE* with itself and the straight line *KH* is equal to the straight line *HE*. We extend *LI* in the direction of *I*, we cut off *IS* equal to *IH* and we join *KI*. So *KI* is parallel to the straight line *HS*, since *IS* is equal to *KH* and is parallel to it; so the plane figure *KHSI* is a parallelogram.

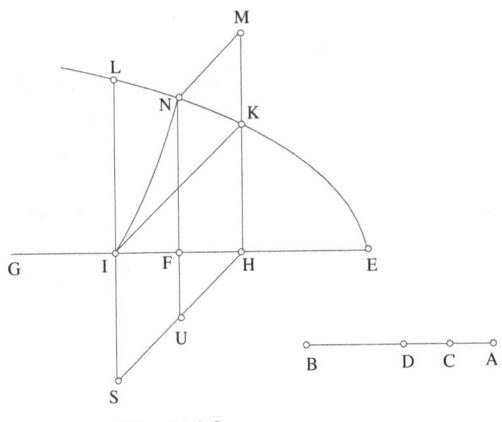

Fig. III.1.2

We draw through the point *I* the hyperbola that does not meet the straight lines *KH*, *HS* as in Proposition 4 of Book II of the *Conics*; let the conic section be *IN*. This conic section cuts the segment *KL*. In fact, the straight line *IL* is parallel to the straight line *HK* which is an asymptote to the conic section, so the straight line *IL* lies inside the hyperbola *IN*. If we extend the straight line *IL* to infinity, it does not meet the conic section *IN* in any point other than the point *I*; in fact, if we extend the two straight lines *HK* and *IL* to infinity in the directions of *K* and *L*, the distance between them always remains the same; and if we extend the conic section *IN* in the direction of *N*, as it is extended further it comes closer to the straight line *HK* and to its extension, as <is shown> in Proposition 14 of Book II of the *Conics*. Since if we extend the straight line *IL* to infinity in the direction of *L*, it will always lie inside the conic section *IN* and the point *K* itself always lies outside the conic section *IN*, because it is on the asymptote, so if we extend the conic section *IN*, it then cuts the segment *KL* of the conic section *EKL*; let it cut it at the point *N*. We extend the straight line *HK* in the direction of *K* and we draw from the point *N* a straight line parallel to the straight line *KI*, let it be *NM*, and we draw the perpendicular *NFU*, it will be parallel to the straight line *LIS*, so the product of *MN* and *NU* is equal to the product of *KI* and *IS*, as has been shown in Proposition 12 of Book II of the *Conics*. So the parallelogram *NH* is equal to the parallelogram *SK*. But the area *NH* is the product of *NU* and *HF*, because *HF* is perpendicular to *NU*, and the area *SK* is equal to the product of *SI* and *IH*, and *SI* is equal to *IH*, and *IH* is equal to *HE*, so the parallelogram *SK* is equal to the square of *EH*.

Now we have shown that the area *SK* is equal to the product of *NU* and *HF*, so the product of *NU* and *HF* is equal to the square of *EH*. We put *FG* equal to *NF*; but *FU* is equal to the straight line *FH* since *SI* is equal to *IH*, so the straight line *HG* is equal to the straight line *NU*, so the product of *GH* and *HF* is equal to the square of *HE*. Similarly, the straight line *NF* is one of the ordinates because it is perpendicular to the axis *EG* and the straight line *EH* is the *latus rectum* of the parabola *EKN*, so the product of *FE* and *EH* is equal to the square of *FN*; but *FN* is equal to *FG*, so the product of *FE* and *EH* is equal to the square of *FG*. But the product of *GH* and *HF* was equal to the square of *HE*. So we divide the straight line *AB* at two points *C*, *D* in the ratio of the straight lines *EH*, *HF*, *FG*, we then have the product of *DA* and *AC* equal to the square of *DB* and the product of *BC* and *CD* equal to the square of *CA*. It remains to show that each of the straight lines *AC* and *DB* is greater than *CD*.

Since the product of *FE* and *EH* is equal to the square of *FG*, *FN* is thus greater than *EH*, so it is greater than *HI* because *HI* is equal to *HE*, so *FN* is much greater than the straight line *HF*. Now *NF* is equal to *FG*, so the straight line *FG* is greater than the straight line *FH*. But *EH* is also greater than *HF* since *EH* is equal to *HI*, so each of the two straight lines *EH* and *FG* is greater than the straight line *HF*. So each of the straight lines *AC*, *DB* is greater than the straight line *CD*, but the straight lines *AC*, *CD*, *DB* are in the same ratio as the straight lines *EH*, *HF*, *FG*. So we have divided the straight line *AB* into <three> straight lines: *AC*, *CD*, *DB* so that the product of *DA* and *AC* is equal to the square of *DB*, the product of *BC* and *CD* is equal to the square of *AC* and each of the two straight lines *AC* and *DB* is greater than the straight line *CD*. That is what we wanted to do.

If we divide the straight line *AB* in this ratio, then we can construct a triangle from the straight lines *AC, CD, DB*; let this triangle be *ECD*, which is the triangle constructed by Archimedes from which he constructed the heptagon. If we construct this triangle, it is possible from it to go on to construct the heptagon by a process different from the process adopted by Archimedes; and we do this by constructing in the circle in which we wish to construct the heptagon a triangle whose angles are equal to the angles of this triangle, then the <major> arc whose chord is the straight line *CD* is a seventh of the circle, the <major> arc whose chord is the straight line *CE* is two sevenths of the circle and the <major> arc whose chord is the straight line *ED* is four sevenths of the circle, because angle *EDC* is twice angle *CED* and angle *ECD* is four times angle *CED*. So if we divide the arc above the straight line *EC* into two equal parts and the arc above the straight line *ED* into four parts and if we put in the chords of the arcs, then what is produced in the circle is a regular heptagon.

It remains for us to show that the angle *EDC* is twice angle *CED* and that the angle *ECD* is four times angle *CED*.

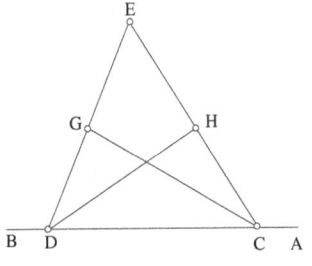

Fig. III.1.3

We divide the angle *CDE* into two equal parts with the straight line *DH* and we divide the angle *ECD* into two equal parts with the straight line *CG*. So the ratio of *EH* to *HC* is equal to the ratio of *ED* to *DC*, which is equal to the ratio of *BD* to *DC*. So by composition, the ratio of *EC* to *CH* is equal to the ratio of *BC* to *CD*. But the ratio of *BC* to *CD* is equal to the ratio of the square of *AC* to the square of *CD*, because the product of *BC* and *CD* is equal to the square of *CA*, so the ratio of *EC* to *CH* is equal to the ratio of the square of *AC* to the square of *CD*, that is to the ratio of the square of *CE* to the square of *CD*, so the ratio of *EC* to *CD* is equal to the ratio of *DC* to *CH*. So the two triangles *DEC* and *CDH* are similar. Then the angle *DHC* is equal to the angle *EDC*, but the angle *DHC* is equal to the sum of the two angles *EDH* and *DEH*, so the angle *DEH* is equal to the angle *HDC*; but the angle *EDC* is twice the angle *HDC*, so the angle *EDC* is twice the angle *DEC*. For the same reason, the ratio of *DG* to *GE* is equal to the ratio of *DC* to *CE*, which is equal to the ratio of *DC* to *CA*. By composition, the ratio of *DE* to *EG* is equal to the ratio of *DA* to *AC*; but the ratio of *DA* to *AC* is equal to the ratio of the square of *BD* to the square of *CA*, so the ratio of *DE* to *EG*[1] is equal to the ratio of the square of *BD* to the square of *CA*, which is equal to the ratio of the square of *DE* to the square of *EC*, so the ratio of *DE* to *EG* is equal to the ratio of the square of *DE* to the square of *EC*, so the ratio of *DE* to *EC* is equal to the ratio of *EC* to *EG*. So the two triangles *ECD* and *ECG* are similar, so the angle *CGE* is equal to the angle *ECD* and the angle *EGC* is equal to the sum of the two angles *GCD* and *GDC*, so the angle *EDC* is equal to the angle *ECG*. But the angle *ECD* is twice the angle *ECG*, so the angle *ECD* is twice the angle *EDC* and the angle *ECD* is four times the angle *CED*.

We have thus shown that the angle *EDC* is twice the angle *CED* and that the angle *ECD* is four times the angle *CED*. So if we construct, in the circle in which we wished to construct the heptagon, a triangle whose angles are equal to the angles of the triangle *ECD*, if we divide the angle *ECD* into two equal parts and each of these halves into two equal parts and if we divide the angle *EDC* into two equal parts, the circle is divided into seven equal parts. So if we draw the straight lines that are chords of these parts, there is produced in the circle a regular heptagon. That is what we wanted to prove.

The chapter on the lemma for the side of the heptagon is completed.
Thanks be given to God alone.

---

[1] The end of manuscript [O].

In the name of God, the Compassionate, the Merciful

TREATISE BY AL-ḤASAN IBN AL-ḤASAN IBN AL-HAYTHAM

On the Construction of the Heptagon in a Circle

One of the geometrical problems – over which geometers enter into rivalry, <problems> in which those who surpass the others take pride, and through which the prowess of those who succeed in solving it is revealed – is the construction of a regular heptagon in a circle. Some ancient and some modern scholars have succeeded in achieving <a solution>, although their achievement included some flaw. Among the ancients, it is Archimedes who constructed the figure; he did indeed write a treatise on finding the side of the heptagon, but in his determination of it he employs a lemma, without presenting a proof of it. We, for our part, have proved the lemma that Archimedes used, in a separate treatise, not in this <present> treatise. From modern scholars, we have received two treatises; in one, Archimedes' lemma has been proved and the construction was based on it; the other treatise is by Abū Sahl Wayjan ibn Rustam al-Kūhī: he determined the side of the heptagon by means of a straight line that he divided into three parts in a particular proportion; this is the straight line that completes Archimedes' lemma. We have not found a treatise that explains sufficiently fully by any of the ancients or by modern scholars, <that is one> in which there are included all the ways in which the construction of the heptagon can be achieved. This being so, we have made a careful study of the construction of the heptagon, and we have given proofs of all the ways in which the construction of the heptagon can be carried out. We have proceeded by analysis and by synthesis.

So we begin our account of the subject by saying: We wish to construct in a given circle a heptagonal figure whose sides and whose angles are <all> equal, inscribed in the circle.

Let the circle be *ABC*; we wish to construct in this circle an inscribed heptagon with equal sides and equal angles.

By the method of analysis:

Let us suppose that this has been carried out, that is that the heptagon is *ADEBCGH*. We join <points to give> the straight lines *CE*, *EB*, *BC*, *CD*, *BD*, *DH*, *BH*, *BA*, *CA*. There are formed in the circle four inscribed triangles, each of whose angles intercepts one or more equal arcs, whose chords are the sides of the heptagon.

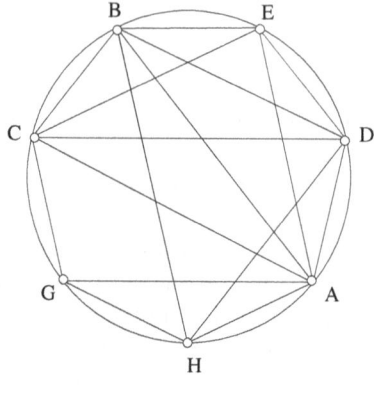

Fig. III.2.1

We say first: we cannot have in the circle an inscribed triangle each of whose angles intercepts one or more arcs of the equal arcs whose chords are the sides of the heptagon, and which is different from one of these triangles. Because in the triangle *ABC* the angle *BAC* intercepts the arc *BC*, which is a seventh of the circle. So the angle *BAC* is one part of the seven parts of the sum of two right angles; the angle *ABC* intercepts *AGC*, which is three sevenths of the circle; it is three parts of the seven parts of the sum of two right angles. Similarly, the angle *ACB* is three parts of the seven parts of the sum of two right angles. In triangle *BDH*, the angle *BDH* is three parts of the seven parts of the sum of two right angles, and each of the angles *DBH* and *DHB* is two parts of the seven parts. In triangle *EBC*, the angle *EBC* is five parts of the seven parts, and each of the angles *BEC* and *BCE* is a single part of the seven parts. In triangle *DBC*, the angle *BDC* is one part of the seven parts, the angle *BCD* is two parts of the seven parts, and the angle *DBC* is four parts of the seven parts.

These triangles are four triangles whose angles are such that each of them is part of the seven parts of the sum of two right <angles>, which has been divided into three parts, using different schemes of division. Seven cannot be divided into three parts except by these four types of division. These are the types we have described in detail, and there are no parts <of the number> seven that are three parts and which do not belong to the set of these four types. No triangle may be inscribed in the circle with angles that intercept the equal arcs whose chords are the sides of the heptagon, apart from these four triangles; if we find a triangle similar to one of these triangles, then we have found the heptagon, because if in the circle we construct a triangle similar to this triangle and if we divide its angles into parts, the circle is divided into seven equal parts; if the angles intercept the arcs, a heptagon is formed which has sides and angles that are equal.

<*First case*>
So let us begin by finding triangles similar to the four triangles whose angles we have described in detail, and let us form the heptagon starting from each of them. Let us make a beginning with the isosceles triangle in which each of the angles at the base is three times the remaining angle. We wish to form the heptagon starting from this triangle.

By the method of analysis:
We suppose that we have found a triangle that shows this property; let the triangle be *ABC*. We put the angle *CBD* equal to the angle *BAC*; so triangle *BCD* is similar to triangle *ABC*, and the angle *BDC* is equal to the angle *ABC*; the angle *ABC* is equal to the angle *ACB*. So the angle *BDC* is equal to the angle *BCD*. So the straight line *BD* is equal to the straight line *BC*. Since the triangle *CBD* is similar to the triangle *ABC*, the ratio of *AC* to *CB* is accordingly equal to the ratio of *BC* to *CD*; so the product of *AC* and *CD* is equal to the square of *BC*. We put the angle *DBE* equal to the angle *BAC*. So the two triangles *ABD* and *DBE* are similar, so the angle *BED* is equal to the angle *ABD*, and the angle *ABD* is two parts of the seven <parts>; so the angle *BEC* is two parts of the seven <parts> and the angle *CBE* is two parts of the seven <parts>. So the straight line *EC* is equal to the straight line *CB*. Since the triangle *DBE* is similar to the triangle *ABD*, the product of *AD* and *DE* is equal to the square of *DB*, and *DB* is equal to *BC*, so the product of *AD* and *DE* is equal to the product of *AC* and *CD*, and *BC* is equal to *CE*; so the product of *AD* and *DE* is equal to the square

of *CE*, and the product of *AC* and *CD* is equal to the square of *CE*. Then on the straight line *EC* we construct a square;[1] let the square be *CEHG*. We extend the two straight lines *CG* and *EH* as far as *I* and *L*. We imagine the hyperbola that has asymptotes *EC* and *CI* and passes through the point *H*; let it be the conic section *HK*. We draw from the point *D* a straight line parallel to the straight line *CI*. It thus meets the conic section; let it meet it at the point *K*. This straight line cuts the straight line *GH*; let it meet it at the point *N*. We cut off <a segment> *HP* equal to *HE* and we draw the two straight lines *PG* and *HC*. The straight line *HC* cuts the straight line *DN*; let it cut it at the point *M*. *CD* is thus equal to *DM* and *DE* is equal to *HN*. We draw *KI* parallel to *DC*. Since the two straight lines *EC* and *CI* are the asymptotes[2] of the conic section *HK*, the product of *KD* and *DC* is equal to the product of *HE* and *EC*, which is equal to the square of *CE*. But the product of *AC* and *CD* is equal to the square of *CE*; the straight line *KD* is thus equal to the straight line *AC*, and *CD* is equal to *DM*. Finally *KM* is equal to *AD*, and the product of *AD* and *DE* is equal to the square of *CE*, so the product of *KM* and *NH* is equal to the square of *EC*, and the ratio of *NH* to *HM* is equal to the ratio of *GH* to *CH*; so the ratio of the product of *KM* and *NH* to the product of *KM* and *MH* is equal to the ratio of *GH* to *HC*, which is equal to the ratio of the square of *GH* to the product of *GH* and *HC*, that is to the product of *HC* and *DN*. Now, the product of *KM* and *NH* is equal to the square of *GH*. So the product of *KM* and *MH* is equal to the product of *HC* and *ND*. We draw *KL* parallel to *MH*. The product of *MK* and *KL* is thus equal to the product of *HP* and *PG*. So the hyperbola that has as asymptotes <the lines> *CH* and *HL* passes through the two points *G* and *K*; let the conic section be *GK*. So if the square *EG* is known in magnitude and position, the two conic sections *GK* and *HK* are known in position, so the point *K* is known, and the point *D* is thus known; it is <starting from this last point> that we construct the <solution to> the problem.

---

[1] Lit.: a square with right angles.

[2] Lit.: does not fall on the conic section. This expression, as we know, is a literal translation of the Greek ἀσύμπωτος, from the verb συμπίπτω, to fall, to meet.

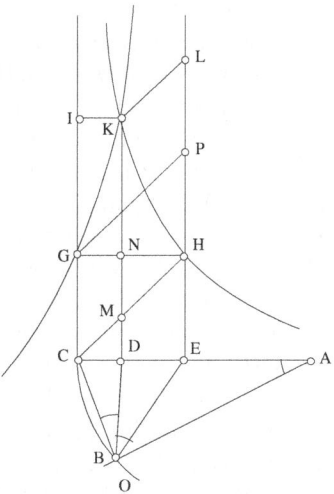

Fig. III.2.2

Let us carry out the synthesis corresponding to this analysis:

Let us suppose we have a known general straight line; let it be *EC*. We construct a square; let it be *EHGC*; we join *CH*, we extend *EH* and *CG*, we cut off <the segment> *HP* equal to *HE* and we join *PG*. We cause to pass through the point *H* a hyperbola that has as asymptotes the straight lines *EC* and *CG*; let the conic section be *HK*. We cause to pass through the point *G* the hyperbola that has as asymptotes the straight lines *CH* and *HP*; this conic section cuts the conic section *HK* because this <former> conic section comes ever closer to the straight line *HL* if we extend *HL*, and the conic section *HK* becomes ever more distant from the straight line *HL* if we extend *HL*. So let the two conic sections cut one another in a point *K*. We draw *KD* parallel to *CG*, *KI* parallel to *CE*, and *KL* parallel to *MH*; we put *CA* equal to *DK*; we take *A* as centre and we draw a circle with distance *AC*; let the circle be *CBO*. We draw *CB* equal to *CE* and we join *AB*, *BD*, *BE*. Since *AC* is equal to *KD*, the product of *AC* and *CD* is equal to the product of *KD* and *DC*, which is equal to the product of *DK* and *KI*, which is equal to the product of *GH* and *HE*, which is equal to the square of *CE*. So the product of *AC* and *CD* is equal to the square of *CE*, by which I mean <that it is equal > to the square of *CB*. Since *KD* is equal to *CA*, and *CD* equal to *DM*, *AD* is equal to *KM*; since the product of *MK* and *KL* is equal to the product of *CG* and *GP*, the product of *KM* and *MH* is equal to the product of *GC* and *CH*, and the ratio of *MH* to *HN* is equal to the ratio of *CH* to *HG*; so the ratio of the product of *KM* and *MH* to the product of *KM* and *HN* is equal to the ratio of the product of *CH* and *HG* to the square of *HG*, which is equal to the ratio of the product of *PG* and *GH* to the square of *GC*. And the product of *KM* and *MH* is equal to the product of *PG* and *GC*; so the product of *KM* and *HN* is equal to the square of *GC*, which is equal to the square of *CE*. *NH* is equal to *DE*, and *KM* is equal to *AD*; thus the product of *AD* and *DE* is equal to the square of *CE*, I mean to say <it is equal> to the square of *CG*. Since the product of *AC* and *CD* is equal to the square of *CB*, the triangle *CBD* is similar to the triangle *ABC*. So the angle *BDC* is equal to the angle *ABC*, and the angle *CBD* is equal to the angle *BAC*; but the angle *ABC* is equal to the angle *ACB*, so the angle *BDC* is equal to the angle *BCD*, so the straight line *BD* is equal to the straight line *BC*, the product of *AD* and *DE* is equal to the square of *DB*, the angle *BED* is equal to the angle *ABD*, and the angle *DBE* is equal to the angle *BAD*. So the angle *DBE* is equal to the angle *CBD*. Since the triangle *ABC* is similar to the triangle *CBD*, the ratio of *AB* to *BC* is equal to the ratio of *BD* to *DC* and *BC* is equal to *BD*, and *BD* is equal to *EC*. So the ratio of *AB* to *BD* is equal to the ratio of *EC* to *CD*, and the ratio of *EC* to *CD* is equal to the ratio of *AC* to *CE*, and is equal to the ratio of *AE*, which remains, to *ED*, which remains. So the ratio of *AB* to *BD* is equal to the ratio of *AE* to *ED*; so the two angles *ABE* and *EBD* are equal; thus the three angles at the point *B* are equal.

So if we take away from the angle *ACB* an angle equal to the angle *CBD* and if we divide the angle that remains into two equal parts, the three angles are equal to the three angles at the point *B*. The angles of the triangle *ABC* will be divided into seven equal angles. So if we construct in the circle a triangle similar to the triangle *ABC*, if we divide the two angles at its base into angles each of which is equal to each of the angles at the point *B* and if we extend the straight lines that divides the two angles as far as the circumference of the circle, the circumference of the circle is divided into seven equal parts; if we put in the chords of the arcs, there is formed in the circle a figure that has seven equal sides and whose angles are equal. In this manner, we can construct in the circle a heptagon with equal sides and equal angles. That is what we wished to construct.

<Second case>

In the same way, we shall consider the isosceles triangle in which each of the angles at the base is two parts, and the angle that remains is three parts, and we shall form the heptagon starting from this triangle.

By the method of analysis:

We suppose that we have found a triangle that shows this property; let it be the triangle *ABC*. Let each of the two angles *B* and *C* be two parts; the angle *A* is three parts. We put the angle *BAD* as two parts. The triangle *ABD* is thus similar to the triangle *ABC*, since the angle *C* is two parts. The product of *CB* and *BD* is thus equal to the square of *BA*. <We extend *CB* to *E*> and we put *BE* equal to *BA*; the product of *CB* and *BD* is equal to the square of *BE*. We join *AE*. The two angles *BAE* and *BEA* are thus equal. So each of them is one single part, because the angle *ABC* is two parts, and the angle *CAD* is one part, since the angle *BAD* is two parts, and the angle *BAC* is three parts; thus the angle *CAD* is equal to the angle *AEC*. So the triangle *ADC* is similar to the triangle *AEC*. So the product of *EC* and *CD* is equal to the square of *AC*, *AC* is equal to *AB*, and *AB* is equal to *BE*; so the product of *EC* and *CD* is equal to the square of *BE*. So the product of *EC* and *CD* is equal to the product of *CB* and *BD*. We erect on the straight line *EB* the perpendicular *EH* and we put *EH* equal to *EB*; we draw from the point *H* the straight line *HI* parallel to the straight line *BE*. We put *HI* equal to *EB*. We join *HB* and *IE*, we extend *HB* on the side of *B*, we erect on the straight line *BE* the perpendicular *BK*, and we put *BK* equal to *BC*. We draw from the point *K* a straight line parallel to the straight line *BC*; let <the straight line> be *KL*. It meets the straight line *HB*; let it meet it at the point *L*. So *LK* is equal to *KB*, because *BE* is equal to *EH*. We draw from the point *D* a straight line parallel to the straight line *BK*; let it be *DG*. It cuts the straight line *BL*; let it cut it at the point *M*. We draw from the point *G* a straight line parallel to the straight line *LH*; let it be *GP*. We put *BN* equal to *BE*, and we draw *NS* parallel to *BK*, and *SO* parallel to *BC*. So <the area> *NO* is equal to the square of *BE* and the product of *BK* and *KG* is equal to the square of *BE*. Hence the product of *DG* and *GK* is equal to the product of *NS* and *SO*. Then the hyperbola that passes through the point *S* and has as asymptotes the two straight lines *DB* and *BO* passes through the point *G*. Let that hyperbola be the conic section *SG*.

Since the ratio of *LB* to *BK*, which is equal to *BC*, is equal to the ratio of *HB* to *BE*, and is equal to the ratio of the whole to the whole,[3] the ratio of *LH* to *EC* is equal to the ratio of *HB* to *BE*, which is equal to the ratio of the product of *HB* and *BE* to the square of *BE*. So the ratio of the product of *LH* and *CD* to the product of *EC* and *CD* is equal to the ratio of the product of *HB* and *BE* to the square of *BE*. But the product of *EC* and *CD* is equal to the square of *BE*. So the product of *LH* and *CD* is equal to the product of *HB* and *BE*, and *CD* is equal to *LG*, *LG* is equal to *HP*, and *LH* is equal to *GP*, so the product of *PG* and *GL* is equal to the product of *HB* and *BE*, that is to say *IE* and *EB*. So the hyperbola that passes through the point *E* and has as asymptotes the two straight lines *LH* and *HI* passes through the point *G*. Let this hyperbola be the conic section *EG*. The point *G* is then the intersection of the two hyperbolas. So if the straight line *BE* is of known magnitude and position, the area *BI* is known in magnitude and shape, and the square *NO* is known in magnitude and shape, so the point *S* will be known; and the two straight lines *KB* and *BC* are known in position, the conic section *SG* is known in position and the two straight lines *HL* and *HI* are known in position. And the point *E* is known <in position>. So the conic section *EG* is known in position and the point *G* is the intersection of two conic sections of known position.

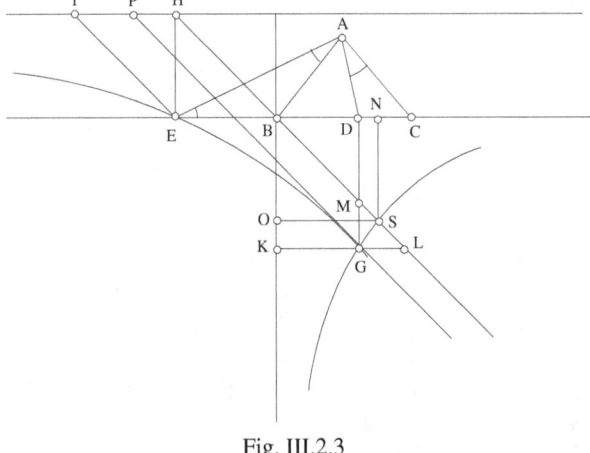

Fig. III.2.3

---

[3] That is to say: the ratio of the sum of *LB* + *BH* to the sum of *BC* + *BE*.

So if we draw from the point *G* the perpendicular *GD*, if we draw the perpendicular *GKL*, and if we put *BC* equal to *LK*, *DC* is then equal to *LG*, hence the product of *CB* and *BD* is equal to the square of *BE*, which is known. And each of the two straight lines *BA* and *AC* is equal to the straight line *BE*, which is known.

Let us carry out the synthesis corresponding to this analysis:

We suppose we have a known straight line; let it be *BE*. And we put *BN* equal to *BE*. We construct on *BN* a square, let it be *BNSO*. We cause to pass through the point *S* the hyperbola that has as asymptotes the two straight lines *NB* and *BO*. Let the conic section be *SG*. We join *BS* and we extend it on both sides, to *H* and to *L*; we draw from the point *E* the perpendicular *EH* and we put it equal to *EB*; we draw *HI* parallel to *BE* and *EI* parallel to *BH*; we cause to pass through the point *E* the hyperbola that has as asymptotes the two straight lines *LH* and *HI*. This conic section cuts the conic section *SG* because it comes ever closer to the straight line *HL*. Let it cut it at the point *G*. We draw from the point *G* the straight line *GD* parallel to the straight line *KB* and we draw *KGL* parallel to the straight line *BD*; we put *DC* equal to *GL*. We have *BC* equal to *KL*, that is to say <equal to> *KB*. So the product of *CB* and *BD* is equal to the square *NO*, which is the square of *BE*. So the product of *PG* and *GL* is equal to the product of *IE* and *EB*; but the ratio of *LB* to *BK*, I mean to say *BC*, is equal to the ratio of *HB* to *HE*, I mean to say *BE*, and is equal to the ratio of *HL* to *EC*; so the ratio of *HB* to *BE*, that is to say the ratio of the product of *IE* and *EB* to the square of *EB*, is equal to the ratio of *HL* to *EC*. So the ratio of the product of *IE* and *EB* to the square of *EB* is equal to the ratio of the product of *HL* and *DC*, to the product of *EC* and *CD*; *CD* is equal to *LG*, and *HL* is equal to *PG*; so the ratio of the product of *PG* and *GL* to the product of *EC* and *CD* is equal to the ratio of the product of *IE* and *EB* to the square of *EB*. But the product of *PG* and *GL* is equal to the product of *IE* and *EB*. So the product of *EC* and *CD* is equal to the square of *BE*. So the product of *EC* and *CD* is equal to the product of *CB* and *BD*; so the ratio of *EC* to *CB* is equal to the ratio of *BD* to *DC*. Now *EC* is greater than *CB*. So *DB* is greater than *DC*. So *BN* is much greater than *DC*. So <the sum> of the two straight lines *BE* and *BN* is much greater than *BC*. So starting from the straight lines *EB*, *BN*, *BC* we can construct a triangle; let this triangle be the triangle *BAC*. Thus each of the two straight lines *BA* and *CA* is equal to the straight line *BE*. So the product of *CB* and *BD* is equal to the square of *BA*.

So the triangle *ABD* is similar to the triangle *ABC*, the angle *BAD* is equal to the angle *ACB* and the angle *ADB* is equal to the angle *BAC*. The product of *EC* and *CD* is equal to the square of *BE*; so it is equal to the square of *CA*. So the triangle *ADC* is similar to the triangle *AEC*, so the angle *CAD* is equal to the angle *AEC* and the angle *ABC* is twice the angle *AEC*, because *AB* is equal to *BE*. So the angle *ABC* is twice the angle *CAD* and the angle *ABC* is equal to the angle *ACD*, so the angle *ACD* is twice the angle *CAD*. So the angle *ADB* is equal to three times the angle *CAD*. Now the angle *ADB* is equal to the angle *BAC*. So the angle *BAC* is equal to three times the angle *CAD*. Now the triangle *ABC* is isosceles; its two equal sides are *AB* and *AC*; so each of the two angles *ABC* and *ACB* is two parts according to the measure that makes the angle *BAC* three parts.

So if we construct in the circle a triangle similar to triangle *ABC*, if we divide each of the two angles at the base into two equal parts, and if we cut off from the angle at the vertex an angle equal to an angle at the base, which we divide into two equal parts, the angles of the triangle are divided into seven equal parts. If we draw the straight lines that cut off the angles as far as the circumference of the circle, the circumference of the circle is divided into seven equal parts. If we put in their chords, there is formed a heptagon with equal sides and equal angles. That is what we wished to construct.

<Third case>

Similarly, we suppose we have an isosceles triangle in which each of the angles at the base is a single part and the angle at the vertex is five parts. We form the heptagon starting from this triangle.

By the method of analysis:

We suppose we have found a triangle that shows this property. Let the triangle be *ABC*. Let each of the two angles *ABC* and *ACB* be a single part. The angle *BAC* is five parts. We put the angle *CAD* equal to the angle *ABC*. We also put the angle *DAE* equal to the angle *ABC*. Since the angle *CAD* is equal to the angle *ABC*, the triangle *ACD* is similar to the triangle *ABC*; hence the ratio of *BC* to *CA* is equal to the ratio of *AC* to *CD*. So the product of *BC* and *CD* is equal to the square of *CA*. But *CA* is equal to *AB*. So the product of *BC* and *CD* is equal to the square of *AB*. Since the angle *DAE* is equal to the angle *ABD*, the triangle *ADE* is similar to the triangle *ABD*. So the product of *BD* and *DE* is equal to the square of *DA*. But *DA* is equal to *DC*, because the angle *CAD* is equal to the angle *ACD*. So the product of *BD* and *DE* is equal to the square of *DC*. Since each of the two angles *CAD* and *DAE* is equal to the angle *ABD*, which is equal to the angle *ACD*, the angle *AEB* is three times the angle *ACB*; the angle *BAC* is five times the angle *ACB* and the angle *EAC* is twice the angle *ACB*. So the angle *BAE* is three times the angle *ACB*. So the angle *BAE* is equal to the angle *AEB*. Hence the straight line *AB* is equal to the straight line *BE*. So the product of *BC* and *CD* is equal to the square of *EB*.

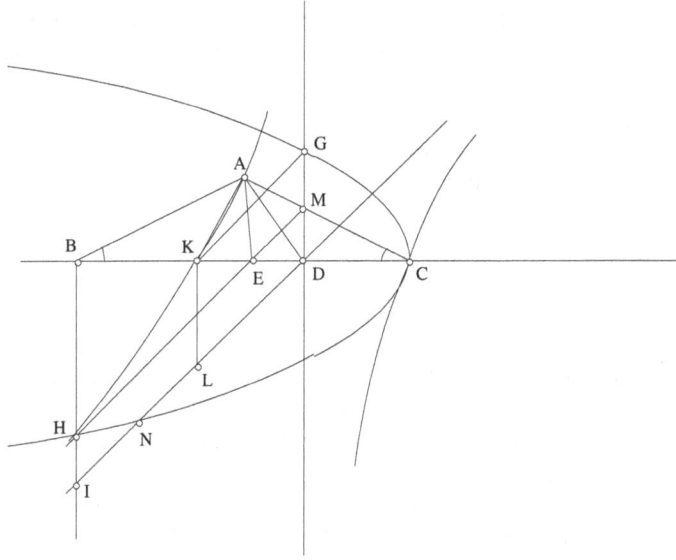

Fig. III.2.4

We put *DK* equal to *DC* and at the point *K* we erect the perpendicular *KL*; we put it equal to *KD*; also, at the point *D* we erect the perpendicular *DG* and we put it equal to *DK*. We join *GK* and *DL* and at the point *B* we erect the perpendicular *BH*; we put it equal to *BE*; we join *HE* and we extend it to *M*. *DM* is then equal to *DE*. We extend the straight line *DL* until it meets the straight line *BH*; let it meet it at the point *I*. Since *HB* is parallel to *MD*, the ratio of *HE* to *EB* is equal to the ratio of *ME* to *ED* and is equal to the ratio of *HM* to *BD* and the ratio of *HE* to *EB* is equal to the ratio of *GK* to *KD*. So the ratio of *HM* to *BD* is equal to the ratio of *GK* to *KD*. Now the ratio of *HM* to *BD* is equal to the ratio of the product of *HM* and *ED* to the product of *BD* and *ED*. So the ratio of the product of *HM* and *ED* to the product of *BD* and *ED* is equal to the ratio of *GK* to *KD*, that is to say to the ratio of *GK* to *KL*, which is equal to the ratio of the product of *GK* and *KL* to the square of *KL*. But the product of *BD* and *ED* is equal to the square of *DC*, which is equal to *KL*. So the product of *HM* and *ED* is equal to the product of *GK* and *KL*. But *ED* is equal to *DM* and *DM* is equal to *HI*. So the product of *HM* and *HI* is equal to the product of *GK* and *KD*. So the hyperbola that passes through the point *K* and has as asymptotes the two straight lines *GD* and *DI* passes through the point *H*. Let this hyperbola be the conic section *KH*. Since the product of *BC* and *CD* is equal to the square of *EB* and *EB* is equal to *BH*, the parabola whose axis is *BC* and *latus rectum DC*, and with vertex at the point *C*, passes through the point *H*. Let this parabola be the conic section *CH*. The point *H* is thus the intersection of these two conic sections. So if *DC* is known, the two conic sections will be known in position, the point *H* will be known, and the two points *E* and *B* will be known.

Let us carry out the synthesis corresponding to this analysis:
We suppose we have a given straight line; let it be *CK*. We divide it into two equal parts at the point *D*, and at the two points *D* and *K* we erect two perpendiculars; let them be *DG* and *KL*. We put each of the <straight lines> *DG* and *KL* equal to *KD*. We join *GK* and *DL* and we extend *DL* to *I*. We cause to pass through the point *K* the hyperbola that has as asymptotes the two straight lines *GD* and *DI*; let it be the conic section *KH*. We extend *DK* on the side of *K*. We draw through the point *C* the parabola whose axis is *CK*, its vertex the point *C*, and with *latus rectum* the straight line *CD*; let the conic section be *CH*. This conic section cuts the straight line *DI* because any straight line that cuts the axis of the parabola cuts the parabola[4] in two points on opposite sides of the axis. So the conic section *CH* cuts the straight line *DI*; if it then goes beyond the straight line *DI*, it becomes more distant from the straight line *DI*, because the straight line drawn from the point of intersection to be a tangent to the conic section cuts the straight line *DI*. If the curve is extended on both sides, it becomes more distant from the straight line *DI*. The conic section <*KH*> cuts the parabola *CH* below the tangent; as it becomes more distant from the point of intersection, it becomes more distant from the straight line *DI*. The conic

---

[4] Lit.: the circumference of the section.

section *KH*, as one extends it, approaches more closely to the straight line *DI*. So it necessarily follows that the two conic sections cut one another; let the two conic sections cut one another at the point *H*. We draw *HB* perpendicular to the axis of the parabola, and we draw, also from the point *H*, a straight line parallel to the straight line *GK*; let it be *HEM*. Each of the two triangles *HBE* and *EDM* is thus similar to the triangle *DKL*. *HB* is accordingly equal to *BE*, and *ED* equal to *DM*. Hence the ratio of *HE* to *EB* is equal to the ratio of *LD* to *DK*, <that is> equal to the ratio of *ME* to *ED* and equal to the ratio of *HM* to *BD*. So the ratio of *HM* to *BD* is equal to the ratio of *LD* to *DK*, that is to the ratio of *GK* to *KL*. So the ratio of the product of *HM* and *ED* to the product of *BD* and *DE* is equal to the ratio of *GK* to *KL*, which is equal to the ratio of the product of *GK* and *KL* to the square of *KL*. But *ED* is equal to *DM* and *DM* is equal to *HI*. So the ratio of the product of *HM* and *HI* to the product of *BD* and *DE* is equal to the ratio of the product of *GK* and *KL* to the square of *KL*. Now the product of *HM* and *HI* is equal to the product of *GK* and *KL*. So the product of *BD* and *DE* is equal to the square of *KL*, that is to the square of *DK*. But *DK* is equal to *DC*. So the product of *BD* and *DE* is equal to the square of *DC*. Since *KC* is twice *CD*, the product of *KC* and *CD* is equal to twice the square of *KL*. So the point *L* lies inside the parabola.[5] So the parabola[6] cuts the straight line *DI* beyond the point *L*. The point *H* is thus situated beyond the point *L*. So the straight line *HB* lies beyond the straight line *KL*. So the straight line *BD* is greater than the straight line *DK*. But the product of *BD* and *DE* is equal to the square of *DK*. So *DE* is smaller than *DK*. So it is smaller than *DC*. Hence *EC* is smaller than twice *DC*. And the product of *BC* and *CD* is equal to the square of *HB*. But *HB* is equal to *BE*. So the product of *BC* and *CD* is equal to the square of *EB*. So the product of *BC* and *CE* is smaller than twice the square of *EB*. So *EC* is smaller than *EB*. So twice *EB* is greater than *BC*. So it is possible to construct on the straight line *BC* an isosceles triangle such that its base is the straight line *BC* and each of the two remaining sides is equal to *BE*. Let this triangle be the triangle *ABC*. We join *AD* and *AE*. Since *AC* is equal to *EB*, the product of *BC* and *CD* is equal to the square of *CA*; so the triangle *ACD* is similar to the triangle *ABC*. So the ratio of *BC* to *CA* is equal to the ratio of *AC* to *CD*. So the angle *CAD* is equal to the angle *ABC* which is equal to the angle *ACB*. So the angle *CAD* is equal to the angle *ACB*. So the straight line *AD* is equal to

---

[5] Lit.: the section.
[6] Lit.: the section.

the straight line *DC*. Thus the product of *BD* and *DE* is equal to the square of *DA*. The triangle *ADE* is accordingly similar to the triangle *ABD*. So the angle *DAE* is equal to the angle *ABD*, which is equal to the angle *ACD*. So, depending on the magnitude of which the angle *ACB* is a single part, the angle *AEB* is three parts. Since *AB* is equal to *BE*, the angle *BAE* is equal to the angle *BEA*. Thus the angle *BAE* is three parts of the magnitude of which the angle *ACB* is a single part. And the angle *CAE* is two parts <measured> in these parts. Thus the angle *BAC* is five parts <measured> in the parts in which each of the two angles *ABC* and *ACB* is a single part. So if we construct in the circle a triangle similar to the triangle *ABC*, and if we cut the angle *BAC* into <five> angles each of which is equal to the angle *ABC*, the angles of the triangle are divided into seven equal parts. If we draw the straight lines <of a length> such that they meet the circumference of the circle, the circle is divided into seven equal parts. If we put in the chords, there is formed in the circle a heptagon with equal sides and equal angles. That is what we wished to construct.

<Fourth case>

Similarly, we suppose <we have a> triangle one of whose angles is a single part, another <angle> two parts, and the remaining one four parts, and we form the heptagon starting from this triangle.

By the method of analysis:

We suppose that we have found a triangle that shows this property; let the triangle be *ABC*. Let its angle *A* be a single part, its angle *B* two parts, its angle *C* four parts. We put the angle *BCD* as a single part; the angle *ACD* will then be three parts and the angle *ADC* is also three parts, because it is equal to the sum of the angles *ABC* and *BCD*. So the triangle *ACD* is the first of the triangles that we have found.[7] So if we find the first triangle, it will be similar to the triangle *ADC*, and if we put the angle *DCB* equal to the angle *CAD*, then the angle *ACB* will be four parts and the angle *ABC* two parts. If we also put the angle *BCE* equal to two parts, the angle *CEB* is then three parts, because the angle *EBC* is two parts. The triangle *BEC* will then be the second of the triangles that we found. If we put the angle *ECA* equal to the angle *ECB*, the angle *ACB* will then be four parts and the angle *CAB* will be a single part. If we also put the angle *ACG* equal to the angle

---

[7] See Fig. 3.67 in this chapter, p. 401.

CAG, then the angle *GCB* will be three parts, and the angle *AGC* will thus be five parts, because it is the sum of the two angles *GCB* and *CBG*. The triangle *AGC* will then be the third of the triangles that we found. If we put the angle *GCD* equal to the angle *CGD*, which is two parts, because it is the sum of the two angles *ACG* and *CAG*, then the angle *ACD* will be three parts.[8] If we next put the angle *DCB* equal to the angle *CAG*, then the angle *ACB* will be four parts and the angle *CAB* will be a single part. So the angle *ABC* will be two parts and the triangle *ABC* reduces to each of the three triangles that we have already established. If we wish to construct the heptagon starting from the triangle one of whose angles is a single part, the second two parts and the third four parts, we find one of the preceding triangles and we increase one of its angles by the extra quantity to which we have just referred. In this way we find the triangle one of whose angles is a single part, the second two parts and the third four parts.

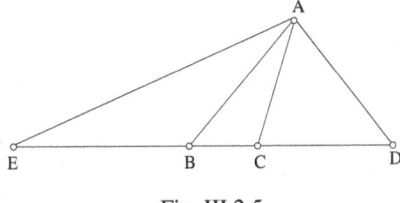

Fig. III.2.5

But it is possible to construct this triangle without reducing it to one of the preceding triangles. So let us return to the triangle. We draw *BC* <produced> in both directions; we put *CD* equal to *CA*, *BE* equal to *BA*, and we join *AE* and *AD* <Fig. III.2.5>.

Since the angle *ACB* is four parts, *ADC* is two parts, the angle *CAD* is two parts and the angle *ABC* is two parts. So the angle *BAD* is three parts, and the two angles *ABD* and *ADB* are equal. So the straight line *AD* is equal to the straight line *AB*. Now *AB* is equal to *BE*. So *AD* is equal to *BE*. Since the angle *ABC* is two parts, the angle *AEB* is a single part, so the angle *AEC* is equal to the angle *BAC*. So the triangle *ABC* is similar to the triangle *AEC*. So the product of *EC* and *CB* is equal to the square of *CA*. But *CA* is equal to *CD*. Hence the product of *EC* and *CB* is equal to the square of *CD*. Since *AC* is equal to *CD*, then the angle *DAC* is equal to the angle *ADC*; but the angle *ACB* is four parts, then <it follows that> the angle

---

[8] See Fig. 3.68 in this chapter, p. 401.

DAC is two parts; but the angle *ABC* is two parts, so the angle *DAC* is equal to the angle *ABC*. So the triangle *ADC* is similar to the triangle *ABD*. So the product of *BD* and *DC* is equal to the square of *DA*. But *DA* is equal to *BE*; so the product of *BD* and *DC* is equal to the square of *BE*. So the straight line *ED* is divided into three parts, the product of *BD* and *DC* is equal to the square of *BE* and the product of *EC* and *CB* is equal to the square of *CD*. The straight line <*DE*> divided in this ratio is the line that completes Archimedes' lemma. This is the straight line that was divided by Abū Sahl al-Kūhī and starting from which he built up the triangle one of whose angles is a single part, the second two parts, and the third four parts, and starting from this he determined the heptagon.

We divide this straight line by a method different from the method by which it was divided by Abū Sahl. We shall prove how it is divided, first by analysis:

We put *CK* equal to *CD*, at the point *K* we erect the perpendicular *KG* and let us put it equal to *KC*. We draw from the point *G* a straight line parallel to *KC*; let it be *GI*. We put *GI* equal to *GK*. Let us join *GC* and *IK*. At the point *C* let us erect the perpendicular *CL* <perpendicular to the straight line *BC*. We draw *BH* perpendicular to *BC*, and we put *BH* equal to *BE*. The straight line *BH* cuts *GC* at the point *M*>. Through the point *D*, we construct the parabola with axis the straight line *DB*, and *latus rectum DC*. <Since *HB* is equal to *EB*, and the square of *EB* is equal to the product of *BD* and *DC*, the square of *HB* is equal to the product of *BD* and *DC*. So the parabola passes through the point *H*>; let it be the conic section *DH*. We draw the hyperbola that passes through the point *K* and has as asymptotes the two straight lines *GC* and *CL*. <Since *KG* is equal to *KC*, *BM* is equal to *BC*; so *HM* is equal to *EC*. But the product of *EC* and *CB* is equal to the square of *CD*. So the product of *MH* and *CB* is equal to the product of *KG* and *KC* and the ratio of *HL* to *BC* is equal to the ratio of *MC* to *BC*, which is equal to the ratio of *GC* to *KC*. So the ratio of the product of *HL* and *MH* to the product of *BC* and *MH* is equal to the ratio of the product of *GC* and *KG* to the product of *KC* and *KG*. So the product of *MH* and *HL* is equal to the product of *GC* and *KG*. So the point *H* lies on the hyperbola>.

*This conic section cuts the conic section *DH* because this conic section, that is the hyperbola, comes ever closer to the straight line *CL*, and the parabola cuts *CL* and then goes beyond it and becomes more distant from it. Let the two conic sections cut one another at the point *H*; so the point *H* lies beyond the straight line *CL*, that is beyond the point *L*, because the hyperbola is always beyond the straight line *CL*.*[9]

We draw from the point *H* the perpendicular *HB* and we draw *HE* parallel to the straight line *GC*. So if the straight line *CD* is known, *CK* will be known in magnitude and in position, the figure *KGI* will be known in magnitude and in shape and the point *K* will be known; then the hyperbola will be known in position. Since *CD* is known in magnitude, the parabola is known in position. So the point *H* is known and the point *B* is known. It is <starting from this point> that the triangle is constructed.

Let us carry out the synthesis corresponding to this analysis:

We suppose we have a known straight line; let it be *KD*. We divide it into two equal parts at the point *C*; at the point *K* we erect the perpendicular *KG* and we put it equal to *KC*. We draw from the point *G* a straight line parallel to the straight line *KC*; let it be *GI*. We put *GI* equal to *KC*. We join *GC* and *IK*. We draw from the point *C* the perpendicular *CL* and we cause to pass through the point *K* the hyperbola that has as asymptotes the two straight lines *GC* and *CL*. We cause to pass through <the point> *D* the parabola with axis *KD* and *latus rectum CD*. This conic section cuts the hyperbola for the reason we have mentioned earlier. Let them cut one another at the point *H*. We draw the perpendicular *HB*, and we draw *HE* parallel to *GC*. We extend *HB* as far as *M*. So *HB* is equal to *BE* and *BM* to *BC*. So *HM* is equal to *EC*, and *HL* to *MC*. So the product of *EC* and *CM* is equal to the product of *HM* and *MC*. But the product of *HM* and *MC* is equal to the product of *IK* and *KC*. Now the ratio of *MC* to *CB* is equal to the ratio of *GC* to *CK*, that is <equal> to the ratio of *IK* to *KC*, which is equal to the ratio of the product of *IK* and *KC* to the square of *KC*. So the ratio of the product of *EC* and *CM* to the product of *EC* and *CB* is equal to the ratio of the product of *IK* and *KC* to the square of *KC*. But the product of *EC* and *CM* is equal to the product of *IK* and *KC*. So the product of *EC* and *CB* is equal to the square of *CK*, that is <to the square> of *CD*. But the product of *BD* and *DC* is equal to the square of *HB*. And *HB* is equal to *BE*; so the product of *BD* and *DC* is equal to the square of *BE*. Thus we have divided the straight line *ED* into three parts such that the product of *EC* and

---

[9] In the preceding paragraph we have, as an exception, used <...> to separate off the sentences that we added when establishing the Arabic text. The paragraph enclosed between *...* should have taken its place in the synthesis, not in the analysis.

CB is equal to the square of *CD*, and the product of *BD* and *DC* is equal to the square of *BE*.

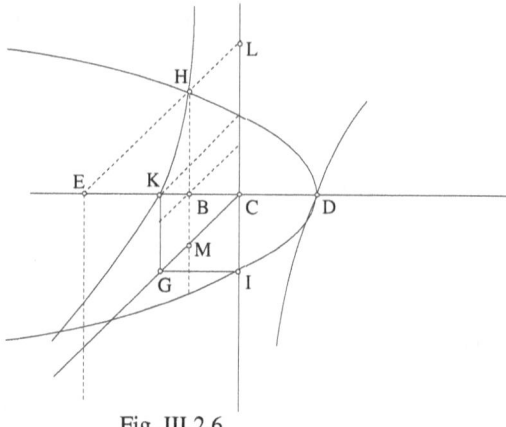

Fig. III.2.6

Since the product of *EC* and *CB* is equal to the square of *CD*, *CD* is greater than *CB*, and *EC*, which is the sum of *EB* and *BC*, is greater than *CD*. Since the product of *BD* and *DC* is equal to the square of *BE*, *BE* is greater than *CD*, and *BD*, which is the sum of *BC* and *CD*, is greater than *BE*. So the sum of any two of the straight lines *EB*, *BC*, *CD* is greater than the remaining straight line. So it is possible to construct a triangle from these three straight lines. Let this triangle be the triangle *ABC*. Let *AB* be equal to *BE* and *AC* equal to *CD*. We join *AE* and *AD*. So the product of *EC* and *CB* is equal to the square of *CA*. So the ratio of *EC* to *CA* is equal to the ratio of *AC* to *CB*. So the two triangles *ACB* and *AEC* are similar. So the angle *CAB* is equal to the angle *AEB*, which is half the angle *ABC*. Since the product of *BD* and *DC* is equal to the square of *BA* and *DC* is equal to *CA*, the product of *BC* and *CD* plus the square of *CA* is equal to the square of *BA*. Accordingly, the triangle *ABD* is isosceles; so the straight line *DA* is equal to the straight line *BA*. So the product of *BD* and *DC* is equal to the square of *DA*. The triangle *ADC* is thus similar to the triangle *ABD*; so the angle *DAC* is equal to the angle *ABD*, which is equal to the angle *ADB*. Each of the two angles *ADC* and *CAD* is two parts in the measure by which the angle *BAC* is a single part. The angle *ACB* is four parts, in the measure by which the angle *BAC* is a single part. So if we divide the angle *ACB* into two equal parts, and we divide each half into two equal parts, the angles of the triangle are <thus> divided into seven equal parts. If we draw the straight lines used to divide the angles <extending them to reach> the circumference of the circle, the circumference of the circle is divided into seven equal parts. If we put in their chords, there is formed in the circle a heptagon with equal sides and equal angles.

We have constructed in the circle a heptagon with equal sides and equal angles in all the cases for which we can construct the heptagon. Such was our purpose in this treatise.

In the name of God, the Compassionate, the Merciful
Glory to God

TREATISE BY IBN AL-HAYTHAM,
SHAYKH ABŪ AL-ḤASAN IBN AL-ḤASAN IBN AL-HAYTHAM

On the Division of the Straight Line Used by Archimedes in Book Two of *The Sphere and the Cylinder*

He says: in the fourth proposition of the second book of his treatise on *The Sphere and the Cylinder*, Archimedes used a straight line that he supposed to have been divided in a particular ratio, without showing how to divide the straight line in this ratio. Since the division of the straight line cannot be carried out except by employing conic sections and in his book he makes no use of conic sections, he did not take the decision to incorporate into his book material that did not belong in a work of this type; so he treated the division of the straight line as an assumption, relying on the fact that it is possible. But insofar as we do not carry out the division of the line in the ratio he assumed, the proof of the proposition in which he used it is not complete.

As this is so, we decided to divide this line in the ratio he assumed and to demonstrate the possibility of the division, in order to make it clear that what Archimedes used was correct. The division employed by Archimedes is that he assumed we have a line on which we have DB. He took each of <the straight lines> DB and BG as known. He assumed the ratio of BG to BI was known. He then said: we put the ratio of HG to GI equal to the ratio of the square of BD to the square of DH. We assume we have a line such as he assumed and we set about dividing it.

At the two points D and G we erect two perpendiculars, let them be DA and GC. We put each of them equal to the known straight line BD. We join AC, which is thus perpendicular to the straight line AD. We draw IE parallel to the straight line CG. We construct through the point E the hyperbola that has as asymptotes the two straight lines CG and GD; let the conic section be *UEK*. This conic section cuts the straight line *AD* because on the side of *K* it extends to infinity and does not meet the straight line *GD*; let it cut the straight line *AD* at the point *K*.

We then construct through the point *D* the parabola whose axis is *DA*, its vertex the point *D* and *latus rectum* the straight line *DB*; let the conic section be *DSO*. This conic section cuts the straight line *AC* if we assume the straight line *AC* has been extended, because on the side of *O* the conic section *DSO* extends to infinity and the straight line *AC* is perpendicular to its axis. Let this conic section cut the straight line *AC* in a point *S*.

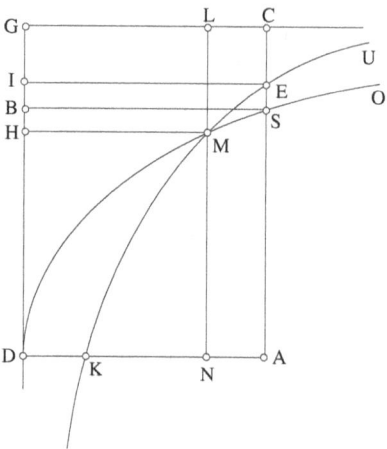

Fig. III.3.1

Since the conic section *DSO* is a parabola with axis *DA* and *latus rectum DB*, the square of *AS* is equal to the product of *DB* and *DA*. Now *DB* is equal to *DA*, so the square of *AS* is equal to the square of *DB*. Thus the straight line *AS* is equal to the straight line *DB*. Now the straight line *AE* is equal to the straight line *DI*, so the straight line *AE* is greater than the straight line *AS*. So the point *E* lies outside the conic section *DSO* and the point *K* lies inside the conic section *DSO*, because it lies on its axis. So part of the hyperbola *UEK* lies outside the parabola *DSO* and part lies inside the conic section *DSO*. So the conic section *DSO* cuts the conic section *KEU*; let it cut it in a point *M*. <From *M*> we draw *MH* perpendicular to the straight line *DB*.

I say that the ratio of HG *to* GI *is equal to the ratio of the square of* BD *to the square of* DH.

Proof: We construct through the point *M* the straight line *LMN* parallel to the straight line *GD*, then *MN* is perpendicular to the straight line *DA*. The straight line *MH* is parallel to the straight line *CG*, because it is perpendicular to the straight line *GD*. So the product of *BD* and *DN* is equal to the square of *MN*, *NM* is equal to *DH* and *ND* is equal to *MH*. The product of *BD* and *MH* is equal to the square of *DH*. So the ratio of *BD* to *DH* is equal to the ratio of *DH* to *MH*. The ratio of *BD* to *MH* is equal to the ratio of the square of *BD* to the square of *DH*. Similarly, since the conic section *UEK* is a hyperbola, and the two straight lines *CG* and *GD* are its two asymptotes, and the two straight lines *EI* and *MH* are parallel to the straight line *CG* and the two straight lines *ML* and *EC* are parallel to the straight line *GD*, <accordingly> the product of *CE* and *EI* is equal to the product of *ML* and *MH*. So the ratio of *ML* to *EC* is equal to the ratio of *EI* to *MH*. Now *ML* is equal to *HG*, *EC* is equal to *GI* and *EI* is equal to *BD*, so the ratio of *HG* to *GI* is equal to the ratio of *BD* to *MH*.

We have shown that the ratio of *BD* to *MH* is equal to the ratio of the square of *BD* to the square of *DH*. So the ratio of *HG* to *GI* is equal to the ratio of the square of *BD* to the square of *DH*. That is what we wanted to prove.

In the name of God, the Compassionate, the Merciful
Glory to God

TREATISE BY AL-ḤASAN IBN AL-ḤASAN IBN AL-HAYTHAM

On a Solid Numerical Problem

We wish to divide up a known number into two parts such that one of them is the cube of the other.

Let the known number be *AB*; we wish to divide up *AB* into two parts such that one of the two parts is the cube of the other. So let us find four magnitudes in continued proportion such that the ratio of the amount by which the greatest of them exceeds the one that follows it, to the smallest, is equal to the ratio of the cube of the number *AB* to the number *AB*.

Let the four magnitudes be the magnitudes *CD*, *DE*, *DG*, *DH* and let the ratio of *CE* to *HD* be equal to the ratio of the cube of *AB* to *AB*, which is a known ratio, because *AB* and its cube are each known.

As for the manner of finding these magnitudes, we shall demonstrate that later. If we find the magnitudes that are in these ratios, then we are dividing up the number *AB* at the point *I* so that the ratio of *AI* to *IB* is equal to the ratio of *CE* to *ED*.

I say that AI *is then the cube of* IB.

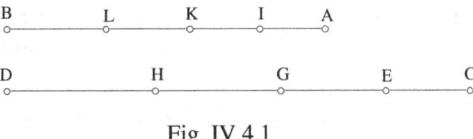

Fig. IV.4.1

Proof: We put the ratio of *IB* to *BK* equal to the ratio of *ED* to *DG* and the ratio of *KB* to *BL* equal to the ratio of *GD* to *DH*. But the ratio of *AI* to *IB* was equal to the ratio of *CE* to *ED*, so, by composition, the ratio of *AB* to BI is equal to the ratio of CD to DE. So we have AB, BI, BK, BL in the same ratios as CD, DE, DG, DH; but CD, DE, DG, DH are in continued proportion, so the numbers AB, BI, BK, BL are in continued proportion. The ratio of AB to BL is equal to the ratio of AB to BI, multiplied by itself three times, and the ratio of AB to BI, multiplied by itself three times, is the ratio of the cube of AB to the cube of BI. But the ratio of AI to IB is equal to the ratio of CE to ED and the ratio of IB to BL is equal to the ratio of ED to DH, so, by <this> equality, the ratio of AI to LB is equal to the ratio of CE to HD. But the ratio of CE to HD is equal to the ratio of the cube of AB to the number AB. So the ratio of the cube of AB to the <number> AB is equal to the ratio of AI to LB; by permutation, the ratio of the cube of AB to AI is equal to the ratio of AB to LB. But it has been shown that the ratio of AB to LB is equal to the ratio of the cube of AB to the cube of BI, so the ratio of the cube of AB to AI is equal to the ratio of the cube of AB to the cube of BI; so AI is the cube of BI.

So we have divided a number AB into two parts such that one of the two parts is the cube of the other, let them be AI and IB. That is what we wished to do.

So as for the manner of finding four magnitudes in continued proportion such that the ratio of the amount by which the greatest of them exceeds the one that follows it to the smallest, is equal to a known ratio, this <procedure> will be as we shall describe.

We suppose <we have> a known straight line AB and at its endpoint we erect a perpendicular; let it be BC. We put the ratio of AB to BD equal to the known ratio and we construct at the point D a straight line parallel to the straight line AB; let it be DE. We draw <the straight line> AE parallel to the straight line BC. The area $ABDE$ is rectangular. We construct through the point E the hyperbola that does not meet the straight lines AB and BC, as has been shown in the fourth proposition of the second book of Apollonius' *Conics*; let the conic section be EGH. We extend the straight line CB in the direction of B and we cut off from it the straight line BN equal to the straight line BD. We construct at the point N a straight line parallel to the straight line AB, let it be NU; we extend it in the other direction as far as M. So the angle BNU is a right angle. Let us construct through the point N the parabola whose axis is NU and *latus rectum NB*, as has been shown in Proposition 52 of Book I of Apollonius' *Conics*; let the conic section be NO. The conic section NO cuts the conic section EGH, since the further we extend the conic section NO in the direction of O, the more distant it becomes from the axis *NU*; so it becomes more distant from the straight line *AB*; and the further we extend the conic section *EGH* in the direction of *E*, the closer it becomes to the straight line *BA*, as has been shown in Proposition 14 of Book II of the *Conics*. So the conic section *NO* cuts the conic section *EGH*; let it cut it at the point *G*.

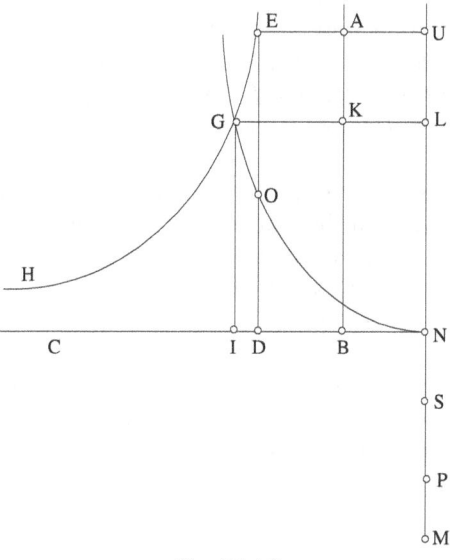

Fig. IV.4.2

We draw from the point *G* the straight line *GI* parallel to the straight line *AB* and the straight line *GKL* parallel to the straight line *CBN*. The area enclosed by the two straight lines *IG* and *GK* is equal to the area enclosed by the two straight lines *DE* and *EA*, as has been shown in Proposition 12 of Book II of the *Conics*. So the ratio of *AB* to *GI* is equal to the ratio of *GK* to *BD*. But *BD* is equal to *BN* and *BN* is equal to *KL*, so the ratio of *AB* to *GI* is equal to the ratio of *GK* to *KL*. We extend *EA* to *U*, so *UN* is equal to *AB*. We put *NM* equal to *GI*, the ratio of *AB* to *GI* is equal to the ratio of *UN* to *NM*. But the ratio of *AB* to *GI* is equal to the ratio of *GK* to *KL*, so the ratio of *UN* to *NM* is equal to the ratio of *GK* to *LK*. But given that the straight line *GL* is parallel to the straight line *IN*, we have that the straight line *GL* is perpendicular to the straight line *NU*. The area enclosed by the straight line *NL* and the straight line *LK*, which is equal to the straight line *NB*, is equal to the square of *LG*, as has been shown in Proposition 12 of

Book I of the *Conics*.[1] So the ratio of *GL* to *LN* is equal to the ratio of *LK* to *LG*. But it has been shown that the ratio of *UM* to *MN* is equal to the ratio of *GL* to *LK*, so the ratio of *UM* to *MN* is equal to the ratio of *NL* to *LG* and to the ratio of *GL* to *LK*. But *UM* is greater than *LN*, *NL* is equal to *GI* and *GI* is equal to *NM*; so *NM* is equal to *NL*, so *NM* is greater than *LG*. We cut off *MS* equal to *LG* and we cut off *MP* equal to *LK*. So the ratio of *UM* to *MN* is equal to the ratio of *MN* to *MS* and equal to the ratio of *MS* to *MP*. So the magnitudes *UM*, *MN*, *MS*, *MP* are in continued proportion; *UN* is equal to *BA* and *MP* is equal to *LK*, which is equal to *BN*, which is equal to *BD*. So the ratio of *UN* to *PM* is equal to the ratio of *AB* to *BD* which is the known ratio.

The magnitudes *UM*, *MN*, *MS*, *MP* are four magnitudes in continued proportion and the ratio of the amount by which the greatest, which is *UM*, exceeds the one that follows it, which is *MN* – by this excess amount I mean *UN* – to *PM*, which is the smallest magnitude, is the known ratio. That is what we wished to do.

The treatise on the solid numerical problem is completed.

Thanks be given to God, Lord of all the worlds, and may a blessing be upon Muḥammad and those that are his.

---

[1] This is Proposition 11 in Heiberg's edition.

CHAPTER IV

PRACTICAL GEOMETRY: MEASUREMENT

4.1. INTRODUCTION

Long before Ibn al-Haytham, al-Fārābī had proposed a classification of sciences that could not be reduced to the classifications known before him.[1] This classification was clearly intended to give an account of the knowledge of the period, to devise a coherent representation of it, and, above all, to explain the new relationships between disciplines. Speaking in terms of the *quadrivium* could not provide what was required. Perhaps that is why al-Fārābī's successors, among them even Avicenna, adopted this new classification.[2] One of its distinguishing characteristics, as indeed of all the later classifications it inspired, is that it includes a complex grouping of several disciplines, whose designation is the significant term: *'ilm al-ḥiyal* ('science of ingenious procedures'). We have here a complex of disciplines that, for the most part, belong among what would be known, much later, as the 'mixed sciences' or 'mixed mathematics', which are 'mixed' in the sense that in them mathematics is combined with elements that relate to physical matter. The concept underlying these disciplines is that they involve 'knowledge' (*'ilm*) and 'action' (*'amal*); that is 'science' and 'art', the categories being inclusive rather than exclusive. On one hand, it is possible to introduce the 'rules of the art' together with its instruments, when we are concerned with defining the specific subject under investigation; on the other hand, this body of knowledge can be applied in the study of objects that lie outside it. Thus if

---

[1] Al-Fārābī, *Iḥsā' al-'ulūm*, ed. 'U. Amīn, 3rd ed., Cairo, 1968.

[2] Ibn Sīnā, 'Épître sur les parties des sciences intellectuelles', transl. R. Mimoune, in *Études sur Avicenne*, dirigées par J. Jolivet et R. Rashed, Collection Sciences et philosophie arabes - Études et reprises, Paris, 1984, pp. 143–51 and R. Rashed, 'Mathématiques et philosophie chez Avicenne', in *Études sur Avicenne*, pp. 29–39; reprinted in *Optique et Mathématiques: Recherches sur l'histoire de la pensée scientifique en arabe*, Variorum reprints, Aldershot, 1992, XV.

knowledge aims at action, action must in turn be based on knowledge. Again in accord with this new representation, a body of knowledge can henceforth take on the status of a science without conforming to either the Aristotelian or the Euclidean model. This new relationship between knowledge and action, between science and art, removed the lines of demarcation that a certain kind of Aristotelianism had, at least in theory, established between the two, and conferred a recognized status on the application of mathematics and the sciences.

Embedded in this complex of the 'science of ingenious procedures' we find many disciplines, among them the art of measurement or mensuration. Of course, this art of measurement, which had been studied well before the time of al-Fārābī, had already given rise to a very rich literature. Thus, in his *Fihrist*, al-Nadīm mentions a number of names, such as those of al-Ḥasan ibn al-Ṣabāḥ, Ibn Nājiya, Ibn Barza and others. We can add those of al-Khwārizmī himself, al-Kindī, Abū Kāmil, al-Qabīṣī, Sinān ibn al-Fatḥ and others, and, a little later, 'Abd al-Qāhir al-Baghdādī, al-Isfizārī, al-Karajī and his successors, to mention only a few. So the list is a rather long one and we should require at least one substantial volume if we were to examine all these writings on mensuration. Here we shall simply describe the various types of works into which the literature may be divided.

Euclid's *Optics*[3] provides the earliest example of the first type of investigation, that of a three-dimensional (stereometric) problem. The matter was taken up by al-Kindī,[4] and, later on, by Sinān ibn al-Fatḥ, under the revealing title *al-misāḥāt al-manāẓiriyya* ('optical mensuration'). The problem is then studied by al-Qabīṣī,[5] and no doubt one could find many others. Ibn al-Haytham is concerned with it in an independent text, a short essay called *On Knowing the Height of Upright Objects, on the Altitude of Mountains and the Height of Clouds*.[6] Another example of the same type, but with no proof, is the

---

[3] Euclide, *L'Optique et la Catoptrique*, Works translated for the first time from Greek into French by Paul Ver Eecke, Paris, 1959, Propositions 18 to 20.

[4] *Sur la rectification des erreurs* [On the Correction of Errors], in R. Rashed, *L'Optique et la Catoptrique d'al-Kindī*, Leiden, 1996, Propositions 19 to 22.

[5] A. Anbouba, 'Un mémoire d'al-Qabīṣī (4ᵉ siècle H.) sur certaines sommations numériques', *Journal for the History of Arabic Science*, 6, 1982, pp. 181–208, esp. pp. 188–9.

[6] R. Rashed, *Ibn al-Haytham and Analytical Mathematics*, London, 2012, p. 412, no. 62, and see below.

text attributed to Ibn al-Haytham: *On the Determination of the Height of Mountains*.

The second type of text on mensuration is also old; it first appears as a chapter in books on algebra, such as that by al-Khwārizmī, and more generally in treatises on *Ḥisāb* (the science of calculation), such as those by al-Būzjānī, al-Karajī, al-Baghdādī, al-Fārisī and others. Most of the texts that deal with mensuration belong to this type, which, as far as we know, seems not to have interested Ibn al-Haytham.

The third type is specific to geometers. One example is a treatise by Thābit ibn Qurra. This is a book without proofs in which the author gives formulae for finding areas of plane figures with straight and curved edges, together with volumes of some solids, including the cube as well as the sphere.[7]

The fourth and final type is represented by Ibn al-Haytham's book *On the Principles of Measurement (Fī uṣūl al-misāḥa)*. In what is in fact a manual, the author is concerned to lay down solid foundations to underpin his practice. This treatise is of great importance, both historical and epistemic: for we are in fact confronted by the work of a theoretical geometer, and no minor one, one of the greatest, who is writing for surveyors. Ibn al-Haytham is certainly not the first to have taken this course. Indeed his predecessor Ibn Sinān had written a practical manual for the artisans who constructed sundials. But, as far as I know, this type of text is not to be found among the writings of any of the great geometers of the Hellenistic period. The new spirit is manifest in the mathematician's dual approach, his two closely linked intentions: to set out geometrical foundations for a practical art, here that of measurement; to provide practitioners with the rules they need to apply. Here again Ibn al-Haytham appears as bringing a tradition to its fulfilment, a tradition that included Thābit ibn Qurra, Ibrāhīm ibn Sinān and doubtless many others. But, as for making a detailed reconstruction of this tradition, scholars have hardly begun to do the work required. Here we shall go no further than to examine the writings of Ibn al-Haytham. There are four of them: the two we have already mentioned (which have come down to us); a third whose title is recorded by old biobibliographers: *On a Problem of Measurement (Fī mas'ala*

---

[7] *Fī misāḥat al-ashkāl al-musaṭṭaḥa wa-al-mujassama*, edition, French translation and commentary in R. Rashed, 'Thābit ibn Qurra et l'art de la mesure', in *id.* (ed.), *Thābit ibn Qurra. Science and Philosophy in Ninth-Century Baghdad*, Scientia Graeco-Arabica, vol. 4, Berlin/New York, 2009, pp. 173–209.

fī al-misāḥa),[8] wording that suggests we have a study of a specific problem. Finally, we may note that, in his preface to his treatise *Fī uṣūl al-misāḥa*, Ibn al-Haytham himself refers to a piece he wrote when young, on the same subject and with the same title, lost at the same time as 'many original manuscripts' of his writings. It is possible that *Fī mas'ala fī al-misāḥa* is the same as the lost text, whose title may have been corrupted.

4.2. MATHEMATICAL COMMENTARY

4.2.1. *Treatise on the principles of measurement*

The plan of the book is clear and well suited to the purpose of the work: to provide surveyors with a rigorous but elementary geometrical manual of their art. After an introduction which explains the basic notions concerned – measurement, unit of measurement, measurable magnitudes – Ibn al-Haytham provides a short first chapter on measuring lines, only straight lines and arcs of circles, these being the only lines his readers will need to use. His second chapter deals with the measurement of areas: rectangle, triangle (using what is now known as 'Heron's formula'[9]), convex polygon, circle. The third chapter is devoted to the measurement of volumes: polyhedron, cylinder, cone, sphere. The fourth chapter describes an experimental method for finding the height of a general body. Ibn al-Haytham concludes with a reminder of the results and procedures that will be of use.

Let us have a brief look at the contents of the various chapters. Ibn al-Haytham, like the Banū Mūsā long before him,[10] begins by explaining that the unit of measurement is a conventional unit, initially chosen for lengths of lines: we have, for instance, the conventional length of a line segment called a 'cubit'. The units for areas and volumes are derived from this one: the square

---

[8] Cited by al-Qifṭī, Ibn Abī Uṣaybi'a and in the Lahore list, see R. Rashed, *Ibn al-Haytham and Analytical Mathematics*, London, 2012, p. 412, no. 65.

[9] Thābit says of this formula: 'Some have claimed it comes from India and others have reported it as due to the Greeks' (*On the Measurement of Plane and Solid Figures*, ed. R. Rashed in *Thābit ibn Qurra. Science and Philosophy in Ninth-Century Baghdad*, pp. 182–3).

[10] R. Rashed, *Founding Figures and Commentators in Arabic Mathematics. A history of Arabic sciences and mathematics*, vol. 1, Culture and Civilization in the Middle East, London, 2012, Chapter I.

unit of area is the square whose side is the unit of length, and the cubic unit of volume is the cube whose edge is the unit of length. The measurement of a magnitude – length, width, depth – is its expression in terms of the unit associated with it. In other words, the measurement of a magnitude is the number, rational or irrational, that expresses the ratio of the magnitude to the magnitude of the same kind taken as the unit. The meaning of the word 'ratio' as used here is that given in Definitions 1, 2, 3, 7 and 9 of Book V of the *Elements*.

In the chapter on the measurement of lines, Ibn al-Haytham considers only the kinds of line handled by surveyors: straight lines and arcs of circles. We measure a straight line by placing on it, one part after another, 'the complete cubit or certain parts of it'. The measurement obtained, that is, the number by which we multiply the unit, is a rational number. Here Ibn al-Haytham does not mention the case in which this procedure would only yield an approximate value of the measurement, something he will mention later, for instance for arcs.

For the measurement c of a circular arc, say that of the circumference of the circle with diameter d, and the measurement a of an arc whose ratio to the circumference is a number k, Ibn al-Haytham here repeats known results, without proof: $c \approx \dfrac{22}{7}d$ and $a = kc$, and indicates that, for surveyors, the procedure is to measure d and find k. He returns to this question later on, and to the method for finding d and k, in his chapter on the measurement of areas.

In the second chapter, on the measurement of areas, Ibn al-Haytham considers only the ones that will be of concern to surveyors: plane areas. He writes 'spherical, cylindrical or conical surfaces do not come into their art of measurement'. He begins with the measurement of the rectangle and gives a proof for the calculation of the area of the rectangle assuming that its two dimensions have a common aliquot part taken as the unit of length; he does not mention other cases.

Ibn al-Haytham then goes on to consider the area of the triangle; the area of a right-angled triangle follows immediately from that of the rectangle, and we then go on, using *Elements* I.37, to the area S of a general triangle with base b and height h, $S = \dfrac{1}{2}b \cdot h$. The problem for surveyors is thus to find the height h. Ibn al-Haytham then makes a very detailed study of the triangle, taking the lengths a, b, c of the three sides as known, and gives ways of answering the following questions:

1. How to know whether the triangle is right-angled or not?

2. How to calculate the height from a particular vertex? The calculation involves considering a segment of the base opposite the vertex in question called the 'foot of the height'. Three cases are then considered:
- height from a vertex whose angle is obtuse,
- height from a vertex whose angle is acute,
- height above the greatest side – this calculation is valid whatever the angles of the triangle are.

Let a be the greatest side of the triangle, $BC = a$, let h be the height AD and x the foot of the height, $x = BD$, we have

$$x = \frac{a^2 + c^2 - b^2}{2a},$$

$$h^2 = c^2 - x^2 = c^2 - \frac{\left(a^2 + c^2 - b^2\right)^2}{4a^2}.$$

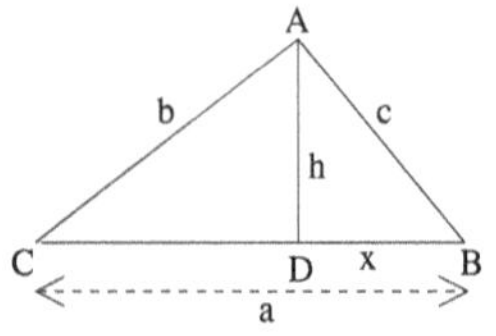

Fig. 4.1

Starting from this calculation of the height we can derive Heron's formula. From the last expression above we have

$$h^2 = \frac{1}{4a^2}(a+b+c)(a+c-b)(b+c-a)(b+a-c);$$

but

$$S^2 = \frac{1}{4}a^2h^2 = \left(\frac{a+b+c}{2}\right)\left(\frac{a+c-b}{2}\right)\left(\frac{b+c-a}{2}\right)\left(\frac{b+a-c}{2}\right).$$

Let us put $p = \frac{a+b+c}{2}$, hence

$$S^2 = p\,(p-b)\,(p-a)\,(p-c).$$

Ibn al-Haytham later derives this formula by a different method, without using the height. His proof involves the centre and radius r of the circle inscribed in the triangle. If $2p$ is the perimeter of the triangle, we have $S = p \cdot r$, where $p^2 r^2 = p(p - a)(p - b)(p - c)$, hence S^2 and thus S. This is in fact the method already used by the Banū Mūsā.[11]

Ibn al-Haytham then considers convex polygons. The idea is as follows: any convex polygon can be dissected into triangles, so its area is the sum of the areas of these triangles, and each of these areas can be expressed in terms of the three sides of the triangle concerned. It is true that every convex n-sided polygon can be dissected into $n - 2$ triangles, by joining any vertex – say A – to each of the others.

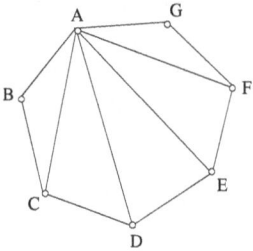

Fig. 4.2

The area of the polygon is equal to the sum of the areas of the $n - 2$ triangles obtained. Ibn al-Haytham says that the division is carried out by means of the chords of the angles of the polygon. His assertion is true for a quadrilateral, a pentagon or a hexagon, in which the $n - 3$ chords will give the $n - 2$ triangles.

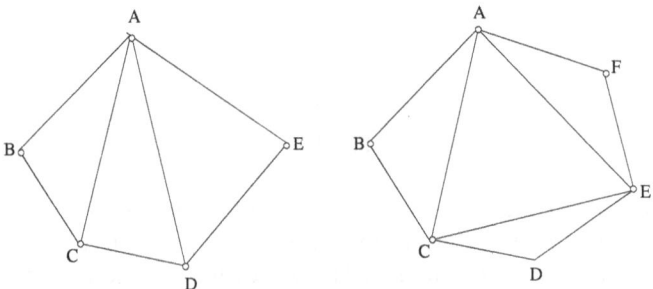

Fig. 4.3

[11] R. Rashed, *Founding Figures and Commentators in Arabic Mathematics*, Proposition 7, p. 46.

But for $n \geq 7$, the division into triangles would require straight lines other than chords. So if $n = 2p$, we would have p chords, the polygon would be dissected into p triangles whose bases are these chords and a polygon having these p chords as sides. If $n = 2p + 1$, we would have p chords, hence p triangles with the chords as their bases and a polygon with $p + 1$ sides: the p chords and one side of the polygon. In both cases, the sides of the initial polygon can be measured and Ibn al-Haytham gives a method of calculating the chords. It then remains to dissect the polygon with p sides or with $p + 1$ sides. The dissection is carried out once we have found the angles of the polygon. For example let us take a polygon $ABCDEFG$ (Fig. 4.2); AD and AE are not chords of the angles of the heptagon. In Fig. 4, we can draw all the chords AC, CE, EG that are the bases of three triangles whose vertices are B, D and F, there then remains a quadrilateral that one can divide into triangles by drawing the diagonal AE. The straight line AE is not the chord of an angle of the heptagon, but it is the chord of the angle G of the triangle AGE. Obtaining a value for the length of this chord is possible either by making an exact measurement, or by the method described later for finding a chord. In fact, when GE is found, the triangle FGE is known, and the angle FGE will also be known; this gives $A\hat{G}E = A\hat{G}F - F\hat{G}E$ and we know the two sides enclosing the angle.

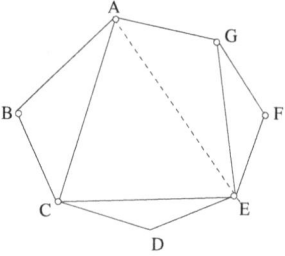

Fig. 4.4

Ibn al-Haytham then gives a procedure for evaluating the chord. Let us consider a triangle BAC; we want the value of AC, the chord of the angle B, starting from the lengths AB and BC. The surveyor must choose a point on BC, say D, so that in triangle BDE, similar to BCA, the length DE can be found by measurement. But we have $\dfrac{AC}{DE} = \dfrac{BC}{BD}$; so if the person making measurements takes BD to be one cubit, we get $AC = DE \cdot BC$.

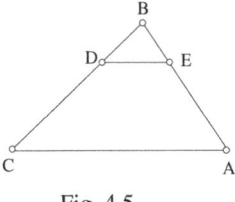

Fig. 4.5

Ibn al-Haytham then returns to the area of the circle and describes Archimedes' proof in *The Measurement of a Circle*.

The area Σ of the circle with diameter d and circumference $2p$ is given by the expression $\Sigma = p\dfrac{d}{2}$.

Ibn al-Haytham also notes that Archimedes found that the ratio $\dfrac{2p}{d}$ has an approximate value $\dfrac{22}{7}$, hence $p \approx \dfrac{11}{7}d$ and $\Sigma \approx \dfrac{11}{14}d^2$ or again $\Sigma = d^2 - \left(\dfrac{1}{7} + \dfrac{1}{14}\right)d^2$.

When we are not given the centre and diameter of the circle, the surveyor needs to be able to find the value of the diameter d. Ibn al-Haytham gives a procedure for finding the length of the diameter.

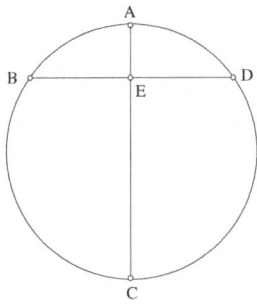

Fig. 4.6

Let DB be a general chord of the circle, let E be its midpoint; the perpendicular bisector of the chord passes through the centre of the circle, it cuts the circle in A and C and AC is thus a diameter; we have

$$BE = ED \text{ and } EA \cdot EC = BE \cdot ED = BE^2,$$

hence

$$EC = \frac{BE^2}{EA} \quad \text{and} \quad AC = EA + \frac{BE^2}{EA}.$$

The length of the diameter is then known if we know the length of the chord BD and that of the sagitta AE.

This procedure is of interest because if the diameter AC is very large, or cannot be found by surveying, we can dispense with measuring its length directly. We choose a chord BD that is small enough for its length and that of its sagitta to be easy to measure.

Ibn al-Haytham says that he summarizes Archimedes' method of finding the area of the circle. In view of the historical interest of this account, we shall give a brief description of it. We require to prove:

The area Σ of a circle of radius R and circumference 2p is equal to the product p $\cdot$ R *or, as we shall see later, we can obtain it 'by multiplying its semi-diameter by its semi-perimeter, that is the number of the multiple of the cubit contained in the semi-diameter and the number of the multiple of the cubit contained in the semi-perimeter'.*

Let us put $U = p \cdot R$.
1. If $U < \Sigma$, then $\Sigma - U = S$, so $S < \Sigma$.

Let E be the centre of the circle, AC and BD two diameters perpendicular to one another, $ABCD$ is thus a square inscribed in the circle. The tangents to the circle at the points A, B, C, D define a square $NMLX$ circumscribed about the circle. The diagonals of this square cut the arcs AD, DC, DB, BA in their midpoints K, I, H, G respectively. The tangent at the point K is parallel to AD. From A and D we draw perpendiculars to AD; we thus construct a rectangle; let its area be s, we have

$$s > \text{area segment } (AKD).$$

We have in succession:

$\bullet$ area $(NMLX) > \Sigma$ and area $(NMLX) = 2$ area $(ABCD)$,

so

$$\text{area } (ABCD) > \frac{1}{2} \Sigma,$$

hence

$$\Sigma - \text{area } (ABCD) < \frac{1}{2}\Sigma.$$

Let r_1 be the first remainder, $r_1 < \frac{1}{2}\Sigma$.

• $s >$ area segment (AKD) and $s = 2$ area triangle (AKD), so

$$\text{area triangle } (AKD) > \frac{1}{2} \text{ area segment } (AKD),$$

hence

$$\text{area segment } (AKD) - \text{area triangle } (AKD) < \frac{1}{2} \text{ area segment } (AKD).$$

We proceed in the same way for the other three segments whose chords are the sides of the square $ABCD$. We obtain the second remainder

$$r_2 = \Sigma - \text{area } (AKDICHBG) < \frac{1}{2} r_1,$$

hence

$$r_2 < \frac{1}{2^2}\Sigma.$$

We repeat the procedure until we obtain a remainder

$$r_n < \frac{1}{2^n}\Sigma < S.$$

Let us suppose that the polygon $AKDICHBG$ has area Σ_n and provides a solution to the problem, that is to say that

$$\Sigma - \Sigma_n < S \text{ or } \Sigma_n > U.$$

We have

$$\text{area } (EAKD) = \text{area triangle } (AED) + \text{area triangle } (AKD)$$
$$= \frac{1}{2}AD \cdot EO + \frac{1}{2}AD \cdot OK = \frac{1}{2}AD \cdot EK;$$

so if the perimeter of *ABCD* is $2p_{n-1}$, we have $\Sigma_n = p_{n-1} \cdot R$. But $\Sigma_n > U$, hence $p_{n-1} \cdot R > p \cdot R$ and $p_{n-1} > p$; which is impossible because $p_{n-1} < p$.

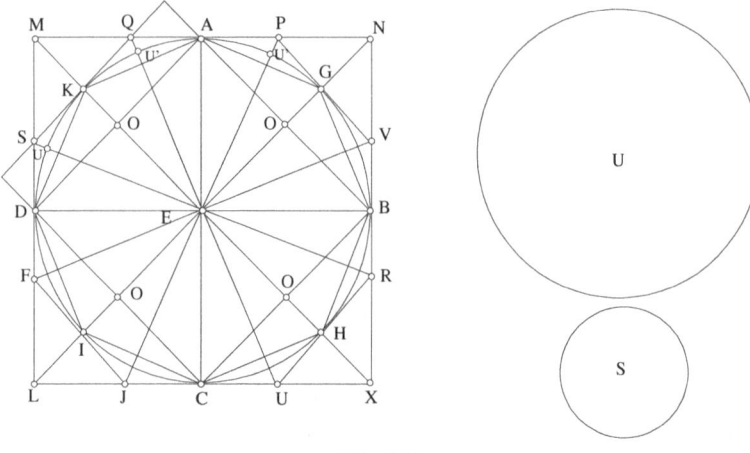

Fig. 4.7

In this first part, Ibn al-Haytham uses the square *NMLX* to prove that area $(ABCD) > \dfrac{1}{2}\Sigma$, so as to be able to apply Proposition X.1 of the *Elements*: we have $\Sigma > S$, we take away more than half of Σ and repeat the procedure.

2. If $U > \Sigma$, we have three cases.
a) $U = \text{area } (NMLX)$ b) $U > \text{area } (NMLX)$ c) $U < \text{area } (NMLX)$.

Let $2p_1$ be the perimeter of the figure *NMLX* circumscribed about the circle, then area $(NMLX) = p_1 \cdot R$.
a) $U = \text{area } (NMLX) \Leftrightarrow p \cdot R = p_1 \cdot R \Leftrightarrow p = p_1$, which is impossible because the arc $AKD = \dfrac{1}{4}p$ and $AM + MD = \dfrac{1}{4}p_1$ and $AM + MD > \overset{\frown}{AKD}$, so $p_1 > p$.

b) $U > \text{area } (NMLX) \Leftrightarrow p \cdot R > p_1 \cdot R \Leftrightarrow p > p_1$, which is impossible because $p_1 > p$.
c) $U < \text{area } (NMLX)$.

We put $U - \Sigma = S$; we then have

$$\text{area } (NMLX) - \Sigma > S,$$

$$\text{area } (NMLX) - \Sigma = r_1 = 4 \text{ [area of the curvilinear triangle } (AKDM)],$$

and

$$r_1 > S.$$

We then consider the tangents at the points K, I, H, G, midpoints of the arcs AD, DC, CB, BA and we obtain the circumscribed octagon. Let us call the midpoints of the arcs U'. We have

$$MQ > QK \text{ and } QK = QA \Rightarrow MQ > QA$$
$$\Rightarrow \text{area } (MKQ) > \text{area triangle } (KQA),$$

hence

$$\text{area } (MKQ) > \text{area portion } (KQAU'),$$

$$\text{area } (QMS) > \text{area [portion } (KQAU') + \text{portion } (KSDU')],$$

consequently

$$\text{area } (QMS) > \frac{1}{2} \text{ area curvilinear triangle } (AKDM),$$

so

$$4 \text{ area } (QMS) > \frac{1}{2} r_1,$$

hence

$$r_1 - 4 \text{ area } (QMS) < \frac{1}{2} r_1.$$

Let us put $r_2 = $ area circumscribed octagon $- \Sigma$; we have

$$r_2 < \frac{1}{2} r_1.$$

If we again double the number of sides of the circumscribed polygon by drawing the tangents to the circle at each of the points U', and if we repeat the same procedure, we take away more than half of r_2 and we obtain

$$r_3 < \frac{1}{2}\, r_2 \Rightarrow r_3 < \frac{1}{2^2}\, r_1;$$

so successive remainders decrease. We repeat the procedure until we obtain

$$r_n < \frac{1}{2^{n-1}}\, r_1 < S.$$

Let Σ_n be the area of the polygon that gives a solution to the problem; if $2p_n$ is its perimeter, we have

$$\Sigma_n = p_n \cdot R;$$

we have

$$r_n = \Sigma_n - \Sigma < S \Rightarrow p_n \cdot R < U \Rightarrow p_n \cdot R < p \cdot R,$$

which is impossible because $p_n > p$.

Notes: In this second part of the proof, we make the hypothesis that $U = p \cdot R > \Sigma$. The square $NMLX$ is circumscribed about the circle, so if its perimeter is $2p_1$, we have $p < p_1$ and area $(NMLX) = p_1 \cdot R$.

So we must have $p \cdot R < p_1 \cdot R$, that is $U <$ area $(NMLX)$; this is what Ibn al-Haytham establishes in a) and b) by showing that $U \geq$ area $(NMLX)$ is impossible.

Now if $U <$ area $(NMLX)$ and $U - \Sigma = S$, we have

$$\text{area } (NMLX) - \Sigma > S.$$

We have

$$\text{area } (NMLX) - \Sigma = r_1 \text{ (shaded area)},$$

which satisfies $r_1 > S$. From this area r_1 we remove a part equal to $4\,[\text{area } (MQS)] > \frac{1}{2}\, r_1$. So we fulfil the conditions for applying *Elements* X.1.

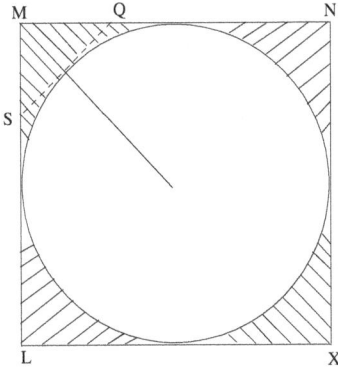

Fig. 4.8

A simple comparison with Archimedes' *Measurement of a Circle*, in the Greek or the Arabic versions, shows that Ibn al-Haytham is reformulating Archimedes' procedure in the language of his own period. But that is not what we are concerned with here.

Ibn al-Haytham then goes on to consider the area S of a sector of a circle, and shows that if l is the length of the circular arc that defines the sector, we have from *Elements* VI.31

$$\frac{S}{\Sigma} = \frac{l}{2p} \Rightarrow S = \frac{l}{2} \cdot \frac{d}{2}.$$

Next he considers the area of a segment of a circle. If the segment is less than a semicircle, we can find its area from that of the corresponding sector by subtracting from the area of the sector the area of the triangle whose vertex is the centre of the circle and whose base is the chord of the arc. So we need to know the length l of the arc that defines the sector or the segment. This length l is known if we know the ratio k of the arc in question to the complete circumference.

Ibn al-Haytham then gives a method for finding an approximate value of the ratio k that is as precise as possible. This method is based on constructing an auxiliary arc equal to a quarter of a circle whose radius is the chord of the arc in question. The actual construction of a ratio equal to the ratio k is carried out on the arc by using compasses whose opening is chosen by repeated trial.

Let ABC be the arc we wish to measure, AC its chord and EB its sagitta. Let us draw the circle with centre A and radius AC; it cuts the line AB in H and the perpendicular to AC at A in D; CHD is a quadrant of a circle; we have

$$\frac{\overparen{CH}}{\overparen{CD}} = \frac{\hat{CAH}}{1 \text{ right angle}}.$$

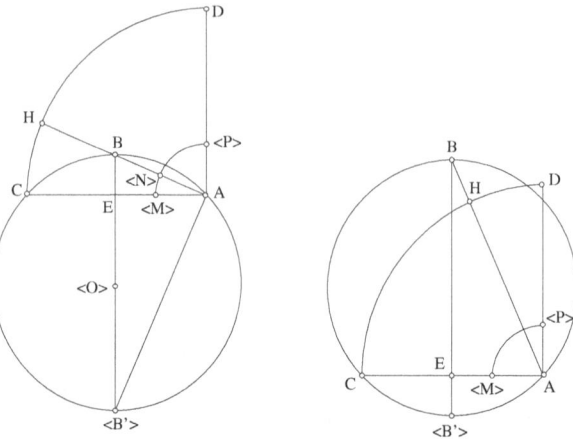

Fig. 4.9

Let BB' be the diameter through B; the arc BCB' is a semicircle cut off by the inscribed angle $BAB' = 1$ right angle; the arc BC is cut off by the angle BAC. We have

$$\frac{\overparen{BC}}{\overparen{BCB'}} = \frac{\hat{BAC}}{1 \text{ right angle}} = \frac{\hat{CAH}}{1 \text{ right angle}};$$

from which we get

$$\frac{\overparen{ABC}}{\text{circumference}} = \frac{\overparen{BC}}{\overparen{BCB'}} = \frac{\overparen{CH}}{\overparen{CD}}.$$

If the arc to be measured is very large, that is subtended by a very long chord, we take a point M on AC and we replace the quadrant of the circle CHD by the quadrant of the circle MNP; we then have

(*) $$\frac{\overparen{CH}}{\overparen{CD}} = \frac{\overparen{MN}}{\overparen{MP}};$$

the measurements can then be made on a smaller figure.

The choice of a point M on AC defines a homothety with centre A and ratio $\dfrac{AM}{AC}$. The arc CHD that defines the ratio $\dfrac{\widehat{CH}}{\widehat{CD}}$ has as its homothetic counterpart the arc MNP, and we have (*). We may note that Ibn al-Haytham states very clearly that he wants to subject the arc CHD to a reduction and proposes to take a part AM to stand for AC. So he is giving himself the homothety $(A, \dfrac{AM}{AC})$.

So Ibn al-Haytham's method reduces to measuring, with compasses, two arcs CH and HD (or MN and NP) whose sum is a quadrant of a circle. Ibn al-Haytham assumes that the size of the opening of the compasses is chosen so that one can apply it (that is step the compasses) an integral number of times between C and H (or M and N) on the one hand and between H and D (or N and P) on the other; let it be m times and n times respectively. Then we have m equal chords between C and H and $m + n$ equal chords between C and D. There are equal arcs that correspond to these equal chords, so

$$\frac{\widehat{CH}}{\widehat{CD}} = \frac{m}{m+n}.$$

It is clear that this method can find the ratio between the two arcs only with some degree of approximation, which can always be improved.

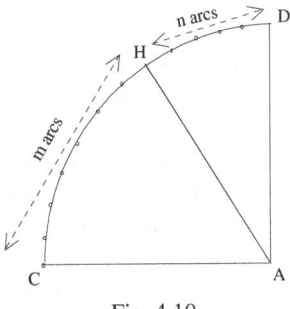

Fig. 4.10

Ibn al-Haytham then explains in great detail that the required ratio may be irrational. But the method recommended leads to a ratio k that is rational. So the surveyor must find the number k to an approximation so close that the difference between the value he finds for k and the exact ratio is so small that it has no effect on the results in which the required ratio plays a part.

After studying the areas of surfaces, Ibn al-Haytham comes to solids, but only those that engage the attention of surveyors, that is, solids bounded by plane faces – polyhedra – spheres, cylinders and cones.

The first polyhedron he investigates is the rectangular parallelepiped. For the calculation of its volume, Ibn al-Haytham gives a proof for the case where all three dimensions have a common aliquot part that is taken as the unit of length; he does not mention other cases. The reasoning assumes that each of the dimensions of the rectangular parallelepiped is an integer number of cubits, that is that each of the dimensions is a multiple of a length taken as a unit, and that the unit of volume is the cube whose edge is that unit. But we know from *Elements* XI.32 that if two parallelepipeds have the same height, their volumes are proportional to their bases; so, whether or not the measurements of the three dimensions in terms of the unit are integer numbers, we can make use of that property.

Ibn al-Haytham considers only the volume of the rectangular parallelepiped and does not indicate how to go on to find the volume of the general parallelepiped, nor that of the upright prism and the oblique prism. He does, however, refer to the result concerning these solids and indicates that, from *Elements* XII.7, the volume of the pyramid is one third of the volume of the prism on the same base and with the same height.

Ibn al-Haytham then explains that, by dividing up the faces of any polyhedron along the chords of their angles, we can dissect the polyhedron into pyramids and the volume of the polyhedron is the sum of the volumes of these pyramids. So for the surveyor, the problem consists of knowing how to find the *base* and *height* of a pyramid. Now, the base of a pyramid is a triangle or a polygon. If it is a triangle, we merely need to measure the three sides to calculate its area. If it is a polygon, its area will be the sum of the areas of the triangles made by the chords of the angles of the polygon, as we saw earlier. In any case, we need to be able to find the measurement of these chords, even if the base of the solid is on the ground. Ibn al-Haytham thus proposes the following procedure for finding the chord of an angle of the base of a solid.

The procedure consists of constructing, in the plane of the base of the solid, but outside the actual solid, an angle xBy equal to the angle xOz of the base. To do this, we apply along one of the sides of the angle a ruler with parallel edges $By \parallel Oz$, and along the side Ox a ruler whose edge, aligned with Ox, cuts By at the point B. The angle B is equal to the angle under investigation (angles with parallel sides).

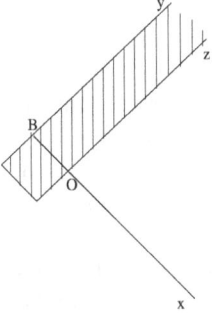

Fig. 4.11

Once angle *xBy* is found, we mark off on the sides enclosing that angle the measured lengths of *BE* and *BD*, the two sides of the angle of the base of the solid.

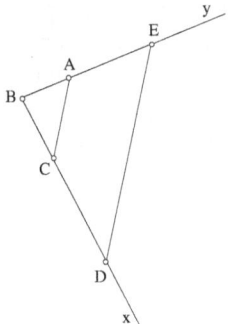

Fig. 4.12

We then return to the method described earlier for finding the length of a chord. If $BA = 1$ and $BC = \dfrac{BD}{BE}$, we have

$$\frac{BD}{BC} = \frac{BE}{BA};$$

the triangles *ABC* and *EBD* are similar. We measure *AC* and from it we find

$$DE = AC \cdot \frac{BE}{BA},$$

hence

$$DE = AC \cdot BE.$$

If the base is on a smooth surface, that is, one that is uniform, without roughness, we can extend the sides of the base to obtain an angle opposed at the vertex to the angle of the base, and thus equal to the angle in question.

Thus we have two methods of obtaining an angle equal to the angle under investigation:

a) Using two rulers, the first of which has parallel edges and a certain width, we construct an angle whose sides are parallel to the sides of the base.

b) We extend the sides of the angle in question along the ground, if that is possible; we obtain an angle that is opposite at the vertex to the required angle of the base.

Whether the base of the pyramid is a triangle or a general polygon, the other faces are triangular. To measure the height, Ibn al-Haytham makes use of points and segments that lie inside the polygon of the base. He then uses a plane auxiliary figure on which he can carry out the constructions needed to obtain the segments to be measured. Thus he considers two faces with a common edge such as AC. The planes ACB, ACD, BCD form the trihedral angle at the vertex C, which is called 'the angle of the pyramid'.

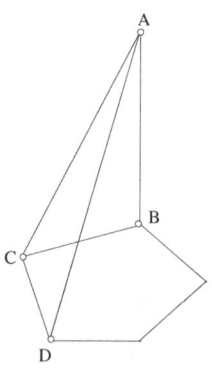

Fig. 4.13

Ibn al-Haytham then uses constructions in the plane of the base of the pyramid to find a value for the height.

Let *AE* and *AG* be the heights of the triangles *ABC* and *ACD*. Let *LIK* be a triangle such that $L\hat{I}K = B\hat{C}D$, *LI* = *CE* and *IK* = *CG*. We have

a) If $I\hat{L}K$ = 1 right angle, then $C\hat{E}G$ = 1 right angle. But $C\hat{E}A$ = 1 right angle, so *CE* is perpendicular to the plane *EAG*. Let *GO* ∥ *CE*, then *GO* is perpendicular to the plane *EAG*, so *GO* ⊥ *GA*, but *GA* ⊥ *DC*, so *GA* is perpendicular to the two lines *GO* and *GC*, and thus to their plane, that is to the plane *BCD*; so *AG* is the height of the pyramid.

Similarly, if $I\hat{K}L$ = 1 right angle, then *AE* is the height of the pyramid.

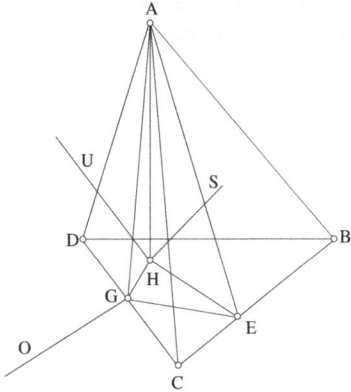

Fig. 4.14

b) If neither of the angles *ILK*, *IKL* is a right angle, we construct at *L* and *K* perpendiculars to the lines enclosing the angle *LIK*. The perpendiculars must intersect, say in a point *M*. If the two angles at *L* and *K* are acute, *M* lies inside the angle *LIK*; if one of the two angles at *L* or *K* is obtuse, *M* lies outside angle *LIK*.

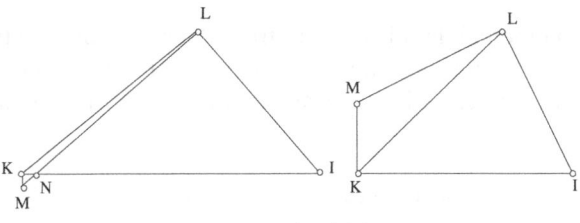

Fig. 4.15

At *G* and *E* we construct perpendiculars to the lines *CG* and *CE*; the perpendiculars intersect in *H*, a point that corresponds to point *M*.

Let $HS \parallel CE$ and $HU \parallel CG$. By hypothesis we have $C\hat{E}A = 1$ right angle, $C\hat{E}H = 1$ right angle, so CE is perpendicular to the plane AHE and $HS \perp (AHE)$, hence $HS \perp AH$. Similarly, $C\hat{G}A = 1$ right angle, $C\hat{G}H = 1$ right angle, so $CG \perp (AHG)$ and consequently $HU \perp (AHG)$, hence $HU \perp AH$. So the straight line AH is perpendicular to the plane SHU, that is to the plane BCD. So the straight line AH is the height of the pyramid. To find the value of the height AH, we consider the right-angled triangle AEH, and we have $AH^2 = AE^2 - EH^2$. But $EH = LM$, a length that can be measured in the plane figure, and AE is a length that can be measured on the face ABC of the pyramid, and $AH^2 = AE^2 - LM^2$, so AH can be calculated.

Ibn al-Haytham next considers the volumes of the cylinder and cone on a circular base. If S is the area of the base and h the height of a right cylinder, the volume, V, is given by $V = S \cdot h$.

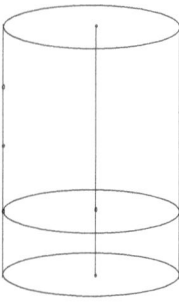

Fig. 4.16

Let there be a straight line joining the centres of the circles that form the bases; this line is the height of the cylinder and it is equal to the length of a generator. Let h be the value of the height, V the required volume and S the area of the base.

We cut off on the height a length, up from the base, equal to the unit, and through the point we obtain we draw a plane parallel to the plane of the base. We thus obtain a cylinder with base S and unit height; let its volume be v. We have

$$\frac{S}{\text{unit of area}} = \frac{v}{\text{unit of volume}}.$$

If we choose uniform units, the area of the base and the volume of the small cylinder are numerically equal: $S = v$.

Ibn al-Haytham then imagines the height being divided into h parts, each equal to the unit. By drawing planes parallel to the base we obtain a number h of cylinders, each equal to the first small cylinder, so

$$V = h \cdot v = h \cdot S.$$

We may note that, from *Elements* XII.14, we have $\dfrac{V}{v} = \dfrac{h}{1}$, hence the above equality holds good for any number h that measures the height (integer or not).

As for the volume of the oblique cylinder, we know that it is equal to that of the right cylinder with the same base and the same height.

Ibn al-Haytham then comes to the volume of a right cone and of an oblique cone and notes that it is equal to one third of the volume of the cylinder that has the same base and the same height as it does, so $V = \dfrac{1}{3} S \cdot h$.

To find the area S of a circular base, we measure its circumference $2p$, from which we find the diameter $d \approx 2p \cdot \dfrac{7}{22}$ and $S = \dfrac{p \cdot d}{2}$.

We shall see later how to find the height h.

Ibn al-Haytham ends his study of volumes of solids with the volume of the sphere. Not only did he know the works of his predecessors on this topic, notably Archimedes and the Banū Mūsā, but he himself had also worked out the volume of the sphere in a treatise[12] composed before the one on *Measurement*. So he had established that

$$V = \text{area of a great circle} \times \frac{2}{3} \text{ of the diameter.}$$

Ibn al-Haytham explains how to construct in a plane a circle equal to the great circle of the sphere and find its diameter. He proposes the following procedure. Using compasses opened so that the distance between the points is e, we draw on the sphere a circle with pole L. We then take two general points X and Y on this circle. The midpoints of the arcs XY, the points G and I, are then found by approximation: we vary the opening of the compasses until it allows us to find the point G such that $GX = GY$, so then $\overset{\frown}{GX} = \overset{\frown}{GY}$ and we then find the point I such that $IX = IY$, so that $\overset{\frown}{IX} = \overset{\frown}{IY}$. The line segment IG is the

[12] *Qawl fī misāḥat al-kura*, in R. Rashed, *Les Mathématiques infinitésimales*, vol. 2, pp. 295–323; English translation in *Ibn al-Haytham and Analytical Mathematics*.

diameter of the circle we drew. If K is the midpoint of the segment IG, the straight line LK passes through the centre of the sphere and gives us LM, the diameter of the sphere. The length of GI can be measured by an opening of the compasses; but the same is not true for the lengths LK and LM, which lie inside the sphere. Ibn al-Haytham then tells us how to use the known lengths, $LI = LG = e$ and $IG = 2\ IK = 2h$, to construct a plane figure the same as the figure $ILGM$. We draw a line segment AB with midpoint C and $AB = IG$, so $CA = IK = h$; on its perpendicular bisector we take the point D such that $AD = LG = e$. The perpendicular to AD at A cuts DC in E.

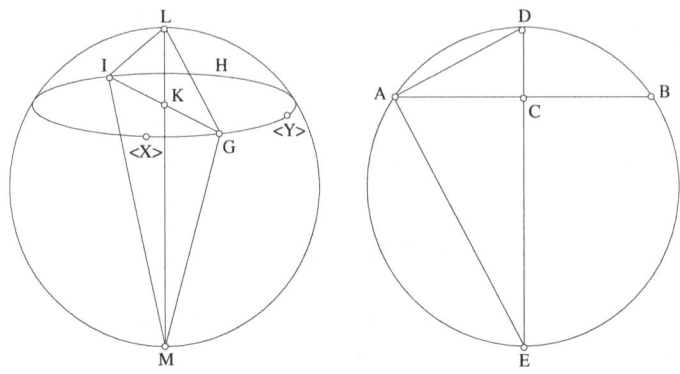

Fig. 4.17

The right-angled triangle DAE in the plane figure is equal to the triangle LGM in the three-dimensional figure, because they are both similar to a right-angled triangle with hypotenuse e ($LG = AD = e$) and side h ($IK = AC = h$) and they have one side equal, $LG = AD$.

Thus we can measure the diameter DE in the plane figure, or calculate DE from the known lengths e and h; $AD = e$, $AC = h$. We have

$$DC^2 = AD^2 - AC^2 = e^2 - h^2,$$

hence

$$DC = \sqrt{e^2 - h^2}.$$

Moreover

$$AD^2 = DC \cdot DE,$$

hence

$$DE = \frac{e^2}{\sqrt{e^2 - h^2}}.$$

We can also write

$$AC^2 = DC \cdot CE,$$

hence

$$CE = \frac{h^2}{\sqrt{e^2 - h^2}},$$

hence

$$DE = \sqrt{e^2 - h^2} + \frac{h^2}{\sqrt{e^2 - h^2}}.$$

We put $DE = d$; the area of the great circle is

$$S = \pi \frac{d^2}{4} \approx \frac{11}{14} d^2 = d^2 - \frac{1}{7} d^2 - \frac{1}{14} d^2$$

and the volume of the sphere is $V = S \times \frac{2}{3} d$.

We may note that the important thing about the plane figure is that it allows us to measure the lengths $DC = LK$ and $DE = LM$. But the calculation of these lengths from the known ones $GL = LI = e$ and $KI = h$ can be carried out using triangle LGM in the three-dimensional figure.

We may also note that the opening of the compasses – a distance $GL = LI = e$ between the points – that is used to draw a circle with pole L on the sphere is arbitrary. The precision of the result obtained by this procedure depends on how careful the practitioner is in finding the points I and G, the midpoints of the two arcs of the circle with pole L, because these points can only be obtained by trial and error, adjusting the opening of the compasses.

The last chapter of Ibn al-Haytham's treatise deals with a problem that is of capital importance for surveyors: to find a height by experimental means – the height of a pyramid, a cylinder, a cone or a solid body standing on the ground. The required height is the length of the perpendicular from the highest point of the solid body to the plane of the base. The procedure is particularly useful when one or both of the two points, the vertex or the foot of the perpendicular, is inaccessible.

The procedure makes use of a rod and a plumb line, both of a length greater than the height of the observer. On the rod, whose tip is X, we carve a circular mark round the shaft at a distance XY equal to one cubit, the chosen unit of measurement. The observer then uses the plumb line to find the height h of his eye above the ground: using a finger he holds the line up to his eye; then he moves the string up or down by sliding it round his finger, until the lead weight touches the ground. The length of string between the finger and the weight is then the required height h. He records this length on the rod, starting from the point Y, and he obtains the point Z, he then carves a circular mark round the rod at this point; we have

$$XY = 1 \text{ and } YZ = h.$$

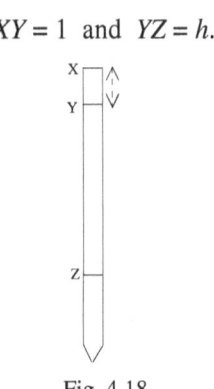

Fig. 4.18

The remainder of the rod may end in a point that allows it to be easily driven into the ground.

Now let AU be the height we want to measure; if we are considering a wall, or a mountain, the ground is represented by the line BC. We have $AU \perp BC$.

The rod XYZ is driven into the ground in the position DGE where $DE \perp BC$, $DG = XY$ and $GE = YZ$. The observer finds the position from which he can take a simultaneous sighting of the points D, the tip of the rod, and A, the vertex of the body whose height it is required to measure; let that position be HI; H represents the eye and I the midpoint of the foot. We have $HI \parallel DE$. He makes a hole in the ground at the point I, and notes the distance $IE = l_1$. He draws the straight line *IE* that passes through the point *U*. He then pulls the rod out of the ground, and moves it to the position *MNP* closer to the required height *AU*, say to point *P* on *EU* where $MP \perp EU$, $MN = XY$ and $NP = YZ$. The observer then finds the position from which he can take a simultaneous sighting of the point *M* and the vertex *A*. Let this position be *KL*, where *K* is the eye and *L* the midpoint of the foot, *L* lies on the straight line *IU*. He makes a hole in the ground at the point *L* and notes the distances $LP = l_2$ and $LI = d$.

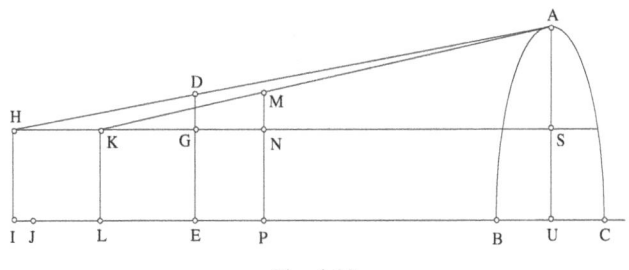

Fig. 4.19

The plane defined by *HI* and *DE* is perpendicular to the plane of the ground and contains the straight line *AU*. The points *I, L, P, U* lie on the straight line *IE* which the observer has drawn. The points *H, K, G, N* lie on a line parallel to *IE* which cuts *AU* in *S*.

So the known lengths are:

$HI = KL = GE = MP = h$, $DG = MN = 1$, which are given.
$IE = l_1$, $LP = l_2$, $IL = d$, which are measured along the straight line *IU*.

The triangles *HDG* and *HAS* are right-angled and similar; we have

$$\frac{HG}{GD} = \frac{HS}{SA}.$$

In the same way, triangles *KMN* and *KAS* are right-angled and similar; we have

$$\frac{MN}{NK} = \frac{AS}{KS}.$$

But $MN = GD$, hence

$$\frac{HG}{NK} = \frac{HS}{KS}.$$

But we have $HS > KS$, hence $HG > NK$ and consequently $IE > LP$.

Let J be a point on the segment IE such that $EJ = LP = NK$, so we have

$$\frac{EI}{EJ} = \frac{HS}{KS},$$

hence

$$\frac{EI}{IJ} = \frac{HS}{HK}.$$

But $HK = IL$, so

$$\frac{HS}{IE} = \frac{IL}{IJ},$$

and

(1) $HS \cdot IJ = EI \cdot IL;$

from the equality $\dfrac{HG}{GD} = \dfrac{HS}{SA},$ where $GD = 1$, we get

(2) $HS = HG \cdot SA.$

From (1) and (2) we get

$$HG \cdot SA \cdot IJ = EI \cdot IL.$$

But $HG = EI$, hence $SA \cdot IJ = IL$ and consequently $SA = \dfrac{IL}{IJ}$. So the height AU is equal to

$$AS + SU = \frac{IL}{IJ} + GE,$$

which can be expressed in terms of the lengths measured by the practitioner

$$AU = \frac{d}{l_1 - l_2} + h,$$

where h is the height of the eye from the ground and the lengths l_1, l_2 and d are measured along a straight line drawn on the ground, all these lengths being accessible, which explains the amount of trouble taken to find the points that define these lengths.

Ibn al-Haytham ends his book with a 'surveyor's repertory', in which he repeats all the results and procedures for measurement, without the proofs, no doubt to make it easier for the practitioner to find the formula he needs. In its own right and because of its position in the book, this repertory fulfils the purpose Ibn al-Haytham had set himself.

We may note that, in the course of this book, Ibn al-Haytham reduces the measurement of curved lines to that of straight ones, and the measurement of surface areas to that of planes, so that in the end everything reduces to linear measurements.

These measurements are referred to an arbitrary unit; so they are expressed by numbers, rational or otherwise. Ibn al-Haytham is in effect introducing a numerical concept of the ratios of magnitudes.

As elsewhere in his work, Ibn al-Haytham's procedure is orderly; the properties and proofs in stereometry are modelled as closely as possible on the analogous properties and proofs of plane geometry. Thus, in this treatise, as in his treatise on figures with equal perimeters or equal areas, polygons are dissected into triangles and, by an analogous process, polyhedra are dissected into pyramids.

Ibn al-Haytham's chief concern in this treatise is to provide rigorous proofs as a foundation for practical geometry. Perhaps that is why he pays no attention to the problem of errors in measurements.

4.2.2. A stereometric problem

The manuscript tradition identifies Ibn al-Haytham as the author of two short essays on a problem in stereometry. The first is called *On Knowing the Height of Upright Objects, on the Altitude of Mountains and the Height of Clouds* (*Fī ma'rifat irtifā' al-ashkāl al-qā'ima wa-'amidat al-jibāl wa-irtifā' al-ghuyūm*). The second essay has the title *On the Determination of the Height of Mountains* (*Fī istikhrāj 'amidat al-jibāl*). These two essays give different treatments of the same problem. The problem, as we have said,[13] belongs to an

---

[13] See above, p. 470.

ancient tradition that in some respects goes back to Euclid's *Optics*; it was considered by al-Kindī, Sinān ibn al-Fatḥ, al-Qabīṣī, and no doubt many others, and Ibn al-Haytham himself had solved it in his manual. If we compare the two solutions, we can see at once that the basic idea is the same, although the method employed in the manual is more subtle, and also more convenient. It may be that the essay presents a first attempt at a solution that Ibn al-Haytham returned to when writing his manual, which would mean this latter was written later. In that case, it seems very likely that the former text, the essay, would have been of no more than merely historical interest to Ibn al-Haytham himself.

The problem is to provide a method of calculating the height AB of an object that is inaccessible to direct measurement, that is, whose foot is at a distance that cannot be measured. Ibn al-Haytham makes use of a rod of given height DE, which he sets up parallel to the height AB, in two different positions, DE and GH, in succession, and he finds the position of the eye that allows one to see the vertex A and the tip of the rod at the same time (that is, one directly in front of the other). Ibn al-Haytham's method reduces to measuring three distances: two distances from the eye to the foot of the rod, say CD in the first case and KH in the second; and the distance between the two positions of the eye.

Let the height to be measured be AB above the horizon Bx. Let us consider a rod DE of given length that can be placed in any position, but such that $DE \perp Bx$. By trial and error we find the position C of the eye such that the sight line CA passes through E. The rod is them moved closer to the height AB; let its new position be GH and let K be the new position of the eye, where K, G, A are aligned. We have

(1) $$\frac{AB}{BC} = \frac{DE}{DC}$$

and

(2) $$\frac{AB}{KB} = \frac{HG}{KH} = \frac{DE}{KH}.$$

We have $BK < BC$, hence $KH < DC$. Let I, a point on CD, be such that $DI = HK$. We have

(3) $$\frac{ID}{DE} = \frac{KH}{HG} = \frac{KB}{AB}.$$

From (1) and (3) we get

$$\frac{CD}{DI} = \frac{CB}{BK} \Rightarrow \frac{CI}{ID} = \frac{CK}{KB};$$

from this last relation and (3), we get

$$\frac{CI}{DE} = \frac{CK}{AB} \Rightarrow AB = \frac{DE \cdot CK}{CI} = \frac{DE \cdot CK}{CD - KH}.$$

We know the length *DE* and we can measure the distances *CK* and *CI* along the horizontal line *Bx*, so we can calculate *AB*.

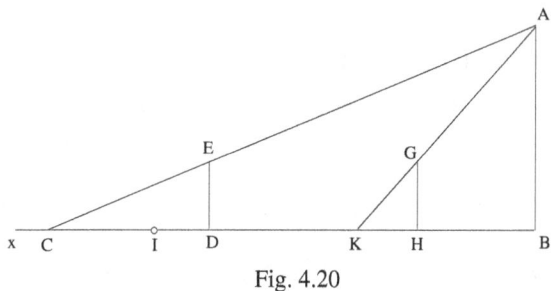

Fig. 4.20

All in all, in this essay Ibn al-Haytham offers surveyors a simple method of calculation, even if the practical procedure is not convenient. For instance, how could one place the eye on the horizontal line *Bx*? This fault will be remedied in his manual.

The second essay – *On the Determination of the Height of Mountains* – differs from the first one not only in its purpose but also in its method and its style. In this second essay, the author is attempting to respond directly to the requirements of surveyors by providing them with a quantitative rule that can be applied immediately, some would say a recipe, whereas in the first essay he makes a point of supplying a proof of the proposed rule. Further, in this second essay, the author gives a fixed numerical value to one parameter, which reinforces the impression that we do indeed have a recipe. Moreover, whereas in the first essay the author employs a rod that he moves in order to carry out the two sightings that are required, this time he keeps the rod fixed and puts a marks on it for making the two observations. In this treatise, contrary to Ibn al-Haytham's habits, there is no explicit statement of his assumptions. To understand how he may have derived his rule, let us draw a figure to take the place of the one that is missing from the manuscript that has come down to us, where the copyist has left a space for it, but has not filled it in.

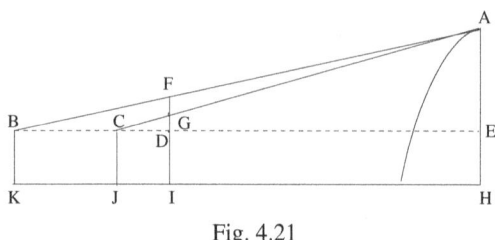

Fig. 4.21

IF is the rod, length $5\frac{1}{2}$ cubits, *ID* is the observer's height, $3\frac{1}{2}$ cubits (an assumption the author does not make explicitly), $DG = GF = 1$ cubit. At *G* we put a mark on the rod. The observer carries out a first sighting in which *B*, *F*, *A* are aligned (*AH* is the altitude to be measured, *B* is the eye of the observer). He then makes a second sighting in which *C*, *G*, *A* are aligned (*C* is the eye of the observer).

Let $l_1 = d(K, J)$ be the distance between the two observation points; $l_2 = d(J, I)$ the distance between the second observation point and the rod; $h = DI = CJ = BK = EH = 3\frac{1}{2}$ cubits, x the unknown height HA, $s = d(H, K)$. Immediately, we have

- $l_1 > l_2$,

- $x = \dfrac{2l_1}{l_1 - l_2} + h,$

- $s = \dfrac{l_1(l_1 + l_2)}{l_1 - l_2}.$

To set these essays in the context of the period, let us compare them with two other essays, by Sinān ibn al-Fatḥ and Abū Ṣaqr al-Qabīṣī respectively.

In a text called *Optical Mensuration* (*al-Misāḥāt al-manāẓiriyya*),[14] Sinān ibn al-Fatḥ addresses the same problem: measuring the height OJ of a mountain.

Let the eye, at G, look towards the mountain peak O. We take a rod EG perpendicular to GO at G. Through B, an arbitrary point on GE, we draw $BA \perp EO$. The right-angled triangles OGE and BAE have a common angle E, so

$$\frac{EA}{BA} = \frac{EG}{GO},$$

hence

$$GO = \frac{BA \cdot EG}{EA}.$$

We then use a rod HI of known height that is placed so as to be perpendicular to EJ in a position such that G, H and O are aligned. We have

$$\frac{GH}{HI} = \frac{GO}{OJ},$$

hence

$$OJ = \frac{GO \cdot HI}{GH}.$$

We assume that we can measure GH.

We may note that Sinān ibn al-Fatḥ first supposes that the eye is at G, and then takes it to be at the point E on the straight line perpendicular to the sight line GO at G. Then, from a point B on EG, he draws BA perpendicular to the sight line EO. However, he does not explain how to draw perpendiculars to straight lines defined by a line of sight, nor how to measure the segments BA, EG, EA and GH, which either lie on a sight line, like segments GH and EA, or on a perpendicular to a sight line, like segments EG and BA. One can certainly imagine a system of hollow rods to give material form to the straight lines and the segments, but Sinān ibn al-Fatḥ does not propose anything like this. Unlike those of Ibn al-Haytham, his procedure is somewhat conceptual.

---

[14] See Appendix II.

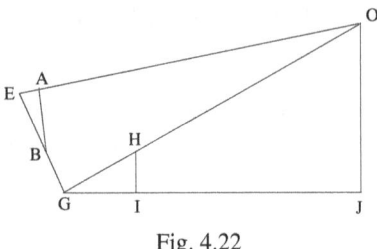

Fig. 4.22

Abū Ṣaqr al-Qabīṣī's solution is different again. In order to find a height *AB* that is inaccessible to direct measurement, he employs an astrolabe to find the height of the point *A* above the horizon for two positions *C* and *D* of the eye of the observer. He then makes use of the sines of the measured angles and of their complements. The calculations are carried out by means of right-angled triangles.

Let $A\hat{C}B = \alpha$ and $A\hat{D}B = \beta$; then

$$\sin A\hat{C}B = \sin\alpha,\ \sin B\hat{A}C = \cos\alpha,$$
$$\sin A\hat{D}B = \sin\beta,\ \sin B\hat{A}D = \cos\beta.$$

Al-Qabīṣī then carries out the following calculation:

$$\cos\beta - \frac{\cos\alpha\,\sin\beta}{\sin\alpha} = \frac{\sin\alpha\,\cos\beta - \cos\alpha\,\sin\beta}{\sin\alpha} = \frac{\sin(\alpha - \beta)}{\sin\alpha}.$$

Let $CD = d$, then $AB = \dfrac{d\,\sin\beta\,\sin\alpha}{\sin(\alpha - \beta)} = h$.

For a general triangle *ADC*, we have

$$\frac{AC}{\sin\beta} = \frac{DC}{\sin D\hat{A}C};$$

now $D\hat{A}C = \alpha - \beta$, hence

$$AC = \frac{d\,\sin\beta}{\sin(\alpha - \beta)}.$$

In triangle ABC, we have $AB = h = AC \sin \alpha$, hence

$$h = d\,\frac{\sin \alpha \sin \beta}{\sin(\alpha - \beta)},$$

which is the result given by al-Qabīṣī.

So the method requires the use of an astrolabe and recourse to a trigonometric table for the calculations. The only length to be measured is the distance CD between the two positions of the eye.

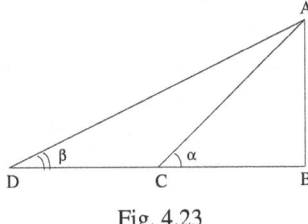

Fig. 4.23

Unlike al-Qabīṣī, Ibn al-Haytham is addressing himself to surveyors and taking into account the means they use. They do not need to employ an astrolabe, or to consult a trigonometric table; they merely need to measure two or three distances.

4.3. HISTORY OF THE TEXTS

4.3.1. *On the Principles of Measurement*

Ibn al-Haytham's treatise *On the Principles of Measurement* (*Fī uṣūl al-misāḥa*) only narrowly escaped suffering the same fate as other writings that are now lost. For this text, contrary to what is affirmed by some modern historians and biobibliographers, we have in fact no complete manuscript, but only fragments. The only fragment that has so far been printed, and that without scholarly editing, is in the India Office Library in London. It is this fragment that has, against all the evidence, been taken to be the whole text. Until now, no other fragment has been examined.

In the course of our research on this treatise, we have been able to obtain four fragments, which luckily complement one another, and thanks to which we have been able to reconstruct the whole of the treatise. We give here the *editio princeps* together with the first translation and study of the work.

1. The first fragment, by far the most substantial, is in St Petersburg, in the Library of the Oriental Institute, no. B 2139. It forms part of a collection that begins with *al-Fawā'id al-bahā'iya*, by Ibn al-Khawwām al-Baghdādī, followed by the piece by Ibn al-Haytham, then the treatise of *al-Ḥisāb* by al-Karajī.

Ibn al-Haytham's text occupies fols 100^r–139^v, each folio is 22.3 × 13.4 cm. The text is transcribed in a rectangle 18.8 × 10.4 cm, drawn in red; each folio has 15 lines, and each line contains about 13 words. The writing is in *naskhī*, in black ink; titles are underlined in red ink, and the geometrical figures are drawn in the same red ink. Unfortunately, the outer edges of the folios have been affected by damp, sometimes making reading very difficult.

These texts were copied by a certain Abū Bakr ibn Khalīl al-Tājir (?), as is indicated in the colophon of the treatise by Ibn al-Khawwām:

وفرغ من تحريرها العبد الضعيف المحتاج إلى رحمة ربه الجليل أبو بكر بن خليل التاحر (لعلها التاجر) ...

We know nothing about this copyist, except that he had a somewhat mediocre grasp of orthography. The number of mistakes in fact suggests that Abū Bakr did not belong to the learned classes. This text is designated by L.

2. The second fragment, the one we referred to above, is in the India Office Library in London, no. 1270, fols 28^v–32^v, designated by I. We have spoken about this manuscript several times, since it contains numerous treatises by Ibn al-Haytham.[15] The treatise *On the Principles of Measurement*, like the treatise that precedes it in the manuscript – *On a Proposition of the Banū Mūsā* – has suffered an significant accident, pointed out by a reader long ago. While the latter text is cut off towards the end, the former is cut off at the beginning, and the two treatises are presented continuously as one single work. Everything seems to point to this accident having happened in the manuscript from which this same manuscript was copied. The accident did not fail to attract the attention of a reader, who wrote in the left-hand margin:

قد فات من هنا آخر رسالة بني موسى

---

[15] See for example *Ibn al-Haytham and Analytical Mathematics*, p. 33; and here, Chapter III.

'missing here <is> the end of the treatise on a proposition of the Banū Mūsā';
and in the right-hand margin he wrote:

من هنا رسالة المساحة التي فات أولها

'From here <we have> the treatise on measurement, whose beginning is missing'.

And this lacuna is rather substantial.

3. The third fragment is in Istanbul, Süleymaniye, Fātiḥ 3439, fols 103ᵛ–104ᵛ, designated by F. We have already told the little we know about this manuscript.[16]

4. The fourth fragment is part of a manuscript in the National Library in St Petersburg, no. fyrk Arabic 143, fols 13ᵛ–15ᵛ. In fact we have a collection of mathematical writings in which the first treatise is the *Book on Measurement* by Abū Bakr al-Māristānī. The copyist gives the date of the transcription in the colophon of this first treatise:

في العشر الأخير من ربيع الأول سنة اثنتي عشرة وستمائة

that is to say, during the month of July 1215. The writing is in *nasta'līq*, and the treatise is designated here by D. We note that this manuscript has been damaged by damp, which sometimes makes reading difficult.

Investigation of variants and omissions shows that these four manuscripts certainly belong to four different families: each one is lacking words and sentences in an individual way.

The editions, translations and studies do not amount to much. As we have indicated, only the fragment in the India Office has been read: E. Wiedemann has made a German translation of its final *Memento*, that is, pages 31ʳ to 32ᵛ only, and not of the fragment as a whole, as some historians and bibliographers have recently suggested.[17] This rather free partial translation is not accompanied by a study. It is this same fragment that was published later,

---

[16] *Ibn al-Haytham and Analytical Mathematics*, p. 33.
[17] E. Wiedemann, *Aufsätze zur arabischen Wissenschafts-Geschichte*, Hildesheim/ New York, 1970, vol. 1, pp. 534–42.

without any critical examination of the text;[18] in any case, it represents only a little more than a third of the *Treatise*.

4.3.2. *On Knowing the Height of Upright Objects, the Altitude of Mountains and the Height of Clouds*

This essay – *Fī ma'rifat irtifā' al-ashkāl al-qā'ima wa-'amidat al-jibāl wa-irtifā' al-ghuyūm* – does not appear in the list of the writings of Ibn al-Haytham established by the old biobibliographers. The case is in no way unique, and does not necessarily cast doubt on the authenticity of the text. It might at most encourage us to check whether this essay was perhaps an extract from another treatise by Ibn al-Haytham. But this is not so: the method employed here is different from those applied in the two other treatises that we know. Further, this essay has come down to us in four manuscripts, duly attributed to Abū 'Alī ibn al-Haytham:

The first manuscript forms part of an important collection in the Columbia University Library, New York, Smith Or. 45/12, fols 243ᵛ–244ʳ, here designated by K.

We have already had recourse to this manuscript to establish texts of writings by al-Khayyām, by Sharaf al-Dīn al-Ṭūsī and by al-Sijzī, and we have shown each time that it was the single original from which the copyist of the collection Leiden Or. 14 had transcribed these texts. On this occasion, Ibn al-Haytham's treatise is again not an exception. This manuscript probably dates to the thirteenth century, it has been copied in *naskhī* script and all the figures are drawn by the copyist. Moreover, examination of the manuscript shows that several parts have been lost, probably pages torn out, and even a whole treatise by al-Qūhī, whose title has been recorded with the other titles that are included, on the first page (fol. 1ʳ). The work in question is *The Art of the Astrolabe by Demonstration*.

The second manuscript forms part of the important collection Leiden Or. 14, fols 236–237, here designated by L. We have just referred to this collection and noted that for several treatises contained in it, its single original was the manuscript Smith Or. 45. Let us return once more to its history.

We know from R.P.A. Dozy, according to the catalogue he compiled in 1851, that Golius, the seventeenth-century mathematician and Arabist, while

---

[18] Ibn al-Haytham, *Majmū' al-rasā'il*, Osmānia Oriental Publications Bureau, Hyderabad, 1938–1939.

travelling in the Orient, at Aleppo had employed al-Darwish Aḥmad to transcribe the last three books of Apollonius' *Conics*. That copy was completed on 15 Dhū al-ḥijja 1036, that is, on 27 August 1627. It was at about this date that Golius had had Or. 14 transcribed. Dozy writes about this:

> Opera, a Nicolao in usum Golii descripta, continentur Codice 14. Pleraque eorum mathematici sunt argumenti, quumque inter ea inveniantur quae unica sunt in Europa, Golio Codicem unum pluresve commodatos esse ab Orientali quodam viro suspicor, quos, quum venales non essent, in Orientem remisit.[19]

So, according to what Dozy conjectures, it was the Arab Nicholas, then resident in Amsterdam, who copied this manuscript for Golius, who then sent the manuscripts used for the copy back to the Orient, one of these copies being Smith Or. 45.

In our previous writings, we have followed Dozy. Recently, J. J. Witkam has rejected Dozy's conjecture and maintained that Or. 14 was copied at Aleppo at the same time, and the end of the 1620s. He writes:

> De codex Or. 14 in de Leidse Universiteitsbibliotheek is zo'n verzameling van afschriften, die voor Golius in Aleppo gemaakt is, duidelijk op kanselarij-papier, en door een Aleppijnse schrijver geschreven. De figuren in de wiskundige tractaten werden door Golius later met de hand bijgetekend in daarvoor aangebrachte uitsparingen in de tekst.[20]

More explicitly, J. J. Witkam writes in connection with this manuscript Or. 14, in a note he kindly communicated to us:

> Collective volume with texts in Arabic, and one in Persian (no. 22), European paper (from the Dutch consular chancellery in Aleppo?), 501 pp, and blanks. Copies made by al-Darwish Aḥmad (his colophon on p. 163, but he is the copyist of the entire volume) in Aleppo for Jacobus Golius. The drawings in No. 1 were made by Golius himself, who occasionally wrote *ḥawāshī* as well. The other drawings and figures appear to be by the copyist. Golius had most texts in this volume copied because he apparently could not acquire the originals. The exception to this is, of course, the first text in the volume, which is a working copy of the MS in his private collection.

In turn we can show that this collection includes manuscripts transcribed from various sources, the most important of which is the collection Smith Or.

---

[19] *Catalogus Codicum Orientalium Bibliothecae Academiae Lugduno Batavae*, Leiden, 1851, p. xv.

[20] *Jacobius Golius (1596-1667) en zijn handschriften*, Oosters Genootschap in Nederland 10, Leiden, 1980, p. 53.

45. Of 26 treatises included in the Leiden manuscript, 12 have as their only original the manuscript at Columbia. They are the following treatises:

1. *Maqāla fī al-Jabr wa-al-muqābala* by al-Khayyām,[21]

2. *Iḍāḥ al-Burhān 'alā ḥisāb al-khaṭṭayn* by Abū Sa'īd al-Ṣābi',

3. Glosses by Abū al-Futūḥ ibn al-Sari on the preceding treatise,

4. *Muqaddima li-ṣan'at āla tu'rafu bihā al-ab'ād* by al-Sijzī,

5. Treatise on the asymptotes to an equilateral hyperbola by al-Sijzī,[22]

6. Treatise on the asymptotes to an equilateral hyperbola by al-Qummī,

7. *Maqāla fī ma'rifat irtifā' al-ashkāl al-qā'ima wa-'amidat al-jibāl wa-irtifā' al-ghuyūm* by Ibn al-Haytham,

8. *Mas'ala dhakarahā Abū Naṣr al-Fārābī fī al-maqāla al-ūla min al-fann al-awwal fī al-mūsīqī,*

9. *Min kalām Abī al-Futūḥ ibn al-Sarī,*

10. *Maqāla fī istikhrāj al-quṭb 'alā ghāyat al-taḥqīq* of Ibn al-Haytham,

11. *Kitāb ṣan'at al-asṭurlāb bi-al-burhān* by al-Qūhī, followed by the commentary of Abū al-'Alā' ibn Sahl,[23]

12. *Mas'ala sa'alahā Shams al-Dīn Amīr al-Umarā' al-Niẓāmiyya ilā Sharaf al-Dīn al-Ṭūsī.*

The third manuscript forms part of the collection in the Malik Library in Teheran, no. 3433, fols 1^v–2^r, designated by A.

This manuscript was transcribed at the Niẓāmiyya school in Baghdad, in the middle of the month of Rabī' al-awwal, in the year 557 of the Hegira, that is in March 1162. The writing is in careful *naskhī* and the text has no glosses or crossing out. However, there is nothing to indicate that the copyist revised his copy by comparing it with the original. The single correction concerns a

---

[21] Edition, French translation and commentary in R. Rashed and B. Vahabzadeh, *Al-Khayyām mathématicien*, Paris, 1999; English version: *Omar Khayyam. The Mathematician*, Persian Heritage Series no. 40, New York, 2000.

[22] Edition, French translation and commentary in R. Rashed, 'Al-Sijzī et Maïmonide: Commentaire mathématique et philosophique de la proposition II–14 des *Coniques* d'Apollonius', *Archives internationales d'histoire des sciences,* 119, 37, 1987, pp. 263–96; repr. in *Optique et Mathématiques: Recherches sur l'histoire de la pensée scientifique en arabe*, Variorum reprints, Aldershot, 1992, XIII.

[23] See R. Rashed, *Géométrie et dioptrique au Xe siècle: Ibn Sahl, al-Qūhī et Ibn al-Haytham*, Paris, 1993; English transl. *Geometry and Dioptrics in Classical Islam*, London, 2005.

gloss that was added after the text and which is found in [K], it refers to a method attributed to Sa'd al-Dīn ibn As'ad ibn Sa'īd al-Hamadhānī.

The fourth manuscript belongs to a collection in the library of Majlis Shūrā in Teheran, no. 2773/2, fols 19–20; here it is designated by I. This collection also contains the commentary on this text of Ibn al-Haytham by Ibn Aḥmad al-Ḥusaynī Muḥammad al-Lāhjānī (fols 1–17). This commentary was completed on Sunday 25 Dhū al-qa'da 1105, that is 18 July 1694. The text by Ibn al-Haytham is in the same hand; so it to was copied at about that date. The copyist has transcribed the final gloss by al-Hamadhānī, but without naming its author and reversing 'the correction' by the copyist of [A]. Thus he writes in the text 'the correction' of this last and in the margin the corrected word; this seems to indicate that [A] was the original from which this copy was made.

So we have two families of manuscripts: the family [K, L], where L is a copy of K and of it only; and the family [A, I], where I is a copy of A and of it only.

The two families have a common origin that goes back to the manuscript whose copyist added the gloss by al-Hamadhānī, that is, to before the mid twelfth century.

4.3.3. On the Determination of the Height of Mountains

Unlike the preceding one, this essay appears under its title in the list of Ibn Abī Uṣaybi'a (*Fī istikhrāj 'amidat al-jibal*). But at the moment we know of only a single manuscript. The attribution to al-Ḥasan ibn al-Ḥasan ibn al-Haytham is explicit; the vocabulary is that of Ibn al-Haytham; the style, to some extent concerned with experimental matters, is that of Ibn al-Haytham the natural philosopher. However, it remains to point out some traits that do not belong to the style of Ibn al-Haytham as a mathematician: the absence of a proof in a properly rigorous form; the presence of a hypothesis that is not made explicit (the height of the observer being $3\frac{1}{2}$ cubits); a rather peculiar step for finding s (he multiplies by $l_1 + l_2$ then he divides by the same quantity instead of, as he usually does, appealing to the fact that the two triangles AEB and FDB are similar). Perhaps we have a version of Ibn al-Haytham's initial text made by someone else, but there is nothing to authorize such a conjecture.

The only manuscript of this text is Arch. Seld. A 32 in the Bodleian Library in Oxford, fols 187ʳ–188ʳ. The copyist has left a space for a figure, which he did not draw when revising the copy against the original. This latter must have had a figure, or also an empty space. The writing is *naskhī*, and the colophon is in *nasta'līq*. The copyist explicitly states that he has revised the copy against the original; he added in the margin a correction and an omission.

Like the preceding ones, this text has not been critically established, or translated or studied before now.

4.4. TRANSLATED TEXTS

Al-Ḥasan ibn al-Haytham

In the Name of God, the Compassionate, the Merciful

TREATISE BY AL-ḤASAN IBN AL-ḤASAN IBN AL-HAYTHAM

On the Principles of Measurement

In my youth I had composed a book on the principles of measurement. Then events took place that completely destroyed many of the original copies of my writings: this book was among the group of things that were lost to me. After some time had elapsed, a friend to the sciences, moved by a virtuous inclination, asked me to compose something on measurement for him. I accordingly returned to this book in order to help him; it contains the principles of everything that is used in <making> measurements. I do, however, suspect that, in some of its expressions and proofs, there may be differences from the expressions in the first book and its proofs. I have, moreover, added complementary material that does not appear in the first book. If it so happens that some people who study this science come across two copies of this book in which the expressions differ, they should note that this is for the reason just given. I now begin my explanation of measurement.

The measurement of magnitudes is their evaluation in terms of the magnitude given as the measure. The magnitude given as the measure is a straight line whose magnitude is a matter of agreement, so that all magnitudes are evaluated in terms of it,[1] such as the bushel that is agreed upon to evaluate all measures of bulk[2] and like the *mithqāl* and the *arṭāl* in terms of which we evaluate things that are weighed. The line that the surveyor calls a cubit or magnitude depends on a suggestion from someone who began by choosing to adopt it.

Magnitudes that can be measured may be divided into three kinds, which are lines, surfaces and solid bodies. The measurable lines that we need to measure are the sizes of distances, the lengths of surfaces of solid bodies and their widths, and the heights of tall solids. Measurable surfaces

---

[1] It is assumed that these are linear magnitudes.

[2] This seems to be concerned with an instrument for measuring dry materials.

are the surfaces of solid bodies. Measurable solid bodies are all the bodies of the kind we seek to measure.

As for the lines, they can be divided into five kinds which are straight lines, circular ones and the three sections that are the sections of cones; in their art surveyors use only straight lines and circular ones.[3]

Surfaces may be divided into three kinds which are plane, convex and concave; in their art surveyors use only plane surfaces. As for convex and concave surfaces, they are surfaces <that are> spherical, cylindrical, conical or surfaces compounded from these and they (such surfaces) play no part in their art of measurement although these surfaces can be reduced to plane surfaces, because for each of them we can determine a ratio to the set <of figures> that it contains.[4] This has been shown in books by geometers and in this manner the measurement of all surfaces is reducible to the measurement of these plane surfaces.

As for bodies, they are of only one kind, that of everything that has a length, a width and a depth, except that their shapes are different.

The measurement of lines is their evaluation in terms of the same cubit, the measurement of surfaces is their evaluation in terms of the square of the cubit and the measurement of bodies is their evaluation in terms of the cube of the cubit. The quantity of the measurement of lines is the number of the multiple of the cubit they contain. The quantity of the measurement of surfaces is the number of the multiple of the square of the cubit that they contain. The quantity of the measurement of bodies is the number of the multiple of the cube of the cubit that they contain.

Procedure for measuring lines: if they are straight lines, we carry out their measurement by applying the cubit, to part after part, on the lines until it exhausts them, either the whole straight line, or some of its parts; for circular <lines>, that is to say the circumference of a circle, the procedure of measuring it consists of measuring the diameter of the circle, so as to then multiply the number of the multiple of the cubit contained in the diameter by three and a seventh; what we obtain is the quantity of the measurement of the circumference of the circle. Archimedes has shown that the circumference of the circle is <the product of> its diameter and three and a seventh, to an extremely close approximation. It is in this manner that we learn the quantity of the measurement of the circumference of the circle. As for an arc of a circle, the quantity of its measurement is

---

[3] This provides a glimpse of the related problem of the rectification of the three conic sections. We may note that Ibn al-Haytham had addressed this problem in the text that was lost.

[4] He is envisaging the measurement of curved surfaces by the inscription and circumscription of plane figures.

known when we know its ratio to the complete circumference. We shall
show later how to find this ratio, when we discuss the measurement of a
sector of a circle.

Procedure for measuring surfaces in general: this will be done by
measuring their lengths and their widths and multiplying the ones by the
others, as we shall show later.

Procedure for measuring bodies in general: this will be done by
measuring their bases and their heights and multiplying the ones by the
others, as we shall show later.

The procedure for measuring surfaces by a detailed account in terms of
the art will be as we shall describe it: it has been shown that all surfaces
can be reduced to plane surfaces and that the plane surfaces with which the
surveyor's art is concerned are those bounded by straight lines and circular
lines. And among the plane surfaces bounded by straight lines, some have
parallel sides and right angles and the others do not.

For a surface with parallel sides and right angles,[5] we carry out its
measurement by measuring one of its lengths and one of its widths and then
multiplying the number of the multiple of the cubit in the length by the
number of the multiple of the cubit in the width; what we obtain is the
number of the multiple of the square of the cubit in the surface.

Example: The surface $ABCD$ has parallel sides and right angles. Let the
cubit given as the measure go into AB four times and go into BC three
times; we multiply four by three, we have twelve.

I say that the quantity of the measure of the surface AC *is twelve
cubits, that is to say the multiple twelve of the square of the cubit given as
the measure.*

Proof: We divide up AB into equal parts: each of them is equal to one
cubit; we shall have four parts, let them be AE, EG, GH, HB. We also
divide up BC into equal parts; we shall have three parts, let them be BI, IK,
KC. We draw from the points E, G, H straight lines parallel to the two
straight lines BC and AD, let the straight lines be EV, GM and HL. We
draw from the two points I and K straight lines parallel to the two straight
lines AB and DC; let the two straight lines be IU and KP. Let the straight
line IU cut the straight lines HL, GM and EV at the points O, S and Q; the
surfaces AQ, ES, GO and HI are equal quadrilaterals whose sides are equal
and whose angles are right angles. They are equal because their bases –
which are AE, EG, GH, HB – are equal, and they are between two parallel
straight lines which are AB and UI. Their sides are equal because each of
their widths, which are AU, EQ, GS and HO, is equal to the straight line BI

---

[5] That is, a rectangle.

and *BI* is equal to each of the <straight lines> *BH, HG, GE, EA*, which are the lengths of these quadrilaterals. They are at right angles since the angle *HBI* is a right angle and the straight line *IO* is parallel to the straight line *BH*, so <the sum of> the angles *HBI* and *OIB* is equal to two right angles and the angle *HBI* is a right angle; so the angle *OIB* is a right angle and the two angles *BHO* and *HOI* are opposite the two angles *OIB* and *HBI*; so they are equal to them, so the quadrilateral *BHOI* is right-angled. In the same way, we show that each of the quadrilaterals *OG, GQ, QA* is right-angled, and that the quadrilaterals *AQ, ES, GO, HI* thus have equal sides and right angles;[6] the number of these squares is the number of the straight lines *AE, EG, GH, HB*, which is the number of the multiple of the cubit in *AB*.

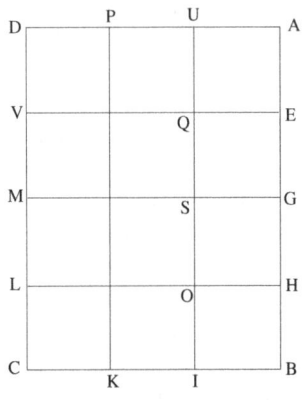

Fig. IV.1.1

In the same way, the surfaces *AI, UK* and *PC* are equal since their bases – which are *BI, IK* and *KC* – are equal and they are between the parallel straight lines *AD* and *BC*; and the number of these surfaces is equal <to the number> of the straight lines *BI, IK, KC*, which is the number of the multiple of the cubit in *BC*. If we multiply the number of the squares which are in the surface *AI* by the number of the surfaces *AI, UK, PC*, what we obtain is the number of everything the surface *AC* contains by way of equal squares, each of which is equal to the square *HI*, which is the square of the cubit given as the measure. But the number of the squares contained in the surface *AI* is the number of the straight lines *AE, EG, GH* and *HB*, which is the number of the multiple of the cubit in *AB*. The number of the surfaces *AI, UK* and *PC* is the number of the straight lines *BI, IK* and *KC*, which is the number of the multiple of the cubit in *BC*. So if we multiply the number

---

[6] They are thus squares.

of the multiple of the cubit in *AB* by the number of the multiple of the cubit in *BC*, we have <that> the number of the multiple of the square of the cubit contained in the surface *AC* with parallel sides and right angles. That is what we wanted to prove.[7]

Among the rectilinear surfaces that are not right-angled, some are bounded by three straight lines and others are bounded by more than three straight lines. Those that are bounded by more than three straight lines can all be divided into triangles. The general method for measuring all rectilinear surfaces is the method for measuring triangles; those bounded by three straight lines are triangles and we measure them as we measure triangles, and surfaces bounded by more than three straight lines are divided into triangles each of which will be measured separately; we then find the sum of the measures of all the triangles into which the surface has been divided; what we obtain is the measure of the complete surface.

Among triangles, some have a right angle, some have an obtuse angle and some have acute angles. We measure each of them by finding the perpendicular drawn from its vertex to its base, and then multiply that perpendicular by half the base; what we obtain is the measure of the triangle, that is to say that we multiply the number of the multiple of the cubit in the perpendicular by the number of the multiple of the cubit in half the base; what we obtain is the number of the multiple of the square of the cubit in the triangle.

Example: *ABC* is a triangle; first let it be right-angled. Let the right angle be the angle *ABC*, the point *A* its vertex; so its perpendicular <height> is the straight line *AB* and its base is the straight line *BC*.

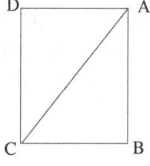

Fig. IV.1.2

I say that its measure is what we obtain from the product of AB *and half* BC.

Proof: We draw from the point *C* a straight line parallel to the straight line *AB*, let it be *CD*; we have *CD* perpendicular to *BC*. We draw from the point *A* a straight line parallel to the straight line *BC*, let it be *AD*; then

---

[7] We may note that the only geometrical result used is the angle property of parallel lines (*Elements*, I.29).

ABCD is a parallelogram with right angles. So the quantity of its measure is what we obtain from the product of the multiple of the cubit in *AB* and the multiple of the cubit in *BC*, as has been shown in the preceding proposition. But the triangle *ABC* is half the surface *ABCD*,[8] so the quantity of its measure is half what we obtain from the product of *AB* and *BC*. But the product of *AB* and half *BC* is half the product of *AB* and *BC*, so what we obtain from the product of *AB* and half *BC* is the quantity of the measure of the triangle *ABC*. That is what we wanted to prove.

Let the triangle *ABC* have an obtuse angle or have acute angles. Let us draw from the point at its vertex, which is *A*, the perpendicular *AD*.

I say that the quantity of the measure of the triangle ABC *is what we obtain from the product of* AD *and half* BC.

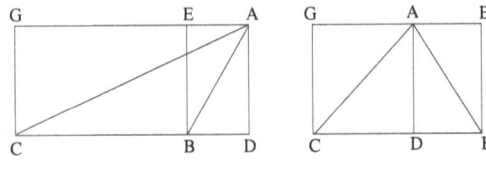

Fig. IV.1.3

Proof: We draw from the points *B* and *C* two straight lines parallel to the straight line *AD*, let it be *BE* and *CG*. They are perpendicular to the base *BC*. We draw from the point *A* a straight line parallel to the straight line *BC*. Let it meet the two straight lines *BE* and *CG* at the points *E* and *G*; then the surface *EBCG* is a parallelogram with right angles, so the quantity of the measure of this surface is what we obtain from the product of *EB* and *BC*. But the triangle *ABC* is half of the surface *EBCG*, because they are on the same base and between two parallel straight lines.[9] So the quantity of the measure of the triangle *ABC* is half the quantity of the measure of the surface *EBCG*; so the product of *EB* and half *BC* is the quantity of the measure of the triangle *ABC*. But *EB* is equal to *AD*, because the surface *ADEB* is a parallelogram, so the quantity of the measure of the triangle *ABC* is what we obtain from the product of *AD*, which is the perpendicular, and half *BC*, which is the base of the triangle. That is what we wanted to prove.

It remains for us to show how we know that a triangle has a right angle or an obtuse angle or acute angles, and how to find the perpendicular if the triangle has an obtuse angle or acute angles.

[8] Euclid, I.34.
[9] Euclid, I.36 and 37.

The method for finding out the nature of the triangle consists of multiplying its greatest side by itself, that is to say the number of the multiple of the cubit in the greatest side <is multiplied> by itself; we note <this square>, we then multiply each of the other two sides by itself, we add up the <results> and we compare <the sum> with the first square. If the sum of the two squares is equal to the first square, then the triangle has a right angle, as has been shown at the end of the first book of the work of Euclid[10] and we shall obtain its measure by multiplying half one of the two smaller sides by the other; what we obtain is the quantity of its measure, as has already been shown.

If what we obtain from the squares of the two smaller sides is smaller than the first square, then the triangle has an obtuse angle and if what we obtain from the squares of the two smaller sides is greater than the first square, then the triangle has acute angles.

To find the perpendicular <height> for a triangle with an obtuse angle, we take away from the square of the greatest side the sum of the squares of the two smaller sides, we take half of what remains, we then divide this half by the base of the triangle, that is to say that we divide the number which is in that half by the number of the multiple of the cubit in the base, and we note the quotient, which we call the foot of the perpendicular;[11] if we find its perpendicular in this way, the base of the triangle with an obtuse angle will be one of the two smaller sides; let one or the other be taken as the base, then the foot of the perpendicular will be in the extension of that side, because from each of the angles of the triangle we can draw a perpendicular to the side opposite it. If we obtain the foot of the perpendicular, we multiply it by itself, we take away its square from the square of the smaller side that is on the side of the vertex of the triangle, that is to say the smaller side that has not been considered to be the base; we take the root of what remains of the square of this side, this is the perpendicular.

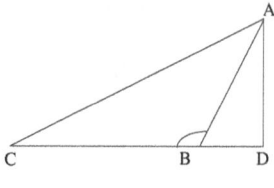

Fig. IV.1.4[12]

---

[10] Euclid, *Elements*, I.47.

[11] Here, this expression refers to the segment *BD*.

[12] This figure does not appear in the manuscript.

Example: Let there be a triangle with an obtuse angle as before, that is to say *ABC*; its angle *ABC* is obtuse. It has been shown in the twelfth proposition of the second book of Euclid's work that the square of *AC* is equal to the square of *AB*, plus the square of *BC*, plus twice the surface enclosed by *CB* and *BD*. If we subtract the square of *AB* and the square of *BC*, added together, from the square of *AC*, what remains is twice the product of *CB* and *BD*. If we take half of it, we obtain the product of *CB* and *BD*. But for any product obtained by multiplying one of the two numbers by the other, if it is divided by one of these two numbers, the quotient we obtain is the other number. That is why if we divide half what remains of the square of *AC* by the straight line *BC*, the quotient we obtain is the straight line *BD* which is called the foot of the perpendicular. But the triangle *ADB* has a right angle, so the square of *AB* is equal to the square of *AD*, plus the square of *DB*. If we subtract the square of *DB* from the square of *AB*, what remains will be the square of *AD*. If we take its root, what we obtain is the straight line *AD*, that is to say the number of the multiple of the cubit in *AD*. But the straight line *AD* is the perpendicular of the triangle *ABC*, which has an obtuse angle, and the perpendicular lies outside the triangle. It is in this way that we find the perpendicular of a triangle that has an obtuse angle.

As for the triangle with acute angles, each of its angles is thus acute and each of its sides can be considered to be the base because, for each of its sides, a perpendicular can be drawn to the side from the opposite angle. To find the perpendicular for a triangle with acute angles, we suppose one of its sides is the base and we multiply one of the two remaining sides by itself; we note <the square>, we then multiply the remaining side by itself, we multiply the base by itself, we add up these two squares and from <their sum> we then subtract the first square that we noted; we take half of what remains, we then divide this half by the base; the quotient we obtain is the foot of the perpendicular. If we obtain the foot of the perpendicular, we multiply it by itself, we then take away <its square> from the square of the remaining side – the side whose square was added to the square of the base – we take the root of what remains, which will be the perpendicular.

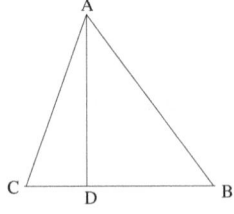

Fig. IV.1.5

Example: Let there be the preceding triangle with acute angles, which is the triangle *ABC* whose angle *ABC* is acute and whose base is *BC*. It has been shown in the thirteenth proposition of the second book of Euclid's work that the square of *AC* is less than the sum of the square of *AB* and the square of *BC* by twice the surface enclosed by the two straight lines *CB* and *BD*. So if we subtract the square of *AC*, which is one of the two remaining sides (other than the base), from the sum of the squares of *AB* and *BC*, what remains is twice the product of *CB* and *BD*; if we take half of this, we have the product of *CB* and *BD*; if we divide that by *BC*, the quotient we obtain is *BD*, and *BD* is called the foot of the perpendicular. But the triangle *ABD* is right-angled, and the angle *ADB* is a right angle because *AD* is perpendicular to *BC*; so the square of *AB* is equal to the sum of the square of *BD* and the square of *AD*. If we subtract the square of *BD* from the square of *AB*, what remains is the square of *AD*. If we take its root, we have the straight line *AD* which is the perpendicular of the triangle. This is the way in which we find the perpendicular of a triangle with acute angles.

We can find the measure of all triangles by a single method which is the method <used> for a triangle with acute angles, because in every triangle there are two acute angles, and the remaining angle is different. So if in every triangle there are two acute angles, then we can find its perpendiculars and its measure by a single method which consists of finding the perpendicular of a triangle with acute angles, and this <is done> by putting the greatest side of the triangle as the base of the triangle if the triangle is scalene; if it is isosceles, we suppose that one of its sides, which is not the smallest of its sides, is a base, and if it is equilateral, we suppose one of its sides is a base; we then multiply one of the two remaining sides by itself, we note <the square>, we multiply the remaining side of the two remaining sides by itself, similarly we multiply the base by itself, we add up the last two squares, from which we subtract the square we noted. The sum of these two squares cannot but be greater than the square we noted – because the square of the base alone is not smaller than the square we noted; if we subtract the first square we noted, from the two squares added up, and if we take half of what remains and we divide it by the base, the quotient we obtain is the foot of the perpendicular; to complete finding the perpendicular, we proceed as described before so as to find the perpendicular of the triangle. If we obtain the perpendicular, we multiply it by half the base and what we obtain is the measure of the triangle.

Proof of this procedure: In any triangle that has a right angle or an obtuse angle, the greatest side is the one opposite the right angle or the obtuse angle; if we put the greatest side of the triangle as the base, the two remaining sides are opposite two acute angles. If we square one of the two remaining sides and we note <the square>, what we note will be the square of the side opposite an acute angle and the two squares we added up will be the squares of the two sides that enclose an acute angle, in this way the perpendicular we find is the perpendicular drawn from the right angle or the obtuse angle to the base of the triangle which is the chord of this angle. The method for finding the perpendicular of a triangle with a right angle or an obtuse one and for finding its measure will be, in this way, the method for finding the perpendicular of a triangle with acute angles and its measure.

If there is not a side in the triangle that is the greatest side, the triangle cannot but have acute angles, because in a triangle with a right angle or an obtuse one, the side that is the chord of the right angle or the obtuse angle is always greater than each of the two remaining sides.

We can find the measure of all triangles by a single general method in which we do not need to find the perpendicular; and this <is done> by adding up the sides of the triangle and taking half of the sum; we then multiply this half by the amount by which it exceeds one of the sides of the triangle, we multiply what we obtain by the amount by which this half exceeds another of the sides of the triangle and we multiply what we obtain by the amount by which this half exceeds the remaining side among the sides of the triangle; we take the root of the product; what we obtain is the measure of the triangle.

<a> *Example*: Let there be a triangle whose sides are ten, eight and six. We add up the three, we have twenty-four; we take half, we have twelve, which we multiply by the amount by which twelve exceeds six, which is six; we have seventy-two; we then multiply seventy-two by the amount by which twelve exceeds eight, which is four, we have two hundred and eighty-eight; we then multiply two hundred and eighty-eight by the amount by which twelve exceeds ten, which is two, we have five hundred and seventy-six, of which we take the root; we have twenty-four, which is the measure of the triangle; this triangle has a right angle, because the square of ten is equal to the square of eight and the square of six, added together. The angle enclosed by eight and six is a right angle, the measure of the triangle is the product of eight and half of six, which is three, which is twenty-four.

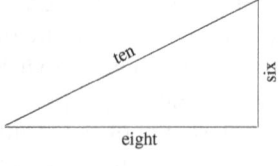

Fig. IV.6

<b> *Proof of the general method we mentioned*: We assume we have the triangle *ABC*, a triangle of any kind, we divide its angle *ABC* into two equal parts with the straight line *BD*, we divide its angle *ACB* into two equal parts with the straight line *CD* and from the point *D* we draw the perpendiculars *DE*, *DH* and *DG*. Since the angle *HBD* is equal to the angle *EBD*, and the angles at the two points *H* and *E* are right angles, and the straight line *BD* is common to the two triangles *HBD* and *EBD*, accordingly the two triangles are equal and have equal sides, the perpendicular *DH* is equal to the perpendicular *DE* and the side *HB* is equal to the side *BE*. Similarly, we show that the perpendicular *DG* is equal to the perpendicular *DE*[13] and the side *EC* is equal to the side *CG*, so the perpendiculars *DH*, *DE* and *DG* are equal and the product of *DG* and half of *AC* is the measure of the triangle *ADC*. So the product of the perpendicular *DE* and half of the perimeter of the triangle is <the sum> of the measures of the triangles *ADB*, *BDC*, *CDA*. But <the sum of> these three triangles is the complete triangle *ABC*, so the product of half the perimeter of *ABC* and the perpendicular *DE* is the measure of the triangle *ABC*. So the product of the square of the number of the multiple of the cubit in *DE* and the square of the number of the multiple of the cubit in half of the perimeter of the triangle is the square of the measure of the triangle, because for two numbers one of which is multiplied by the other, if the product is then multiplied by itself, what we obtain is equal to what we get by multiplying the square of one of the two numbers by the square of the other. But the product of the square of one of the two numbers and the square of the other is the product of one of the two numbers and itself and <the product> of what we obtain and the square of the other. But the product of one of the two numbers and itself and then <the product> of what we obtain and the square of the other number is the product of one of the two numbers and the square of the other and then <the product> of what we obtain and the first one, because what we obtain from the product of the numbers one by another, by commuting, is always the same. So the product of the square of the semiperimeter of the triangle and the square of *DE* is the product of the semiperimeter and the square of *DE* and <the product> of what we obtain and the semiperimeter. So the product of the number of the multiple of the cubit in the semiperimeter of the triangle *ABC* and the square of the number of the multiple of the cubit in *DE* and then <the product> of what we obtain and the number of th multiple of the cubit in the semiperimeter, is the square of the area of the triangle *ABC*.

---

[13] Ibn al-Haytham does not say that *ED* is the radius of the circle inscribed in the triangle.

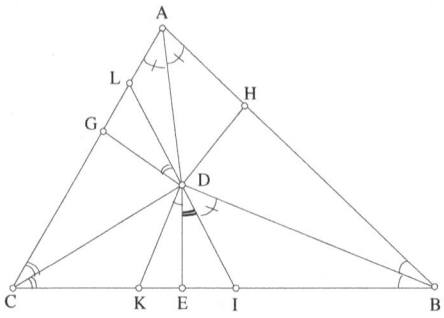

Fig. IV.1.7

<c> Moreover, the square of *AD* is equal to the <sum of the> squares of *AH* and *HD*, because the angle at *H* is a right angle; similarly the square of *AD* is equal to the <sum of the> squares of *AG* and *GD*, so the squares of *AH* and *HD* are equal to the <sum of the> squares of *AG* and *DG*; but the square of *HD* is equal to the square of *GD*, because we have shown that *DH* is equal to *DG*; it remains <that> the square of *AH* is equal to the square of *AG*, so *AH* is equal to *AG*, *HD* is equal to *GD* and *AD* is common; so the two triangles *AHD* and *AGD* have sides and angles equal and the angle *HAD* is equal to the angle *GAD*. Moreover, given that the angle *BED* is a right angle and the angle *BDE* is acute, we make the angle *BDK* a right angle; similarly the angle *CDE* is acute, we make the angle *CDI* a right angle, so it is equal to the angle *DEI* and the angle *DIE* is common to the two triangles *DIE* and *DIC*; it remains <that> the angle *IDE* is equal to the angle *DCI*. But the angle *DCI* is equal to the angle *DCL* and the angle *DCL* is equal to the angle *LDG* because the two triangles *LDG* and *DCL* are similar; so the angles *IDE* and *LDG* are equal, the angles *DEI* and *DGL* are equal because they are right angles and the straight line *DE* is equal to the straight line *DG*; so the two triangles *DIE* and *DLG* have equal sides and angles, the straight line *IE* is equal to the straight line *LG* and the angle *DIE* is equal to the angle *DLG*. Similarly, the angle *DBC* is half the angle *ABC*, the angle *DCB* is half the angle *ACB*, the angle *DAC* is half the angle *BAC* and <the sum> of the angles *DBC*, *DCB* and *DAC* is half that of the angles of the triangle; but <the sum> of the three angles of the triangle is equal to two right angles, so the sum of the angles *DBC*, *DCB* and *DAC* is equal to a right angle. But the angle *DBC* is equal to the angle *KDE* because the triangle *EDK* is similar to the triangle *BDK*, since the angle *BDK* is a right angle; similarly the angle *EDI* is equal to the angle *DCB*, so the sum of the two angles *KDE* and *EDI*, which is equal to the angle *KDI*,

is equal to the angle that remains after the angle *DAG* is subtracted from a right angle. But the angle *KDI* plus the angle *IDB* makes a right angle, so the angle *IDB* is equal to the angle *DAG*. We have shown that the angle *DIE* is equal to the angle *DLG*, so the angle *DIB* is equal to the angle *ALD* and there remains the angle *DBI* <which is> equal to the angle *LDA*; so the two triangles *ADL* and *DBI* are similar; so the ratio of *BI* to *ID* is equal to the ratio of *LD* to *LA* and the ratio of *DI* to *IE* is equal to the ratio of *DL* to *LG*; the ratio compounded from the ratio of *BI* to *ID* and the ratio of *ID* to *IE*, which is the ratio of *BI* to *IE*, is the ratio compounded from the ratio of *DL* to *LA* and the ratio of *DL* to *LG*, and the ratio compounded from these two ratios is the ratio of the square of *DL* to the product of *AL* and *LG*; so the ratio of *BI* to *IE* is the ratio of the square of *DL* to the product of *AL* and *LG*. But the square of *DL* is the product of *CL* and *LG*, because the right-angled triangle *CLD* is similar to the triangle *DLG*; so the ratio of *CL* to *LD* is equal to the ratio of *DL* to *LG*, so the ratio of *BI* to *IE* is equal to the ratio of the product of *CL* and *LG* to the product of *AL* and *LG*; but the ratio of the product of *CL* and *LG* to the product of *AL* and *LG* is the ratio of *CL* to *LA*, since we take away *LG* from both; so the ratio of *BI* to *IE* is equal to the ratio of *CL* to *LA*. By composition, the ratio of *BE* to *EI* is equal to the ratio of *CA* to *AL*. But we have shown that *EI* is equal to *LG*, so the ratio of *CA* to *AL* is equal to the ratio of *BE* to *LG* and is equal to the ratio of the whole to the whole, so the ratio of the sum of *AC* and *BE* to *AG* is equal to the ratio of *BE* to *EI*; now we have shown that *AG* is equal to *AH*, that *CG* is equal to *CE* and that *BE* is equal to *BH*, so the sum of *AC* and *BE* is the semiperimeter of the triangle *ABC*; so the ratio of the semiperimeter of the triangle *ABC* to *AG* is equal to the ratio of *BE* to *EI* and it is compounded from the ratio of *BE* to *ED* and the ratio of *DE* to *EI*; but the ratio of *DE* to *EI* is equal to the ratio of *CE* to *ED*, because the two triangles *CED* and *DEI* are similar, so the ratio of *BE* to *EI* is compounded from the ratio of *BE* to *ED* and the ratio of *CE* to *ED*; but the ratio compounded from these two ratios is the ratio of the product of *BE* and *EC* to the square of *ED*, so the ratio of *BE* to *EI* is equal to the ratio of the product of *BE* and *EC* to the square of *ED*. But we have shown that the ratio of *BE* to *EI* is equal to the ratio of the semiperimeter of the triangle *ABC* to the straight line *AG*, so the ratio of the semiperimeter of the triangle *ABC* to the straight line *AG* is the ratio of the product of *BE* and *EC* to the square of *ED*; the product of the semiperimeter of the triangle and the square of *ED* is thus equal to the product of *BE* and *EC* and <the product> of what we obtain and *AG*. If we once again multiply the whole by the semiperimeter, the two <products> are also equal. The product of the semiperimeter and the square of *DE*, then <the product> of what we obtain and the semiperimeter, is equal to the product of *BE* and *EC*, then <the product> of what we obtain and *AG*, then <the product> of what we obtain and the semiperimeter. But the product of the semiperimeter and the square of *DE*, then <the product> of what we obtain and the semiperimeter, is the product of the square of the semiperimeter and the square of *DE*, as we have mentioned for the product of numbers one by another: if we commute, it is still the same. The product of the square of the semiperimeter and the square of *DE*, which we have shown to be equal to the square of the measure of the triangle, is equal to the product of *BE* and *EC*, then <the product> of what we obtain and *AG*, then <the product> of what we obtain and the semiperimeter. But the product of *BE* and *EC*, then <the product> of what we obtain and *AG*, then <the product> of what we obtain and the semiperimeter, is equal to the product of the semiperimeter and *AG*, then <the product> of what we obtain and *EC*, then <the product> of what we obtain and *BE*, because the product of numbers one by another remains the same if we commute. So the product of the semiperimeter and *AG*, then <the product> of what we obtain and *EC*, then <the product> of what we obtain and *BE*, is equal to the square of the measure of the triangle. Now we have shown that *AC* plus *BE* is the semiperimeter, similarly *AB* plus *CE* is the semiperimeter, since *HB* is equal to *BE*, and *HA* is equal to *AG* and *EC* is equal to *CG*. Similarly *BC* plus *AG* is the semiperimeter, so the straight line *AG* is the amount by which the semiperimeter exceeds the side *BC*, *CE* is the amount by which the semiperimeter exceeds the side *AB* and *BE* is the amount by which the semiperimeter exceeds the side *AC*. But we have shown that if we multiply the semiperimeter by *AG*, then <multiply> what we obtain by *CE*, then <multiply> what we obtain by *BE*, then the product is the square of the measure of the triangle. So if we multiply the semiperimeter by the amount by which it exceeds the side *BC*, which is *AG*, then <multiply> what we obtain by the amount by which the semi<perimeter> exceeds the side *AB*, which is *EC*, then <multiply> what we obtain by the amount by which the semi<perimeter> exceeds the side *AC*, which is *BE*, the product is the square of the measure of the triangle. Now <we have shown> earlier that the product of numbers one by another, remains the same if we commute; and this is allowed whatever the exceeding amounts we commute. So if we multiply the semiperimeter of the triangle by the amount by which it exceeds one of the sides of the triangle, whichever side this is, if we then multiply what we obtain by the amount by which the semi<perimeter> exceeds one of the other sides of the triangle, then <multiply> what we obtain by the amount by which the semi<perimeter> exceeds the remaining side, the product is the square of the measure of the triangle. If we take the root of the product, we have the measure of the triangle; I mean what we obtain from what I have mentioned and <again> draw attention to concerning the product of the numbers that are homonyms for each of the numbers of the multiples of the cubit contained by the straight lines and <the product> of the numbers that are homonyms for each of the numbers of the multiples of the square of the cubit contained by the surfaces. That is what we wanted to prove.

We have thus given a complete treatment of the measurement of triangles.

All rectilinear surfaces can be divided into triangles, so the measure of all rectilinear surfaces can be reduced to the measure of triangles by dividing each of these surfaces into triangles, measuring each of these triangles separately and then adding up all the measures; what we obtain is the measure of <each of> these surfaces. Now we have shown above that we measure a rectangle by multiplying one of its sides by the side that makes a right angle with the former one. However, there is no method of finding out that the angles of a surface[14] are right angles except to draw the two diagonals, which then divide it up into four triangles, each of them (the diagonals) dividing it (the surface) into two triangles. We then test each of the four triangles, and do this by testing its sides. If in each of them (the triangles) there is a right angle and if the right angles are those that are opposite the diagonals, then the surface is a rectangle; and if that is not so, it is not a rectangle.

But if there is no method for measuring a surface which does not have right angles, except by dividing it into triangles and finding the sides of the triangles, and if <the sum> of the measures of the triangles is the measure of the surface, then <knowing> the measure of the triangles into which this surface is divided dispenses us from testing the surface.[15] The principle used as the basis for measuring all rectilinear surfaces is the measurement of triangles. It remains to show how to divide surfaces into triangles. The division of surfaces into triangles may be carried out by determining the chords <on which> the angles of the surface <stand>, but one cannot, for every surface, measure out the chords that divide it up, because the unevennesses in certain surfaces give rise to obstacles and impediments that make it impossible to measure out the chords. But we can determine the chords that divide up rectilinear surfaces without measuring out the

---

[14] The surface under consideration is a convex quadrilateral.

[15] That is from knowing, in the case of a convex quadrilateral, whether the angles are right angles.

chords. The method for determining the chord of any angle enclosed by two straight lines is the one we shall describe:

We measure out the two straight lines that enclose the angle, then we cut off a cubit on one of the two; we divide the other straight line by the former straight line,[16] and we cut off on the other side – the one that is on the other side of the angle – a magnitude equal to the quotient. We then draw a straight line from the first point of division to the second point of division, we multiply the quantity of its magnitude, which we have obtained, by the first side from which we cut off a single cubit; what we obtain is the chord that joins the ends of the two straight lines that enclose the angle.

Example: The two straight lines AB and BC enclose an angle ABC. We wish to know the magnitude of the chord AC; we know AB and the magnitude of BC. On BC we cut off a single cubit, that is BD. We then divide AB by BC; we cut off on the other side, <that is> on AB, a magnitude equal to the quotient, let it be BE. We draw a straight line from D to E, which is not difficult because it is close by and small, and we find the value of the straight line DE; what we obtain we <then> multiply by BC and what we obtain is the magnitude AC.

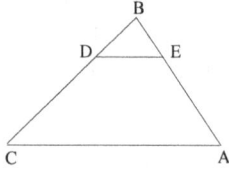

Fig. IV.1.8

Proof: We have divided AB by BC, we have obtained BE; so the product of BC and BE is the magnitude AB; but the product of AB and BD is the magnitude AB, because BD is equal to unity, so the product of AB and BD is equal to the product of CB and BE, the four magnitudes AB, BC, BD and BE are magnitudes in proportion, the ratio of AB to BE is equal to the ratio of BC to BD and the straight line AC is parallel to the straight line ED, as was proved in book six of the work of Euclid.[17] But the two triangles ABC and EBD are similar; so the ratio of AC to ED is equal to the ratio of CB to BD and the product of AC and DB is equal to the product of

[16] See following page.
[17] Euclid, *Elements*, VI.2.

CB and BE; but the product of AC and DB is AC, because DB is unity,[18] so the product of CB and BE is AC. That is what we wanted to prove.[19]

Thus this method enables us to divide all rectilinear surfaces into triangles and dividing them will be simple and easy.

As for the circle, we obtain its measure by multiplying its semidiameter by its semiperimeter, that is to say the number of the multiple of the cubit contained in the semidiameter <is multiplied> by the number of the multiple of the cubit contained in the semiperimeter; what we obtain is the measure of the circle, that is to say the number of the multiple of the square of the cubit contained in its surface.

This was understood clearly by Archimedes thanks to a proof he set out in connection with this notion. Here we give an extremely brief summary of the proof.

Let there be a circle ABCD with centre E.

I say that the product of its semidiameter and its semiperimeter is equal to its measure.

Proof: It cannot be otherwise. If it were possible, let the product of its semidiameter and its semiperimeter be greater or smaller than its measure.

<1> Let the product of the semidiameter of the circle and its semiperimeter be equal to <the measure> of the figure U, first let the latter be smaller than the measure of the circle and let the amount by which the circle exceeds the figure U be the magnitude of the figure S. We draw in the circle two diameters that cut one another at right angles, let these be the two diameters AEC and BED; we join the straight lines AB, BC, CD, DA. The figure ABCD will be a quadrilateral with equal sides and right angles. We cause to pass through the points A, B, C and D straight lines that are tangents to the circle; let the straight lines be NAM, MDL, LCX, XBN; so the figure NMLX is a quadrilateral with equal sides and right angles, because the sides are parallel to the diameters that cut one another at right angles and are equal to them. The square ABCD is half of the square NMLX, so the square ABCD is greater than the semicircle.[20] We join the

---

[18] The operations Ibn al-Haytham mentions throughout this part of his work apply to the numbers that measure the segments with the cubit taken as the unit.

[19] The segments are represented by their numerical measures, once the unit of length has been chosen. This allows a segment to be identified with its product by the unit; an analogous procedure is to be found, in a different context, in 'Umar al-Khayyām, in his *Algebra*; see R. Rashed and B. Vahabzadeh, *Al-Khayyām mathématicien*, Paris, 1999.

[20] Without saying so, Ibn al-Haytham assumes here that the square NMLX circumscribed about the circle has an area greater than that of the circle (Archimedes, *The Sphere and the Cylinder*, Postulate 4).

straight lines $EOGN$, $EOKM$, $EOIL$, $EOHX$,[21] thus the angles AEB, AED, DEC, CEB have been divided, each of them into two equal parts, because the straight line AE is equal to the straight line EB, the straight line EN is common and the base AN is equal to the base BN; so the angle AEG is equal to the angle BEG; similarly for the remaining angles. So each of the arcs AD, DC, CB and BA has been divided into two equal parts, at the points K, I, H, G. We join the straight lines AK, KD, DI, IC, CH, HB, BG and GA. The triangle AKD will be greater than half the segment AKD, because if we draw at the point K a straight line that is a tangent to the circle, it will be parallel to the straight line AD, because the tangent is perpendicular to the diameter EK and the diameter EK is perpendicular to the straight line AD. If from the points A and D we draw to the tangent two <lines> perpendicular to the straight line AD, a parallelogram is formed that is greater than the segment AKD[22] and the triangle AKD is half of this parallelogram; so the triangle AKD is greater than half the segment AKD. Similarly for all the remaining triangles that are homologous to the triangle AKD, each of them is greater than half the segment AKD. Similarly for all the remaining triangles that are homologous to the triangle AKD, each of them is greater than half the segment in which it is inscribed. If we also divide the arcs AK, KD, DI, IC, CH, HB, BG and GA, each of them into two equal parts, and if we put in the straight lines that are their chords, we generate triangles that are (each) greater than half the segments in which they are inscribed. If we continue to proceed in this way, it follows that we cut off from the circle <an area> more than half of it and <then> from what remains <cut off> more than half of that. But the magnitude S is the amount by which the circle exceeds the magnitude U which is smaller than it is; so S is smaller than the circle.

But for two unequal magnitudes, if we cut off from the greater of them more than half of it, and from what remains more than half of it, and if we continue to proceed in this way, it necessarily follows that there remains a magnitude smaller than the smaller magnitude.[23]

---

[21] The midpoints of the four chords AB, BC, CD, DA are all given the same letter, O.

[22] Archimedes, *The Sphere and the Cylinder*, Postulate 4.

[23] Ibn al-Haytham is referring to Euclid, *Elements*, X.1; see also his own proposition, in R. Rashed, *Les Mathématiques infinitésimales*, vol. 2, pp. 495–7 and the critique by Ibn al-Sarī pp. 498–510; English translation in *Ibn al-Haytham and Analytical Mathematics*.

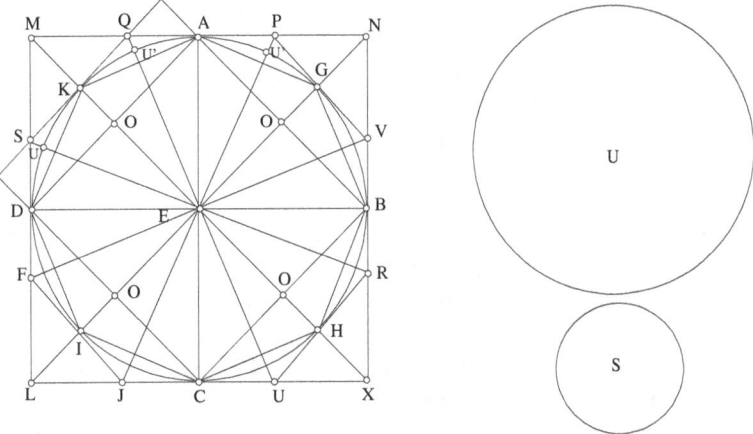

Fig. IV.1.9

Let the excess amounts that remain from the circle, whose sum is smaller than the figure *S*, be the segments *AK*, *KD*, *DI*, *IC*, *CH*, *HB*, *BG* and *GA*. So the figure *AKDICHBGA* is greater than the figure *U*. But since the straight line *KE* is perpendicular to the straight line *AD*, the product of *EO* and half of *AD* is equal to the measure of the triangle *AED*. Similarly, the product of *KO* and half of *AD* is the measure of the triangle *AKD*, so the product of *EK*, which is the semidiameter of the circle, and half of *AD*, is the measure of the figure *AEDK*. Similarly, the product of the semidiameter and half of *AB* is the measure of the figure *AEBG*; similarly for the part that remains. So the product of the semidiameter of the circle and the semi-perimeter of the figure *ABCD* is the measure of the figure *AKDICHBG*, which is greater than the figure *U*, which is obtained from the product of the semidiameter of the circle and half its perimeter. The product of the semidiameter of the circle and half the perimeter of the figure *ABCD* is thus greater than the product of the semidiameter of the circle and its semiperimeter, so the semiperimeter of *ABCD* is greater than the semi-perimeter of the circle and the perimeter of that figure is greater than the perimeter of the circle; so the straight line *AD*, which is a part of the figure[24] as the arc *AKD* is a part of the circumference of the circle, is greater than the arc *AKD*. But a straight line is the shortest line joining two points, and something that comes closer to it is shorter than something that becomes more distant from it, so the straight line *AD* is shorter than the arc *AB*; however it is greater than it, which is impossible. So the figure *U* is not smaller than the circle.

---

[24] That is to say a part of the perimeter *ABCD*.

<2> I say that the figure U is not, either, greater than the circle. If that were possible, let it be greater than it; if it is greater than the circle, it is either equal to the figure $NMLX$, or smaller than it, or greater than it. But the product of EA, which is the semidiameter of the circle, and half AM, is equal to the measure of the triangle MEA; similarly the product of ED, which is the semidiameter of the circle, and half DM is equal to the measure of the triangle DEM; similarly for the triangles DEL, LEC, XEC, XEB, NEB and NEA; so the product of the semidiameter of the circle and the semiperimeter of the figure $NMLX$ is the measure of the figure $NMLX$.

<a> If the figure U is equal to the figure $NMLX$, then the product of the semidiameter of the circle and its semiperimeter is equal to the product of the semidiameter of the circle and the semiperimeter of the figure $NMLX$, so the perimeter of the figure $NMLX$ is equal to the perimeter of the circle; but the line AMD is a part of the perimeter of this figure, as the arc AKD is a part of the perimeter of the circle, so the line AMD is equal to the arc AKD. But the arc AKD is shorter than the line AMD, because the arc is closer to the straight line AD than the line AMD is, so the arc AKD is shorter than the line AMD; but it is equal to it, which is impossible. So the figure $NMLX$ is not equal to the figure U.

<b> If the figure $NMLX$ were smaller than the figure U, we show as before that the perimeter of the figure $NMLX$ would be smaller than the perimeter of the circle; which is impossible.

<c> If the figure $NMLX$ is greater than the figure U, let the amount by which the figure U exceeds the circle be the magnitude of the figure S. Since the figure $NMLX$ is greater than the figure U, the amount by which it exceeds the circle will be greater than the magnitude S. The excesses have the magnitude of the figures by which the figure $NMLX$ exceeds the circle and <their sum> is greater than the magnitude S. We cause to pass through the points H, I, K, G straight lines that are tangents to the circle, let them be the straight lines QKS, FIJ, UHR, VGP, and we join the straight lines EQ, ES, EF, EJ, EU, ER, EV, EP; let these straight lines cut the circumference of the circle at the points U'.[25] Since KS is a tangent to the circle and EK is a semidiameter, EK is perpendicular to QS, so the angle QKM is a right angle, so the angle MQK is acute and the straight line MQ is greater than the straight line QK. But the straight line QK is equal to the straight line QA, because they are tangents drawn to the circle from the same point; so the straight line MQ is greater than the straight line QA and the triangle QMK is greater than the triangle KQA; but the triangle QKA is greater than the part <of it> enclosed by the two straight lines KQ, QA and the arc $AU'K$; so the triangle MQK is greater than the part $KQAU'K$. Similarly the

---

[25] The same letter, U', is used for several points.

triangle *MKS* is greater than the part adjacent to it, so the triangle *QMS* is greater than the part adjacent to it and the triangle *QMS* is greater than half of the part *AMDKA*; similarly the triangles *PNV*, *RXU*, *JLF* are <each> greater than half the part in which they are inscribed. If we cause to pass through the points *U*²⁶ straight lines that are tangents to the circle, they cut off parts that are again triangles that are greater than half the former ones; we show this as we showed it for the triangle *QMS*. If we continue to proceed in such a way, we cut off these excess amounts, <the amounts> by which the figure *NMLX* exceeds the circle, <parts> greater than half of them and also more than half of what remains. These excess amounts are greater than the magnitude *S*. But for two unequal magnitudes, if we cut off from the greater of them more than half of it, and from what remains more than half of that, and if we continue to repeat this procedure, it necessarily follows that there remains a magnitude smaller than the smaller magnitude. Let the excess amounts at the points *Q*, *S*, *F*, *J*, *U*, *R*, *V* and *P* be smaller than the magnitude *S*; but the magnitude *S* is the amount by which the figure *U* exceeds the circle; the excess amounts found at the points *Q*, *S*, *F*, *J*, *U*, *R*, *V* and *P* on the circle are smaller than the magnitude *U*. So the figure *QSFJURVP* is smaller than the magnitude *U*. But the straight line *EK* which is the semidiameter of the circle is perpendicular to the straight line *QS*, so the product of *EK* and half *QS* is the measure of the triangle *EQS*; similarly for the remaining triangles. The product of the semidiameter of the circle and the semiperimeter of the figure *QSFJURVP* is the measure of that figure, but that figure is smaller than the figure *U*; now <the measure of> the figure *U* is the product of the semidiameter of the circle and the semiperimeter of the circle, so <the measure of> the figure *QSFJURVP* is smaller than the product of the semidiameter of the circle and its semiperimeter and the perimeter of the figure *QSFJURVP* is smaller than the perimeter of the circle; now, for this figure, the perimeter of the circle is smaller than its perimeter, as has been proved earlier; which is impossible. So the figure *U* is not greater than the circle nor smaller than it, so it is equal to it.

But <the measure of> the figure *U* is the product of the semidiameter of the circle and its semiperimeter; consequently the product of the semidiameter of the circle and its semiperimeter is equal to the measure of the circle. That is what we wanted to prove.

Archimedes then determined the ratio of the diameter of the circle to its circumference to the closest possible approximation, because the diameter

²⁶ The same letter, *U′*, is used for the midpoints of each of the arcs *AK*, *KD*, *AG* and *GB*.

does not have an exact ratio to ratio to the circumference,[27] because they do not belong to the same genus; which is why, for determining this ratio, he employed a form of approximation. If he adopted this method, it is because his purpose in doing so was to determine the measure of the circle; the approximation in the ratio of the diameter to the circumference does not change the measure of the circle and the discrepancy in the former (the ratio) does not have a noticeable effect on the measure of the circle; he then found <that> the ratio of the diameter of the circle to its circumference is equal to the ratio of one to three and a seventh. He <wrote> a separate treatise on this notion which is available for people to read. So if anyone wishes to measure a circle, he measures its diameter, which he then multiplies by three and a seventh and what he obtains is the perimeter of the circle. He then takes half of this perimeter and half the diameter, he multiplies one by the other; what he obtains is the measure of the circle. If he so wishes, he multiplies the whole diameter by a quarter of the perimeter, and if he wishes, he multiplies the whole perimeter by a quarter of the diameter; what he obtains from all this is the same.

We can determine the measure of the circle in another way also by multiplying the diameter by itself, from the result we <then> subtract a seventh of it, plus half of a seventh; thus what remains is the measure of the circle, and that is in agreement with the first procedure.

Proof: The perimeter is equal to three times the diameter plus a seventh, so the perimeter is twenty-two sevenths of the diameter and a quarter of the perimeter is five sevenths of the diameter plus half a seventh of it; the product of the diameter and a quarter of the perimeter is the measure of the circle, so the product of the diameter and five sevenths of it, plus half of a seventh of it, is the measure of the circle. But five sevenths of the diameter plus half of a seventh of it, is less than the complete diameter by a seventh of it, plus half of a seventh of it; the product of the diameter and five sevenths of it, plus half of a seventh of it, is less than the product of the diameter with itself, by the product of the diameter and a seventh of it plus half of a seventh of it; but the product of the diameter with itself is the square of the diameter and its product with five sevenths of itself, plus half of a seventh of it, is less than its square by the magnitude of its product with a seventh of itself, plus half a seventh of it, which is a seventh of the

---

[27] This means that the ratio is not that of an integer to an integer. In his work on squaring the circle, Ibn al-Haytham states, on the contrary, that the ratio exists even if we cannot know it. Here, he seems to be making a concession to a common belief by taking curves and straight lines to belong to two different categories. This belief was no doubt more familiar to the readers of this treatise.

square plus half a seventh of it. So if we multiply the diameter by itself and if from that we subtract a seventh of it, plus half of a seventh of it, what remains will be the product of the diameter and five sevenths of it, plus half of a seventh of it, <a product> that has been shown to be equal to the measure of the circle. So if we multiply the diameter of the circle by itself and if we subtract from that one seventh of it, plus half of one seventh of it, what remains will be the measure of the circle. That is what we wanted to prove.

We shall now show how to determine the diameter of the circle, because for a circle, we have not, in general, <been given> its diameter and, in general, the centre of a circle is not <given> in a way that allows us to cause to pass through it a straight line that will be the diameter.

The method of determining the diameter of the circle is to draw in it an arbitrary chord which we then divide into two equal parts; we draw from its midpoint a straight line at right angles that ends on the circumference of the segment cut off by the chord, we then measure the chord, we measure the perpendicular, we multiply half the chord by itself, we divide the product by the perpendicular and we add the perpendicular to the quotient; what we obtain is the diameter of the circle.

Example: Let there be the circle *ABCD* whose diameter we wish to know. In the circle we draw an arbitrary chord, let it be the straight line *BD*, which we divide into two equal parts at the point *E*. We draw from the point *E* the straight line *EA* perpendicular to the straight line *BD*, which we extend on the side of *E* until it meets the circle; let that straight line be *AEC*.

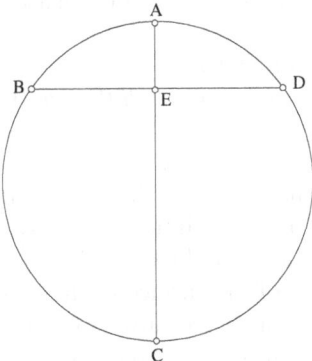

Fig. IV.1.10

Proof: Since the straight line *BD* is a chord in the circle, and has been divided into two equal parts at the point *E* and we have drawn from the point *E* the perpendicular *AEC*, the centre of the circle will lie on the straight line *AEC*, as has been shown in the first proposition of the third book of the work of Euclid. So the straight line *AC* is the diameter of the circle. Since the straight lines *AC* and *BD* intersect inside the circle, the product of *AE* and *EC* is equal to the product of *BE* and *ED*, as has also been shown in the third book.[28] But the product of *BE* and *ED* is equal to the square of *BE*, since *BE* is equal to *ED*, so the square of *BE* is equal to the product of *AE* and *EC*. So if we divide the square of *BE* by the straight line *EA*, the quotient will be the straight line *EC* to which we add the perpendicular, which is the perpendicular *EA*; we obtain *CA*; so *CA* is the diameter of the circle.

So if in the circle we draw an arbitrary chord which we divide into two equal parts and if we draw at its midpoint a perpendicular to <reach as far as> the circumference of the segment cut off by the chord, if we then multiply half the chord by itself and if we divide that (the result) by the perpendicular, if we add the quotient to the perpendicular, then the sum is the diameter of the circle. That is what we wanted to prove.

How to measure a sector of a circle. We measure the sector by multiplying the side that is the semidiameter of the circle by half the arc of the sector; what we obtain is the measure of the sector.

As for a segment of a circle, it is the complement of the triangle of the sector. We measure the sector, we then measure the excess triangle and we subtract it from the measure of the sector, what remains is the measure of the segment of the circle.

Example: Let the sector be *ABC*.

I say that the product of AB *and half the arc* BEC *is the measure of the sector.*

Proof: We complete the circle, let it be *BDC*; so the point which is the vertex of the sector is the centre of the circle, since a sector is a figure whose vertex is the centre of a circle and its base an arc of the circumference of the circle. So the ratio of the arc *BEC* to the circumference of the circle is equal to the ratio of the surface of the sector *ABEC* to the surface of the complete circle, as has been shown by a proof analogous to the one Euclid presented in the final proposition of the sixth book. The ratio of half the arc *BEC* to half the circumference of the circle is thus equal to the ratio of the sector to the complete circle; but the ratio of half the arc *BEC* to half

---

[28] Euclid, *Elements*, III.35.

the circumference of the circle is equal to the ratio of the product of the straight line *AB*, which is the semidiameter of the circle, and half the arc *BEC*, to the product of *AB* and half the perimeter of the circle which is the measure of the surface of the circle. The ratio of the product of *AB* and half the arc *BEC* to the measure of the surface of the circle is equal to the ratio of the measure of the surface of the sector *ABEC* to the measure of the surface of the circle. So the product of the straight line *AB*, which is the semidiameter of the circle, and half the arc *BEC*, is the measure of the surface of the sector *ABEC*. That is what we wanted to prove.[29]

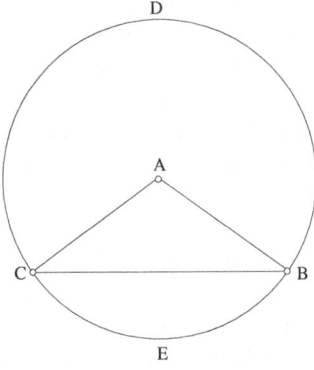

Fig. IV.1.11

If we measure the triangle *ABC* and take it away from the measure of the sector, what remains is the measure of the segment *BEC*, because the sector is the sum[30] of the triangle *ABC* and the segment *BEC*.

It remains to show how to find the magnitude of the arc of the sector and of the segment, and this is because for an arbitrary sector, its arc is not known and, for a segment of an arbitrary circle, its arc is not known. A known arc is one whose ratio to the circumference of the circle is a known ratio. In fact if the ratio of the arc of the sector or the segment to the circumference of the circle is not known, then there is no way to find the measure of the sector or the segment, so we cannot find them until we know the ratio of the arc to the circumference of the circle.

The method for finding the ratio of any arc to the circumference of its circle is to divide the chord of the arc into two equal parts, to draw from its

---

[29] The result is valid, irrespective of whether the sector is smaller or greater than a semicircle.

[30] This corresponds to the case in which the segment in question is smaller than a semicircle. Ibn al-Haytham does not consider the case of a segment greater than a semicircle.

midpoint a perpendicular that ends on the arc and to join the end of the chord to the end of the perpendicular by a straight line, to erect at the end of the chord a perpendicular to the chord, to take the end of the chord as centre and, with the distance to the other end of the chord, draw an arc of a circle that cuts[31] the perpendicular erected at the end of the chord; this arc is a quarter of a circle because it subtends a right angle at the centre of its circle. We extend the straight line that joins the end of the chord to the end of the first perpendicular until it reaches the arc which is a quarter of a circle, then we find the size of the arc that lies between this straight line and the end of the chord through which the quarter circle passes, by means of compasses, so that the compasses give a size for this arc, either once or several times, and we also find the size of the complete quarter circle;[32] from this we obtain the ratio of the arc, which lies between the chord and the straight line that cuts the arc, <that is its ratio> to the quarter circle; this ratio is the ratio of the first arc, whose ratio we are looking for, to the circumference of its circle.

Example: Let the arc be *ABC*. We wish to know its ratio to the circumference of its circle; so we draw the chord *AC* which we divide into two equal parts at the point *E*. We draw the perpendicular *EB*, we join *AB* and at the point *A* on the straight line *AC* we erect a perpendicular to the straight line *AC*; let it be *AD*. We take *A* as centre and with distance *AC* we draw an arc of a circle; let it be *CHD*. So the arc *CD* is a quarter circle, because the angle *CAD* is a right angle. We extend the straight line *AB* until it meets the arc *CD*; let it meet it at the point *H*.[33] So if we evaluate the arc *CH* in terms of a magnitude by which we <also> evaluate the complete arc *CHD*, these <values> give us the ratio of the arc *CH* to the arc *CD*.[34]

I say that the ratio of the arc CH *to the arc* CD *is the ratio of the arc* ABC *to the circumference of the whole circle.*

---

[31] Lit.: separates.

[32] See note 34 below.

[33] If the arc *ABC* is less than a semicircle, *H* lies on the extension of *AB*. If $\overset{\frown}{ABC}$ is greater than a semicircle, *H* lies between *A* and *B*.

[34] Ibn al-Haytham is recommending a system of trial and error to find an opening of the compasses that fits into the arc a whole number of times. This procedure, known in English as 'stepping', was used by craftsmen to make toothed wheels. See note 48 below. [note J. V. Field]

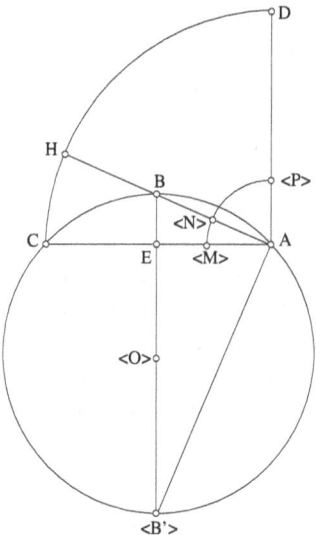

Fig. IV.1.12

Proof: The angle *CAH* is at the centre of the circle *CHD* and it is inscribed in the circumference of the circle *ABC*. So the arc *CB* is twice <the arc> similar to the arc *CH*, and the arc *ABC* is twice the arc *BC*; so the arc *ABC* is four times <the arc> similar to the arc *CH*. Thus the ratio of the arc *CH* to a quarter of its circle is equal to the ratio of the arc *ABC* to its complete circle. But the arc *CHD* is a quarter circle, so the ratio of the arc *CH* to the arc *CD* is equal to the ratio of the arc *ABC* to its complete circle. That is what we wanted to prove.

If we take a point on the chord *AC*, in any position on it that we wish, and if with *A* as centre and the distance of this point we draw an arc of a circle that cuts the straight line *AC*, this arc is a quarter circle; we cut off on this arc an arc similar to the arc *CHD*. This notion is necessary if the arc *ABC* is of a considerable magnitude. From this stage onwards we make a reduction in the arc *CHD* and instead of *AC* we use a part of the straight line *AC*, which thus takes the place of *AC*[35] and from it there follows the determination of the ratio; it is in this way that we shall determine the ratio of the arc to the circumference of its circle.

However, for a general arc, its ratio to its circle is not necessarily a numerical ratio[36] because, since for two arbitrary arcs the ratio of one to the

---

[35] See above, pp. 484–5.

[36] That is to say rational.

other is not necessarily a numerical ratio, so for the arc *CH* its ratio to the arc *CD* is not necessarily a numerical ratio. But if the ratio of these two arcs one to the other is not a numerical ratio, then it is not possible that one common magnitude, that we can specify, can evaluate each of them. So if the ratio of the arc *CH* to the arc *CD* is not a numerical ratio, then the ratio of the arc *ABC* to the circumference of its circle is not a numerical ratio. If the arc is of this kind, then we cannot express it; if this happens in regard to measure, then the ratio of something inexpressible leads to something we can express by means of a ratio such that there is no perceptible discrepancy between it and the true ratio. Similarly, if a magnitude is inexpressible (irrational), then it leads to something we can express by way of a magnitude such that the discrepancy between it and that magnitude is not perceptible, because a practical art cannot carry out real operations without making use of a kind of approximation in cases where one cannot achieve extreme accuracy.

This method is the one Archimedes adopted for determining the ratio of the diameter of the circle to its circumference. The method by which we determine the ratio of the arc *CH* to the arc *CD* is to decrease the opening of the compasses[37] until it is as small as possible; we use this (the small compass opening) to measure out the arc *CH* and the arc *CD*; so if we decrease the opening of the compasses, we necessarily arrive at a magnitude that measures out the two arcs, even if the ratio is not a numerical ratio; in fact if the parts become smaller and smaller, the difference by which one of the two arcs exceeds an arc that has a <numerical> ratio with the other arc becomes smaller and smaller until this difference has become imperceptible.

It is in this way that we can determine the ratio of the arc *CH* to the arc *CD*, which is equal to the ratio of the arc *ABC* to the circumference of its circle.

We have completed the explanation of the procedures for measuring all surfaces whose measures are used, which are rectilinear surfaces and the circle.

Procedure for measuring solids, a detailed technical exposition: among solids there are those that are enclosed by plane surfaces, those that are enclosed by surfaces that are not plane and those that are enclosed by some plane surfaces and some that are not plane. Among the solids enclosed by

---

[37] That is to say the separation of the points of the compasses. The method of trial and error that follows is akin to 'stepping' used to divide an arc, see note 34 above and note 48 below. [note J. V. Field]

plane surfaces, there are those that have parallel faces and those that have faces that are not parallel. Among those with parallel faces, there are some that have perpendicular faces and some that do not have perpendicular faces.

The measure of a solid with parallel faces and faces that are perpendicular[38] is obtained by multiplying the length of its base by its width, then <multiplying> what we obtain by the height of the solid, that is to say that we multiply the number of the multiple of the cubit that is in the length of its base by the number of the multiple of the cubit that is in the width of the base, then we multiply the product by the number of the multiple of the cubit that is in the height. Any one of its faces can be taken as the base and any one of its sides can be taken as the height, since all its sides are perpendicular to its faces.

Example: Let there be the solid *ABCDEGHI* with parallel faces and with all the faces perpendicular.[39]

I say that its measure is the product of the length of its base and its width, then <the product of> what we obtain and the height, that is to say that the number of the multiple of the cube of the cubit in the solid BI *is the product of the number of the multiple of the cubit in* AB, *which is the length of the base, and the number of the multiple of the cubit in* BC, *which is the width of the base, and the number of the multiple of the cubit in* BG, *which is the height of the solid.*

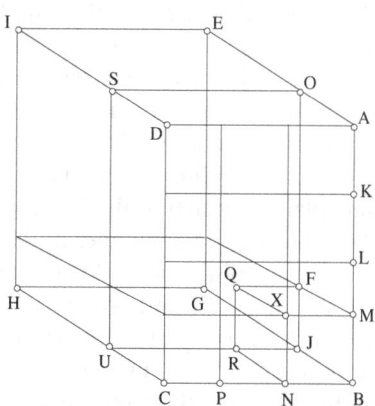

Fig. IV.1.13

---

[38] If the solid is a rectangular parallelepiped, each face is perpendicular to the four faces that surround it.

[39] See preceding note.

Proof: We divide up *AB* into cubits, let them be *AK*, *KL*, *LM* and *MB*; we also divide up *BC* into cubits, let them be *BN*, *NP* and *PC*, and we draw through the dividing points straight lines parallel to the sides of the base. We thus divide up the base into right-angled quadrilaterals whose sides are equal and perpendicular; each is equal to the square *BMXN* which is the square of the cubit, as we showed in the first proposition of this book. We then divide up *BG*, the height of the solid; let us cut off a single cubit, that is *BJ*. We also cut off a single cubit from each of the <straight lines> *AE*, *DI*, *CH*, at the points *O*, *S*, *U*. We join the straight lines *JO*, *OS*, *SU* and *UJ*, then the quadrilateral *JOSU* is equal to the quadrilateral *ABCD*; it is parallel to it, because the perpendiculars *BJ*, *AO*, *DS* and *CU* are equal. We draw from the parallel and perpendicular straight lines in the base planes perpendicular to the base; they then divide up the quadrilateral *OJUS* into equal squares <that are> also equal to the squares into which the base has been divided and <the planes> divide up the solid *BS* into equal cubes, each of which is equal to the cube *BMNXJQFR*; the cube *BMNXJQFR* is the cube of *BM* which is the cubit. Thus the solid *BS* is divided up by the planes perpendicular to the base into equal cubes each of which is equal to the cube of the cubit; the number of these cubes is the number of the squares into which the base was divided, because the base of each of these cubes is one of the squares in the base, and the number of the squares into which the base was divided is the number obtained from the product of the number of the multiple of the cubit that is in the length of the base and the number of the multiple of the cubit that is in the width of the base. So the number of cubes in the solid *BS* is the number obtained from the product of the number of the multiple of the cubit that is in the length of the base and the number of the multiple of the cubit that is in the width of the base. If we then divide up the height of the solid into cubits, <and> if we then draw from the dividing points planes parallel to the base, solids are formed each of which is equal to the solid *BS*, so the number of the multiple of the solid *BS* in the solid *BI* is the number of the multiple of the cubit in the height of the solid *BI*. But each of the solids into which the solid *BI* is divided contains cubes whose <number> is equal to the number of the cubes of the cubit contained by the solid *BS*, so the number of the multiple of the cube of the cubit contained by the solid *BI* is the number we obtain from the product of the number of the multiple of the cubit in the length of the base and the number of the multiple of the cubit in the width of the base and the number of the multiple of the cubit in the height of the solid.

The quantity of the measure of any solid with parallel faces and right angles is what we obtain from the product of the length of the base and its width, and then from the product of what we obtain and the height of the solid, I mean that the number of the multiple of the cube of the cubit that the solid includes is the product of the number of the multiple of the cubit in the length of the base and the number of the multiple of the cubit in the width of the base and the number of the multiple of the cubit that is in the height of the solid. That is what we wanted to prove.

But solids enclosed by planes do not all have parallel faces and those among them that have parallel faces do not all have perpendicular faces. The method for measuring solids that do not have parallel faces and solids with parallel faces that are not at right angles is to divide them up into pyramids, then to measure each of these pyramids separately and add up their measures; the sum we obtain is then the measure of the complete solid.

Moreover, for the solid with parallel faces at right angles, we do not know it has right angles until we have checked its angles, and there is no method of checking its angles except by dividing up all its faces into triangles. But if we divide up all its faces into triangles, then we are dividing the solid into pyramids. And if we divide the solid into pyramids, we can measure each of these pyramids separately, and if we add up <the measures>, what we obtain is the measure of the whole solid. The general method for measuring all solids enclosed by plane surfaces is to measure the pyramid.

The method for measuring the pyramid is to measure its base, then to multiply what we obtain by a third of the height; what we obtain is its measure, because it has been shown in book twelve of the work of Euclid[40] that any rectilinear pyramid is one third of a prism enclosed by faces with parallel sides and two opposed equal and parallel bases, and that the measure of such a prism is the product of its base and its whole height; so the measure of the pyramid is the product of its base and a third of its height. We have shown how to measure the base of the pyramid, because the base of the pyramid is a surface with rectilinear sides and we have shown above how to measure surfaces with rectilinear sides; so it remains to show how to determine the height of a pyramid. It also remains to show how to divide up the surfaces of solids into triangles so as, by this means, to divide up the solid into pyramids.

We carry out the division of the faces of the solid into triangles by means of determining chords of the angles. But this can be carried out on all the faces of a solid with rectilinear sides by the method we demonstrated for finding the chords of angles, in proposition six of this treatise, except

[40] Euclid, *Elements*, Book XII, Proposition 7 and corollary.

for the base of the solid, because it is not possible to work inside the solid; so let us employ the procedure described below.

We shall now show how to find the chords of the angles of the base.

We construct two rulers, one of which has a certain width; we attach the one that has a certain width to one of the two sides that surround the angle of the base whose chord we wish to determine. We place the ruler so that a part of it goes beyond the angle, then we attach the other ruler to the other side of the base that, with the first side, encloses the angle in question. We fit the end of this second ruler to the end of the first ruler, thus in the plane of the first ruler, <the one> that has a certain width, there is formed an angle equal to the angle of the base, because the outer edge of the ruler that has a certain width is parallel to the edge of it attached to the side of the solid that is the side of the base and the straight line drawn against the edge of the second ruler lies along the other side of the base; so the angle enclosed between the straight line drawn in the plane of the ruler and the edge of the first ruler is equal to the angle of the base of the solid. If we obtain this angle, we fix the ruler that has a certain width in a plane, we draw against the edge that is adjacent to the solid a straight line in the plane, then we fit the second ruler onto the straight line drawn in the plane of the first ruler, <the one that> has a certain width, and we draw against the edge of this ruler a straight line in the plane; it then forms in the plane an angle equal to the angle in the plane of the ruler, because the first straight line is parallel to the edge of the ruler that has a certain width which lies alongside the angle which is in the plane of the ruler and the second straight line lies alongside the straight line drawn in the plane of the ruler. If we obtain this angle in the plane, on one of these two straight lines we cut off a cubit, we divide the number of the multiple of the cubit that is in the side of the solid homologous to the other straight line of the angle by the number of the multiple of the cubit in the side of the solid homologous to the first straight line, we take away a number equal to the quotient of the straight line that remains from the two straight lines of the angle drawn in the plane, and we join the point of intersection and the end of the first cubit with a straight line, then we evaluate this straight line, we multiply what we obtain from evaluating it by the number of the multiple of the cubit that is in the side of the solid homologous to the straight line from which we cut off a single cubit; what we obtain is the chord of the angle of the base of the solid, a chord that cuts off a triangle from the base of the solid.

Example: The angle *ABC* is the angle equal to the angle of the base of the solid. Let *AB* be a single cubit; let *BC* be equal to the quotient of the side of the solid homologous to the straight line *CB* <divided> by the side homologous to the straight line *BA*. We join *AC* and find its value. We multiply the magnitude we found for it by the side of the solid homologous to the straight line *AB*.

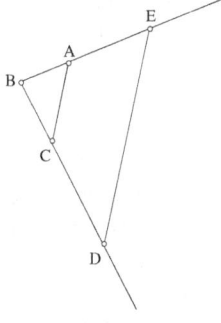

Fig. IV.1.14

I say that what we obtain is the chord that joins the two ends of the sides of the base.

Proof: If we extend the two straight lines *BA* and *BC* as the straight lines *BE* and *BD*, if we put *BE* equal to the side of the solid homologous to the straight line *BA*, if we put *BD* equal to the side of the solid homologous to the straight line *BC* and if we join *ED*, then *ED* will be equal to the chord that joins the two ends of the two sides of the solid. The two sides *EB* and *BD* are equal to the two sides of the solid and the angle *EBD* is equal to the angle of the base of the solid, so the base *ED*[41] is equal to the chord that is opposite the angle of the base of the solid. But it has been shown in the sixth proposition of this treatise that the product of *AC* and *BE* is equal to the magnitude of the straight line *DE*, so the product of *AC* and *BE* is equal to the chord that divides up the base of the solid in a plane. That is what we wanted to prove.

If the base of the solid lies in a single plane, we extend the two sides of the base; thus there is formed, outside the solid, an angle equal to the angle of the base of the solid; we construct in this angle the analogue of what has been constructed in the angle *ABC*. So by this means we obtain the required chord.

How to determine the height of the pyramid. This is done as follows. We determine the two perpendiculars of two triangles among the triangles seen on the outside of the pyramid, which go to the same angle among the angles of the pyramid, and the base of the pyramid may be a triangle or a

---

[41] The base *ED* of the triangle *EBD*.

polygon, provided the two vertices of the triangles of the pyramid are the same point. We determine the two feet of the <perpendiculars in the> two triangles and we draw in a plane an angle equal to the angle of the base of the pyramid that <forms> with the two angles of the two triangles the angle of the pyramid, as we have shown above. We then cut off on the two straight lines of this angle drawn in the plane two straight lines equal to the two feet of the two triangles; we join their endpoints, this forms two angles and a triangle. We then examine the two angles: if one is a right angle, then the height of the triangle that cuts off the foot for which the angle at its endpoint is not a right angle is the height of the pyramid. We carry out our examination of the angles by multiplying each of the sides of the triangle by itself and adding up the results two by two to compare <the sum> to the third. So if two of them <have a sum> equal to the square of the third, then the angle enclosed by the two is a right angle, if their sum is smaller, then the angle enclosed by the two is greater than a right angle and if their sum is greater, then the angle enclosed by the two is smaller than a right angle. And if one of the two angles is not a right angle, then at the endpoint of each of these two straight lines we erect a perpendicular and we extend them; they meet one another; when they meet, we find the value of one of them and we multiply it by itself, then we multiply by itself the height of the triangle that cuts off in the triangle the foot at the end of which we erected the straight line whose value we found, then we take away the square of the straight line from the square of that height, we take the root of what remains; what we obtain is the height of the pyramid.

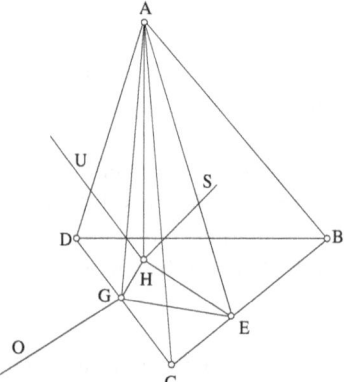

Fig. IV.1.15

Example: Let the pyramid be *ABCD*, with base *BCD*, vertex the point *A*, the remaining triangles *ABC, ABD* and *ACD*. We wish to know its height.

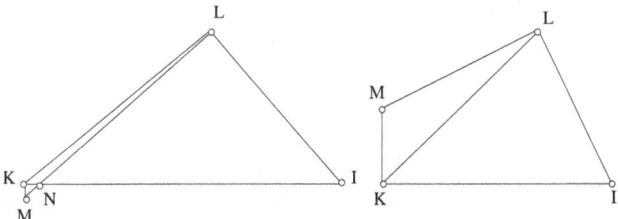

Fig. IV.1.16

We find the heights of the triangles *ABC* and *ACD*; let the heights be *AE* and *AG*. The two straight lines *EC* and *CG* are thus their feet. Let us draw in a plane an angle equal to the angle *ECD*; let the angle be *LIK*. We cut off *LI* equal to *EC* and *IK* equal to *CG*. We join *LK*. We then consider two angles equal to the angles *ILK* and *LKI*. If one of these two angles is a right angle, then the height of the triangle whose foot is the other side of the angle is the height of the pyramid, that is to say that if the angle *ILK* is a right angle, then the perpendicular *AG* is the height of the pyramid and if the angle *IKL* is a right angle, then the perpendicular *AE* is the height of the pyramid. And if neither of these two angles is a right angle, then we erect at the points *L* and *K* perpendiculars to the two straight lines *LI* and *KI*; let them be *LM* and *KM*. We extend them, they meet one another; let them meet one another at the point *M*. But if the two angles *ILK* and *IKL* are acute, then if the straight lines meet one another, that will be inside the angle *LIK*, that is to say that the straight line drawn from the angle to the point where they meet cuts the angle *LIK*. And if one of the two angles *ILK, IKL* is obtuse, the meeting <point> is outside the angle, that is to say that the straight line drawn from the angle to the meeting point falls outside the angle. If the angle *ILK* is obtuse, then the angle *INL* is acute, so the angle *KNM* is acute and the angle *NKM* is a right angle, and the two straight lines meet one another outside the angle. When the two straight lines meet one another, we find the value of one of the two straight lines that are *LM* and *KM*; for example let this be *LM*. We multiply it by itself and we subtract its square from the square of *AE*; we take the root of what remains; what we obtain is the height of the pyramid.

The proof of all this consists of imagining that in the base of the pyramid we have a straight line that joins the two points *E* and *G*; let the straight line be *EG*. So we have *ECG* equal to the triangle *LIK*. So if the angle *ILK* is a right angle, then the angle *CEG* is a right angle.

Thus I say that the perpendicular AG *is the height of the pyramid.*

Proof: We draw from the point *G* a straight line parallel to the straight line *CE*, in the plane of the base; let it be the straight line *GO*. Since the angle *CEG* is a right angle and the angle *CEA* is a right angle, the straight line *CE* is perpendicular to the plane *AEG*. Since the straight line *GO* is parallel to the straight line *CE*, accordingly the straight line *GO* will also be perpendicular to the plane *AEG*, so the angle *OGA* is a right angle. But the angle *CGA* is a right angle, so the straight line *AG* is perpendicular to the plane that contains the straight lines *CG* and *GO*. But the plane that contains these two straight lines is the plane of the base of the pyramid, so the straight line *AG* is perpendicular to the plane of the base of the pyramid; so it is the height of the pyramid.

Similarly, if the angle *IKL* is a right angle, we show that the perpendicular *AE* is the height of the pyramid.

But if neither of the two angles *ILK*, *IKL* is a right angle and if the two straight lines meet one another at the point *M*, then we imagine we have two perpendiculars drawn from the points *E* and *G* to the straight lines *CE* and *CG*; so they meet one another as the two straight lines *LM* and *KM* met one another, because the triangle *ECG* is equal to the triangle *ILK* and is similar to it. Let the two perpendiculars meet one another at the point *H*. We join *AH* and we cause to pass through *H* a straight line parallel to the straight line *CE*, in the plane of the base of the pyramid, let it be *HS*, and a straight line parallel to the straight line *CG*, also in the plane of the base, let it be *HU*. Since the angle *CEH* is a right angle and the angle *CEA* is a right angle, the straight line *CE* is perpendicular to the plane of the triangle *AEH*; but since the straight line *HS* is parallel to the straight line *CE*, the straight line *HS* will also be perpendicular to the plane of the triangle *AEH*, so the angle *SHA* is a right angle. But since the angle *CGH* is a right angle and the angle *CGA* is a right angle, the straight line *CG* is perpendicular to the plane of the triangle *AGH*; but the straight line *HU* is parallel to the straight line *CG*, so the straight line *HU* is perpendicular to the plane of the triangle *AGH*, so the angle *UHA* is a right angle; now the angle *SHA* is a right angle, so the straight line *AH* is perpendicular to the two straight lines *SH* and *UH*, so it is perpendicular to the plane that contains these two straight lines; but these two straight lines are in the plane of the base of the pyramid, so the straight line *AH* is perpendicular to the plane of the base of the pyramid, so it is the height of the pyramid. But since *AH* is perpendicular to the plane of the base, it makes a right angle with every straight line in the plane of the base, the angle *AHE* is a right angle, so the triangle *AEH* is right-angled, so the square of *AE* is equal to the square of *AH* plus the square of *HE*; so if we subtract the square of *HE* from the square of *AE*, what remains is the square of *AH*. But *HE* is equal to *LM*; so if we subtract the square of *LM* from the square of *AE*, what remains is the square of *AH*, if we take its root, we then have *AH*, which is the height of the pyramid. That is what we wanted to prove.

We can also find the heights of pyramids by a method different from this; it is a method by which we can find the heights of all tall bodies and by which we can find the heights of mountains and of tall things; we demonstrate that later.

If we take one of the faces of the pyramid as a base of the pyramid, <a face> other than its natural base, which is the base of the solid, and if we find the perpendicular dropped to that base from the angle of the pyramid that is opposite to it,[42] by means of the first procedure that we have set out for determining the height of the pyramid and if we multiply the measure of this base by a third of this height, what we obtain is the measure of the pyramid.

All the bodies on which surveyors carry out measurements are bodies with plane faces, cylinders on a circular base, cones on a circular base, or spheres, and nothing apart from these belongs to the art of measurement. We have shown <in the present work> how to measure all bodies with plane faces.

The cylinder on a circular base is one whose two bases are two equal and parallel circles and which is enclosed by a single rounded surface. The method for measuring this cylinder is to measure its base and to measure its height, then to multiply the measure of the base by the height; what we obtain is the measure of the cylinder.

If the cylinder is at right angles to its base, then this is clear, because the length of the cylinder is its height. So if we consider a piece of the cylinder one cubit in height and if from the end of the cubit we draw a plane parallel to the base, then a part is cut off from the cylinder, <a part> whose base is the base of the cylinder and its height a single cubit. The number of the multiple of the cube of the cubit in this part is the number of the multiple of the square of the cubit in the base, because we have erected on every part of the base a part of the solid of revolution. So the ratio of the part of the base to the base as a whole is equal to the ratio of the part of the solid to the solid as a whole. The number of the multiple of the cube of the cubit in the solid which is a part of the cylinder is thus equal to the number of the multiple of the square of the cubit in the base. Then when we

---

[42] This comment only applies to a pyramid with a triangular base. If a pyramid with apex *S* has as its base a quadrilateral *ABCD*, if we take *SAB* as base, we can no longer regard the solid as a pyramid.

imagine the height of the cylinder has been divided into cubits and we cause to pass through each dividing point a plane parallel to the base, the cylinder will be divided into parts whose number is equal to the number of the multiple of the cubit in the height of the cylinder; each of these parts will be equal to the first part whose height is a single cubit. So the number of the multiple of the cube of the cubit that is in the whole cylinder is the product of the number of the cube of the cubit that is in the single first part and the number of the multiple of the cubit that is in the height of the cylinder. But the number of the multiple of the cube of the cubit that is in the first part is the number of the multiple of the square of the cubit that is in the base. So if we multiply the measure of the base by the height of the cylinder, the product is thus equal to the measure of the cylinder.

As for the oblique cylinder, it is equal to the right cylinder whose base is the base of the oblique cylinder and whose height is equal to its height. We show this by the proofs in book twelve of the work of Euclid.[43] So the method for measuring the oblique cylinder is to measure its base then determine its height and multiply the measure of the base by the height; what we obtain is the measure of the oblique cylinder.

The method for measuring the cone[44] consists of measuring its base and multiply <the measure> by a third of the height; thus what we obtain is the measure of the cone if <the axis of the> cone is perpendicular to its base or inclined, because it has been shown in book twelve of the work of Euclid[45] that any cone whose base is a circle is a third of the cylinder whose base is its base and whose height is its height.

How to measure the bases of cylinders and cones. This is done by finding the circumference of its base; what we obtain for the magnitude of the circumference is divided by three and a seventh; the quotient we obtain

---

[43] Euclid, *Elements*, Book XII, Propositions 10 to 15; the cones and cylinders investigated are cones and cylinders of revolution (*Les Œuvres d'Euclide*, translated literally by F. Peyrard, Paris, 1966, proofs given on p. 397). Ibn al-Haytham perhaps means to imply that the reasoning for going from the right cylinder to the oblique cylinder should be deduced from the reasoning Euclid presents in the last part of Proposition 31 of Book XI for going from the right parallelepiped to the oblique parallelepiped. See *The Thirteen Books of Euclid's Elements*, translated with introduction and commentary by Th. L. Heath, Cambridge, 1908; repr. New York, Dover Publications, 1956.

[44] Lit.: the rounded cone, which we shall henceforth translate as 'cone'. By using this expression 'rounded cone', Ibn al-Haytham wished to distinguish between pyramid (*makhrūṭ*) and cone (*makhrūṭ mustadīr*).

[45] Euclid, *Elements*, XII.10, for the right cone.

is the diameter of the base; when we obtain the diameter of the base and its circumference, we then find the measure of the circle by the method we have set out above in connection with measuring the circle.

As for the procedure for determining the heights of inclined cylinders and inclined cones, we show that later.

The method for measuring the sphere is to measure a great circle of the sphere, then multiply the measure of the circle by two thirds of the diameter of the circle, which is the diameter of the sphere; what we obtain is the measure of the sphere. In fact, the sphere is two thirds of the cylinder whose base is <equal to> a great circle of the sphere and whose height is equal to the diameter of the sphere. Geometers[46] have proved this in their books and their books are available, and we too have proved it in a separate treatise.

How to find a great circle on the sphere. This is done as we shall describe. We open the compasses by an arbitrary amount, then we put <the tip of> one of its two legs on a point of the sphere, then with the other leg we draw a circle on the surface of the sphere, then we remove the compasses while maintaining their adjustment[47] and we mark two points on the circumference of the circle, that is on the sphere; the circle is thus divided into two arcs, we divide each of the two arcs into two equal parts using another pair of compasses with which we step out one of the two arcs. We increase and decrease the opening of the compasses until we can step out the arc in two moves, so the arc is divided into two equal parts;[48] we make a mark at its midpoint, then we step out the other arc in the same

---

[46] Specifically the Banū Mūsā, *On the Knowledge of the Measurement of Plane and Spherical Figures*, Proposition 15 (see *Founding Figures and Commentators in Arabic Mathematics*, Chapter I).

[47] That is to say while preserving the opening of the instrument.

[48] This is an example of stepping being used to divide an arc into equal parts. In this case we have simple bisection, and stepping is used only because, for practical reasons, operations cannot be carried out inside the circle. However, the method can also be employed to obtain any number of equal divisions on a circle and there is every reason to suppose it was well known to craftsmen in Ibn al-Haytham's time (and indeed almost certainly since Antiquity). Some such method would have been convenient for laying out the teeth on the gear wheels of the 'Box of the Moon' described by al-Bīrūnī, a calendrical device that includes wheels with 7, 19 and 59 teeth (see Donald R. Hill, 'Al-Bīrunī's mechanical calendar', *Annals of Science*, 42, 1985, pp. 139–63; reprinted in J. V. Field, D. R. Hill and M. T. Wright, *Byzantine and Arabic Mathematical Gearing*, London, 1985). Ibn al-Haytham's testimony is of historical interest because artisan practices are rarely the subject of written descriptions. Moreover, it seems to be among the earliest known accounts of stepping. [Note J. V. Field]

way until it is divided into two equal parts and we make a mark at its midpoint. If we obtain these two points, then they divide the circumference of the circle into two equal parts. The straight line that we imagine joining these two points is the diameter of the circle. We open the second pair of compasses and we put <the tip of> one of its two legs at one of these two points that divide the circumference of the circle into two equal parts and we open the compasses until <the tip of> the other leg reaches the other point. So if <the tips of> the two legs of the compasses reach the two opposite points, then the opening of the compasses will be equal to the diameter of the circle drawn on the surface of the sphere; we then position the two <tips of> the legs of these compasses on a plane surface so that <the tips of> its legs leave a mark on the plane surface, then we put a ruler across these two points and we join the two points with a straight line, so this straight line will be equal to the diameter of the circle drawn on the surface of the sphere. We divide this straight line into two equal parts and we draw from its midpoint a perpendicular to the straight line; we then take the first compasses, we put one of the <tips of its> two legs on the end of the straight line we divided and we cause the other leg of the compasses to move until it meets the perpendicular we erected; it will necessarily meet the perpendicular, because the opening of the first compasses is greater than the semidiameter of the circle that they drew on the sphere, since the position of the <tip of the> second leg of the first compasses is the pole of the circle that they drew on the sphere, and any straight line drawn from the pole of a circle on the sphere to its circumference is greater than the semidiameter of the circle; that can be shown from the book of the *Sphærica* of Theodosius. So if the <tip> of the leg of the compasses meets the perpendicular to the straight line, at the position where they meet we mark a point, we join this point to the end of the straight line on which the <tip of the> leg of the compasses stands with a straight line, then we extend the perpendicular in the other direction and at the end of the straight line drawn from the end of the straight line we divided we erect, as far as the perpendicular, a straight line at right angles that we extend until it meets the perpendicular. The straight line cut off on the perpendicular between this straight line and the first straight line is the diameter of the sphere.

If we wish, we may find the value of half the straight line that is equal to the diameter of the circle drawn on the sphere and we find the value of what is cut off from the perpendicular, then we multiply the result of the evaluation of half the straight line by itself, we divide what we obtain by the magnitude of what has been cut off from the perpendicular, to what we obtain we add the perpendicular, the sum will be the diameter of the sphere.

If we multiply it by itself and from that subtract a seventh and half of a seventh, what remains will be the great circle of the sphere. So if we multiply the measure of this circle by two thirds of the diameter, what we obtain is the measure of the sphere.

To prove that it is the diameter of the sphere, we put the straight line equal to the diameter of the circle drawn on the sphere, the straight line *AB*, which we divide into two equal parts at the point *C*; we draw from the point *C* the straight line *CD* perpendicular to the straight line *AB*; let the point *D* be that found by the <tip of the> leg of the first compasses. We join *AD* and we erect on *AD* a straight line at a right angle; let it be *AE*. We extend *DC* until it meets *AE*; it is necessary that it meets it because the angle *CAE* is acute and the angle *ACE* is a right angle, let them meet one another at the point *E*.

I say that DE *is equal to the diameter of the sphere.*

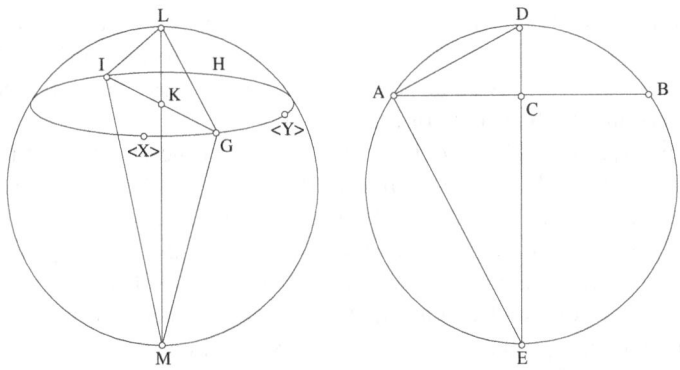

Fig. IV.1.17

Proof: We imagine the circle drawn on the sphere <to be> the circle *GHI*; let the value of its diameter found by the compasses be the straight line *GI* and let *L* be its pole. We divide the straight line *GI* into two equal parts at the point *K*, so the point *K* is the centre of the circle. We join *LK*, *LK* will be perpendicular to the plane of the circle because any straight line drawn from the point *L* to the circumference of the circle is equal to the straight line *LG*, and any straight line drawn from the point *K* to the circumference of the circle is equal to the straight line *KG*, because the point *K* is the centre of the circle; two straight lines drawn from the points *L* and *K* to a point on the circumference of the circle are thus equal to the straight lines *LG* and *GK*. But the straight line *LK* is common to all the triangles; so all the triangles formed are equal to the triangle *LKG* and their angles at the point K are equal to the right angle LKG, so the straight line LK makes a right angle with any straight line drawn from the point K to the circumference of the circle; so the straight line LK is perpendicular to the plane of the circle. But any straight line drawn from the centre of the circle and perpendicular to its plane passes through the centre of the sphere; this has been proved in Theodosius' book the *Sphærica*. So we imagine the straight line LK <to be> extended until it ends on the surface of the sphere. It meets the surface of the sphere at the point M, so the straight line LM is a diameter of the sphere. We join GM; it forms the triangle LGM. We imagine the plane of the triangle LGM cutting the sphere, on its surface, it forms a circle whose centre is the centre of the sphere; this too has been shown in Theodosius' book the *Sphærica*. Let this circle be the circle $GLIM$, thus this circle lies on the surface of the sphere and its centre is the centre of the sphere. But if its centre is the centre of the sphere, then it is a great circle of the sphere and its centre lies on the straight line LM. If the centre of the circle $GLIM$ lies on the straight line LM, the straight line LM is a diameter of the circle and the arc LGM is a semicircle, so the angle LGM is a right angle, the triangle GLM is similar to the triangle GLK; so the ratio of ML to LG is equal to the ratio of GL to LK. The product of ML and LK is equal to the square of LG.

Similarly, the angle EAD is a right angle and the angle ACD is a right angle, so the triangle ADE is similar to the triangle ADC. So the product of ED and DC is equal to the square of AD. But AD is equal to GL, AC is equal to GK, the square of AD is equal to <the sum> of the squares of AC and CD and the square of GL is equal to <the sum> of the squares of GK and KL, so the square of CD is equal to the square of KL, so CD is equal to KL. But since the product of ED and DC is equal to the square of AD, and AD is equal to GL, the product of ED and DC is equal to the square of GL and the product of ML and LK is equal to the square of GL, so the product of ED and DC is equal to the product of ML and LK; but DC is equal to LK, so the straight line DE is equal to the straight line LM; but LM is the diameter of the sphere, so the straight line DE is equal to the diameter of the sphere. That is what we wanted to prove.

But since the angle DAE is a right angle, and AC is perpendicular to DE, the product of EC and CD is equal to the square of AC. So if we divide the square of AC by the straight line CD, the quotient is the straight line CE. So if we add to it the straight line CD, the sum will be the straight line DE, which is equal to the diameter of the sphere.

What we have explained is the method for measuring all the bodies that are used in the art of the surveyor.[49]

It remains for us to show how to determine the heights of bodies if their height is unknown, whether such bodies are cylinders on a circular base or bodies with straight edges or walls or buildings or mountains such that one cannot reach either their peak or the point at the foot of their perpendicular height. The method for this is to take a straight rod whose length is not less than five cubits, then to measure from its endpoint a single cubit, by means of the cubit used for measurement,[50] then to mark at the end of the cubit a distinctive sign on the rod, circularly around the rod; we then take a thread at the end of which there is heavy lead weight. The observer holds onto a part of the thread, and remains upright, he puts the thread against one of his eyes and lets the lead weight go, increasing or decreasing the length of the thread until the end of the lead weight is on the surface of the ground. He then marks a sign on place on the thread where it was at his eye, then he puts this thread against the straight rod and puts the sign that is on the thread on the sign that is on the rod and which is the end of the cubit measured out on the rod. Then he pays out the thread on the side with the plumb bob and lays it along the rod, he holds the weight in his other hand and pays out the thread along the length of the rod, then, at the position on the rod reached by the end of the plumb bob, he marks a distinctive sign which cannot be removed, circularly around the rod. Thus there will be a part of the rod remaining, because the height of a man and a cubit add up to less than five cubits. If the observer wishes to determine the height of any body or the height of a mountain, let him stand upright on the ground opposite the body whose height he wishes to determine; he then drives the rod into the earth and arranges that the measured cubit is in the upper part of the rod, he drives the rod into the ground until all that remains of the rod beyond the measured parts has disappeared and he adjusts the rod until it is standing vertically in the surface of the ground, without leaning sideways. When the rod has been set up and is standing upright, the observer positions himself behind it, looks at the body whose height he seeks to know and selects a particular place on it – if it is a pyramid, that will be the point at its apex, if it is a wall or a cylinder or a mountain, that will be a particular place – then he moves forwards and moves backwards, he leans right and left and, in all these cases, he looks at the top of the rod and the place he selected until he can see them at the same time; if he sees them at the same time, then he covers one of his two eyes and looks with

[49] That is to say all the things surveyors measure.

[50] The cubit used for the measurement probably refers to a standard rod carried by the surveyor. [Note J. V. Field]

the other eye and fixes his gaze on the top of the rod. If he fixes his gaze on the top of the rod, he must see the body whose height he wishes to know, because it is behind the rod and in the same direction as the rod. So if he sees the tall body, then let him direct his glance to the top of the rod; he bends to right and left, moves forwards and backwards, and adjusts his posture as much as possible until he sees the place on the body that he selected at the same time as the top of the rod to which he is directing his glance and sees at the same time only the top of the rod and in the direction of the top of the rod only this place, and so as he sees both of them with one of his two eyes. When that happens, he keeps his leg still, <that is the leg> on the side of the eye with which he is looking, then sits down and puts his finger on the point on the surface of the ground below the midpoint of his foot on the side of the eye with which he was looking, then he takes his leg away from that position and marks this point with a distinctive sign that cannot be effaced, either by means of a small rod that he drives in at that point or by a small hole dug at that point. If he is proceeding in this way, at this stage he draws a straight line on the surface of the ground from the position of the sign to the foot of the rod that was set up, then he measures out that straight line by means of the cubit used for measurement – the cubit being divided into parts that are as small as possible – then he notes the magnitude of the straight line and records it. He then pulls up the rod from its place in the ground and extends the straight line drawn on the earth in the direction of the body whose height he seeks to know, then he puts a mark on a point of that straight line, sets the rod at that point, drives it into the earth deep enough for the magnitude of the remainder in the lower part of the rod to be hidden from view; he adjusts how it stands until it is upright and vertical. Then he positions himself behind the rod and places his foot on the straight line drawn on the surface of the earth and looks at the place he selected on the tall body, he moves forwards and back, leans right and left, covers the eye he covered the first time and looks with the eye he looked with first, and directs his glance to the top of the rod until he sees the top of the rod and the place he selected on the tall body at the same time. If he sees them at the same time, he keeps his foot still, <that is the foot> on the side of the eye with which he was looking, he sits down and, at the midpoint of his foot, he makes on the surface of the ground a distinctive mark that cannot be effaced, then he measures out the straight line between this mark and the position of the rod, using the cubit used for measurement, notes the magnitude and records it. If he obtains the two magnitudes we mentioned, he also measures out the distance between the position of his foot during the first sighting and the position of his foot during the second sighting, he notes that magnitude also and records it, then he subtracts the second magnitude from the first magnitude. The second magnitude cannot but be smaller than the first, we shall show that later. If he subtracts the second from the first, there will be something remaining from the first one, he then notes this remainder, then he divides the magnitude of the distance between the two positions of his foot by this remainder, he adds to the quotient the magnitude of the rod measured with the plumb line; what he obtains is the height of the body whose height we seek, whether that is a mountain or anything else.

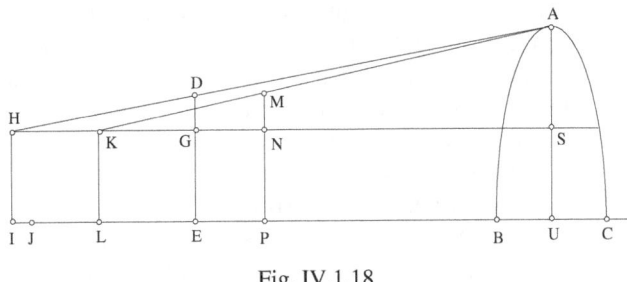

Fig. IV.1.18

Proof of this procedure: let the mountain or cylinder or pyramid or <other> body whose height we wish to know be called *ABC*; let the rod that we have set up on the surface of the ground the first time be the straight line *DE*, let the cubit measured on this be *DG*, let the measurement of the rod by the plumb line be *GE* and let the remainder of the rod be driven into the ground. Let the height of the observer be *HI* and let the point *H* be the position of the eye through which he makes sightings and the point *I* be the midpoint of his foot, and let the place selected on the tall body be the point *A*. We draw the visual ray[51] from the point *H* that passes through the end of the rod, that is the point *D*, and through the point *A*, which is the place selected on the body, let the ray be *HDA*; so *HDA* is a straight line because the visual ray is propagated only in a straight line; that has been shown in the book of *Optics*. Let the straight line *IE* be the straight line drawn on the surface of the ground, and let the rod, in the second case, be the straight line *MP*; let the cubit measured on it be *MN*, so *NP* is what is measured on it by the plumb line. Let the observer, in the second case, be *KL*; we draw the ray *KMA*, it will be a straight line. Since *HI, KL, GE, NP* are perpendicular to the surface of the ground, they are all parallel, I mean by these perpendiculars the perpendiculars to the straight lines that join the

---

[51] In his reform of optics, Ibn al-Haytham rejected all the various theories of visual rays. The conventional term he uses here has the sense of 'line of sight'.

midpoints of the positions we mentioned.[52] Since they are erected on the same straight line, they are all in the same plane; since they are all measured by the plumb line, they are all equal, so the line that passes through the points *H, K, G, N* is a single straight line parallel to the straight line *IP*. Let us draw this straight line, let it be the straight line *HKGN*, and let us imagine a straight line starting from the point *A* parallel to the straight lines *HI, KL, GE, NP*, which are parallel; let the straight line be *AU*. So this straight line is perpendicular to the surface of the ground because it is parallel to the straight lines we mentioned that are perpendicular to the surface of the ground. This straight line meets the straight lines *HN* and *IP* if we extend them, because the straight line *AU* is parallel to the two straight lines *HI* and *DE* and starts from the point *A* which belongs to the straight line *HA* which is in the plane of the two straight lines *HI* and *DE*; so the straight line *AU* is in the plane of the two straight lines *HI* and *DE*, which are parallel, and the two straight lines *HN* and *IP* are in the plane of these two parallel straight lines; so the straight line *AU* meets the two straight lines *HN* and *IP* if we extend them. Let us imagine <that> the two straight lines *HN* and *IP* are extended, let the straight line *AU* meet them, let the meeting of the straight line *AU* with the straight line *HN* be at the point *S* and let its meeting with the straight line *IP* be at the point *U*. Since the straight line *AS* is parallel to the straight line *DG*, the ratio of *HG* to *GD* is equal to the ratio of *HS* to *SA*, because the triangles *HGD* and *HSA* are similar. Since the straight line *AS* is parallel to the straight line *MN*, the ratio of *MN* to *NK* is equal to the ratio of *AS* to *SK*; but *MN* is equal to *DG* because each is a single cubit, so the ratio of *DG* to *NK* is equal to the ratio of *AS* to *SK*; so, by the ratio of equality, the ratio of *HG* to *NK* is equal to the ratio of *HS* to *SK*. But *HS* is greater than *SK*, so the straight line *HG* is greater than the straight line *NK*; but *HG* is equal to *IE*, because the area *HIEG* is a rectangle; but the straight line *KN* is equal to the straight line *LP*, so the straight line *IE* is greater than the straight line *LP*; which is what we stated earlier that we were going to prove.

We put *JE* equal to *LP*. So the ratio of *IE* to *EJ* is equal to the ratio of *HG* to *NK*; but the ratio of *HG* to *NK* is equal to the ratio of *HS* to *SK*, so the ratio of *IE* to *EJ* is equal to the ratio of *HS* to *SK*. By inversion (of the ratios) and composition,[53] the ratio of *SH* to *HK* is equal to the ratio of *EI* to *IJ*. If we permute, we have <that> the ratio of *SH* to *IE* is equal to the ratio

---

[52] By hypothesis, the points *I, L, E, P* all lie on the straight line *IU*; this straight line joins the points *I* and *L*, which represent the midpoint of the observer's foot in each of the two positions that are mentioned.

[53] After inverting the two ratios, it would be necessary to subtract unity from the two members, and then to invert again to obtain the result.

of *HK* to *IJ*. But *HK* is equal to *IL*, because the area *HILK* is a rectangle, so the ratio of *SH* to *IE* is equal to the ratio of *LI* to *IJ*, so the product of *EI* and *IL* is equal to the product of *SH* and *IJ*.

Similarly, the ratio of *HG* to *GD* is equal to the ratio of *HS* to *SA*, so the product of *HS* and *GD* is equal to the product of *AS* and *H G*; but the product of *HS* and *GD* is *HS* because *GD* is unity, so the product of *AS* and *HG* is equal to the magnitude *HS*, and the product of *HS* and *IJ* is equal to the product of *EI* and *IL*; the product of *AS* and *HG*, then <the product of> what we obtain and *IJ*, is thus equal to the product of *EI* and *IL*. But *HG* is equal to *IE*, so the product of *AS* and *IE*, then <the product of> what we obtain and *IJ*, is equal to the product of *EI* and *IL*. But the product of numbers one with another remains equal if we commute them, so the product of *AS* and *IJ*, then <the product of> what we obtain and *IE*, is equal to the product of *LI* and *IE*; so if the magnitude *IE* is taken away from both, we have <that> the product of *AS* and *IJ* is equal to the magnitude *IL* and if we divide *IL* by the magnitude *IJ*, we have the quotient *AS*; but *SU* is equal to *GE*, the magnitude of the plumb line, *EJ* is equal to *LP* which is the second magnitude and *EI* is the first magnitude, so the straight line *IJ* is the remainder that is the amount by which the first magnitude exceeds the second magnitude, and *IL* is the magnitude of the distance between the two positions of the foot of the observer. So if we divide the magnitude *IL*, which is the magnitude of the distance between the two positions of the foot of the observer, by *IJ*, which is the amount by which the first magnitude exceeds the second magnitude, and if we add to the quotient the magnitude *GE*, which is the magnitude of the plumb line, the sum will be *A U* which is the height of the body *ABC* whose height we seek, because *AU* is perpendicular to the surface of the ground. That is what we wanted to prove.

We have completed the explanation of the procedures for all the measurements of magnitudes used in the art of surveyors, by means of their proofs and their causes. This is the matter to which we intended to draw attention in this treatise. But since what is used in all that we have considered of the art of measurement is only the practical procedure and since surveyors do not make use of any proofs when making their measurements, we need to excerpt from the sum of what we have explained in this treatise the practical procedures we have mentioned, so that they may be accessible and easy for someone who wishes to acquire instruction in the art of measurement and make use of the practical procedures.

An Account of the Procedures for Measurement set out in this Treatise

All the plane figures whose measurement is undertaken by the surveyor are rectilinear figures, circles or pieces of them. All the solid figures whose measurement is undertaken by the surveyor are rectilinear solids, circular cylinders, cones and spheres.

The measurement of all plane rectilinear figures can be reduced to the measurement of triangles and to the determination of the chords of the angles that divide the surfaces into triangles. We carry out the measurement of all triangles by adding up the sides of the triangle and taking half the sum, then multiplying that half by the amount by which it exceeds one of the sides of the triangle, then multiplying what we obtain by the amount by which that half exceeds one of the other sides of the triangle, then multiplying what we obtain by the amount by which that half exceeds the remaining side of the triangle; we take the square root of what we obtain, that will be the measure of the triangle.

To determine the chords of the angles, we cut off one cubit from one of the two sides that enclose the angles, then we divide the magnitude of the other side by the magnitude of the first side; we cut off on the other side <a magnitude> equal to the quotient and we join the two cutting <points> by a straight line, we find the value of that straight line, and we multiply what we obtain as its value by the magnitude of the first side; what we then obtain is the chord.

We carry out the measurement of circles by determining the diameter of the circle and multiplying the diameter by itself and subtracting from its square a seventh of the square and half of a seventh of it; what remains is then the measure of the circle.

We determine the diameter of the circle, if the diameter is unknown, by drawing an arbitrary chord in the circle, dividing it into two equal parts and from its midpoint drawing a perpendicular as far as the arc the chord has cut off. We then find the value of half the chord and we find the value of the perpendicular, then we multiply the magnitude of half the chord by itself, we divide <the square> by the magnitude of the perpendicular and we add the perpendicular to the quotient; <the sum> is the diameter of the circle.

The measure of a sector of a circle is the product of its side and half its arc. We carry out the measurement of a segment of a circle by completing it to give us a sector and measuring the sector, then measuring the triangle of the sector and subtracting it from the measure of the sector; what remains is the measure of the segment.

We determine the ratio of the arc to the circumference of the circle by drawing the chord of the arc and dividing the chord into two equal parts, drawing a perpendicular from its midpoint as far as the arc and joining the endpoint of the chord to the endpoint of the perpendicular by a straight line that we extend; then erecting, at the endpoint of the chord from which we drew the straight line, a straight line at a right angle. We take this endpoint as centre and, with the distance to the other endpoint of the chord or with a part of that distance on the chord, we draw an arc of a circle so that this arc cuts the two straight lines drawn from the endpoint of the chord. We then find the value of the arc cut off by the straight line from the midpoint in terms of a magnitude that gives a value for the whole arc that is a quarter of a circle, from that we obtain the ratio of the small arc to a quarter of a circle and it will thus be the ratio of the first arc to the circumference of its circle.

The measurement of all rectilinear bodies reduces to measuring pyramids. We carry out the measurement of a pyramid by measuring its base and multiplying it by a third of its height, what we obtain is its measure.

The measurement of the base of the pyramid, if the base is a triangle, is as the measurement of triangles; and if the base is a polygon, we carry out <the measurement> by dividing the base into triangles; their (polygons') division into triangles is done by determining the chords of the angles.

We determine the chords of the angles of the base of a solid, whether it is a pyramid or something else, by finding an angle equal to the angle of the base in a plane, and this <is done> by considering two rulers; we attach one of them to one of the two sides of the base <around the angle> and we position the end of that ruler beyond the angle, then we attach the second ruler to the other side enclosing the angle, then we draw against the edge of that ruler a straight line in the plane of the first ruler, then we put the first ruler in a plane and we fit the second ruler to the straight line drawn against the first ruler, then we draw against the two edges of the rulers, that is to say the two inner edges, two straight lines; we thus form in the plane an angle equal to the angle of the base of the solid; we find the chord of that angle by the method <given> above for finding the chords of angles. So this chord is the chord of the angle of the base of the solid. If the base of the solid is in a continuous plane, we extend the two sides of the base, we thus form outside the solid an angle equal to the angle of the base of the solid. We proceed as we proceeded for the angle we mentioned before; we then obtain the chord we wanted.

We carry out the measurement of a circular cylinder by measuring its base and multiplying it by the height; if the cylinder is set on its base at right angles, then its height is its length; if it is inclined, we shall show later how to find its height.

We determine the measure of its base by finding the value of the circumference of its base, we divide the magnitude we obtain by three and a seventh; what we obtain is its diameter. When we have obtained the diameter <of the base>, we determine its measure as we have set out above.

We carry out the measurement of a cone by measuring its base, then multiplying the measure of its base by a third of its height; what we obtain is its measure.

We carry out the measurement of a sphere by determining the measure of the greatest circle that can be drawn on it (a great circle), then multiplying the measure of that circle by two thirds of its diameter; what we obtain is the measure of the sphere.

We determine the diameter of the sphere by drawing on the surface of the sphere an arbitrary circle by means of compasses <the point of> one of whose legs we place on the surface of the sphere; with <the point of> the other leg we draw a circle on the surface of the sphere; we then mark on the circumference of this circle two points, the circle is thus cut into two arcs each of which we divide into two equal parts[54] using another pair of compasses with which we step out the circumference of this circle.[55] If each of the two arcs is divided into two equal parts, then the circumference has been divided into two equal parts; we then put <the tip of> one of the legs of a second pair of compasses on one of the two opposite points and we move the other leg until <its tip> reaches to the point opposite that one. We then place <the tips of> the two legs of these compasses on a plane and with <the tips of> its two legs we make two marks that we then join by a straight line. We draw from the midpoint of this straight line a perpendicular to the line, then we put <the tip of> one of the legs of the first compasses at the endpoint of the straight line we divided and we cause the other leg to move until it meets the perpendicular. We then mark the position of <the tip of> this leg on the perpendicular. We then find the value of half the straight line we divided and we find the value of what has been cut off from the perpendicular. We multiply the magnitude of half the straight line by itself and we divide <the square> by the magnitude of the perpendicular; we add the perpendicular to what we obtain, so the sum is the diameter of the sphere. When we obtain the diameter of the sphere, we multiply it by itself, we subtract <from the square> a seventh and half a

---

[54] The points that divide the arcs are at opposite ends of a diameter.

[55] On stepping, see note 48 above.

seventh of it; what remains is the measure of the greatest circle that can be drawn on the sphere. We then multiply the value of the measure of the circle by two thirds of its diameter; what we obtain is then the measure of the sphere.

To determine the heights of pyramids, cylinders, mountains, walls and all tall bodies, we consider a straight rod whose length is not less than five cubits, in terms of the cubit used for measurement. We then measure out on this rod a single cubit by means of the cubit used for measurement. At the end of the cubit we make a distinctive mark around the top of the rod.

The observer next takes a thread at the end of which there is a lead weight, he then holds the thread in his hand and remains standing, then he applies a point on the thread to one of his eyes, then he releases the thread, increases and decreases <its length> until the lead weight touches the surface of the ground; when that happens he marks with a sign the place on the thread applied to his eye, then he applies the thread to the straight rod and superimposes the sign on the thread on the sign that is on the rod and which is at the end of the cubit. With his other hand, he pays out the thread until the end of the lead weight reaches a point on the rod. When this happens, he puts a mark on the place on the rod which is at the end of the thread; part of the rod will remain over because the sum of <the lengths of> the plumb line and the cubit ls less than five cubits. If the observer wishes to know the height of one of the bodies, let him stand facing the body, then drive the rod into a place on the ground, between himself and the tall body, and arrange matters so that the cubit measured out on the rod is on the side of the upper part of the rod; he drives the rod into the earth until the part of it that remained over disappears. He adjusts the rod until it is at right angles to the surface of the ground, then he places himself behind the rod and looks at the top of the rod and the top of the thing whose height he wishes to know and directs his glance to a particular place on the top of the thing, if the top of it is not a point; he then covers one of his eyes and looks with the other and looks at the top of the rod, then he moves forwards and back, bends to the right and bends to the left until he can see at the same time both the top of the rod and the place he had selected on the top of the thing; when this happens he sits down and puts his finger on the place on the ground that is beneath the midpoint of his foot, <that is the foot> on the same side as the eye he was looking with, and puts a mark at that place; then he draws a straight line from that mark to the foot of the rod; he then measures that straight line with the cubit used for measurement – let that cubit be divided into parts as small as possible. He notes the magnitude of the straight line and records it. He then pulls up the rod from its position and extends the straight line drawn on the surface of the ground in the direction of the thing, he then makes a mark on this straight line and puts the rod on this mark and arranges matters so that the cubit measured out on the rod is on the side of the upper part of the rod; he drives the rod into the earth until the part that remained over disappears. The observer places himself behind the rod and covers the eye he covered <before> and looks with the other eye and puts his foot, <that is the foot> on the side of the eye he is looking with, on the straight line drawn on the surface of the ground and looks at the top of the rod until he sees the top of the rod at the same time as the place that he selected at the top of the thing; when that happens he sits down and makes a mark at the place beneath the midpoint of his foot. He measures out the straight line that lies between the mark and the foot of the rod and subtracts this magnitude from the first magnitude; what remains from the straight line is the part that becomes the divisor. He then measures out the straight line that runs between the position of his foot in the first case and the position of his foot in the second case, and he divides what he obtains by the remainder he recorded. We add to the quotient the magnitude of the rod measured by the plumb line; the sum is the height of the thing whose height we sought.

These are the practical procedures that surveyors need in their art. Here we complete this treatise.

In the name of God, the Compassionate, the Merciful

TREATISE BY THE SHAYKH ABŪ ʿALĪ IBN AL-HAYTHAM

On Knowing the Height of Upright Objects, the Altitude of Mountains and the Height of Clouds

We take *AB* as the height of a mountain or of an object and we wish to know what it is.

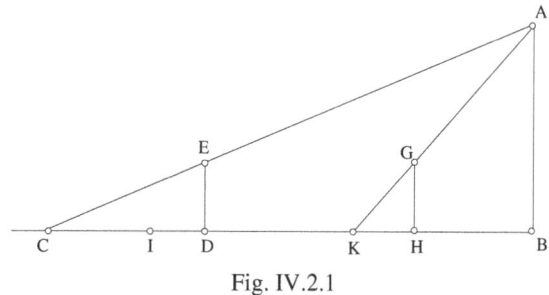

Fig. IV.2.1

We set up an object on the surface of the ground, such as *DE*; let the observer move forwards and back until he sees the top of the object at the same time as the top of the mountain; under these conditions let the eye be shown as the point *C*, the object as *DE* and the line of sight as *CA*. We imagine the straight line *BDC* on the surface of the ground. So the ratio of *CB* to *BA* is equal to the ratio of *CD* to *DE*. We now remove the object *DE* and place it in a position closer to the mountain, such as *HG*. Let the observer move forwards and back until he sees the top of the standing object at the same time as the top of the mountain, as he did in the first case; in this second case let the eye be shown as *K* and the line of sight as *KGA*, then the ratio of *AB* to *BK* is equal to the ratio of *GH* to *HK*. But since the straight line *BK* is smaller than the straight line *BC*, the straight line *HK* is smaller than the straight line *CD*. We cut off from *CD* <a piece> equal to *HK*, let it be *ID*. The ratio of *ID* to *DE* is equal to the ratio of *KH* to *HG*, but the ratio of *CB* to *BA* is equal to the ratio of *CD* to *DE* and the ratio of *AB* to *BK* is equal to the ratio of *ED*, which is equal to *GH*, to *DI*, which is equal to *HK*. So, by the ratio of equality, the ratio of *CD* to *DI* is equal to the ratio of *CB* to *BK*; by separation, the ratio of *CI* to *ID* is equal to the ratio of *CK* to *KB*. But the ratio of *CI* to *ID* is equal to the ratio of *CK* to *KB* and the ratio of *ID* to *DE* is equal to the ratio of *KB* to *BA*, thus, by the ratio of equality, the ratio of *CI* to *DE* is equal to the ratio of *CK* to *AB*. So the product of *KC* and *DE* is equal to the product of *CI* and *AB*. So if we multiply *CK*, <which is> known, by *DE*, <which is> known – because it is the object – and if we divide by *CI*, then the quotient is *AB*. That is what we wanted to prove.

The treatise is completed. Thanks be given to God, Lord of the worlds, may blessing be upon His prophet Muḥammad and those that are his, the pure.

TREATISE BY AL-ḤASAN IBN AL-ḤASAN IBN AL-HAYTHAM

On the Determination of the Height of Mountains

If someone wishes to know the height of a mountain, of a high building, or of one of the bodies that have a height, he should employ a straight rod, whose length is five cubits and a half. Let him measure off a single cubit from one of its two ends, and make a distinctive black mark, round, around the rod, at the end of the cubit. Then he measures out on the rod another cubit, following on from that cubit, and at the end of it makes a round black mark, again <one that is> easy to see. The measurement can be done by means of a short straight ruler, whose length is a single cubit. Let this ruler be divided into sixty equal parts, so that we can make use of its parts in what follows. If he wishes to know the height of a mountain or of a body that has a height, he goes to a place close to the mountain, without being extremely close to it. He relies on finding a piece of flat ground, or almost flat; he drives the rod into the earth and sets it up in the ground so that it is vertical; he pushes it into the earth until the second mark drawn on the rod is level with his eye, then he consolidates <the supports for> the base of the rod, so that it cannot lean and stays vertical. Then he moves back and covers one of his two eyes; he looks at the mountain and the top of the rod; he takes a sighting of a place at the top of the mountain that is distinct either because there is a protruding part on the mountain, or there is a rock, or sufficiently vivid colour, to enable him to recognize this place if he looks at it later. He moves backwards and forwards until he sees this place on the mountain at the same time as the top of the rod, that is to say that he sees them in the same direction; and in such a way that the sighting is carried out by only one of the two eyes. If this place is determined, then let him sit down and make a mark on the earth at the midpoint of the position of his foot, that is to say the foot that is on the <same> side as the eye with which he saw the mountain and the rod at the same time. He drives in a small rod at this place so that the mark does not disappear.

Then, he takes a little wax, <a piece> about the size of a walnut, which he sticks on one or the other of the faces of the upright rod, on the position of the higher circular mark, that is to say the first mark. Then he moves back and covers one of his two eyes, and he looks at the thing stuck onto the rod and the chosen place at the top of the mountain, until he sees the chosen place at the top of the mountain and the piece of wax at the same time, that is to say in the same direction. When this place is determined, he again make a mark under the mild point of his foot and there too drives in a small rod. When he has done that, he measures out the distance between the first mark on the ground and the second mark; this measurement is carried out by means of the divided ruler, and he determines the measurement by <using> the parts of the <cubit> ruler because in the majority of cases the distance will not simply be a whole number of cubits, but often <will be> cubits and parts of a cubit, and perhaps only parts of a cubit, without any whole cubits. Then we measure out the second distance, <the distance> that is between the second mark on the ground and the upright rod, and we determine this distance with extreme accuracy. If we determine the two distances, then you will always find <that> the first distance, that is to say the one between the two marks on the ground, is greater than the distance between the second mark and the rod.

You subtract the second distance from the first distance, and you note what remains; then you add the two distances, you multiply the result by the first distance, you divide the product by <the quantity> that was noted, you double the quotient, you divide the result by the sum of the two distances and you add three and a half cubits to the result. What we obtain is the height of the mountain or of a body whose height was sought. And what we obtain from the first division before doubling, is the distance there is between the first mark on the ground and the foot of the perpendicular <dropped from the top> of the mountain. That is what we wanted to show.

The comparison with the original copy is completed.

Finished by the grace of God and His blessings. Blessed be the best of His created beings, the prophet Muḥammad, and those that are his. Let there be peace.

A RESEARCH TRADITION: THE REGULAR HEPTAGON

1. HISTORY OF THE TEXTS

We cannot come to a deep understanding of the unprecedented flowering of research on the regular heptagon in the final third of the tenth century if we confine ourselves to recounting events and describing construction procedures – work that has, moreover, already been carried out satisfactorily by A. Anbouba.[1] Understanding on that level comes through undertaking two tasks that are inseparably linked. First of all we must carry out exploratory work, so as to shed light on the position of this research in relation to its own theoretical and technical aims. As we have shown, once such work has been carried out, the research on the heptagon is seen to be part of a much larger enterprise on which geometers were then engaged, but along with them also algebraists, an enterprise dedicated to investigating geometrical constructions, notably those that employed conic sections. But to follow the course of development of this enterprise, at least in part, and to uncover the exact significance of this research, we needed to establish texts of all available works, assess the various testimonies and evaluate the documents; establishing this basic framework was made all the more necessary by the fact that the construction of the heptagon was the subject of controversy so impassioned that it attracted attention, and a little censure, from contemporaries of the events concerned, such as al-Bīrūnī,[2]

---

[1] *'Tasbī' al-Dā'ira* (La construction de l'heptagone régulier)', *Journal for the History of Arabic Science*, 1, 2, 1977, pp. 352–84.

[2] 'The centuries have passed, writes al-Bīrūnī, and we have arrived at our own time, a time rich in things <that are> surprising, unprecedented and amazing, and <a time> that combines contraries; I mean the richness of sources of knowledge in this time and the natural disposition of its men to recognize the near-perfection and completeness in every branch of learning, the increase of distinction among them as well as the capacity to discover amazing things that had defeated the illustrious Ancients; and side by side with all this, the appearance of behaviour that runs counter to

(*Cont. on next page*)

and later, at least indirectly, from Ibn al-Haytham.[3] So we needed to establish texts of all the writings that have come down to us from Ibn al-Haytham's predecessors and his immediate successors. That is the task to which this bulky appendix is dedicated.[4]

So what we are giving here is the English translation of these writings, but not of the summary versions that have sometimes been made from them.

A single occurrence is not necessarily significant: three collections combine the majority of the writings – Cairo, Dār al-Kutub, 41; Paris, Bibliothèque Nationale, 4821; Istanbul, Süleymaniye, Aya Sofya 4832. The remaining texts are found in other collections; we shall first look at these collections.

(*Cont.*) what we have just said, in contradiction with it, on the part of the majority of those among them who are in rivalry; they are animated by jealousy, conflicts dominate and they are so opinionated that some appropriate things that belong to others and pride themselves on what is not theirs; some pillage others' learning and make it theirs to sell it for gain; they ensure that others are blind to what they are doing themselves, but vent all their anger upon anyone who has seen through them and bear him hatred and enmity; this is what has happened to a group of eminent mathematicians of our time, regarding the regular heptagon and, with it, trisection of an angle; regarding duplication of the cube, and other things [...].'

إلى أن طال الأمد ، وانتهت المدّة إلى زماننا هذا ، ذي العجائب والبدائع والغرائب ، الجامع بين الأضداد . أعني بذلك غزارة ينابيع العلوم فيه ، وتهيّؤ طبائع أهله لقبول ما يكاد أن يكون الكمال والنهاية في كل علم ، وانتشار الفضل فيهم والقدرة على استنباط العجائب المعجزة جل القدماء ، مع ظهور أخلاق منهم تضادّ ما ذكرناه ومناقضة ‹له› من عموم المتنافسين ، والتحاسد إيّاهم ، واحتواذ التنازع والتعاند عليهم ، حتى يغير بعضهم على بعض وافتخر بما ليس له . ويسلب بعضهم بعضًا علمه وينسبه إلى نفسه ، متكسّبًا به ، ويكلّف الناس التعامي عن فعله ، بل يصرف عنان قوّته الغضبيّة إلى من فطن بحاله وينطوي على عداوة وبغضاء له . كما وقع بين جماعة من أفاضل عصرنا في تسبيع الدائرة وفي تثليث الزاوية بالسواء وفي تضعيف المكعّب وغير ذلك....

This passage from *Kitāb maqālīd 'ilm al-hay'a* (ed. M.-T. Debarnot, Damascus, 1985, p. 95) is valuable testimony from one of the members of the mathematical community (that is indeed what it is), concerning some of the behaviour during controversies that, as we see, relate in particular to the classical problems of geometrical construction.

[3] See the introduction to his treatise on the heptagon, p. 441.

[4] We gave the *editio princeps* of more or less all these writings – only one of which has already been the subject of a critical edition, and that edition was unsatisfactory; the work in question was the treatise by al-Sijzī (see below) – as well as the French translation of the edited texts, in *Mathématiques infinitésimales*, vol. III.

I. The manuscript collection 41 Dār al-Kutub (Cairo) includes 32 treatises and short works. It is one of the most important collections of scientific manuscripts – in mathematics and astronomy – that is now known. It includes some rare texts, and some unique ones, for instance a copy of Ibn Abī Jarrāda's version of the *Sections of the Cylinder and its Lateral Surface* by Thābit ibn Qurra,[5] a copy of the *Book on the Synthesis of the Problems Analysed by Abū Sa'd al-'Alā' ibn Sahl,*[6] as well as others by Ibn Qurra, by al-Qūhī, by al-Bīrūnī, and so on. It also contains five treatises on the regular heptagon, composed in the period that we are concerned with here: a piece by Abū al-Jūd, the one by al-Sijzī, the treatise by al-Shannī and an anonymous text, in addition to the one attributed to Archimedes. Apart from the middle folios 87ᵛ to 94ᵛ, it is in a single hand. It was indeed copied relatively recently, in the eighteenth century, by the famous Muṣṭafā Ṣidqī, in careful *naskhī* script. We have already several times come across this copyist, who is well educated and familiar with the mathematical sciences.[7] The copy is, on the whole, without glosses or additions. But Muṣṭafā Ṣidqī does sometimes intervene, and give his version rather than a true copy. This is what he did in a treatise by al-Qūhī on the volume of the paraboloid[8] in this same collection. He also intervened in the text on the regular heptagon attributed to Archimedes.

The different treatises brought together in this collection were transcribed between 1146 and 1153, that is, between 1733 and 1740. However, most of the copies were completed in 1740. Being all by one hand, this collection does not raise any problems relating to specific texts. For the moment we shall not touch on the question of its sources. The case is very different for the collection in Paris.

II. The collection 4821 of the Bibliothèque Nationale in Paris is no less important than the preceding one. It is moreover much older. It too contains rare texts, sometimes unique ones; five pieces on the heptagon: one by Abū al-Jūd, the one by al-Sijzī, the one by al-Ṣāghānī, and two treatises by al-Qūhī. But as several hands are at work in the transcription of this

---

[5] R. Rashed, *Founding Figures and Commentators in Arabic Mathematics*. A history of Arabic sciences and mathematics, vol. 1, Culture and Civilization in the Middle East, London, 2012, pp. 381–458.

[6] R. Rashed, *Geometry and Dioptrics in Classical Islam*, London, 2005, pp. 444–89.

[7] See, for example, *Geometry and Dioptrics*, pp. 10, 22, 27; *Founding Figures and Commentators in Arabic Mathematics*, pp. 126, 127, 129, 464, 585, 586.

[8] *Founding Figures and Commentators in Arabic Mathematics*, pp. 583–7.

collection, we need to give a once and for all description of it, so as not to have to return to the matter.[9]

In its present state, the volume has 86 paper leaves (230 × 150 mm). It was copied in Iran, but is to be found in Turkey, at Istanbul, in the fifteenth century (if we are to believe the seals and ownership indications on fol. 1^r).[10] Originally, this collection included eighteen treatises, three of which have been missing for a long time. The table of contents transcribed on fol. 1^r is evidence of this. It is written in black ink, and states that the manuscript contains:

– *The Second Book of Euclid with the Additions of al-Qūhī*,

– *The Sphericity of the Earth by al-Khāzin*,

– *The Book of Abī Saʿd al-ʿAlāʾ ibn Sahl on the Critical Examination of the Book by Ptolemy on Optics* (incomplete).[11]

A later hand has noted in red ink, above the titles in the contents, the number of the folio on which the work appears; in the case of the three lost treatises, their absence is noted.

Let us now turn to the treatises that are preserved, in the order:

1. *Treatise on the Construction of the Side of the Regular Heptagon in the Circle by Abū Sahl al-Qūhī*, fols 1^v–8^r. As a rule, each folio has 16 lines, sometimes 17, with about eight words in each. The copyist (hand A) notes in the colophon that he has compared the copy with its original; and, in fact, a marginal note on fol. 3^v is in his hand, and is probably to be explained within the framework of that revision. Hand B has put in two marginal notes, on fols 1^v and 6^v.

Copyist A did not give his name in the colophon. His handwriting, *naskhī*, has a certain similarity (but no more) to that of al-Ḥusayn Muḥammad ibn ʿAlī, who transcribed fols 29^v–36^v of the collection. We

---

[9] The summary description given by G. Vajda [*Index général des manuscrits arabes musulmans de la Bibliothèque nationale de Paris*, Publications de l'Institut de recherche et d'histoire des textes, Paris, 1953], and his unpublished description of the manuscript preserved in the Department of oriental manuscripts of the BnF, are not adequate for the task of disentangling the complicated historical network, textual and conceptual, within which this fundamentally important manuscript plays its part.

[10] On fol. 1^r we read that this collection was the property of ʿAbd al-Raḥmān ibn ʿAlī [...] at Constantinople in 891/1486, then that of Alī ibn Amr Allāh ibn Muḥammad in Rajab 970/1563. Then it belonged to Sulaymān ibn Yūsuf in Ramaḍān 1077/1677, and then to Ḥunayn Ḥalabī (or Celebi) in 1089/1678-1679. Later it was taken into the Bibliothèque Nationale in Paris.

[11] Perhaps this incomplete text has survived under the title *Proof that the Celestial Sphere is not of Extreme Transparency*. See *Geometry and Dioptrics*, pp. 144–9.

may also observe that on page 8ʳ, well below the colophon, there is a line of Persian verse in the hand of Abū Isḥāq b. ʿAbd Allāh al-Kūbnānī:

خوشا اندر شبی روشن چراغی

کتابی واز (کذا) همه عالَم فراغی.

Next to it, a third hand confirms the identity of the last one:

خط أُستاذ أكابر الرياضيين محيي مراسم علوم الحكماء الماضيين المولى أبو إسحاق الكوبناني مد الله أظلاله وحقق آماله.

'Writing of the master of great mathematicians, who revived the traces of the sciences of past philosophers, his Excellency Abū Isḥāq ibn ʿAbd Allāh al-Kūbnānī [...]'.

This very probably identifies him as an owner of the collection. Again in another hand, we read, under the name of al-Kūbnānī: 'Commentator on the *Fawāʾid Bahāʾiyya* composed by Khawwām Baghdādī', followed by the signature of Sayyid Jalāl al-Dīn Tehrānī,[12] with the place – Paris – and the date – 1936.

2. *Al-Ghurba al-gharbiyya by al-Suhrawardī*, fols 8ᵛ–9ʳ. We have a short treatise on philosophy transcribed much later, on Tuesday 17 Jumādā al-ūlā of the year 744 (7 October 1343), on two folios that had been left blank.

3. *On the Construction of the Side of the Heptagon by al-Sijzī*, fols 10ᵛ–16ᵛ. Each folio has 16 to 18 lines, with about 10 to 11 words. The script is *naskhī*, in the same hand as that of the preceding mathematical text (number 1). At the end of the copy, we read: 'Transcribed from a damaged copy, with which it (the transcription) has been compared; praise be to God (*nuqila min nuskha saqīma wa-qūbila bihā wa-Allāh al-ḥamd*)'. The transcribed copy further includes three additions that are of significance for the history of the text. The first is on fol. 11ʳ (*allādhī*), the second, fol. 13ᵛ (*ka-khaṭṭ AD*); the third, a complete sentence, is on fol. 15ᵛ: *quṭruhu al-mujānib DB*, above this has been written the letter *ẓ*, an abbreviation conventionally used by copyists for the word *al-ẓāhir* – which means that this sentence was difficult to read in the original. These additions are in the same anonymous hand B in which most of the treatises of the collection are transcribed, notably the writings on the regular heptagon. So, as in the case of Text 1, copyist B must have had access to the original used by copyist A. From this point on, it does not seem overly bold to suppose there was some

---

[12] Jalāl al-Dīn Tehrānī is a scholar and a great collector of manuscripts, who has recently died.

degree of collaboration between the two men. This hypothesis will, moreover, be confirmed by the fact that copyist B is known, this time as a matter of certainty, to have collaborate with the third copyist of the collection, al-Ḥusayn Muḥammad ibn 'Alī.

4. *On the Construction of the Regular Heptagon Inscribed in a Given Circle by al-Qūhī*, fols 17^v–23^r. Each folio has 17–19 lines, each with about 11 words. This text was copied in *naskhī* by copyist B. Everyone who has spoken about this manuscript has confused the two hands.

The colophon tells us that B transcribed this text at 'Kushk Hamadhān (at the fort of Hamadhān) on Thursday 13 Rajab 544 AH', that is 16 November 1149, from an authoritative original – no less than an autograph text by al-Sijzī. It is a remarkable fact that several texts copied by B were taken from this autograph.

The copy shows two important crossings-out, on 18^v and 22^r, corresponding to repetitions that the copyist himself has struck out; there is a single addition, on 21r, used to indicate his place in the text.

5. *Treatise by al-Ṣāghānī for 'Aḍud al-Dawla,* fols 23^v–29^r. These pages are in the same hand as Text 4 and show the same characteristics. The copy was completed two days after the previous item, in the same place, and again from al-Sijzī's autograph. It contains neither additions nor glosses, and the two crossings-out (fol. 23^v) are again there to remove repetitions in the transcription.

6. *The Construction of a Regular Pentagon in a Known Square by al-Qūhī*, fols 29^v–33^v. Each folio has 18 or 19 lines, each of about 10 words. The writing, *naskhī*, is according to the colophon, in the hand of al-Ḥusayn Muḥammad ibn 'Alī. There are only two added words, on 30^v and 32^v, in the hand of the copyist. He completed the transcription on Tuesday 15 Ramaḍān 544 AH, that is 16 January 1150.

We know that al-Ḥusayn and B were collaborating in their copying because here, as in the following treatise, which was copied by the same person, it is B who took on the task of drawing the geometrical figures. Since al-Ḥusayn begins his copying on the verso of fol. 29, whose recto had been copied by B, we are compelled to conclude that the two men worked together, at the same time (the beginning of winter 1149–1150).

7. *On the Knowledge of the Magnitude of the Distance Between the Centre of the Earth and the Position of a Shooting Star, by al-Qūhī*, fols 34^r–36^v. Each folio has 18 lines, each with about 11 words. This text is in

naskhī script, in the hand of al-Ḥusayn Muḥammad ibn ʿAlī, transcribed, like the preceding one, on Tuesday 15 Ramaḍān 544 AH. We see no additions or glosses. After the colophon, we read, in another hand, in Persian, that the copy was completed on 15 Ramaḍān l'an 544, 'at Asad, God knows it'.[13]

8. *Letter of Abū al-Jūd to Abū Muḥammad ʿAbd Allāh ibn ʿAlī al-Ḥāsib*, fols 37ᵛ–46ʳ. Each folio has 18 lines, each with about 11 words. The text, like all those that now follow in the manuscript, was copied by B, again from the autograph by al-Sijzī and again at the same place, Kushk (Fort) of Hamadhān, on Wednesday 12 Rajab 544 AH, that is the day before the copy of the fourth text, on 15 November 1149. In all there are two words added in the margin by the scribe, words omitted during the copying (37ᵛ–39ᵛ).

9. *Commentary on the First Book of the* Almagest *by Abū Jaʿfar al-Khāzin*, fols 47ᵛ–67ᵛ. It is in the same hand as the preceding text, and is presented in the same form. The additions and crossings-out are by the copyist and, as for the preceding treatises, were made at the time of the transcription itself and not when comparing the copy with the original. We have already established a text of this work, translated it and written comments on it.[14]

10. *The Surface of any Circle is Greater than the Surface of any Regular Polygon with the Same Perimeter*, by al-Sumaysāṭī. This text – fol. 68 – presented here as anonymous, is that by al-Sumaysāṭī. It is in the same hand as the preceding text, and, like that one, has no colophon. We have already established a text of this work also, translated it and written comments on it.[15]

11. *Glosses by Maslama ibn Aḥmad al-Andalusī on the Planisphere of Ptolemy*, fols 69ᵛ–75ᵛ. The same characteristics. The colophon tells us that

---

[13] See *Geometry and Dioptrics*, pp. 1008–17.

[14] *Mathématiques infinitésimales*, vol. I, pp. 737–829; English translation in *Founding Figures and Commentators*, pp. 551–76. We have, like everyone else, confused this copyist with al-Ḥusayn Muḥammad ibn ʿAlī (*Mathématiques infinitésimales*, vol. I, p. 740; *Founding Figures and Commentators*, p. 507) being misled by our confidence in the catalogue of the manuscripts. We ask readers to accept our apologies.

[15] *Mathématiques infinitésimales*, vol. I, p. 830–3; English translation in *Founding Figures and Commentators*, pp. 577–8.

the copy was produced at Asadābād, a town close to Hamadhān, on Wednesday 11 Sha'bān 544, that is 14 December 1149.

12. *A Chapter that does not Belong in the Book (the Planisphere) of the Sayings of Maslama ibn Aḥmad*, fols 76ʳ–81ᵛ. Same characteristics.

13. *Copy of an Autograph text by 'Abd Allāh ibn al-Ḥasan al-Qūmasī, pupil of Yaḥyā ibn 'Adī*, fol. 82. This text, a translation from Syriac attributed in the manuscript to Ibn 'Adī, is also copied by B.

14. *On the Generation of the Climate by al-Nayrīzī*, fols 82ᵛ–86ʳ. By the same anonymous hand, the copy was completed, according to the colophon, at Hamadhān, on 13 Sha'bān 544, that is two days after Text 11, which was copied at Asadābād.

So we see that three different scribes took part in the copying of this collection: copyist A, whose original was in the hands of the anonymous B – who transcribed the majority of the treatises – and, finally, al-Ḥusayn Muḥammad ibn 'Alī. These texts were copied, in 1149–1150, in the immediate environs of Hamadhān and Asadābād – this is certain for B and al-Ḥusayn, very probable for A. So everything suggests that the three copyists collaborated in the transcription of these mathematical texts (which are on the highest level), even if we should perhaps see B as the master in charge of the project. The texts on the heptagon copied by B are all taken from an original that is an autograph by al-Sijzī [4, 5, 8].

Let us now turn to the treatises whose texts are translated here.
1. *Book of the Construction of the Circle Divided into Seven Equal Parts by Archimedes (Kitāb 'amal al-dā'ira al-maqsūma bi-sab'at aqsām mutasāwiya li-Arshimīdis)*
There exists only one copy of this text, the one transcribed by Muṣṭafā Ṣidqī, ms. 41 in Dār al-Kutub, fols 105ʳ–110ʳ. The transcription was completed 'on Sunday the seventh day of Jumādā al-ūlā of the year one thousand one hundred and fifty-three', that is on 31 July 1740. Moreover, there is nothing to indicate that Muṣṭafā Ṣidqī compared the copy with his original. We have already noted that often what we have is more like a rewritten version than a copy, and moreover it is not unusual for Muṣṭafā Ṣidqī to incorporate into his version proofs that come from mathematicians of the end of the tenth century, such as al-Ḥubūbī (transcribed 'al-Juyūbī') and al-Shannī. This has, as it were, the advantage of sketching a more

precise portrait of the *faqīh* and mathematician, al-Ḥubūbī,[16] of whom we still know very little.

The text of this book has never been established before. It is, however, widely available and well known since the German translation was made of it by C. Schoy in 1927,[17] and, more recently, the Russian translation by B. Rosenfeld.[18]

Abū al-Jūd wrote three treatises, the first of which, as we have explained before, has been lost.[19] We still have the following treatises:

---

[16] Abū ʿAlī al-Ḥasan ibn Ḥārith al-Ḥubūbī is a correspondent of Abū al-Wafāʾ al-Būzjānī. As his contemporaries indicate, he is a *faqīh* (a jurist in Islamic law) and a mathematician. Thus we learn from the manuscript Bodleian, Thurston 3, that Abū al-Wafāʾ al-Būzjānī wrote a treatise titled 'The reply of Abū al-Wafāʾ Muḥammad ibn Muḥammad al-Būzjānī to what he was asked by the *faqīh* (jurist) Abū ʿAlī al-Ḥasan ibn Ḥārith al-Ḥubūbī: the determination of the area of triangles without determining the perpendiculars or their feet' (fol. 3):

جواب أبي الوفاء محمد بن محمد البوزجاني عما سأله الفقيه أبو علي الحسن بن حارث الحبوبي، وهو البرهان على مساحة المثلثات من غير استخراج العمود ومسقط الحجر.

We also meet him in the correspondence exchanged between Abū Naṣr ibn ʿIrāq and al-Bīrūnī. The latter writes:

... إلى أن ورد كتاب شيخنا أبي الوفاء محمد بن محمد البوزجاني على الفقيه أبي علي الحبوبي يذكر فيه أنه تأمل أكثر كتابي في السموت...

'Until the book of our Master Abū al-Wafāʾ Muḥammad ibn Muḥammad al-Būzjānī reached the *faqīh* (jurist) Abū ʿAlī al-Ḥubūbī: <the book> in which he mentions that anyone who ponders on my book on *Azimuths*...' (*Risāla fī maʿrifat al-qusiy al-falakiyya*, in *Rasāʾil Abī Naṣr Manṣūr ibn ʿIrāq ilā al-Bīrūnī*, Hyderabad, 1948, 8th treatise, p. 2). See F. Sezgin, *Geschichte des arabischen Schrifttums*, Leiden, 1974, V, p. 336.

See also the treatise titled *Fī tashrīḥ al-kura*, very probably by al-Ḥubūbī, in ms. Cairo, Dār al-Kutub, no. 1202.

[17] C. Schoy began by translating Propositions 17 and 18 in 'Graeco-Arabische Studien nach mathematischen Handschriftender Viceköniglichen Bibliothek zu Kairo', *Isis*, 8, 1926, pp. 21–40; and then went on to translate the whole treatise in *Die trigonometrischen Lehren des persischen Astronomen Abūʾl Raihān Muḥammad Ibn Aḥmad al-Bīrūnī*, Hanover, 1927.

[18] This translation was published in I. N. Weselowskii, *Archimed-socinenja*, Moscou, 1962.

[19] It seems that this text by Abū al-Jūd was still in circulation at the beginning of the thirteenth century. In fact, the copyist of the manuscript in the Bodleian (Thurston no. 3, fol. 129^r) wrote in the margin beside the lemma given by al-Sijzī, that deals with the division D_2 by Abū al-Jūd, that Abū al-Jūd had given it in this first essay:

(Cont. on next page)

2.1. *Treatise by Abū al-Jūd on the Construction of the Heptagon in the Circle* (*Kitāb 'amal al-musabba' fī al-dā'ira li-Abī al-Jūd Muḥammad ibn al-Layth*)

This treatise belongs to the same collection as the preceding one, fols 117ᵛ–120ʳ, and is thus still in the hand of Muṣṭafā Ṣidqī, who completed his copy of it on Wednesday 10 Jumādā al-ūlā 1153 AH, that is 3 August 1740. Like the preceding one, this text contains no glosses or additions, and there is nothing to encourage us to suppose that Muṣṭafā Ṣidqī revised it by comparing it with his original. The text of the treatise has never been established or translated before.

2.2. *Treatise by Abū al-Jūd on the Account of the Two Methods of al-Qūhī and al-Ṣāghānī* (*Risālat Abī al-Jūd Muḥammad ibn al-Layth ilā Abī Muḥammad 'Abd Allāh ibn 'Alī al-Ḥāsib fī ṭarīqay Abī Sahl al-Qūhī wa-shaykhihi Abī Ḥāmid al-Ṣāghānī fī 'amal al-musabba' al-mutasāwī al-aḍlā' fī al-dā'ira*)

This treatise has come down to us in a single copy, number 8 in the collection 4821 described before, in the hand of the copyist B and not that of al-Ḥusayn Muḥammad ibn 'Alī as has been claimed.

(*Cont.*)

وجدت هذه في رسالة أبي الجود محمد بن الليث إلى الشيخ عبد الله بن أحمد بن الحسين، وذكر فيها : 'وما أعلم أن أحداً سبقني إلى هذا العمل على ما اعترف به المهندسون ونطقت كتبهم إلى هذه الغاية، وهي أواخر سنة ثمان وخمسين وثلاثمائة هجرية' ، وذكر فيها أبي جعفر الخازن .

'I found this (that is the division D₂) in a letter by Abū al-Jūd Muḥammad ibn al-Layth <addressed> to the Master 'Abd Allāh ibn Aḥmad ibn al-Ḥusayn, and he remarked: "I know no one who has preceded me in this construction, according to what geometers have recognized and what their books have stated up to this point in time, that is the end of the year three hundred and fifty-eight of the Hegira", and he mentioned Abū Ja'far al-Khāzin'.

This testimony, although very brief, is nevertheless very important. It is the only one that has come down to us from someone who has read this essay by Abū al-Jūd, without being committed to one side or the other in the controversy. It confirms certain statements made by both Abū al-Jūd and his detractors concerning the division D₂, and discreetly raises the veil over Abū al-Jūd's interests, by reminding that he mentions the name of al-Khāzin – who is one of the first mathematicians who tried to solve a cubic equation with the help of the intersection of the curves of two conics. Now this is precisely one of the chief preoccupations of Abū al-Jūd, according to the testimony of his successor al-Khayyām.

2.3. An abridged version of the preceding treatise is in Oxford, Bodleian Library, Thurston 3, fols 133^r–134^r. Part of the date of transcription of this version has been effaced and all we have is 'Friday 2 Sha'bān of the year six hundred ...' (fol. 134^r). On fol. 136^r, the date written by the copyist is effaced in the same way. On the other hand, on fols 69^r and 92^v respectively we read the dates: 'the end of the month of Rajab in the year six hundred and seventy-five' and 'seven Rajab in the year six hundred and seventy-five'. So we can assuredly fill in the effaced part to read '2 Sha'bān six hundred and seventy-five', that is, 8 January 1277.

A relatively recent copy of Thurston 3 is in Oxford, Bodleian Library, Marsh 720, fols 261^r–264^r.

We gave the *editio princeps* of the two texts 2.2 and 2.3, but the translation only of the first, 2.2, in *Mathématiques infinitésimales*, vol. III.

3.1. *Book by al-Sijzī on the Construction of the Heptagon* (*Kitāb Aḥmad ibn Muḥammad ibn 'Abd al-Jalīl al-Sijzī fī 'amal al-musabba' fī al-dā'ira wa-qismat al-zāwiya al-mustaqīma al-khaṭṭayn bi-thalātha aqsām mutasāwiya*)

This book exists in three manuscripts. We find it, written in the hand of Muṣṭafā Ṣidqī, in the collection 41 of Dār al-Kutub, Cairo, referred to above, fols 113^v–115^v. This copy, designated by Q, was transcribed on Tuesday 9 Jumādā al-ūlā 1153 AH, that is 2 August 1740.

The second manuscript is number 4821 in the Bibliothèque Nationale, Paris, fols 10^v–16^v, in the hand of al-Ḥusayn Muḥammad ibn 'Alī. It was very probably copied at Hamadhān or at Asadābād about 1149–1150. We have designated it by B.

The third manuscript belongs to the collection Reshit 1191 in the Süleymaniye Library of Istanbul, fols 80^v–83^r, designated by T. We have a collection of works by al-Sijzī, copied in *nasta'līq* by the same hand.

Now, examining omissions as well as other accidents of copying shows that Q and T are related. Their common ancestor must very probably have been in Istanbul, and it is a copy of that manuscript that Muṣṭafā Ṣidqī transcribed. In 1926, C. Schoy translated the text of Q into German, but without giving a critical edition of it.[20] An edition and a translation of the text have recently been published.[21]

---

[20] C. Schoy, 'Graeco-Arabische Studien nach mathematischen Handschriften...'.

[21] See J. P. Hogendijk, 'Greek and Arabic Constructions of Regular Heptagon', *Archive for History of Exact Sciences*, 30, 1984, pp. 197–330, on pp. 292–316. The edition (see apparatus criticus in *Mathématiques infinitésimales*, vol. III, pp. 739–57) and the English translation remain completely unsatisfactory, despite the notable effort.

3.2. As was the case for the *Treatise* by Abū al-Jūd, the collection Thurston 3 in the Bodleian Library, fol. 129, includes an abridged version of the preceding book. The collection Marsh 720, fols 267^v–268^r, also includes a late copy of this latter.

This abridged version deliberately omits all the beginning of the book, a little more than three pages in our edition, as well as all the historical and polemical references that appear in the body of the text.

It is possible that this abridged version was made from a copy in the tradition of manuscript B. If we look at the example of the omission (*quṭruhu al-mujānib BD*) in other manuscripts, the omission is repaired in B and is found in the abridged version.

We give the *editio princeps* of this version, but without translating it; the translation of the complete version is enough.

4.1. *Solution by al-Qūhī for the Construction of the Regular Heptagon in a Given Circle (Istikhrāj Wayjan ibn Rustum al-ma'rūf bi-Abī Sahl al-Qūhī fī 'amal al-musabba' al-mutasāwī al-aḍlā' fī dā'ira ma'lūma)*

This treatise has come down to us in five known manuscripts:

Fols 222^v–225^r, in Dar al-Kutub 40, also in the hand of Muṣṭafā Ṣidqī, who completed the transcription on Monday 29 Dhū al-qaʿda 1159, that is, 13 December 1746; designated here by Q. We have shown that this manuscript has the same ancestor as the important manuscript 4832 in Aya Sofya, Istanbul.[22]

Fols 17^v–23^v, of the collection 4821 in the Bibliothèque Nationale, Paris, in the hand of the anonymous copyist we have called B (and not that of al-Ḥusayn Muḥammad ibn 'Alī), who transcribed it from the autograph by al-Sijzī, designated here by B.

Fols 145^v–147^v, of the collection Aya Sofia 4832 in the Süleymaniye Library, in Istanbul. We have described this collection, and shown that it was transcribed at the latest in the sixth century of the Hegira (twelfth century).[23] This manuscript was transcribed from Abū 'Alī al-Ṣūfī; designated by A.

Fols 65^v–67^r, of the collection 1751 in the University of Tehran; designated by D.

Fols 215^v–219^v, of the collection 5648 in the Ẓāhiriyya Library of Damascus. We have here a recent copy of the manuscript 40 of Dār al-Kutub, as we have shown more than once.[24] So we have not taken this Damascus manuscript into account in establishing the text.

---

[22] *Founding Figures and Commentators*, pp. 126 and 464.
[23] *Ibid.*, pp. 124–5.
[24] *Ibid.*, p. 126

We note immediately an important difference that separates these manuscripts into two families, B on one side, and A, D and Q on the other: the introductory section in B is notably different from that in A, D and Q. Moreover, a synthesis is presented differently in each of the traditions. This difference compelled us to establish texts of the two versions in parallel.

A more careful examination of the accidents of copying allows us to refine our analysis, and eventually to establish the following stemma:

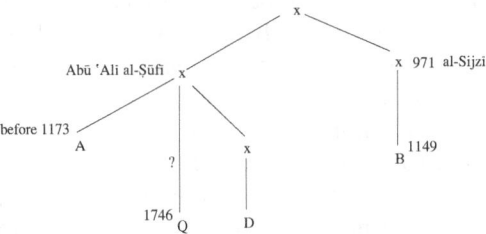

The text of this treatise has not been established before. There is a German translation[25] from the copy by Muṣṭafā Ṣidqī.

4.2. The collection Thurston 3, fol. 130 includes an abridged version of this text; again we have established the text here without translating it.

4.3. *Treatise on the Construction of the Side of the Regular Heptagon Inscribed in the Circle, by al-Qūhī (Risāla fī 'amal ḍil' al-musabba' al-mutasāwī al-aḍlā' fī al-dā'ira li-Abī Sahl al-Qūhī)*

There exist two manuscripts of this treatise. The first belongs to the collection 4821 in the Bibliothèque Nationale, Paris, fols 1ᵛ–8ʳ, designated by B. As we have already said, this copy is in the hand A, and no doubt dates from the years 1149–1150.[26]

The second manuscript belongs to the collection in the India Office, London, no. 461 (Loth 767), fols 182ᵛ–189ʳ, designated by I. We have set out all we know about this collection, which was copied in India around 1784.[27] We have shown that a substantial part of this collection (the treatise by Sharaf al-Dīn al-Ṭūsī on *The Equations*) was taken from a single original, a manuscript that today is in India (Khuda Bakhsh, no. 2928), which itself was copied in 696 AH, that is in 1297.

[25] Y. Samplonius, 'Die Konstruktion des regelmässigen Sibeneckes nach Abū Sahl al-Qūhī Waiǧan ibn Rustam', *Janus*, 50, 1963, pp. 227–49.

[26] For the history of the manuscript tradition, see above p. 571 and *Mathématiques infinitésimales*, vol. III, p. 657.

[27] R. Rashed, *Sharaf al-Dīn al-Ṭūsī, Œuvres mathématiques,* Paris, 1986, vol. I, pp. XLII–XLVII.

So it is from these two different manuscript traditions that this text will be established (for the first time).

5. *Treatise by al-Ṣāghānī for 'Aḍud al-Dawla <on the Regular Heptagon>* (*Risālat Aḥmad ibn Muḥammad ibn al-Ḥusayn al-Ṣāghānī ilā al-Malik al-jalīl 'Aḍud al-Dawla ibn Abī 'Alī Rukn al-Dawla*)

This treatise has come down to us in a single version, which is part of the collection 4821 in the Bibliothèque Nationale, Paris, fols 23ᵛ–29ʳ, in the hand of the anonymous copyist B (and not that of al-Ḥusayn Muḥammad ibn 'Alī), which was copied from the autograph by al-Sijzī. The copy by al-Sijzī is itself dated 12 Shawwal 360, that is, 7 August 971, and the transcription of the anonymous copyist is dated 15 Rajab 544, at Hamadhān, that is 18 November 1149. It has crossings-out (fol. 23ᵛ), made in the course of the transcription, removing words or phrases that were repeated, but there are no glosses or additions. This text has never been established or translated before.

6. *Book on the Discovery of the Deceit of Abū al-Jūd, by al-Shannī* (*Kitāb tamwīh Abī al-Jūd fī mā qaddamahu min al-muqaddimatayn li-'amal al-musabba' bi-za'mihi li-Abī 'Abd Allāh Muḥammad ibn Aḥmad al-Shannī*)

This book has come down to us in complete form in only one manuscript, Dār al-Kutub 41, Cairo, in the hand of Muṣṭafā Ṣidqī, fols 129ᵛ–134ᵛ, whose copy was completed on Sunday 21 Jumādā al-ūlā 1153 AH, that is 14 August 1740; it is designated by Q. In addition to this there are some fragments in two other manuscripts.

The first belongs to the Université St Joseph in Beirut, no. 223, fols 16–19, designated by L. This fragment lacks three words and two sentences that are present in Q.

The second is part of the collection T-S Ar. 41.64 in the University Library, Cambridge. This fragment – designated by C – consists of two pages. The fragment lacks a word and a sentence that are present in Q.

This text, which is difficult to read in places, has not been established or translated before.

7. *On the Determination of the Chord of the Heptagon, by Naṣr ibn 'Abd Allāh* (*Risāla Naṣr ibn 'Abd Allāh fī istikhrāj watar al-musabba'*)

This text belongs to the collection Thurston 3, fol. 131, recopied in Marsh, fols 266ʳ–267ʳ, in the Bodleian Library, Oxford. There cannot be any doubt that here, as for the preceding texts, we have an abridged version of an original that today still cannot be found, if it is not definitively lost.

As we have repeated several times, the collection Thurston 3 was copied in 675 AH, that is in 1277.

This text has never been established or translated before.

8. *Synthesis for the Analysis of the Lemma on the Regular Heptagon Inscribed in the Circle, Anonymous* (*Tarkīb li-taḥlīl muqaddimat al-musabbaʻ al-mutasāwī al-aḍlāʻ fī al-dāʼira*)

This text is part of the collection Dār al-Kutub, Cairo, 41, fols 100ᵛ–101ᵛ. It is in the hand of Muṣṭafā Ṣidqī, who completed the copy on Monday Jumādā al-ūlā 1153 AH, that is, 15 August 1740.

Like the others, this text has not been established or translated before.

9. *Treatise on the Proof for the Lemma Neglected by Archimedes, by Kamāl al-Dīn ibn Yūnus* (*Risālat al-mawlā Kamāl al-Dīn ibn Yūnus ilā khādimihi Muḥammad ibn al-Ḥusayn fī al-burhān ʻalā ījād al-muqaddima allatī ahmalahā Arshimīdis fī kitābihi fī tasbīʻ al-dāʼira wa-kayfiyyat dhālika*)

This *treatise* has come down to us in two versions, which have repeatedly been confused: one is complete, the other is abridged. Let us begin with the first one.

This version exists in two collections: one in Kuwait, the other in Istanbul. The first is in Dār al-āthār al-islāmiyya, LNS 67, fols 138ᵛ–140ʳ, copied by the mathematician ʻAbd al-ʻAzīz al-Khilāṭī.[28] He did not give his name in the colophon, but he had done so in the colophon of the treatise that precedes the one by Kamāl al-Dīn ibn Yūnus. The end of this treatise in fact immediately precedes the beginning of the following one, on the same page. As the copy of the first was completed on 11 Dhū al-Qaʻda 630 AH, the one by Kamāl al-Dīn ibn Yūnus was copied shortly after that date, that is after 19 August 1233. The writing is in *naskhī*, the figures have been drawn in red ink. We designate this manuscript by the letter K.

The second manuscript of the complete version appears in the collection Aḥmet III, no. 3342, on three unnumbered folios, in the Topkapi Saray Museum in Istanbul. The script is *naskhī*, and we shall designate this collection by I.

We should note that K and I are not independent. Comparing with K, I is lacking the name of the addressee of the treatise, the introductory passage (three lines), four words, which are all to be found in the abridged version – which, on the whole, permits us to conclude that they were not added by K. On the other hand, K and I have seven errors in common, while I has six errors of its own and K two errors. Is manuscript I a

---

[28] R. Rashed, *Sharaf al-Dīn al-Ṭūsī, Œuvres mathématiques,* vol. I, pp. XXXVI–XXXVII.

descendant of K, or do they have the same immediate ancestor? It is difficult to pass judgement on the basis of three pages, but we can say that they are definitely related.

The abridged version has also come down to us in two manuscripts, if we do not count Marsh 720.

The first is part of the collection Thurston 3, fols 128^v–129^r, dated 675 AH (see fols 69^v, 92^v), designated by O. To that we may add the seventeenth-century copy of that collection: Marsh 720, fols 257^r–258^v, a copy we shall not consider in establishing the text.

The second manuscript is part of the collection Genel 1706/8, fols 184^v–186^r, in Manisa, in Turkey,[29] designated by C.

In this abridged version, the introductory paragraph (3 lines) and the name of the addressee are absent. A group of terms has been eliminated, such as *khaṭṭ, saṭh, nisba* ... . But, although it is abridged, this version includes three additional sentences, one of which appears to be due to the scribe omitting some words by accidentally skipping to the same word later in the text. Our edition gives these sentences in italics. Finally, the last paragraph is different from what it is in the complete version, and, further, it includes many incoherencies. Let us conclude by noting that C is not a copy of O and that the latter is not a copy of the first either: a sentence is lacking in O but is found in C and in the complete version; but on the other hand, four words of O are absent from C.

In the circumstances, we have given the *editio princeps* of both versions, as well as a French translation of the text established for the complete version in *Les Mathématiques infinitésimales*, vol. III. Here, we are giving the English translation of the first version.

---

[29] This collection was copied in Tabrīz in 699 AH (fol. 210^r), see also above, pp. 33ff.

1.2. TRANSLATED TEXTS

In the name of God, the Compassionate, the Merciful

Book of the Construction of the Circle Divided into Seven Equal Parts by Archimedes

TRANSLATED BY
ABŪ AL-ḤASAN THĀBIT IBN QURRA OF ḤARRĀN

A Single Book in Eighteen Propositions

I say, after the praise of God and <invoking> his blessing upon his Prophet and chosen one and on his family, his companions and his friends: when I wished to transcribe this book, I had obtained only a copy <that was> damaged and suffered from the ill effects of the ignorance of the copyist and his lack of understanding. I did all I could by way of making sure I checked the problems, carried out the synthesis for the analyses and presented the propositions in easy and accessible terms; and I have introduced some proofs by modern scholars – with the support and help of God.

Propositions

– 1 – We draw *AB* and we mark on it two points *C* and *D* such that the square of *CD* is equal to the sum[1] of the squares of *AC* and *DB*.

I say that the square of *AB* is equal to twice the product of *AD* and *CB*.

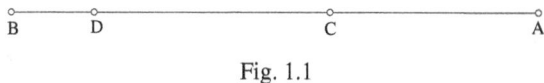

Fig. 1.1

---

[1] In such expressions we have added the term 'sum' in order to conform with English usage.

And that is because the sum of the squares of *AC* and *DB* is equal to the square of *CD*; so if, on one side and on the other, we add the square of *CD* plus twice the product of *AC* and *CD*, we shall have the sum of the three squares of *AC, CD, DB* and twice the product of *AC* and *CD* equal to twice the square of *CD* plus twice the product of *AC* and *CD*. But twice the square of *CD* plus twice the product of *AC* and *CD* is equal to twice the product of *AD* and *DC*; so the sum of the three squares of *AC, CD* and *DB* and twice the product of *AC* and *CD* is equal to twice the product of *AD* and *DC*. Since the square of *AD* is equal to the sum of the squares of *AC*, *CD* and twice the product of *AC* and *CD*, we have the sum of the squares of *AD* and *DB* equal to twice the product of *AD* and *DC*. We add on one side and on the other twice the product of *DB* and *AD*, we have the sum of the squares of *AD* and *DB* and twice the product of *AD* and *DB*, that is to say the square of *AB* equal to twice the product of *AD* and *DC* and <that> of *AD* and *DB*, that is to say twice the product of *AD* and *CB*. That is what we wanted.

– **2** – In another way: since the square of *AB* is equal to the sum of the three squares of *AC, CD* and *DB* plus the three <terms that are> twice the products of *AC* and *CD*, of *AC* and *DB* and of *CD* and *DB,* and since the square of *CD* is equal to the sum of the squares of *AC* and *DB*, the square of *AB* is equal to twice the square of *CD* plus the three <terms that are> twice the products of *AC* and *CD*, of *AC* and *DB* and of *CD* and *DB*. But Since twice the product of *AB* and *CD* is equal to twice the square of *CD* plus twice the products of *AC* and *CD* and of *DB* and *CD*, we have the square of *AB* equal to the sum of twice the products of *AB* and *CD* and of *AC* and *DB*. But twice the product of *AB* and *CD* is equal to the sum of twice the products of *CB* and *CD* and of *AC* and *CD*, and the sum of twice the products of *AC* and *CD* and of *AC* and *DB* is equal to twice the product of *AC* and *CB*. The square of *AB* is thus equal to the sum of twice the products of *CB* and *CD* and of *AC* and *CB*, that is to say to twice the product of *AD* and *CB*. That is what we wanted.

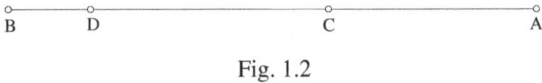

Fig. 1.2

– **3** – For every right-angled triangle, twice the product of the sum of one of the sides that enclose the right angle and the hypotenuse, considered as a single straight line, and the sum of the other side and the hypotenuse, considered as a single straight line, is equal to the square of the whole perimeter considered as a single straight line.

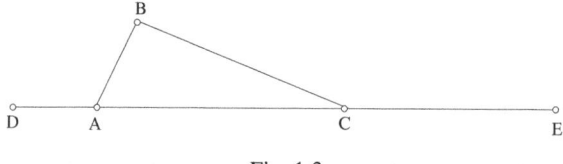

Fig. 1.3

Let the triangle be ABC with its right angle at B; let us extend AC on one side and on the other and let us put AD equal to AB and CE equal to CB. It is clear that DC is equal to AB plus AC, that AE is equal to AC plus CB, that DE is equal to the perimeter of the triangle, and that the square of AC is equal to the sum of the squares of AB and BC, that is to say of AD and CE; twice the product of DC and AE is thus equal to the square of ED. That is what we wanted.

I say, in another way that comes from Abū 'Alī al-Ḥubūbī: since the product of DC and AE is equal to the square of AC plus the three products of AC and CE, of AC and AD and of AD and CE – but the square of AC is equal to the sum of the squares of DA and CE – twice the product of DC and AE is equal to the sum of the three squares of DA, AC and CE and the three <terms that are> twice the products of AC and CE, of AC and AD and of AD and CE; but this sum is equal to the square of DE, consequently twice the product of DC and AE is equal to the square of DE. That is what we wanted.

– **4** – For every right-angled triangle, if from its right angle we drop a perpendicular to the hypotenuse, then the square of the perimeter, considered as a single straight line, is equal to twice the product of the hypotenuse and the sum of the perimeter and the perpendicular <that was> dropped, considered as a single straight line.

Let the triangle be ABC, the right angle B and the perpendicular we draw BD. Let us extend AC on one side and on the other and let us cut off EA equal to AB, CG equal to BC and GH equal to BD.

I say that the square of EG is equal to twice the product of AC and EH.

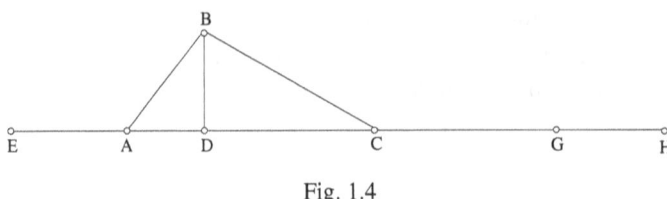

Fig. 1.4

And this is because the ratio of *GH*, that is to say *BD*, to *CG*, that is to say *BC*, is equal to the ratio of *EA*, that is to say *AB*, to *AC*. So by composition, the ratio of *HC* to *CG* is equal to the ratio of *EC* to *AC*. By permutation, the ratio of *CH* to *EC* is equal to the ratio of *CG* to *AC*. By composition, the ratio of *EH* to *EC* is equal to the ratio of *AG* to *AC*, so the product of *EH* and *AC* is equal to the product of *EC* and *AG*, and twice the product of *EH* and *AC* is equal to twice the product of *EC* and *AG*; but twice the product of *EC* and *AG* is equal to the square of *EG*, consequently the square of *EG* is equal to twice the product of *AC* and *EH*. That is what we wanted.

– **5** – In another way: let us return to the triangle and its perpendicular, let us extend *BC* on one side and on the other and let us cut off *CG* equal to *AC*, *BE* equal to *AB* and *EH* equal to *BD*.
    I say that the square of *EG* is equal to twice the product of *GC* and *GH*.

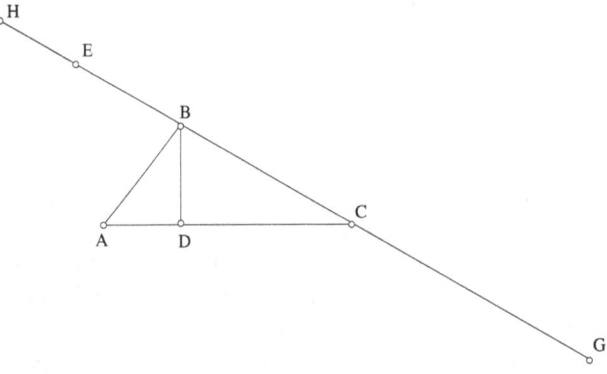

Fig. 1.5

And this is because the product of *BE*, that is to say *AB*, and *BC*, is equal to the product of *EH*, that is to say *BD*, and *GC*, that is to say *AC*; we have twice the product of *EB* and *BC* equal to twice the product of *EH* and *CG*. The sum of the squares of *EB* and *BC* is equal to the square of *CG*. But the square of *EC* is equal to twice the product of *EB* and *BC* plus the sum of the squares of *EB* and *BC*, so the square of *EC* is equal to the sum of twice the product of *EH* and *CG* and the square of *CG*. We add to the two sides the square of *CG*, we have the sum of the squares of *EC* and *CG* equal to twice the product of *EH* and *CG* plus twice the square of *CG*. But since the square of *EG* is equal to twice the product of *GC* and *CE* plus the squares of *GC* and *CE*, that is to say equal to twice the square of *GC* plus twice the <sum of the> products of *GC* and *EH* and of *GC* and *CE*, which is equal to twice the product of *GC* and *CE* plus twice the square of *CG*, that is to say twice the product of *CG* and *GH*, then the square of *EG* is equal to twice the product of *GC* and *GH*. That is what we wanted.

I say, in another way that comes from Abū ʿAlī al-Ḥubūbī: since the square of *GE* is equal to the sum of the three squares of *EB*, *BC* and *GC* plus twice the products of *GC* and *CE* and of *BC* and *BE*, and since twice the product of *GC* and *GH* is equal to twice the square of *GC* plus twice the product of *GC* and *CH* – but the square of *GC* is equal to the sum of the squares of *BE* and *BC* – and since twice the product of *GC* and *CH* is equal to the sum of twice the products of *GC* and *EH* and of *GC* and *CE* – but twice the product of *GC* and *EH* is equal to twice the product of *EB* and *BC* – so twice the product of *GC* and *GH* is equal to the sum of the three squares of *EB*, *BC* and *GC* and twice the products of *GC* and *CE* and of *BC* and *BE*, consequently the square of *GE* is equal to twice the product of *GC* and *GH*. That is what we wanted.

In another way that comes from Abū ʿAbd Allāh al-Shannī: since the square of *GE* is equal to the sum of the two squares of *GC* and *CE* plus twice the product of *GC* and *CE*, and since the square of *CE* is equal to the sum of the squares of *CB* and *BE* – that is to say the square of *GC* – plus twice the product of *CB* and *BE*, the square of *GE* is equal to twice the square of *GC* plus twice the products of *GC* and *CE* and of *CB* and *BE*. But twice the square of *GC* plus twice the product of *GC* and *CE* is equal to twice the product of *GC* and *GE*, and twice the product of *CB* and *BE* is equal to twice the product of *GC* and *EH*, so the sum of twice the products of *GC* and *GE* and of *GC* and *EH* is twice the product of *GC* and *GH* and in consequence the square of *GE* is equal to twice the product of *GC* and *GH*. That is what we wanted.

– 6 – For every right-angled triangle with unequal sides, the sum of the square of the perimeter, considered as a single straight line, and the square of the difference between the two sides that enclose the right <angle> is equal to the square of the sum of the hypotenuse and one of the sides, considered as a single straight line, plus the square of the sum of the hypotenuse and the other side, considered as a single straight line.

Let the triangle be *ABC* with its right angle at *B*. Let us cut off from *BC* the straight line *BD* equal to *AB*.

I say that the square of the perimeter, considered as a single straight line, plus the square of *DC* is equal to the square of the sum of *AB* and *AC*, considered as a single straight line, plus the square of the sum of *AC* and *BC* considered as a single straight line.

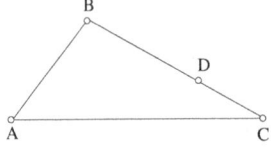

Fig. 1.6

And this is because the sum of the squares of *BC* and *BD* is equal to the square of *CD* plus twice the product of *BC* and *BD*. Now *BD* is equal to *AB*, so the square of *AC* is equal to the square of *CD* plus twice the product of *AB* and *BC*. We add to the two sides the square of *AB* plus twice the product of *AB* and *AC*, the result is the sum of the squares of *AB* and *AC* plus twice the product of *AB* and *AC*, that is to say the square of the sum of *AB* and *AC*, considered as a single straight line, is equal to the sum of the squares of *AB* and *CD* plus twice the product of *AB* and the sum of *AC* and *CB*, considered as a single straight line. Let us also add to the two sides the square of the sum of *AC* and *BC*, considered as a single straight line, so the square of the sum of *AC* and *BC*, considered as a single straight line, plus the square of the sum of *AB* and *AC*, considered as a single straight line, is equal to the sum of the squares of *AB* and *CD* plus the square of the sum of *AC* and *BC*, considered as a single straight line, plus twice the product of *AB* and the sum of *AC* and *BC*, considered as a single straight line. But the square of *AB* plus the square of the sum of *AC* and *BC*, considered as a single straight line, plus twice the product of AB and the sum of AC and BC, considered as a single straight line, is equal to the square of the perimeter of the triangle. The square of the sum of AC and BC, considered as a single straight line, plus the square of the sum of AB and AC, considered as a single straight line, is consequently equal to the square of the perimeter of the triangle, considered as a single straight line, plus the square of CD. That is what we wanted.

– **7** – In another way: let the triangle be ABC and the right angle B; let us extend AC on one side and on the other and let us make AE equal to AB, CG equal to CB and GH equal to the amount by which BC exceeds AB, then EC is equal to the sum of AB and AC, AG is equal to the sum of AC and BC, and EG as a whole is equal to the perimeter of the triangle.

I say that the sum of the squares of EG and GH is equal to the sum of the squares of EC and AG.

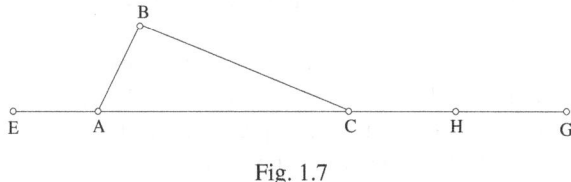

Fig. 1.7

And this is because the square of AC is equal to the sum of the squares of AB and BC, that is to say of CH and CG; but the sum of the squares of CH and CG is equal to the square of GH plus twice the product of CH and CG, so the square of AC is equal to the square of GH plus twice the product of CH and CG. But CH is equal to AE, so the square of AC is equal to the square of GH plus twice the product of AE and CG. We add to the two sides twice the product of EA and AC, then the square of AC plus twice the product of EA and AC is equal to the square of GH plus twice the product of EA and AG. We also add to the two sides the square of EA, then the square of EC is equal to the sum of the squares of EA and GH and twice the product of EA and AG. We next add the square of AG to the two sides, then the sum of the squares of EC and AG is equal to the sum of the three squares of EA, AG and GH and twice the product of EA and AG. But the square of EG is equal to the sum of the squares of EA and AG and twice the product of EA and AG, so the sum of the squares of EG and GH is equal to the sum of the squares of EC and AG. That is what we wanted.

I say, in another way that comes from Abū ʿAlī al-Ḥubūbī: since the sum of the squares of *EC* and *AG* is equal to the sum of the squares of *EA* and *GC*, twice the square of *AC* and the sum of twice the two products of *EA* and *AC* and of *CG* and *AC*, and since the sum of the squares of *EG* and *GH* is equal to the sum of the four squares of *EA*, *AC*, *CG* and *GH* and twice the three products of *EA* and *CG*, of *EA* and *AC* and of *AC* and *CG* – but twice the product of *EA* and *CG*, that is to say of *CH* and *CG*, is equal to twice the square of *CH*, plus twice the product of *GH* and *CH* – accordingly if we add to the two sides the square of *GH*, the square of *GH* plus twice the product of *EA* and *CG* is equal to the sum of the squares of *CH* and *CG*, that is to say <equal to> the sum of the squares of *EA* and *CG*. But the sum of the squares of *EA* and *CG* is equal to the square of *AC*, so the sum of the squares of *EG* and *GH* is equal to the sum of the squares of *EA* and *CG*, twice the square of *AC*, and twice the products of *EA* and *AC* and of *AC* and *CG*. Consequently, the sum of the squares of *EG* and *GH* is equal to the sum of the squares of *EC* and *AG*. That is what we wanted.

In another way that comes from Abū ʿAbd Allāh al-Shannī: since the square of *EC* is equal to the sum of the squares of *EA* and *AC* and twice the product of *EA* and *AC*, and since the square of *AG* is equal to the sum of the squares of *AC* and *CG* and twice the product of *AC* and *CG* – but the sum of the squares of *EA* and *CG* is equal to the square of *AC* – accordingly the sum of the squares of *EC* and *AG* is equal to three times the square of *AC* plus twice the products of *EA* and *AC* and of *CG* and *AC*. But since the sum of twice the square of *AC*, twice the products of *EA* and *AC* and of *CG* and *AC* and twice the product of *EA* and *CG* is equal to the square of *EG*, then if we add up and subtract what is common <to the terms>, there remains the sum of the squares of *EC* and *AG* plus twice the product of *EA* and *CG* <which is> equal to the sum of the squares of *EG* and *AC*. We add to the two sides the square of *GH* and we suppose the two straight lines *EA* and *CH* as being a single straight line divided into two equal parts to which has been added the excess *HG*, then the square of the sum of *EA* and *CG* and the square of *HG* is equal to the sum of the squares of *EA* and *CG*. But twice the product of *EA* and *CG* plus the square of *GH* is equal to the sum of the squares of *EA* and *CG*, that is to say <equal to> the square of *AC*. So if we subtract that, there remains the sum of the squares of *EG* and *GH* <which is> equal to the sum of the squares of *EC* and *AG*. That is what we wanted.

– **8** – We draw *AB* and on it we mark two points *C* and *D* such that the product of *CD* and *AB* is equal to the product of *AC* and *DB*.

I say that twice the product of *AB* and *CD* is equal to the product of *AD* and *CB*.

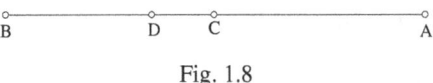

Fig. 1.8

And this is because the product of *AC* and *DB* is equal to the product of *CD* and *AB*. But the product of *CD* and *AB* is equal to the sum of the products of *AC* and *CD* and of *CB* and *CD*, so the product of *AC* and *DB* is equal to the sum of the products of *AC* and *CD* and of *CB* and *CD*; so twice the product of *AC* and *DB* is equal to the sum of the three products of *AC* and *CD*, of *BC* and *CD* and of *AC* and *DB*, so twice the product of *AB* and *CD* is equal to the sum of the products we mentioned. But the sum of the products of *AC* and *CD* and of *AC* and *DB* is equal to the product of *AC* and *CB*, so twice the product of *AB* and *CD* is equal to the sum of the products of *AC* and *CB* and of *BC* and *CD*; but the sum of the products of *AC* and *CB* and of *BC* and *CD* is equal to the product of *AD* and *CB*, so twice the product of *AB* and *CD* is equal to the product of *AD* and *CB*. That is what we wanted.

– **9** – Let there be a right-angled triangle *ABC* – its right angle is *B* – circumscribed about the circle *DEG*; we join *DE*, we extend it and we extend *CB* until they meet one another in a point *H*.

I say that *BH* is equal to *AD*.

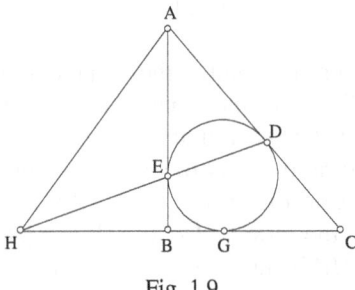

Fig. 1.9

Proof: We join *AH*. Since *DA* is equal to *AE*, since *DE* has been extended and since *AH* meets it, the product of *HD* and *HE* plus the square of *AE* is equal to the square of *AH*.[2] But the square of *AH* is equal to the sum of the squares of *AB* and *BH*, so the sum of the product of *HD* and *HE* and the square of *AE* is equal to the sum of the squares of *AB* and *BH*. But the product of *HD* and *HE* is equal to the square of *HG*, so the sum of the squares of *AB* and *BH* is equal to the sum of the squares of *AE* and *HG*. We subtract the square of *BH* from the two sides, there remains the square of *AB* <which is> equal to the sum of twice the product of *HB* and *BG* and the squares of *BG* and *AE*. We subtract the square of *AE* from the two sides, there remains the sum of twice the product of *AE* and *EB* and the square of *EB* <which is> equal to the sum of twice the product of *HB* and *BG* and the square of *BG*. But the square of *EB* is equal to the square of *BG*, so there remains twice the product of *AE* and *EB* <which is> equal to the product of *HB* and *BG*; but *EB* is equal to *BG*, so *AE*, that is to say *AD*, is equal to *BH*. That is what we wanted.

     – **10** – Let us return to the figure in its <present> form and let us say: the ratio of *CH* to *HB* is equal to the ratio of *DC* to *EB*.

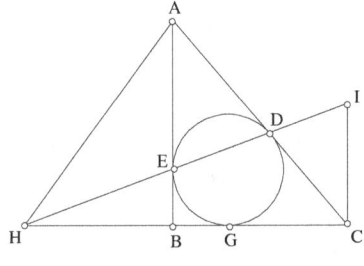

Fig. 1.10

     Proof: We draw from the point *C* the perpendicular *CI*, we extend *HD* and we extend both <lines> to meet one another in *I*. Because *CI* and *AE* are parallel, the triangle *ICD* is similar to the triangle *ADE* and the ratio of *AE* to *CI* is equal to the ratio of *AD* to *CD*. By permutation, the ratio of *AE* to *AD* is equal to the ratio of *CI* to *CD*. But *AE* is equal to *AD*, so *CI* is equal to *CD*. But since the ratio of *CH* to *HB* is equal to the ratio of *CI* to *EB*, the ratio of *CH* to *HB* is thus equal to the ratio of *CI*, that is to say *CD*, to *EB*. That is what we wanted.

---

[2] See Mathematical commentary.

– **11** – Let us return to the figure in its <present> form and let us say: the product of *AD* and *DC* is equal to the area of the triangle.

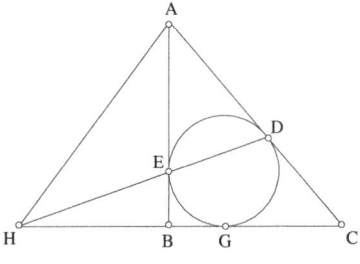

Fig. 1.11

Proof: Since the ratio of *HC* to *HB* is equal to the ratio of *CG* to *GB*, accordingly the product of *HC* and *BG* is equal to the product of *HB* and *CG*. Since the product of *HG* and *BC* is equal to twice the product of *AD* and *DC*[3] – but *HB* is equal to *AE* and *BG* is equal to *BE* and the whole of the straight line *AB* is equal to the straight line *HG* – the product of *AB* and *BC* is equal to twice the product of *AD* and *DC*; but the product of *AB* and *BC* is equal to twice the area of the triangle, so the product of *AD* and *DC* is equal to the area of the triangle. That is what we wanted.

– **12** – In another way: let us suppose <we have> a right-angled triangle *ABC* with its right angle at *B*, circumscribed about the circle *DEG*.

---

[3] Comment: It seems possible that some intermediate steps have been omitted in the Arabic text.

We have $\frac{HC}{HB} = \frac{GC}{GB}$ (because *GC* = *CD* and *GB* = *EB*), so

$$HC \cdot GB = HB \cdot GC = HB \cdot CD = AD \cdot CD,$$

because from Proposition 9, *HB* = *AD*.

On the other hand

$$HG \cdot BC = (HB + BG)(CG + GB)$$
$$= HB \cdot CG + HB \cdot BG + BG \cdot CG + GB^2$$
$$= HB \cdot CG + BG(HB + BG + GC)$$
$$= HB \cdot CG + HC \cdot BG = 2AD \cdot CD.$$

But from Proposition 9, *HG* = *AB*, so *AB* · *BC* = 2*AD* · *CD*, hence *AD* · *CD* = area (*ABC*).

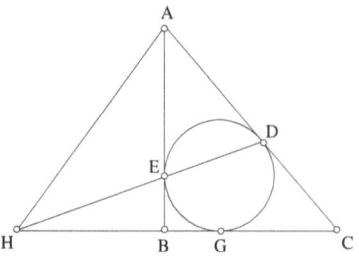

Fig. 1.12

I say that the product of *AD* and *DC* is equal to the area of the triangle.

And this is because the straight line *AD* is equal to *AE* and since *CD* is equal to *CG*, thus the square of *AC* is equal to the sum of the squares of *AE* and *CG* and twice the product of *AD* and *DC*. But the square of *AC* is equal to the sum of the squares of *AB* and *BC*, so the sum of the squares of *AB* and *BC* is equal to the sum of the squares of *AE* and *CG* and twice the product of *AD* and *DC*. If we subtract the sum of the squares of *AE* and *GC* from the two sides, there remains the sum of the squares of *EB* and *BG* and twice the products of *EB* and *AE* and of *CG* and *GB* <which is> equal to twice the product of *AD* and *DC*. But the square of *EB* is equal to the square of *BG* and the sum of the square of *BG* and the product of *CG* and *GB* is equal to the product of *CB* and *BG*, so the sum of twice the products of *EB* and *AE* and of *CB* and *BG* is equal to twice the product of *AD* and *DC*. But *EB* is equal to *BG*, so the sum of twice the products of *GB* and *AE* and of *EB* and *CB* is equal to twice the product of *AD* and *DC*. But since twice the product of *AE* and *BC* is equal to the sum of twice the products of *GB* and *AE* and of *CG* and *AE*, that is to say <equal to> twice the product of *AD* and *DC*, accordingly if we add each of the two sides to its homologue in the other two and if we subtract the common <element> twice the product of *GB* and *AE*, there remains the sum of twice the products of *AE* and *CB* and of *EB* and *CB* <which is> equal to four times the product of *AD* and *DC*. So twice the product of *AD* and *DC* is equal to the sum of the products of *AE* and *BC* and of *EB* and *BC*, that its to say of *AB* and *BC*; but the product of *AB* and *BC* is twice the area of the triangle, consequently the product of *AD* and *DC* is equal to the area of the triangle. That is what we wanted.

– **13** – In another way: let us put each of the straight lines *DH* and *EI* equal to *CD*,[4] so *BI* is equal to *CB*. Since the square of *AC* is equal to the sum of the square of *AD* and *DH* and twice the product of *AD* and *DH*, and since it is also equal to the sum of the squares of *AB* and *BI*, accordingly the sum of the squares of *AD* and *DH* and twice the product of *AD* and *DH* is equal to the sum of the squares of *AB* and *BI*. But the sum of the squares of *AD* and *DH* and twice the product of *AD* and *DH* is equal to the sum of the square of *AH* and four times the product of *AD* and *DH*, so the sum of the squares of *AB* and *BI* is equal to the sum of the square of *AH* and four times the product of *AD* and *DH*. But the sum of the square of *AH* and four times the product of *AD* and *DH* is equal to the sum of the square of *AI* and twice the product of *AB* and *BI*. So if we subtract the squares of *AH* and *AI* which are equal to one another, there remains four times the product of *AD* and *DH* <which is> equal to twice the product of *AB* and *BI*. If we take half of this, we thus find <the area of the triangle>. That is what we wanted.

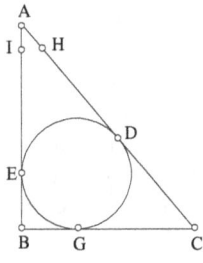

Fig. 1.13

– **14** – Let *ABC* be a right-angled triangle and *B* the right angle, and let *AD* be equal to *AB* and *EC* equal to *BC*.

I say that the product of *ED* and the perimeter of the triangle is equal to four times the area of the triangle.[5]

---

[4] This concerns a different figure, the points *H* and *I* are different from those that were used earlier. Here we need to take *H* on *AD* and *I* on *AE*, with *DH* = *EI* = *CD* = *CG*.

[5] The result follows immediately:

$$ED \cdot \text{perimeter} = (AB + BC - AC)(AB + BC + CA) = (AB + BC)^2 - CA^2;$$

but $AB^2 + BC^2 = CA^2$, hence

$$ED \cdot \text{perimeter} = 2\,AB \cdot BC = 4 \cdot \text{area }(ABC).$$

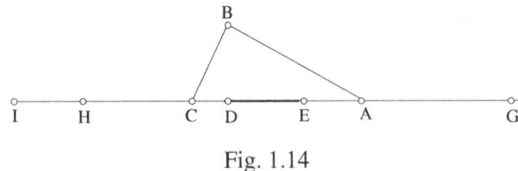

Fig. 1.14

Let us extend *AC* on one side and on the other and let us put *GA* equal to *AB*, *CH* equal to *CB* and *HI* equal to *ED*; the sum *GH* is equal to the perimeter of the triangle. Since the sum of the straight lines *CA* and *HI* is equal to the sum of *AB* and *BC*, the sum of twice the product of *AC* and *GH* and the product of *HI* and *GH* is equal to the square of *GH*. But the square of *GH* is equal to the sum of the three squares of *GA*, *AC* and *CH* and twice the three products of *GA* and *AC*, of *GA* and *CH* and of *AC* and *CH*. Then the sum of twice the product of *AC* and *GH* and the product of *HI* and *GH* is equal to the sum of the three squares of *GA*, *AC* and *CH* and twice the three products of *GA* and *AC*, of *GA* and *CH* and of *AC* and *CH*. But the product of *AC* and *GH* is equal to the sum of the square of *AC* and the products of *GA* and *AC* and of *CH* and *AC*; but the square of *AC* is equal to the sum of the squares of *GA* and *CH*, so twice the product of *AC* and *GH* is equal to the sum of the three squares of *GA*, *AC* and *CH* and twice the products of *GA* and *AC* and of *CH* and *AC*. If we subtract from the two sides twice the product of *AC* and *GH*, <which is> common <to them>, there remains the product of *HI* and *GH* <which is> equal to twice the product of *GA* and *CH*, that is to say twice the product of *AB* and *BC*. But the product of *AB* and *BC* is equal to twice the area of the triangle, consequently the product of *HI* and *GH*, that is to say *ED*, and the perimeter of the triangle, is four times the area of the triangle. That is what we wanted.

– **15** – Let there be a semicircle *ACDB* with centre *G* and containing the chord *AC*. We divide the arc *BC* into two equal parts at *D*, we join *DB* and we put *AE* equal to *AC*.

I say that the product of *GB* and *BE* is equal to the square of *DB*.

Let us join *DC*, *DA*, *DG*, *DE*; because the arcs *CD* and *DB* are equal, the two angles *CAD* and *DAB* are equal. Now *AC* is equal to *AE* and *AD* is common, so *DE* is equal to *CD*, that is to say to *DB*, and the angle *DEB* is equal to the angle *DBE*, that is to say to *BDG*, so the ratio of *EB* to *BD* is equal to the ratio of *DB* to *BG*, and the product of *GB* and *BE* is equal to the square of *DB*. That is what we wanted.

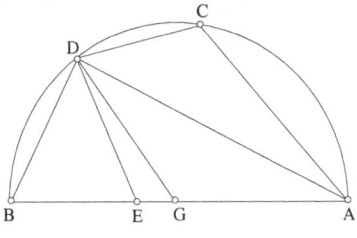

Fig. 1.15

– **16** – Let us return to the preceding figure. We say that the product of the semidiameter and *AC* plus the square of *DB* is equal to twice the square of the semidiameter.

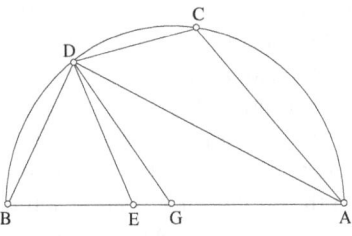

Fig. 1.16

Twice the square of *GB*, that is to say the product of *AB* and *GB*, is indeed equal to the product of *GB* and *AE*, that is to say of *GB* and *AC*, that is to say of the semidiameter and *AC*, plus the product of *GB* and *EB*. Now from what went before, the product of *GB* and *EB* is equal to the square of *DB*, so twice the square of *GB* is equal to the product of the semidiameter and *AC* plus the square of *DB*. That is what we wanted.

– **17** – Let us suppose <we have> a square *ABCD*. We extend the side *AB* in the direction of *A* as far as *E*; we join the diagonal *BC* and we put one end of the ruler at the point *D* and the other end on the straight line *EA* <which is> such that it cuts *EA* at the point *G* and makes the triangle *GAH* equal to the triangle *CID*. We draw from the point *I* the straight line *KIL* parallel to *AC*.

I say that the product of *AB* and *KB* is equal to the square of *GA*, that the product of *GK* and *AK* is equal to the square of *KB* and that each of the straight lines *BK* and *GA* is longer than the straight line *AK*.

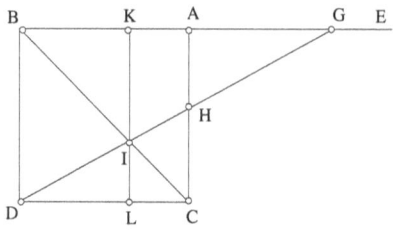

Fig. 1.17

And this is because the product of *CD* and *IL* is equal to the product of *GA* and *AH*, thus the ratio of the straight line *CD*, that is to say *AB*, to *GA* is equal to the ratio of *AH* to *IL*. Since each of the triangles *GAH* and *GKI* is similar to the triangle *ILD*, accordingly the ratio of *AH* to *IL* is equal to the ratio of *GA* to *LD*, that is to say *KB*, so the ratio of *AB* to *GA* is equal to the ratio of *GA* to *KB* and the ratio of *IL*, that is to say *AK*, to *KI*, that is to say *KB*, is itself also equal to the ratio of *LD*, that is to say *KB*, to *GK*. The product of *AB* and *KB* is thus equal to the square of *GA* and the product of *GK* and *AK* is equal to the square of *KB* and each of the straight lines *GA* and *KB* is longer than the straight line *AK*. That is what we wanted.

– **18** – We wish to construct a circle divided into seven equal parts.

Let us draw *AB* with known endpoints and on it let us mark two points *C* and *D* such that the product of *AD* and *CD* is equal to the square of *DB*, that the product of *CB* and *DB* is equal to the square of *AC*[6] and each of the straight lines *AC* and *DB* is longer than *CD*, by the preceding construction. From the straight lines *AC*, *CD* and *DB* we construct a triangle *CED* such that the side *CE* is equal to the straight line *AC* and the side *DE* is equal to the straight line *DB*. We join *AE* and *EB*, we circumscribe about the triangle *AEB* a circle *AEBHG* and we extend the straight lines *EC* and *ED* to <meet> the circumference; let them meet it at the points *G* and *H*. We

[6] This assumes we know how to construct the points *C* and *D* that give the range (*A*, *C*, *D*, *B*), of type I, construction by means of conic sections, which the author does not discuss.

join *BG* and let us draw from the point *C* the straight line *CI* to the <point of> intersection (of *EH* and *BG*).

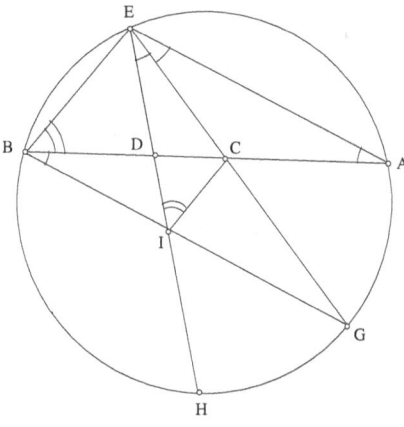

Fig. 1.18

Because the two sides of the triangle *ACE*, *AC* and *CE*, are equal, the angle *EAC* is equal to the angle *AEC* and the arc *AG* is equal to the arc *EB*. But since the product of *AD* and *CD* is equal to the square of *DB*, that is to say *DE*, accordingly the triangle *AED* is similar to the triangle *CED*, the angle *DAE* is equal to the angle *CED* and the arc *GH* is equal to the arc *EB*. The three arcs *EB*, *AG* and *GH* are equal to one another, *GB* is parallel to *AE* and the angle *CAE*, that is to say *CED*, is equal to the angle *DBI*. But given that the angle *CED* is equal to the angle *DBI*, that the angle *CDE* is equal to the angle *IDB* and the straight line *ED* is equal to the straight line *DB*, then *CD* is equal to *DI*, *CE* is equal to *IB* and the four points *B*, *E*, *C*, *I* <all> lie on the same circle. Since the product of *CB* and *DB* is equal to the square of *AC*, that is to say *EC*, and since the straight line *CB* is equal to <the straight line> *IE* and since *DB* was equal to the straight line *DE*, accordingly the product of *IE* and *ED* is equal to the square of *EC* and the triangle *IEC* is similar to the triangle *CED*, so the angle *DCE* is equal to the angle *EIC*. But the angle *DCE* is twice the angle *CAE*, so the angle *CIE* is twice the angle *CAE*. But the angle *CID* is equal to the angle *DBE*, so the angle *DBE* is twice the angle *CAE* and the arc *AE* is twice the arc *EB*. But since the angle *DEB* is equal to the angle *DBE*, accordingly the arc *HB* is also twice the arc *EB*. So we divide each of the arcs *AE* and *BH* into two equal parts <that are also> equal to the arc EB, consequently the circle $AEBHG$ is divided into seven equal parts. That is what we wanted.

Praise be to God alone, and blessing upon him who is the last of the Prophets. The correction and drafting of this honourable copy has been accomplished by the pen of the man who made the corrections, humble before God the All-Highest al-Ḥajj Muṣṭafā Ṣidqī ibn Ṣāliḥ – may God pardon him and pardon all Muslims.

Sunday, the seventh day of Jumādā al-ūlā of the year one thousand one hundred and fifty-three.

In the name of God, the Compassionate, the Merciful

TREATISE BY ABŪ AL-JŪD MUḤAMMAD IBN AL-LAYTH

On the Construction of the Heptagon in the Circle, which he sent to Abū al-Ḥasan Aḥmad ibn Muḥammad ibn Isḥāq al-Ghādī; composed using the two procedures by which he distinguished himself

He said: knowing your concern for gaining instruction, the authentic nature of your talent in matters of geometry and your inclination to learn more, I shall enable you to profit in this respect through something that has become clear to me among matters that others found difficult – unless others have found this without our knowing about it, and without our seeing any trace of it.

I had proceeded by analysis of this proposition, that is to say on the heptagon, as far as an isosceles triangle in which each of the two angles at the base is three times the third angle, so that the angles of this triangle are seven times the small angle and this small angle is a seventh of the sum of the angles of the triangle, which is equal to two right angles, so that, if the triangle is fitted into an arc of a general circle, its two sides cut off a seventh from the circumference. I then carried out analysis on this triangle as far as <obtaining> a straight line with known endpoints, divided up into two parts such that the product of the whole straight line and one of the parts is equal to the square of a straight line whose ratio to the other part is equal to the ratio of the whole straight line to the sum of the latter and the other part. Then I constructed the said triangle, as a whole, by analogy with the construction of the pentagon starting from an isosceles triangle in which each of the angles at the base is twice the third angle; the sum of the angles of the triangle is thus equal to five times the small angle, and this latter is a fifth of two right angles.

I learned that a certain geometer arbitrarily attributed this construction to Abū Sahl al-Qūhī, then that he changed part of it and claimed it for his own, as I was told, without, as an act of will, engaging so far as to discover a similar construction, and without his animal passions allowing him to think of any propositions at all. Some years, and no small number of them, after what I did, Abū Sahl al-Qūhī composed a treatise on this figure, in which he relies upon Archimedes' lemmas, <to be found> in a treatise in which the latter seeks to determine the chord of a seventh <part of a circle> and in which he assumed a proposition that he did not prove, which he did not refer to in any book and which is: we suppose for this that we have a square *ABCD*; we draw the diagonal *AC* and from the point *D* we draw a straight line that cuts the diagonal *AC* in *E* and the side *BC* in *G* and which meets the extension of *AB* in *H*, <a line> such that the triangle *CED* is equal to the triangle *BGH*.

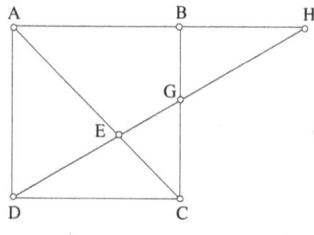

Fig. 2.1

Abū Sahl al-Qūhī avoided mentioning the square and by using two conic sections he divided a straight line into parts like those of the straight line *DEGH*, and <in> their ratio, and he found the chord of a seventh <of a circle>.

His treatise indicates that I was innovative in what I did and that I stood out from the others by the route I followed, and in which I was ahead of everyone else.

Abū Ḥāmid al-Ṣāghānī later composed a treatise on this figure, in which he addressed himself to this square and drew the straight line *DEGH* in accordance with the condition we mentioned, that very condition, and for that he made use of three hyperbolas, two opposite conic sections and a third one, in a long construction and using many figures and straight lines. I myself proceeded to carry out the analysis for this square by means of the straight line I mentioned, it (the square) is subjected to analysis, giving us something simpler than that, more obvious, more correct and more lucid and by means of this I found what was sought, in a single proposition.

<Proposition>: Let us suppose <that> for a square *ABCD* with equal sides and angles, in which is drawn the diagonal *AC*, we extend *BC* to *I* to make *CI* equal to *BC*, we construct a parabola whose vertex is the point *I*, whose axis is *IB* and whose *latus rectum* is *CD*, as has been shown in proposition fifty-six[1] of the first book of the *Conics*; let the conic section be *IKL*. We construct a hyperbola whose vertex is the point *C*, its transverse diameter twice *AC*, and whose *latus rectum* is equal to the transverse diameter, as has been shown in proposition fifty-eight[2] of the book we mentioned; let the conic section be *CL*. It necessarily cuts the parabola *IKL*; let it cut it in *L*. We draw from the point *L* <the straight line> *LM* perpendicular to *CB*; we draw from the point *M* the straight line *MD* which cuts *AC* in *N* and we extend it to meet the extension of *AB* in *S*; from the point *N* we drop the perpendicular *NE* to *AB* and we construct on the base *BE* a triangle *BGE* whose two sides *BG* and *EG* are equal and each equal to *SB*.

I say that the angle *BGE* is a seventh of two right angles and that if it is fitted into an arc of a general circle, its sides cut off one seventh from the circumference of the circle.

Proof: We extend *DC*, which meets the parabola at the point *K*, we complete the rectangle *AL* and we add to *AB*, <the line> *AH* equal to *BS*. We draw the straight line *GH* and from the point *G* the perpendicular *PG* to the midpoint of *BE*, and starting from the straight line *EH* the perpendicular *UO* to the mid point of *EH*, and we join *UE*. The two straight lines *AD* and *AB* do not meet the hyperbola *CL*, because the tangent at its vertex between the hyperbola and one of the asymptotes is equal to *AC* and because the square *AC* is a quarter of the area applied to the transverse diameter and to the *latus rectum*, from what has been shown in the first proposition of the second book of the *Conics*, and the square *AC* will be equal to the rectangle *AL*, from what has been shown in the eighth[3] proposition of the book we

[1] Proposition 52 in Heiberg's edition.

[2] Proposition 54 in Heiberg's edition.

[3] Proposition 12 in Heiberg's edition.

mentioned. We take away the common rectangle *AM*, there remains the rectangle *MD* <which is> equal to the rectangle *BL*, so the ratio of *CM* to *BM* is equal to the ratio of *LM* to *CD*. But *CK* is equal to *CD*, because the square of the perpendicular dropped from the parabola onto its axis is equal to the rectangle of the part cut off from the axis, which is *CI*, and its *latus rectum* which is *CD*, which are equal, from what has been shown in proposition fourteen[4] of the first book of the *Conics*. So the ratio of *CM* to *BM* is equal to the ratio of *LM* to *CK* and the ratio of the square of *CM* to the square of *BM* is equal to the ratio of the square of *LM* to the square of *CK*. But the ratio of the square of *LM* to the square of *CK* is equal to the ratio of *IM* to *IC*, cut off from the axis by these straight lines, as has been shown in proposition nineteen[5] of the book we mentioned. The ratio of the square of *CM* to the square of *BM* is thus equal to the ratio of *IM* to *IC*.

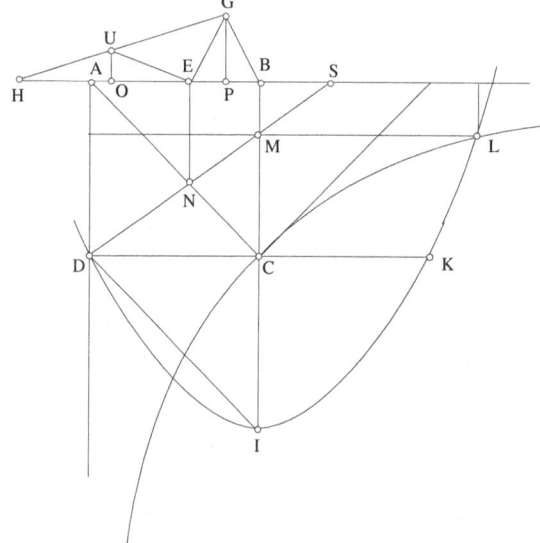

Fig. 2.2

We join *I*, *D* with a straight line. It has been shown that the ratio of *IM* to *IC* is equal to the ratio of *DM* to *DN* because the two straight lines *ID*

---

[4] Proposition 11 in Heiberg's edition.

[5] Proposition 20 in Heiberg's edition.

and *CN* are parallel. So the ratio of the square of *CM* to the square of *BM* is equal to the ratio of *DM* to *DN*. As for the ratio of the square of *CM* to the square of *BM*, it is equal to the ratio of *CM* to *BM* repeated twice,[6] that is to say the ratio of *DM* to *SM* repeated twice, so the ratio of *DM* to *SM*, the ratio that is repeated twice, is equal to the ratio of the product of *DM* and *CM* to the product of *SM* and *BM*.[7] As for the ratio of *DM* to *DN*, it is equal to the ratio of the product of *DM* and *CM* to the product of *DN* and *CM*, so the ratio of the product of *DM* and *CM* to the product of *SM* and *BM* is equal to the ratio of the product of *DM* and *CM* to the product of *DN* and *CM*. The product of *DN* and *CM* is thus equal to the product of *SM* and *BM*. So the ratio of *DN* to *SM* is equal to the ratio of *BM* to *CM* by proportionality. But the ratio of *BM* to *CM* is equal to the ratio of *SB* to *CD*, so the ratio of *DN* to *SM* is equal to the ratio of *SB* to *CD*. On the one hand, *CD* is equal to *AB* and on the other hand the ratio of *DN* to *SM* is equal to the ratio of *AE* to *SB*, so the ratio of *SB* to *AB* is equal to the ratio of *AE* to *SB*, so the product of *AB* and *AE* is equal to the square of *SB*. But *AH* is equal to *SB*, so the product of *AB* and *AE* is equal to the square of *AH*. Similarly, the ratio of *AS* to *AD* is equal to the ratio of *ES* to *EN*. On one side *AD* is equal to *AB* and on the other side *EN* is equal to *AE*, so the ratio of *AS* to *AB* is equal to the ratio of *ES* to *AE*. By permutation, the ratio of *AS* to *ES* is equal to the ratio of *AB* to *AE* and, by inversion, the ratio of *AS* to *AE* is equal to the ratio of *AB* to *BE*.[8] So the product of *AS* and *BE* is equal to the product of *AE* and *AB*. On the one hand *AS* is equal to *BH* and, on the other hand, it has been shown when we first stated that *AB* times *AE* is equal to the square of *AH*, so the product of *BH* and *BE* is equal to the square of *AH*, that is to say <equal> to the square of *BG*. So the ratio of *BH* to *BG* is equal to the ratio of *BG* to *BE*; but the angle *GBE* is common to the two triangles *BGE* and *BGH*, so the two triangles *BGE* and *BGH* are similar and the side *GE* is equal to the side *BG*; so the side *GH* is equal to the side *BH*. Similarly, the ratio of *AB* to *AH* is equal to the ratio of *AH* to *AE*; by composition, the ratio of *BH* to *AH* is equal to the ratio of *EH* to *AE* and, by permutation, the ratio of *BH* to *EH* is equal to the ratio of *AH* to *AE*; again by composition, the ratio of *BH* to half the sum of *BH* and *EH* is

---

[6] That is to say *CM* repeated twice to *BM* repeated twice.

[7] We have $\dfrac{DM}{SM} = \dfrac{CM}{BM}$, hence $\left(\dfrac{DM}{SM}\right)^2 = \dfrac{DM \cdot CM}{SM \cdot BM}$.

[8] $\dfrac{AS}{AB} = \dfrac{ES}{AE} \Rightarrow \dfrac{AS}{ES} = \dfrac{AB}{AE} \Rightarrow \dfrac{AS}{ES - AS} = \dfrac{AB}{AB - AE} \Rightarrow \dfrac{AS}{AE} = \dfrac{AB}{BE}$.

equal to the ratio of *AH* to half of *EH*. On the one hand half of the sum of *BH* and *EH* is equal to *PH* and on the other hand half of *EH* is *OH*; now *BH* is equal to *GH*, so the ratio of *GH* to *PH* is equal to the ratio of *AH* to *OH*. But the ratio of *GH* to *PH* is equal to the ratio of *UH* to *OH*; but it is equal to the ratio of *AH* to *OH*, so *UH* is equal to *AH* and *AH* is equal to *GE*, and *UE* is equal to *UH*, so the straight lines *UH*, *UE* and *EG* are equal; that is why the angle *EUG* is twice the angle *H* and the angle *GEB* is equal to <the sum of> the two angles *EGH* and *AHG*, so the angle *GEB* is three times the angle *H*, and it is equal to the angle *EBG*, so the angle *EBG* is three times the angle *H*. Similarly, the angle *BGH* is three times the angle *H*, the sum of the two angles *GBH* and *BGH* is six times the angle *H*, the sum of the angles of the triangle *BGH* is seven times the angle *H*, and the sum of the three angles of the triangle *BGH*, that is to say seven times the angle *H*, is equal to two right angles; but the angle *BGE* is equal to the angle *H*, because the two triangles *BGE* and *BGH* are similar, so the angle *BGE* is also a seventh of two right angles. If one of these two angles is fitted into an arc of a general circle, the two sides enclosing the angle cut off a seventh from the circumference. That is what we wanted to construct.

As for my former letter about the construction of the heptagon, in which I was ahead of everyone and I distinguished myself from others by the route I followed, I repeat it here for you in its entirety in a single proposition proved with the aid of God and His assistance.

For this purpose let us suppose <we have> a straight line *AB* whose two endpoints are known. We add to it <the line> *BI* equal to *AB*, we construct on it the square *BIKL* and we construct a parabola whose vertex[10] is the point *A*, its axis *AI*, its *latus rectum AB*, as has been shown in proposition fifty-six[9] of the first book of the *Conics*; let the conic section be *AM*. We construct a hyperbola whose vertex[10] is the point *B*, its transverse diameter twice the diagonal of the square *BIKL*, and whose *latus rectum* is equal to the transverse diameter, as has been shown in proposition fifty-eight[11] of the book we mentioned; it necessarily cuts the parabola *AM*; let it cut it in *M*, let the conic section be *BM*. From the point *M* we drop the

[9] Proposition 52 in Heiberg's edition.

[10] Lit.: beginning.

[11] Proposition 54 in Heiberg's edition.

perpendicular *MD* onto *AB* and let us construct on the base *AD* a triangle *ACD* whose sides *AC* and *CD* are equal, each being equal to the perpendicular *MD*.

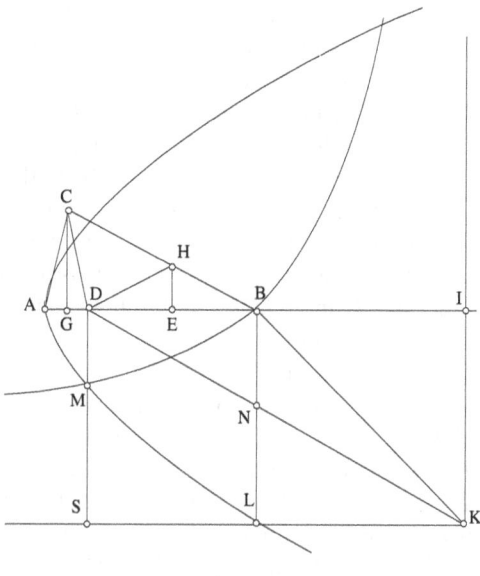

Fig. 2.3

I say that the angle *ACD* is a seventh of two right angles and that the angle *ABC* is also a seventh of two right angles.

If this is so, then it is clear that if one of the two angles is fitted into an arc of a general circle, its two sides cut off a seventh from the circumference of the circle.

Proof: We join *K* and *D* with a straight line; let it cut the side *LB* in *N*. We extend *DM* and *KL*; let them meet at the point *S*. Let us draw from the midpoints of *AD* and *BD* two perpendiculars to *AB*, *GC* and *EH*; we join *DH*. Now the straight lines *KI* and *KS* do not meet the hyperbola *MB*, from what has been shown in the first proposition of the second book of the *Conics*; and the ratio of *BI* to *SM*, <lines> which meet the conic section <and are> parallel to the asymptotes, is equal to the ratio of *KS* to *KI*, <lines> cut off on the two asymptotes of the conic section by the two straight lines we mentioned, as has been shown in the eighth[12] proposition of the book we mentioned and other propositions of the book of the *Conics*. Now *BI* is equal to *KL* and *KI* is equal to *SD*, so the ratio of *KL* to *SM* is equal to the ratio of *KS* to *SD*. But the ratio of *KS* to *SD* is equal to the ratio of *KL* to *LN*, so the ratio of *KL* to *LN* is equal to the ratio of *KL* to *SM*; *LN* is equal to *SM*, lastly *BN* is equal to *MD* and the ratio of *BN* to *BD* is equal to the ratio of *IK* to *ID*. Now *BN* is equal to *MD* and *IK* is equal to *AB*, so the ratio of *MD* to *BD* is equal to the ratio of *AB* to *ID*. Now we have put *AC* equal to *MD*, so the ratio of *AC* to *BD* is equal to the ratio of *AB* to *ID*. The square of *MD* – the perpendicular dropped from the parabola *AM* to its axis *AI* – is also equal to the product of *AB* – its *latus rectum* – and *AD*, cut off from the axis by the perpendicular *MD*, from what has been shown in proposition fourteen[13] of the first book of the *Conics*. But *AC* is equal to *MD*, so the product of *AB* and *AD* is equal to the square of *AC*, and the ratio of *AB* to *AC* is equal to the ratio of *AC* to *AD*; but the angle *CAD* is common to the two triangles *ABC* and *ACD*, so they are similar; but *CD* is equal to *AC*, so *BC* is equal to *AB*. But it has been shown in our first statement that the ratio of *AC* to *BD* is equal to the ratio of *AB* to *ID*; so the ratio of *AC* to *BD* is equal to the ratio of *BC* to *ID*; but *BE* is half of *BD* and *BG* is half of *ID*; so the ratio of *AC* to *BE* is equal to the ratio of *BC* to *BG*. But the ratio of *BC* to *BG* is equal to the ratio of *BH* to *BE*; so the ratio of *BH* to *BE* is equal to the ratio of *AC* to *BE*, *BH* is equal to *AC* and *DH* is equal to *BH*; so *DH* is equal to *AC*. So the straight lines *BH*, *DH* and *CD* are equal; that is why the angle *DHC* is twice the angle *ABC* and the angle *DCH* is equal to the angle *DHC*, so the angle *DCH* is also twice the angle *ABC*; but the angle *ADC* is equal to the sum of the two angles *DCB* and *ABC*, so the angle *ADC* is three times the angle *ABC*. Similarly, the angle *ACB* is three times the angle *ABC*, so <the sum> of the two angles *ACB* and *CAB* is six times the angle *ABC* and the sum of the angles of the triangle *ABC* is seven times the angle *ABC*; the angle *ABC* is thus a seventh <of the sum> of the angles of the triangle *ABC*, that is to say a seventh of two right angles. But the angle *ACD* is equal to the angle *ABC* because the two triangles *ACD* and *ABC* are similar, so the angle *ACD* is a seventh of two right angles.

---

[12] Proposition 12 in Heiberg's edition.

[13] Proposition 11 in Heiberg's edition.

If one of these two angles *ABC* and *ACD* is fitted into an arc of a general circle, its two sides cut off a seventh from the circumference. That is what we wanted to prove.

God has made it easier for us, thanks be to Him, to find the chord of a seventh of the circle, before everyone else and after them by two methods which are peculiar to us, which is more evident, clearer and more lucid than what others have done since the first construction. Thanks be given to God, all powerful and great, for His assistance and His support and may His blessings and His peace be upon Muḥammad and those that are his.

Completed on Wednesday 10 of Jumādā al-ūlā, the year one thousand one hundred and fifty-three.

In the name of God, the Compassionate, the Merciful

TREATISE BY ABŪ AL-JŪD MUḤAMMAD IBN AL-LAYTH
ADDRESSED TO THE EMINENT MASTER
ABŪ MUḤAMMAD 'ABD ALLĀH IBN 'ALĪ AL-ḤĀSIB

**On the Account of the Two Methods of the Master Abū Sahl al-Qūhī
the Geometer, and of his Own Master Abū Ḥāmid al-Ṣāghānī,
and on the Route he himself Took to Construct
the Regular Heptagon in the Circle**

I have received the treatise of the Master, my Lord – may God continue
to help him – including the two treatises composed by the pre-eminent
Master Abū Sahl al-Qūhī and our Master the geometer Abū Ḥāmid al-
Ṣāghānī – may God sustain them – on finding the chord of a seventh of the
circle, which were brought to him from Baghdad. Accordingly, I gave
thanks for the favour he had done me in sending them to me; may God
reward him for having sent them and give him his due <reward>. I show
the route each of them took in his construction, as well as the route I
followed for this, and which made me stand out in finding it (the chord); as
well as the state of doubt engendered in the construction of our Master Abū
Ḥāmid – may God sustain him – because of an error that occurred, perhaps
when the copyist was making the transcription, so that the Master – may
God allow him to continue to exercise his powers – may know from my
treatise what the three methods are, and how much is known by the author
of each of them.

I say that each of the two geometers we mentioned intended to <prove>
the proposition that Archimedes introduced in his treatise on the
construction of the heptagon, <a proposition> that he assumed without
carrying the construction or proving it in that treatise – unless he had made
a correction to it somewhere else, and this proposition had been handed

down to some, not to others – <a proposition> he continued to believe was true. God knows <the truth of> this. Each of these two pre-eminent Masters sought to correct the proposition and to prove it.

Let there be a square *ABCD*; if we draw its diagonal *BC*, and we extend its side *BD* indefinitely and we draw from the angle *A* a straight line that cuts the diagonal in *E* and the side *CD* in *G* and meets the extension of the straight line *BD* in *H*, it forms two equal triangles *ACE* and *DGH*, inside the square and outside it.

The Master Abū Sahl, by his mastery of the art and his skill in geometry avoided all mention of this square and the two equal triangles inside it and outside it, instead he ignored all of them and went on to why they had been drawn and the reason for their being constructed, namely for the division of a given straight line into three parts such that the product of the sum of the first and the second parts and the first one is equal to the square of the third part and the product of the sum of the second and third parts and the second one is equal to the square of the first part; from these three parts he constructed a triangle and he showed that one of its angles is twice the second angle and four times the third angle, that is to say its three angles are successive terms in double proportion, so that the sum of its angles is once, plus twice, plus four times, that is to say seven times the smallest angle, and the smallest angle is a seventh of two right angles, so that if he fitted it into the circumference of a general circle, its two sides cut off a seventh.

This is a triangle known to Archimedes; and all those who have constructed the heptagon by using motion and an instrument, have constructed the heptagon with the help of this triangle.

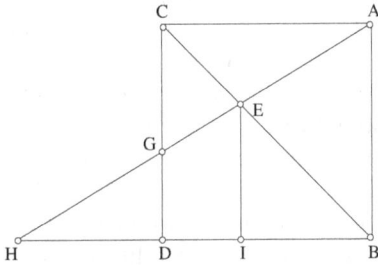

Fig. 3.1

The straight line divided into these parts at the points E and G is the straight line AH, because if the triangle ACE is equal to the triangle DGH and if the two angles CAE and DHA are equal because they are alternate internal angles, their sides are inversely proportional: the ratio of AE to GH is equal to the ratio of DH to AC. But the ratio of DH to AC – since the triangles ACG and DHG are similar – is equal to the ratio of GH to AG, so the ratio of AE to GH is equal to the ratio of GH to AG, the product of AG and AE is equal to the square of GH, and the product of the sum of the first and second parts and the first is equal to the square of the third part. Similarly, if we draw EI perpendicular to BD, the ratio of IH to IE, that is to say IB, is equal to the ratio of BH to AB, that is to say BD. So if we permute, the ratio of IH to BH is equal to the ratio of IB to BD, but if we separate <ratios>, the ratio of IH to IB is equal to the ratio of IB to ID, and the product of IH and ID is equal to the square of IB. But since the parts of AH are in the same ratio as the parts of BH, accordingly the product of the sum of the second and third parts of AH, which is EH, and EG, the second part, is equal to the square of AE, the first part, and the parts of BH, which are BI, ID and DH, are in this same ratio.

The Master Abū Sahl divided the straight line into three parts in these ratios, without however mentioning the square or the two equal triangles, because of the superiority of his knowledge and the penetration of his intelligence – using two intersecting conic sections, a hyperbola and a parabola, and starting from them he constructed the triangle we mentioned; he showed that its angles were in a series of double ratios, and in this way he obtained that one of the angles was a seventh of two right <angles>; he then constructed the heptagon by fixed geometry in the treatise that bears his name.

As regards our Master Abū Ḥāmid – may God sustain him – he sought <to prove> the same proposition that was introduced by Archimedes, that is to say <concerning> the square with diagonal BC, and the straight line AH.

I say that the triangle ACE is equal to the triangle DGH; he set about his analysis with the help of three hyperbolas, two opposite conic sections and a third one that cuts one of the <other> two. He then proceeded by synthesis, by which he established the construction with the help of a straight line divided into three parts in the ratios we mentioned; he then completed his treatise as others completed theirs. Perhaps the doubts about his treatise arise from an error that is due to the copyist when making the transcription of the original, I shall remove them and I shall correct what is defective.

For this I set up a square $ABCD$ with diagonal BC and side BD indefinite in the direction of D. He wished to draw from the point A a straight line to cut the straight line BC at the point E, CD at the point G, to meet BD in H and to be such that the triangle ACE formed inside the square is equal to the triangle DGH formed outside it. He increased AC by CI which is equal to it and constructed two opposite conic sections which pass through the points A and I and which have as asymptotes the two straight lines BC and DC – that is to say, if we extend the two straight lines BC and DC in the direction of C, for example so that the opposite angles they make at C are obtuse when the two straight lines in question are extended; we are concerned with the conic sections KL and PN; and he constructed a third hyperbola which passes through the point C and has as asymptotes the straight lines AB and BD; let the conic section be MN. This conic section cuts the conic section PN, because if we extend the conic section PN it meets the straight line BD and if we draw the conic section MN, it does not meet it; let the two conic sections PN and MN cut one another at the point N. Then from N he dropped a perpendicular NH to BD, and he joined AH, which cuts BC in E and CD in G. He said that he had constructed what he sought and had divided up AH into three parts in E and G in the ratios in question.

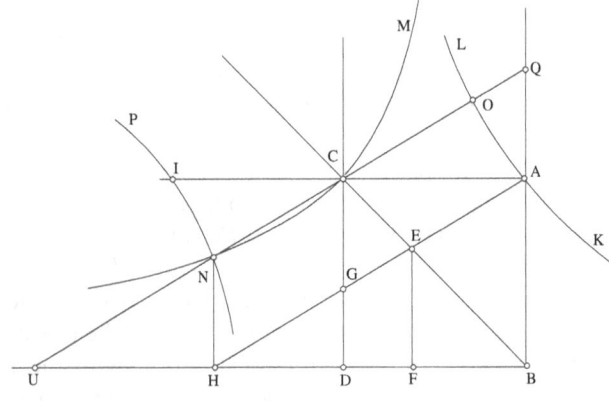

Fig. 3.2

Proof: He drew the straight line CN and extended it on one side and on the other to the extended straight lines AB and BD; it met them at the points U and Q. It cut the conic section KL at the point O. The two straight lines CQ and UN were separated off between the conic section MN and its two asymptotes, so they are equal, as Apollonius showed in the sixth proposition of the second book of the *Conics*. But the two angles A and H are right <angles> and UH is parallel to AC and is equal to it,[1] so CU is parallel to AH. But the square of CO – drawn from the angle of the conic section KL towards it – is equal to the product of AG, which is parallel to it and intercepts the angle supplementary to the angle of the conic section, and AE which is the excess of the latter, between the conic section and the asymptote, from what Apollonius showed in the seventh proposition of the book we mentioned. But CO is equal to CN, because they lie between the two opposite conic sections and their angles, from what Apollonius has shown in the thirty-first proposition of the first book of the *Conics*. But CN is equal to GH, because they are parallel; consequently the product of AG and AE is equal to the square of GH. He drew the perpendicular EF to BD. From the preceding we have shown that the product of FH and FD is equal to the square of BF; similarly the product of EH and EG is equal to the square of AE. Similarly BH is divided <as required> because the ratio between its parts is equal to the ratio between the parts of AH, so the product of BD and BF is equal to the square of DH and the product of FH and FD is equal to the square of BF.

He then constructed a triangle one of whose sides is equal to BF, the second equal to FD and the third equal to DH. He extended the side equal to FD on one side and on the other so that each of the two added parts is equal to the one of the two remaining sides of the triangle that is on its side. We obtain the well known triangle constructed by Archimedes and others who have sought to construct the heptagon using an instrument and movement, with the help of the lemma he had assumed, because the angles of this triangle follow one another in double ratio, that is to say in the ratio of one to two and of two to four, and the sum of them all is seven, and the unit is one seventh of seven. One of its angles is one seventh of two right <angles>. He fitted it into the circumference of the circle so as to cut off a seventh of it between the two sides <of the angle>, and this is obvious.

---

[1] The two right-angled triangles QCA and NHA are similar and their hypotenuses CQ and UN are equal, so we have $UH = AC$ and $NH = QA$; but $NH \parallel QA$, so $CU \parallel AH$.

For my part – may God sustain my Master and Lord – it was because of the modest extent of my knowledge, the lack of depth in my art, that I wished to approach something far from my usual concerns and to conquer something difficult; so I followed the route laid down by Euclid in the early part of his book on the *Elements* for inscribing the pentagon in the circle, the place where he introduces an isosceles triangle in which each of the angles at the base is twice the remaining angle, so that the sum of its three angles is equal to five times its smallest angle and the latter is a fifth of the sum of the three angles which is equal to two right <angles>. He fitted this small angle into the circumference of a circle and he extended its sides, which cut off a fifth of it (*sc.* of the circumference of the circle). I then knew that, if I construct an isosceles triangle such that each of the two angles at the base is three times the remaining angle, then the sum of its three angles is equal to seven times its smallest angle and the latter is a seventh of the sum of the three angles, which is equal to two right <angles>; so that if I fit it into the circumference of a general circle, it cuts off, with its two sides, a seventh <of the circumference>.

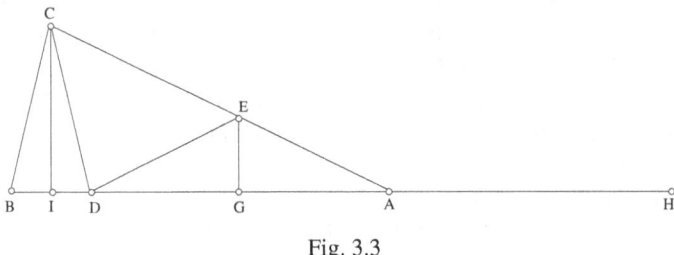

Fig. 3.3

For the analysis I supposed that the triangle *ABC* was like this: the two sides *AB* and *AC* are equal, each of the angles *ABC* and *ACB* is three times the angle *BAC*, and I proceeded by analysis. So I cut off from the angle *ACB* an angle *BCD* equal to the angle *BAC* and I cut off from the angle *ADC* an angle *ADE* equal to the angle *BAC*, so the side *AE* is equal to the side *ED*, because of the equality of the angles *DAE* and *EDA*, that is why the angle *DEC* is equal to twice the angle *DAE*. Since the angle *ACB* is equal to three times the angle *BAC*, and since we cut off from it the angle *BCD* equal to the angle *EDA* and equal to the angle *BAC*, we are left with the angle *ECD* <which is> equal to twice the angle *DAE*, that is to say equal to the angle *DEC*, so the side *DC* is equal to the side *DE*; but since the angle *B* is common to the two triangles *ABC* and *DBC*, and since the angle *BCD* is equal to the angle *BAC*, accordingly the isosceles triangle *ABC* is similar to the triangle *BCD*; so this latter is also isosceles, so the side *BC* is equal to the side *CD*. The straight lines *BC*, *CD*, *DE* and *EA* are equal. But because the triangles *ABC* and *BCD* are similar, the ratio of *AB* to *BC* is equal to the ratio of *BC* to *BD*, so the product of *AB* and *BD* is equal to the square of *BC*. But *AE* is equal to *BC*, so the product of *AB* and *BD* is equal to the square of *AE*. From the points *C* and *E* I dropped the two perpendiculars *CI* and *EG* to the straight line *AB*; they are thus parallel and the ratio of *AE* to *AG* is equal to the ratio of *AC*, that is to say *AB*, to *AI*. To *AB* I added *AH*, equal to *AD*; the whole of *BH* will thus be equal to twice *AI*. But *AD* is twice *AG*, that is why the ratio of *AE* to *AD* is equal to the ratio of *AB* to *BH*. So we have divided *AB*, for example at the point *D*, into two parts such that the product of *AB* and one of them, which is *BD*, is equal to the square of *AE* and the ratio of *AE* to the other part of *AB*, which is *AD*, is equal to the ratio of the whole of *AB* to *BH*. I found *BH* equal to the sum of the two straight lines *AB* and *AD*, I then knew that if I divide a given straight line into two parts such that the product of the whole straight line and one of the two parts is equal to the square of a straight line whose ratio to the other part is equal to the ratio of the whole straight line to the sum of this latter and the other part, I shall obtain an isosceles triangle such that the sum of its angles is seven times its small angle, and from that there follows, from what I have shown earlier, the construction of the regular heptagon in a circle, because I have fitted its small angle into its circumference and a seventh has been cut off from the circumference by its two sides.

I have omitted describing the analysis to avoid excessive length and to be sparing of weighty argument; I have assumed we have a straight line *AB*, I have sought to divide it into two parts in the ratio we mentioned, that is to say so that the product of *AB* and one of its two parts is equal to the square of a straight line whose ratio to the second part is equal to the ratio of *AB* to the sum of the latter and the other part. It was not possible for me to do this except by using two intersecting conic sections, a hyperbola and a parabola. I divided it using them, I constructed the triangle on which I had carried out the analysis and by using this latter I cut off a seventh of the circumference of the circle. I composed the treatise that bears my name in the year three hundred and fifty-eight from the Hegira addressed to the Eminent Shaykh Abū al-Ḥusayn ʿUbayd Allāh ibn Aḥmad, may God give him long life. In this same year I had described the first draft of this treatise to the Master, my Lord, may God cause his power to endure.

Anyone who has examined my construction of the heptagon and the constructions <put forward> by others will know that I stood out on account of the route I followed, and that at the time I was the one who came closest (*sc.* to solving the problem). The pre-eminent Master Abū Sahl al-Qūhī and our excellent Master the geometer Abū Ḥāmid al-Ṣāghānī, may God sustain them, followed Archimedes' route, corrected the lemma he stated, and built on the foundation he had provided. And they did well. But dividing the given straight line into two parts, as I did it, is more immediate than dividing it into three parts, as they did; and the analogy I employed in constructing the isosceles triangle in which each of the angles at the base is three times the remaining angle is valid for all polygons that have an odd number of sides, but their analogy is not valid for all polygons: it is in fact possible to find an isosceles triangle such that each of the angles at the base is five times the remaining angle, we then obtain the polygon with eleven equal sides inscribed in the circle; and there does not exist a triangle whose three angles are successively in general ratios of doubling, <a triangle> such that we obtain a polygon with eleven sides inscribed in the circle. The same is true for the majority of regular polygons whose number of sides is odd. We also know that the parabola is more accessible than the hyperbola. Our Master Abū Ḥāmid, may God sustain him, instead of it used two hyperbolas, which is why his construction of that and of everything else is less accessible.

I recognize the advance that Master Abū Sahl – may God give him long health and pre-eminence – made over me and those like me; I also recognize that he is unique in his time as regards the art of geometry; and I recognize that our Master Abū Ḥāmid – may God sustain him – is capable of constructing the heptagon and other interesting geometrical figures. He has worked on them in the perfect manner, and practiced it with great skill.

State duties have prevented me from undertaking <the study of> these constructions, and high responsibilities have caused me to turn away from this art, <responsibilities> that I did not seek. Indeed for several years some of them took me away from studying them (geometrical problems) and from teaching them. So there are geometers who deny <my having> the least knowledge and <object to> the slightness of my work, and foster the illusion that I borrowed it and did not create it. That is why I asked the person whom one consults, the Master, my Lord – may God cause his power to endure – him who is moderate, pre-eminent, and learned in the sciences; the just witness and the truthful judge of my earlier book – <I asked him> to inquire of the Master geometers present at the court – may God load it with his blessings and may he sustain them (the geometers) – whether anyone had constructed the heptagon using a single conic section and if anyone, to their knowledge, acknowledged having constructed the regular hendecagon in a circle; and for him to make the reply known to me, so that, if I send my construction of the two figures concerned, they do not desist from denouncing it, as they have done many times, in denouncing others, and in attributing them (the constructions) to others than myself. It is from God that aid and assistance come to us, and it is from him there is strength and power. God is all we need, and it is from him that there comes the best of our help. It is to him that we render thanks for inspiring us so that we should know it (the construction); and it is from him that we ask support to grasp that of which we are ignorant and in which we are lacking.

I show the analysis for what I constructed before in connection with the heptagon, in preparation for addressing to the Master, my Lord – may God cause his power to endure – my treatise devoted to this construction. Thus he will know that it is more accessible and easier than what others have done, and than what I myself did before.

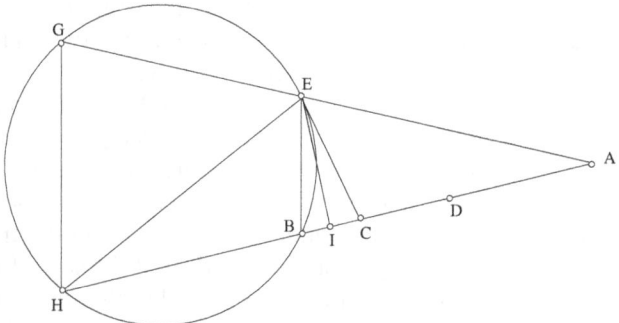

Fig. 3.4

For this let us draw a circle *BEG* circumscribed about a quadrilateral *BEGH*; let the sides *EG*, *GH*, *BH* be equal because the three arcs *EG*, *GH*, *BH* are equal; let each of them be twice the arc *BE*, so their sum is six times the arc *BE* and the circumference of the whole circle is seven times the arc *BE*. Seven times the arc *BE* is thus the circumference of the circle *BEG* and the chord *BE* is the side of the heptagon inscribed in the circle *BEG*. If we extend *GE* and *HB*, they meet one another; let them meet in *A*. We draw *EH*. So the angle *GEH* is equal to twice the angle *EHB*, because the arc *GH* has been assumed to be twice the arc *BE*. But the angle *GEH* is equal to <the sum> of the two angles *EAH* and *AHE*, so <the sum> of the two angles *EAH* and *AHE* is twice the angle *AHE*; the angle *EAH* is equal to the angle *AHE*, the side *AE* is equal to the side *EH* and *AE* is equal to *AB* because *EG* is equal to *BH*; and the angle *AGH* is equal to the angle *AHG* because the two arcs *EG* and *BH* are equal. From the point *E* we drop the perpendicular *EI* to *AB*; so *AI* is equal to *IH*; we cut off from *AI* <the line> *IC* equal to *IB*, so that leaves *AC* <which is> equal to *BH*; but *BH* is longer than *BE*, *AC* is longer than *BE*. We cut off from *AC* <the line> *AD*, equal to *BE*. Since the ratio of *BE* to *AB* is equal to the ratio of *GH* to *AH*, because *BE* and *GH* are parallel, and since *AD* is equal to *BE* and *AC* equal to *BH*, that is to say <equal to> *GH*, accordingly the ratio of *AD* to *AB* is equal to the ratio of *AC* to *AH*. By separation, the ratio of *AD* to *DB* is equal to the ratio of *AC* to *CH*. If we permute <the ratios>, then the ratio of *AD* to *AC* is equal to the ratio of *DB* to *CH*. But *CH* is equal to *AB*, because *AC* is equal to *BH* and *CB* is common, so the ratio of *AD* to *AC* is equal to the ratio of *DB* to *AB*. By separation, the ratio of *AD* to *DC* is equal to the ratio of *DB* to *AD*, so the product of *DB* and *DC* is equal to the product of *AD* by itself. Similarly, we join *CE*; but *CI* is equal to *IB* and the angle *I* is a right angle, so *CE* is equal to *BE*; but the angle *B* is common to the two triangles *ABE* and *CBE*, so they are similar and the ratio of *AB* to *BE* is equal to the ratio of *BE* to *BC*. But *BE* is equal to *AD*, so the ratio of *AB* to *AD* is equal to the ratio of *AD* to *BC*, and the product of *AB* and *BC* is equal to the square of *AD*. Now the product of *BD* and *DC* was also equal to the square of *AD*.

This analysis has thus led to dividing a given straight line into three parts such that the product of the whole straight line and the third part is equal to the square of the first part and the product of the sum of the second and third parts and the second <part> is also equal to the square of the first part. This is more accessible and easier than dividing a straight line into three parts such that the product of the sum of the first and the second parts and the first <part> is equal to the square of the third part and the product of the sum of the second and the third parts and the second <part> is equal to the square of the first part, as was proposed by Archimedes and as has been constructed by the Master Abū Sahl and our Master Abū Ḥāmid – may God sustain them – in order to construct the heptagon. It is also easier than dividing the straight line into two parts such that the product of the complete straight line and one of them is equal to the square of a straight line whose ratio to the other part is equal to the ratio of the whole straight line to the sum of the straight line and this other part, which is what I did before, also in order to construct the heptagon. It is easy to divide the straight line in the ratio concerned by all the procedures described above, but I divide it by means of a single conic section, whereas all the other procedures divided it either by using two conic sections, a parabola and a hyperbola, or by using three hyperbolic conic sections.

I also mention the synthesis, but without the proof of the corrected lemma to be found in the treatise devoted to this construction.

Let there be a given straight line *AB* divided at the points *C* and *D* such that the product of *AB* and *BC* is equal to the square of *AD* and similarly the product of *BD* and *DC* is equal to the square of *AD*.[2] We construct the triangle *ABE*, with *AE* equal to *AB* and *BE* equal to *AD*. We extend *AB* and *AE* to *H* and *G* so that each <of the straight lines> *BH* and *EG* becomes equal to *AC*. We join *GH* and we draw a circle *BEG* circumscribed about the quadrilateral *BEGH*; this is easy.

I say that the three sides *EG*, *GH*, *BH* are equal, that each of the three arcs that they cut off is twice the arc *BE*, that the arc *BE* is a seventh of the circumference of the circle *BEG* and that the chord *BE* is the side of the regular heptagon inscribed in the circle *BEG*.

Proof: The product of *DB* and *DC* is equal to the square of *AD*, the ratio of *AD* to *DC* is equal to the ratio of *DB* to *AD*. By composition, the ratio of *AD* to *CA* is equal to the ratio of *DB* to *AB*. But *AB* is equal to *CH*, because *AC* is equal to *BH* and *CB* is common, so the ratio of *AD* to *AC* is equal to the ratio of *DB* to *CH*. By permutation, the ratio of *AD* to *DB* is equal to the ratio of *AC* to *CH*. By composition, the ratio of *AD* to *AB* is equal to the ratio of *AC* to *AH*. But *AD* is equal to *BE* and *AC* is equal to *BH*, so the ratio of *BE* to *AB* is equal to the ratio of *BH* to *AH*. But the ratio of *BE* to *AB* is equal to the ratio of *GH* to *AH* because *GH* and *BE* are parallel, so the ratio of *GH* to *AH* is equal to the ratio of *BH* to *AH*, and *GH* is equal to *BH*; so it is equal to *EG*. The straight lines *EG*, *GH*, *BH* are thus equal and the three arcs *EG*, *GH*, *BH* are also equal.

Similarly, we join *EC*, we divide up *CB* into two equal parts at the point *I* and we join *EI*. Since the product of *AB* and *BC* is equal to the square of *AD* and since *BE* is equal to *AD*, the product of *AB* and *BC* is

---

[2] The author does not indicate here how to construct this division (see the construction by Abū al-Jūd, in Text 2.2, p. 605).

equal to the square of *BE*, so the ratio of *AB* to *BE* is equal to the ratio of *BE* to *BC*. But the angle *ABE* is common to the triangles *ABE* and *CBE*, so they are similar. But *AB* is equal to *AE*, so *CE* is equal to *BE*, *EI* is perpendicular to *CB* and *AI* is equal to *IH*, because *AC* is equal to *BH* and *CI* is equal to *IB*; so *AE* is equal to *EH* and the angle *A* is equal to the angle *EHB*. But the external angle *GEH* is equal to <the sum> of the angles *A* and *EHB* which are equal, so the angle *GEH* is twice the angle *EHB*, so the arc *GH* is twice the arc *BE*. The same holds for each of the arcs *EG* and *BH* which is twice the arc *BE*, and the sum of the arcs *EG*, *GH*, *HB* is six times the arc *BE*, so the circumference of the circle *BEG*, as a whole, is seven times the arc *BE*, the arc *BE* is a seventh of the circumference of the circle *BEG* and the chord *BE* is the side of the regular heptagon inscribed in the circle *BEG*. That is what we wanted to prove.

The figure for this proposition is given above.

If I address the treatise devoted to this construction to the Master, my Lord – may God continue to sustain him – once he has given his approval to the allusions I make in it, he will know the proof, by means of a single conic section, of the lemma to which I have referred – if God so wills.

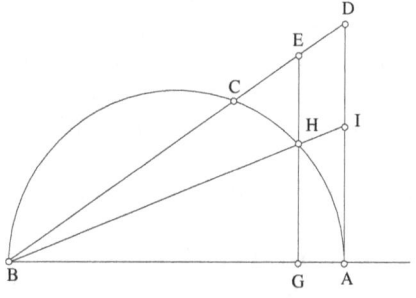

Fig. 3.5

I used – may God sustain the Master, my Lord – with a single section, <one of> the conic sections, in what I constructed before, two lemmas belonging to the book of the *Elements*. The first: if we draw from the point *B* of the straight line *AB*, the diameter, a straight line that cuts the circle *ACB* in *C* and if we draw from the point *A* a perpendicular to *AB* to meet the extension of *BC* in *D*, <we wish to know> how to draw starting from the straight line *CD* a straight line, for example the straight line *EG*, parallel to

AD, which cuts the circumference in *H* and is such that the ratio of *EH* to *GH* is equal to a given ratio.

This is easy if we divide *AD* at the point *I* in the given ratio, by drawing *BI* which necessarily cuts the circumference; let it cut it in *H*. We cause to pass through this point <the line> *EG* parallel to *AD*, then the ratio of *EH* to *GH* is equal to the given ratio, and this is clear.

Second lemma: We draw *EG* parallel to *AD*, the perpendicular, so that it is equal to the straight line joining *A* and *H*. This too is not inaccessible.

I never sent the treatise about the construction of the lemma I had introduced earlier. It is from God that there comes assistance for the truth by his goodness.

Finished thanks be to God and transcribed from a copy made by Aḥmad ibn Muḥammad ibn 'Abd al-Jalīl al-Sijzī. Finished at the fort of Hamadhān, on Wednesday 12.7.544 – may the blessing of God be upon His prophet Muḥammad and those that are his.

In the name of God, the Compassionate, the Merciful

BOOK BY
AḤMAD IBN MUḤAMMAD IBN ʿABD AL-JALĪL AL-SIJZĪ

On the Construction of the Heptagon in the Circle and the Trisection of a Rectilinear Angle

He said: we are astonished to come across someone who seeks <to acquire> the art of geometry and gives himself to it and, when he borrows from eminent ancients, sees in them weaknesses and shortcomings; in particular when it is a beginner and a pupil, who knows little of geometry, so that he starts to imagine that after some minimal effort there will come to him things which he considers easy of access and within the grasp of the understanding, whereas they are past the understanding of those who are experienced in this art and have been trained in it.

Would that I could know by what power, by what intuition, by what skill and by what insight, he thought so well of himself as to find the heptagon by starting from the preliminaries in the manner of someone who reads part of the introductory book, that is to say Euclid's book on the *Elements*, without being either skilful or practiced, and that he belittles those who have distinguished themselves in this art.

And there is surely no need to believe in weakness on the part of the eminent Archimedes, despite the fact that in geometry he is in advance of all other geometers: indeed his attainments in geometry were such that the Greeks called him 'the Geometer' – it is Archimedes, and no other among the ancients or the moderns, who received that name[1] – on account of his eminence in the art of geometry. He was extremely diligent in designing useful things, and thanks to his great intellect he succeeded in making instruments, machines and mechanical devices, he established the lemmas

---

[1] Lit.: his name.

for the heptagon and he took the right route; it is through the power of his intellect that we have a grasp of the heptagon, and Heron owed his understanding of machines to the power of his (Archimedes') intellect, because of his care and his diligence in matters of mathematics. Despite his eminence, his primacy and his rank in the art of geometry, this miserable misguided man taxes Archimedes with inadequacy and proposes the first of his <own> preliminaries, bad, corrupt, far from the road to truth, and incapable of leading to the construction of the heptagon. This deceit which misled no one but himself and by which he expected to deceive someone else, would only work on someone who does nothing good in geometry, not even as a beginner. And further, he then attributes to Archimedes things that would be an insult to anyone endowed with the least understanding, not to mention geometers; and he claims that the lemma introduced by Archimedes is more difficult than what he seeks, and he finds fault with his method of proceeding and taxes him with imitation. What Archimedes did is very beautiful, in the proof he obtained from the lemmas for the heptagon, and in what he wrote in his book, so that he who is not worthy cannot draw profit from them, as is the case for this pitiable man.

I too, after having borrowed from the learned work of Archimedes and from the preliminaries in Apollonius, and in particular from modern scholars such as al-'Alā' ibn Sahl, I retained this notable and admirable proposition as something precious, as something I was able to make use of, in a very small step, <a proposition> concerning the trisection of a rectilinear angle with the help of the first book of Apollonius' work on the *Conics*.

Now, I shall describe how matters stand and I begin with what is said by the self-deceiver, so that it may be a lesson for beginners, and I show how thoroughly suspect his remarks are and his construction fallacious; I shall go on to the lemmas for the heptagon, and I shall continue by <considering> the construction of the heptagon. I shall finish this book with the trisection of a rectilinear angle, and I pray to God for assistance.

This is the beginning of his book, and the order of these lemmas.

He[2] said: among the many lemmas that he introduced for the division of the circle into seven equal parts, Archimedes imitated[3] a lemma whose construction he has not shown and which he has not proved; and it perhaps

---

[2] Abū al-Jūd.

[3] Here the translation follows the sense as understood by al-Sijzī, cf. above, p. 305.

involves a more difficult construction and a less accessible proof than the matter for which he introduced it. It is the following:

He[4] said that he – he means Archimedes – said: Let us draw the diagonal of the square $ABCD$, say AC. Let us extend AB to E, without an endpoint, and let us draw from a point of BE, say E, a straight line to the corner of the square, at the point D, which cuts the diagonal AC at the point G and the side BC at the point H, so that the triangle BHE outside the square is equal to the triangle CDG.

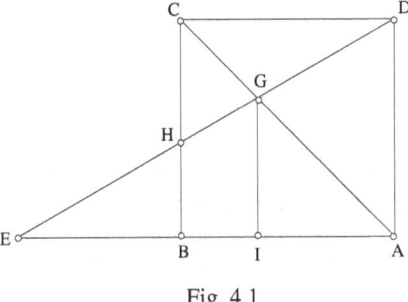

Fig. 4.1

Archimedes wished – from what he imitated in this matter – to draw a perpendicular GI to AB to divide AB in I so that the product of AI and the whole of AE is equal to the product of AB and IE and the product of AB and AI is equal to the square of BE. But the division of AB in accordance with this condition is easier to carry out and to prove than the act of drawing a straight line ED in accordance with the condition mentioned by Archimedes. Perhaps this latter is not possible without dividing the straight line AB in accordance with the condition mentioned and perhaps the division of AB is similarly more difficult than the division of the circle into seven equal parts.

In what has come into my mind for this chapter, I engage, on a different basis, upon a shorter route, a simpler construction and lemmas that are less numerous and easier.

The first is this one: If we draw a circle with distance of a perpendicular to a straight line, then it touches the straight line on which the perpendicular was erected, as the straight line AB is perpendicular to

[4] This is still Abū al-Jūd, quoted by al-Sijzī.

CD at the point *B*, and if we draw with distance *AB* a circle *BEG*, I say that it touches the straight line *CD*, the proof of this is easy.

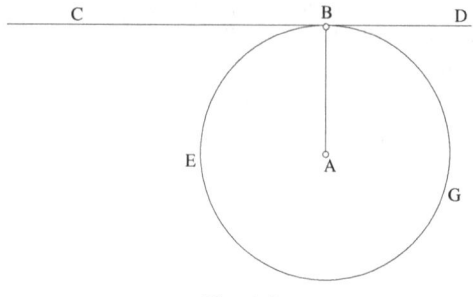

Fig. 4.2

Second lemma: We wish to draw from one of the sides of a given triangle, say the side *AB* of the triangle *ABC*, to the second side, which is *BC*, a straight line, equal to what it cuts off from it outside the small triangle, and parallel to the third side, which is *AC*.

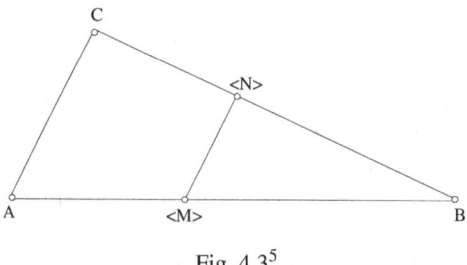

Fig. 4.3[5]

Third lemma: We wish to draw a straight line whose ratio to a known straight line, say the straight line *AB*, is a known ratio, say the ratio of *C* to *D*.

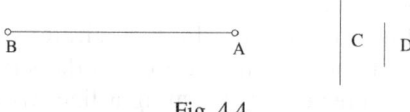

Fig. 4.4

[5] In the figure in manuscript [B], al-Sijzī (or the copyist?) draws *DI* – here *NM* – without later using it; in manuscript [T], we have *DE* instead of *DI*.

Fourth lemma: We wish to divide up a known straight line, say the straight line *AB*, into two parts such that the product of the whole straight line and one of the two parts is equal to the square of a straight line whose ratio to the other part is a known ratio; let the ratio be <that of> *C* to *D*.

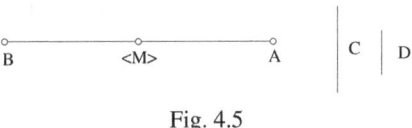

Fig. 4.5

These lemmas are those on which he relied when constructing the heptagon. Then, for the construction of the heptagon, he ordered that a given straight line be divided into two parts such that the product of the whole straight line and one of the two parts is equal to the square of a straight line whose ratio to the other part is equal to the ratio of the whole straight line to the sum of the whole straight line and this part. He gave the ratio in the fourth proposition and for the construction of the heptagon he used another ratio, different from the one he introduced in his lemma ; and he believed it is possible to construct that by means of the lemma in the fourth proposition. But the construction is not possible except by means of conic sections, and, for someone who, in geometry, knows neither the cone nor the sections, it would be by means of the lemmas presented in the books of the ancients, thanks to which it is possible to construct the heptagon <that is> for someone who adds his lemmas to them. But by means of his lemmas, and by others analogous to them, it is difficult to find the hexagon inscribed in the circle – which carpenters make on the lids of cauldrons by means of a single opening of the compasses – *a fortiori* if it is a matter of finding the heptagon. This is his mistake and his deception in regard to the lemmas for the heptagon and its construction.

Let us now begin on what we have found out in connection with the heptagon, its lemmas and the trisection of a rectilinear angle.

Lemma: We wish to divide a straight line *AB* into two parts at <a point> *C*, for example, such that the ratio of the straight line equal in the power to *AB* times *BC* to the straight line *AC* is equal to the ratio of *AB* to the sum of *AB* and *AC*, considered as a single straight line.

We extend *BA* to *D* such that *AD* is equal to *AB*; we apply to *AD* the square *ADEG* and we construct through the point *A* a hyperbola such that the two straight lines *EG* and *ED* do not meet it, but continually come closer to it, from <proposition> four of <book> two of the *Conics* of the eminent Apollonius, and from <proposition> one in the translation of Isḥāq; let it be *KAH*. We construct on the axis *BD* a parabola such that its *latus rectum* is *AB*, let it be *LBH*. We draw from the intersection of the two conic sections, which is the point *H*, the perpendicular *HC* to the straight line *AB*.

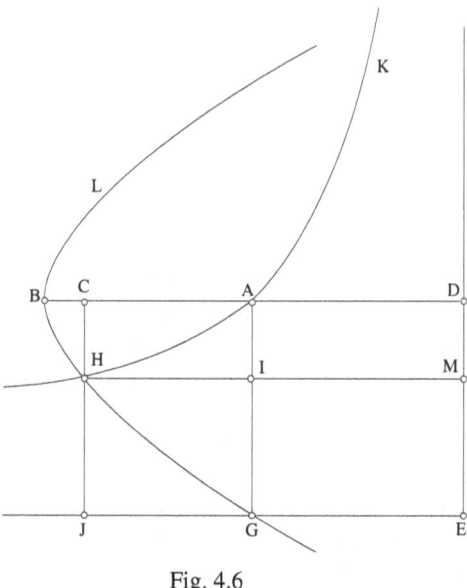

Fig. 4.6

I say that we have divided the straight line *AB* at the point *C* in the way that we wished.

Proof: We extend *EG* and *CH* until they meet one another in *J*; we draw *HIM* parallel to *JE* and *AIG* parallel to *CJ*, since the rectangle *MJ* is equal to the square *GD*, accordingly *JI* will be equal to *ID*. We take the common rectangle *IC*, then the rectangle *JA* is equal to the rectangle *HD*. But the rectangle *HD* is <the product> of the straight line *CH* and the straight line *CD* and *JA* is *CA* times *AG*, that is to say *AB*, so *AB* times *AC* is equal to *CH* times *CD*. So the ratio of *CH* to *AC* is equal to the ratio of *AB* to *CD*, but *CH* is equal in the power to *AB* times *BC* since *AB* is the *latus rectum* of the parabola *LBH*; but *CD* is *AB* plus *CA*, so the ratio of the straight line which in the power is equal to *AB* times *BC* to the straight line

CA is equal to the ratio of the straight line *BA* to *BA* and *AC* considered as a single straight line. So we have constructed what we wished. That is what we wanted to prove.

Abū Saʻd al-ʻAlāʼ ibn Sahl has proved this proposition by using the method of analysis and our synthesis is a part of his analysis.

Here is another lemma: We wish to construct on a straight line *AB* an isosceles triangle such that each of its angles at the base is three times the remaining angle.

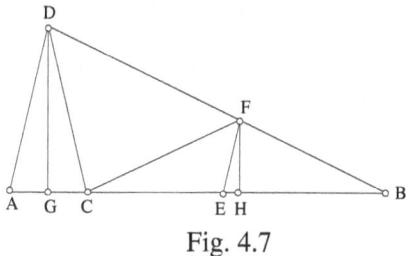

Fig. 4.7

Let us divide *AB* into two parts at *C*, <a point> such that a straight line, say the straight line *AD*, is equal in the power to *AB* times *AC*; so its ratio to *CB* is equal to the ratio of *AB* to *AB* and *BC* considered as a single straight line, from the preceding construction. We draw *BD* equal to *AB* and to enclose with *AD* an angle *D*.

I say that the triangle *ABD* is what is sought and that each of the two angles *A* and *D* is three times the angle *B*.

Proof: We join *DC*; since the ratio of *AD* to *CB* is equal to the ratio of *AB* to *AB* and *BC*, considered as a single straight line, and since *AB* is smaller than *AB* and *BC* combined, *AD* is consequently smaller than *CB*. We take from *CB*, *BE* <which is> equal to *AD*; we draw *EF* parallel to *AD* and we draw *FH* and *DG* perpendicular to *AB*. Since the product of *AB* and *AC* is equal to the square of *AD*, the ratio of *AB* to *AD* in the triangle *ABD* is equal to the ratio of *AD* to *AC* in the triangle *ADC* and the angle *A* of the two triangles is common, so the triangle *ADC* is similar to the triangle *ABD*, so the straight line *DC* is equal to the straight line *AD* and *AG* is equal to *GC*. But since *GC* is half of *AC* and *CB* is half of twice *CB*, we have that *GB* is half of *AB* plus half of *CB* also, and the ratio of *EB*, which is equal to *AD*, to half of *CB* is equal to the ratio of *AB* to half of *AB* and of

CB, that is to say *GB*. But the ratio of *FB*, which is equal to *EB*, to *HB*, is equal to the ratio of *DB*, which is equal to *AB*, to *GB*. So the ratio of *EB* to *HB* is equal to the ratio of *AB* to *GB*, that is to say to half <the sum of> *AB* and *BC*; but the ratio of *EB* to half *CB* is also equal to the ratio of *AB* to *BG*. Consequently half of *CB* is equal to *HB*, so the straight line *FC* is equal to the straight line *FB*, that is to say *EB*. So the straight lines *AD*, *DC*, *FC*, *FB* are all equal, but the angle *ACD* is equal to the <sum> of the two angles *CDF* and *DBC* and the angle *DFC* is twice the angle *B*, the angle *ACD*, that is to say the angle *CAD*, is thus three times the angle *B*. Each of the two angles *A* and *D* in the triangle *ABD* is three times the angle *B*. So we have constructed what we wished. That is what we wanted to construct.

<Proposition>: We wish to construct in a circle *ABC* a heptagon with equal sides and equal angles.

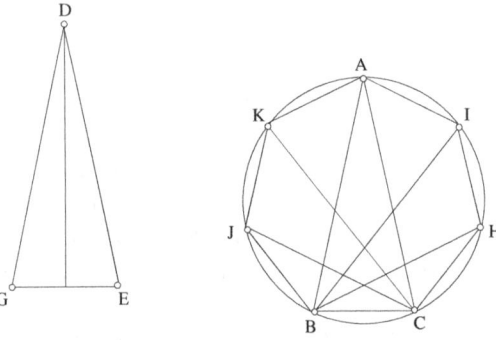

Fig. 4.8

We construct an isosceles triangle *EGD* such that each of the two angles *E* and *G* is three times the angle *D* and we construct in the circle *ABCH* a triangle whose angles are equal to the angles of the triangle *EGD*, that is the triangle *ABC*; we draw *BH* such that the angle *CBH* is equal to the angle *BAC* and let us divide up the angle *HBA* into two equal parts with the straight line *BI*. It is clear that the three angles *CBH*, *HBI*, *IBA* are equal. We do to the angle *BCA* what we did to the angle *CBA* and we draw the straight lines *CJ*, *CK*. Since the six angles which are at the <points> *B* and *C* are equal and equal to the angle *A*, their arcs, which are *CH*, *HI*, *IA*, *BJ*, *JK*, *KA*, are equal and equal to the arc *BC*; so we have constructed in the circle *ABCH* a regular heptagon. That is what we wanted to construct.

Here is a lemma for the trisection of a rectilinear angle.

The semicircle ACB is given, the straight line AG is known in position, the diameter is AB and the centre is D. We wish to find on the diameter AB a point, say the point E, such that if we draw from this point to the circumference of the semicircle ACB a straight line parallel to the straight line AG, say the straight line EC, its square, that is to say <the square of> EC, is equal to the straight line BE times ED. Let us construct on the diameter DB a hyperbola such that its transverse diameter is DB and its *latus rectum* is equal to the straight line DB and such that the ordinates are at an angle equal to the angle GAB; that is the conic section DCH which meets the circumference of the semicircle at the point C. We draw CE parallel to AG.

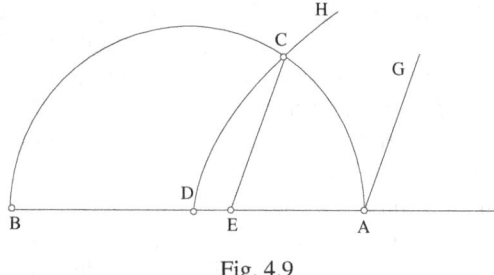

Fig. 4.9

I say that EB times ED is equal to the square of EC.

Proof: The ratio of EB times ED to the square of EC is equal to the ratio of DB to the *latus rectum*. But DB is equal to the *latus rectum*, BE times ED is consequently equal to the square of EC. So we have constructed what we wished. That is what we wanted to construct.

Once we have introduced the preceding <result>, trisection of a rectilinear angle becomes easy.

Let the angle BAC be given, we wish to divide it into three equal parts. Let us extend BA to D to whatever length we want and let us construct on the diameter AD a semicircle AGD with centre E. Let us draw HG parallel to AC and such that HD times HE is equal to the square of HG using the preceding construction. Let us join GD and GE and let us draw AI parallel to EG.

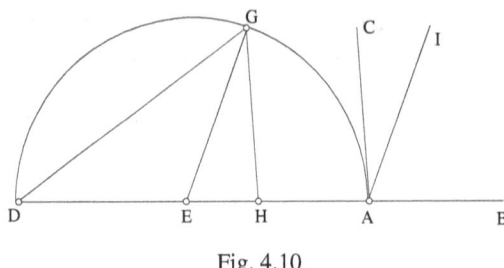

Fig. 4.10

I say that the angle *BAI* is twice the angle *IAC*.

Proof: Since the product of *HD* and *HE* is equal to the square of *HG*, we have <that> the ratio of *HD* to *HG* in the triangle *HDG* is equal to the ratio of *HG* to *HE* in the triangle *HGE*; but the angle *H* is common to the triangles, so the two triangles are similar and the angle *HDG* is equal to the angle *HGE*; consequently, the angle *HGE* is equal to the angle *EGD*, but the exterior angle *HEG* is equal to twice the angle *EGD*, and since *EG* is equal to *ED*, the angle *HEG* is twice the angle *EGH*; but the angle *AHG*, which is equal to the angle *BAC*, is equal to the sum of the two angles *HGE* and *GEH* <which are> interior <angles> of the triangle; now the angle *BAI* is equal to the angle *HEG*; and finally the angle *IAC* <which is> equal to the angle *EGH*, so the angle *BAI* is twice the angle *IAC*. We divide the angle *BAI* into two equal parts, so we have divided the angle *BAC* into three equal parts. That is what we wanted to prove.

The book of Aḥmad ibn Muḥammad ibn 'Abd al-Jalil on the heptagon and the trisection of the rectilinear angle is completed.

In the name of God,
the Compassionate, the Merciful

In the name of God,
the Compassionate, the Merciful

Solution by Wayjan ibn Rustam Known by the Name Abū Sahl al-Qūhī for the Construction of the Regular Heptagon in a Given Circle

Treatise by Abū Sahl al-Qūhī Wayjan ibn Rustam on the Determination of the Side of the Heptagon

There have appeared in the time of our Lord the Great King the victorious 'Aḍud al-Dawla – may God give him long life and cause his power to endure – many noble sciences, of belles lettres, of refined arts, of ingenious constructions; a large political outlook, a high-minded life, justice dispensed widely, prosperity in every region, security for the people, <these have appeared> in his reign and at the time of his glory.

There have also appeared many geometrical propositions which had not appeared in the time of any King, when they (mathematicians) had the intention of bringing them out and had done everything to

God, Thanks be rendered to him, has made to appear in the time of Our Lord the Great King the victorious and triumphant, 'Aḍud al-Dawla, may God give him long life and cause his glory to endure, <together with> his eminence, his authority, his power and his rule, many branches of the sciences and belles lettres, and many kinds of research and inquiry, which were hidden and did not reveal themselves come into the open; as strangers, they did not make themselves available; being haughty, they did not lower themselves or submit; being distant, they were neither easily met with nor accessible. There also appeared, thanks to the good fortune of his reign and the happiness of its spirit, many subtle geometrical propositions, which required much preparation before one could approach them and which were so difficult to envisage for the Ancients that they bequeathed <the

resolve them, because they knew that this family of mathematical sciences, as astronomy, numbers, weights, centres of gravity and similar matters of philosophical mathematics and comparable sciences that indeed belong among the true sciences which are not susceptible to corruption, or to change, or to criticism, or to attack, as others are susceptible to them – because their premises are necessarily true, their arguments are true and their proofs satisfy all the premises and syllogisms. The easiest part among all these propositions that have appeared in the course of these blessed times is a proposition that the Ancients who are mentioned as associated with it worked actively to resolve, without any one of them succeeding. God, powerful and great, achieved it, in the reign of our Lord, the Great King the victorious 'Aḍud al-Dawla, may God give him long life and cause his power to endure by the hand of the King's servant; this is the construction of the regular heptagon in the circle.

task of> examining them to the Moderns, so difficult were these <propositions> for the most distinguished among them, and arduous for the most eminent. Defeated, they turned away from <seeking> solutions, and they fled from <attempting to> resolve them, shedding their perspicacity and their power, and, when faced with these <propositions>, they saw their vigour and their energy melt away. But, all this, once they had worn out all their efforts to find solutions, and exhausted all their means of deducing them, confident as they were in what their good faith had promised them: the survival of geometry for all time, its development beyond their life spans, its permanence as a beautiful unchanging memory and a treasure huge and inexhaustible, and moreover the benefit it offers to the mathematical sciences, in particular <those concerning> numbers, sounds, heavenly bodies, weights, and everything that is like them, is connected with them, goes with them and is close to them; even more, to all theoretical sciences in general, it being given that the science of geometry is a model that sets the standard for what is true, a guide one follows to find truth, because its foundation is well established, its reasoning is stable and continuous, not susceptible to damage: it is not subject to any weakness nor can any corruption, any blow, reach to it; no rejection, no refusal <to believe> changes it; it is incomparable in regard to truth, and there is nothing similar in regard to nobility. The simplest among those <propositions of which proofs are> still sought, is to know the side of the regular heptagon in the circle. The ingenuity of the eminent geometers we have mentioned, and notably Archimedes, who worked on it, without any of them being so fortunate as to succeed. The attention of the subject of our Lord the Great King, victorious and triumphant, 'Aḍud al-Dawla, may God give him long life, cause his power to endure and sustain his victory, <his attention> turned to it, and he found <a solution>; he pays him (the King) homage by making the approach to the matter simple, and in giving a proof of it, hoping that it will meet with a kindly reception from him, if God so wishes. It is He who is all I need and he is the best support.

– **1** – We wish to construct in a known circle ABC the side of the regular heptagon.

By analysis, we suppose that the straight line BC is the side of the regular heptagon inscribed in the circle ABC and that the arc AB is twice the arc BC which is a seventh of the circumference of the circle; let us join the two straight lines AB and AC; so the angle ACB is twice the angle BAC, because the ratio of the arc to the arc is equal to the ratio of the angle to the angle if the two angles are on the circumference or at the centre; so we have that the arc ADC is four times the arc BC and twice the arc AB, because we have supposed the arc BC is a seventh of the circumference of the circle. The angle ABC is thus also four times the angle BAC. But the angle BCA is twice the angle CAB, so the angle ABC is twice the angle BCA. So in the rectilinear triangle ABC, the angle ABC is twice the angle ACB and four times the remaining angle BAC.

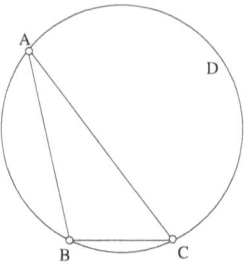

Fig. 5.1

So the analysis has led to the construction of a rectilinear triangle in which one of the angles is twice one of the two remaining angles and four times the last angle.

– **2** – We wish to construct a rectilinear triangle in which one of the angles is twice one of the two remaining angles and four times the last angle.

By analysis, we suppose that one of the angles of the triangle ABC, which is the angle ABC, is twice the angle BCA and four times the remaining angle BAC.

We put the straight line CD equal to the straight line AC on the extension BC, and we join the straight line AD. The angle BCA is twice the angle ADC, because it is exterior to the triangle ACD and it is equal to <the sum> of the two angles CAD and CDA which are equal. But the angle ABC is equal to twice the angle ACB, so the angle ABC is four times the angle ADB. But the angle ABC is also four times the angle BAC, so the angle ADB of the triangle ADB is equal to the angle CAB of the triangle CAB. But the angle ABC is common to the two triangles, so the remaining angle of one of the two triangles is equal to the remaining angle of the other triangle. So the two triangles ABD and ABC are similar and the ratio of DB to BA is equal to the ratio of BA to BC. The product of DB and BC is thus equal to the square of BA.

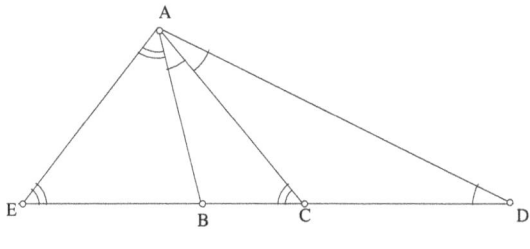

Fig. 5.2

We put the straight line BE equal to the straight line AB on the extension of the straight line BD, and we join the straight line AE. So the product of DB and BC is equal to the square of BE. Similarly, since the angle ABC is exterior to the triangle ABE, it is accordingly equal to the sum of the two angles BAE and AEB which are equal, so the angle ABC is twice the angle EAB. But the angle ABC is twice the angle ACB, so the angle ACB of the triangle ACE is equal to the angle BAE of the triangle ABE. So if we take the common angle to be AEB, then the remaining angle in one of the two triangles is equal to the remaining angle in the other triangle; so the two triangles ACE and ABE are similar, so the ratio of CE to EA is equal to the ratio of AE to EB and the product of CE and EB is equal to the square of EA; but the straight line EA is equal to the straight line AC because the angle ACE is equal to the angle AEC. But the straight line AC is equal to the straight line CD, so the straight line CD is equal to the straight line AE, the square of the straight line CD is equal to the square of the straight line AE, and thus the product of CE and EB is equal to the square of the straight line CD. But it has been shown that the product of DB and BC is equal to the square of BE. So the straight line ED is divided at the two points B and C such that the product of CE and EB is equal to the square of CD and the product of DB and BC is also equal to the square of BE.

The construction of the triangle as we have described has led to finding a straight line, let it be ED, divided at the points B and C such that the product of DB and BC is equal to the square of BE and the product of CE and EB is equal to the square of CD.

– **3** – We wish to find this by analysis and we suppose that AB is a straight line and is divided at the points C and D such that the product of BC and CD is equal to the square of AC and the product of DA and AC is equal to the square of DB.

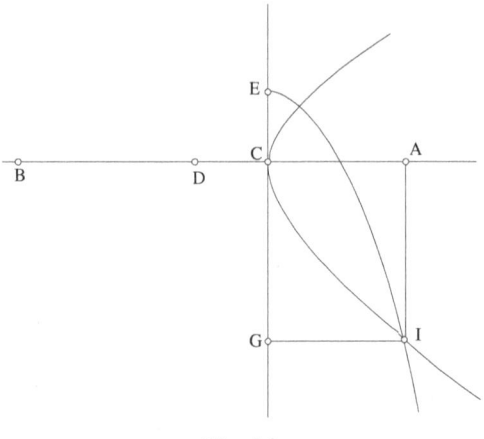

Fig. 5.3

We draw the straight line *EC* perpendicular to the straight line *AB* and equal to the straight line *CD*, the straight line *CG* is equal to the straight line *DB* on the extension of the straight line *EC*, and each of the straight lines *GI* and *AI* is parallel, respectively, to one of the straight lines *AC* and *CG*. Since the straight line *GC* is equal to the straight line *BD* and the straight line *CE* is equal to the straight line *DC*, the product of *GE* and *EC* is equal to the product of *BC* and *CD*. But the product of *BC* and *CD* is equal to the square of *AC*, so the product of *GE* and *EC* is equal to the square of *AC*; but the square of *AC* is equal to the square of *GI* and the straight line *EC* is equal to the straight line *CD*, so the product of *CD* and *EG* is equal to the square of *GI*. So the point *I* lies on the outline of the parabola with axis the straight line *EG*, vertex the point *E* and *latus rectum* the straight line *CD*. Similarly, since the product of *DA* and *AC* is equal to the square of *DB*, and since the square of *DB* is equal to the square of *CG*, accordingly the product of *DA* and *AC* is equal to the square of *GC*; but the square of *GC* is equal to the square of *AI* because the area *ACGI* is a parallelogram, so the product of *DA* and *AC* is equal to the square of *AI*. So the point *I* also lies on the outline of the hyperbola with vertex the point *C*, with axis and *latus rectum* both equal to the straight line *DC*; in fact the ratio of the product of *DA* and *AC* to the square of *AI* is equal to the ratio of the transverse diameter to the *latus rectum* for the hyperbola. Thus if we suppose we have the straight line *CD*, which is the axis of the hyperbola and which is equal to the *latera recta* of the two conic sections, known in position and magnitude, the outlines of the hyperbola and the parabola will be known in position. The point *I*, which is their point of intersection, is thus known, the straight line *IA* is known because it is perpendicular to the straight line *AB*, <which is> known in position, at the known point *I*. So the straight line *DB* is known, because it is equal to the straight line *AI* and each of the straight lines *AC*, *CD* and *DB* is known, so the straight line *AB* is divided at the two points *C* and *D* such that the product of *BC* and *CD* is equal to the square of *AC* and the product of *DA* and *AC* is equal to the square of *DB*. That is what we wanted to construct.

– **4** – We wish to show that if the straight line *AB* is divided at the points *C* and *D* such that the product of *AD* and *DC* is equal to the square of *DB* and the product of *CB* and *BD* is equal to the square of *AC*, as we have described, then <the sum of> two of the parts is greater than the remaining part.

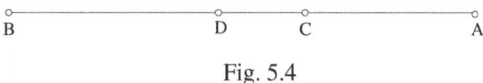

Fig. 5.4

Proof: Since the product of *AD* and *DC* is equal to the square of *DB*, the straight line *DB* is the mean proportional between the two straight lines *AD* and *DC*. But *AD* is <the sum of> two of the parts of the straight line *AB* and it is greater than the remaining part *DB*, because *AD* is greater than *DC* and *DB* is greater than *DC*. Similarly, since the product of *CB* and *BD* is equal to the square of *AC*, the straight line *AC* is the mean proportional between the two straight lines *CB* and *BD* and the straight line *CB*, which is <the sum of> two of the parts of the straight line *AB*, is greater than the remaining straight line *AC*, because *BC*, the first <part>, is greater than *CD*, the third. Similarly, since <the sum> of the two straight lines *AC* and *DB* is greater than the straight line *DB* and since the straight line *DB* is greater than the straight line *CD*, the sum of the two straight lines *AC* and *DB* is greater than the remaining straight line *CD*. So <the sum> of two parts of the straight line *AB*, if this latter is divided as we have described, is greater than the remaining part. That is what we wanted to prove.

– **5** – We wish to find a straight line divided as we have described.

By synthesis, we take it that each of the two lines *AB* and *AC* is straight, that they are equal and of known magnitude and that they enclose a right angle; we extend each of them and we construct in the plane of the two straight lines *AB* and *AC* a parabola with *latus rectum* the straight line *AC*, with vertex the point *B* and axis the straight line *BA*, let the conic section be *BDE*. We also put in the same plane a hyperbola with axis the straight line *AC*, equal to its *latus rectum*, and with vertex the point *A*, let the conic section be *DAE*. We take each of the straight lines *EG* and *EH* parallel to each of the straight lines *AH* and *AG* and we put the straight line *IC* equal to one of the straight lines *AH* or *GE*.

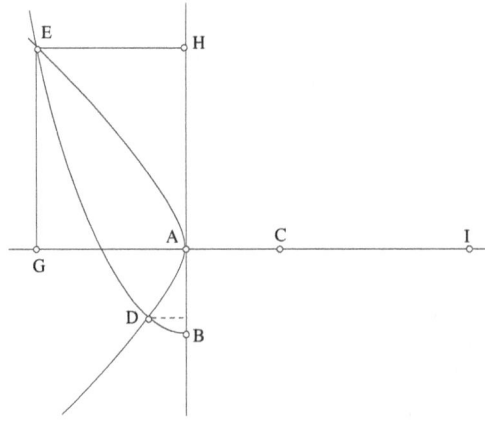

Fig. 5.5

I say that the straight line *GI* is divided at the points *C* and *A* such that the product of *IA* and *AC* is equal to the square of *AG* and the product of *CG* and *AG* is equal to the square of *IC*.

Proof: The straight line *IC* is equal to the straight line *AH* and the straight line *CA* is equal to the straight line *AB*, so the straight line *HB* is equal to the straight line *AI* and the product of *IA* and *AC* is equal to the product of *HB* and *AC*; but the product of *HB* and *AC* is equal to the square of *EH*, because *AC* is equal to the *latus rectum* of the parabola *BDE* and the straight line *EH* is an ordinate; so the product of *IA* and *AC* is equal to the square of *EH*; but the square of *EH* is equal to the square of *AG*, because the area *AHEG* is a parallelogram, so the product of *IA* and *AC* is equal to the square of *AG*.

Similarly, since the conic section *DAE* is a hyperbola, since its axis is the straight line *AC* which is equal to its *latus rectum*, as we have assumed, and since the straight line *EG* is an ordinate, accordingly the product of *CG* and *GA* is equal to the square of *EG*, because the ratio of the product of *CG* and *GA* to the square of *EG* is equal to the ratio of the diameter of the hyperbola to its *latus rectum*, as Apollonius showed in Proposition 20[1] of Book I of the *Conics*. So the product of *CG* and *GA* is equal to the square of *EG*; but the square of *EG* is equal to the square of *IC*, so the product of *CG* and *GA* is equal to the square of the straight line *IC*. But it has been shown that the product of *IA* and *AC* is equal to the square of the straight line *GA*, so the straight line *IG* has been divided at the points *C* and *A* such that the product of *IA* and *AC* is equal to the square of the straight line *AG* and the product of *CG* and *GA* is equal to the square of the straight line *IC*. That is what we wanted to prove.

– **6** – We wish to construct a triangle with rectilinear sides as we have described.

By synthesis, we find a straight line *AB* divided at the points *C* and *D* such that the product of *BC* and *CD* is equal to the square of *AC* and the product of *DA* and *AC* is equal to the square of *DB*. From three straight lines equal to the straight lines *AC*, *CD* and *DB*, we construct a triangle; this construction is easy and accessible because <the sum of> two of the straight lines is greater than the remaining one as we have shown; let the triangle be *CDE* such that *DE* is equal to *DB* and *CE* equal to *AC*.

I say that the angle *ECD* of the triangle *ECD* is twice the angle

<6> We wish to construct a triangle with rectilinear sides as we have described.

By synthesis, we find a straight line *AB* divided at the points *C* and *D* such that the product of *AD* and *CD* is equal to the square of *DB* and the product of *BC* and *DB* <is> equal to the square of *AC*. From three straight lines equal to the straight lines *AC*, *CD*, *DB*, we construct a triangle; his construction is easy and accessible, because <the sum of> two of the straight lines is greater than the remaining one, as we have shown; let the triangle be *CDE* such that *DE* is equal to *DB* and *CE* equal to *AC*.

I say that the angle *EDC* of the triangle *ECD* is twice the angle *ECD*

[1] This is Proposition 21 in Heiberg's edition and 21 in the translation of the Banū Mūsā.

EDC and that it is four times the angle *CED*.

and four times the angle *CED*.

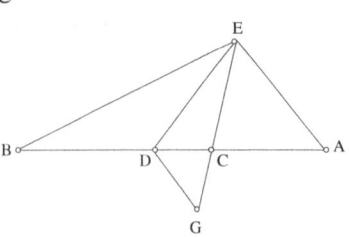

Fig. 5.6.1

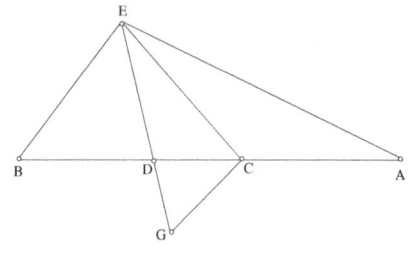

Fig. 5.6.2

Proof: We put the straight line *CG* equal to the straight line *CD* and let us join the straight lines *DG*, *EA*, *EB*. Since the product of *BC* and *CD* is equal to the square of *AC* and the square of *AC* is equal to the square of *CE*, accordingly the product of *BC* and *CD* is equal to the square of *CE*, so the ratio of *BC* to *CE* is equal to the ratio of *CE* to *CD*. But the angle *BCE* is common to the two triangles *BCE* and *DCE*, so the two triangles *BCE* and *DCE* are similar. The angle *CED* of one of the two triangles is thus equal to the angle *EBD* of the other triangle. But the angle *EDC* <which is> exterior to the triangle *EDB* is equal to <the sum of the> two equal interior angles *DEB* and *DBE* because the straight line *DB* is equal to the straight line *DE*; so the angle *EDC* is twice the angle *EBD*. But the angle *EBD* is equal to the angle *CED*, so the angle *EDC* of the triangle *EDC* is twice the angle *CED*. Similarly since the angle *ECD* is <an>

Proof: We put the straight line *DG* equal to the straight line *CD* and let us join the straight lines *CG*, *EA* and *EB*. Since the product of *AD* and *DC* is equal to the square of *BD* and the square of *BD* is equal to the square of *DE*, the product of *AD* and *DC* is equal to the square of *DE*, so the ratio of *AD* to *DE* is equal to the ratio of *DE* to *CD*. But the angle *ADE* is common to the two triangles *ADE* and *DCE*, so the two triangles *ADE* and *DCE* are similar. The angle *CED* of one of the two triangles is equal to the angle *EAD* of the other triangle. But the angle *ECD* <which is> exterior to the triangle *ECA* is equal to <the sum of the> two equal interior *CEA* and *CAE*, because the straight line *CA* is equal to the straight line *CE*; so the angle *ECD* is twice the angle *EAC*. But the angle *EAC* is equal to the angle *CED*, so the angle *ECD* of the triangle *EDC* is twice the angle *CED*. Similarly, since the angle *EDC* is <an> exterior <angle> of the triangle *CGD*, it is thus equal to <the

exterior <angle> of the triangle *CGD*, it is thus equal to <the sum of the> two equal interior angles *CDG* and *CGD*, because the straight line *CG* is equal to the straight line *CD*, thus the angle *ECD* is twice the angle *DGC*. Similarly, since the straight line *CE* is equal to the straight line *AC* and the straight line *GE* is equal to the straight line *DA* – in fact the straight line *CG* is equal to the straight line *CD* – the product of *GE* and *EC* is equal to the product of *DA* and *AC*. But the product of *DA* and *AC* is equal to the square of *DB*, so the product of *GE* and *EC* is equal to the square of *DB*; but the square of *DB* is equal to the square of *DE*, because the straight line *DE* is equal to the straight line *DB*, so the product of *GE* and *EC* is equal to the square of *DE*. So the ratio of *GE* to *ED* is equal to the ratio of *ED* to *EC*. But the angle *GED* is common to the triangles *GED* and *CED*, so the triangle *GED* is similar to the triangle *CED* and the angle *EDC* is equal to the angle *EGD*. But the angle *ECD* is twice the angle *DGE*, so the angle *ECD* is twice the angle *EDC*. But it has been shown that the angle *EDC* is twice the angle *CED*, so the angle *ECD* is four times the angle *CED*. Consequently one of the two angles of the triangle *ECD* is twice one of the remaining angles and four times the remaining angle. That is what we wanted to prove.

sum of the> two equal interior angles *DCG* and *CGD*, because the straight line *DG* is equal to the straight line *CD*, thus the angle *EDC* is twice the angle *DGC*. Similarly, since the straight line *DE* is equal to the straight line *BD* and the straight line *GE* equal to the straight line *CB* – in fact the straight line *DG* is equal to the straight line *CD* – thus the product of *GE* and *ED* is equal to the product of *CB* and *BD*. But the product of *BC* and *BD* is equal to the square of *CA*, so the product of *GE* and *ED* is equal to the square of *AC*; but the square of *AC* is equal to the square of *CE*, because the straight line *CE* is equal to the straight line *AC*; so the product of *GE* and *ED* is equal to the square of *CE*. So the ratio of *GE* to *EC* is equal to the ratio of *EC* to *ED*. But the angle *GEC* is common to the triangles *GEC* and *CED*, so the triangle *GEC* is similar to the triangle *CED* and the angle *ECD* is equal to the angle *EGC*. But the angle *EDC* is twice the angle *CGE*, so the angle *EDC* is twice the angle *ECD*. But it has been shown that the angle *ECD* is twice the angle *CED*, so the angle *EDC* is four times the angle *CED*. Consequently, one of the two angles of the triangle *ECD* is twice one of the two remaining angles and four times the third angle. That is what we wanted to prove.

– **7** – We wish to find the side of the regular heptagon inscribed in a given circle *ABC*.

By synthesis, we construct a rectilinear triangle one of whose angles is twice one of the two remaining angles and four times the other angle; let it be the triangle *DEG* such that the angle *DEG* is twice the angle *EGD* and four times the other remaining angle; we construct in the circle *ABC* the triangle *ABC*, let each of its angles be equal to one of the angles of the triangle *DEG*: the angle *ABC* to the angle *DEG*, the angle *ACB* to the angle *DGE* and the remaining angle to the remaining angle.

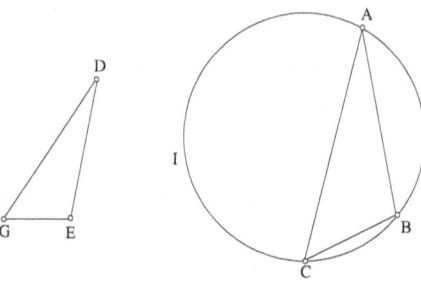

Fig. 5.7

I say that the straight line *BC* is the side of the regular heptagon inscribed in the circle *ABC*.

Proof: The angle *DEG* is twice the angle *DGE* and four times the angle *EDG*, so the angle *ABC* is twice the angle *ACB* and four times the angle *BAC*. So the arc *AIC* is twice the arc *AB* and four times the arc *BC* because the ratio of the arc to the arc in the same circle is equal to the ratio of the angle to the angle, whether these angles be on the circumference or at the centre. But the arc *AB* is twice the arc *BC*, so the whole arc *CIAB* is six times the arc *BC*, so the circumference of the circle is seven times the arc *BC*, so the straight line *BC* is the side of the regular heptagon inscribed in the given circle *ABC*. So we have constructed in the given circle *ABC* the side of the regular heptagon. That is what we wanted to prove.

The treatise is completed.

Treatise on the Construction of the Side of the Regular Heptagon Inscribed in the Circle by Abū Sahl al-Qūhī

Mathematicians agree in recognizing Archimedes is eminent and holds first place among the Ancients, because they have taken note of how many beautiful and advanced things he discovered, and <how many> difficult and abstruse propositions, in the highly valued demonstrative sciences; clear testimony is to be found in his surviving books, such as the book *On Centres of Gravity*, the book *On the Sphere and the Cylinder*, and the other books, each of which represents a summit beyond which nothing more is to be found. Thus they believed that what he had trouble in solving, and what he did not succeed in completing, <were matters> in which no one would ever find a way to achieve a solution, and no one else would find a route that led to the result – this is what they believed regarding the construction of the side of the regular heptagon inscribed in the circle, on the basis of what could be deduced from the book he had composed on the subject. It is a subtle book whose purpose he did not fulfil nor did he achieve his aim in finding the heptagon by a single method, not to speak of more than one; as God has given it to be accomplished by the subject of our Lord, Abū al-Fawāris ibn ʿAḍud al-Dawla, and by his servant, Wayjan ibn Rustam.

– **1** – We wish to find in a known circle *ABCD* the side of the inscribed regular heptagon.

By analysis, we suppose that each of the straight lines *AB* and *BC* is a side of the regular heptagon inscribed in the circle *ABCD*. Each of the arcs *AB* and *BC* is a seventh of the circumference of the circle *ABCD*. If we divide it up, each of the arcs *AB* and *BC* is a fifth of the arc *ADC*, so the arc *ADC* is five times each of the arcs *AB* and *BC* and the angle *ABC* is five times each of the angles *BAC* and *BCA*, because the ratio of the arc to the arc in the circle is equal to the ratio of the angle to the angle, whether the

angles are on the circumference or whether they are at the centre. The
triangle *ABC* is isosceles and the angle *ABC* is five times each of the
remaining angles *BAC* and *BCA*. So we have reduced this <problem> to the
construction of an isosceles triangle in which one of the angles is five times
each of the two remaining angles.

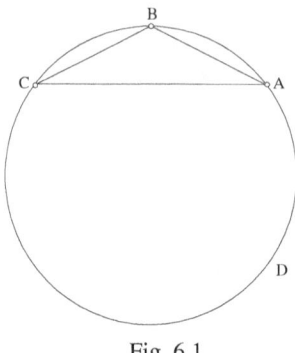

Fig. 6.1

<2> Thus we wish to construct an isosceles triangle in which one of the
angles is five times each of the two remaining angles.

By analysis, we suppose we have an isosceles triangle *ABC* in which
the angle *ABC* is five times each of the angles *BAC* and *BCA*; we suppose
that the line *CBDE* is straight, that the angle *BAD* is equal to each of the
angles *BAC* and *BCA* and that the straight line *DE* is equal to the straight
line *DA*.

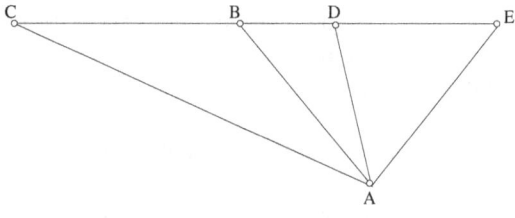

Fig. 6.2

Since the angle *ACD* is equal to the angle *BAD* and the angle *ADB* is
common to the triangles *ACD* and *ABD*, the remaining angle is equal to the
remaining angle and the triangle *ACD* is similar to the triangle *ABD*. So the
ratio of the straight line *CD* to the straight line *DA* is equal to the ratio of

the straight line *DA* to the straight line *DB*, so the product of *CD* and *DB* is equal to the square of the straight line *DA*. But the straight line *DA* is equal to the straight line *DE*, so the product of *CD* and *DB* is equal to the square of the straight line *DE*.

Similarly, since the angle *ABC* is five times the angle *BCA* and the angle *BCA* is equal to the angle *BAD*, the angle *ABC* is equal to five times the angle *BAD*. But the angle *ABC* is equal to the sum of the angles *BAD* and *BDA*, so the sum of the angles *BDA* and *BAD* is five times the angle *BAD*. If we separate <ratios>, then the angle *BDA* will be four times the angle *BAD*; but the angle *ABD* is twice the angle *BCA*, because the triangle *ABC* is isosceles and the angle *ABD* is an exterior <angle> of the triangle *ABC*. But the angle *BCA* is equal to the angle *BAD*, so the angle *ABD* is twice the angle *BAD* and the angle *ADB* is four times the angle *BAD*. That is why the angle *ADB* is twice the angle *ABD* and the angle *ADB* is twice each of the angles *DAE* and *DEA*, because the triangle *ADE* is isosceles and the angle *ADB* is an exterior <angle> of it; so the angle *ABD* is equal to each of the angles *DAE* and *DEA*. But since the angle *ABD* is equal to the angle *DAE* and the angle *AED* is common to the two triangles *ABE* and *ADE*, accordingly the remaining angle is equal to the remaining angle and the two triangles *AEB* and *ADE* are similar; so the ratio of *BE* to *EA* is equal to the ratio of the straight line *EA* to the straight line *ED* and the product of *BE* and *ED* is equal to the square of the straight line *EA*. But the straight line *EA* is equal to the straight line *AB*, because the angle *AEB* is equal to the angle *ABE* and the straight line *AB* is equal to the straight line *CB* because the triangle *ABC* is isosceles; so the product of *BE* and *ED* is equal to the square of the straight line *BC*. But the product of *CD* and *DB* is equal to the square of the straight line *DE*. So there must be a certain straight line divided in the ratio of *CB*, *BD* and *DE* so that the product of *CD* and *DB* is equal to the square of *DE* and the product of *BE* and *ED* is equal to the square of *BC*. So we have reduced <the problem> to finding a straight line divided in this ratio.

<3> We wish to find a straight line divided in this ratio.

By analysis, we suppose that the straight lines *AB*, *BC*, *CD* are in this ratio, that is to say that the product of *AC* and *CB* is equal to the square of *CD* and the product of *BD* and *DC* is equal to the square of *AB*. We suppose that the straight line *CE* is equal to the straight line *CD*, the straight line *BG* is equal to the straight line *AB*, they are parallel and make a known angle with the straight line *CA*, the angle *ABG* is divided into two equal parts by the straight line *KBI* and the line *ECI* is straight. Since the product of the straight line *AC* and the straight line *CB* is equal to the square of *CD* and the square of the straight line *CD* is equal to the square of the straight line *CE*, the product of *AC* and *CB* is equal to the square of the straight line *CE*. But the angle *ECB* is known, so the point *E* lies on the outline of the hyperbola – which is *BE* – whose transverse diameter is the straight line *AB*, and whose *latus rectum* is equal to its transverse diameter – which is the straight line *AB* – and the angle of the ordinates is the angle *ECB*.

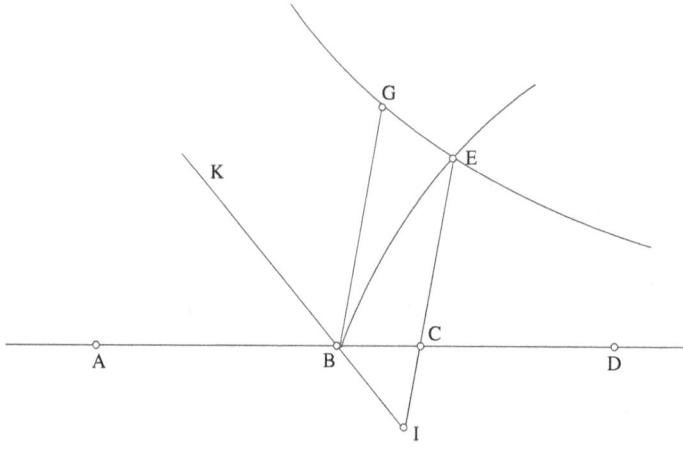

Fig. 6.3

Similarly, the straight line *BC* is equal to the straight line *CI*, because the angle *CBI* is equal to the angle *BIC* and the straight line *CD* is equal to the straight line *CE*; so the product of *IE* and *EC* is equal to the square of *BA*, that is to say to the square of *BG*, because the two straight lines *AB* and *BG* are equal. So the product of *IE* and *EC* is equal to the square of the straight line *GB*. The point *E* also lies on the outline of the hyperbola which is *GE*, whose asymptotes are the two straight lines *BD* and *BK* and which passes through the point *G*. If we take *AB* of known magnitude and known in position, all these straight lines will be known in position, the point *G* is known and each of the conic sections *BE* and *GE* is known in position, so the point *E* is known, and the point *C* is known, because the angle *EBC* is known; so each of the points A, B, C, D is known. That is what we wished to know.

<4> If the straight line AB is divided at the points C and D as we have described, that is to say points such that the product of the straight line BC and the straight line CD is equal to the square of the straight line AC and the product of the straight line DA and the straight line AC is equal to the square of the straight line DB, I say that the sum of two of the straight lines AC, CD and DB is greater than the remaining straight line.

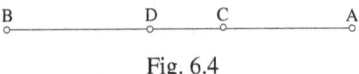

Fig. 6.4

Proof: Since the product of BC and CD is equal to the square of AC, the ratio of BC to CA is equal to the ratio of CA to CD. But the straight line BC is greater than the straight line CD, so the straight line BC is greater than the straight line CA, and since the straight line CA is the mean proportional between the straight lines BC and CD, the straight line CA is accordingly greater than the straight line CD; but the straight line BC is two parts of the three parts, that is to say the sum[1] of the straight lines BD and DC, so it is greater than the remaining part, which is AC.

Moreover, since the straight line AC is greater than the straight line DC, it will be, with <the addition of> the straight line DB, much greater than the straight line CD, so the sum of the two straight lines AC and DB is greater than the remaining straight line CD. Similarly, since the product of the straight line DA and the straight line AC is equal to the square of DB, the straight line DB is accordingly the mean proportional between the two straight lines DA and AC. It is for this reason that the straight line DA is greater than the straight line DB; but the straight line DA is two parts of the three parts, so the sum of the two parts AC and CD is greater than the remaining part DB. The straight line AB being divided into parts, as we have described, then the sum of two of the parts is greater than the remaining part. That is what we wanted to prove.

---

[1] We have added the word 'sum' to conform with English usage.

<5> We wish to find a certain straight line that can be divided in this ratio.

By synthesis, we take the straight line *AB*, known in position and in magnitude, and the straight line *AC*, equal to the straight line *AB*, making a known angle: we divide the angle *CAB* into two equal parts by means of the straight line *DAE*, and we draw *BAG* to be a straight line. We construct through the point *A* a hyperbola whose transverse diameter is the straight line *AB* and whose *latus rectum* is equal to the straight line *AB*; so the angle of its ordinates is equal to the angle *CAB*; let *AI* be the outline of the conic section. Again, we construct through the point *C* a hyperbola whose asymptotes are the straight lines *AD* and *AG*; let the conic section be *CI*; so the conic sections cut one another at the point *I*. We draw the straight line *IKE*, its ordinate, and we put the straight line *KG* equal to the straight line *KI*.

I say that the parts of the straight line *BG* <divided> at the points *A* and *K* are as we wished, that is to say that the product of the straight line *BK* and *KA* is equal to the square of *KG* and the product of the straight line *AG* and *GK* is equal to the square of the straight line *AB*.

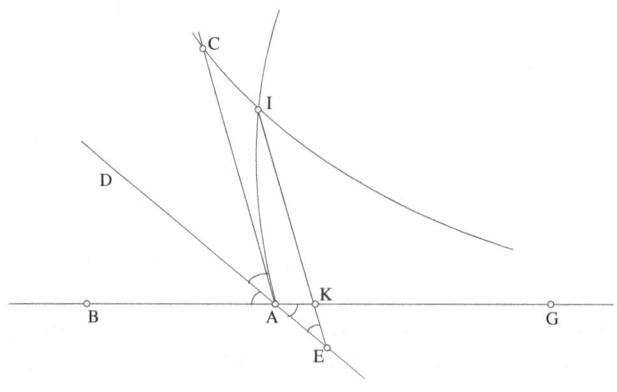

Fig. 6.5

Proof: Since the conic section *AI* is a hyperbola, its transverse diameter the straight line *AB*, which is equal to its *latus rectum*, and the ratio of the product of *BK* and *KA* to the square of *KI* is equal to the ratio of its transverse diameter to its *latus rectum*, which are equal, accordingly the product of *BK* and *KA* is equal to the square of *KI*. But the square of *KI* is equal to the square of *KG*, because the two straight lines *KI* and *KG* are equal. So the product of *BK* and *KA* is equal to the square of *KG*.

Similarly, the straight line *KA* is equal to the straight line *KE* because the angle *KEA* is equal to the angle *KAE* and the straight line *KG* is equal to the straight line *KI*; so the straight line *AG* is equal to the straight line *EI*. So the product of *AG* and *GK* is equal to the product of *EI* and *IK*; but the product of *EI* and *IK* is equal to the square of *AC*, since the two straight lines *AD* and *AG* do not meet the hyperbola *CI* that passes through the point *C* and since the straight line *CA* is parallel to the straight line *IKE*; so the product of *AG* and *GK* is equal to the square of *AC*; but the square of *AC* is equal to the square of *AB* because the lines are equal. It is for this reason that the product of *AG* and *GK* is equal to the square of *AB*. But the product of *BK* and *KA* is equal to the square of *KG*. So we have found the straight line *BG* divided at the points *K* and *A* such that the product of *BK* and *KA* is equal to the square of *KG* and the product of *AG* and *GK* is equal to the square of *AB*, as we described. That is what we wanted to prove.

<6> We wish to construct an isosceles triangle in which one of the angles is five times each of the remaining angles, as we have said in the analysis.

By synthesis, we find a straight line *AB* divided at the points *C* and *D* such that the product of *BC* and *CD* is equal to the square of *CA* and the product of *DA* and *AC* is equal to the square of *DB*, <a straight line> whose construction we have shown above.

Since the sum of two of the straight lines *AC*, *CD*, *DB* is greater than the remaining straight line, accordingly we construct a triangle from three straight lines equal to the <straight lines> *AC*, *CD*, *DB*, let it be the triangle *CDE*, and we join the straight line *BE*.

I say that the triangle *BDE* is isosceles and the angle *BDE* is five times each of the angles *DEB* and *DBE*.

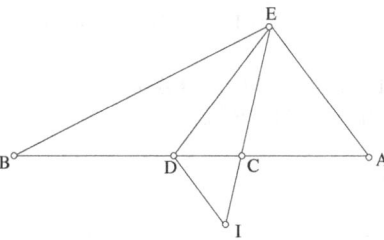

Fig. 6.6

Proof: We join the straight line *AE* and we make the straight line *DI* parallel to the straight line *AE*, and the straight line *ECI*, so the straight line *CI* is equal to the straight line *CD* – because the triangle *ACE* is isosceles – and the product of *IE* and *EC* is equal to the product of the straight line *DA* and *AC*. But the product of the straight line *DA* and *AC* is equal to the square of *DB*, that is to say to the square of the straight line *ED,* because the two straight lines *DB* and *DE* are equal. So the product of *IE* and *EC* is equal to the square of *ED*, thus the ratio of *IE* to *ED* is equal to the ratio of *ED* to *EC*; but the angle *DEC* is common, so the angle *EDC* is equal to the angle *EID*. But the angle *ECD* is twice the angle *CID*, because the triangle *CID* is isosceles. So the angle *ECD* is twice the angle *EDC* and the angle *EDC* is twice the angle *EBD*, because it is <an> exterior <angle> of the isosceles triangle *EDB*. But the angle *EBD* is equal to the angle *DEC*,[2] so the angle *EDC* is twice the angle *DEC* and the angle *ECD* is four times the angle *CED*. If we combine, the sum of the angles *ECD* and *CED* is five times the angle *CED*; but the angle *EDB*, an exterior <angle> of the triangle *CED*, is equal to the sum of the angles *ECD* and *DEC*, and the angle *CED* is equal to the angle *EBD*, so the angle *EDB* is five times the angle *EBD*. But the angle *EBD* is equal to the angle *DEB*, because the triangle *EDB* is isosceles. So the triangle *EDB* is isosceles and one of its angles, which is the angle *EDB*, is five times each of the angles *DEB* and *DBE*. So we have constructed an isosceles triangle in which one of the angles is five times each of the two remaining angles; it is the triangle *EDB*. That is what we wanted to prove.

<7> We wish to construct in the known circle *ABCD* the side of a regular heptagon.

By synthesis, we construct the isosceles triangle *EGI* in which the angle *EGI* is five times each of the remaining angles *GEI* and *EIG*, as we have shown above. In the circle *ABCD* we inscribe a triangle *ABC* similar to the triangle *EGI* and such that the straight line *AC* is homologous to the straight line *EI*.

I say that each of the straight lines *AB* and *BC* is a side of the regular heptagon inscribed in the circle *ABCD*.

---

[2] The equality $E\hat{B}D = D\hat{E}C$ is a consequence of the hypothesis $BC \cdot CD = AD^2 = EC^2$ which means that the triangles *BCE* and *ECD* are similar.

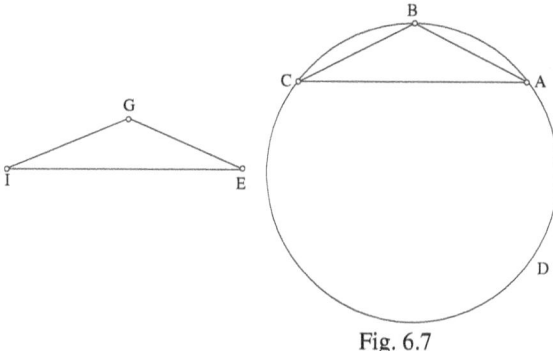

Fig. 6.7

Proof: Since the angle *EGI* is five times each of the angles *GEI* and *GIE*, the angle *ABC* is five times each of the angles *BAC* and *BCA*. So the arc *ADC* is five times each of the arcs *AB* and *BC*, because the ratio of the arc to the arc in the circle is equal to the ratio of the angle to the angle, whether these angles are on the circumference or at the centre. If we combine the arcs, the whole circumference of the circle *ABCD* is seven times each of the arcs *AB* and *BC*. So each of the arcs *AB* and *BC* is a seventh of the circumference of the circle *ABCD*. It is for this reason that each of the straight lines *AB* and *BC* is the side of the regular heptagon inscribed in the circle *ABCD*. So we have constructed in the circle *ABCD* the side of the regular heptagon, which is *AB* or *BC*. That is what we wanted to prove.

The treatise on the construction of the side of the regular heptagon inscribed in the circle is completed.

In the Name of God, the Compassionate, the Merciful

TREATISE BY AḤMAD IBN MUḤAMMAD IBN AL-ḤUSAYN AL-ṢĀGHĀNĪ FOR HIS MAJESTY THE KING ʿAḌUD AL-DAWLA IBN ABĪ ʿALĪ RUKN AL-DAWLA

Since the Good is something one seeks ardently for its own sake, the Good is thus what one truly seeks, and he who seeks it is blessed in the absolute sense. Among the qualities of the man who is blessed are good conduct and perfection in his actions. If this is as we have said, if this defines the man who does good and the man who is blessed, then our Lord the victorious Great King, ʿAḍud al-Dawla, may God give him long life, is in truth a man who does good and is blessed, he whose high actions and virtuous life is certainly known to all. Eyes are drawn to him, and hearts agree in obeying him.

Among the blessings of kings and rulers there is the emergence of sciences that take shape in their time. The determination of the chord of the heptagon resisted the efforts of geometers. Archimedes had stated a lemma whose proof would have made it possible to find the chord of the heptagon. And this is the way the problem was passed down to our own time. Then a solution to it, using fixed geometry, came to Aḥmad ibn Muḥammad ibn al-Ḥusayn al-Ṣāghānī, who accomplished this in the reign of the victorious Great King, ʿAḍud al-Dawla. May God give him long life and heap happiness upon His Highness; his are the times we are proud to live in, and we are proud of the virtues that characterize him and for which he is celebrated. I pray to God that he may lengthen his days and that he may continue his blessings for ever; may he fulfil the hopes that scholars have of him; may he help them to gain their due reward for what service they can render, each according to what he can do.

When I was living in Rayy, I had sent this problem to his library, which prospers thanks to the favourable effect of his rank and benefits by his good fortune. Now, I have given it another form in which I have shown how to reduce the problem to the lemma, then I have recast it as synthesis; and through this <essay> have rendered a service to the King, my great victorious Lord – may God give him long life. It is from God that we beg for assistance, and it is from him that we ask for help. It is he who brings us fulfilment, and we entrust ourselves to him, our best support.

– **1** – The circle *ABC* is known: how should we proceed to inscribe in it a regular heptagon?

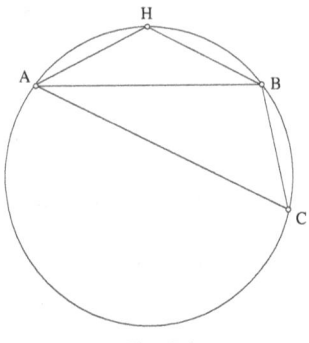

Fig. 7.1

Let us suppose, by analysis, that this has been done. Let the straight lines *AH*, *BH*, *BC* be some of the sides and let us imagine *AB* and *AC* to have been joined. Since the arc *AB* is twice the arc *BC*, the angle *ACB* is twice the angle *BAC*, and since the arc *AC* is twice the arc *AB*, the angle *ABC* is twice the angle *ACB*. So if we construct a triangle whose angles are in proportion with the ratio of doubling, then the problem is solved.

– **2** – We suppose, by analysis, that in the triangle *ADU* the angle *AUD* is twice the angle *ADU* and the angle *ADU* is twice the angle *DAU*. Let us imagine that we have extended *DU* to *H* and to *B*, so that *DH* is equal to *AD* and *UB* equal to *AU*, and *AH* and *AB* are joined. It is clear that the angle *AUD* is twice the angle *ABU*, so the angle *BAU* is equal to the angle *ADU* and the angle *B* is common, so the triangle *ADB* is similar to the triangle *AUB*, and the ratio of *DB* to *AB* is equal to the ratio of *AB* to *BU*. But *AB* is equal to *AD*, because the angle *D* is equal to the angle *B*; but *AD* is equal to *DH*, so the ratio of *BD* to *DH* is equal to the ratio of *DH* to *BU*. Similarly, since the angle *ADU* is twice each of the angles *DAU* and *AHU*, the angle *AHU* is equal to the angle *DAU*. But the angle *AUD* is common, so the triangle AUD is similar to the triangle AUH, and the ratio of UH to AU, that is to say to BU, is equal to the ratio of AU, that is to say BU, to UD.

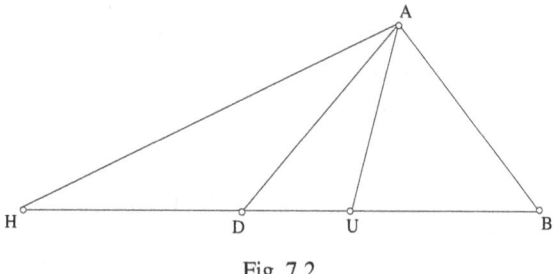

Fig. 7.2

The analysis of this problem has led us to find a straight line <divided> into these parts.

– 3 – Let us take the straight line AB on which there are the two points D and E; let us suppose that the ratio of AD to DB is equal to the ratio of DB to AE and the ratio of BE to AE is equal to the ratio of AE to ED. We construct on AD the square AI, we join the diagonal AI and we join FB.

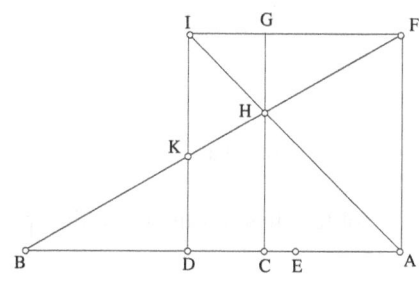

Fig. 7.3

I say that if we draw from the point H a perpendicular to AB, then it ends at the point E.

If it were otherwise, let it end at the point C. We extend it to G. It is clear that GI is equal to GH, similarly CH is equal to AC, so the ratio of BC to CH, that is to say to AC, is equal to the ratio of FG, that is to say AC, to GH, that is to say to GI, that is to say to CD, and the ratio of BC to AC is equal to the ratio of *AC* to *DC*. By composition and permutation, we have that the ratio of *AB* to *AD* is equal to the ratio of *AC* to *CD*. But since we have supposed that the ratio of *BE* to *AE* is equal to the ratio of *AE* to *ED*, by composition and permutation, we have that the ratio of *AB* to *AD*, which was equal to the ratio of *AC* to *CD*, is equal to the ratio of *AE* to *ED*, so the ratio of *AC* to *CD* is equal to the ratio of *AE* to *ED*; which is absurd. Consequently, the perpendicular drawn from *H* ends at *E*. That is what we wanted to prove.

– **4** – Let us return to the same figure; let the perpendicular be *GHE*. Since we have supposed that the ratio of *AD* to *BD*, that is to say the ratio of *FK* to *KB*, is equal to the ratio of *BD* to *AE*, that is to say to the ratio of *BK* to *FH*, the ratio of *FK* to *KB* is consequently equal to the ratio of *KB* to *FH*. But the ratio of *FK* to *KB* is equal to the ratio of *FI* to *DB*, so the ratio of *KB* to *FH* is equal to the ratio of *FI* to *DB*. But the angle *IFH* is equal to the angle *B*, so the triangle *BDK* is equal to the triangle *FIH*.

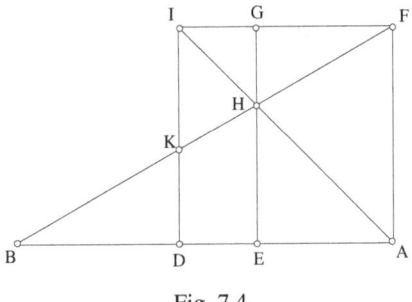

Fig. 7.4

The analysis of this problem has led us to another proposition, which is:

– **5** – Let there be the square *ABCD*, we extend the straight line *BD* on the side of the point *D* and we join the diagonal *BC*; we wish to draw a straight line from the point *A*, say the straight line *AH*, so that the triangle *AEC* is equal to the triangle *GDH*.

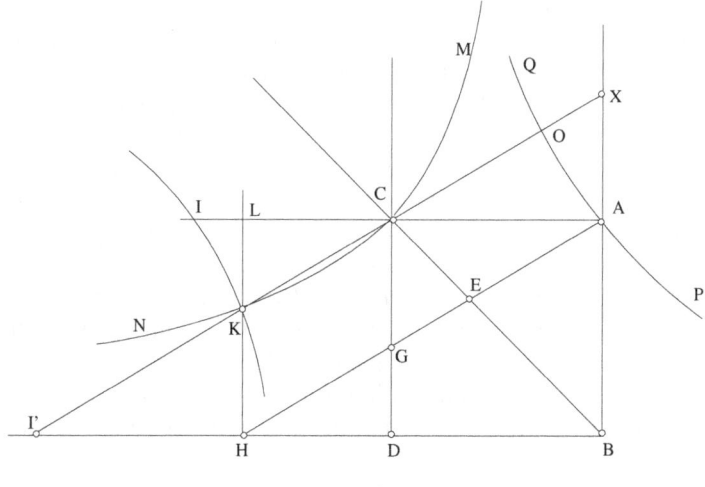

Fig. 7.5

Let us suppose by analysis that this has been done and that the straight line *AH* solves the problem.[1] Let us imagine that we have completed the rectangular area *DL* and let us imagine the straight line *XI′* drawn parallel to the straight line *AH*; *AC* will thus be equal to *I′H* and *XC* will be equal to *KI′* because the two triangles *AXC* and *KI′H* are similar. Since the triangle *AEC* is equal to the triangle *GDH* and the angle *EAC* is equal to the angle *GHD*, their sides are inversely proportional, and the ratio of *GH* to *AE* is equal to the ratio of *AC* to *DH*. But the ratio of *AC* to *DH* is equal to the ratio of *AG* to *GH*, so the ratio of *AG* to *GH* is equal to the ratio of *GH* to *AE*, and the product of *AG* and *AE* is equal to the square of *GH*. Now the product of *AG* and *AE* is smaller than the square of *AC*, because the perpendicular drawn from the point *C* to *AG* ends at a point between the two points *E* and *G*; so *AC* is longer than *GH*, that is to say <longer than> *CK*; but *XC* is longer than *AC*, consequently *XC* is much longer than *CK*. Let us imagine we have *OC* equal to *KC*, consequently the product of *AG* and *AE* will be equal to the square of *OC*. So if we construct a hyperbola that passes through the point *A* and is such that the asymptotes are *CB* and *CD*, that is the conic section *PQ*, it passes through the point *O*, as Apollonius showed in the seventh proposition of the second book of the

---

[1] Lit.: makes the problem.

work[2] the *Conics*, and the conic section *PQ* will be known in position. But since we have shown that *XC* is equal to *KI'*, if we construct a hyperbola that passes through the point *C* and has asymptotes *XB* and *BI'*, it passes through the point *K*, as Apollonius showed in the sixth proposition of the second book of the work the *Conics*[3] and it is known in position. Let this conic section be *MN*. We have also shown that *AC* is longer than *GH*; so it is longer than *CL*. We make *CI* equal to *AC*. If we construct through the point *I* the hyperbola that is opposite to the conic section *PQ*, it passes through the point *K*, as Apollonius showed in the thirty-first proposition of the first book of the work of the *Conics*,[4] and it too is known in position. Let *IK* be this conic section; so the conic section *IK* is known in position; now the conic section *MN* was too; so the point *K* is known and the point *C* also. So the straight line *CK* is known in position, the point *A* is known and *AH* is parallel to *CK*; so *AH* is known in position. That is what we wanted to prove.

– **6** – Synthesis corresponding to that analysis: We take the square *ABCD*, we extend *BD*, we join *BC* and we take *CI* equal to *AC*. We construct two opposite hyperbolas[5] that pass through the points *A* and *I* and are such that their asymptotes are *BC* and *CD*; let them (the conic sections) be *PQ* and *KI*. We construct a hyperbola that passes through the point *C* and is such that the straight lines *AB* and *BD* are its asymptotes; let the conic section be *MN*. The two conic sections *MN* and *IK* necessarily cut one another in a point between *AI* and *BH*, because, indeed, if it is extended the straight line *BD* cuts the conic section *IK*; let them cut one another at the point *K*. We join *KC*, we draw the perpendicular *KH* to *BD* and we join *HA*.

I say that the triangle *GDH* is equal to the triangle *AEC*.

Proof: We extend the straight line *KC* to the points *X* and *I'*. It is clear that *XC* is equal to *KI'*[6] and that the triangle *AXC* is similar to the triangle *KI'H*; so *I'H* is equal to *AC* and they are parallel; so *AH* is parallel to *XI'*. But since the points *O* and *K* lie on the two opposite conic sections *PQ* and

---

[2] This is Proposition 11 of Book II.

[3] This is Proposition 8 of Book II.

[4] This is Proposition 30 of Book I.

[5] Lit.: two hyperbolic sections.

[6] From Proposition 8 of Book II; the author does not give a reference.

IK, accordingly *OC* will be equal to *KC*, as Apollonius showed in the thirty-first proposition of the first book of the work the *Conics*.[7] And since the two straight lines *CB* and *CD* do not meet the conic section *PQ* and we have drawn *AG* and *OC* to be parallel, accordingly the product of *GA* and *AE* is equal to the square of *OC*, as Apollonius showed in the seventh proposition of the second book of the work the *Conics*,[8] that is to say the square of *GH*; so the product of *AG* and *AE* is equal to the square of *GH*. So the ratio of *AG* to *GH* is equal to the ratio of *GH* to *AE*. But the ratio of *AG* to *GH* is equal to the ratio of *AC* to *DH*, because the triangles *AGC* and *GHD* are similar, so the ratio of *AC* to *DH* is equal to the ratio of *GH* to *AE*; the two angles *EAC* and *GHD* are equal, and the triangle *AEC* is thus equal to the triangle *GDH*. That is what we wanted to prove.

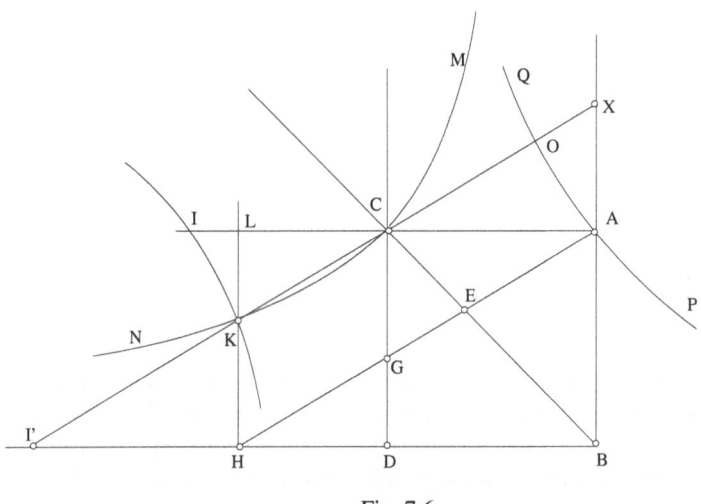

Fig. 7.6

– 7 – Once that has been proved, we take the square *ABCD*, together with the straight line *AH*, and we draw the perpendicular *OEU*.

Since the angle *OCE* is half a right angle and the angle *O* is a right angle, *OC* is equal to *OE* and similarly *EU* is equal to *BU*. But since we have shown that the ratio of *AG* to *GH* is equal to the ratio of *GH* to *AE*,

[7] This is Proposition 30 of Book I.
[8] This is Proposition 11 of Book II.

accordingly the ratio of *BD* to *DH* is equal to the ratio of *DH* to *BU*. Similarly, the ratio of *HU* to *UE*, that is to say to *UB*, is equal to the ratio of *AO* to *OE* because the triangles *AOE* and *EUH* are similar. But *OE* is equal to *OC*, that is to say to *UD*, so the ratio of *UH* to *UB* is equal to the ratio of *UB* to *UD*. But since we have shown in our last proposition that *BD* is longer than *DH*,[9] the sum of *BU* and *UD* is longer than *DH*. Similarly we have shown here that *BU* is the mean proportional between *HU* and *UD*; so *BU* is smaller than the sum of *DU* and *DH*; but *BU* is longer than *UD*, so the sum of *BU* and *DH* is longer than *DU*, and <the sum> of two of the three straight lines *BU*, *UD* and *DH* is longer than the third; so they can be used to construct a triangle. Similarly, we can construct a triangle whose sides are equal to the straight lines *AE*, *EG*, *GH* starting from the straight line *AH* and the points *E* and *G*.[10]

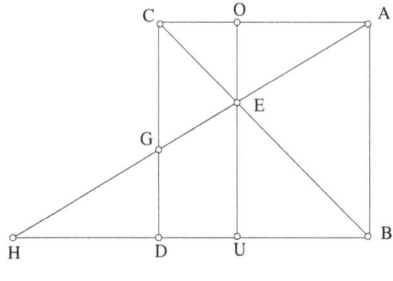

Fig. 7.7

We take the straight line *BH* with the points *D* and *U* and we construct the triangle *AUD* such that *BU* is equal to *AU* and *AD* equal to *DH*.

I say that the angles of the triangle *AUD* are in proportion in double ratio, that is to say that the angle *UAD* is half the angle *ADU* and the angle *ADU* is half the angle *AUD*.

Proof: We join *AB* and *AH* and we extend *AU* to <the point> *G* such that *GU* is equal to *DU*. It is clear that the angle *DAH* is equal to the angle *DHA*; so the angle *UDA* is twice the angle *DAH*. But since the ratio of *UH* to *UB*, that is to say to *UA*, is equal to the ratio of *UA* to *UD*, the triangle *AUH* is similar to the triangle *AUD* and the angle *UAD* is thus equal to the angle *AHD*. But the angle *ADU* is twice the angle *AHD*, so the angle *ADU*

---

[9] It has been shown that *AC* > *GH*; but *AC* = *BD* and *GH* > *DH*, so *BD* > *DH*.

[10] The ranges (*B*, *U*, *D*, *H*) and (*A*, *E*, *G*, *H*) are similar because *AB* ∥ *EU* ∥ *GD*.

is twice the angle *DAU*. Similarly, the ratio of *BD* to *DH*, that is to say of *AG* to *AD*, is equal to the ratio of *DH*, that is to say *AD*, to *BU*, that is to say to *AU*; so the triangle *ADG* is similar to the triangle *AUD* and the angle *ADG* is equal to the angle *AUD*; but the angle *AUD* is equal to the sum of the angles *AGD* and *UDG*, and since *UD* is equal to *UG*, the angle *ADU* is equal to the angle *AGD*, and the angle *AUD* is thus twice the angle *ADU*. That is what we wanted to prove.

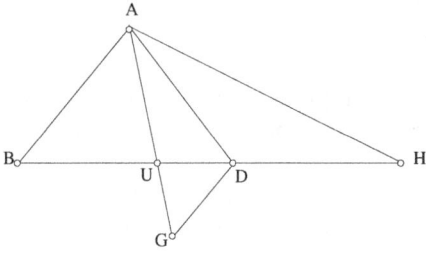

Fig. 7.8

The same is true for the angles of the triangle *AUD* and those of the triangle *AUH*.

– **8** – Once these lemmas have been proved, we take the circle *ABC* and in it we wish to inscribe a regular heptagon.

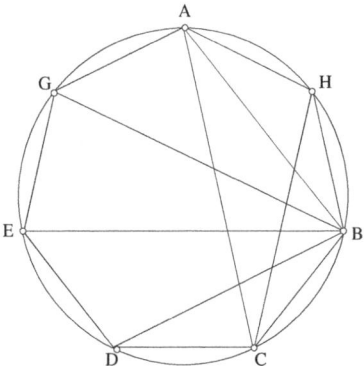

Fig. 7.9

We inscribe in it a triangle similar to the triangle *AUD*, which is the triangle *ABC*, such that the angle *ABC* is the homologue of the angle *AUD* and the angle *ACB* is the homologue of the angle *ADU*. We divide up the angle *ABC* into four equal parts by the straight lines *BG*, *BE*, *BD*; we divide up the angle *ACB* into two equal parts by the straight line *CH* and we join the straight lines *AH*, *BH*, *AG*, *GE*, *ED* and *DC*. Since the angles *ACH*, *BCH*, *ABG*, *GBE*, *EBD*, *DBC* are equal and equal to the angle *BAC*, the arcs are equal and the heptagon *AGEDCBH* is regular. That is what we wanted to construct.

The problem is completed. Thanks be rendered to God. May the blessing and the grace of God be upon Muḥammad and those that are his.

I solved this problem on Saturday the 12th day of the month of Shawwāl in the year 360; the 27th day of the 5th month <340 in the era of Yazdegerd>.

<Copy> completed at the fortress of Hamadhān, on Saturday 15.7.544 of the Hegira, from a copy in the hand of Aḥmad ibn Muḥammad ibn ʿAbd al-Jalīl al-Sijzī.

Thanks be rendered to God. May the blessing and the grace of God be upon our master Muḥammad and those that are his.

In the Name of God, the Compassionate, the Merciful

Book on the Discovery of the Deceit of Abū al-Jūd concerning the Two Lemmas he Introduced in order, as he Claimed, to Construct the Heptagon
by

ABŪ ʿABD ALLĀH MUḤAMMAD IBN AḤMAD AL-SHANNĪ

Since geometry holds the highest rank among the sciences, those who distinguish themselves in it are very few in number, though one loses count of those who make the attempt, so great is their number. It has been said that those truly learned in this art, that is to say in the science of geometry, are three: Euclid, Archimedes and Menelaus. Euclid, for his part, was the first to bring together the fundamentals of geometry; he put them in order, and marked out the routes to get to them; he made them accessible, to such a point that it is from his time onwards that this science has developed. As regards Archimedes – thanks to his zeal in this science and his working on abstruse notions like the study of mechanics and the instruments required – he attained such a peak of achievement that the Greeks called him 'the Geometer'. No other, neither among the ancients nor among the moderns, has deserved that name, such is his superiority and his pre-eminence. He set out a proposition and introduced it for constructing the regular heptagon in the circle; but, since he was not able to complete it from the fundamentals of geometry, he left it as it was, while nevertheless showing that, if it is granted, then, from that fact, we <can> construct the heptagon; in this he followed the example set by Euclid, when the latter was not able, using the fundamentals he had assembled, to find the chord of the heptagon in the circle, nor to find a third of an angle with straight sides, which, if it were found, would allow one to find the chord of the heptagon in the circle; he did not mention it, and did not introduce any statement (of a proposition). However, it is far from true that either he or Archimedes was unable to do this, or that <in doing as they did> they were following some other example.[1] Archimedes did the same in his book on *The Sphere and the Cylinder*, where he wished to divide the sphere in a given ratio by the plane of a circle; he needed to divide the diameter of the sphere in the said ratio – this is Proposition 4 of the second book of the work – something he could not carry out using the fundamentals from Euclid; he then gave the ratio, but left out the construction, but later Eutocius of Askelon wrote a commentary on this book and divided the diameter in this ratio with the help of two intersecting conic sections, a hyperbola and a parabola.

The proposition he (Archimedes) introduced to construct the heptagon is the following:

Let there be a square *ABCD*, we draw its diagonal *AC*, we extend the side *AB* to infinity on the side of *B*: how do we draw from the point *D* a straight line, for instance the straight line *DGHE*, so that the triangle *DGC* is equal to the triangle *HBE*?

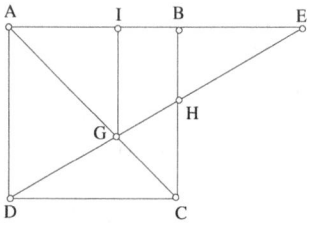

Fig. 8.1

Archimedes in fact wished to drop the perpendicular *GI* to *AB* and in this way divide the straight line *AE* at the points *I* and *B* so that the product of *AB* and *AI* is equal to the square of *BE* and the product of *EI* and *IB* is equal to the square of *AI*.

Proof: In the triangle *DGC* which is equal to the triangle *BHE* the angle *GDC* is equal to the angle *E* of the triangle *BHE*. The sides which enclose the two equal angles are inversely proportional: so the ratio of *BE* to *DC* is equal to the ratio of *DG* to *EH*. But the ratio of *BE* to *DC* is equal to the ratio of *EH* to *DH*, so the ratio of *EH* to *HD* is equal to the ratio of *DG* to *HE*. But since the parts of the straight line *DE* are in the ratio of the parts of

---

[1] Lit.: to imitate each of them in a thing. The Arabic sentence is ambiguous. See the note on the verb *qallada*.

the straight line AE,[2] accordingly the ratio of EB to BA is equal to the ratio of AI to EB and the product of AB and AI is thus equal to the square of BE; similarly, the ratio of EI to IG, that is to say AI, is equal to the ratio of AE to AD, that is to say AB. By permutation, the ratio of EI to AE is equal to the ratio of AI to AB and, by separation, the ratio of EI to AI is equal to the ratio of AI to IB, so the product of EI and IB is equal to the square of AI.

But since the straight line BE is the mean proportional between the straight lines AB and AI, the straight line BE is smaller than the sum of AI and IB. The straight line AI is also a mean proportional between the straight lines IE and IB, so AI is smaller than the sum of the straight lines IB and BE. But the straight line IB is smaller than each of the straight lines AI and BE, thus the sum of the straight lines AI and BE is greater than the straight line IB. So it is possible to construct a triangle from these three parts. Let us construct it; let the triangle be ABC, and let the side AB be equal to the side AI, the side BC equal to <the side> IB and the side AC equal to <the side> BE. So it is clear that the angles of the triangle ABC are successively in double ratio, that is to say the angle B is twice the angle C and the angle C is twice the angle A.

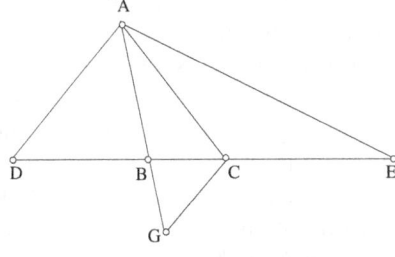

Fig. 8.2

Proof: We extend the straight line BC in both directions to the points D and E in such a way that DB is equal to AB and EC is equal to AC; we extend AB in the direction of B as far as G in such a way that BG is equal to BC. We join AD, AE and CG. It is clear that the angle CAE is equal to the angle CEA, so the angle BCA is twice the angle CAE. But since the ratio of EB to BD, that is to say to BA, is equal to the ratio of AB to BC, the triangle AEB is similar to the triangle ABC, so the angle BAC of the triangle ABC is

---

[2] That is to say that the ranges (D, G, H, E) and (A, I, B, E) are similar; we in fact have $AD \parallel IG \parallel BH$ (see al-Ṣāghānī, p. 668).

equal to the angle *AEC*, that is to say to the angle *EAC* of the triangle *AEB*. The angle *ACB* is thus equal to the angle *EAB*, that is to say to twice the angle *CAB*. Similarly, the ratio of *CD* to *CE*, that is to say the ratio of *AG* to *AC*, is equal to the ratio of *CE*, that is to say *AC*, to *BD*, that is to say *AB*. So the triangle *ACG* is similar to the triangle *ABC*; the angle *ACB* of the triangle *ABC* is thus equal to the angle *AGC* of the triangle *AGC*. But *CB* is equal to *BG*, the angle *ACB* is thus equal to the angle *BCG*. But the angle *ABC* is twice the angle *BCG*, that is to say twice the angle *ACB*. Now it has been shown that the angle *BCA* is twice the angle *CAB*; thus the angle *CAB* is a seventh of the sum of the angles of the triangle *ABC*. We fit the angle *CAB* into the circumference of the circle, the two sides *AC* and *AB* then cut off one seventh of it.

This proposition remained in this state until it was possible for Abū Sahl Wayjan ibn Rustum al-Qūhī and for Abū Ḥāmid Aḥmad ibn Muḥammad ibn al-Ḥusayn al-Ṣāghānī, for each of them, to carry out what was required by using conic sections; both of them being among those to whom is accorded primacy, skill and distinction in this art and in particular Abū Sahl al-Qūhī who was unique in his time both for his mastery and for his skill, avoided mentioning the square and the two equal triangles, and he went beyond it to arrive at why they had been drawn and the reason for which they had been constructed, that is the division of the straight line into three parts such that the product of the sum of the first and the second parts and the first is equal to the square of the third part, and the product of the sum of the second and the third parts and the second is equal to the square of the first part. He proceeded to his analysis with the help of two intersecting sections, from among the conic sections, a hyperbola and a parabola, then he proceeded to the synthesis and on it he based <the construction of> the heptagon.

As for Abū Ḥāmid, he intended <to prove> the proposition, that is to say <including> the square and the two equal triangles. He proceeded to his analysis by means of three hyperbolas: two opposite sections and another which cuts one of them; he next proceeded to the synthesis and on that he based <the construction of> the heptagon.

All that I have said about the precedence accorded to Archimedes and about his pre-eminence – even if it is too well known to require description – and then about what followed from these two talented men, Abū Sahl al-Qūhī and Abū Ḥāmid al-Ṣāghānī, and that they had recognized Archimedes' pre-eminence, that they confirmed what he said, adopted the basis he had established, corrected what he had been able to say and to introduce as a lemma, all that has led <us> to Abū al-Jūd Muḥammad ibn al-Layth, who, by being audacious and unjust together with lacking a store of experience in science, reached the point of supposing that Archimedes, in the matter of the lemma he had introduced, had stooped to imitation – such is the admiration he conceives for his own brutish intelligence; and he claimed for himself the construction of the heptagon using simple lemmas, from the book of the *Elements*, <lemmas> that are easily found and acquired.

The first is the following: If we construct a circle with the distance of a perpendicular to a straight line, then it touches the straight line on which the perpendicular was erected.

The second: To draw from one of the sides of a given triangle to its second side a straight line parallel to the third side and equal to what it cuts off from the second side outside the smaller triangle.

The third: To find a straight line whose ratio to a known straight line is a known ratio.

The fourth: To divide a known straight line into two parts such that the product of the whole straight line and one of its two parts is equal to the square of a straight line whose ratio to the other part is equal to a given ratio.

He noted this ratio and later in the construction of the heptagon he used another ratio, different from the one he had introduced, and which is the division of a straight line into two parts such that the product of the whole straight line and one of its two parts is equal to the square of a straight line whose ratio to the other part is equal to the ratio of the whole straight line to the sum of this straight line and the other part. He believed, because of his ignorance and his having learned so little, that the ratio of this straight line to the sum of this straight line and this part was equal to a known ratio; thus he provided the same proof for that and based <the construction of the> heptagon on it, although in fact this division is the same as the one that was presented by Archimedes, as I shall show at the end of this book.

Thus Abū al-Jūd wished to divide a straight line, for example *AB*, in this ratio at a point *C* such that the product of *AB* and *AC* is equal to the square of a straight line whose ratio to the straight line *CB* is equal to the ratio of *AB* to the sum of *AB* and *BC* and the square of this straight line is that of a straight line smaller than the straight line *BC*, since its ratio to *BC* is equal to the ratio of *AB* to the sum of *AB* and *BC*. Let this straight line be

BE; it is possible for us to construct on the straight line *AC* an isosceles triangle *ADC*, with *AD* and *DC* <equal>; let each of them be equal to the straight line *BE*. We join *BD*, we draw *EF* parallel to *AD*, we draw *FH*, *DG* perpendicular to *AB* and we join *FC*. Since the product of *AB* and *AC* is equal to the square of *EB*, that is to say to the square of *AD*, the ratio of *AB* to *AD* of the triangle *ABD* is equal to the ratio of *AD* to *AC* of the triangle *ADC*; but the angle *A* is common to the two triangles, which are thus similar, and *DB* will be equal to *AB*. But since *GC* is half of *AC* and *BC* is half of twice *BC*, accordingly *GB* is half of the sum of *AB* and *BC*. The ratio of *EB*, which is equal to *AD*, to half of *BC*, is equal to the ratio of *AB* to half of <the sum of> *AB* and *BC*, that is to say *GB*. But the ratio of *FB*, which is equal to *EB*, to *BH*, is equal to the ratio of *DB*, which is equal to *AB*, to *GB*, that is to say to half of <the sum of> *AB* and *BC*. So the ratio of *EB* to half of *BC* and to *BH* is the same. So *BH* is half of *BC*, the straight line *FC* is consequently equal to *FB*, that is to say to *EB*; so the straight lines *AD*, *DC*, *FC*, *FB* are all equal. But the angle *ACD* is equal to <the sum> of the angles *CDF* and *FBC*, and the angle *DFC* is twice the angle *B*, so the angle *ACD*, that is to say the angle *CAD*, is three times the angle *B*, and the sum of the angles of the triangle *ABD* is the seven times the angle *B*.

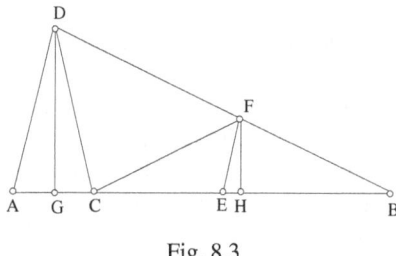

Fig. 8.3

This treatise came into the hands of Abū Saʿīd Aḥmad ibn Muḥammad ibn ʿAbd al-Jalīl al-Sijzī; the unrighteousness of what it said and the error in its <proposed> construction were clearly apparent to him. Abū Saʿīd al-Sijzī desired to divide the straight line in the ratio prescribed by Abū al-Jūd for constructing the heptagon, but he found that could not be done. Accordingly he wrote to Abū Saʿd al-ʿAlāʾ ibn Sahl the geometer and asked him for the <method of> dividing the straight line in the ratio concerned. It was possible for al-ʿAlāʾ ibn Sahl to divide up the straight line in this ratio by using two intersecting sections, conic sections, a hyperbola and a parabola. He carried out his analysis and he dispatched it to Abū Saʿīd al-Sijzī. When this reached Abū Saʿīd al-Sijzī, he carried out the synthesis on which he based <the construction of> the heptagon and he claimed it as his own. Here is his synthesis:

We wish to divide a straight line *AB* in the ratio we mentioned. We extend *BA* to *D*, so as to make *AD* equal to *AB*; let us apply to *AD* the square *ADEG*; we construct through the point *A* a hyperbola such that the straight lines *EG* and *ED* do not meet it, which is the conic section *AHK*; we also construct on the axis *BD* a parabola such that its *latus rectum* is equal to *AB*, which is the conic section *BHL*. We draw from the <point of> intersection of the two conic sections, which is the point *H*, the <straight line> *HC* perpendicular to *AB*.

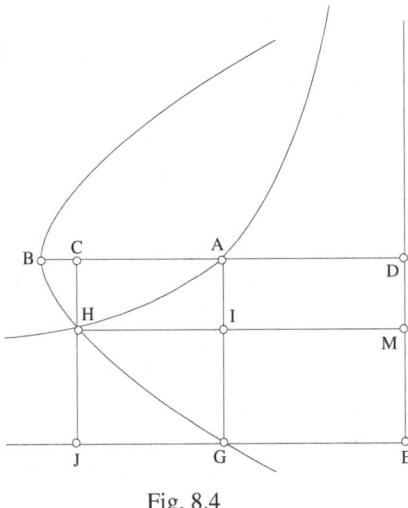

Fig. 8.4

I say that we have divided the straight line *BA* at the point *C*, in the way we wished.

Proof: We extend *EG* and *CH* to meet one another in *J*. We draw *HIM* parallel to *JE* and *AIG* parallel to *CJ*. Since the rectangle *MJ* is equal to the square *GD*, accordingly <the area> *JI* will be equal to <the area> *ID*. The rectangle *IC* is added to both[3] (*JI* and *ID*), so the rectangle *JA* is equal to the rectangle *CM*. But the rectangle *CM* is the product of *CH* and *CD*, and

---

[3] Lit.: we take the surface *IC* in common.

JA is <the product of> *CA* and *AG*, that is to say *AB*. So <the product of> *AB* and *AC* is equal to <the product of> *CH* and *CD*, so the ratio of *CH* to *AC* is equal to the ratio of *AB* to *CD*. But *CH* in the power is equal to <the product of> *AB* and *BC* since *AB* is the *latus rectum* of the parabola *BHL*. But *CD* is *AB* plus *CA*, so the ratio of the straight line that is equal in the power to <the product of> *AB* and *BC* to the straight line *CA* is equal to the ratio of the straight line *BA* to *BA* plus *AC*, considered as a single straight line. So we have constructed what we wanted.

News of what al-'Alā' ibn Sahl had constructed in order to divide the straight line in this ratio later reached Abū al-Jūd. He changed it very little, and that in the sense of completely avoiding any mention of the ratio and going on to what can be constructed. He recognized that the ratio of *CH* to *AC*, that is to say of *IA* to *AC*, is equal to the ratio of *AB*, that is to say *DE*, to *DC*. So he drew in the rectangle *CE* the diagonal that of necessity passes through the point *I*, and showed that *AI* is equal to *CH*; but he did not draw the straight line *HM* and then he based <the construction of> the heptagon on that, and claimed it as his own as he had claimed what Abū Sahl had constructed for dividing the straight line, which he needed to construct the heptagon we have already mentioned.

Indeed, in what he wrote to Abū Saʿīd al-Sijzī in reply to the question concerning the division of the straight line that was mentioned above,[4] al-'Alā' ibn Sahl mentioned the analysis of a proposition which he had also asked him about, which is the following:

ABCD is a parallelogram in which there has been drawn a diagonal which is *BC*; the side *CD* has been extended indefinitely in the direction of *D*. How may we draw a straight line, say the straight line *AEGH*, in such a way that the ratio of the triangle *BEG* to the triangle *GDH* is a given ratio?

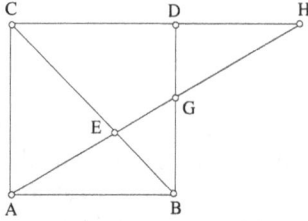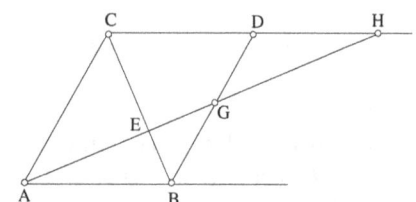

Fig. 8.5

---

[4] This is a range of type II.

He said, in long and emphatic statements, at the end of his analysis of this proposition: as for the ratio between the two triangles *AEB* and *GDH*, there is no method for arriving at this, and if we had found a way of achieving it, we should have adopted it. I do not understand how he came to find it so difficult that he abandoned it and how he had so good an opinion of himself in what he said, because there is a certain proportion between the two problems and it is possible to find it, because if the area *ABCD* is a square and if the triangle *AEB* is equal to the triangle *GDH*, then it is the proposition introduced by Archimedes for constructing the heptagon, and for which Abū Sahl al-Qūhī adopted the approach of dividing the straight line, in the ratio found there (i.e. in Archimedes' lemma). Here is his synthesis:

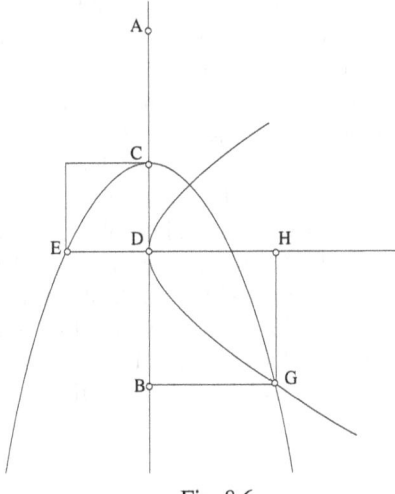

Fig. 8.6

Abū Sahl al-Qūhī said in his treatise: to find a straight line divided into three parts such that the product of the sum of the first and the second parts and the first is equal to the square of the third part, and the product of the sum of the second and third parts and the second is equal to the square of the first part. He supposed he had the two straight lines *CD* and *DE*, equal and each of them perpendicular to the other. He constructed a parabola whose vertex is the point *C*, whose *latus rectum* is equal to *CD* and whose axis lies on the extension of *CD*; let the conic section be *CG*; he constructed a hyperbola with vertex the point *D*, a transverse diameter that is its axis *DE* and is equal to its *latus rectum*; it necessarily cuts the parabola, say at the point *G*. He drew *GB* perpendicular to *CD* and *GH* parallel and equal to *DB*; he added to *CB*, *AC* equal to *GB* and he showed that the straight line *AB* had been divided at the two points *C* and *D* in the ratio we mentioned.

Abū al-Jūd later said in his collection entitled *Geometrical* *<Collections>*, after quoting what al-ʿAlāʾ ibn Sahl said about this: I have myself found how to do what al-ʿAlāʾ ibn Sahl said was impossible; by which he means to give the ratio between the two triangles *AEB* and *GDH* of the preceding proposition.

He then said: here is his lemma, and then came the figure constructed by Abū Sahl, the very same figure, except that, knowing that the ratio of the square of *HG* to the area enclosed by *HE* and *HD* is equal to the ratio of the *latus rectum* of the hyperbola to its transverse diameter, *<a latus rectum>* which has been supposed to be *<equal to>* any straight line, such as the straight line *K*; he put its ratio to the straight line *DE* as the given ratio,[5] then he constructed the conic section *DG* such that its *latus rectum* is *K* and its transverse diameter *<is equal to> DE*. So the straight line *AB* will be divided at the two points *C* and *D*, the product of *BC* and *CD* is equal to the square of *AC* and the ratio of the square of *BD* to the area enclosed by *DA* and *AC* is equal to the given ratio. He took a parallelogram, say *AEGB*, in it he drew the diagonal *AG* and divided its side *AB* in this ratio at the points *C* and *D*; he drew *CI* parallel to *AE* and joined *EI*, *ID* which he extended, as he did the side *GB*, as far as where they met at the point *H*. He then showed that the line *EDH* is a straight line[6] and that the ratio of the triangle *EIA* to the triangle *BDH* is equal to the given ratio.

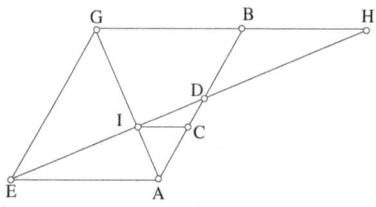

Fig. 8.7

[5] See *Livre sur la synthèse des problèmes analysés par Abū Saʿd al-ʿAlāʾ ibn Sahl*, in R. Rashed, *Géométrie et dioptrique au Xᵉ siècle*, Paris, 1993, Appendix I, p. 187.

[6] Al-Shannī does not show that *E*, *I*, *D* are collinear (see *Géométrie et dioptrique au Xᵉ siècle*, Appendix I, page CII).

He also laid claim to Menæchmus' construction for finding two straight lines between two straight lines so that the four are in continued proportion, a construction he incorporated into this book that he named the *Geometrical <Collections>*, after having mentioned Thābit's construction for this.

He then said: what I have done, myself, is in terms of the most accessible and the most lucid. The book by Eutocius in which he assembled the statements made by the Ancients in regard to finding two straight lines between two straight lines so that the four are in continued proportion, attests to this. In this book Eutocius reported on two methods due to Menæchmus; in one of them he used two conic sections, a hyperbola and a parabola, and in the other, two parabolic sections. It is <as follows>.

Since we wish to shed light on his case, I shall give an account of it here together with what he himself changed and the construction carried out by Menæchmus.

Menæchmus said: we take two straight lines *AB* and *BC*, one being perpendicular to the other; we extend each of them to infinity and we construct a parabola such that its axis is *BC* and its *latus rectum BC*; and another parabola such that its axis is *AB* and its *latus rectum AB*. These two parabolas cut one another at the point *G*. We draw *GE* and *GD* parallel to the straight lines *AB* and *BC*; they will be means in proportion between the straight lines *AB* and *BC*.

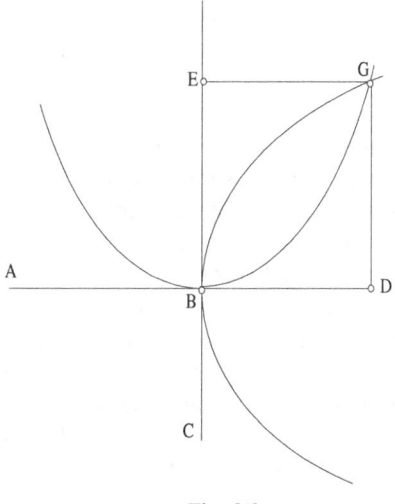

Fig. 8.8

As for Abū al-Jūd, he took the two straight lines *AB* and *BC* <which are> known in that one is perpendicular to the other, and constructed a parabola with axis *AB* and *latus rectum AB* and another parabola with axis *BC* and *latus rectum BC*. They cut one another at the point *G*, he drew *GD* and *GE* parallel to *AB* and *BC* and showed that they are means in proportion between *AB* and *BC*.

I cannot permit myself to impute this to his having chosen to be brief, as has happened to others in many procedures, because of what we know about his case and because of the deceit <we know he practised> in many of his procedures.

When Abū Saʿīd al-Sijzī learned of what Abū al-Jūd had done in connection with this proposition that was established by al-ʿAlāʾ ibn Sahl, in claiming it for himself, he was excessively insulting and critical, in revealing his (Abū al-Jūd's) case and his portrait, and he stated all this in a treatise. But this poisonous creature[7] did not draw back in the face of what had been heaped on him, that is the shame that attaches to someone who gets everything muddled up; and under the violent, heavy blow, he did not lower his eyes and he put his honour at risk by what he endured. Abū al-Jūd then wrote to Abū Muḥammad ʿAbd Allāh ibn ʿAlī al-Ḥāsib, claiming to have constructed the heptagon. He began by giving an indication of the two methods of the two masters, Abū Sahl al-Qūhī and Abū Ḥāmid al-Ṣāghānī, while disparaging their construction, and maintaining that both of them were intending to prove the proposition that Archimedes introduced in his treatise on the construction of the heptagon, and which he assumed without carrying out the construction or proof in this treatise; and <said> both of them desired to correct it and prove it. For my part, <he said>, I have made the inaccessible approachable and smoothed out what was difficult, and I have done this and that … He then set out the construction we mentioned before.

He then said: by what I constructed previously, I have distinguished myself and I am in advance of all of them since the analysis resulted in dividing a given straight line into three parts such that the product of the whole straight line and the third part is equal to the square of the first part, and the product of the sum of the second and third parts and the second is equal to the square of the first part.

And he said: this is much easier than the division of the straight line into three parts such that the product of the sum of the two parts, the first and the second, and the first, is equal to the square of the third part, and the product of the sum of the two parts, the second and the third, and the second, is equal to the square of the first part, <the division that> was

---

[7] This refers to Abū al-Jūd.

carried out by Abū Sahl and Abū Ḥāmid; and this is also easier than the division of the straight line into two parts such that the product of the whole straight line and one of its two parts is equal to the square of a straight line whose ratio to the other part is equal to the ratio of that straight line to the sum of the latter and the part already mentioned, <a division> I myself carried out earlier.

He then said: it being given that all these constructions are carried out either by means of intersecting conic sections, a hyperbola and a parabola, or by means of three hyperbolas, as for me, I divided the straight line into its three parts by means of a single conic section. However, I did not send you the construction and the treatise that is devoted to it, until you ask the geometers of the prestigious court if one of them had constructed the heptagon by means of a single conic section, so that, if I send my construction, their behaviour towards me will not be as bad as it was on several occasions when they attacked me and attributed to other people things I had discovered; he means the construction of al-'Alā' ibn Sahl which he had earlier ascribed to himself, and other similar impostures. If this person who deceives himself, that is to say Abū al-Jūd, had mentioned this construction in this book, in which he taxes these two eminent men with incompetence and copying, that would have been the better for him and for this passage in his book. If he did not do so, the unhappy man <nevertheless> pursued his purpose by saying at the end of his treatise: in what I constructed earlier I used, together with the single conic section, two lemmas from the *Elements*.

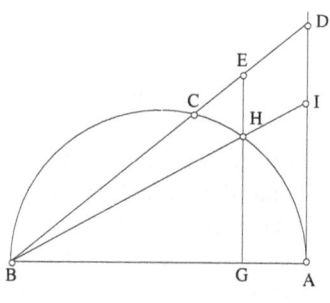

Fig. 8.9

The first: if we draw from point *B* of a straight line *AB*, the diameter, a straight line that cuts the circle *ACB* in *C*; if we draw from the point *A* a perpendicular to *AB* so that it meets the straight line *BC* produced in *D*,

<we wish to know> how to draw starting from the straight line *CD* a straight line, such as the straight line *EG*, parallel to *AD* and which cuts the circumference <of the circle> in <the point> *H* such that the ratio of *EH* to *GH* is equal to a given ratio.

He said: this is in fact easy if we divide *AD* at the point *I* in the given ratio by drawing *BI*, which necessarily cuts the circumference; let it cut it in *H*. We cause to pass through this point <the line> *EG* parallel to *AD*, then the ratio of *EH* to *GH* is equal to the given ratio.

Second lemma: Let us draw *EG* parallel to *AD*, the perpendicular, so as to make it equal to the straight line joining *A* and *H*.

He said: This too is not inaccessible. We extend *AB* to *I*; we put the ratio of *AB* to *BI* equal to the ratio of *AD* to *AB*; we put the product of *AB* and *BI* equal to <the product of> *IG* and *GB*, we draw from the point *G* a perpendicular to the diameter *AB*, such as the straight line *GHE*, which cuts the circumference in *H*, and we join *AH*. He claimed, without proving it, that *AH* is equal to *EG*.

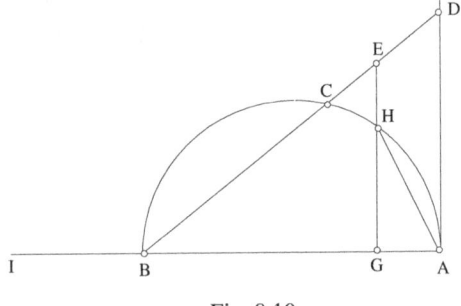

Fig. 8.10

I wished, for my part, to set out a proof of <the proposition> he had stated, I examined it closely and thus found that he had made a mistake, but he had observed it worked that way if the perpendicular *AD* is equal to the diameter *AB*. Whether he had had that idea or not, he imagined – being ignorant and negligent – that this led to what he wanted and <served> his purpose <even> if *AD* is longer or shorter than *AB*, and he accordingly supplied a single proof of this <proposition>, or perhaps he knew about it and let himself be blinded by incompetence and in this way intended to commit a fraud; or perhaps he was looking for <no more than> a proposition that satisfied him and those like him, people who are on a par with him in his shallowness and mediocrity of understanding.

I shall set out the proof that there are defects in what he did in this proposition, and I shall abandon it after having introduced a proposition I need for the notion I have in mind, with God's aid – he is sufficient my most excellent help.

I say: for any straight line divided in a ratio that has a mean and two extremes, if we add to its smaller part a straight line that causes the smaller part to become equal to the greater one, then the smaller part plus the straight line that has been added to it is divided in a ratio that has a mean and two extremes and its smaller part is the straight line that has been added to it.

I show this by <considering> an example: Let the straight line AB be divided at the point C in a ratio that has a mean and two extremes; let the smaller part be AC, let us add to AC, AD to make the complete line CD equal to CB.

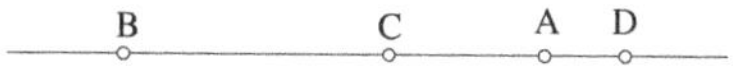

Fig. 8.11

I say that the straight line DC is divided in a ratio which has a mean and two extremes at the point A and that its smaller part is AD.

Proof: Since the product of AB and AC is equal to the square of BC, that is to say the square of AC plus the product of AC and CB, that is to say DA and AC plus twice the square of AC, and since DC is equal to CB and its square is equal to the square of CB, that is to say the squares of DA and AC plus twice the product of DA and AC, we take away the product of DA and AC, a single time, from both sides and the square of AC, a single time, from both sides, it remains <that> the product of DA and AC plus the square of DA, that is to say the product of DC and DA, is equal to the square of AC. The straight line DC is thus divided at the point A in a ratio that has a mean and two extremes and its smaller part is AD. That is what we wanted to prove.

Having introduced that, we come back to the problem and return to the figure. We say: if AD is equal to AB and if we put the ratio of AB to BI equal to the ratio of AD to AB, then we have BI equal to AB. We put the product of AB and BI, that is to say the square of BI, equal to the product of IG and GB, so the straight line AB is divided at the point G in a ratio that has a mean and two extremes and its smaller part is AG, from what we have shown above. We draw the perpendicular *EHG*, which cuts the circumference at the point *H*, and we join *AH*. Since the product of *AB* and *AG* is equal to the square of *AH*, because the triangles inscribed in the semicircle and the triangle *AGH* are similar – but the product of *AB* and *AG* is equal to the square of *GB* – so the straight line *GB* is equal to the straight line *AH*. But *AD* is equal to *AB* and *EG* is parallel to *AD*, so *EG* is equal to *GB*, that is to say to *AH*. That is what we wanted to prove.

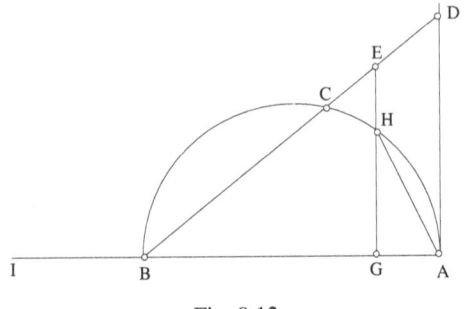

Fig. 8.12

I say that if the perpendicular *AD* is longer or shorter than the straight line *AB*; if we put the ratio of *AB* to *BI* equal to the ratio of *AD* to *AB*; if we put the product of *IG* and *GB* equal to the product of *AB* and *BI*; if we draw *GE* parallel to *AD* and if we join *AH*, then *AH* will never be equal to *EG*.

Proof: This cannot be so; if it were possible, let *AH* be equal to *EG*. Since the ratio of *AD* to *AB* is equal to the ratio of *AB* to *BI* and since the product of *AB* and *BI* is equal to the product of *IG* and *GB*, the ratio of *AB* to *BG* is equal to the ratio of *GI* to *IB*. By separation, the ratio of *AG* to *GB* is equal to the ratio of *GB* to *BI*; but the ratio of *AD* to *AB* is equal to the ratio of *EG* to *GB*, so the ratio of *EG* to *GB* is equal to the ratio of *AB* to *BI*. By permutation, we have <that> the ratio of *EG*, that is to say *AH*, to *AB*, is equal to the ratio of *AB* to *BI*, that is to say the ratio of *AG* to *GB*, from the above; so the ratio of *AH* to *AB* is equal to the ratio of *AG* to *GB*. By inversion, the ratio of *AB* to *AH*, that is to say the ratio of *AH* to *AG*, is equal to the ratio of *BG* to *AG*, so *AH*, that is to say *EG*, is equal to *GB*, so *AD* is equal to *AB*; now we have taken it to be longer or shorter; this is absurd and cannot be so.

Given that I have shown that what he constructed in this proposition is defective, it has thus been shown that what he constructed from it is defective, even though that has not reached me.

Abū al-Jūd wished to divide the straight line in the same ratio that is to be found in the first division itself, if he could succeed in this, then construct the heptagon as he had constructed it by this procedure, and make it clear that this is another ratio, distinct from the one constructed by al-Alā' ibn Sahl, although this division is also the division introduced by Archimedes for constructing the heptagon. He relies on it (the division), as I shall now show.

Let there be the straight line *AB* divided at the point *C* such that the product of *AB* and *AC* is equal to the square of a straight line, say *BE*, and let the ratio of *BE* to *BC* be equal to the ratio of *AB* to the sum of *AB* and *BC*.

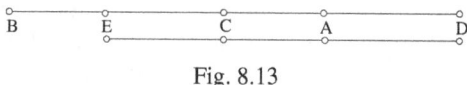

Fig. 8.13

I say that the straight line *AB* has also been divided at the points *C* and *E* such that the product of *AB* and *AC* is equal to the square of *BE* and the product of *AE* and *EC* is also equal to the square of *BE*.

Proof: The ratio of *AB* to the sum of *AB* and *BC* is equal to the ratio of *BE* to *BC*. By separation, the ratio of *AB* to *CB* is equal to the ratio of *BE* to *CE*. By permutation, the ratio of *AB* to *BE* is equal to the ratio of *CB* to *CE*. By separation, the ratio of *AE* to *EB* is equal to the ratio of *EB* to *EC*, so the product of *AE* and *EC* is equal to the square of *BE*. That is what we wanted to prove.

Next we extend *AB* to *D* in the direction of *A* so that *AD* is equal to *BE*.

I say that the straight line *ED* has been divided in the ratio ascribed to Archimedes, that is to say that the product of *AE* and *EC* is equal to the square of *AD* and the product of *DC* and *CA* is equal to the square of *EC*.

Proof: The product of *DE* and *AC* is equal to the product of *AE* and *EC*, the ratio of *ED* to *EA* will be equal to the ratio of *EC* to *CA*. By permutation, the ratio of *ED* to *EC* will be equal to the ratio of *EA* to *AC*. By separation, the ratio of *DC* to *EC* will be equal to the ratio of *EC* to *CA*, so the product of *DC* and *CA* is equal to the square of *EC*; but the product of *AE* and *EC* was equal to the square of *AD*. That is what we wanted to prove.

What is surprising in this man is not that he made a mistake in what he did, or that he laid claim to work done by someone else, but what is surprising is what occurred to his imagination, and that he believed it; and that he should have had so good an opinion of himself, notably in what he alleged and claimed, in regard to dividing a straight line in this ratio by means of a single conic section, although he knew this was not <found> possible by any of the Moderns such as al-ʿAlāʾ ibn Sahl, Abū Sahl al-Qūhī and Abū Ḥāmid al-Ṣāghānī, despite their being the leaders in this science, for their practice and their superiority over all their contemporary colleagues, except by using two sections from among the conic sections. And we may note, also, that he rejects their constructions and ascribes their procedure and their results to imitation. May God preserve us from pretending to know things we do not know, and we ask him for success so that we may give thanks for what we understand; this is in his hands. There is no strength or power, save from Him.

The treatise is completed – thanks be rendered to God alone, blessing and peace be upon him who is the last of the prophets.

Sunday 21 Jumādā al-ūlā, in the year one thousand one hundred and fifty-three.

TREATISE BY NAṢR IBN ʿABD ALLĀH

On the Determination of the Chord of the Heptagon

He said: the circle *ABC* is given; we wish to construct in it the chord of the regular heptagon.

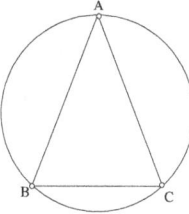

Fig. 9.1

Let us suppose, by the method of analysis, that we have constructed it and that it is *BC*. Let us divide <the arc> *BAC* into two equal parts in *A* and let us join *BA* and *CA*; they are equal. Since the arc *BC* is a seventh of the circle, each of the arcs *BA* and *AC* is three sevenths of it and each of them is three times the arc *BC*. From what has been shown by Euclid, each of the angles *B* and *C* is three times the angle *A*.

Thus analysis has led to constructing an isosceles triangle in which each of the angles at the base is three times the angle at the vertex.

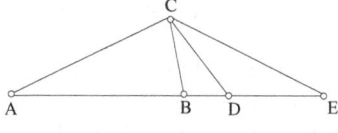

Fig. 9.2

Let us suppose, by the method of analysis, that we have found it; let it be the triangle *ABC* in which each <of the angles> *B* and *C* is three times <angle> *A*. We construct <the angle> *BCD* equal to *A*; now <the angle>

CDA is common to the two triangles *ACD* and *BCD*, so *AD* to *DC* is equal to *DC* to *DB*, and *AD* by *DB* is equal to the square of *DC*. Let us put *DE* equal to *DC*; so *AD* by *DB* is equal to the square of *DE*. Let us join *CE*; since *CBA* is equal to *BDC* plus *BCD* and it is three times <angle> *A*, that is to say <three times> *BCD*, accordingly *BDC* is twice *BCD*; but it is also twice each of the <angles> *DEC* and *DCE*; now *E* is common, so *AE* to *EC* is equal to *EC* to *ED*, and *AE* by *ED* is equal to the square of *EC*, that is to say of *CA*, because the angles *E* and *A* are equal, say *BA*;[1] so *AE* by *ED* is equal to the square of *AB*; but *AD* by *DB* was equal to the square of *DE*.

So the analysis has led to finding three straight lines such as *AB*, *BD* and *DE*, put together to make a single straight line, such that the product of the whole straight line and the part that is at one of the ends is equal to the square of the part that is at the other end and the product of the sum of the part on the end[2] and the part in the middle and <the part> in the middle is equal to the square of the part at the other end.[3]

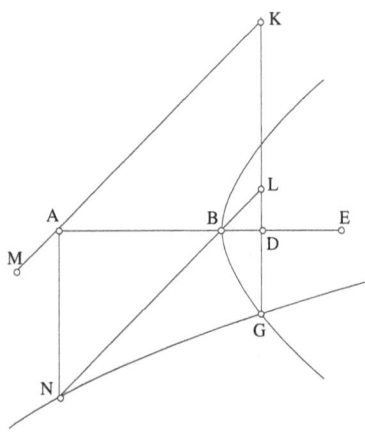

Fig. 9.3

Let *AB* be known in position and magnitude.

We suppose by the method of analysis that we have found the two remaining <straight lines>, let them be *BD* and *DE* subject to the condition that *AE* by *ED* is equal to the square of *AB*, and *AD* by *DB* is equal to the

---

[1] By hypothesis we have *CA* = *BA*.
[2] That is on this second end.
[3] That is on the first end.

square of *DE*. We draw from *A* the perpendicular *AN* equal to *AB* and we draw from *D* the perpendicular *DG* equal to *DE*. We join *NB* and we extend it to *L*; it is known in position. We extend it on the other side to *K* and from *A* <we draw> *MAK* parallel to *NBL*; it is also known in position, so *GD* to *LK* is equal to *DE* to *BA*, so *KG* by *GD* is equal to the square of *AB*, that is to say of *AN*; but it is parallel to *GK* and *MA*, *AE* are known in position, so *N* and *G* are on the outline of the hyperbola of which *MA* and *AE* are the asymptotes, and *N* is known. The conic section that passes through *N* and *G* is thus known in position and magnitude. Similarly, *AD* by *DB* is equal to the square of *DE*, that is to say of *DG*, so *B* and *G* are on the outline of a hyperbola with transverse diameter and *latus rectum AB* and its axis *BE*, which is extended. But *AB* is known in position and magnitude and *B* is known; so the conic section that passes through *B* and *G* is known in position and magnitude. The point of intersection of the two conic sections, which is *G*, is known and *GDK*, which is known in position, because it is at a known angle, meets *ADE*, known in position, and *MAK*, known in position; so *K* is known; *D* and *L* are likewise known and each <of the straight lines> *GD* and *DL* is known; the same is true for each <of the straight lines> *DE* and *DB*, so *D* and *E* are likewise known. That is what we wanted.

The synthesis of this problem is like this: the straight line *AB* is known in position and magnitude. We wish to find two straight lines put together to make a single straight line as we have described it.

We draw from *A* the perpendicular *AN* equal to *AB*; we join *NB*, and from *A* <we draw the straight line> *MAK* parallel to *NB*, and we cause to pass through *N* the hyperbola *NG*, such that *MA* and *AB* are its asymptotes; <and> through *B* the hyperbola *BG* such that its transverse diameter and its *latus rectum* are *AB* and its axis is *BE* extended. We draw from *G*, a point of intersection of the two conic sections, the <line> *GD* perpendicular to *AB* and we extend it to *K*; we put *DE* equal to *DG*.

I say that *BD* and *DE* are the lines we required.

Proof: Since *NG* is a hyperbola and *MA* and *AE* are its asymptotes, and from the centre to the conic section we have drawn <the straight line> *AN* and *KLDG* which is parallel to it, and which cuts the angle which is next to the angle between the asymptotes,[4] accordingly *KG* by *GD*, that is to say

---

[4] That is the angle formed by an asymptote and the extension of the other one.

AE by *ED*, is equal to the square of *AN*, that is to say to the square of *AB*. But since *BG* is a hyperbola whose transverse diameter and *latus rectum* are *AB* and whose axis is *BE* extended, we have *AD* by *DB* equal to the square of *DG*, that is to say to the square of *DE*. That is what we wanted.

We wish to construct a triangle of the form we have described.

We find three straight lines put together as we have shown and we construct with <centre> *A* and distance *AB* a circle *KC*; and with <centre> *D* and distance *DE* a circle *EC*; we join *AC*.

I say that the triangle *ABC* is as we required.

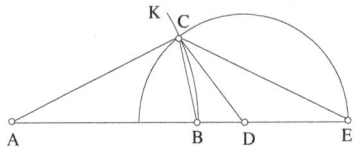

Fig. 9.4

Proof: We join *DC* and *EC*. Since *AD* by *DB* is equal to the square of *DE*, that is to say to the square of *DC*, accordingly *AD* to *DC* is equal to *DC* to *DB*; so the two triangles *ADC* and *BDC* are similar, and *BCD* is equal to *A*. But since *AE* by *ED* is equal to the square of *EC*,[5] we have <that> *A* is equal to *ECD*, so *ECD* is equal to *BCD*; but *BDC* is twice *ECD*, so it is twice *BCD*; but *ABC* is equal to *BCD* plus *BDC*, so *ABC*, which is equal to *ACB*, is three times *BCD*, that is to say <three times> *BAC*. So we have constructed a triangle as we required.

We wish to construct in a circle known in position and magnitude the chord of the regular heptagon.

We construct a triangle of the form we have explained and we draw in the circle a triangle that is similar to it. Since each of the angles at the base is three times the angle at the vertex, each of the arcs of the angles at the base is three times the arc of the angle at the vertex; so the arc on which the angle at the vertex stands is a seventh of the circumference and its chord is the side of the regular heptagon. That is what we wanted to prove. From God, strength and power.

---

[5] We have $DE \cdot AE = AB^2$, but it has not been shown that $EC = AB$.

In the name of God, the Compassionate, the Merciful

Synthesis for the Analysis of the Lemma on the Regular Heptagon Inscribed in the Circle

(ANONYMOUS)

We wish to divide a straight line into two parts such that the product of the whole straight line and the sum of the straight line and one of the parts is equal to the square of a known straight line whose ratio to the whole straight line is equal to the ratio of the part we mentioned to the part that remains.

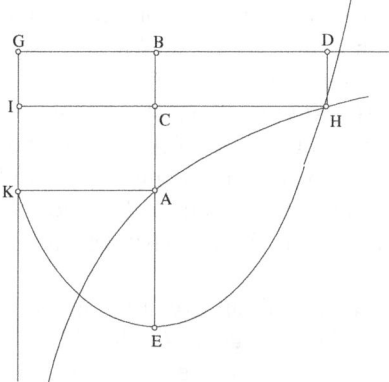

Fig. 10.1

We suppose <we have> a straight line say *AB* and the perpendicular *GB*, which is equal to it. We complete the square *AKGB* and we cause to pass through the point *A* a hyperbola such that the straight lines *KG* and *GB* are its asymptotes, a construction Apollonius shows in Proposition 4 of Book II of the work the *Conics*; let the conic section be *AH*. We extend the straight line *AB* on the side of *A* to the point *E* such that *AE* is equal to *AB*; we extend the straight line *GB* indefinitely on the side of *B*. We construct a parabola with vertex the point *E*, whose transverse diameter, which is an axis, is the straight line *EB* extended, whose *latus rectum* is equal to the straight line *EA* and the angle of the ordinate is a right angle; it (the parabola) necessarily cuts the hyperbola – let it cut it at the point *H*; this is the conic section *EH*. We draw from the point *H* the <line> *HC* perpendicular to *AB*.

I say that the point *C* is <the point> we seek.

Proof: Let us draw the <line> *HD* perpendicular to the straight line *BG* and let us extend *HC* until it meets the straight line *KG* at the point *I*. But since the two points *A* and *H* are on the hyperbola and we have drawn from them the two straight lines *AK* and *HI* and the two straight lines *AB* and *DH* as far as the two asymptotes and parallel to them, accordingly, from what has been shown in Proposition 12 of Book II of the work the *Conics*, the square *AG*, that is to say the square of *AB*, that is to say the product of *AB* and *AC* and <the product of *AB*> and *BC*, is equal to the rectangle *HG*, that is to say <equal> to the product of *CB* and *GB* and <the product of *CB*> and *BD*; but *GB* is equal to *AB*, so we take away the product of *AB* and *BC*, that is to say the product of *GB* and *BC*, since they are equal; so it remains <that> the product of *AB* and *AC* is equal to the product of *BD* and *BC* and thus the ratio of *BD*, that is to say *HC*, which is equal to it, to *AB*, is equal to the ratio of *AC* to *BC*. But since the product of the straight line *HC* and itself is equal to the product of *EC* and *EA* because we have a parabola – but *EA* is equal to *AB* – the square of *HC* is equal to the product of the sum of *AB* and *AC* and *AB*. Now we have shown that the ratio of the straight line *HC* to *AB* is equal to the ratio of *AC* to *CB*. So we have divided the straight line *AB* into two parts at the point *C*; the product of the sum of *AB* and *AC* and *AB* is equal to the square of *HC* and the ratio of *HC* to *AB* is equal to the ratio of *AC* to *CB*. That is what we wanted to construct.

Once this has been carried out, we suppose we have a square *ABCD* whose diagonal *BC* has been drawn and whose side *CD* has been extended indefinitely in the direction of *D*; we divide the side *BD* at the point *G* such that the product of *BD* and the sum of *BD* and *BG* is equal to the square of a straight line; let it be the straight line *F*, such that the ratio of the straight line *F* to the straight line *BGD* is equal to the ratio of *BG* to *GD*. We join the straight line *AG* and we extend it until it meets the side *CD* extended, at the point *H*.

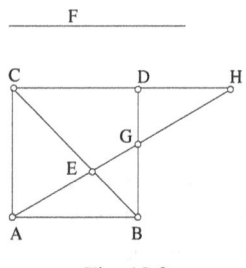

Fig. 10.2

I say that the triangle *AEB* is equal to the triangle *DGH*.

Proof: The ratio of the straight line *F* to *BD* is equal to the ratio of *BG* to *GD*, similarly the ratio of the square of the straight line *F* to the square of the straight line BD is equal to the ratio of the square of the straight line BG to the square of the straight line GD. But the ratio of the square of the straight line F to the square of BD is equal to the ratio of the sum of BD and BG to BD, it being given that the product of the sum of BD and BG and BD is equal to the square of the straight line F. Now if we have three magnitudes in proportion, the ratio of the first to the third is equal to the ratio of the square of the second to the square of the third. But BD is equal to AB and the ratio of the square of the straight line F to the square of BD is equal to the ratio of the square of BG to the square of DG; the ratio of the square of BG to the square of DG is accordingly equal to the ratio of the sum of BG and AB to AB. But the ratio of the sum of BG and AB to AB is equal to the ratio of AG to AE, because the straight line BE has divided up the angle ABG into two equal parts and that has been shown in Proposition <3> of Book VI of the *Elements*. The ratio of the square of BG to the square of DG is equal to the ratio of AG to AE. But the ratio of the square of BG to the square of DG is composed of the ratio of BG to DG and the ratio of AG to GH. Now the ratio composed of the ratio of BG to DG and the ratio of AG to GH is the ratio of the product of AG and BG to the product of GH and DG. But the ratio of AG to AE, if we multiply throughout by BG, is equal to the ratio of the product of AG and BG to the product of AE and BG, so the ratio of the product of AG and BG to the product of GH and GD is equal to the ratio of the product of AG and BG to the product of AE and BG. So the product of AE and BG is equal to the product of GH and GD, and the ratio of AE to GH is equal to the ratio of DG to BG, that is to say to the ratio of DH to AB; so the ratio of AE to GH is equal to the ratio of DH to AB and, in the triangle AEB, the angle BAE is equal to the angle DHG of the triangle DGH and the sides that enclose the two equal angles are inversely proportional; so the triangle AEB is equal to the triangle DGH. That is what we wanted to construct.

The treatise is completed, with the assistance of God and with his help.

Monday twenty-two Jumādā al-ūlā, the year one thousand one hundred and fifty-three.

In the name of God, the Compassionate, the Merciful
May God assist us

TREATISE BY THE LORD KAMĀL AL-DĪN IBN YŪNUS

— may God give him enduring eminence —

FOR HIS SERVANT MUḤAMMAD IBN AL-ḤUSAYN

On the Proof for the Lemma[1] Neglected by Archimedes in his Book on the Construction of the Heptagon Inscribed in the Circle, and How that is Done

He said – may God give him lasting glory: you have confided to me – may God give you enduring eminence – the question of the lemma neglected by Archimedes in his book on the heptagon inscribed in the circle, and for which he explained neither the construction nor the proof. It is the following:

Let there be the square *ABCD*; we extend the straight line *AB* to *E*, and in it (the square) we draw the diagonal *BC*. We wish to draw from the point *D* a straight line, such as the straight line *DGIE*, so that the triangle *CGD* is equal to the triangle *EAI*.

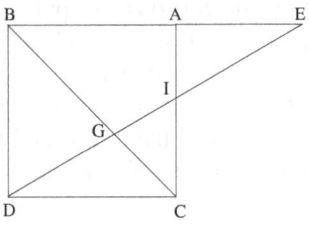

Fig. 11.1

---

[1] Lit.: on the proof of finding the lemma.

You referred to the fact that Aḥmad ibn Muḥammad ibn ʿAbd al-Jalīl al-Sijzī attached such importance to the question of this lemma that at the beginning of his book on the regular heptagon inscribed in the circle he discussed remarks made by the man who affirmed: 'and it perhaps involves a more difficult construction and a less accessible proof than the matter for which he introduced it; and perhaps these latter are not possible'.[2] He tells us that he proceeded to his synthesis by starting from the analysis of al-ʿAlāʾ ibn Sahl.

That Archimedes, with all the majesty of his greatness and the eminence of his position in this science, should have appealed, to prove what he sought, to something whose proof requires it to be preceded by what is sought, that is truly too much. You wanted to know the truth of the matter. I am responding to your request.

When I examined this, the proof of how to find the lemma revealed itself to me in several ways; and it appeared to me clearly while I was pondering the possibility of a proof of the existence of this lemma, that it was possible to set up a proof of the existence of the following lemma, without there being any need for it to be preceded by the latter one.[3] Anyone who examines what I have set out, and handles it correctly, will succeed in this. Moreover, this is possible starting from the very lemma introduced by Aḥmad <ibn Muḥammad ibn ʿAbd al-Jalīl al-Sijzī> whose synthesis he carried out starting from the analysis of al-ʿAlāʾ ibn Sahl, as we shall show, even if the former did not realize this was so. I now begin by giving the proof, in a way that does not imply that Archimedes' proof is circular, contrary to the opinion of the man who spoke about it in al-Sijzī's company.

I say: Let us return to the preceding figure. We extend CD to J and we put DJ equal to CD; we extend BA to H, we put AH equal to AB and on AH we construct a square, let it be the square $AHKL$. We construct through the point K a hyperbola having as asymptotes the straight lines AL and AH, let the conic section be KMO. We construct a hyperbola whose *latus rectum* and transverse diameter are each equal to DJ, and whose vertex is the point D and its axis is along the extension of CD; let the conic section be DM; it

---

[2] The comments concerned are ones al-Sijzī attributed to Abū al-Jūd. He means that the construction in the lemma and its proof may not be possible.

[3] This refers to the equality of the two triangles AIE and CGD.

cuts the conic section *KM*, on the side of *M*:[4] Given that if we extend *HK* until it meets *CD* at <the point> *N*, *CN* is equal to *CD* and *NK* equal to *ND*, thus it[5] is smaller than the ordinate drawn from the point *N* <to the second hyperbola>, since the square of the ordinate we mentioned is equal to the product of *JN* and *ND*, which is greater than the square of *NK*; so the ordinate drawn from *N* meets the conic section beyond the point *K*, after having cut the conic section *KMO*; and this <is so> because the straight line *NK* does not meet the conic section *KMO* in any point other than the point *K*, as was shown by Apollonius in the *Conics*.[6] We draw from the point *M* a perpendicular to *HB*, let the perpendicular be *ME*. We join *ED*, it cuts *BC* at <the point> *G* and *AC* at <the point> *I*.

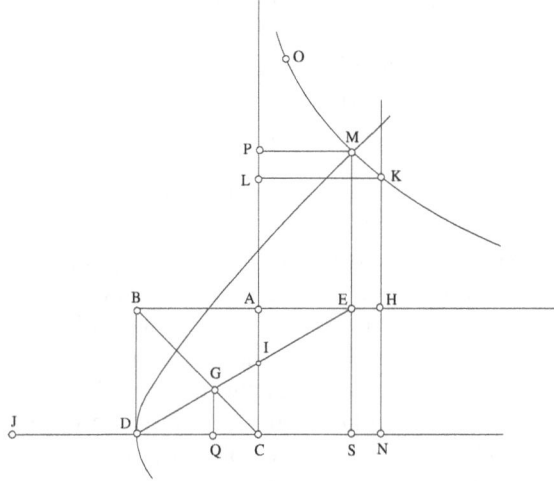

Fig. 11.2

I say that the triangle *CGD* is equal to the triangle *EAI*.

Proof: We extend *ME*, it meets *ND* in *S*; we draw from the point *G* the <line> *GQ* perpendicular to *CD*; since the two triangles *ESD*, *GQD* are similar, accordingly the ratio of *ES*, that is to say *DJ*, to *SD* is equal to the ratio of *GQ*, that is to say *CQ*, to *QD*. By composition, we have <that> the

---

[4] The point *K* divides the conic section *KM* into two parts and *KM* is the part that approaches the asymptote *KL*.

[5] That is *NK*.

[6] Apollonius, *Conics*, II.13.

ratio of *JS* to *SD* is equal to the ratio of *CD* to *DQ*. But the ratio of *JS* to *SD* is equal to the ratio of the square of *MS* to the square of *SD*, because *DM* is a hyperbola. But the ratio of the square of *MS* to the square of *SD* is equal to the ratio of the square of *ES* to the square of *SC*, given that we complete the area *AEMP*, which is equal to the square *KA*, that is to say the square *AD*, because *KMO* is a hyperbola. We take the common rectangle *EC*; the rectangle *MC* is equal to the rectangle *ED*. So the ratio of *MS* to *SD* is equal to the ratio of *ES* to *SC*, their squares are also in proportion, so the ratio of *CD* to *DQ* is equal to the ratio of the square of *ES*, that is to say *CD*, to the square of *SC*, so the ratio of *CD* to *SC*, that is to say *EA*, is equal to the ratio of *SC*, that is to say *EA*, to *DQ*. But the ratio of *EA* to *DQ* is equal to the ratio of *AI* to *GQ* because the triangles *AEI* and *GDQ* are similar, so the ratio of *CD* to *EA* is equal to the ratio of *AI* to *GQ*, and the product of *CD* and *GQ*, that is to say twice the triangle *CDG*, is equal to the product of *AE* and *AI*, that is to say twice the triangle *EAI*, so the two triangles *CGD* and *EIA* are equal. That is what we wanted to prove.

Since we have completed what we wanted <to do>, let us now turn our attention to what Aḥmad ibn ʿAbd al-Jalīl al-Sijzī left out in the lemma of his on which he based his construction of the heptagon; and let us show that we can divide a given straight line in accordance with the condition that Archimedes relied on in his book on the question of the heptagon, by the smallest of steps, and not as he had thought in his comments: 'Perhaps his division of that is a more difficult matter than the division of the circle into seven parts.' We have explained this, because it escaped him and he found it difficult, whereas it is easy to obtain it starting from his own lemma. For our part, we have carried out the division by several methods.

Let us construct the entire figure that he introduced to show his lemma and on <the straight line> *AC* let us cut off *CN*, equal to *CH* and *BP* which is also equal to it.

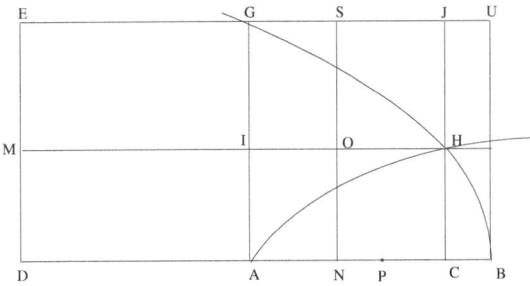

Fig. 11.3

I say that *AB* has been divided in the required manner at the points *P* and *N*.

Proof: We draw *NS* parallel to *AG*; it cuts *HI* at the point *O* and *JG* at the point *S*. Since the ratio of *CH*, that is to say *CN*, to *CA* is equal to the ratio of *AB* to <the sum of> *BA* and *AC*, from what he has said in his lemma; by separation, the ratio of *CN* to *NA*, that is to say *AI* to *IO*, or *GS*, is equal to the ratio of *AB* to *AC*, that is to say *AG* to *GJ*, and equal to the ratio of the remainder to the remainder, that is to say of *GI* to *JS*, that is to say *CN*; so the product of *AI* and *CN* which is equal to it, that is to say the square of *BP*, is equal to the product of *GI* and *IO*, that is to say <the product of> *PA* and *AN*. Since the rectangle *HG* is equal to the rectangle *ID*, that is to say to the rectangle *BI*, we take the rectangle *HU* as common; the rectangle *UI* is equal to the sum of the rectangles *CU* and *CI*. If we subtract <from that sum> the rectangle *CU*, and <also subtract it> from the rectangle *UI*, which is equal to it (the sum), the rectangle *OG* which is equal to *CU*, because they are equal to the square of *CH* – for *CU*, it is because of the parabola and for *GI*, it is from the preceding – there remains *CI*, or *NG*, <which is> equal to *UO*. The ratio of *AG*, or *AB*, to *US*, or *BN*, is equal to the ratio of *SO*, or *AP*, to *AN*. By separation, the ratio of *AN* to *NB* is equal to the ratio of *PN* to *NA*, so the product of *BN* and *PN* is equal to the square of *NA*. That is what we wanted to prove.

SINĀN IBN AL-FATḤ AND AL-QABĪṢĪ:
OPTICAL MENSURATION

Sinān ibn al-Fatḥ is not an unknown figure; al-Nadīm and, following him, al-Qifṭī each dedicate a short article to him. This is what the former writes:

> He belongs to the peoples of Ḥarrān, he was pre-eminent in the art of arithmetic and of numbers, there are among his books: the book on the dust board (*al-takht*) in Indian arithmetic, the book on addition and subtraction, the book on wills, the book on calculation of cubes, the book of commentary on algebra and *al-muqābala* of al-Khwārizmī[1].

On the other hand, al-Nadīm does not mention Sinān ibn al-Fatḥ's dates, or what he did. All one can say is that he lived after al-Khwārizmī and, of course, before al-Nadīm himself. In fact, in the *Book on the Calculation of Cubes*,[2] Sinān ibn al-Fatḥ mentions his commentary (*tafsīr*) on the *Algebra* of al-Khwārizmī.

[1] Al-Nadīm, *Kitāb al-fihrist*, ed. R. Tajaddud, Teheran, 1971, pp. 339–40.

[2] This book has come down to us in a manuscript in Cairo, Dār al-Kutub, Riyāḍa 260, fols 94^v–105^v. See R. Rashed, *Entre arithmétique et algèbre. Recherches sur l'histoire des mathématiques arabes,* Paris, 1984, pp. 21–2, and n. 11; English transl. *The Development of Arabic Mathematics Between Arithmetic and Algebra*, Boston Studies in Philosophy of Science 156, Dordrecht/Boston/London, 1994.

SINĀN IBN AL-FATḤ

Extracts from Optical Mensuration[3]

Sinān ibn al-Fatḥ says:

<1> If you want to know the length[4] of the straight line DE, <as seen> from the position D, let[5] the height be the straight line DA; we draw the line of sight from the point A to the point E. On the straight line DA you take any magnitude you wish; let it be the straight line AC. Then from the point C you draw a straight line to the line of sight AE; let the straight line be CB. The visual ray goes out from the point A to the <point> B and to the <point> E. You know that the ratio of the straight line AC to the straight line CB is equal to the ratio of the straight line AD to the straight line DE, because angle ACB is equal to angle ADE and all the angles of the triangle ADE are equal to those of the triangle ACB, so the sides of the triangle ACB are proportional to those of the triangle ADE.

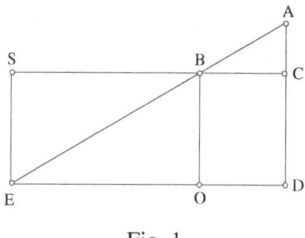

Fig. 1

But if the ratio of the first to the second is equal to the ratio of the third to the fourth, then the product of the first and the fourth divided by the second has as quotient the third; similarly, if you divide by the third, the result is the second; similarly, if you multiply the second by the third and you divide the product by the fourth, the result is the first and if you divide

---

[3] The short treatise of Sinān ibn al-Fatḥ on *Optical Mensuration* has come down to us in the collection Riyāḍa 260, fols 91ᵛ–94ʳ in the Dār al-Kutub, Cairo.

[4] Lit.: distance; this is an Arabism, but one that does not introduce any ambiguity.

[5] Lit.: as if, which we translate this way in what follows.

it by the first, the result is the fourth. So if the magnitude of *AC* compared to *CB* is equal to the magnitude of *AD* compared to *DE* and if you know the magnitude of *AC*, of *CB* and of *AD* – these three magnitudes – then you know the magnitude of *DE*, because you multiply the straight line *CB* by the straight line *AD*, you divide the product by the straight line *AC*, the result is the straight line *DE*. This <is> in order to know the length without measuring it <directly>.

<2> Similarly, to know a height without measuring it: Let the height of a wall be the straight line *DE*; let the visual ray go out from the point *A*; you erect a rod of any magnitude <at a distance less than> the distance *AD*; let there be a perpendicular *BC*; the magnitude of *CB* will be known to you, the magnitude of *AC* will be known to you and the magnitude of *AD* will be known to you; so the height *DE* will be known, from what we have described in regard to multiplication and division.

<3> Similarly, to know the depth of something without measuring it: Let the depth be *BO*, the width of the head of the well *BS* which is equal to *OE*, the rim of the well the point *B*; let us move from the point *B* to the point *C* and let the size be *CA*. The visual ray goes out from *A* towards *B* and *E* and the ratio of *AC* to *CB* is equal to the ratio of *BO* to *OE*. But *OE* is equal to *BS* and *AC*, *CB* and *BS* are three straight lines that you know. So the straight line *BO* is known, from what we have set out in regard to multiplication and division.

<4> To know the height of a mountain without measuring it: Let the height of the mountain be like <the length of> the straight line *OJ*. In relation to it you place yourself in the position of the point *G* from which there goes out the visual ray *GO*. Then from the point *G* you draw a perpendicular touching the straight line *GO*, let it be *GE*.

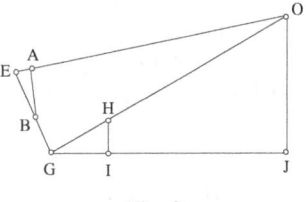

Fig. 2

The visual ray goes out from the position *E* toward the point *O*; let it be the straight line *EO*. You then take on the length of *EG* a length as you choose, let it be *EB*; you draw the straight line *BA* as far as the ray <and perpendicular to> *EO*, then the magnitude *EA* compared to *BA* is as the magnitude *EG* compared to *GO*. But the magnitudes of *EB*, *BA* and *EG* are three magnitudes that are known to you, so the magnitude of *GO* is known to you, from what we have set out in regard to multiplication and division.

If you then draw the straight line *IH* such that the magnitude of *GH* compared to *HI* is equal to the magnitude of *GO* compared to *OJ*, the magnitudes of *HI*, *HG* and *GO* are three magnitudes known to you, so the magnitude of *OJ* is known to you, from what we have set out in regard to multiplication and division.

These four kinds of proportionality that I have just mentioned enable one to obtain any distance, any height or any depth <in the> plane or not <in the> plane; understand it – if God the All-Highest so wishes.

<5> To know the width of a thing that you cannot measure, situated in a sea: Let *BA*, *DE* be a sea whose edge is *BE*;[6] you want to know the distance *AD*. Starting from the position *B* you know the distance *BA*, from what we have described above. You set up <the instrument> through which, at the position *B*, you take a sighting of the point *A* and <you turn it> until it (the line of sight) reaches the point *E*, then you measure <the distance> from *B* to *E*.

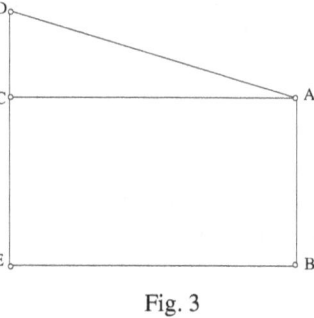

Fig. 3

Then you put <the instrument> through which you are taking sightings at the point *E* <and turn it> until it (the line of sight) reaches *D*. Then you find the distance *ED*, in the way we have described above. If the distance *AB* is equal to the distance *DE*, then the distance *AD* is equal to the distance *BE*, and if one of the distances is greater than the other, then subtract the smaller from the greater, multiply the result by itself, multiply the distance *BE* by itself, add the two products and take the root of the sum; what you obtain is the distance *AD*; understand it – if God the All-Highest so wishes.

The treatise of Sinān ibn al-Fatḥ on optical mensuration is completed. Praise be to God, Lord of the worlds.

---

[6] *BA* and *DE* are perpendicular to *BE*.

AL-QABĪṢĪ

Fragment on Optical Mensuration[7]

To know the height of something <that stands> on the surface of the ground, if we cannot reach its foot, that is as for knowing the height of mountains.

If you want <to do> this, by measurement with an astrolabe take the height of the summit of the mountain above the plane of the ground as when you take the height of a heavenly body, then you move back from this position by a distance great enough to change the height by several degrees; then take its height a second time from that position.

<For> the first height put a sine, that is the first sine. Then subtract the height from 90° and put the result as a sine, that is the second sine. Do the same for the second height; you obtain the third and fourth sines. Next multiply the second sine by the third sine, divide by the first sine, you subtract your result from the fourth sine and you keep what is left. Next you multiply the number of cubits between the two positions from which you have taken the heights by the third sine and you divide by what you were left with <at the end of the previous step>; what you obtain is the height of the mountain or the height of the thing whose height you want to know.

If you want to know how many <cubits there are> between the position from which you took the first height and the foot of the perpendicular from the <summit of the> mountain to the plane of the ground, then multiply the quotient formed by the division <carried out> before the subtraction of the fourth sine by the <number of> cubits between the two positions and divide it also by what you had left over from what remained; what you obtain is <the distance> between the first position from which you took the height

---

[7] An extract from the treatise by Abū Ṣaqr 'Abd al-'Azīz ibn 'Uthmān al-Qabīṣī, *Fī anwā' min al-a'dād wa-ṭarā'if min al-a'māl*, MS Istanbul, Aya Sofya 4832, fols 85ᵛ–88ʳ. A. Anbouba has published a critical edition of the whole of this treatise in 'Un mémoire d'al-Qabīṣī (4ᵉ siècle H.) sur certaines sommations numériques', *Journal for the History of Arabic Science*, vol. 6, nos 1 and 2, 1982, pp. 181–208, to pp. 188–9.

and the foot of the perpendicular from the <summit of the> mountain to the plane of the ground.

If you want to know the distance between your eye at the position from which you took the first height and the summit of the mountain, then multiply the distance between the position and the foot of the perpendicular from <the summit of> the mountain by itself, and multiply the perpendicular from <the summit of> the mountain by itself; you add the two products and you then take the root of this sum, this is what <distance> there is between your eye and the summit of the mountain. That is what we wanted to know.

SUPPLEMENTARY NOTES

I. *ON THE COMPLETION OF THE* CONICS

[**1**, Proposition 1, p. 175] In propositions 1 to 11, Ibn al-Haytham considers only half the conic section, in the case of the parabola the curve is cut off by its axis, in the case of the hyperbola we have half of one branch and we have one quadrant of the ellipse.

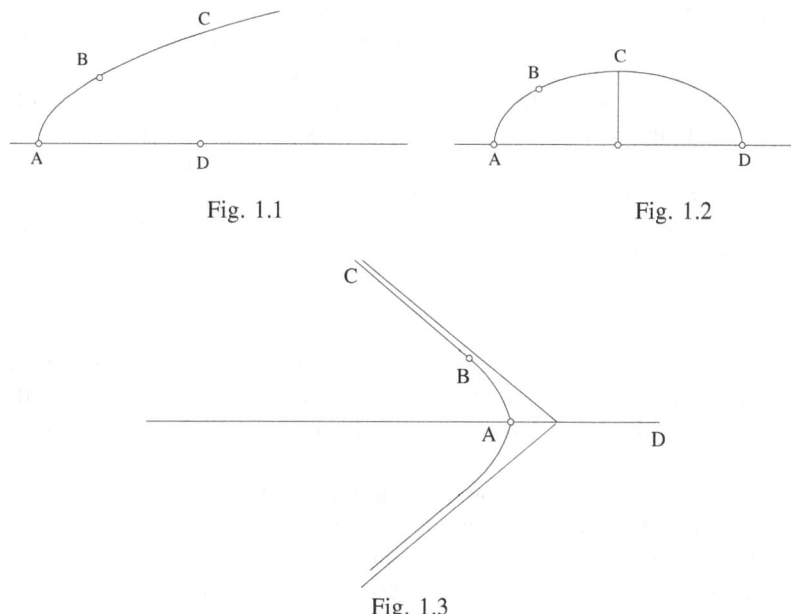

Fig. 1.1   Fig. 1.2

Fig. 1.3

[**2**, Proposition 3, page 178] We have explained (p. 42, note 42) what Apollonius means by this notion of a 'homologous straight line' or the straight line having an analogous ratio, which he introduces in Book VII of the *Conics*; see VII.2 ff. Thus, for example, for a hyperbola with transverse axis $A\Gamma$, of length d and *latus rectum* of length c, if B is a general point on the hyperbola, BE the perpendicular whose foot on the transverse axis is E, and N is a point on the axis such that $\dfrac{\Gamma N}{AN} = \dfrac{d}{c}$, the homologous straight line

is AN. The term 'homologous', which is the translation of the Arabic *shabīh al-nisba* (literally 'of analogous ratio'), requires no explanation. In VII.3, Apollonius introduces the notion of a homologous straight line for the ellipse. We give here Apollonius' text for the case of the hyperbola (VII.2):

> If we produce the axis of a hyperbola in such a way that the part that lies outside the section is the transverse diameter; if we cut off a straight line, starting from one of the ends of the transverse diameter, in such a way that the transverse diameter is divided into two parts whose ratio one to the other is equal to the ratio of the transverse diameter to the *latus rectum*, with the straight line that has been cut off corresponding to the *latus rectum*; if we draw from the end of the transverse diameter – the end that is also the end of the straight line that has been cut away – a straight line that meets the section in a general point, and if we draw from this point a perpendicular to the axis, then the ratio of the square of the straight line drawn from the end of the transverse diameter to the rectangle enclosed by the two straight lines that lie between the foot of the perpendicular and the two ends of the straight line that has been cut off, is equal to the ratio of the transverse diameter to the length by which it exceeds the straight line that has been cut off. Let us call the straight line that has been cut off <the straight line> having an analogous ratio.[1]

[**3**, Proposition 7, page 189] In this paragraph, Ibn al-Haytham considers the half NX of the parabola $\mathcal{P}$ and the half VX of the branch $\mathcal{H}_V$ of the hyperbola, and when he refers to asymptotes he means the half-lines wx and wy.

To prove that X is the only common point of the two arcs thus identified, he employs an argument of *reductio ad absurdum*.

If there existed a second common point Y:

1) the straight line XY that cuts $\mathcal{H}_V$ would cut wx and wy (*Conics*, II.8);

2) the straight line XY that cuts the arc NX of $\mathcal{P}$ would come out from inside $\mathcal{P}$, and would cut the asymptote wx, and would go on to cut the axis of $\mathcal{P}$ beyond N and thus could not cut the half-line xy.

The two arcs considered have only a single common point.

---

[1] *Apollonius: Les Coniques*, tome 4: *Livres VI et VII*, edition, translation and commentary by R. Rashed, Berlin/New York, 2009, see pp. 354–77; Arabic pp. 353, 11–355, 2.

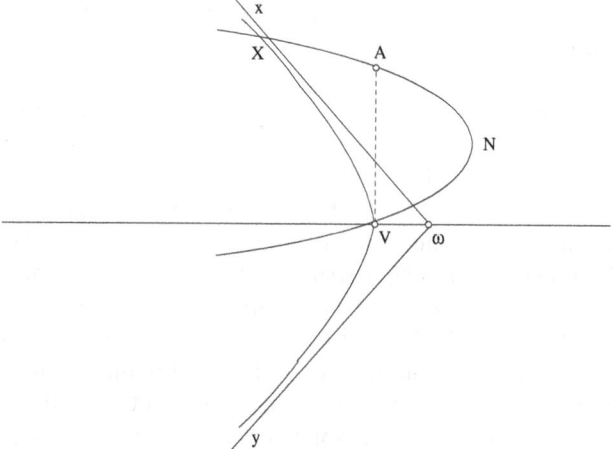

Fig. 2.1

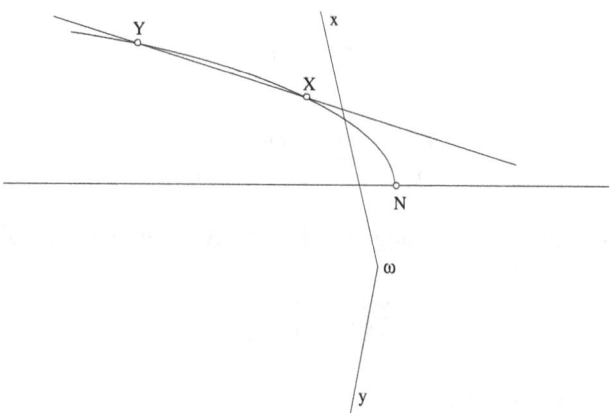

Fig. 2.2

II. A *NEUSIS* TO DIVIDE THE STRAIGHT LINE USED BY ARCHIMEDES

The manuscript tradition of Ibn al-Haytham's *On the Division of the Straight Line Used by Archimedes* includes after his text a different solution proposed by another mathematician.[2] The attribution to an anonymous author is explicit, as can be seen in the first sentence. We have a *neusis*.

The author gives a mechanical construction to solve Archimedes' problem. He draws ΔA and ZE perpendicular to ΔZ, in the same half-plane; he marks $\Delta A = \Delta B$ and $Z\Gamma = Z\Theta$ on ΔZ produced. He then imagines three movable straight lines AX, XE and $E\Gamma$, the first of which can turn about the fixed point A and the last can turn about the fixed point Γ while remaining parallel to the first one; the pivot points at X and E are constrained to move on ΔZ and ZE respectively. The solution is found when angle AXE is a right angle.

Angles ΔXA and $Z\Gamma E$ are equal, since XA and ΓE are parallel. So the right-angled triangles $A\Delta X$ and $XE\Gamma$ are similar, hence there is the proportional relationship

$$\frac{A\Delta}{\Delta X} = \frac{XE}{E\Gamma}.$$

Thus

$$\frac{B\Delta^2}{\Delta X^2} = \frac{A\Delta^2}{\Delta X^2} = \frac{XE^2}{E\Gamma^2} = \frac{XZ^2}{ZE^2},$$

because the right-angled triangles $XE\Gamma$ and XZE are similar. Since $ZE^2 = XZ \cdot Z\Gamma$, we get

$$\frac{B\Delta^2}{\Delta X^2} = \frac{XZ}{Z\Gamma} = \frac{XZ}{Z\Theta},$$

which is what we wanted.

This construction is reminiscent of the one for the two proportional means that Eutocius attributes to Plato[3] and the analogous construction

---

[2] Our edition of this text is based on the following manuscripts: Istanbul, Beşiraga 440, fols 275ᵛ–276ʳ [called B]; Istanbul, Süleymaniye, Atif 1712, fol. 147ᵛ [called O]; Istanbul, Süleymaniye, Carullah 1502, fol. 223ʳ [called C]; Leiden, Or. 14, fols 499–500 [called L]. For the history of the texts, see above, Chapter III. Comments on this text by F. Woepcke appear in *L'Algèbre d'Omar Alkhayyæmî*, publiée, traduite et accompagnée d'extraits de manuscrits inédits, Paris, 1851, pp. 93 ff.

[3] *Commentaire du livre II d'Archimède* Sur la sphère et le cylindre, ed. Mugler, pp. 45–6.

found in the Banū Mūsā.[4] The difference is that, in these latter constructions, the two right angles are fixed; one of the vertices runs along an axis and the solution is found when the second vertex lies on another axis. In contrast, in the present construction two vertices run along fixed axes and we have found the solution when the angles are right angles.

'In a different way, due to someone else, if we make the straight line move: Let the straight line be ΔZ[5] on which we mark the two points B and Θ; we want to divide the straight line ΔB at the point X such that the ratio of XZ to $Z\Theta$ is equal to the ratio of the square of $B\Delta$ to the square of ΔX. Let us draw from the points Δ and Z perpendiculars ΔA and ZE in the same direction; let us cut off ΔA equal to ΔB, let us produce ΔZ to Γ and let us cut off $Z\Gamma$ equal to $Z\Theta$. Let us imagine that at the two points A and X there are two movable straight lines in two different directions, say the lines AX and XE, such that AX cuts ΔB and XE cuts ZE and that when the lines move they are always parallel,[6] with the condition that the straight line joining the two points of intersection, that is, points X and E, encloses a right angle with each of the lines AX and XE; let the line be EX. It is then that angle AXE is a right angle, as is angle $XE\Gamma$.

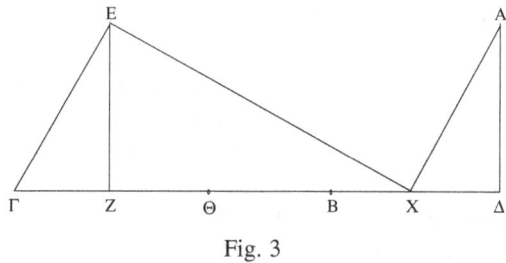

Fig. 3

I say that the ratio of XZ to $Z\Theta$ is equal to the ratio of the square of $B\Delta$ to the square of ΔX because the two angles $A\Delta X$ and $XE\Gamma$ are right angles and angle $AX\Delta$ is equal to angle Γ because the two lines are parallel; the remaining angle, angle A, is equal to angle EXZ. So the two right-angled triangles $A\Delta X$ and $XE\Gamma$ are similar, the ratio of $A\Delta$ to ΔX is equal to the ratio of XE to $E\Gamma$, and the ratio of the square of $A\Delta$, that is, the square of $B\Delta$, to the square of ΔX is equal to the ratio of the square of EX to the square of $E\Gamma$, that is, the ratio of the square of XZ to the square of ZE. But the ratio of the square of XZ to the square of ZE is equal to the ratio of XZ

[4] *Mathématiques infinitésimales*, vol. I, pp. 52–4.
[5] We have deliberately adopted here the lettering used by Archimedes.
[6] He means that AX must be parallel to $E\Gamma$.

to $Z\Gamma$, because ZE is a mean in proportion between the lines XZ and $Z\Gamma$; but the line $Z\Gamma$ is equal to the line $Z\Theta$, so the ratio of XZ to $Z\Theta$ is equal to the ratio of the square of $B\Delta$ to the square of ΔX; that is what we wanted.'

III. AL-QŪHĪ: *THE LEMMA TO ARCHIMEDES' DIVISION OF A STRAIGHT LINE*

Lemma: *Let there be two segments,* AB *and* C; *to find on the straight line* AB, *between* A *and* B *or beyond* B, *a point* D *such that* $\dfrac{AD}{C} = \dfrac{C^2}{BD^2}$, *or* $AD \cdot BD^2 = C^3$.

Here we recognize the problem raised by Archimedes in *The Sphere and the Cylinder* II.4, with the sole difference that Archimedes considers a segment C and an area Γ unrelated to C, whereas here we have $\Gamma = C^2$. Al-Qūhī's equation is no less general, since we can always construct a segment C' such that $C \cdot \Gamma = C'^3$,[3] if we know how to insert two proportional means between two given magnitudes. Note that the form adopted by al-Qūhī enables him to express the problem's conditions of solvability by bounding above the segment C', whereas the commentary of Eutocius gave it by bounding above the volume $C \cdot \Gamma$.

We suppose $BE = C$, with E, and D on the same side of B, and we complete the square $BEGH$. Let $\mathscr{P}$ be the parabola of vertex A, with axis AB and *latus rectum* C; and let $\mathscr{H}$ be the hyperbola of vertex G, having BE and BH as asymptotes.

Let I be common to $\mathscr{H}$ and $\mathscr{P}$, $ID \perp AB$, $IK \perp BH$.

(1) $I \in \mathscr{P}$, whence $ID^2 = C \cdot AD$;
(2) $I \in \mathscr{H}$, whence $BE \cdot EG = BK \cdot KI = ID \cdot DB = C^2$.

From (1) and (2), we deduce

$$\frac{ID}{C} = \frac{AD}{ID} = \frac{C}{DB};$$

hence

$$\frac{AD}{C} = \frac{C^2}{DB^2},$$

and hence the result.

Remark on the existence of I:

In the case of Fig. 4.2, point G, vertex of the hyperbola, is inside the parabola, for $AE = AB + C$; and if F is the point of the parabola that is projected upon AB at E, we have $EF^2 = (AB + C)C$; therefore $EF > C$, and $\mathscr{P}$ cuts $\mathscr{H}$ at a point I of the arc AF.

However, in the case of Fig. 4.1, a discussion is necessary, which the author does not undertake; nevertheless, he does give the condition, which allows us to suppose that he had envisaged it.

In fact, we may have three cases:

1) The first case is that of the manuscript's Fig. 4.1. In this case, the parabola and the hyperbola cut one another at two points, I and I'; whence there are two solutions, D and D'.

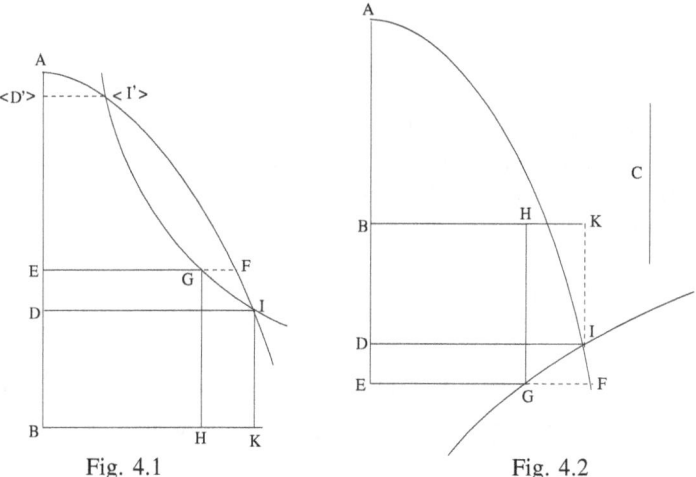

Fig. 4.1   Fig. 4.2

2) The case of Fig. 4.3. In this case, the parabola and the hyperbola are tangent, and to the point of contact I there corresponds a point D.

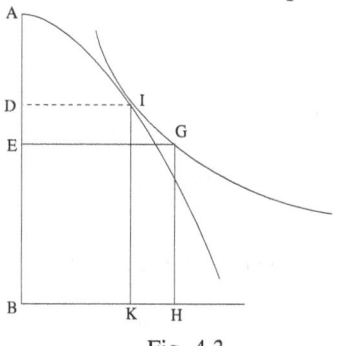

Fig. 4.3

3) The case of Fig. 4.4. In this case, the parabola and the hyperbola do not cut one another, and there is no point D between A and B.

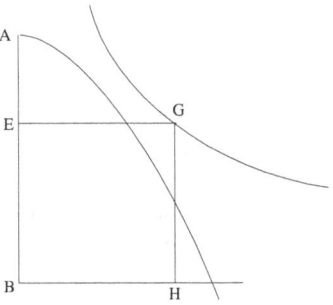

Fig. 4.4

The discussion of this problem can be reduced to that of a cubic equation. We take the half-line AB as axis, and we suppose $AB = a > 0$, $AD = x > 0$, and $BD = a - x$. We designate by c the length of the given segment C, and we have

$$\frac{AD}{c} = \frac{c^2}{BD^2} \iff AD \cdot BD^2 = c^3 \iff x(a-x)^2 = c^3.$$

We suppose $f(x) = x(a - x)^2$. This function has a maximum at point $x = \frac{a}{3}$, and we have $M = f\left(\frac{a}{3}\right) = \frac{4a^3}{27}$. The equation is written $f(x) = c^3$, and we have:

1) if $0 < c^3 < \frac{4a^3}{27}$, the equation has three roots:

$$0 < x_1 < \frac{a}{3} < x_2 < a < x_3.$$

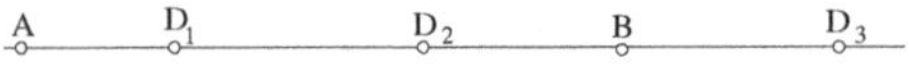

A D_1 D_2 B D_3

Fig. 4.5

2) If $c^3 = \frac{4a^3}{27}$, the equation has two roots, one double and one simple:

$$x_1 = x_2 = \frac{a}{3}, \text{ and } x_3 > a.$$

Fig. 4.6

3) If $c^3 > \dfrac{4a^3}{27}$, the equation has a single root:

$$x_3 > a.$$

Fig. 4.7

Al-Qūhī's procedure amounts to taking two points A and B as given, and showing that the function $f(D) = AD \cdot BD^2$, $D \in [A, B]$, presents a maximum for D verifying $AD = \dfrac{AB}{3}$. He did not consider D on AB produced; nevertheless, he affirms without the slightest ambiguity that in order to have D between A and B, it is necessary that $c^3 \leq \dfrac{4a^3}{27}$.

Now that this lemma is established, we have therefore found point D such that

(1) $AD \cdot BD^2 = C^3,$

with

(2) $AD = AB \pm BD.$

Al-Qūhī indicates that in the case where $AD = AB - BD$, it is necessary that $c^3 \leq \dfrac{4}{27} AB^3$ – a condition that he had mentioned at the beginning of the text.

From (2), we deduce

$$AD \cdot BD^2 = AB \cdot BD^2 \pm BD^3,$$

and, according to (1), we have

$$C^3 = AB \cdot BD^2 \pm BD^3.$$

As AB and C are two known segments, we know BD according to the lemma. But $AB \cdot BD^2$ is the volume of a parallelepiped (P) of altitude AB, whose base is a square with sides equal to BD; in addition, BD^3 is the volume of a cube constructed on this same square. If, then, C^3 is the volume of a given solid, we know how to solve the problem: construct on AB a parallelepiped of square base, such that if a cube of the same base is added to it or subtracted from it, the solid obtained has a known volume.

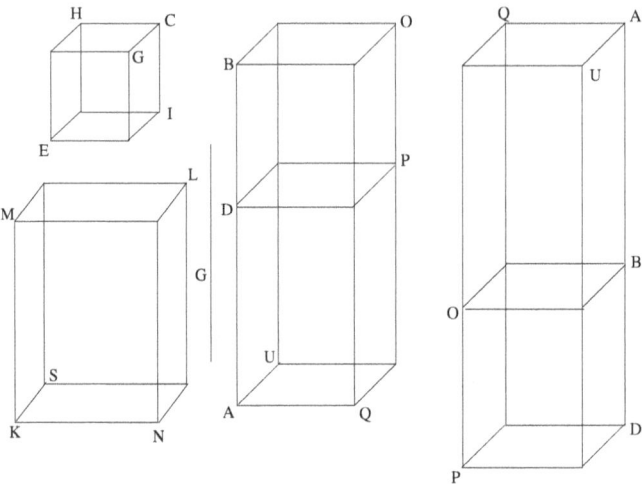

Fig. 4.8

To generalize the problem, al-Qūhī takes as given two points A and B and a parallelepiped (CE) of known form, and replaces the cube of volume C^3 by a parallelepiped similar to (CE) whose volume is known; let there be the solid (KL), of which KM is an edge. The trihedral angles (C, HIG) and (K, MNS) are equal. We suppose $(KL) = V$; on the straight line AB we consider a point D such that $\dfrac{AD}{KM} = \dfrac{KM^2}{BD^2}$. We construct on BD (in both cases) the solid (DO), similar to (CE) and to (KL); from its base (DP) we construct the solid (AP) with edge AD. We show that

$$\mathrm{vol.}(AP) = V.$$

Demonstration: Let G be a segment such that $\dfrac{AD}{G} = \dfrac{G}{KM}$; we then have[7]

$$\left(\frac{AD}{G}\right)^2 = \left(\frac{G}{KM}\right)^2 = \left(\frac{AD}{G}\right)\left(\frac{G}{KM}\right) = \frac{AD}{KM}.$$

[7] The letter G has two different meanings.

But

$$\frac{AD}{KM} = \frac{KM^2}{BD^2},$$

and therefore

$$\frac{AD}{G} = \frac{G}{KM} = \frac{KM}{BD};$$

from which we deduce

$$\left(\frac{KM}{BD}\right)^3 = \left(\frac{AD}{G}\right)\left(\frac{G}{KM}\right)\left(\frac{KM}{BD}\right) = \frac{AD}{BD}.$$

In addition,

$$\frac{AD}{BD} = \frac{\text{vol.}(AP)}{\text{vol.}(DO)},$$

for the two solids have the same base, and

$$\left(\frac{KM}{BD}\right)^3 = \frac{\text{vol.}(KL)}{\text{vol.}(DO)},$$

for these two solids are similar; hence the conclusion

$$\text{vol.}(AP) = \text{vol.}(KL) = V.$$

The solid constructed on AB is (AO), and we have

$$V = \text{vol.}(AO) \pm \text{vol.}(DO), \text{ or vol.}(AO) = V \mp \text{vol.}(DO).$$

Al-Qūhī proposes another, simpler variant of this demonstration. He takes as given the segments AB and C; let D be on AB or on its extension, such that $\frac{AD}{C} = \frac{C^2}{BD^2}$ (two possible cases).

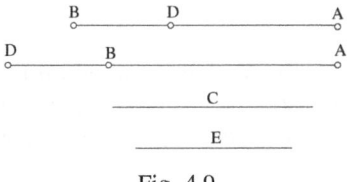

Fig. 4.9

On BD, we construct a parallelepiped P_1; let $v(BD)$ be its volume. And on AD we construct a parallelepiped P_2 with the same base as P_1; let $v(AD)$ be its volume. If we construct on C a parallelepiped P similar to P_1, and if V is its volume, then $v(AD) = V$.

Demonstration: Let E be the segment defined by $\dfrac{AD}{E} = \dfrac{E}{C}$; then

$$\frac{AD}{C} = \left(\frac{AD}{E}\right)^2 = \left(\frac{E}{C}\right)^2.$$

But we know that

$$\frac{AD}{C} = \frac{C^2}{BD^2};$$

we therefore have

$$\frac{AD}{E} = \frac{E}{C} = \frac{C}{BD},$$

hence

$$\frac{AD}{BD} = \left(\frac{C}{BD}\right)^3.$$

Yet, since P_1 and P_2 have the same base, we have

$$\frac{AD}{BD} = \frac{v(AD)}{v(BD)}.$$

In addition, the solids P and P_1 are similar; therefore

$$\frac{V}{v(BD)} = \left(\frac{C}{BD}\right)^3,$$

and therefore $V = \text{vol.}(AD)$, or else $V = \text{vol.}(AB) \pm \text{vol.}(BD)$.

We see that the extension of Archimedes' problem proposed by al-Qūhī may be interpreted as an 'application of volumes' analogous in three dimensions to the application of areas dealt with in Book VI of Euclid's *Elements*: A segment AB being given, we seek to apply along this segment a given volume V, with a deficiency or excess similar to a given parallelepiped.

The *editio princeps* of the text[8] was edited on the basis of the Leiden manuscript Or. 168/8, fols. 80^v–84^v. F. Woepcke has analysed this text; see the 'Additions' to his *Algèbre d'Omar Alkhayyāmī*, Paris, 1851; reprinted in

---

[8] See *Les Mathématiques infinitésimales*, vol. III, pp. 919–35.

Études sur les mathématiques arabo-islamiques, Frankfurt am Main, 1986, vol. I, Appendice B, pp. 96–102.

Finally, let us note that the title of this epistle does not appear in any of the lists of al-Qūhī's works drawn up by the ancient bio-bibliographers. As we know, these lists are often incomplete, and the absence of a title is no indication with regard to its authenticity. Al-Qūhī himself does not come to our aid, since he does not mention this epistle in any of his other writings. Yet this argument *ex silentio* has no more validity, especially in view of the affirmation at the end of the epistle: 'This lemma was established by the Master Abū Sahl al-Qūhī, may God be satisfied with him. I gave a copy of it to the Master Abū al-Jūd ...'. This testimony is solid, and we have no argument to doubt it. With regard to al-Qūhī's correspondent, who transmitted a copy to Abū al-Jūd, nothing enables us to identify him, at least for the time being. We know that al-Qūhī carried out some scientific correspondence, like the famous one he exchanged with al-Ṣābi'.

Al-Qūhī and the Lemma to the Division of the Straight Line
by Archimedes

In the Name of God, the Merciful, the Compassionate
On Him I rely

I took pause, O my brother, over what you mentioned of the statements of the geometer Abū ʿAbd Allāh al-Māhānī in a treatise intended to comment upon the second book of Archimedes' work on *The Cylinder, the Sphere, and the Cone*, according to which he was able to deal with eight chapters of the total of nine chapters of this work, and he was incapable of rectifying the fourth chapter, which deals with the division of the sphere into two parts according to a given ratio, because of the difficulty of a lemma which he needed;[9] he[10] tried to solve it by algebra, which led him to equalize the cube, the squares, and the number.[11] Now, these terms are not proportional. Thus, this is the same as the application to a given straight line of a parallelepiped that is deficient by a cube with regard to it. You asked me to explain this lemma. I then needed to introduce another lemma, which facilitates the approach to it, and which is as follows.[12]

The two straight lines *AB* and *C* being given, we wish to divide *AB* at *D* such that the ratio of *AD* to *C* is equal to the ratio of the square of *C* to the square of *BD*. This is what we need for the lemma which al-Māhānī found difficult.

---

[9] Archimedes.

[10] That is, al-Māhānī.

[11] In the manuscript, we read: *muʿādalat al-mukaʿʿab wa-al-amwāl ʿadadan* – that is, $x^3 + ax^2 = c^3$; this, as we shall see later, is not the equation under consideration. F. Woepcke thought this was the result of a *lapsus calami*, which is quite possible. It is also probable that there was an error as early as the transcription, and that the initial sentence read: *muʿādalat al-mukaʿʿab wa-al-amwāl wa-al-ʿadad*, 'to equalize the cube, the squares, and the number'. In order to decide, we must find a text of this treatise from a family other than the only one we know. In fact, what is at issue is the equation $x^3 + c^3 = ax^2$, whose terms are not proportional; for we move from affinity $x \to (a - x)$ to the equation $x^3 + a^2x = c^3 + 2ax^2$.

[12] Here, al-Qūhī does not take up Archimedes' lemma in the latter's terms (see the commentary above).

This is possible only if straight line *C* is not longer than the straight line that is capable of the solid having as its edge[13] one-third of *AB*, and deficient by a cube[14] whose side is two-thirds of *AB* – that is, the straight line capable of four-ninths of one-third of the cube of *AB*. But we want it to be the latter, which is more general than the former. We then consider *AB* according to two cases: in one, we wish to separate *BD*, and in the other we add *BD*, so that the ratio of *AD* to *C* is equal to the ratio of the square of *C* to the square of *BD*.

We suppose *BE* to be equal to *C*; we complete the square *BEGH*, and we construct a parabola of vertex *A*, diameter *AB*, and *latus rectum* the straight line *C*; let it be the section *AI*. We construct a hyperbola that passes through point *G*, and such that the straight lines *BE* and *BH* are its asymptotes; let it be the section *GI*. The two sections necessarily cut one another; let them cut one another at *I*. From point *I*, we drop a perpendicular to *AB*; let it fall at *D*. Apollonius has shown in his book on *Conics* that the square of the perpendicular lowered from the parabola to its diameter is equal to the product of what it separates from the diameter, starting with the vertex of the section, by the *latus rectum*;[15] the product of *AD* by *C* is therefore equal to the square of *ID*, and the ratio of *AD* to *ID* is equal to the ratio of *ID* to *C*.

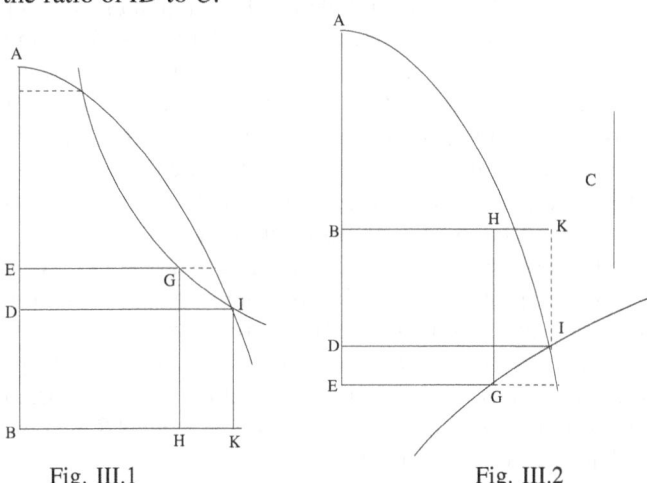

Fig. III.1   Fig. III.2

---

[13] This expression renders the Arabic *al-muḍāf ilayhi*, which translates literally as 'applied to'. Here the meaning is clear.

[14] The text reads 'square' instead of 'cube'; this is probably a *lapsus calami* brought about by analogy with the text of Euclid.

[15] *Conics*, I.11.

Likewise, from point *I* we draw the straight line *IK* parallel to *BD*, and we extend straight line *BH* until it meets it at *K*. Straight line *EG* therefore falls on the hyperbola from the asymptote in a manner parallel to the other asymptote; similarly for the straight lines *GH*, *KI*, and *ID*. According to what Apollonius has shown,[16] the product of *BE* by *EG* is equal to the product of *BK* by *KI*. But each of the <straight lines> *BE* and *EG* is equal to *C*, *BK* is equal to *ID*, and *KI* is equal to *BD*; therefore, the product of *ID* by *BD* is equal to the square of *C*, and the ratio of *ID* to *C* is equal to the ratio of *C* to *BD*. We have thus shown that the ratio of *AD* to *ID* is equal to the ratio of *ID* to *C*, and is equal to the ratio of *C* to *BD*; therefore, the ratio of *AD*, the first, to *C*, the third, is equal to the ratio of the square of *C*, the third, to the square of *BD*, the fourth. That is what we wanted to prove.

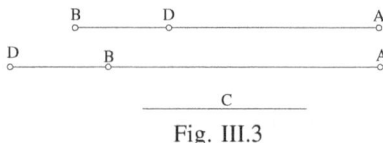

Fig. III.3

Once this lemma has been introduced, let there be *AB* for us in two cases: we wish to apply to it a parallelepiped equal to a given solid, to which we add or from which we subtract a cube;[17] let the straight line *C* be the side of a cube equal to the given solid. In one case, we separate *BD* from *AB*, and in the other we add *BD* to it, so that the ratio of *AD* to *C* is equal to the ratio of the square of *C* to the square of *BD*. Straight line *C* is limitless when we add it, but it is necessarily limited when we subtract it; in other words, straight line *C* cannot be longer than the straight line that is capable of a solid[18] that is four-ninths of one-third of the cube of *AB*, whose edge is one-third of *AB*, to which we have added a cube whose side is two-thirds of *AB*. The product of *AD* by the square of *BD* is a parallelepiped surrounded twice by the square of *BD*, and four times by the surface *AD* by *DB*; the product of *C* by the square of *C* is the cube equal to the given solid. The solid applied to *AB* in the first case is deficient with regard to *AB* by a cube whose side is *BD*, and in the other case exceeds *AB* also by a cube of side *BD*. That is what we wanted to prove.

This having been constructed, I would like to make this proposition universal – that is, that the solid that is applied to *AB* should be equal to a

---

[16] *Conics*, II.12.

[17] The cube is added to or subtracted from the given parallelepiped.

[18] Lit.: a cube.

given parallelepiped, augmented or diminished by a solid similar to a parallelepiped of known form.

Let the solid of known form be the solid *CE*; its angle is that which is limited by the straight lines *CG*, *CH*, and *CI*. We construct a solid similar to the given solid, like the solid *KL*, whose angle, equal to angle *C* of solid *CE*, is that which is limited by the straight lines *KM*, *KN*, and *KS*. We subtract *BD* from the given *AB*, in one case, and we add it in the other case, so that the ratio of *AD* to *KM* is equal to the ratio of the square of *KM* to the square of *BD*, according to the condition mentioned in the lemma, as we have shown in the first proposition of this treatise. On *BD*, we construct a solid similar to solid *CE*; let it be the solid *DO*. We complete the solid *AP* of length *AD*, and of the width and depth <of solid> *DO*; let the angle which is equal to angle *C* and equal to angle *D* of the two solids *CE* and *DO* be the angle limited by the straight lines *AD*, *AU*, and *AQ*; I say that solid *AP* is equal to solid *KL*.

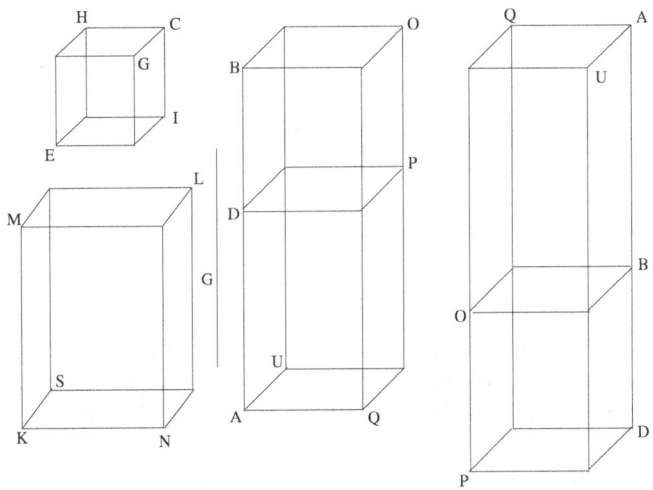

Fig. III.4

Demonstration: Between the two straight lines *AD* and *KM*, we draw the straight line *G*, a proportional mean; we have the ratio of *AD* to *KM* equal to the ratio of *AD* to straight line *G* repeated twice. Now, the ratio of *AD* to *KM* is equal to the ratio of the square of *KM* to the square of *BD*, and the ratio of *AD* to *G* is equal to the ratio of *KM* to *BD*; therefore, the ratio of *AD* to *G* is equal to the ratio of *G* to *KM*, and is equal to the ratio of *KM* to *BD*. The ratio of *AD* to *BD* is therefore equal to the ratio of *KM* to *BD*, repeated three times. But the ratio of *AD* to *BD* is equal to the ratio of solid *AP* to solid *DO*, for they are in accordance with the length of a single straight line, and of the same width and depth. But the ratio of *KM* to *BD*, repeated three times, is equal to the ratio of solid *KL* to solid *DO*, for they are similar; therefore, the ratios of solid *AP* and of solid *KL* to solid *DO* are the same. Solid *AP* is therefore equal to solid *KL*, which is equal to the given solid. Yet we have applied it to *AB*, deficient in one case and excessive in the other by a solid *DO*, similar to the solid *CE* of known form; this is what we wished to construct.

More accessible than this, and with fewer straight lines: if we suppose the two straight lines *AB* and *C*, if we separate *BD* from *AB* in one case, if we add *BD* to it in the other, if we suppose the ratio of *AD* to straight line *C* to be equal to the ratio of the square of straight line *C* to the square of straight line *BD*, and if we construct on the side *BD* a parallelepiped of known form, and on straight line *C* a solid that is similar to it, then the latter is equal to the solid constructed on *AD* with the same width, the same depth, and angles equal to those of the solid of known form constructed on *BD*.

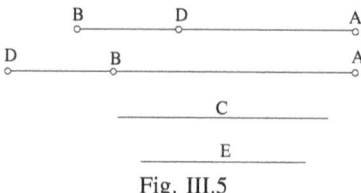

Fig. III.5

Demonstration: Between *AD* and *C* we draw the straight line *E*, a proportional mean. We have the ratio of *AD* to *C* equal to the ratio of *AD* to *E* repeated twice, and the ratio of the square of *C* to the square of *BD* is equal to the ratio of *C* to *BD*, repeated twice; therefore, the ratio of *AD* to *E* is equal to the ratio of *C* to *BD*, and the ratio of *AD* to *E* is equal to the ratio of *E* to *C*. Therefore, the four straight lines *AD*, *E*, *C*, and *BD* are in continuous proportion. The ratio of *AD* to *BD* is therefore equal to the ratio of *C* to *BD*, repeated three times. But the ratio of *AD* to *BD* is equal to the ratio of the solid constructed on *AD* with width, depth, and angles equal to those of the solid of known form constructed on *BD*, to the solid constructed on *BD*, since they are of the same width and same depth; and the ratio of *C* to *BD*, repeated three times, is equal to the ratio of the solid constructed on straight line *C* to the solid constructed on *BD*, since they are similar. Therefore, the similar <solid> constructed on *C* is equal to the solid constructed on the side *AD*, with width, depth, and angles equal to those of the <solid> constructed on *BD*; it[19] is deficient with regard to *AB* in one of the two cases, and it exceeds it in the other case, by the solid of known form, one of whose sides is *BD*. That is what we wanted.

If we suppose the straight line *C* to be the side of any parallelepiped equal to a given solid, and similar to a solid of known form, it will be equal to the one we mentioned applied to *AB*, which is deficient or excessive with regard to it by a solid similar to a solid of known form. The side of the solid such that you wish to apply to *AB* an equal solid, deficient by a similar solid to a known solid, must not be greater than that whose edge is one-third of *AB*, which is the side of the similar solid, deficient by a cube of two-thirds of *AB*.

This lemma was established by the Master Abū Sahl al-Qūhī – may God be satisfied with him. I have given a copy of it to Master Abū al-Jūd – may God have pity upon him.

---

[19] That is, the solid constructed on the side *AD*.

ADDENDA (VOL. 2)

AL-ḤASAN IBN AL-HAYTHAM AND MUḤAMMAD IBN AL-HAYTHAM: THE MATHEMATICIAN AND THE PHILOSOPHER

It was in a chapter with the same title, in the preceding volume, that we pointed out a confusion that certainly dates back to the work of the biobibliographer Ibn Abī Uṣaybiʿa, and perhaps to earlier authors, and has been perpetuated by all modern historians. The confusion is between two figures who were contemporaries of one another: Abū ʿAlī al-Ḥasan ibn al-Ḥasan ibn al-Haytham and Muḥammad ibn al-Ḥasan ibn al-Haytham. In our earlier account we put forward many arguments, some of which appeared to us to be irrefutable, and still appear to be so. Here we shall add three pieces of evidence that have come to light since we wrote the earlier account. They have not been noticed until now, and all of them lend support to our proof.

1. Highly informative testimony is provided by the famous philosopher Fakhr al-Dīn al-Rāzī. He mentions the name of *Abū 'Alī (al-Ḥasan) ibn al-Haytham* as well as that of *Muḥammad ibn al-Haytham*. But whereas he refers to the former only in connection with mathematics and optics, he discusses the latter only in a context of theology and philosophy. Thus, at no time does al-Rāzī seem to confuse the two figures or their respective areas of activity.

In his own works, al-Rāzī cites several treatises that he expressly attributes to Abū ʿAlī (al-Ḥasan) ibn al-Haytham. These are *Optics, On the Resolution of Doubts on the Book of the* Elements *of Euclid*, the treatise *On Place*, a treatise on Proposition X.1 of the *Elements* and the treatise *On Errors in the Method of Making Observations*. Now these titles correspond closely with those of surviving works by al-Ḥasan ibn al-Haytham. Thus, in his book *al-Mulakhkhaṣ*, al-Rāzī cites *On the Resolution of Doubts*[1] as well

---

[1] *Die Erkenntnislehre des 'Aḍudaddin al-Īcī*, Übersetzung und Kommentar des ersten Buches seiner Mawāqif von Josef van Ess, Akademie der Wissenschaften und der Literatur Veröffentlichungen der Orientalischen Kommission, vol. XXII, Wiesbaden, 1966, p. 175.

as the treatise *On Place*.[2] In his treatise called *Higher Researches* (*al-Maṭālib al-ʿāliyya*), of 605 AH (1208/1209) he writes:

إن لأبي علي بن الهيثم رسالة في بيان أن كل مقدار يفصل منه جزء من أجزائه، ويفصل من الباقي
جزء نسبته إلى الجزء الأول مثل نسبة الجزء الأول إلى الكل، ويفعل ذلك دائمًا، فإن جميع تلك
الأجزاء المأخوذة على تلك النسبة إلى غير النهاية، إذا جمعت فليس تبلغ جملتها إلى الجزء الذي كان
أعظم من الجزء الأول.

There is a treatise by Abū ʿAlī al-Ḥasan ibn al-Haytham that proves that if we cut off one part of a magnitude, and from the remainder cut off a part whose ratio to the first part is equal to the ratio of the first part to the whole, and if we continue to proceed in this way, then if we add up the parts we have taken, in this ratio, to infinity, the sum will not be so great as the part that was greater than the first part.[3]

In other words, let A be a magnitude, αA a part of this magnitude, where $0 < \alpha < 1$, and let there be a sequence of equal ratios $0 < \alpha_i < 1$ (for $i = 1, 2, 3 \ldots$), we have for $\alpha_i = \alpha$ for all values of i, $\lim_{n \to \infty}(1 - \alpha)^n = 0$. This is exactly what Abū ʿAlī al-Ḥasan ibn al-Haytham proved in his treatise *On the Division of Two Different Magnitudes*[4] and which he to some extent took up again in his *Commentary on Euclid's* Elements.[5]

So there is absolutely no doubt which source was consulted by al-Rāzī, or about who wrote it.

A little later in the same book, al-Rāzī again cites the mathematician, in these terms:

إن أبا علي بن الهيثم بيّن في كتاب حل شكوك أقليدس.

Abū ʿAlī ibn al-Haytham has proved in his book *On the Resolution of Doubts concerning Euclid* [...].[6]

---

[2] Al-Rāzī uses Ibn al-Haytham's own terms in taking up the latter's criticism of the traditional account of place as the surface enclosing the body. See *al-Mulakhkhaṣ*, ms. Majlis Shūrā, no. 827, fols 92–93. See also *Les Mathématiques infinitésimales du IXe au XIe siècle*, vol. IV: *Méthodes géométriques, transformations ponctuelles et philosophie des mathématiques*, London, 2002, which gives al-Rāzī's text with a translation into French.

[3] Fakhr al-Dīn al-Rāzī, *al-Maṭālib al-ʿāliyya*, ed. Aḥmad Ḥijāzī al-Saqqā, Beirut, 1987, VI, pp. 81–2.

[4] See R. Rashed, *Les Mathématiques infinitésimales du IXe au XIe siècle*, vol. II: *Ibn al-Haytham*, London, 1993.

[5] *Ibid*.

[6] Fakhr al-Dīn al-Rāzī, *al-Maṭālib al-ʿāliyya*, p. 165.

In the eighth volume of the same book (p. 155), al-Rāzī writes:

إن الشيخ أبا علي بن الهيثم صنف رسالة في أنواع الخلل الواقع في آلات الرصد ، وعد منها قريباً من
ثلاثين وجهاً من الوجوه التي لا يمكن الاحتراز عنها .

The Master Abū ʿAlī ibn al-Haytham has written a treatise on the kinds of error that occur with observational instruments and he has counted about thirty kinds, which cannot be avoided.

Citations in al-Rāzī's two treatises, *al-Mulakhkhaṣ* and *Higher Researches*, show that the author was familiar with some works by the mathematician, whose names he gives without the slightest ambiguity: Abū ʿAlī ibn al-Haytham.

Let us now turn to al-Rāzī's *Opus Magnum*, his *Great Commentary* (*al-Tafsīr al-Kabīr*) on the Koran. In this work, he cites both Abū ʿAlī ibn al-Haytham and Muḥammad ibn al-Haytham. In Volume XIII of his *Commentary*, in a discussion of morning twilight, al-Rāzī writes:[7]

فإن قالوا : لم لا يجوز أن يقال : الشمس حين كونها تحت الأرض توجب إضاءة ذلك الهواء المقابل
له (الضمير يعود على قرص الشمس)، ثم ذلك الهواء المقابل (مقابل) للهواء الواقف فوق الأرض،
فيصير[ه] ضوء الهواء الواقف تحت الأرض سبباً لضوء الهواء الواقف فوق الأرض، ثم لا يزال يسري
ذلك الضوء من هواء إلى هواء آخر ملاصق له حتى يصل إلى الهواء المحيط بنا ؛ هذا هو الوجه الذي
عول عليه أبو علي بن الهيثم في تقدير هذا المعنى في كتابه الذي سماه بالمناظر [الكثه؟].

If they say: why is it not permitted to say that the Sun, when it is below the Earth, necessarily illuminates the air that faces it, and then that air facing it will illuminate the air that is above the Earth, and thus the light of the air that is below the Earth will be the cause of the light of the air that is above the Earth, then that light continues to propagate from one air to another air that is near it until it reaches the air that surrounds us? This is the method on which Abū ʿAlī ibn al-Haytham relied to establish this idea in his book which he called *Optics*.

Having summarized the theory, al-Rāzī sets about criticizing it, and in the process demonstrates his knowledge of Abū ʿAlī ibn al-Haytham's optics. On the other hand, in Volume XIV al-Rāzī develops the following philosophico-theological thesis: God cannot be located either in a place or in

---

[7] Fakhr al-Dīn al-Rāzī, *al-Tafsīr al-Kabīr*, 3rd ed., Beirut, n.d., vol. XIII, pp. 95–6.

a direction. Among other arguments there is one concerning the infinite difference between God and the Universe:[8]

فإن قيل : أليس أنه تعالى متقدم على العالم من الأزل إلى الأبد ، فتقدمه على العالم محصور بين حاصرين

ومحدود بين حدين وطرفين، أحدهما : الأزل، والثاني : أول وجود العالم، ولم يلزم من كون هذا التقدم

محصوراً بين حاصرين أن يكون لهذا التقدم أول وبداية. فكذا ههنا. وهذا هو الذي عول عليه محمد بن

الهيثم في دفع هذا الإشكال عن هذا القسم.

If we say: Is it not so that the Most High existed before the Universe from time eternal in the past and will exist to time eternal that is to come? His existence before the Universe is thus confined between two markers, and delimited by two limits and end points: one is eternity in the past and the other the beginning of the existence of the Universe; and it does not necessarily follow from the fact that his anteriority is confined between two markers that it has a beginning and a starting point. It is so here. This is what Muḥammad ibn al-Haytham relies on to remove in this section of the book.

Now, if we examine the list of writings by Muḥammad ibn al-Haytham, there are several that could be the source used by al-Rāzī. For example, we might suggest his treatise *On the Universe Regarding its Beginning, its Nature and its Perfection* (*Maqāla fī al-'ālam min jihat mabda'ihi wa-ṭabī'atihi wa-kamālihi*).

It is clear that the difference in context of these citations is highly significant, particularly since – as we have seen – al-Rāzī was a careful reader of the mathematical writings of Abū 'Alī ibn al-Haytham. In short, whether it is a matter of mathematics and optics on the one hand, or one of philosophy and theology[9] on the other, al-Rāzī never makes a mistake, either about the title or about the author, and refers explicitly to Abū 'Alī ibn al-Haytham, that is al-Ḥasan, in the former case, and to Muḥammad ibn al-Haytham in the latter. Unless we simply reject the evidence, there are in fact two figures, whom al-Rāzī distinguishes.

---

[8] *Ibid.*, vol. XIV, pp. 110–11.

[9] An independent piece of information provided by Ibn Abī Uṣaybi'a gives us a glimpse of the social context and the direction of Muḥammad ibn al-Haytham's interests: he is said to have made two replies to Ibn Fasānjus in the course of a controversy in which the latter criticized the opinions of astrologers (556). Ibn Fasānjus is a very instructive figure. As al-Najāshī (372/982.983–450/1058.1059) reports in his *Rijāl*, art. 704, published in Qum, Ibn Fasānjus is a literary scholar who wrote on history, poetry and also philosophy, and from whom we have a book criticizing astrologers. He is not known to have written anything on mathematics, astronomy or optics.

2. Pursuing the same train of thought, we may also mention the evidence provided by ʿAbd al-Laṭīf al-Baghdādī (d. 629 / 1231-2). Not only does he cite al-Ḥasan ibn al-Haytham, he also composed a critical commentary on the latter's treatise *On Place*. Now, physician–philosopher as he was, al-Baghdādī, whenever he speaks of al-Ḥasan ibn al-Haytham, sees him only as a mathematician, an expert on optics and an astronomer, in no case as purporting to be a physician or a philosopher. This is what al-Baghdādī actually says:

غرضي في هذه المقالة أن أبحث عن ماهية المكان بحسب رأي ابن الهيثم. وهذا الرجل فاضل في العلوم

الرياضية، واسع الدسيعة في أنواعها، طويل الباع في علم الهيئة وعلم المناظر، وهو من أهل مصر

معاصر ابن رضوان الطبيب.

My purpose in this treatise is to examine the essential nature of place according to the opinion of Ibn al-Haytham. This author is eminent in mathematics, greatly gifted in its various branches and well versed in astronomy and optics; he comes from Egypt and is a contemporary of the physician Ibn Riḍwān.[10]

In the course of his critical commentary on Ibn al-Haytham's treatise *On Place*, al-Baghdādī censures him for knowing so little about logic, and, indirectly, for his ignorance of the writings of Aristotle. Thus he complains about 'his (Ibn al-Haytham's) lack of skill in the art of logic (*qillat riyāḍatihi bi-ṣināʿat al-manṭiq*)'; or 'neglect of the art of logic (*al-ihmāl li-ṣināʿat al-manṭiq*)'. This criticism is the more significant because al-Baghdādī seems to be well informed about Ibn al-Haytham's writings. In this same book, he refers to the treatise Ibn al-Haytham composed on astronomy, called *Fī ḥarakat al-iltifāf* (*On the Winding Movement*).

In short, al-Baghdādī, who was a famous philosopher and physician, a pupil of Ibn Nāʾilī, and a man we find in Mosul in the company of the mathematician Kamāl al-Dīn ibn Yūnus, in Damascus, in Jerusalem, in Acre (587/1190), in Egypt (where he met Maimonides), and who knew the writings of philosophers, physicians and mathematicians, such as al-Samawʾal (according to Ibn Abī Uṣaybiʿa), al-Baghdādī sees in Ibn al-Haytham only a mathematician who is ignorant in the matter of logic, that is, in philosophy. Now, we know from Ibn Abī Uṣaybiʿa that Muḥammad ibn al-Haytham had made a summary of Porphyry's *Isagoge* and Aristotle's *Organon* (his writings on logic), as well as his *On the Soul*, the *Physics* and

---

[10] ʿAbd al-Laṭīf al-Baghdādī, *Maqāla fī al-makān*, ms. Bursa, Çelebi 323, fols 23–52 (text, French translation and commentary in R. Rashed, *Les Mathématiques infinitésimales*, vol. IV).

On the Heavens. We also know that he had been a physician, and had made summaries of about thirty of the books by Galen.[11]

In view of such comments, if al-Ḥasan and Muḥammad were one and the same, it seems impossible that al-Baghdādī, who was well informed about the learned writings of the time, and specifically about those of al-Ḥasan ibn al-Haytham, could have failed to mention the medical and philosophical works by this author (supposedly one single author), or at least have referred to the fact that he was a physician.

3. Finally, two titles are explicitly attributed to Muḥammad ibn al-Haytham by al-Baghdādī (the bibliographer), in the supplement to his bibliographical work *Kashf al-Zunūn*. We read:

Proofs of Prophecy (Fī ithbāt al-nubuwāt)[12] – cited also by Ibn Abī Uṣaybiʿa, in the list of the works of Muḥammad ibn al-Haytham.

The Superiority of al-Ahwāz over Baghdad in Regard to Natural Factors (Tafḍīl Ahwāz ʿalā Baghdād min jihat al-umūr al-ṭabīʿiyya)[13] – again cited by Ibn Abī Uṣaybiʿa, in the list for Muḥammad.

So it seems probable that these two texts by Muḥammad were still in circulation under his own name when al-Baghdādī wrote his Supplement.

Taken together with the evidence we put forward in the previous volume, it seems to me that these further indications must be seem as settling the argument. They show us that Muḥammad ibn al-Haytham still had an existence as a historical figure at least a century after his death, and that the error made by one biobibliographer was not shared by all the philosophers and scholars of his time.

[11] Ibn Abī Uṣaybiʿa, *ʿUyūn al-anbāʾ fī ṭabaqāt al-aṭibbāʾ*, ed. N. Riḍā, Beirut, 1965.

[12] Ismāʿīl al-Baghdādī, *Hadiyyat al-ʿĀrifīn,* Istanbul, 1955, vol. 1, p. 23.

[13] Ibn Abī Uṣaybiʿa, *ʿUyūn al-anbāʾ fī ṭabaqāt al-aṭibbāʾ*, p. 311.

BIBLIOGRAPHY

1.1. MANUSCRIPTS

Anonymous treatise
  Tarkīb li-taḥlil muqaddimat al-musabbaʿ al-mutasāwī al-aḍlāʿ fī al-dāʾira
  Cairo, Dār al-Kutub, Riyāḍa 41, fols 100^v–101^v.

[Archimedes]
  Kitāb ʿamal al-dāʾira al-maqsūma bi-sabʿat aqsām mutasāwiya li-Arshimīdis -
  Tarjamat Abī al-Ḥasan Thābit ibn Qurra al-Ḥarrānī
  Cairo, Dār al-Kutub, Riyāḍa 41, fols 105^r–110^r.

Abū al-Jūd
  Kitāb ʿamal al-musabbaʿ fī al-dāʾira li-Abī al-Jūd Muḥammad ibn al-Layth;
  arsalahu ilā Abī al-Ḥasan ibn Muḥammad ibn Isḥāq al-Ghādī, wa-huwa ʿalā al-
  wajhayn alladhayni tafarrada bi-himā
  Cairo, Dār al-Kutub, Riyāḍa 41, fols 117^v–120^r.

  Risālat Abī al-Jūd Muḥammad ibn al-Layth ilā al-ustādh al-fāḍil Abī Muḥammad
  ʿAbd Allāh ibn ʿAlī al-Ḥāsib fī al-dalāla ʿalā ṭarīqay al-ustādh Abī Sahl al-Qūhī
  al-Muhandis wa-shaykhihi Abī Ḥāmid al-Ṣāghānī wa tarīqihi allatī salakahā fī
  ʿamal al-musabbaʿ al-mutasāwī al-aḍlāʿ fī al-dāʾira
  Paris, Bibliothèque nationale 4821, fols 37^v–46^r.

  Risālat Muḥammad ibn al-Layth ilā Abī Muḥammad ʿAbd Allāh ibn ʿAlī al-Ḥāsib
  fī ṭarīqay Abī Sahl al-Qūhī wa-shaykhihi Abī Ḥāmid al-Ṣāghānī fī ʿamal al-
  musabbaʿ al-mutasāwī al-aḍlāʿ fī al-dāʾira
  Oxford, Bodleian Library, Thurston 3, fols 133^r–134^r.
  Oxford, Bodleian Library, Marsh 720, fols 261^r–264^r.

Ibn ʿAbd Allāh, Naṣr
  Risālat Naṣr ibn ʿAbd Allāh fī istikhrāj watar al-musabbaʿ
  Oxford, Bodleian Library, Thurston 3, fol. 131^{r-v}.
  Oxford, Bodleian Library, Marsh 720, fols 266^r–267^r.

Ibn al-Fatḥ, Sinān
  Fī al-Misāḥāt al-manāẓiriyya
  Cairo, Dār al-Kutub, Riyāḍa 260, fols 94^v–105^v.

Ibn al-Haytham, al-Ḥasan
  Fī ʿamal al-musabbaʿ fī al-dāʾira

Istanbul, Süleymaniye, 'Āṭif 1714/ 19, fols 200^v–210^r.
Istanbul, Askari Müze 3025 (without numbering).

Fī istikhrāj 'amidat al-jibal
Oxford, Bodleian Library, Seld. A.32, fols 187^r–188^r.

Fī ma'rifat irtifā' al-ashkāl al-qā'ima wa 'amidat al-jibāl wa irtifā' al-ghuyūm
Leiden, Universiteitsbibliotheek, Or. 14/8.
New York, University of Columbia Library, Smith Or. 45/12.
Teheran, Majlis Shūrā 2773/2, fols 19–20.
Teheran, Malik 3433, fols 1^v–2^r.

Fī mas'ala 'adadiyya mujassama
London, India Office 1270 (= Loth 734), fols 118^v–119^r.

Fī muqaddimat ḍil' al-musabba'
Aligarh, University Library, 'Abd al-Ḥayy.
London, India Office 1270 (= Loth 734), fols 122^r–123^v.
Oxford, Bodleian Library, Thurston 3, fol. 131^{r-v}.

Fī qismat al-khaṭṭ alladhī ista'maluhu Arshimīdis
Istanbul, Beshiraga 440, fol. 275^v.
Istanbul, Haci Selimaga 743, fols 135^r–136^v.
Istanbul, Süleymaniye, 'Āṭif 1712/17, fol. 147^{r-v}.
Istanbul, Süleymaniye, Carullah 1502, fols 222^v–223^r.
Istanbul, Topkapi Sarayi, Ahmet III 3453/16, fol. 179^v.
Istanbul, Topkapi Sarayi, Ahmet III 3456/18, fols 81^v–82^r.
Leiden, Universiteitsbibliotheek, Or. 14/16, fols 498–499.
London, India Office 1270/18, fol. 119^v.

Fī shakl Banī Mūsā
Aligarh, University Library, no. I, fols 28–38.
Istanbul, Askari Müze 3025 (without numbering).
Istanbul, Süleymaniye, 'Āṭif 1174, fols 149^r–157^r.
London, British Museum, Add. 14332/2, fols 42–61.
London, India Office 1270 (= Loth 734), fols 28^r–28^v.

Fī tamān Kitāb al-Makhrūṭāt
Manisa Genel 1706, fols 1^v–25^r.

Fī uṣūl al-misāḥa
St Petersburg, Oriental Institute, B 2139, fols 100^r–139^v.
London, India Office 1270, fols 28^v–32^v.
Istanbul, Süleymaniye, Fātiḥ 3459, fols 103^v–104^v.
St Petersburg, National Library 143, fols 13^v–15^v.

Ibn Yūnus, Kamāl al-Dīn

Risālat al-mawlā Kamāl al-Dīn ibn Yūnus ilā khādimihi Muḥammad ibn al-Ḥusayn fī al-burhān 'alā ijād al-muqaddima allatī ahmalahā Arshimīdis fī kitābihi fī tasbī' al-dā'ira wa-kayfiyyat dhālika

Koweit, Dār al-athār al-islāmiyya, LNS 67, fols 138ᵛ–140ʳ.
Istanbul, Topkapi Sarayi, Ahmet III, no. 3342 (without numbering).

Risāla li-mawlānā Kamāl al-Dīn Abī al-ma'ālā Mūsā ibn Yūnus fī al-burhān 'alā ijād al-muqaddima allatī ahmalahā Arshimīdis fī tasbī' al-dā'ira wa-kayfiyyat dhālika

Manisa Genel 1706/8, fols 184ᵛ–185ᵛ.
Oxford, Bodleian Library, Thurston 3, fols 128ᵛ–129ʳ.
Oxford, Bodleian Library, Marsh 720, 257ʳ–258ᵛ.

Al-Qabīṣī

Fī anwā' min al-a'dād wa ṭarā'if min al-a'māl
Istanbul, Süleymaniye, Aya Sofya 4832, fols 85ᵛ–88ʳ.

Al-Qūhī

Istikhrāj Wayjan ibn Rustum al-ma'arūf bi-Abī Sahl al-Qūhī fī 'amal al-musabba' al-mutasāwī al-aḍlā' fī dā'ira ma'alūma

Istanbul, Süleymaniye, Aya Sofya 4832/27, fols 145ᵛ–147ᵛ.
Cairo, Dār al-Kutub, Riyāḍa 40, fols 222ᵛ–225ʳ.
Damascus, al-Ẓāhiriyya 5648, fols 215ᵛ–219ᵛ.
Teheran, Danishka 1751, fols 65ᵛ–67ʳ.

Risāla li-Abī Sahl al-Qūhī fī istikhrāj ḍil' al-musabba' al-mutasāwī al-aḍlā' fī al-dā'ira

Oxford, Bodleian Library, Thurston 3, fol. 130ʳ⁻ᵛ.

Risāla fī 'amal ḍil' al-musabba' al-mutasāwī al-aḍlā' fī al-dā'ira li-Abī Sahl al-Qūhī

Paris, Bibliothèque nationale 4821, fols 1ᵛ–18ʳ.
London, India Office, Loth 767, fols 182ᵛ–189ʳ.

Lemme à la division de la droite
Leiden, Universiteitsbibliotheek, Or. 168/8, fols 80ᵛ–84ᵛ.

Al-Ṣāghānī

Risālat Aḥmad ibn Muḥammad ibn al-Ḥusayn al-Ṣāghānī ilā al-Malik al-jalīl 'Aḍud al-Dawla ibn Abī 'Alī Rukn al-Dawla

Paris, Bibliothèque nationale 4821, fols 23ᵛ–29ʳ.

Al-Shannī

Kitāb tamwīh Abī al-Jūd fī mā qaddamahu min al-muqaddimatayn li-'amal al-musabba'

Cairo, Dār al-Kutub, Riyāḍa 41, fols 129ᵛ–134ᵛ.

Cambridge, University Library, T–S Ar. 41.64 (fragment).
Liban, St Joseph 223, fols 16–19 (fragment).

Al-Sijzī

Kitāb Aḥmad ibn Muḥammad ibn 'Abd al-Jalīl al-Sijzī fī 'amal al-musabba' fī al-dā'ira wa qismat al-zāwiya al-mustaqīma al-khaṭṭayn bi-thalāthat aqsām mutasāwiya
Cairo, Dār al-Kutub, Riyāḍa 41, fols 113^v–115^v.
Istanbul, Reshit 1191, fols 80^v–83^r.
Paris, Bibliothèque nationale 4821, fols 10^v–16^v.

Maqāla li-Aḥmad ibn 'Abd al-Jalīl al-Sijzī fī 'amal al-musabba' fī al-dā'ira wa qismat al-zāwiya al-mustaqīmat al-khaṭṭayn
Oxford, Bodleian Library, Thurston 3, fols 129$^{r–v}$.
Oxfrod, Bodleian Library, Marsh 720, fols 267^v–268^r.

1.2.   OTHER MANUSCRIPTS CONSULTED FOR THE ANALYSIS AND THE SUPPLEMENTARY NOTES

Apollonius
   Kitāb al-Makhrūṭāt
   Istanbul, Aya Sofya 2762 (photographic reproduction by M. Nazim Terzioğlu, Publications of the Mathematical Research Institute 4, Istanbul, 1981).

Al-Baghdādī, 'Abd al-Laṭīf
   Maqāla fī al-makān
   Bursa, Çelebi 323, fols 23–52.

Banū Mūsā
   Muqaddimāt kitāb al-Makhrūṭāt
   Istanbul, Süleymaniye, Aya Sofya 4832, fols 223^v–226^v.

Ibn Abī al-Shukr al-Maghribī
   Sharḥ Kitāb Abulūniyūs fī al-Makhrūṭāt
   Teheran, Sepahsalar 556.

Al-Iṣfahānī
   Talkhīṣ al-Makhrūṭāt
   Istanbul, Aya Sofya 2724.

Al-Rāzī, Fakhr al-Dīn
   al-Mulakhkhaṣ
   Teheran, Majlis Shūrā, no. 827.

Al-Shīrāzī, Abū al-Ḥusāyn 'Abd al-Malik ibn Muḥammad
  Kitāb Taṣaffuḥ al-Makhrūṭāt,
  Istanbul, Süleymaniye, Carullah 1507.
  Istanbul, Topkapi Sarayi, Ahmed III, 3463.
  Istanbul, Yeni Cami 803.

Al-Sijzī
  Jawāb Aḥmad ibn Muḥammad ibn 'Abd al-Jalīl 'an masā'il handasiyya sa'ala
  'anhā ahlu Khurāsān
  Dublin, Chester Beatty Library 3652.
  Istanbul, Reshit 1191.

Thābit ibn Qurra
  Fī misāḥat al-ashkāl al-musaṭṭaḥa wa al-mujassama
  Istanbul, Aya Sofya 4832, fols 41^r–44^r.

Al-Ṭūsī, Naṣīr al-Dīn
  Taḥrīr kitāb al-Makhrūṭāt
  Dublin, Chester Beatty Library 3076.
  London, India Office 924.

2. BOOKS AND ARTICLES

M. Abdulkabirov, *Matematika i astronomiya v trudakh Ibn Sina, yego*
  sovrenrennikov i posledovatelei, Tashkent, FAN, 1981.

A. Anbouba
'Tasbī' al-Dā'ira (La construction de l'heptagone régulier)', *Journal for the History of*
  Arabic Science, vol. 1, no. 2, 1977, pp. 352–84; French abstract under the title
  'La construction de l'heptagone régulier', *ibid.*, vol. 2, no. 2, pp. 264–9.

'Un mémoire d'al-Qabīṣī (4e siècle H.) sur certaines sommations numériques', *Journal*
  for the History of Arabic Science, vol. 6, nos 1 and 2, 1982, pp. 181–208.

Apollonius
Apollonius Pergaeus, ed. J. L. Heiberg, 2 vols, Leipzig, 1891–1893; repr. Stuttgart,
  1974.

Les Coniques d'Apollonius de Perge, French transl. by Paul Ver Eecke, Paris, 1959.

Apollonius, *Les Coniques (I–VII)*, edition, translation and commentary by R. Rashed,
  Berlin/New York, Walter de Gruyter, 2008–2010.

See also T. L. Heath

Archimède, Commentaires d'Eutocius et fragments, Texte établi et traduit par Charles
  Mugler, Collection des Universités de France, Paris, 1972.

O. Becker
Grundlagen der Mathematik, 2nd ed., Munich, 1964.

Das mathematische Denken der Antike, Göttingen, 1966.

Al-Bīrūnī
al-Qanūn al-Mas'ūdī, Oriental Publications Bureau, Hyderabad, 1954.

*Kitāb maqālīd 'ilm al-hay'a. La Trigonométrie sphérique chez les Arabes de l'Est à la
      fin du Xe siècle*, edition and translation by M.Th. Debarnot, Damascus, Institut
      français de Damascus, 1985.

M. Clagett, *Archimedes in the Middle Ages*, vol. V: *Quasi-Archimedean Geometry in
      the Thirteenth Century*, Philadelphia, 1984.

M. Decorps-Foulquier, *Recherches sur les* Coniques *d'Apollonios de Pergè*, Paris,
      Klincksieck, 2000.

E.J. Dijksterhuis, *Archimedes*, translated by C. Dickshoorn with a new bibliographic
      essay by Wilbur Knorr, Princeton, 1987.

Euclid
The Thirteen Books of Euclid's Elements, translated with introduction and commentary
      by T. L. Heath, Cambridge, 1908; repr. New York, Dover Publications, 1956.

L'Optique et la Catoptrique, Œuvres traduites pour la première fois du grec en français
      avec une introduction et des notes par Paul Ver Eecke, Nouveau Tirage, Paris,
      Librairie Albert Blanchard, 1959.

Les Œuvres d'Euclide, traduites littéralement par F. Peyrard, Paris, 1819; Nouveau
      tirage, augmenté d'une importante Introduction par M. Jean Itard, Paris, Librairie
      A. Blanchard, 1966.

Al-Fārābī
Iḥsā' al-'ulūm, ed. 'Uthmān Amīn, 3rd ed., Cairo, 1968.

Kitāb al-Mūsīqā al-kabīr, edited and expounded by Ghattas Abd-el-Malek Khashaba,
      revised and introduced by Dr. Mahmoud Ahmed El Hefny, Cairo, The Arab
      Writer-Publishers & Printers, s.d.

T. L. Heath
The Works of Archimedes, Cambridge, 1897; repr. New York, Dover Publications,
      1953.

A Manual of Greek Mathematics, repr. New York, Dover Publications, 1963.

Apollonius of Perga. Treatise on Conic Sections, Cambridge, 1896; repr. 1961.

A History of Greek Mathematics, 2 vols, Oxford, 1921; repr. Oxford, 1965.

Héron, *Metrica*, ed. E.M. Bruins, *Codex Constantinopolitanus Palatii Veteris n. 1*,
      Part two [Greek Text], Leiden, 1964.

Donald R. Hill, 'Al-Bīrunī's mechanical calendar', *Annals of Science*, 42, 1985, pp. 139–63; reprinted in J. V. Field, D. R. Hill and M. T. Wright, *Byzantine and Arabic Mathematical Gearing*, London, 1985.

J. P. Hogendijk
'Greek and Arabic constructions of regular heptagon', *Archive for History of Exact Sciences*, 30,1984, pp. 197–330.

Ibn al-Haytham's Completion of the Conics, Sources in the History of Mathematics and Physical Sciences 7, New York/Berlin/Heidelberg/Tokyo, Springer-Verlag, 1985.

Ibn Abī Uṣaybi'a, *'Uyūn al-anbā' fī ṭabaqāt al-aṭibbā'*, ed. N. Riḍā, Beirut, 1965.

Ibn al-Haytham, *Majmū' al-rasā'il*, Osmānia Oriental Publications Bureau, Hyderabad, 1938–1939.

Ibn 'Irāq, *Rasā'il Abī Naṣr Manṣūr ibn 'Irāq ilā al-Bīrūnī*, Osmānia Oriental Publications Bureau, Hyderabad, 1948.

Khalīfa, Ḥajjī, *Kashf al-ẓunūn*, ed. Yatkaya, Istanbul, 1943.

Ch. Mugler, *Dictionnaire historique de la terminologie géométrique des Grecs*, Études et commentaires, XXVIII, Paris, Librairie C. Klincksieck, 1958.

Al-Nadīm, *Kitāb al-fihrist*, ed. R. Tajaddud, Teheran, 1971.

Al-Najāshī, Abū al-'Abbās, *Rijāl al-Najāshī,* Qum, Mu'assasa al-nashr al-islāmī, 372–450 H.

Pappus of Alexandria
Pappi Alexandrini Collectionis quae supersunt e libris manu scriptis edidit latina interpretation et commentariis instruxit F. Hultsch, 3 vols, Berlin, 1876–1878.

La Collection mathématique, Œuvre traduite pour la première fois du grec en français, avec une introduction et des notes par Paul Ver Eecke, 2 vols, Paris/Bruges, 1933; Nouveau tirage Paris, 1982.

Pappus of Alexandria, Book 7 of the Collection. Part 1. *Introduction, Text, and Translation*; Part 2. *Commentary, Index, and Figures*, edited with translation and commentary by Alexander Jones, Sources in the History of Mathematics and Physical Sciences 8, New York/Berlin/Heidelberg/Tokyo, Springer-Verlag, 1986.

Al-Qifṭī, *Ta'rikh al-ḥukamā'*, ed. J. Lippert, Leipzig, 1903.

R. Rashed
'La construction de l'heptagone régulier par Ibn al-Haytham', *Journal for the History of Arabic Science*, vol. 3, no. 2, 1979, pp. 309–87.

'Mathématiques et philosophie chez Avicenne', in J. Jolivet and R. Rashed (eds), *Études sur Avicenne*, Collection Sciences et philosophie arabes – Études et reprises, Paris, Les Belles Lettres, 1984, pp. 29–39; reprinted in *Optique et Mathématiques: Recherches sur l'histoire de la pensée scientifique en arabe*, Variorum reprints, Aldershot, 1992, XV.

Sharaf al-Dīn al-Ṭūsī, Œuvres mathématiques, 2 vols, Paris, Les Belles Lettres, 1986.

'Al-Sijzī et Maïmonide: Commentaire mathématique et philosophique de la proposition II-14 des *Coniques* d'Apollonius', *Archives Internationales d'Histoire des Sciences*, no. 119, vol. 37, 1987, pp. 263–6; reprinted in *Optique et Mathématiques: Recherches sur l'histoire de la pensée scientifique en arabe*, Variorum reprints, Aldershot, 1992, XIII.

'La philosophie mathématique d'Ibn al-Haytham. I: L'analyse et la synthèse', *Mélanges de l'Institut Dominicain d'Études Orientales du Caire (MIDEO)*, 20, 1991, pp. 31-231.

Optique et Mathématiques: Recherches sur l'histoire de la pensée scientifique en arabe, Variorum reprints, Aldershot, 1992.

'La philosophie mathématique d'Ibn al-Haytham. II: Les Connus', *MIDEO*, 21, 1993, pp. 87-275.

Les Mathématiques infinitésimales du IXe au XIe siècle, vol. I: *Fondateurs et commentateurs: Banū Mūsā, Thābit ibn Qurra, Ibn Sinān, al-Khāzin, al-Qūhī, Ibn al-Samḥ, Ibn Hūd*, London, al-Furqān, 1996; English translation: *Founding Figures and Commentators in Arabic Mathematics*. A History of Arabic Sciences and Mathematics, vol. 1, Culture and Civilization in the Middle East, London, Centre for Arab Unity Studies/Routledge, 2012.

Les Mathématiques infinitésimales du IXe au XIe siècle, vol. II: *Ibn al-Haytham*, London, al-Furqān, 1993. English translation: *Ibn al-Haytham and Analytical Mathematics*. A History of Arabic Sciences and Mathematics, vol. 2, Culture and Civilization in the Middle East, London, Centre for Arab Unity Studies/Routledge, 2012.

Géométrie et dioptrique au Xe siècle. Ibn Sahl, al-Qūhī et Ibn al-Haytham, Paris, Les Belles Lettres, 1993.

Les Œuvres scientifiques et philosophiques d'al-Kindī. Vol. I: *L'Optique et la Catoptrique d'al-Kindī*, Leiden, E.J. Brill, 1996.

'Algebra', in R. Rashed (ed.), *Encyclopedia of the History of Arabic Science*, London/New York, Routledge, 1996; French version *Histoire des sciences arabes*, 3 vols, Paris, Le Seuil, 1997, vol. II.

'L'histoire des sciences entre épistémologie et histoire', *Historia scientiarum*, 7.1, 1997, pp. 1–10.

Les Catoptriciens grecs. I: *Les miroirs ardents*, edition, French translation and with a commentary, Collection des Universités de France, Paris, Les Belles Lettres, 2000.

Les Mathématiques infinitésimales du IXe au XIe siècle, vol. IV: *Méthodes géométriques, transformations ponctuelles et philosophie des mathématiques*, London, 2002.

Geometry and Dioptrics in Classical Islam, London, al-Furqān, 2005.

'Thābit ibn Qurra et l'art de la mesure', in R. Rashed, *Thābit ibn Qurra. Science and Philosophy in Ninth-Century Baghdad*, Scientia Graeco-Arabica, vol. 4, Berlin/New York, Walter de Gruyter, 2009, pp. 173–209.

D'al-Khwārizmī à Descartes. Études sur l'histoire des mathématiques classiques, Paris, Hermann, 2011.

Abū Kāmil. Algèbre et analyse diophantienne, edition, translation and commentary, Berlin, Walter de Gruyter, 2012.

R. Rashed and H. Bellosta, *Ibrāhīm ibn Sinān: Logique et géométrie au Xe siècle*, Leiden, E.J. Brill, 2000.

R. Rashed and B. Vahabzadeh, *Al-Khayyām mathématicien*, Paris, Blanchard, 1999. English translation: *Omar Khayyam. The Mathematician*, Persian Heritage Series no. 40, New York, Bibliotheca Persica Press, 2000.

Fakhr al-Dīn al-Rāzī
Al-Maṭālib al-'āliyya, ed. Aḥmad Ḥijāzī al-Saqqā, Beirut, 1987.

Al-Tafsīr al-Kabīr, 3rd ed., Beirut, Dār Iḥyā' al-turāth al-'arabī, s.d., t. XIII.

Kh. Samir, 'Une correspondance islamo-chrétienne entre Ibn al-Munağğim, Ḥunayn ibn Isḥāq et Qusṭā ibn Lūqā', Introduction, edition, notes and index by Khalil Samir; introduction, translation and notes by Paul Nwyia in F. Graffin, *Patrologia Orientalis*, t. 40, fasc. 4, no. 185, Turnhout, 1981.

Y. Samplonius, 'Die Konstruktion des regelmässigen Sibeneckes nach Abū Sahl al-Qūhī Waiğan ibn Rustam', *Janus*, 50, 1963, pp. 227–49.

C. Schoy
'Graeco-Arabische Studien nach mathematischen Handschriften der Vieköniglichen Bibliothek zu Kairo', *Isis*, 8, 1926, pp. 21–40.

Die trigonometrischen Lehren des persischen Astronomen Abū'l Raiḥān Muḥammad Ibn Aḥmad al-Bīrūnī, Hanover, 1927.

J. Sesiano, 'Mémoire d'Ibn al-Haytham sur un problème arithmétique solide', *Centaurus*, 20.3, 1976, pp. 189–95.

F. Sezgin, *Geschichte des arabischen Schrifttums*. Band V: *Mathematik*, Leiden, E.J. Brill, 1974.

N. Terzioğlu, *Das Achte Buch zu den* Conica *des Apollonius von Perge. Rekonstruiert von Ibn al-Haysam*, Herausgegeben und eingeleitet von N. Terzioğlu, Istanbul, 1974.

G. Vajda, *Index général des manuscrits arabes musulmans de la Bibliothèque nationale de Paris*, Publications de l'Institut de recherche et d'histoire des textes, Paris, ed. CNRS, 1953.

J. Van Ess, *Die Erkenntnislehre des 'Aḍudaddīn al-Īcī*, Übersetzung und Kommentar des ersten Buches seiner Mawāqif, Akademie der Wissenschaften und der Literatur. Veröffentlichungen der Orientalischen Kommission, Band XXII, Wiesbaden, Franz Steiner Verlag GMBH, 1966.

E. Wiedemann, *Aufsätze zur arabischen Wissenschafts-Geschichte*, Hildesheim/New York, 1970, vol. I.

J.J. Witkam, *Jacobius Golius (1596-1667) en zijn handschriften*, Oosters Genootschap in Nederland, 10, Leiden, E.J. Brill, 1980.

F. Woepcke, *L'Algèbre d'Omar Alkhayyāmī*, publiée, traduite et accompagnée d'extraits de manuscrits inédits, Paris, Benjamin Duprat, 1851; reprinted in *Études sur les mathématiques arabo-islamiques*, herausgegeben von Fuat Sezgin, Frankfurt am Main, 1986.

INDEX OF NAMES

SUBJECT INDEX

INDEX OF WORKS

INDEX OF MANUSCRIPTS